Infinitely Divisible Point Processes

Infinitely Divisible Point Processes

Professor Dr. Klaus Matthes

Akademie der Wissenschaften der DDR
Zentralinstitut für Mathematik und Mechanik

Professor Dr. Johannes Kerstan

Professor Dr. Joseph Mecke

Friedrich-Schiller-Universität Jena
Sektion Mathematik

JOHN WILEY & SONS

Chichester · New York · Brisbane · Toronto

English translation: B. Simon

Library of Congress Cataloging in Publication Data

Kerstan, Johannes.
Infinitely divisible point processes.

(Wiley series in probability and mathematical statistics)
Translation of unbegrenzt teilbare Punktprozesse.
In the 1974 German ed. J. Kerstan's name appeared first on t. p.
Bibliography: p.
Includes indexes.
1. Point processes. I. Matthes, Klaus, joint author. II. Mecke, Joseph, joint author. III. Title.
QA274.42.K4713 519.2 76-50627

ISBN 0 471 99460 X

Printed in GDR

To

Boris Vladimirovich Gnedenko

Preface to the German Edition

During the past decade, the theory of point processes has developed into a special discipline of probability theory. Starting from problems concerning special classes of stochastic models, especially of those of queueing and reliability theory, but also by the generalization and more precise formulation of theoretical concepts and ideas such as they are encountered in the theory of branching processes, an appropriate system of concepts and propositions has been created. This book is devoted to a coherent presentation of an essential part of this system. The authors considered that the present state of the theory of point processes not only permits, but also requires such a systematic presentation, in order to establish generally accepted, secure conceptional foundations for further investigations. While reference was made to the existing literature to the best of our knowledge and belief, it should be noted here that in our presentation emphasis was laid on the contents rather than the precise determination of the authorship of single statements. Therefore, wherever in this book references are lacking, this does not mean any claim for priority put in by the authors.

In accordance with to the scope of their own scientific work, the concept of the infinitely divisible point process served the authors as a guideline and a criterion of choice. In all cases they endeavoured to achieve an appropriate degree of generality. For the sake of better readability they renounced, however, the idea of using arbitrary locally compact Abelian topological groups satisfying the second axiom of countability as phase spaces in the theory of stationary point processes, rather confining themselves to the Euclidean spaces R^s. For similar reasons they also desisted from a detailed presentation of the theory of marked point processes. In designing the contents of this book, the authors were discouraged by the fact that at that time the use of marks was yet hardly common in the international literature.

During their work on this book, the authors were frequently faced with the necessity to incorporate new results, thus exceeding the initially given scope. Whenever possible, this was done until proof reading was under way. Thus in several cases redundancies had to be put up with.

The authors wish to thank all those of their colleagues who contributed to an enrichment of this presentation by sending preprints of their own publications. Here we are particularly indebted to D. Daley and O. Kallenberg. Further thanks are due to Mrs. B. Röder as well as to P. Franken and U. Prehn who enabled the completion of this book by valuable suggestions or direct co-operation in producing the manuscript. Our thanks are likewise due to Miss R. Helle and R. Höppner of the Akademie-Verlag, Berlin, for their judicious patience and assistance in the complicated cooperation with the authors.

The problem of studying limit distributions of arbitrary infinitesimal triangular arrays of point processes, i.e. of arbitrary infinitely divisible point processes, was formulated by B. Gnedenko in a paper read jointly with J. Belyaev and I. Kovalenko before the Sixth All-Union Congress on Probability Theory and Mathematical Statistics, held at Vilnius in 1960. The friendly dedication of this book to B. Gnedenko is, however, not primarily due to this fact, but expresses the authors' appreciation for his pioneering role, which made him, by his teaching activities in Berlin early in the fifties, the originator of the research in probability theory in the German Democratic Republic.

Berlin and Jena, January, 1974

The authors

Preface to the English Edition

This translation is based on a revised German edition.

To begin with, the classification has been altered in order to arrive at a more systematic, and consequently more comprehensible, arrangement. The sequential dependence of each chapter on its predecessor has been replaced by the interdependence table referred to on page XII, and the contents of the first chapter have been somewhat enhanced in order to give a better orientation for some of the results which follow. This process has, however, led to a certain amount of overlap between different parts of this edition. The theory of Campbell and Palm measure has been developed for stochastic marked point processes, as was originally envisaged for the German edition. Furthermore, new material has been included at many points. This applies particularly to the estimates of the variation distance of Poisson distributions, the characterization of the g_ν-property in terms of the reduced Campbell measure, and the general limit theorem for sequences of substochastic translations. The discussion of the theory of spatially homogeneous branching processes has been extensively revised, and in particular the Kallenberg method of backward clusters and some of its applications have been incorporated. A detailed treatment of the subcritical case has also been included for the first time.

Lack of time did not make it possible to tackle the problem of a fundamental rationalisation of all proofs. Similarly, it was not possible to take into account the results of the theory of stochastic point processes associated with the terms of the Gibbs distribution and conditional intensity. The principal courses to which the authors will direct their attention in their future work on this book are thus indicated.

The authors would like to thank a number of colleagues who assisted them while this new manuscript was being drawn up. This applies particularly to Dr. E. Warmuth and Dr. K. Fleischmann. This English edition would have been unthinkable without their continued support. Many improvements to the contents are the result of their contributions. We owe a number of valuable remarks to Dr. J. Grandell, Dr. P. M. Lee, Dr. F. Liese and to Professor K. Krickeberg, who together with Dr. K. H. Fichtner, rendered assistance in proof reading. Last, but not least, we would like to thank particularly Professor D. G. Kendall upon whose initiative this translation was produced.

Berlin and Jena 1977 *The authors*

Contents

Jnterdependence Table

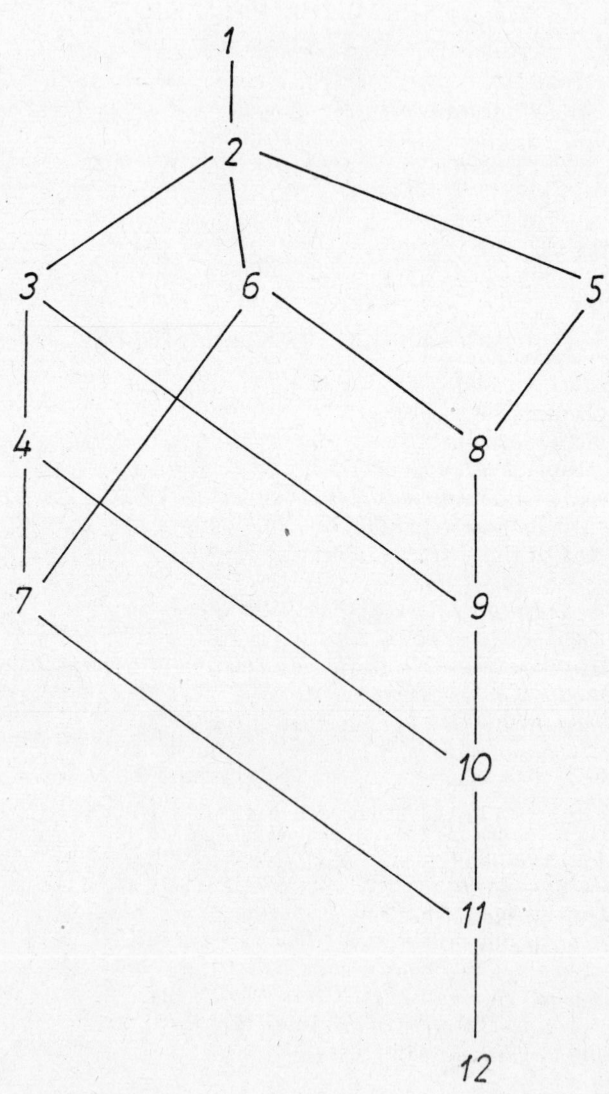

1. Basic Concepts

1.0. Introduction

In introducing the concept of a point process, we find our guidance in the concept that certain events occur at the points of a random subset Z of R^s ($s = 1, 2, \ldots$), which has no accumulation points. Thus for instance a cathode emits electrons at random times. Calls arrive at a telephone exchange at random times. As a result of a microbiological experiment, a random population of micro-organisms develops at the surface of a nutrient solution. Subject to certain assumptions, for almost all realizations of a random real-valued function defined on the real axis, the set of zeros is well-defined and has no accumulation points.

The above mentioned examples suggest that the concept of a point process should be formalized in the following way: Let $\boldsymbol{Z}$ denote the family of those subsets of R^s, the intersections of which with each bounded subset of R^s are finite. Consequently each set Z in $\boldsymbol{Z}$ is at most countable. Further let $\mathfrak{Z}$ denote the smallest σ-algebra of subsets of $\boldsymbol{Z}$ such that for all bounded sets X in the σ-algebra $\mathfrak{R}^s$ of the Borel sets of R^s the mapping

$$Z \curvearrowright \text{number of elements of } (Z \cap X) \qquad (Z \in \boldsymbol{Z})$$

is measurable with respect to $\mathfrak{Z}$. We shall further denote the number of elements of any set X by the symbol card X. Then we may conceive a point process to be a random element of the measurable space $[\boldsymbol{Z}, \mathfrak{Z}]$, i.e. a probability space $[\boldsymbol{Z}, \mathfrak{Z}, P]$. This formally simple definition of the concept of a point process is unsatisfactory in the following respect: It is not possible to express the fact that points having multiplicities different from unity occur. However, such multiplicities commonly occur in simple mathematical models, as for example in renewal theory.

To any point z in R^s, we can assign in a one-to-one way the Dirac distribution δ_z on $\mathfrak{R}^s$ so that for all X in $\mathfrak{R}^s$ the relation $\delta_z(X) = k_X(z)$ holds, where k_X denotes the indicator function of X. Now, if Z is any set in $\mathfrak{Z}$, then $\Phi = \sum_{z \in Z} \delta_z$ provides a measure Φ on $\mathfrak{R}^s$ which assigns the value $\Phi(X) = \text{card}\,(Z \cap X)$ to each X in $\mathfrak{R}^s$. Let M denote the set of integer-valued measures Φ on $\mathfrak{R}^s$, the values of which are finite on all bounded X in $\mathfrak{R}^s$.

Then the correspondence $Z \curvearrowright \Phi = \sum_{z \in Z} \delta_z$ provides a one-to-one mapping of $\boldsymbol{Z}$ onto the set of those Φ in M which satisfy the condition $\Phi(\{x\}) \leqq 1$ for all x in R^s. Such measures Φ in M are said to be simple. The corresponding Z in $\mathbf{Z}$ is the set $\{z : z \in R^s, \Phi(\{z\}) > 0\}$. Each element Φ of M can be expressed as

$$\Phi = \sum_{z \in Z} \Phi(\{z\})\, \delta_z$$

where $Z = \{z : z \in R^s, \Phi(\{z\}) > 0\}$ is an element of $\mathbf{Z}$. Therefore we may think of Φ as picking out the points in $Z = \{z : z \in R^s, \Phi(\{z\}) > 0\}$ and giving each such z the multiplicity $\Phi(\{z\})$. Accordingly the value of $\Phi(X)$ can still be interpreted as the number of "the points of Φ lying in X". For that reason, the elements of M are also called counting measures.

The above statements provide a justification for replacing the set $\mathbf{Z}$ by the more powerful set M. Now instead of $\mathfrak{Z}$ we have the smallest σ-algebra $\mathfrak{M}$ of subsets of M, such that for all bounded X in $\mathfrak{R}^s$ the mapping

$$\Phi \curvearrowright \Phi(X) \qquad (\Phi \in M)$$

is measurable with respect to $\mathfrak{M}$. We conceive of a point process with state space R^s as a random element of the measurable space $[M, \mathfrak{M}]$, i.e. a probability field $[M, \mathfrak{M}, P]$. In this way we introduce not only multiplicities of points, but also gain advantages in performing mathematical operations, because calculating with the measures in M is less difficult than with the sets in $\mathbf{Z}$.

In contrast to the theory of stochastic processes, where complicated problems of measurability are encountered, the concept of the σ-algebra $\mathfrak{M}$ is sufficient for all problems touched on below: All "reasonable" sets and mappings are measurable, too. In particular the set of all simple Φ in M is belongs to $\mathfrak{M}$, and moreover the correspondence

$$Z \curvearrowright \sum_{z \in Z} \delta_z \qquad (Z \in \mathbf{Z})$$

provides a one-to-one mapping of $\mathbf{Z}$ into M which is measurable with respect to the σ-algebras $\mathfrak{Z}$ and $\mathfrak{M}$ in both directions. A probability distribution P on $\mathfrak{M}$ is said to be simple if the condition P (Φ is not simple) $= 0$ is satisfied, i.e. if P can be considered to be a probability distribution on $\mathfrak{Z}$.

In the present exposition we do not confine ourselves to phase spaces of the type R^s, but allow arbitrary complete separable metric spaces $[A, \varrho_A]$ to be considered. This extends the range of application of the theory without introducing any essential, additional, conceptual or technical difficulties. At the same time the theory becomes more complete in itself. Thus for instance we can now consider the random m-dimensional vectors with non-negative integer coordinates to be point processes with phase space $A = \{1, \ldots, m\}$, for the correspondence

$$[k_1, \ldots, k_m] \curvearrowright \sum_{i=1}^{m} k_i \delta_i$$

provides a one-to-one mapping of $\{0, 1, \ldots\}^m$ onto the set M belonging to the phase space A. In this case we think of $A = \{1, \ldots, m\}$ as being endowed with the metric

$$\varrho_A(x, y) = \begin{cases} 0 & \text{if} \quad x = y \\ 1 & \phantom{\text{if}} \quad x \neq y \end{cases}$$

to form a complete separable metric space. Generally the "natural metric" is meant whenever the phase space $[A, \varrho_A]$ is denoted by A only.

For all finite sequences $X_1, \ldots, X_m$ of bounded sets in the σ-algebra $\mathfrak{A}$ which consists of the Borel sets of the phase space $[A, \varrho_A]$, the correspondence

$$\Phi \curvearrowright [\Phi(X_1), \ldots, \Phi(X_m)] \qquad (\Phi \in M)$$

provides a measurable mapping of M into $\{0, 1, \ldots\}^m$. In this way each probability distribution P on $\mathfrak{M}$ is transformed into a probability distribution $P_{X_1,\ldots,X_m}$ of a random m-dimensional vector with non-negative integer coordinates. The probability distributions $P_{X_1,\ldots,X_m}$ are called the finite-dimensional distributions of P. It is easily shown (Proposition 1.3.2.) that P is uniquely determined by the family of finite-dimensional distributions. For simple distributions a more stringent statement (Theorem 1.4.9.) is true: If two distributions P, Q on $\mathfrak{M}$ satisfy the condition $P_X = Q_X$ for all bounded sets X in $\mathfrak{A}$, i.e. if their one-dimensional distributions are identical, then, if Q is simple, it can be concluded that $P = Q$.

The proof of the analogue of Kolmogorov's existence theorem (Theorem 1.3.5.) requires considerable conceptual efforts and hence is somewhat beyond the scope of demonstration of this chapter. There is, however, no difficulty in following the text if it is omitted at first reading.

Together with Φ_1 and Φ_2, the sum $\Phi_1 + \Phi_2$ also belongs to M. It can be interpreted as the superposition of the realizations Φ_1 and Φ_2. The superposition of independent random elements Φ_1 and Φ_2 of $[M, \mathfrak{M}]$ leads to a new random element $\Phi = \Phi_1 + \Phi_2$ of this measurable space, since the transition from $[\Phi_1, \Phi_2]$ to $\Phi_1 + \Phi_2$ is measurable with respect to the σ-algebras $\mathfrak{M} \otimes \mathfrak{M}$ and $\mathfrak{M}$. The probability distribution of the superposition Φ is called the convolution of P_1 and P_2 and is denoted by $P_1 * P_2$. Consequently the following relation holds:

$$(P_1 * P_2)(Y) = (P_1 \times P_2)\big((\Phi_1 + \Phi_1) \in Y\big) \qquad (Y \in \mathfrak{M}).$$

We use this formula in defining the convolution $*$ for the larger vector space $\boldsymbol{E}$ of all finite complex-valued measures on $\mathfrak{M}$. In this way $\boldsymbol{E}$ may be regarded as a commutative and associative algebra over the field of complex numbers. Moreover this algebra is free from divisors of zero and has a unity element, namely the probability distribution δ_o. In this case the symbol "o" denotes the zero measure on $\mathfrak{A}$, i.e. the "empty" realization. On the other hand the zero element of the vector space $\boldsymbol{E}$, i.e. the zero measure on $\mathfrak{M}$, is denoted by 0.

The convolution of distributions P_1, P_2 on $\mathfrak{M}$ represents a generalization of the convolution of distributions of random m-dimensional vectors with non-negative integer coordinates, and moreover, for all finite sequences $X_1, \ldots, X_m$ of bounded sets we have

$$(P_1 * P_2)_{X_1,\ldots,X_m} = (P_1)_{X_1,\ldots,X_m} * (P_2)_{X_1,\ldots,X_m}.$$

If we had introduced the point process by means of the measurable space $[\boldsymbol{Z}, \mathfrak{Z}]$, and accordingly defined the superposition of Z_1 and Z_2 as a union

$Z_1 \cup Z_2$, then the above relation would be valid only subject to the validity of an additional assumption: The condition $Z_1 \cap Z_2 \cap (X_1 \cup \cdots \cup X_m) = \emptyset$ has to be satisfied almost everywhere with respect to the distribution $P_1 \times P_2$.

For each natural number n and each distribution P on $\mathfrak{M}$, there exists at most one n-th convolution root $\sqrt[n]{P}$ in the set $\boldsymbol{P}$ of all probability distributions on $\mathfrak{M}$. A distribution P in $\boldsymbol{P}$ is said to be infinitely divisible, if all of the convolution roots $\sqrt[n]{P}$ $(n = 1, 2, \ldots)$ exist. For each finite measure E on $\mathfrak{M}$ we introduce the following mixture $\mathscr{U}_E$ of convolution powers E^n of E:

$$\mathscr{U}_E = e^{-E(M)} \sum_{n=0}^{\infty} \frac{1}{n!} E^n.$$

In this way we obtain an infinitely divisible distribution $\mathscr{U}_E$ where $\mathscr{U}_{E_1} * \mathscr{U}_{E_2} = \mathscr{U}_{E_1+E_2}$. The correspondence $E \rightsquigarrow \mathscr{U}_E$ provides a one-to-one mapping of the set of all finite measures E on $\mathfrak{M}$ satisfying the condition $E(\{o\}) = 0$ onto the set of those infinitely divisible distributions P in $\boldsymbol{P}$, for which $P(\{o\}) > 0$ holds (Theorem 1.10.1.).

Every finite measure E on $\mathfrak{M}$ is of the form $E = qQ$ with $q \geqq 0, Q \in \boldsymbol{P}$. If E is different from zero, there is only one representation of this type, namely $q = E(M)$, $Q = \big(E(M)\big)^{-1}E$. On the other hand, if $E(M) = 0$, then we have to set $q = 0$, while Q can be chosen arbitrarily. The distribution $\mathscr{U}_{qQ}$ corresponds to a random mechanism proceeding in two stages. At first a non-negative integer n is selected at random according to the Poisson distribution π_q having the expectation q. Subsequently n measures $\Phi_1, \ldots, \Phi_n$ distributed according to Q are realized independently of each other. Then their superposition $\Phi = \Phi_1 + \cdots + \Phi_n$ will be distributed according to $\mathscr{U}_{qQ}$. In this way we derived a model conception for all infinitely divisible P in $\boldsymbol{P}$, which satisfy $P(\{o\}) > 0$; this model will be further developed in section 4.5.

Any infinitely divisible P in $\boldsymbol{P}$ which has the property $P(\Phi \text{ finite}) = 1$, satisfies the inequality $P(\{o\}) > 0$ (Proposition 1.10.2.). For bounded phase spaces $[A, \varrho_A]$, consequently all infinitely divisible distributions on $\mathfrak{M}$ are of the form $\mathscr{U}_E$. In particular this is also true for finite phase spaces $A = \{1, \ldots, m\}$, so that we obtain the "classical" representation of the infinitely divisible distributions of random m-dimensional vectors having non-negative integer coordinates.

A triangular array $(P_{n,j})_{\substack{n=1,2\ldots\\ 1 \leqq j \leqq m_n}}$ of distributions on $\mathfrak{M}$ is said to be infinitesimal if for all bounded X in $\mathfrak{A}$ the convergence $\max\limits_{1 \leqq j \leqq m_n} P_{n,j}\big(\Phi(X) > 0\big) \xrightarrow[n\to\infty]{} 0$ takes place. It will be shown in section 3.4. that a distribution on $\mathfrak{M}$ is infinitely divisible if and only if for an infinitesimal triangular array $(P_{n,j})$ the sequence $\mathop{\ast}\limits_{1 \leqq j \leqq m_n} P_{n,j}$ tends to P as n increases in the sense of the weak convergence introduced in chapter 3, viz. a random element Φ of $[M, \mathfrak{M}]$ can be approximated by means of an infinitely divisible distribution, if it can be represented as a superposition of uniformly "thin" independent random elements $\Phi_1, \ldots, \Phi_s$. It is just this fact which motivated the investigation of the infinitely divisible distributions on $\mathfrak{M}$.

The proofs set forth in section 3.4. are based on the "classical" statements about infinitesimal triangular arrays of distributions of random m-dimensional vectors having non-negative integer coordinates, i.e. they are based on the previous study of the special case $A = \{1, \ldots, m\}$. For finite phase spaces the weak convergence of distributions on $\mathfrak{M}$ coincides with the convergence in variation distance. In section 1.11., for arbitrary phase spaces $[A, \varrho_A]$, triangular arrays with the property $\max_{1 \leqq j \leqq m_n} P_{n,j}\big(\Phi(A) > 0\big) \xrightarrow[n\to\infty]{} 0$ will be investigated with respect to the convergence behaviour of the sequence $\underset{1 \leqq s \leqq m_n}{\ast} P_{n,j}$ in the sense of variation distance. For a bounded phase space $[A, \varrho_A]$ a well-rounded theory results, the proofs being hardly more difficult than in the case of finite phase spaces. Thus we have not only integrated the "classical" theory into the conceptual framework of this chapter, but also introduced the method of calculating with variation distance, which will find many further applications.

Non-normalized finite measures on $\mathfrak{M}$, which are introduced by the representation $P = \mathscr{U}_E$, also play a part in the theory of infinitely divisible distributions on $\mathfrak{M}$. In the following chapter we are even led to infinite measures on $\mathfrak{M}$. For this reason a number of definitions set forth in this chapter are from the outset concerned with arbitrary measures on $\mathfrak{M}$. In particular we define the intensity measure ϱ_H of such a measure H by means of the set up

$$\varrho_H(X) = \int \Phi(X)\, H(\mathrm{d}\Phi) \qquad (X \in \mathfrak{A})$$

where H is said to be of finite intensity if for all bounded X in $\mathfrak{A}$ the value $\varrho_H(X)$ is finite.

A distribution P on $\mathfrak{M}$ is said to be free from after-effects if, for all pairwise disjoint $X_1, \ldots, X_m$ in $\mathfrak{A}$ the finite sequence $\Phi(X_1), \ldots, \Phi(X_m)$ is independent with respect to P. Consequently in the case $A = \{1, \ldots, n\}$ this property is encountered if and only if the "coordinates" $\Phi(\{1\}), \ldots, \Phi(\{n\})$ of Φ are independent with respect to P. If for a measure ν on $\mathfrak{A}$ the value of $\nu(X)$ is finite for all bounded X in $\mathfrak{A}$, then there exists one and only one distribution P_ν on $\mathfrak{M}$ being free from after-effects, with the intensity measure ν, such that all of the one-dimensional distributions $(P_\nu)_X$ are Poisson distributions of mean $\nu(X)$. We call P_ν the Poisson distribution having the intensity measure ν. Then, according to 1.7.6., a Poisson distribution P_ν is simple if and only if P_ν is continuous, i.e. if P_ν satisfies the condition $P_\nu\big(\Phi(\{a\}) > 0\big) = 0$ for all a in A. Thus the well-known theorem of Rényi (Theorem 1.7.8.), which states that a continuous distribution P on $\mathfrak{M}$ is certainly of Poisson type if this is true for all one-dimensional distributions P_X, is derived immediately from the above mentioned Theorem 1.4.9. As a counterpart to Theorem 1.7.8., the following characterization of the continuous Poisson distributions is derived in Theorem 1.11.8.: A continuous distribution P on $\mathfrak{M}$ is Poisson if and only if it is both free from after-effects and simple (cf. Khinchin [3]). Later on we shall become acquainted with a number of further characterizations of classes of Poisson distributions.

The relation $P_\nu * P_\gamma = P_{\nu+\gamma}$ is derived immediately from our introduction of the Poisson distributions. Thus in particular the relation $\sqrt[n]{P_\nu} = P_{\nu/n}$ holds. Consequently all P_ν are infinitely divisible; however, the condition $P_\nu(\{o\}) > 0$ is satisfied only if the intensity measure ν is finite. With respect to their structures, the Poisson distributions are the simplest infinitely divisible distributions on $\mathfrak{M}$. In Section 4.5. it will be observed that all of the infinitely divisible distributions in $\boldsymbol{P}$ can be derived from Poisson distributions by means of a simple stochastic model.

Any measure ν on $\mathfrak{A}$ can be transplanted to $\mathfrak{M}$ by means of the measurable mapping $a \curvearrowright \delta_a$. In this way we obtain a measure Q_ν on $\mathfrak{M}$ with the property $Q_\nu\big(\Phi(A) \neq 1\big) = 0$. Let γ be a distribution, then Q_γ can be interpreted as a distribution of a point selected at random according to γ, and consequently $(Q_\gamma)^n$ $(n = 1, 2, \ldots)$ can be interpreted as the distribution of a random ensemble of exactly n random points realized independently of each other according to γ. The Poisson distributions P_ν with finite intensity measure ν lead one to consider distributions of the form Q_γ, for by 1.7.4., we have $P_\nu = \mathscr{U}_{Q_\nu}$, i.e. in the non-trivial case $\nu \neq o$, $P_\nu = e^{-q} \sum\limits_{n=0}^{\infty} \frac{q^n}{n!} (Q_\gamma)^n$, where $q = \nu(A)$ and $\gamma = \big(\nu(A)\big)^{-1}\nu$. This relation provides not only an insight into the structure of P_ν, but can also be employed in order to construct, for finite ν, the distribution P_ν without resorting to the analogue of Kolmogorov's existence theorem.

To each Φ in M, and to each X in $\mathfrak{A}$, we assign the measure ${}_X\Phi = \Phi\big((.) \cap X\big)$ in M so that all points of Φ which are not contained in X, are cancelled. The mapping $\Phi \curvearrowright {}_X\Phi$ of M into itself is measurable with respect to $\mathfrak{M}$, and hence transforms any measure H on $\mathfrak{M}$ into a measure ${}_XH$ on $\mathfrak{M}$. Now the relation ${}_X(P_\nu) = P_{\nu((.)\cap X)}$ can be inferred immediately from the above definitions. Therefore, if $\nu(X)$ is finite, we obtain ${}_X(P_\nu) = \mathscr{U}_{Q_{\nu((.)\cap X)}}$.

Let ν be a measure on $\mathfrak{A}$, the values $\nu(X)$ of which are finite for all bounded sets X in $\mathfrak{A}$. A distribution P on $\mathfrak{M}$ is said to have the property g_ν, if for all bounded X in $\mathfrak{A}$

1. $P\big(\Phi(X) > 0\big) > 0$ implies $\nu(X) > 0$, and
2. the relation $({}_XP)\big((.) \mid \Phi(X) = k\big) = (Q_{\nu((.)|X)})^k$ holds for all k having the property $P\big(\Phi(X) = k\big) > 0$.

Here $\nu\big((.) \mid X\big)$ denotes the distribution $\big(\nu(X)\big)^{-1} \nu\big((.) \cap X\big)$. In consideration of the underlying meaning of the distributions of form Q_ν, the property g_ν means that, as referred to the measure $\nu\big((.) \cap X\big)$, P is "purely random" on each bounded X in $\mathfrak{A}$.

According to the above statements, P_ν has always the property g_ν. For $\nu = o$ only the distribution $P_o = \delta_o$ has the property g_ν. However, if ν is finite and different from zero, then any P in $\boldsymbol{P}$ satisfies the condition g_ν if and only if it can be represented as a mixture of the powers $(Q_{(\nu(.)|A)})^k$ $(k = 0, 1, \ldots)$ (Proposition 1.14.11.). Finally, generalizing a theorem derived by Nawrotzki (cf. Nawrotzki [1]), we can show for infinite ν (Theorem

1.14.8.) that a distribution P on $\mathfrak{M}$ has the property g_ν if and only if it can be represented as a mixture $P = \int P_{l\nu}(.)\, \sigma(dl)$ of Poisson distributions $P_{l\nu}, l \geqq 0$. For infinite ν, the Poisson distributions P_ν can be characterized by the fact that they have the property g_ν and that for any monotone increasing sequence (X_n) of bounded sets in $\mathfrak{A}$ having the two properties $\bigcup_{n=1}^{\infty} X_n = A$, $\nu(X_n) > 0$ for $n = 1, 2, \ldots$, the sequence of $(\nu(X_n))^{-1}\, \Phi(X_n)$ tends to unity in the sense of convergence in probability (Theorem 1.14.9.).

In the introduction to the concept of a point process, the examples selected by us suggest an interpretation of a point a of the phase space $[A, \varrho_A]$ as a position where events can occur. Nothing was said about the nature of these events, i.e. they were tacitly considered to be of the same kind. However, this practice is not sufficient for many problems encountered. Thus for instance it may prove necessary to record the velocities of the electrons emitted by a cathode. The calls arriving at a telephone exchange have virtual durations. And finally we can investigate random populations where different species of micro-organisms occur.

Therefore it is appropriate in many models that the phase point a should be considered to be an ordered pair $a = [x, y]$ consisting of a position x and a mark y. In this case y describes the quality of the event. Accordingly we shall assume that two complete separable metric spaces $[O, \varrho_O]$ and $[K, \varrho_K]$ are given, which are called the position space and the mark space, respectively. Then, if we introduce the phase space $[A, \varrho_A]$ as a direct product of the position space and the mark space, i.e. if we set

$$A = O \times K, \qquad \varrho_A([x, y], [x', y']) = \varrho_O(x, x') + \varrho_K(y, y'),$$

then we can apply all hitherto introduced concepts and constructions and obtain the concept of a marked point process.

As a rule it is assumed that the mark space $[K, \varrho_K]$ be bounded. (If this is not the case, we can help ourselves by introducing the topologically equivalent metric $\varrho_K(1 + \varrho_K)^{-1}$.) In this way we can ensure that the inequality $\Phi(X \times K) < +\infty$ is satisfied for any bounded Borel subset X of the position space and any Φ in M, and consequently only a finite number of points of Φ have positions in X.

Each "multivariate" point process with phase space $[A, \varrho_A]$, i.e. each random m-tuple $[\Phi_1, \ldots, \Phi_m]$ of counting measures in M, can be considered as a marked point process. Indeed, if we set $[K, \varrho_K] = \{1, \ldots, m\}$, $[A', \varrho_{A'}] = [A \times K, \varrho_{A \times K}]$ then $[\Phi_1, \ldots, \Phi_m] \curvearrowright \sum_{i=1}^{m} \Phi_i \times \delta_i$ provides a measurable one-to-one mapping of M^n onto the space M' of counting measures belonging to the phase space $[A', \varrho_{A'}]$, wich is measurable in both directions. In other words the points in Φ_i are given the mark i.

The introduction of marked point processes leads to a refinement and unification of the general theory, as we shall see in particular in the chapters 5, 6, 8 and 9.

In section 1.15., providing the set M with a distance function ϱ_M which permits an intuitive interpretation, we shall establish a bounded complete separable metric space $[M, \varrho_M]$. A sequence (Φ_n) of counting measures in M converges to Φ with respect to ϱ_M if and only if, for all bounded real functions f which are continuous on A and identically zero outside of suitable bounded subsets, the convergence relation

$$\int f(a)\, \Phi_n(\mathrm{d}a) \xrightarrow[n\to\infty]{} \int f(a)\, \Phi(\mathrm{d}a)$$

is valid. In this way the σ-algebra $\mathfrak{M}$ is given a topological interpretation, since it coincides with the σ-algebra of the Borel sets of $[M, \varrho_M]$.

For reasons of abbreviation we use in the sequel the term "measure" in the sense of "non-negative measure" and "distribution" in the sense of "probability measure".

1.1. Continuous and Discrete Distributions

Let $[A, \varrho_A]$ be a complete separable metric space. We denote by $\mathfrak{A}$ the σ-algebra of the Borel subsets of A, i.e. the smallest σ-algebra containing all open subsets of A. Further let $\mathfrak{B}$ be the ring of all bounded X in $\mathfrak{A}$.

A subring $\mathfrak{H}$ of $\mathfrak{B}$ is called a σ-subring of $\mathfrak{B}$ if the two following conditions are satisfied: For each sequence (H_n) of sets in $\mathfrak{H}$, $\bigcap_{n=1}^{\infty} H_n$ belongs to $\mathfrak{H}$, and for each sequence (H_n') of sets in $\mathfrak{H}$ with the property that $\bigcup_{n=1}^{\infty} H_n' \in \mathfrak{B}$, we have $\bigcup_{n=1}^{\infty} H_n' \in \mathfrak{H}$. A subsystem $\mathfrak{S}$ of $\mathfrak{B}$ is said to generate $\mathfrak{B}$ if the smallest σ-subring of $\mathfrak{B}$ which includes $\mathfrak{S}$, coincides with $\mathfrak{B}$. A subsystem $\mathfrak{S}$ of $\mathfrak{B}$ is called monotone closed in $\mathfrak{B}$ if it includes the intersection of each monotone decreasing sequence of sets in $\mathfrak{S}$ as well as the union of each monotone increasing sequence (S_n) of sets in $\mathfrak{S}$ satisfying the condition $\bigcup_{n=1}^{\infty} S_n \in \mathfrak{B}$.

In what follows, we shall frequently use the following modification of the monotone class theorem (see, for instance, Halmos [1], Chapt. I, Par. 6):

1.1.1. *A subsystem $\mathfrak{S}$ of $\mathfrak{B}$ coincides with $\mathfrak{B}$ if it is monotone closed in $\mathfrak{B}$ and includes a subring $\mathfrak{H}$ which generates $\mathfrak{B}$.*

Proof. Let $\mathfrak{V}$ denote the smallest set system which includes $\mathfrak{H}$ and is monotone closed in $\mathfrak{B}$. To each V in $\mathfrak{V}$, we associate the system $\mathfrak{K}(V)$ of those E in $\mathfrak{V}$ for which $E \setminus V$, $V \setminus E$, and $V \cup E$ are in $\mathfrak{V}$. Then obviously $E \in \mathfrak{K}(V)$ is equivalent to $V \in \mathfrak{K}(E)$.

For each monotone decreasing sequence (E_n) of sets in $\mathfrak{K}(V)$, we obtain

$$\left(\bigcap_{n=1}^{\infty} E_n\right) \setminus V = \bigcap_{n=1}^{\infty} (E_n \setminus V) \in \mathfrak{V},$$

$$V \setminus \left(\bigcap_{n=1}^{\infty} E_n\right) = \bigcup_{n=1}^{\infty} (V \setminus E_n) \in \mathfrak{V},$$

$$V \cup \left(\bigcap_{n=1}^{\infty} E_n\right) = \bigcap_{n=1}^{\infty} (V \cup E_n) \in \mathfrak{V},$$

i.e. we have

$$\bigcap_{n=1}^{\infty} E_n \in \mathfrak{K}(V).$$

Correspondingly, for each monotone increasing sequence (E_n') of sets in $\mathfrak{K}(V)$ with the property that $\bigcup_{n=1}^{\infty} E_n' \in \mathfrak{B}$, we obtain the relation $\bigcup_{n=1}^{\infty} E_n' \in \mathfrak{K}(V)$. Consequently, for all V in $\mathfrak{B}$, the set system $\mathfrak{K}(V)$ is monotone closed in $\mathfrak{B}$.

If V is in $\mathfrak{H}$, then the relation $E \in \mathfrak{K}(V)$ holds for all E in $\mathfrak{H}$. Thus we obtain $\mathfrak{H} \subseteq \mathfrak{K}(V)$, and hence $V \subseteq \mathfrak{K}(V)$. Consequently, for all V in $\mathfrak{H}$ and all E in $\mathfrak{B}$, we have $V \in \mathfrak{K}(E)$, from which we may further conclude that $V \subseteq \mathfrak{K}(E)$. Therefore the set system $\mathfrak{B}$ is a subring of $\mathfrak{B}$ which is monotone closed in $\mathfrak{B}$, and hence a σ-subring of $\mathfrak{B}$ which includes $\mathfrak{H}$, i.e. $\mathfrak{B} = \mathfrak{B}$, and hence also $\mathfrak{S} = \mathfrak{B}$. ∎

Let N be the set of those measures on $\mathfrak{A}$ which are finite on $\mathfrak{B}$. By to the relation

$$\delta_a(X) = k_X(a) \qquad (X \in \mathfrak{A})$$

a distribution δ_a in N corresponds in a one-to-one way to any a in A, the indicator function of X being denoted by k_X in the above formula as well as below.

A measure ν in N is said to be integer-valued if for all X in $\mathfrak{B}$ the relation $\nu(X) \in \{0, 1, \ldots\}$ is valid. Let M be the set of all integer-valued Φ in N. Obviously a ν in N belongs to M if it can be represented as a sum of an at most countable number of δ_a, $a \in A$. In this way the set M will already be exhausted; for we have:

1.1.2. Proposition. *For all Φ in M, the set $\{a : a \in A,\ \Phi(\{a\}) > 0\}$ is at most countable, and*

$$\Phi = \sum_{a \in A, \Phi(\{a\})>0} \Phi(\{a\})\, \delta_a .$$

Proof. For each X in $\mathfrak{B}$, the number of elements of $X \cap \{a : \Phi(\{a\}) > 0\}$ is at most equal to $\Phi(X)$, and hence finite, and since A can be covered by a countable number of X_n in $\mathfrak{B}$, this proves the first statement on $\{a : \Phi(\{a\}) > 0\}$. Now, if we set

$$\Phi' = \Phi - \sum_{a \in A, \Phi(\{a\})>0} \Phi(\{a\})\, \delta_a ,$$

then $\Phi' \in M$, and $\Phi'(\{a\}) = 0$ for all a in A. Let us suppose that $\Phi' \neq o$. Since A can be covered by a countable number of open balls $S_1(b_n)$ with the radius 1, there exists an a_1 in A with the property $\Phi'(S_1(a_1)) > 0$. If we employ a countable covering of $S_1(a_1)$ by open balls with the radius $1/2$, then it can be concluded that there exists an a_2 in A with the property $\Phi'(S_1(a_1) \cap S_{2^{-1}}(a_2)) > 0$. Now we cover $S_1(a_1) \cap S_{2^{-1}}(a_2)$ by a countable number of open balls with the radius 2^{-2}, and thus can infer the existence of an a_3 in A with the property $\Phi'(S_1(a_1) \cap S_{2^{-1}}(a_2) \cap S_{2^{-2}}(a_3)) > 0$. By repeating this procedure, we see: There exists a sequence (a_n) of elements of A, which for $n = 1, 2, \ldots$ satisfies the inequality:

$$\Phi'(S_1(a_1) \cap S_{2^{-1}}(a_2) \cap \cdots \cap S_{2^{-n}}(a_{n+1})) > 0 .$$

The (a_n) form a Cauchy sequence, since for $n, m = 1, 2, \ldots$

$$\begin{aligned}\varrho_A(a_n, a_{n+m}) &\leqq \varrho(a_n, a_{n+1}) + \cdots + \varrho(a_{n+m-1}, a_{n+m}) \\ &\leqq \frac{2}{2^{n-1}} + \cdots + \frac{2}{2^{n+m-2}} < \frac{1}{2^{n-3}}.\end{aligned}$$

Consequently (a_n) tends to an a in A. For $n = 1, 2, \ldots$ we have $S_{2^{-n}}(a_{n+1}) \subseteqq S_{2^{-n+3}}(a)$, which implies that $\Phi'\big(S_{1/n}(a)\big) > 0$ and therefore $\Phi'\big(S_{1/n}(a)\big) \geqq 1$. But this leads to $\Phi'(\{a\}) = \inf_n \Phi'\big(S_{1/n}(a)\big) \geqq 1$, in contradiction to the above. ∎

On the basis of 1.1.2., any Φ in M can be considered to describe the following situation: In A, the at most countable set $\{a : \Phi(\{a\}) > 0\}$ is distinguished, every point a in this set occurring with the multiplicity $\Phi(\{a\})$. Then for all X in $\mathfrak{A}$, the measure $\Phi(X)$ indicates the number of distinguished points in X, determined with consideration of the respective multiplicities. Obviously $\{a : \Phi(\{a\}) > 0\}$ dependent on $\Phi \in M$ runs over the entire family of those subsets of A the intersections of which with each bounded set are finite. The empty set $\emptyset$ corresponds to the zero measure o.

Let $\mathfrak{N}$ denote the smallest σ-algebra of subsets of N with respect to which the real-valued function

$$\nu \rightsquigarrow \nu(X) \qquad (\nu \in N)$$

is measurable for all X in $\mathfrak{B}$.

For all measures ν on $\mathfrak{A}$ and all X in $\mathfrak{A}$, let ${}_X\nu$ denote the measure $\nu\big((.) \cap X\big)$, so that the total mass lying outside of X is eliminated. Obviously $\nu \rightsquigarrow {}_X\nu$ provides a measurable mapping of $[N, \mathfrak{N}]$ into itself for all X in $\mathfrak{A}$.

We define a *random measure with phase space* $[A, \varrho_A]$ to be a random element of the measurable space $[N, \mathfrak{N}]$, i.e. a probability field $[N, \mathfrak{N}, P]$.

1.1.3. *M is an element of $\mathfrak{N}$.*

Proof. Let $\mathfrak{H}$ be an at most countable base of $[A, \varrho_A]$ consisting of bounded open subsets, and $\mathfrak{B}'$ the subring of $\mathfrak{B}$ generated by $\mathfrak{H}$. This subring is at most countable, too. Thus we have

$$\big\{\nu : \nu \in N, \nu(X) \in \{0, 1, \ldots\} \text{ for all } X \in \mathfrak{B}'\big\} \in \mathfrak{N}.$$

The family $\mathfrak{S}$ of all Z in $\mathfrak{B}$ with the property

$$\big\{\nu : \nu \in N, \nu(Z) \in \{0, 1, \ldots\}\big\} \supseteqq \big\{\nu : \nu \in N, \nu(X) \in \{0, 1, \ldots\} \text{ for all } X \in \mathfrak{B}'\big\}$$

is monotone closed in $\mathfrak{B}$ and includes $\mathfrak{B}'$. On the other hand, the σ-subring of $\mathfrak{B}$ generated by $\mathfrak{H}$ coincides with $\mathfrak{B}$. Consequently $\mathfrak{S} = \mathfrak{B}$, and we have

$$\begin{aligned}M &= \big\{\nu : \nu \in N, \nu(X) \in \{0, 1, \ldots\} \text{ for all } X \in \mathfrak{B}\big\} \\ &= \big\{\nu : \nu \in N, \nu(X) \in \{0, 1, \ldots\} \text{ for all } X \in \mathfrak{B}'\big\} \in \mathfrak{N}. \ ∎\end{aligned}$$

We set $\mathfrak{M} = M \cap \mathfrak{N}$, and thus obtain the smallest σ-algebra of subsets of M, with respect to which, for all X in $\mathfrak{B}$, the mapping

$$\Phi \rightsquigarrow \Phi(X)$$

of M into $\{0, 1, \ldots\}$ is measurable. Then obviously, for all X in $\mathfrak{A}$, the mapping $\Phi \curvearrowright \Phi(X)$ of M into $\{0, 1, \ldots, +\infty\}$ is measurable, too. In particular the set $\{\Phi : \Phi \in M, \Phi(A) = 1\}$ of all δ_a, $a \in A$, belongs to in $\mathfrak{M}$.

1.1.4. *The correspondence $a \curvearrowright \delta_a$ provides a one-to-one mapping of $[A, \mathfrak{A}]$ into $[M, \mathfrak{M}]$, which is measurable in both directions.*

Proof. Let $\mathfrak{S}$ denote the family of all inverse images of sets in $\mathfrak{M}$, with respect to the mapping $a \curvearrowright \delta_a$. Obviously $\mathfrak{A} \subseteq \mathfrak{S}$ since, for all X in $\mathfrak{A}$, X is the inverse image of $\{\Phi : \Phi \in M, \Phi(X) = 1\}$. For all Z in $\mathfrak{B}$, the inverse image of $\{\Phi : \Phi \in M, \Phi(Z) = n\}$ coincides with Z for $n = 1$, with the complement $\bar{Z}$ for $n = 0$, and with the empty set $\emptyset$ for all other values of n. Thus also conversely the inverse image of each Y in $\mathfrak{M}$ belongs to $\mathfrak{A}$, and we obtain $\mathfrak{S} = \mathfrak{A}$ from which our proposition can be immediately inferred. ∎

For any X in A we denote by ${}_X\mathfrak{M}$ the σ-algebra generated by the family of all sets $\{\Phi : \Phi \in M, \Phi(Z) = k\}$ with Z in $X \cap \mathfrak{B}$ and $k \in \{0, 1, \ldots\}$; thus the σ-subalgebra ${}_X\mathfrak{M}$ comprises all statements concerning the behaviour of Φ on the set X. Obviously $\Phi \curvearrowright {}_X\Phi$ provides a measurable mapping of $[M, \mathfrak{M}]$ into itself for all X in $\mathfrak{A}$, ${}_X\mathfrak{M}$ being the inverse image of $\mathfrak{M}$ with respect to this mapping. Thus this mapping transforms each measure L on $\mathfrak{M}$ into a new measure on $\mathfrak{M}$, which is denoted by ${}_XL$.

We define a *point process with phase space* $[A, \varrho_A]$ to be a random element of the measurable space $[M, \mathfrak{M}]$, i.e. a probability field $[M, \mathfrak{M}, P]$.

According to 1.1.3. we can conceive a point process with phase space $[A, \varrho_A]$ to be a random measure $[N, \mathfrak{N}, V]$ having the property $V(M) = 1$.

Let $\boldsymbol{P}$ denote the set of all probability distributions on $\mathfrak{M}$, and $\boldsymbol{\mathfrak{P}}$ the smallest σ-algebra of subsets of $\boldsymbol{P}$ with respect to which all real-valued functions

$$P \curvearrowright P(Y) \qquad (P \in \boldsymbol{P})$$

are measurable for all $Y \in \mathfrak{M}$. Obviously $\Phi \curvearrowright \delta_\Phi$ is a measurable mapping of $[M, \mathfrak{M}]$ into $[\boldsymbol{P}, \boldsymbol{\mathfrak{P}}]$, and hence $a \curvearrowright \delta_{\delta_a}$ is a measurable mapping of $[A, \mathfrak{A}]$ into $[\boldsymbol{P}, \boldsymbol{\mathfrak{P}}]$.

A measure L on $\mathfrak{M}$ is called *continuous*, if for all a in A the relation $L\big(\Phi(\{a\}) > 0\big) = 0$ is valid. Thus in particular δ_Φ is continuous if and only if $\Phi = o$.

1.1.5. Proposition. *For all P in $\boldsymbol{P}$, the set $\big\{a : a \in A,\ P\big(\Phi(\{a\}) > 0\big) > 0\big\}$ is at most countable.*

Proof. Suppose that the set $\big\{a : a \in A, P\big(\Phi(\{a\}) > 0\big) > 0\big\}$ be uncountable. Since A can be covered by a sequence (X_n) of sets in $\mathfrak{B}$, there exists an X in $\mathfrak{B}$ for which the set $X \cap \big\{a : a \in A, P\big(\Phi(\{a\}) > 0\big) > 0\big\}$ is uncountable. Further we can conclude that there exist a $c > 0$ as well as a sequence (a_n) of pairwise distinct points in X such that all sets $H_{a_n} = \big\{\Phi : \Phi(\{a_n\}) > 0\big\}$ satisfy the condition $P(H_{a_n}) \geqq c$. Let Y_k denote the set of those Φ in M which are at least $(k + 1)$-fold covered by the sequence (H_{a_n}). The Y_k form a monotone decreasing sequence of sets in $\mathfrak{M}$.

Now, if for any k the probability $P(Y_k) = d_k$ were less than c, then for all n we would have the inequality $P(H_{a_n} \setminus Y_k) \geqq c - d_k > 0$. On the other hand, each Φ in M would be at most k-fold covered by the sequence $(H_{a_n} \setminus Y_k)$, which would imply that

$$1 \geqq P\left(\bigcup_{n=1}^{\infty} (H_{a_n} \setminus Y_k)\right) \geqq \frac{1}{k} \sum_{n=1}^{\infty} P(H_{a_n} \setminus Y_k) = +\infty .$$

So we conclude that $P(Y_k) \geqq c$ for $k = 1, 2, \ldots$, and consequently $P\left(\bigcap_{k=1}^{\infty} Y_k\right) \geqq c$. However, $\bigcap_{k=1}^{\infty} Y_k$ coincides with the set of all Φ in M having the property $\Phi(\{a_1, a_2, \ldots\}) = +\infty$. This gives $P(\Phi(X) = +\infty) \geqq c$, in contradiction to the definition of M. ∎

According to 1.1.5., for each P in $\boldsymbol{P}$ the set $X_P = \{a : a \in A, P(\Phi(\{a\}) > 0) = 0\}$ belongs to $\mathfrak{A}$. We call

$$P_c = {}_{X_P}P$$

the *continuous part* of P.

A probability distribution P on $\mathfrak{M}$ is called *discrete*, if $P_c = \delta_o$, i.e. if an at most countable set Z with the property ${}_ZP = P$ exists. The smallest possible set of this kind is $\overline{X_P} = A \setminus X_P$. For arbitrary P in $\boldsymbol{P}$ we call $P_d = {}_{\overline{X_P}}P$ the *discrete part* of P. Any P in $\boldsymbol{P}$ is continuous if and only if $P_d = \delta_o$ is valid. Obviously δ_o is the only distribution being both continuous and discrete at the same time.

Let Z be an arbitrary, at most countable subset of A. Then to each family $(y_z)_{z \in Z}$ of non-negative integers we can assign a measure on $\mathfrak{A}$ by means of $\Phi = \sum_{z \in Z} y_z \delta_z$. Now, if z_0 is any of the points in A, then Φ is in M if and only if the condition

$$(k) \qquad \sum_{z \in Z \cap S_n(z_o)} y_z < +\infty \qquad (n = 1, 2, \ldots)$$

is satisfied. Conversely, a Φ in M can be represented as an image of a family $(y_z)_{z \in Z}$ if and only if Φ satisfies the condition $\Phi(\overline{Z}) = 0$, i.e. if Φ coincides with ${}_Z\Phi$.

Obviously the set Y of all $(y_z)_{z \in Z}$ satisfying the condition (k) is included in the Kolmogorov σ-algebra $\left(\mathfrak{P}(\{0, 1, \ldots\})\right)^Z$. The correspondence

$$(y_z)_{z \in Z} \curvearrowright \Phi = \sum_{z \in Z} y_z \delta_z \qquad \left((y_z)_{z \in Z} \in Y\right)$$

is measurable in both directions. For, due to $y_z = \Phi(\{z\})$, each y_z, and hence also the family $(y_z)_{z \in Z}$, is a measurable function of Φ. Conversely, $\Phi(X) = \sum_{z \in Z \cap X} y_z$ is a measurable function of $(y_z)_{z \in Z}$ for all X in $\mathfrak{B}$, and consequently also Φ itself depends measurably on $(y_z)_{z \in Z}$. The discrete probability distributions on $\mathfrak{M}$ have therefore the following simple structure:

1.1.6. *For each at most countable subset Z of A, the mapping $(y_z)_{z \in Z} \curvearrowright \sum_{z \in Z} y_z \delta_z$ from $\{0, 1, \ldots\}^Z$ into M provides a one-to-one correspondence between the* *distributions p of families $(\eta_z)_{z\in Z}$ of random non-negative integers which have the property*

$$p\left(\sum_{z\in Z\cap S_n(z_0)} \eta_z < +\infty\right) = 1 \qquad (n = 1, 2, \ldots; \qquad z_0 \text{ arbitrary, fixed})$$

and the distributions P on $\mathfrak{M}$ which have the property $P = {}_ZP$.

In the special case $A = \{1, \ldots, m\}$, the correspondence

$$\Phi \curvearrowright [\Phi(\{1\}), \ldots, \Phi(\{m\})]$$

provides a one-to-one mapping of M onto the countable set of all m-dimensional row vectors with non-negative integer coordinates. In what follows, we shall always identify Φ with the corresponding vector, i.e. $\sum_{k=1}^{m} y_k\delta_k$ with $[y_1, \ldots, y_m]$. *In particular each random m-dimensional vector with non-negative integer coordinates will accordingly be interpreted as a point process with phase space* $\{1, \ldots, m\}$.

1.2. The Intensity Measure

To each measure L on $\mathfrak{M}$ we assign a measure ϱ_L on $\mathfrak{A}$ by means of the relation

$$\varrho_L(X) = \int \Phi(X)\, L(\mathrm{d}\Phi) \qquad (X \in \mathfrak{A}).$$

We call ϱ_L the *intensity measure* of L, and say that L is *of finite intensity* if ϱ_L is in N.

Obviously the relation $\varrho_{\delta_\Phi} = \Phi$ is always valid. Any P in $\boldsymbol{P}$ satisfies the condition $\varrho_P = o$ if and only if P is equal to δ_o. In the special case $A = \{1, \ldots, m\}$, ϱ_P is equal to $\sum_{i=1}^{m} \left(\int \Phi(\{i\})\, P(\mathrm{d}\Phi)\right) \delta_i$, and consequently P is of finite intensity if and only if all coordinates of the corresponding random m-dimensional vector have finite expectations.

A distribution P on $\mathfrak{M}$ is continuous if and only if ϱ_P is *diffuse* (*non-atomic*), i.e. if $\varrho_P(\{a\}) = 0$ is satisfied for all a in A.

By virtue of 1.1.4., the mapping $a \curvearrowright \delta_a$ bi-uniquely transforms the measures ν on $\mathfrak{A}$ into the measures L on $\mathfrak{M}$ with the property $L\big(\Phi(A) \neq 1\big) = 0$. Then obviously $\varrho_L = \nu$. Consequently L is of finite intensity if and only if ν is in N. In this case we denote L by Q_ν. The correspondence $\nu \curvearrowright Q_\nu$ thus provides a one-to-one mapping of N onto the set of those measures L on $\mathfrak{M}$ which are of finite intensity and satisfy the condition $L\big(\Phi(A) \neq 1\big) = 0$, the corresponding relation being

$$\varrho_{Q_\nu} = \nu.$$

Then Q_ν is in $\boldsymbol{P}$ if and only if $\nu(A) = 1$, and can in this case be interpreted as a distribution of a point in A with multiplicity one which is distributed according to ν.

For each measurable mapping $c \curvearrowright f_{(c)}$ of a measurable space $[C, \mathfrak{C}]$ into $[\boldsymbol{P}, \mathfrak{P}]$ and for each distribution L on $\mathfrak{C}$, the distribution

$$Y \curvearrowright \int f_{(c)}(Y)\, L(\mathrm{d}c) \qquad (Y \in \mathfrak{M})$$

is denoted by $\int f_{(c)}(.)\, L(\mathrm{d}c)$. Then also $c \curvearrowright {}_X(f_{(c)})$ is measurable for all X in $\mathfrak{A}$, and we have

$${}_X\left(\int f_{(c)}(.)\, L(\mathrm{d}c)\right) = \int {}_X(f_{(c)})(.)\, L(\mathrm{d}c).$$

For each finite sequence $Y_1, \ldots, Y_m$ of sets in $\mathfrak{M}$ and all non-negative real $v_1, \ldots, v_m$, the linear combination $g = \sum_{i=1}^{m} v_i k_{Y_i}$ has the following two properties:

a) The real function $c \curvearrowright \int g(\Phi)\, f_{(c)}(\mathrm{d}\Phi)$ is measurable.
b) $\int \left(\int g(\Phi)\, f_{(c)}(\mathrm{d}\Phi)\right) L(\mathrm{d}c) = \int g(\Phi)\, P(\mathrm{d}\Phi)$,

where we set $P = \int f_{(c)}(.)\, L(\mathrm{d}c)$. These two properties are transferred to all suprema of monotone increasing sequences (g_n), and hence especially to the functions $\Phi \curvearrowright \Phi(X)$, $X \in \mathfrak{A}$. Thus we see the validity of

1.2.1. *Subject to the above-mentioned assumptions, for each X in $\mathfrak{A}$ the mapping $c \curvearrowright \varrho_{f_{(c)}}(X)$ of C into $[0, +\infty]$ is measurable with respect to $\mathfrak{C}$, and*

$$\varrho_{\int f_{(c)}(.) L(\mathrm{d}c)}(X) = \int \varrho_{f_{(c)}}(X)\, L(\mathrm{d}c).$$

The equation defining ϱ_L can be rewritten to

$$\varrho_L(X) = \int \left(\int k_X(a)\, \Phi(\mathrm{d}a)\right) L(\mathrm{d}\Phi) \qquad (X \in \mathfrak{A}).$$

Using the theorem of B. Levi, the following version of the so-called Campbell theorem can be inferred from the above equation:

1.2.2. *For all measures L defined on $\mathfrak{M}$, and for all non-negative real valued functions f which are measurable with respect to $\mathfrak{A}$, the mapping*

$$\Phi \curvearrowright \int f(a)\, \Phi(\mathrm{d}a)$$

of M into $[0, +\infty]$ is measurable, and

$$\int f(a)\, \varrho_L(\mathrm{d}a) = \int \left(\int f(a)\, \Phi(\mathrm{d}a)\right) L(\mathrm{d}\Phi).$$

For all natural numbers n, the set $A' = A^n$ can be complemented by means of the set-up

$$\varrho_{A'}([a_1, \ldots, a_n], [a_1', \ldots, a_n']) = \sum_{i=1}^{n} \varrho_A(a_i, a_i') \qquad (a_1, \ldots, a_n, a_1', \ldots, a_n' \in A)$$

to form a complete separable metric space. Then to each Φ in M we can assign an integer-valued measure Φ' on the σ-algebra $\mathfrak{A}'$ of the Borel subsets of $[A', \varrho_{A'}]$, using the correspondence

$$\Phi \curvearrowright \Phi' = \underbrace{\Phi \times \cdots \times \Phi}_{n-\text{times}}.$$

For all $X_1, \ldots, X_n$ in $\mathfrak{B}$, $\Phi'(X_1 \times \ldots \times X_n) = \prod_{i=1}^{n} \Phi(X_i)$ is finite. For all X in the ring $\mathfrak{B}'$ of the bounded sets in $\mathfrak{A}'$, we thus have $\Phi'(X) < +\infty$, i.e. Φ' is always in the set M' of counting measures on A', associated with the phase space $[A', \varrho_{A'}]$.

For all $X_1, \ldots, X_n$ in $\mathfrak{B}$, $\Phi'(X_1 \times \ldots \times X_n)$ is a measurable function of Φ. Consequently, for all X in $\mathfrak{B}'$, the real-valued function $\Phi \curvearrowright \Phi'(X)$ is measurable, i.e. the mapping $\Phi \curvearrowright \Phi'$ of $[M, \mathfrak{M}]$ into $[M', \mathfrak{M}']$ is measurable. Here $\mathfrak{M}'$ denotes the σ-algebra of subsets of M', which corresponds to the phase space $[A', \varrho_{A'}]$. Hence by means of $\Phi \curvearrowright \Phi'$, to each measure L on $\mathfrak{M}$ a measure $L^{\langle n \rangle}$ on $\mathfrak{M}'$ can be assigned. The intensity measure $\varrho_{L^{\langle n \rangle}}$ on $\mathfrak{A}'$ is called the *n-th moment measure of L*. For $n = 1$ it coincides with ϱ_L. Generally for any $X_1, \ldots, X_n$ in $\mathfrak{B}$,

$$\varrho_{L^{\langle n \rangle}}(X_1 \times \cdots \times X_n) = \int \left(\prod_{i=1}^{n} \Phi(X_i) \right) L(\mathrm{d}\Phi),$$

which justifies our notation.

Here we shall not further develop the theory of the higher-order moment measures; instead we refer to the more detailed presentations in Daley & Vere-Jones [1], Krickeberg [1], Lenard [1], [3] and Belyaev [2].

1.3. The Finite-dimensional Distributions

Let L be a measure on $\mathfrak{M}$ and $X_1, \ldots, X_m$ a finite sequence of sets in $\mathfrak{A}$. If L satisfies the condition $L(\Phi(X_i) = +\infty) = 0$ for $1 \leqq i \leqq m$, which is always satisfied if $X_i \in \mathfrak{B}$, then the measurable mapping

$$\Phi \curvearrowright [\Phi(X_1), \ldots, \Phi(X_m)]$$

of M into $\{0, 1, \ldots, +\infty\}$ transforms the measure L into a measure $L_{X_1,\ldots,X_m}$ on $\mathfrak{P}(\{0, 1, \ldots\}^m)$. Together with $L_{X_1,\ldots,X_m}$, $({}_X L)_{X_1,\ldots,X_m}$ also exists for all X in $\mathfrak{A}$ and is equal to $L_{X \cap X_1,\ldots,X \cap X_m}$.

For all $X_1, \ldots, X_m$ in $\mathfrak{B}$, let $\mathfrak{M}_{X_1,\ldots,X_m}$ denote the σ-algebra generated by the family of all sets $\{\Phi : \Phi \in M, \Phi(X_1) = k_1, \ldots, \Phi(X_m) = k_m\}$ with $k_i \in \{0, 1, \ldots\}$, $i = 1, \ldots, m$, i.e. the smallest σ-algebra of subsets of M with respect to which the mappings

$$\Phi \curvearrowright \Phi(X_i) \qquad (1 \leqq i \leqq m)$$

are measurable.

We have

1.3.1. *Let $\mathfrak{H}$ be a generating semiring of $\mathfrak{B}$. Then the union $\mathfrak{Y}$ of all σ-algebras $\mathfrak{M}_{H_1,\ldots,H_m}$ with $H_1, \ldots, H_m$ pairwise disjoint sets in $\mathfrak{H}$ is a generating subalgebra of $\mathfrak{M}$.*

Proof. For all $H_1, \ldots, H_k, H_{k+1}, \ldots, H_{k+s}$ in $\mathfrak{H}$ there exists a finite sequence $H_1', \ldots, H_m'$ of pairwise disjoint sets in $\mathfrak{H}$ such that

$$\mathfrak{M}_{H_1,\ldots,H_k} \cup \mathfrak{M}_{H_{k+1},\ldots,H_{k+s}} \subseteq \mathfrak{M}_{H_1',\ldots,H_m'}.$$

This implies that $\mathfrak{Y}$ is a subalgebra of $\mathfrak{M}$. Now let $\mathfrak{S}$ denote the family of all sets C in $\mathfrak{B}$ having the property that $\{\Phi : \Phi \in M, \Phi(C) = k\}$ belongs to the σ-algebra $\mathfrak{Y}'$ generated by $\mathfrak{Y}$. Because of the definition of $\mathfrak{Y}$ the family $\mathfrak{S}$ includes $\mathfrak{H}$ as well as the subring $\mathfrak{H}^*$ of $\mathfrak{B}$ which is generated by $\mathfrak{H}$, i.e. the family of all unions of finite numbers of pairwise disjoint sets in $\mathfrak{H}$. Bearing in mind that the Φ in M are integer-valued it is obvious that $\mathfrak{S}$ is monotone closed in $\mathfrak{B}$. Consequently $\mathfrak{S}$ coincides with $\mathfrak{B}$ and hence $\mathfrak{Y}$ generates $\mathfrak{M}$. ∎

If L is in $\boldsymbol{P}$, then the $L_{X_1,\ldots,X_m}$ are called the *finite-dimensional distributions of L*. Bearing in mind the pertinent remarks in section 1.1., we can consider the distribution $L_{X_1,\ldots,X_m}$ on $\mathfrak{P}(\{0, 1, \ldots\}^m)$ to be the distribution of a point process with phase space $\{1, \ldots, m\}$, i.e. $L_{X_1,\ldots,X_m}$ can be interpreted as the distribution of $\Phi' = \sum_{i=1}^{m} \Phi(X_i)\, \delta_i$ which has been established with respect to L.

As an immediate consequence of 1.3.1. we obtain the following uniqueness proposition.

1.3.2. Proposition. *Let $\mathfrak{H}$ be a generating semiring of sets in $\mathfrak{B}$. If the distributions P, Q on $\mathfrak{M}$ satisfy*

$$P_{H_1,\ldots,H_m} = Q_{H_1,\ldots,H_m},$$

for all finite sequences $H_1, \ldots, H_m$ of pairwise disjoint sets in $\mathfrak{H}$, then $P = Q$.

In the special case $[A, \varrho_A] = R^s$ we can choose for $\mathfrak{H}$ the system of all direct products of half-open intervals $[a, b)$ with $-\infty < a \leqq b < +\infty$.

A distribution P on $\mathfrak{M}$ is said to be *free from after-effects* if for all finite sequences $X_1, \ldots, X_m$ of pairwise disjoint sets in $\mathfrak{B}$ the relation $P_{X_1,\ldots,X_m} = P_{X_1} \times \cdots \times P_{X_m}$ is valid, i.e. if the coordinates of the random vector $[\Phi(X_1), \ldots, \Phi(X_m)]$ are independent with respect to P. Obviously, if P is free from after-effects, then this is true also for ${}_XP$ for all X in $\mathfrak{A}$. Consequently, if P is free from after-effects, this is true also for both the continuous and the discrete parts of P. For any at most countable subset Z of A, a distribution P on $\mathfrak{M}$ with the property ${}_ZP = P$ is free from after-effects if and only if the random family $(\eta_z)_{z\in Z}$ constructed according to 1.1.6. is independent.

The definition of freedom from after-effects can be weakened as follows:

1.3.3. *Let $\mathfrak{H}$ be a generating semiring of sets in $\mathfrak{B}$. Then a distribution P on $\mathfrak{M}$ is free from after-effects if and only if for all finite sequences $H_1, \ldots, H_n$ of pairwise disjoint sets in $\mathfrak{H}$ the relation $P_{H_1,\ldots,H_n} = P_{H_1} \times \cdots \times P_{H_n}$ is valid.*

Proof. We denote by $\mathfrak{G}$ the subring of $\mathfrak{B}$ generated by $\mathfrak{H}$. Now let $X_1, \ldots, X_m$ be any finite sequence of pairwise disjoint sets in $\mathfrak{G}$. Each X_i can be represented as a union of a finite number of pairwise disjoint $H_{i,j}$ $(1 \leqq j \leqq m_i)$ in $\mathfrak{H}$. By assumption the family

$$\left(\Phi(H_{i,j})\right)_{\substack{1\leq i\leq m\\ 1\leq j\leq m_i}}$$

is independent with respect to P. Hence also

$$\Big(\sum_{1\leqq j\leqq m_i} \Phi(H_{i,j})\Big)_{1\leqq i\leqq m} = \big(\Phi(X_i)\big)_{1\leqq i\leqq m}$$

is independent, i.e. we have

$$P_{X_1,\ldots,X_m} = P_{X_1}\times\cdots\times P_{X_m}.$$

For a given sequence $X_2, \ldots, X_m$ of pairwise disjoint sets in $\mathfrak{G}$, let $\mathfrak{V}$ denote the family of those C in $\mathfrak{B}$ for which $\Phi\left(C\setminus\bigcup_{i=2}^m X_i\right)$, $\Phi(X_2), \ldots, \Phi(X_m)$ is independent with respect to P. If a sequence (C_n) of sets in $\mathfrak{V}$ tends monotonically to some C in $\mathfrak{M}$, then for all Φ

$$\Phi\left(C_n\setminus\bigcup_{i=2}^m X_i\right) \quad \text{converges toward} \quad \Phi\left(C\setminus\bigcup_{i=2}^m X_i\right),$$

which gives

$$\begin{aligned}
&P\left(\Phi\left(C\setminus\bigcup_{i=2}^m X_i\right) = l_1,\ \Phi(X_2) = l_2,\ \ldots,\ \Phi(X_m) = l_m\right)\\
&= \lim_{n\to\infty} P\left(\Phi\left(C_n\setminus\bigcup_{i=2}^m X_i\right) = l_1,\ \Phi(X_2) = l_2,\ \ldots,\ \Phi(X_m) = l_m\right)\\
&= \lim_{n\to\infty} P\left(\Phi\left(C_n\setminus\bigcup_{i=2}^m X_i\right) = l_1\right)\prod_{i=2}^m P\big(\Phi(X_i) = l_i\big)\\
&= P\left(\Phi\left(C\setminus\bigcup_{i=2}^m X_i\right) = l_1\right)\prod_{i=2}^m P\big(\Phi(X_i) = l_i\big).
\end{aligned}$$

Consequently, $\mathfrak{V}$ is monotone closed in $\mathfrak{B}$. On the other hand $\mathfrak{V}$ includes the ring $\mathfrak{G}$, and we obtain $\mathfrak{V} = \mathfrak{B}$.

By induction on k we observe: For all finite sequences $C_1, \ldots, C_k$ of pairwise disjoint sets in $\mathfrak{B}$ and all finite sequences $X_{k+1}, \ldots, X_m$ of pairwise disjoint sets in $\mathfrak{G}$, the sequence

$$\Phi\left(C_1\setminus\bigcup_{i=k+1}^m X_i\right), \ldots, \Phi\left(C_k\setminus\bigcup_{i=k+1}^m X_i\right),\ \Phi(X_{k+1}), \ldots,\ \Phi(X_m)$$

is independent. For $m = k$, this implies in particular the independence of $\Phi(C_1), \ldots, \Phi(C_k)$. ∎

A distribution P on $\mathfrak{M}$ is free from after-effects if and only if the behaviour of Φ on disjoint sets in $\mathfrak{A}$ is independent:

1.3.4. Proposition. *For all P in $\mathbf{P}$, the following properties are equivalent:*

a) *P is free from after-effects.*
b) *For all finite sequences $X_1, \ldots, X_m$ of pairwise disjoint sets in $\mathfrak{A}$, the sequence ${}_{X_1}\Phi, \ldots, {}_{X_m}\Phi$ is independent with respect to P.*
c) *For all X in $\mathfrak{A}$, ${}_X\Phi$ is independent of ${}_{\bar{X}}\Phi$ with respect to P.*

Proof. 1. a) implies b):

Let $X_1, \ldots, X_m$ be any finite sequence of pairwise disjoint sets in $\mathfrak{A}$. We have to prove that the sequence of σ-algebras ${}_{X_1}\mathfrak{M}, \ldots, {}_{X_m}\mathfrak{M}$ is independent.

If now the union of all $\mathfrak{M}_{Z_1,\ldots,Z_k}$ for pairwise disjoint $Z_1, \ldots, Z_k$ in $\mathfrak{B}$, with $Z_1, \ldots, Z_k \subseteq X_i$, is denoted by ${}_i\mathfrak{M}$, then we obtain subalgebras ${}_i\mathfrak{M}$ of ${}_{X_i}\mathfrak{M}$ for $1 \leqq i \leqq m$. The smallest σ-algebra including ${}_i\mathfrak{M}$ is ${}_{X_i}\mathfrak{M}$. Hence it will suffice to prove the independence of the sequence ${}_1\mathfrak{M}, \ldots, {}_m\mathfrak{M}$. On the basis of the definition of the ${}_i\mathfrak{M}$, it remains only to show that the following independence property is true: If to each i in $\{1, \ldots, m\}$ there corresponds a finite sequence $Z_{i,1}, \ldots, Z_{i,m_i}$ of pairwise disjoint sets in $\mathfrak{B}$ which are included in X_i, then the family $\left([\Phi(Z_{i,1}), \ldots, \Phi(Z_{i,m_i})]\right)_{1 \leqq i \leqq m}$ is independent with respect to P. But this is an immediate consequence of the independence of $\left(\Phi(Z_{i,j})\right)_{\substack{1 \leqq i \leqq m \\ 1 \leqq j \leqq m_i}}$.

2. Obviously c) is a weaker version of b).

3. c) implies a): Let $X_1, \ldots, X_m$ be a finite sequence of disjoint sets in $\mathfrak{B}$. By assumption, for $1 \leqq i \leqq m$ the random number $\Phi(X_i) = ({}_{X_i}\Phi)(X_i)$ is independent of the family

$$\left(\Phi(X_j)\right)_{\substack{1 \leqq j \leqq m \\ j \neq i}} = \left(({}_{\bar{X}_i}\Phi)(X_j)\right)_{\substack{1 \leqq j \leqq m \\ j \neq i}}.$$

Consequently $\left(\Phi(X_i)\right)_{1 \leqq i \leqq m}$ is independent. ∎

In virtue of the uniqueness proposition 1.3.2., a distribution P in $\boldsymbol{P}$ is uniquely determined by its one-dimensional distributions P_X, $X \in \mathfrak{B}$, and the requirement of freedom from after-effects. Generally, $P \in \boldsymbol{P}$ is not uniquely determined by its one-dimensional distributions, as can be seen from the following simple example:

Let p be the direct product $\pi_1 \times \pi_1$ of two Poisson distributions π_1 with unit expectation, and P the corresponding distribution on the σ-algebra $\mathfrak{M}$ corresponding to the phase space $A = \{1, 2\}$. This distribution is free from after-effects. We can find a family of pairwise different distributions q of two-dimensional random vectors with non-negative integer coordinates, such that for the corresponding distributions Q on $\mathfrak{M}$ the relations $P_\emptyset = Q_\emptyset$, $P_{\{1\}} = Q_{\{1\}}$, $P_{\{2\}} = Q_{\{2\}}$, $P_{\{1,2\}} = Q_{\{1,2\}}$ are satisfied; this is accomplished by setting

$$q(\{[0, 1]\}) = e^{-2} + u, \qquad q(\{[1, 0]\}) = e^{-2} - u,$$

$$q(\{[0, 2]\}) = \frac{e^{-2}}{2} - u, \qquad q(\{[2, 0]\}) = \frac{e^{-2}}{2} + u,$$

$$q(\{[1, 2]\}) = \frac{e^{-2}}{2} + u, \qquad q(\{[2, 1]\}) = \frac{e^{-2}}{2} - u,$$

as well as $q(\{[i, j]\}) = p(\{[i, j]\})$ for all other $[i, j]$ with $i, j \geqq 0$, where the parameter u of the family satisfies the inequality $-\frac{e^{-2}}{2} \leqq u \leqq \frac{e^{-2}}{2}$ (cf. Lee [3]).

The family $\left(P_{X_1,\ldots,X_m}\right)$ of the finite-dimensional distributions of any P in $\boldsymbol{P}$ has simple compatibility and continuity properties. The following existence theorem shows that these properties are characteristic, and thus represents a means for constructing distributions on $\mathfrak{M}$.

1.3.5. Theorem.[1]) *Let $\mathfrak{H}$ be a generating semiring of sets in $\mathfrak{B}$ such that each X in $\mathfrak{B}$ can be covered by a finite sequence $H_1, \ldots, H_k$ of sets in $\mathfrak{H}$. To each finite sequence $H_1, \ldots, H_m$ of pairwise disjoint sets in $\mathfrak{H}$, let there be associated the distribution $p^{H_1,\ldots,H_m}$ of a random m-dimensional vector with non-negative integer coordinates. Subject to these assumptions, there exists a P in $\mathbf{P}$ with the property $P_{H_1,\ldots,H_m} = p^{H_1,\ldots,H_m}$ if and only if the following conditions are satisfied:*

a) *For each permutation $i_1, \ldots, i_m$ of $1, \ldots, m$ and all non-negative integer $l_1, \ldots, l_m$:*

$$p^{H_1,\ldots,H_m}(\{[l_1, \ldots, l_m]\}) = p^{H_{i_1},\ldots,H_{i_m}}(\{[l_{i_1}, \ldots, l_{i_m}]\}).$$

b) *For all non-negative $l_1, \ldots, l_{m-1}$:*

$$p^{H_1,\ldots,H_m}(\{[v_1, \ldots, v_m]: v_1 = l_1, \ldots, v_{m-1} = l_{m-1}, \ v_m \geqq 0\})$$
$$= p^{H_1,\ldots,H_{m-1}}(\{[l_1, \ldots, l_{m-1}]\}).$$

c) *If the union H of the sets $H_k, \ldots, H_m$ is in $\mathfrak{H}$, then all non-negative $l_1, \ldots, l_k$ satisfy the relation:*

$$p^{H_1,\ldots,H_{k-1},H}(\{[l_1, \ldots, l_k]\})$$
$$= p^{H_1,\ldots,H_m}(\{[v_1, \ldots, v_m]: v_1 = l_1, \ldots, v_{k-1} = l_{k-1}, v_k + \cdots + v_m = l_k\}).$$

d) *For all sequences $(H_{n,1} \cup \cdots \cup H_{n,m_n})_{n=1,2,\ldots}$ of finite unions of pairwise disjoint sets in $\mathfrak{H}$ which decrease monotonically toward $\emptyset$, the following relation holds:*

$$\lim_{n\to\infty} p^{H_{n,1},\ldots,H_{n,m_n}}(\{[0, \ldots, 0]\}) = 1.$$

Proof. 1. Obviously the finite-dimensional distributions $P_{H_1,\ldots,H_m}$ of a P in $\mathbf{P}$ satisfy the conditions a), b), and c). The corresponding assertion for d) results immediately from the convergence encountered for all Φ

$$\Phi(H_{n,1} \cup \cdots \cup H_{n,m_n}) \xrightarrow[n\to\infty]{} 0.$$

2. Conversely, suppose now that the correspondence $[H_1, \ldots, H_m] \frown p^{H_1,\ldots,H_m}$ has the above-mentioned properties.

Let $X_1, \ldots, X_s$ be a finite sequence of sets in $\mathfrak{H}$ which are not necessarily pairwise disjoint. By assumption there exists a finite sequence $H_1, \ldots, H_m$ of pairwise disjoint sets in $\mathfrak{H}$ such that all X_i can be represented as unions of certain H_j:

$$X_i = \bigcup_{j\in J_i} H_j \qquad (1 \leqq i \leqq s).$$

If we now set for all non-negative integer $l_1, \ldots, l_s$:

$$p^{X_1,\ldots,X_s}(\{[l_1, \ldots, l_s]\})$$
$$= p^{H_1,\ldots,H_m}\left(\left\{[n_1, \ldots, n_m]: \sum_{j\in J_i} n_j = l_i \text{ for } 1 \leqq i \leqq s\right\}\right)$$

---

[1] Cf. Harris [1], [2].

then, in virtue of a), b), and c), this value no longer depends on the selection of the finite sequence $H_1, \ldots, H_m$. If in particular the $X_1, \ldots, X_s$ are pairwise disjoint, then for $H_1, \ldots, H_m$, we can choose the original sequence $X_1, \ldots, X_s$. The correspondence $[X_1, \ldots, X_s] \rightsquigarrow p^{X_1,\ldots,X_s}$ has the properties a), b), c), and d).

3. Using Kolmogorov's existence theorem, a) and b) imply the existence of one and only one distribution Q on the measurable space $\left[\{0, 1, \ldots\}^{\mathfrak{H}}, \left(\mathfrak{P}(\{0, 1, \ldots\})\right)^{\mathfrak{H}}\right]$ with the finite-dimensional distributions $p^{X_1,\ldots,X_s}$.

Let S denote the quotient space of the linear space of all real functions on $\{0, 1, \ldots\}^{\mathfrak{H}}$ which are measurable with respect to $\left(\mathfrak{P}(\{0, 1, \ldots\})\right)^{\mathfrak{H}}$, the quotient space being taken with respect to the subspace of the functions vanishing almost everywhere with respect to Q. For all Z of $\mathfrak{H}$, let $\mathfrak{K}(Z)$ be the residue class in S corresponding to the projection $x = (x_H)_{H \in \mathfrak{H}} \rightsquigarrow x_Z$.

The mapping $H \rightsquigarrow \mathfrak{K}(H)$ of $\mathfrak{H}$ into S is finitely additive. For if the union Z of a finite sequence $H_1, \ldots, H_m$ of pairwise disjoint sets in $\mathfrak{H}$ belongs to $\mathfrak{H}$, let Q' denote the distribution, as defined with respect to Q, of the reduced family $(x_H)_{\substack{H \in \mathfrak{H} \\ H \neq Z}}$. Then in virtue of c), for the mapping $(y_H)_{\substack{H \in \mathfrak{H} \\ H \neq Z}} \rightsquigarrow (x_H)_{H \in \mathfrak{H}}$ defined by

$$x_H = y_H \quad \text{for} \quad H \in \mathfrak{H},\ H \neq Z, \quad \text{and} \quad x_Z = y_{H_1} + \cdots + y_{H_m},$$

Q' induces on $\left(\mathfrak{P}(\{0, 1, \ldots\})\right)^{\mathfrak{H}}$ a distribution with the same finite-dimensional distributions as Q, i.e. the distribution Q itself. Consequently the following relation is true with probability one:

$$x_Z = x_{H_1} + \cdots + x_{H_m},$$

that means

$$\mathfrak{K}(Z) = \mathfrak{K}(H_1) + \cdots + \mathfrak{K}(H_m)$$

Each set X in the ring $\mathfrak{H}^*$ generated by $\mathfrak{H}$ can be represented as a finite union of pairwise disjoint sets $H_1, \ldots, H_m$ in $\mathfrak{H}$, and by means of the set-up

$$\mathfrak{K}(X) = \mathfrak{K}(H_1) + \cdots + \mathfrak{K}(H_m).$$

$\mathfrak{K}$ is extended to a finitely additive mapping of $\mathfrak{H}^*$ into S.

The usual definition "$\xi \leqq \eta$ if and only if $\eta - \xi$ has a non-negative representative" makes S a partially ordered set. Obviously in S, $\xi \leqq \eta$ implies $\xi + \gamma \leqq \eta + \gamma$ for all γ and $r\xi \leqq r\eta$ for all real, positive r, i.e. S represents a partially ordered vector space with respect to $\leqq$.

By virtue of our construction of $\mathfrak{K}$, the relation $\mathfrak{K}(X) \geqq 0$ holds for all X in $\mathfrak{H}^*$, i.e. $\mathfrak{K}$ is a finitely additive finite, S-valued measure on $\mathfrak{H}^*$.

Any ϱ in S is an upper bound of an at most countable subset V of S if and only if the following condition is satisfied:

If r is any representative of ϱ then to each ν in V there exists a fixed representative r_ν such that r is an upper bound of $\{r_\nu\}_{\nu \in V}$. Consequently each non-empty, at most countable subset V of S, which is bounded from above, has a lowest upper bound, namely the residue class of $\sup_{\nu \in V} r_\nu$. In other words: the vector space S being partially ordered by $\leqq$ is a conditionally σ-complete vector lattice.

We define the sum $\sum_{n=1}^{\infty} \xi_n$ of a sequence (ξ_n) of elements of S with the property $\xi_n \geqq 0$ $(n = 1, 2, \ldots)$ to be the supremum of the sequence of partial sums $\sum_{n=1}^{k} \xi_n$, subject to the existence of this supremum.

If a sequence of finite unions $X_n = H_{n,1} \cup \cdots \cup H_{n,m_n}$ of pairwise disjoint sets in $\mathfrak{H}$ decreases monotonically toward $\emptyset$, then by means of d) it can be concluded that

$$Q(x_{H_{n,1}} + \cdots + x_{H_{n,m_n}} > 0) \xrightarrow[n\to\infty]{} 0,$$

i.e. the sequence $(x_{H_{n,1}} + \cdots + x_{H_{n,m_n}})$ is not only almost everywhere monotone decreasing, in virtue of the finite additivity of $\mathcal{H}$, but moreover converges in probability to zero. Consequently, this sequence decreases monotonically toward zero almost everywhere and the monotone decreasing sequence of the $\mathcal{H}(X_n)$ has the infimum 0 in S. Just as in the case of real-valued set functions, this leads to the conclusion that $\mathcal{H}$ is a totally additive, S-valued measure on the ring $\mathfrak{H}^*$.

4. For all ξ in S, we set

$$\|\xi\|_S = \int \frac{|r_\xi(x)|}{1 + |r_\xi(x)|} Q(\mathrm{d}x),$$

where r_ξ denotes a representative of ξ. The inequality $0 \leqq \xi \leqq \eta$ always implies $\|\xi\|_S \leqq \|\eta\|_S$. As it is known, with the introduction of

$$\varrho_S(\xi, \eta) = \|\xi - \eta\|_S$$

S becomes a complete metric space. The convergence with respect to ϱ_S coincides with the convergence in probability. Consequently an infinite series $\sum_{n=1}^{\infty} \xi_n$ of elements of S with the property $\xi_n \geqq 0$ has the sum ξ if and only if the sequence of its partial sums tends to ξ with respect to ϱ_S.

Now suppose that to each natural number n there be associated a monotone decreasing sequence $(\xi_{n,m})_{m=1,2,\ldots}$ of elements of S with the infimum 0. Moreover let the set $\{\xi_{n,m}\}_{n,m=1,2,\ldots}$ be bounded from above within S. Subject to these assumptions, the following relation holds:

$$\bigwedge_{(m_n)} \left(\bigvee_{n=1}^{\infty} \xi_{n,m_n} \right) = 0,$$

where (m_n) runs through all sequences of natural numbers.

Bearing in mind that, for all n, $\|\xi_{n,m}\|_S \xrightarrow[m\to\infty]{} 0$, it follows that for each natural number l there exists a sequence $(m_{n,l})_{n=1,2,\ldots}$ of natural numbers with the property

$$\|\xi_{n,m_{n,l}}\|_S < (l\, 2^n)^{-1}.$$

We can select the $m_{n,l}$ in a suitable way so that for fixed n, $m_{n,l}$ is monotone increasing with respect to l. For all l, there exists the sum $\xi_l = \sum_{n=1}^{\infty} \xi_{n,m_{n,l}}$, and we have

$$\|\xi_l\|_S \leqq \sum_{n=1}^{\infty} \|\xi_{n,m_{n,l}}\|_S < l^{-1}.$$

Consequently the sequence (ξ_l) converges with respect to ϱ_S, i.e. in probability, toward 0, and moreover it is monotone decreasing. Therefore we can conclude that $\bigwedge_{l=1}^{\infty} \xi_l = 0$. On the other hand we have

$$0 \leqq \bigvee_{n=1}^{\infty} \xi_{n,m_{n,l}} \leqq \xi_l .$$

Hence the uncountable family $\left(\bigvee_{n=1}^{\infty} \xi_{n,m_n}\right)_{(m_n)}$ contains a countable subfamily with the infimum 0, and thus has itself the infimum 0, because all of its members are greater than or equal to 0.

5. On the basis of the property of distributivity derived in 4. for the conditionally σ-complete vector lattice S, the classical extension theorem remains valid also for S-valued set-functions (cf. for instance Fremlin [1]). Thus, for each H in $\mathfrak{H}^*$ there exists a well-defined, S-valued finite measure ${}_H\text{Ш}$ on the σ-algebra $H \cap \mathfrak{B}$ of subsets of H generated by $H \cap \mathfrak{H}^*$, which coincides with $\mathfrak{K}$ on $H \cap \mathfrak{H}^*$. For all H_1, H_2 in $\mathfrak{H}^*$ the measures ${}_{H_1}\text{Ш}$, ${}_{H_2}\text{Ш}$ are equal to ${}_{H_1 \cap H_2}\text{Ш}$ on $(H_1 \cap \mathfrak{B}) \cap (H_2 \cap \mathfrak{B}) = (H_1 \cap H_2) \cap \mathfrak{B}$. Consequently, there exists a common extension Ш of all ${}_H\text{Ш}$ on the ring $\mathfrak{B}' = \bigcup_{H \in \mathfrak{H}^*} (H \cap \mathfrak{B})$. By virtue of our assumptions on $\mathfrak{H}$ we have $\mathfrak{B}' = \mathfrak{B}$.

Now, let (X_n) be a sequence of pairwise disjoint sets in $\mathfrak{B}$ such that the union $X = \bigcup_n X_n$ also is in $\mathfrak{B}$.

By assumption there exists an H in $\mathfrak{H}^*$ such that X, and, therefore, all X_n are in $H \cap \mathfrak{B}$, and we obtain

$$\sum_{n=1}^{\infty} \text{Ш}(X_n) = \sum_{n=1}^{\infty} {}_H\text{Ш}(X_n) = {}_H\text{Ш}(X) = \text{Ш}(X),$$

i.e. Ш is the extension of $\mathfrak{K}$ to a finite, S-valued measure on $\mathfrak{B}$.

6. Let $\mathfrak{S}$ denote the system of those Z in $\mathfrak{B}$ for which there exists a representative y^Z of Ш(Z) which takes only values within $\{0, 1, \ldots\}$. Obviously $\mathfrak{S}$ is monotone closed in $\mathfrak{B}$ and includes $\mathfrak{H}$, i.e. $\mathfrak{S} = \mathfrak{B}$.

Now, if we set for all finite sequences $X_1, \ldots, X_m$ of sets in $\mathfrak{B}$:

$$p^{X_1,\ldots,X_m} = Q([y^{X_1}, \ldots, y^{X_m}] \in (.)),$$

then we obtain a continuation of the initially given mapping $[H_1, \ldots, H_m] \curvearrowright p^{H_1,\ldots,H_m}$ which still has the properties a) through d).

Therefore, from now on we can assume in addition that $\mathfrak{H}$ coincides with $\mathfrak{B}$.

7. Let X be an arbitrary bounded, closed subset of A, and $\varepsilon > 0$. To each natural number n we associate a representation $X = \bigcup_{m=1}^{\infty} X_{n,m}$ of X, as a

union of a sequence of pairwise disjoint sets in $\mathfrak{B}$, the diameters of which do not exceed $1/n$. Because of

$$Ш(X) = \sum_{m=1}^{\infty} Ш(X_{n,m})$$

we can select an m_n for each n such that

$$\left\| \sum_{m=m_n+1}^{\infty} Ш(X_{n,m}) \right\|_S < \varepsilon \cdot 2^{-n}$$

is satisfied. Thus we obtain

$$\left\| Ш \left(X \setminus \bigcap_{n=1}^{\infty} \bigcup_{m=1}^{m_n} X_{n,m} \right) \right\|_S \leqq \left\| \sum_{n=1}^{\infty} \sum_{m=m_n+1}^{\infty} Ш(X_{n,m}) \right\|_S$$
$$\leqq \sum_{n=1}^{\infty} \left\| \sum_{m=m_n+1}^{\infty} Ш(X_{n,m}) \right\|_S < \varepsilon .$$

The set $\bigcap_{n=1}^{\infty} \bigcup_{m=1}^{m_n} X_{n,m}$ is totally bounded and hence relatively compact. Consequently its closure B_ε provides a compact subset of X with the property

$$\|Ш(X \setminus B_\varepsilon)\|_S \leqq \varepsilon .$$

If we let now ε run through the sequence $1/n$, we can infer the existence of a monotone increasing sequence $B_{1/n}$ of compact subsets of X with the property

$$\|Ш(X \setminus B_{1/n})\|_S \leqq 1/n .$$

Hence the monotone decreasing sequence $Ш(X \setminus B_{1/n})$ has the infimum 0 in S.

Now, if we successively substitute for X the closed balls of radius r around a fixed point z in A, then we observe: There exists a monotone increasing sequence $(B_n)_{n=1,2,\ldots}$ of compact subsets of A such that for all X in B the relation

$$Ш(X) = \bigvee_{n=1}^{\infty} Ш(X \cap B_n)$$

is satisfied.

8. We can select a sequence of continuous real functions f_n ($n = 1, 2, \ldots$) on $[A, \varrho_A]$, such that all inverse images $f_n^{-1}(I)$ of bounded intervals I are bounded and that the smallest σ-subring of $\mathfrak{B}$ which includes all these inverse images, coincides with $\mathfrak{B}$.

For each natural number s, let $\mathfrak{H}_s$ denote the countable ring of subsets of R^s which is generated by all of the direct products $\mathop{\times}\limits_{i=1}^{s} I_i$ of bounded intervals with rational left and right boundary points. Obviously each set in $\mathfrak{H}_s$ can be represented both as the union of a monotone increasing sequence of closed sets in $\mathfrak{H}_s$ and as the intersection of a monotone decreasing sequence of open sets in $\mathfrak{H}_s$.

To each finite sequence $n_1, \ldots, n_s$ of natural numbers, and to each W in $\mathfrak{H}_s$, we associate the inverse image $W_{n_1,\ldots,n_s}$ of W with respect to the mapping $a \rightsquigarrow [f_{n_1}(a), \ldots, f_{n_s}(a)]$ of A into R^s. Then $\mathfrak{G}_{n_1,\ldots,n_s} = \{W_{n_1,\ldots,n_s}\}_{W\in\mathfrak{H}_s}$ represents a subring of $\mathfrak{B}$ for which each X in $\mathfrak{G}_{n_1,\ldots,n_s}$ can be represented both as the union of a monotone increasing sequence of closed, and as the intersection of a monotone decreasing sequence of open sets in this ring. We denote by $\mathfrak{G}$ the union of the rings $\mathfrak{G}_{n_1,\ldots,n_s}$ which extends over all finite sequences $n_1, \ldots, n_s$, and by $\mathfrak{G}^*$ the subring of $\mathfrak{B}$ generated by $\mathfrak{G}$. The smallest σ-subring of $\mathfrak{B}$ which includes $\mathfrak{G}^*$ coincides with $\mathfrak{B}$. Each X in $\mathfrak{G}^*$ can be represented as a union of a monotone increasing sequence of closed sets in $\mathfrak{G}^*$.

Let $\mathfrak{V}$ denote the smallest subring of $\mathfrak{B}$ which includes both $\mathfrak{G}^*$ and $\{B_n\}_{n=1,2,\ldots}$. For each Z in $\mathfrak{G}^*$, there exists a monotone increasing sequence $(K_{Z,n})_{n=1,2,\ldots}$ of compact sets in $\mathfrak{V}$ with the properties $K_{Z,n} \subseteq Z$ for $n = 1, 2, \ldots$; $\text{III}(Z) = \bigvee_{n=1}^{\infty} \text{III}(K_{Z,n})$. Like $\mathfrak{G}^*$, $\mathfrak{V}$ is also at most countable.

To each $x = (x_H)_{H\in\mathfrak{V}}$ contained in $\{0, 1, \ldots\}^{\mathfrak{V}}$, we associate a set function γ_x on $\mathfrak{V}$ by means of the relation

$$\gamma_x(H) = x_H \qquad (H \in \mathfrak{V}).$$

Then γ_x is an integer-valued, finitely additive measure on $\mathfrak{V}$ almost everywhere with respect to Q.

We suppose that a fixed sequence $(K_{Z,n})_{n=1,2,\ldots}$ with the above-mentioned properties be associated to each Z in $\mathfrak{G}^*$. Then for all Z in $\mathfrak{G}^*$, the following relation holds almost everywhere with respect to Q:

$$\gamma_x(Z) = \sup_{n=1,2,\ldots} \gamma_x(K_{Z,n}).$$

Henceforth we shall permit only such x in $\{0, 1, \ldots\}^{\mathfrak{V}}$ for which all the statements formulated above are valid.

Let (H_m) be a monotone decreasing sequence of sets in $\mathfrak{G}^*$ with intersection $\emptyset$. For a given $\varepsilon > 0$ and for all m, we select an n_m with the property

$$\gamma_x\left(H_m \setminus K_{H_m,n_m}\right) < \varepsilon \cdot 2^{-m}$$

and obtain

$$\inf_{m=1,2,\ldots} \gamma_x(H_m) \leqq \varepsilon + \inf_{m=1,2,\ldots} \gamma_x\left(\bigcap_{l=1}^{m} K_{H_l,n_l}\right).$$

But $\bigcap_{l=1}^{m} K_{H_l,n_l}$ is a monotone decreasing sequence of compact sets with intersection $\emptyset$. Hence $\bigcap_{l=1}^{m} K_{H_l,n_l}$ is empty for all sufficiently large m, and we have

$$\inf_{m=1,2,\ldots} \gamma_x(H_m) \leqq \varepsilon,$$

and therefore

$$\inf_{m=1,2,\ldots} \gamma_x(H_m) = 0.$$

Thus almost everywhere with respect to Q, γ_x is a finite, integer-valued measure on the at most countable subring $\mathfrak{G}^*$ of $\mathfrak{B}$, and hence can be extended to an integer-valued measure ω_x on $\mathfrak{B}$. For all Z in $\mathfrak{G}^*$, and hence also for all Z in $\mathfrak{B}$, the residue class of $\omega_x(Z)$ coincides with Ш(Z). If we now cover A by the sequence $\big(S_n(z_0)\big)$ of open balls around a fixed point z_0 in A, then we see that almost everywhere with respect to Q the measure ω_x is finite on the whole ring $\mathfrak{B}$, and hence can be extended to a well-defined measure Φ_x in M.

For all X in $\mathfrak{G}^*$, the real function

$$x \curvearrowright \Phi_x(X)$$

which is defined almost everywhere, is measurable with respect to the σ-algebra $\big(\mathfrak{P}(\{0, 1, \ldots\})\big)^{\mathfrak{B}}$. From this the measurability for all X in $\mathfrak{B}$ is inferred. Thus in the form of $x \curvearrowright \Phi_x$, we have found a mapping from $\{0, 1, \ldots\}^{\mathfrak{B}}$ into M, which is defined almost everywhere with respect to Q. We denote by P the distribution of Φ_x constructed with respect to Q.

For all finite sequences $H_1, \ldots, H_m$ of pairwise disjoint sets in $\mathfrak{B}$ and for all non-negative, integer $l_1, \ldots, l_m$, we obtain

$$\begin{aligned} P\big(\Phi(H_i) = l_i \text{ for } 1 \leqq i \leqq m\big) &= Q\big(\Phi_x(H_i) = l_i \text{ for } 1 \leqq i \leqq m\big) \\ &= Q(x_{H_i} = l_i \text{ for } 1 \leqq i \leqq m) = p^{H_1,\ldots,H_m}(\{[l_1, \ldots, l_m]\}), \end{aligned}$$

and consequently $P_{H_1,\ldots,H_m} = p^{H_1,\ldots,H_m}$. ∎

If we consider distributions of finite intensity, Theorem 1.3.5. can be simplified in the following manner.

1.3.6. Theorem. *Let $\mathfrak{H}$ be a generating semiring of sets in $\mathfrak{B}$ and ν a measure in N. To each finite sequence $H_1, \ldots, H_m$ of pairwise disjoint sets in $\mathfrak{H}$ let there be associated the distribution $p^{H_1,\ldots,H_m}$ of a random m-dimensional vector with non-negative integer coordinates.*

If now the relation

$$\sum_{l=1}^{\infty} l p^H(\{l\}) = \nu(H)$$

holds for all H in $\mathfrak{H}$, then a distribution P in $\mathbf{P}$ with the property $P_{H_1,\ldots,H_m} = p^{H_1,\ldots,H_m}$ exists if and only if the conditions a), b) and c) of Theorem 1.3.5. *are fulfilled. The distribution P is uniquely determined and has the intensity measure ν.*

Proof. We have only to prove that the conditions are sufficient.

1. In the same way as in the third step of the proof of Theorem 1.3.5. by the aid of the $p^{H_1,\ldots,H_m}$ we define a distribution Q on the σ-algebra $\big(\mathfrak{P}(\{0, 1, \ldots\})\big)^{\mathfrak{H}}$. By assumption, for all H in $\mathfrak{H}$ the expected value with respect to Q of the projection x_H is equal to $\nu(H)$.

All functions of a residue class z in S are equivalent with respect to the existence and the value of the expectation formed according to Q. Therefore we can deal with the expected value $\mathrm{E}_Q(z)$.

Taking into consideration the above remarks, by virtue of the finite additivity of $\mathfrak{K}$ we obtain the relation

$$\mathrm{E}_Q\big(\mathfrak{K}(H)\big) = \nu(H)$$

for all H in $\mathfrak{H}^*$.

2. On the basis of 1. for all sequences $(H_{n,1} \cup \cdots \cup H_{n,m_n})_{n=1,2,\ldots}$ of finite unions of pairwise disjoint sets in $\mathfrak{H}$ which decrease monotonically toward $\emptyset$ we obtain

$$\overline{\lim_{n\to\infty}} \left(1 - p^{H_{n,1},\ldots,H_{n,m_n}}(\{[0, \ldots, 0]\})\right) = \overline{\lim_{n\to\infty}}\, Q\big(x_{H_{n,1}} + \cdots + x_{H_{n,m_n}} > 0\big)$$

$$\leqq \overline{\lim_{n\to\infty}}\, \mathrm{E}_Q\big(x_{H_{n,1}} + \cdots + x_{H_{n,m_n}}\big) = \overline{\lim_{n\to\infty}}\, \nu(H_{n,1} \cup \cdots \cup H_{n,m_n}) = 0.$$

Therefore the condition d) of Theorem 1.3.5. is also valid.

3. In the proof of Theorem 1.3.5. the condition that each X in $\mathfrak{B}$ can be covered by a set H in $\mathfrak{H}$ was* used only in the fifth step of the proof. Now we show how this step can be carried out under the above assumptions.

First we introduce as before the measures ${}_H\text{Ш}$, $H \in \mathfrak{H}^*$, and compose them to an S-valued set-function $\text{Ш}'$, defined on the ring $\mathfrak{B}' = \bigcup_{H\in\mathfrak{H}^*} (H \cap \mathfrak{B})$.

Obviously $\text{Ш}'$ is finitely additive, because for all $X_1, \ldots, X_m$ in $H_1 \cap \mathfrak{B}, \ldots, H_m \cap \mathfrak{B}$ we have $X_1, \ldots, X_m \in (H_1 \cup \cdots \cup H_m) \cap \mathfrak{B}$.

For all H in $\mathfrak{H}^*$ the family of those X in $H \cap \mathfrak{B}$ which satisfy

$$\mathrm{E}_Q\big({}_H\text{Ш}(X)\big) = \nu(X)$$

is monotone closed and includes the algebra $H \cap \mathfrak{H}^*$, and hence coincides with $H \cap \mathfrak{B}$. Therefore, for all X in $\mathfrak{B}'$ the relation

$$\mathrm{E}_Q\big(\text{Ш}'(X)\big) = \nu(X)$$

holds.

Now let Z be some set in $\mathfrak{B}$. According to our assumption on $\mathfrak{H}$ there exists a sequence (H_n) of pairwise disjoint sets in $\mathfrak{H}$ with the property $Z \subseteq \bigcup_{n=1}^{\infty} H_n$. The infinite series $\sum_{n=1}^{\infty} \text{Ш}'(H_n \cap Z)$ converges in S since clearly $\text{Ш}'(H_n \cap Z) \geqq 0$, $n = 1, 2, \ldots$, and, moreover,

$$\sum_{n=1}^{\infty} \mathrm{E}_Q\big(\text{Ш}'(H_n \cap Z)\big) = \sum_{n=1}^{\infty} \nu(H_n \cap Z) = \nu(Z) < +\infty,$$

so that we can apply the Theorem of B. Levi.

Now if (V_m) is another sequence of pairwise disjoint sets in $\mathfrak{H}$ with the property $Z \subseteq \bigcup_{m=1}^{\infty} V_m$, then we obtain

$$\sum_{n=1}^{\infty} \text{Ш}'(H_n \cap Z) = \sum_{n=1}^{\infty} {}_{H_n}\text{Ш}(H_n \cap Z) = \sum_{n=1}^{\infty} \left(\sum_{m=1}^{\infty} {}_{H_n}\text{Ш}(H_n \cap V_m \cap Z) \right)$$

$$= \sum_{n,m=1}^{\infty} {}_{V_m}\text{Ш}(H_n \cap V_m \cap Z) = \sum_{m=1}^{\infty} \left(\sum_{n=1}^{\infty} {}_{V_m}\text{Ш}(H_n \cap V_m \cap Z) \right) = \sum_{m=1}^{\infty} \text{Ш}'(V_m \cap Z).$$

By means of the set-up

$$Ш(Z) = \sum_{n=1}^{\infty} Ш'(H_n \cap Z)$$

we get a well-defined, S-valued set-function on $\mathfrak{B}$, and $Ш(Z) \geqq 0$ for all Z in $\mathfrak{B}$.

Let (X_m) be a sequence of pairwise disjoint sets in $\mathfrak{B}$ with the property $X = \bigcup_m X_m \in \mathfrak{B}$. Then we have

$$\sum_{m=1}^{\infty} Ш(X_m) = \sum_{m=1}^{\infty} \left(\sum_{n=1}^{\infty} Ш'(H_n \cap X_m) \right)$$

$$= \sum_{n=1}^{\infty} \left(\sum_{m=1}^{\infty} {}_{H_n}Ш(H_n \cap X_m) \right) = \sum_{n=1}^{\infty} Ш'(H_n \cap X) = Ш(X),$$

i.e. Ш is an extension of $\mathfrak{K}$ to a finite, S-valued measure on $\mathfrak{B}$.

4. Now we can take over the rest of the proof of Theorem 1.3.5. and derive the existence of a distribution P which has the property

$$P_{H_1,\ldots,H_m} = p^{H_1,\ldots,H_m}.$$

Because of 1.3.2., this distribution is uniquely determined. By assumption, for all H in $\mathfrak{H}$ the relation

$$\varrho_P(H) = \nu(H)$$

holds, and hence ϱ_P coincides with ν. ∎

1.4. Simple Distributions

An element Φ of M is said to be *simple*, if for all a in A the inequality $\Phi(\{a\}) \leqq 1$ is satisfied, i.e. if no point belongs to Φ with a multiplicity greater than one. By virtue of $\Phi \curvearrowright \{a : a \in A,\ \Phi(\{a\}) > 0\}$, we obtain a one-to-one mapping of the set M_s of all simple Φ in M onto the system of the at most countable subsets X of A whose the intersections with each X in $\mathfrak{B}$ are finite. In the special case $A = \{1, \ldots, m\}$, the simple Φ correspond to the m-dimensional vectors, whose coordinates assume only the values zero or one.

To each Φ in M we associate a Φ^* in M_s in the following way

$$\Phi^* = \sum_{a \in A,\, \Phi(\{a\}) > 0} \delta_a .$$

Each point in Φ is given the multiplicity one. In order to make this definition readily applicable by means of a simple limiting process, we introduce the concept of the decomposition of a set X in $\mathfrak{A}$ by which we will mean an at most countable system $\mathfrak{Z}$ of pairwise disjoint sets in $\mathfrak{A}$ which has the union X. Let $\mathrm{d}(\mathfrak{Z})$ denote the supremum of the diameters of the sets in $\mathfrak{Z}$. A sequence $(\mathfrak{Z}_n)$ of decompositions of X is called a *distinguished sequence of decompositions* if $\lim_{n\to\infty} \mathrm{d}(\mathfrak{Z}_n) = 0$.

1.4.1. *For all Φ in M, all X in $\mathfrak{A}$, and all distinguished sequences $(\mathfrak{Z}_n)$ of decompositions of X,*

$$\sum_{Z \in \mathfrak{Z}_n} \min\left(\Phi(Z), 1\right) \xrightarrow[n\to\infty]{} \Phi^*(X).$$

Proof. Let X' be any set in $\mathfrak{B}$ with the property $X' \subseteq X$. The set $\{a : a \in X', \Phi(\{a\}) > 0\}$ is finite, so that there exists a positive minimum distance between the points a (if such points do exist at all). For all sufficiently large n, we thus have:

$$\Phi^*(X') \leqq \sum_{Z \in \mathfrak{Z}_n} \min\left(\Phi(Z), 1\right) \leqq \Phi^*(X),$$

and we obtain

$$\Phi^*(X') \leqq \varliminf_{n\to\infty} \sum_{Z \in \mathfrak{Z}_n} \min\left(\Phi(Z), 1\right) \leqq \varlimsup_{n\to\infty} \sum_{Z \in \mathfrak{Z}_n} \min\left(\Phi(Z), 1\right) \leqq \Phi^*(X),$$

from which because of

$$\Phi^*(X) = \sup_{\substack{X' \in \mathfrak{B} \\ X' \subseteq X}} \Phi^*(X')$$

it can be concluded that

$$\sum_{Z \in \mathfrak{Z}_n} \min\left(\Phi(Z), 1\right) \xrightarrow[n\to\infty]{} \Phi^*(X). \blacksquare$$

As an immediate consequence of 1.4.1., we have:

1.4.2. *The mapping $\Phi \curvearrowright \Phi^*$ of M into itself is measurable with respect to $\mathfrak{M}$.*

Obviously $\Phi \in M_s$ is equivalent to $\Phi = \Phi^*$. Thus we have:

1.4.3. *The set M_s belongs to $\mathfrak{M}$.*

The mapping $\Phi \curvearrowright \Phi^*$ transforms each measure L on $\mathfrak{M}$ into a measure L^* on $\mathfrak{M}$ with the property $L^*(\Phi \notin M_s) = 0$. L is said to be *simple* if $L = L^*$, i.e. if $L(\Phi \neq \Phi^*) = 0$.

Obviously for each covering $\mathfrak{X}$ of A by sets in $\mathfrak{A}$, any Φ in M is simple if and only if for all X in $\mathfrak{X}$ the condition $\Phi(X) = \Phi^*(X)$ is satisfied. We thus observe that the following proposition is valid:

1.4.4. *For each at most countable covering $\mathfrak{X}$ of A by sets in $\mathfrak{B}$, a measure L on $\mathfrak{M}$ is simple if and only if it satisfies, for all X in $\mathfrak{X}$, the condition*

$$L\left(\Phi^*(X) \neq \Phi(X)\right) = 0.$$

The simplicity of a distribution on $\mathfrak{M}$ can be expressed in the following way by means of its finite-dimensional distributions (cf. Karbe [1]):

1.4.5. Proposition. *Let $\mathfrak{H}$ be a generating semiring of sets in $\mathfrak{B}$. Then a distribution P on $\mathfrak{M}$ is simple if and only if for each H in $\mathfrak{H}$, and for each $\varepsilon > 0$, there exists a finite sequence $H_1, \ldots, H_m$ of pairwise disjoint sets in $\mathfrak{H}$ which has the properties*

$$\bigcup_{i=1}^{m} H_i = H, \; P_{H_1,\ldots,H_m}(\{0, 1\}^m) > 1 - \varepsilon.$$

Proof. 1. For each representation $H = H_1 \cup \cdots \cup H_m$ of a set H in $\mathfrak{H}$ as a finite union of pairwise disjoint sets of $\mathfrak{H}$, we have

$$P\big(\Phi(H) \neq \Phi^*(H)\big) \leqq P\big(\Phi(H_i) \notin \{0, 1\} \text{ for at least one } i \text{ in } \{1, \ldots, m\}\big).$$

Consequently, if the condition stated above is satisfied, we can conclude that $P\big(\Phi(H) \neq \Phi^*(H)\big) = 0$. On the basis of our assumptions on $\mathfrak{H}$, each X in $\mathfrak{B}$, and hence also the phase space A, can be covered by a sequence of sets in $\mathfrak{H}$, so that, using 1.4.4., we can conclude that $P = P^*$. This shows that the above-mentioned condition is sufficient.

2. Conversely, let us suppose that the distribution P be simple.

Let H be a set in $\mathfrak{H}$, and ε a positive real number. For all Φ in M_s, there exists a positive minimum distance between the points of the finite set $\{a : a \in H, \Phi(\{a\}) > 0\}$. Therefore, if we cover H by a sequence $(X_n)_{n=1,2,\ldots}$ of pairwise disjoint Borel subsets of H with sufficiently small diameters, we have

$$P\big(\Phi(X_n) > 1 \text{ for at least one } n\big) < \frac{\varepsilon}{4}.$$

If we consider the convergence

$$P\left(\Phi\left(H \setminus \bigcup_{n=1}^{k} X_n\right) > 0\right) \xrightarrow[k\to\infty]{} 0,$$

we can conclude that for sufficiently large k the finite sequence

$$X_1' = X_1, \ldots, X_{k-1}' = X_{k-1}, X_k' = \bigcup_{n=k}^{\infty} X_n$$

of pairwise disjoint sets in $\mathfrak{B}$ satisfies the conditions

$$\bigcup_{n=1}^{k} X_n' = H, \; P_{X_1',\ldots,X_k'}(\{0, 1\}^k) > 1 - \frac{\varepsilon}{2}.$$

3. Let $\mathfrak{S}$ denote the system of those sets X in $\mathfrak{B}$, for which there exist, for each $\eta > 0$, sets $V_{X,\eta}$ in the ring $\mathfrak{H}^*$ generated by $\mathfrak{H}$ such that

$$P\big(\Phi((X \setminus V_{X,\eta}) \cup (V_{X,\eta} \setminus X)) > 0\big) < \eta$$

is satisfied. Obviously $\mathfrak{H}^*$ is included in $\mathfrak{S}$. If $(X_n)_{n=1,2,\ldots}$ is a monotone decreasing sequence of sets in $\mathfrak{S}$ with the intersection X, then for $n = 1, 2, \ldots$ we have

$$P\big(\Phi((X \setminus V_{X_n,\eta}) \cup (V_{X_n,\eta} \setminus X)) > 0\big) \leqq P\big(\Phi(X_n \setminus X) > 0\big)$$
$$+ P\big(\Phi((X_n \setminus V_{X_n,\eta}) \cup (V_{X_n,\eta} \setminus X_n)) > 0\big) < P\big(\Phi(X_n \setminus X) > 0\big) + \eta.$$

If we now note that $P\big(\Phi(X_n \setminus X) > 0\big)$ tends to zero with increasing n, then we see that for sufficiently large n

$$P\big(\Phi((X \setminus V_{X_n,\eta}) \cup (V_{X_n,\eta} \setminus X)) > 0\big) < 2\eta$$

is satisfied, and hence X is an element of $\mathfrak{S}$. In a similar way it can be shown that also the union of each monotone increasing sequence of sets in $\mathfrak{S}$ which is bounded in $\mathfrak{B}$, is again an element of $\mathfrak{S}$. Hence the set system $\mathfrak{S}$ is monotone closed in $\mathfrak{B}$ so that we can conclude that $\mathfrak{S} = \mathfrak{B}$.

Referring now to the results of 2., we associate a set V_i in $\mathfrak{H}^*$ to each set X_i' $(i = 1, \ldots, k)$, so that

$$P\big(\Phi\big((X_i' \setminus V_i) \cup (V_i \setminus X_i')\big) > 0\big) < \frac{\varepsilon}{4k}$$

is satisfied. If in addition we set $V_{k+1} = H \setminus \bigcup_{i=1}^{k} V_i$, then the $V_1, \ldots, V_{k+1}$ provide a finite covering of H by sets in $\mathfrak{H}^*$ with the property

$$\begin{aligned}
&P\big(\Phi(V_n) > 1 \text{ for at least one } n \text{ in } \{1, \ldots, k+1\}\big)\\
&\leqq P\big(\Phi(V_n) > 1 \text{ for at least one } n \text{ in } \{1, \ldots, k\}\big) + P\big(\Phi(V_{k+1}) > 1\big)\\
&\leqq P\big(\Phi(X_n') > 1 \text{ for at least one } n \text{ in } \{1, \ldots, k\}\big)\\
&\quad + P\left(\Phi\left(\bigcup_{n=1}^{k} (V_n \setminus X_n')\right) > 0\right) + P\left(\Phi\left(\bigcup_{n=1}^{k} (X_n' \setminus V_n)\right) > 0\right)\\
&\leqq \frac{\varepsilon}{2} + 2\sum_{n=1}^{k} P\big(\Phi\big((X_n' \setminus V_n) \cup (V_n \setminus X_n')\big) > 0\big) < \frac{\varepsilon}{2} + k\frac{2\varepsilon}{4k} = \varepsilon.
\end{aligned}$$

If we set $G_n = V_n \setminus \bigcup_{i=1}^{n-1} V_i$ $(n = 1, \ldots, k+1)$, then obviously also

$$P_{G_1,\ldots,G_{k+1}}(\{0, 1\}^{k+1}) > 1 - \varepsilon$$

is true.

Now we select a sequence $H_1, \ldots, H_m$ of pairwise disjoint sets in $\mathfrak{H}$ for which $\bigcup_{i=1}^{m} H_i = H$ is satisfied and which allows to represent each $H \cap G_i$ $(i = 1, \ldots, k+1)$ as a union of certain H_i. Then we have

$$P_{H_1,\ldots,H_m}(\{0, 1\}^m) > 1 - \varepsilon.$$

Consequently the above-mentioned condition is also necessary. ∎

As a criterion for verifying the simplicity of given distributions P, the theorem proved above is of little use. Later on we shall introduce a more convenient, but only sufficient simplicity criterion which involves only the one-dimensional distributions P_X.

Now, referring again to 1.4.1., we shall prove (cf. Leadbetter [1]):

1.4.6. Proposition. *For all measures L on $\mathfrak{M}$, all X in $\mathfrak{A}$, and all distinguished sequences $(\mathfrak{Z}_n)$ of decompositions of X,*

$$\varrho_{L^*}(X) = \lim_{n\to\infty} \sum_{Z\in\mathfrak{Z}_n} L\big(\Phi(Z) > 0\big).$$

Proof. In the case $\varrho_{L^*}(X) < +\infty$, the asserted convergence is derived by means of Lebesgue's theorem and of 1.4.1. from

$$\sum_{Z\in\mathfrak{Z}_n} \min\big(\Phi(Z), 1\big) \leqq \Phi^*(X)$$

bearing in mind that

$$\int \sum_{Z \in \mathfrak{Z}_n} \min\left(\Phi(Z), 1\right) L(\mathrm{d}\Phi) = \sum_{Z \in \mathfrak{Z}_n} L\left(\Phi(Z) > 0\right).$$

However, if $\varrho_{L^*}(X)$ is infinite, then, using Fatou's lemma, we obtain

$$+\infty = \varrho_{L^*}(X) \leqq \underline{\lim}_{n\to\infty} \sum_{Z \in \mathfrak{Z}_n} L\left(\Phi(Z) > 0\right),$$

that means

$$\lim_{n\to\infty} \sum_{Z \in \mathfrak{Z}_n} L\left(\Phi(Z) > 0\right) = +\infty. \blacksquare$$

For all measures L on $\mathfrak{M}$, we have $\varrho_{L^*} \leqq \varrho_L$, since for each X in $\mathfrak{A}$ and each Φ in M, $\Phi^*(X) \leqq \Phi(X)$. From $\varrho_{L^*} = \varrho_L$ and $\varrho_L \in N$ it follows that $L^* = L$. For if z is an arbitrarily selected point in A, we obtain for $n = 1, 2, \ldots$

$$\int \Phi^*\left(S_n(z)\right) L(\mathrm{d}\Phi) = \int \Phi\left(S_n(z)\right) L(\mathrm{d}\Phi) < +\infty,$$

and hence

$$\int \left(\Phi\left(S_n(z)\right) - \Phi^*\left(S_n(z)\right)\right) L(\mathrm{d}\Phi) = 0$$

which means that

$$L\left(\Phi\left(S_n(z)\right) \neq \Phi^*\left(S_n(z)\right)\right) = 0,$$

from which according to 1.4.4. it can be concluded that

$$L(\Phi \neq \Phi^*) = 0.$$

Without using the assumption that $\varrho_L \in N$, the equation $L = L^*$ cannot be derived from $\varrho_{L^*} = \varrho_L$. An example illustrating this situation will be constructed in section 6.4.

For distributions P on $\mathfrak{M}$, the value of $\varrho_{P^*}(X)$ can be calculated according to 1.4.6. for all X from the probabilities $P\left(\Phi(X') > 0\right)$, $X' \in \mathfrak{B}$. It turns out now that not only the intensity measure ϱ_{P^*} of P^*, but also P^* itself can be determined by construction from these special probabilities (cf. Kallenberg [1], [2]):

1.4.7. Proposition. *Let $\mathfrak{H}$ be a generating subring of $\mathfrak{B}$. If two distributions P, Q on $\mathfrak{M}$ satisfy the condition $P\left(\Phi(H) = 0\right) = Q\left(\Phi(H) = 0\right)$ for all H in $\mathfrak{H}$, then it follows that $P^* = Q^*$.*

Proof. 1. Let $\mathbf{\mathfrak{L}}$ denote the set of all non-empty finite systems $\mathfrak{L}$ of pairwise disjoint non-empty sets in $\mathfrak{H}$. Then for all distributions V on $\mathfrak{M}$ and all $\mathfrak{L}$ in $\mathbf{\mathfrak{L}}$ and all subsystems $\mathfrak{J}$ of $\mathfrak{L}$

$$\begin{aligned} &V\left(\Phi(L) = 0 \text{ for } L \in \mathfrak{J},\ \Phi(L) > 0 \text{ for } L \in (\mathfrak{L} \setminus \mathfrak{J})\right) \\ &= \sum_{\mathfrak{J} \subseteq \mathfrak{K} \subseteq \mathfrak{L}} (-1)^{\operatorname{card}(\mathfrak{K} \setminus \mathfrak{J})}\, V\left(\Phi\left(\bigcup_{K \in \mathfrak{K}} K\right) = 0\right). \end{aligned}$$

We shall prove the validity of this formula by induction on $m = \operatorname{card}(\mathfrak{L} \setminus \mathfrak{J})$.

For $m = 0$ we have $\mathfrak{L} = \mathfrak{J}$, and our assertion is reduced to

$$V\big(\Phi(L) = 0 \text{ for } L \in \mathfrak{J}\big) = V\left(\Phi\left(\bigcup_{L \in \mathfrak{J}} L\right) = 0\right).$$

Let us now assume that our assertion is valid for $m = m_0$. Let $\mathfrak{L} \in \mathbf{L}$ as well as $\mathfrak{J} \subseteq \mathfrak{L}$ be given such that card $(\mathfrak{L} \setminus \mathfrak{J}) = m_0 + 1$ is satisfied. We select any set L_0 from $\mathfrak{L} \setminus \mathfrak{J}$. If $\mathfrak{L} \neq \{L_0\}$, then our induction hypothesis leads to

$$(*) \qquad V\big(\Phi(L) = 0 \text{ for } L \in \mathfrak{J}, \Phi(L) > 0 \text{ for } L \in \big((\mathfrak{L} \setminus \mathfrak{J}) \setminus \{L_0\}\big)\big) = \sum_{\mathfrak{J} \subseteq \mathfrak{K} \subseteq \mathfrak{L} \setminus \{L_0\}} (-1)^{\operatorname{card}(\mathfrak{K} \setminus \mathfrak{J})} V\left(\Phi\left(\bigcup_{K \in \mathfrak{K}} K\right) = 0\right).$$

In the case $\mathfrak{L} = \{L_0\}$, the relation $(*)$ is reduced to

$$V(M) = V\big(\Phi(\emptyset) = 0\big)$$

and hence remains valid.

Now, if we replace $\mathfrak{J}$ by $\mathfrak{J} \cup \{L_0\}$, then we can again use the induction hypothesis, and obtain

$$(**) \quad P\big(\Phi(L) = 0 \text{ for } L \in \mathfrak{J} \cup \{L_0\}, \Phi(L) > 0 \text{ for } L \in \big((\mathfrak{L} \setminus \mathfrak{J}) \setminus \{L_0\}\big)\big) = \sum_{\mathfrak{J} \cup \{L_0\} \subseteq \mathfrak{K} \subseteq \mathfrak{L}} (-1)^{\operatorname{card}(\mathfrak{K} \setminus \mathfrak{J})+1} P\left(\Phi\left(\bigcup_{K \in \mathfrak{K}} K\right) = 0\right).$$

We have, however:

$$\begin{aligned} &P\big(\Phi(L) = 0 \text{ for } L \in \mathfrak{J}, \Phi(L) > 0 \text{ for } L \in (\mathfrak{L} \setminus \mathfrak{J})\big) \\ &= P\big(\Phi(L) = 0 \text{ for } L \in \mathfrak{J}, \Phi(L) > 0 \text{ for } L \in \big((\mathfrak{L} \setminus \mathfrak{J}) \setminus \{L_0\}\big)\big) \\ &\quad - P\big(\Phi(L) = 0 \text{ for } L \in \mathfrak{J} \cup \{L_0\}, \Phi(L) > 0 \text{ for } L \in \big((\mathfrak{L} \setminus \mathfrak{J}) \setminus \{L_0\}\big)\big). \end{aligned}$$

Using $(*)$ and $(**)$, this equation can be continued as follows:

$$\begin{aligned} &= \sum_{\mathfrak{J} \subseteq \mathfrak{K} \subseteq \mathfrak{L} \setminus \{L_0\}} (-1)^{\operatorname{card}(\mathfrak{K} \setminus \mathfrak{J})} P\left(\Phi\left(\bigcup_{K \in \mathfrak{K}} K\right) = 0\right) \\ &\quad + \sum_{\mathfrak{J} \cup \{L_0\} \subseteq \mathfrak{K} \subseteq \mathfrak{L}} (-1)^{\operatorname{card}(\mathfrak{K} \setminus \mathfrak{J})} P\left(\Phi\left(\bigcup_{K \in \mathfrak{K}} K\right) = 0\right) \\ &= \sum_{\mathfrak{J} \subseteq \mathfrak{K} \subseteq \mathfrak{L}} (-1)^{\operatorname{card}(\mathfrak{K} \setminus \mathfrak{J})} P\left(\Phi\left(\bigcup_{K \in \mathfrak{K}} K\right) = 0\right). \end{aligned}$$

Consequently our assertion remains valid also for $m = m_0 + 1$.

2. From 1. it follows that for each non-empty, finite family $(X_i)_{i \in I}$ of pairwise disjoint sets in $\mathfrak{H}$, and for each subset J of I

$$\begin{aligned} &P\big(\Phi(X_i) \neq 0 \quad \text{for} \quad i \in J, \ \Phi(X_i) = 0 \quad \text{for} \quad i \in I \setminus J\big) \\ &= Q\big(\Phi(X_i) \neq 0 \quad \text{for} \quad i \in J, \ \Phi(X_i) = 0 \quad \text{for} \quad i \in I \setminus J\big). \end{aligned}$$

This is clear if the X_i are non-empty and remains also valid if some of the X_i are empty.

3. Now let $X_1, \ldots, X_m$ be any finite sequence of pairwise disjoint sets in $\mathfrak{H}$ and $\varepsilon > 0$. According to 1.4.5. where we set $P = \frac{1}{2}(P^* + Q^*)$, each X_i can be covered by pairwise disjoint sets $Z_{i,1}, \ldots, Z_{i,m_i}$ in $\mathfrak{H}$ such that

$$P_{Z_{i,1},\ldots,Z_{i,m_i}}(\{0, 1\}^{m_i}) > 1 - \varepsilon.$$

In this way we obtain a covering $(Z_i)_{i \in I}$ of $X = \bigcup_{i=1}^{m} X_i$ by pairwise disjoint sets in $X \cap \mathfrak{H}$. Let $I_s = \{i : i \in I, Z_i \subseteq X_s\}$ for $1 \leqq s \leqq m$.

For all non-negative, integer $l_1, \ldots, l_m$

$$\begin{aligned}
&P^*\big(\Phi(X_s) = l_s \quad \text{for} \quad 1 \leqq s \leqq m, \Phi(Z_i) \leqq 1 \quad \text{for} \quad i \in I\big) \\
&= \sum_{\substack{J \subseteq I \\ \operatorname{card}(J \cap I_s) = l_s \text{for} 1 \leqq s \leqq m}} P^*\big(\Phi(Z_i) \geqq 1 \quad \text{for} \quad i \in J, \Phi(Z_i) = 0 \quad \text{for} \quad i \in I \setminus J\big) \\
&\quad - \sum_{\substack{J \subseteq I \\ \operatorname{card}(J \cap I_s) = l_s \text{for} 1 \leqq s \leqq m}} P^*\big(\Phi(Z_i) \geqq 1 \quad \text{for} \quad i \in J, \Phi(Z_i) = 0 \quad \text{for} \quad i \in I \setminus J, \\
&\qquad\qquad \Phi(Z_i) > 1 \quad \text{for at least one } i \in J\big).
\end{aligned}$$

However, the sets

$$\{\Phi : \Phi \in M, \Phi(Z_i) \geqq 1 \quad \text{for} \quad i \in J, \Phi(Z_i) = 0 \quad \text{for} \quad i \in I \setminus J, \; \Phi(Z_i) > 1 \quad \text{for at least one } i \text{ in } J\}$$

are pairwise disjoint and for $\sum_{i=1}^{m} l_i \neq 0$ they are included in

$$\{\Phi : \Phi \in M, \; \Phi(Z_i) > 1 \quad \text{for at least one } i \text{ in } I\}.$$

Hence we obtain

$$\begin{aligned}
&\Big|P^*\big(\Phi(X_s) = l_s \quad \text{for} \quad 1 \leqq s \leqq m, \Phi(Z_i) \leqq 1 \quad \text{for} \quad i \in I\big) \\
&\quad - \sum_{\substack{J \subseteq I \\ \operatorname{card}(J \cap I_s) = l_s \text{for} 1 \leqq s \leqq m}} P^*\big(\Phi(Z_i) \geqq 1 \quad \text{for} \quad i \in J, \Phi(Z_i) = 0 \quad \text{for} \quad i \in I \setminus J\big)\Big| \\
&\leqq P^*\big(\Phi(Z_i) > 1 \quad \text{for at least one } i \text{ in } I\big),
\end{aligned}$$

and consequently

$$\begin{aligned}
&\Big|P^*\big(\Phi(X_s) = l_s \quad \text{for} \quad 1 \leqq s \leqq m\big) \\
&\quad - \sum_{\substack{J \subseteq I \\ \operatorname{card}(J \cap I_s) = l_s \text{for} 1 \leqq s \leqq m}} P^*\Big(\Phi(Z_i) \geqq 1 \quad \text{for} \quad i \in J, \Phi\Big(\bigcup_{i \in I \setminus J} Z_i\Big) = 0\Big)\Big| \\
&\leqq 2P^*\big(\Phi(Z_i) > 1 \quad \text{for at least one } i \text{ in } I\big).
\end{aligned}$$

A corresponding inequality is valid also for Q, so that, using 2., we can conclude that

$$\begin{aligned}
&|P^*_{X_1,\ldots,X_m}(\{[l_1, \ldots, l_m]\}) - Q^*_{X_1,\ldots,X_m}(\{[l_1, \ldots, l_m]\})| \\
&\leqq 2\big(P^*\big(\Phi(Z_i) > 1 \quad \text{for at least one } i \in I\big) \\
&\qquad + Q^*\big(\Phi(Z_i) > 1 \quad \text{for at least one } i \in I\big)\big) < 4\varepsilon.
\end{aligned}$$

The right-hand side of this inequality can be made arbitrarily small by a suitable selection of $(Z_i)_{i\in I}$. If we now take into consideration that by assumption

$$(P^*)_{X_1,\ldots,X_m}(\{[0,\ldots,0]\}) = P(\Phi(X) = 0)$$
$$= Q(\Phi(X) = 0) = (Q^*)_{X_1,\ldots,X_m}(\{[0,\ldots,0]\})$$

is satisfied, then we obtain

$$(P^*)_{X_1,\ldots,X_m} = (Q^*)_{X_1,\ldots,X_m}$$

from which, using the uniqueness proposition 1.3.2., it can be concluded that $P^* = Q^*$. ∎

The premises of the uniqueness proposition 1.4.7. are not sufficient for concluding $P = Q$ without additional assumptions. If for instance 2Φ is substituted for Φ, then each distribution P on M different from δ_0 is transformed into a distribution P' different from P, although

$$P(\Phi(X) = 0) = P'(\Phi(X) = 0) \qquad (X \in \mathfrak{B})$$

is satisfied. However, the following theorem can be proved (cf. Kallenberg [1], [2]):

1.4.8. Theorem. *Let $\mathfrak{H}$ be a generating semiring of sets in $\mathfrak{B}$ and $\mathfrak{H}^*$ the subring of $\mathfrak{B}$ which is generated by $\mathfrak{H}$. If two distributions P, Q on $\mathfrak{M}$ satisfy the conditions*

$$Q \text{ is simple,}$$
$$P(\Phi(H) = 0) = Q(\Phi(H) = 0) \quad \text{for all} \quad H \text{ in } \mathfrak{H}^*,$$
$$P(\Phi(H) > 1) \leqq Q(\Phi(H) > 1) \quad \text{for all} \quad H \text{ in } \mathfrak{H},$$

then it follows that $P = Q$.

Proof. From 1.4.7. it follows that $P^* = Q^* = Q$. Let H be a set in $\mathfrak{H}$ and ε any positive number. By virtue of 1.4.5., there exists a finite decomposition $\mathfrak{Z}$ of H into pairwise disjoint sets in $\mathfrak{H}$, such that

$$P^*(\Phi(Z) \leqq 1 \quad \text{for all} \quad Z \text{ in } \mathfrak{Z}) = P(\Phi^*(Z) \leqq 1 \quad \text{for all} \quad Z \text{ in } \mathfrak{Z}) \geqq 1 - \varepsilon$$

and therefore also

$$Q(\Phi(Z) \leqq 1 \quad \text{for all} \quad Z \text{ in } \mathfrak{Z}) \geqq 1 - \varepsilon$$

is satisfied. Thus we obtain

$$\begin{aligned}
&P(\Phi(Z) > 1 \quad \text{for at least one } Z \text{ in } \mathfrak{Z}) \\
&\leqq P(\Phi(Z) > 1 \quad \text{for at least one } Z \text{ in } \mathfrak{Z}, \quad \Phi^*(Z) \leqq 1 \quad \text{for all} \quad Z \text{ in } \mathfrak{Z}) + \varepsilon \\
&\leqq \varepsilon + P(\Phi(Z) > \Phi^*(Z) = 1 \quad \text{for at least one } Z \text{ in } \mathfrak{Z}) \\
&\leqq \varepsilon + \sum_{Z\in\mathfrak{Z}} P(\Phi(Z) > 1,\ \Phi^*(Z) \leqq 1) \\
&= \varepsilon + \sum_{Z\in\mathfrak{Z}} \big(P(\Phi(Z) > 1) - P(\Phi^*(Z) > 1)\big) \leqq \varepsilon + \sum_{Z\in\mathfrak{Z}} \big(Q(\Phi(Z) > 1) \\
&\quad - Q(\Phi(Z) > 1)\big) = \varepsilon.
\end{aligned}$$

where we used the fact that, for all $\mathfrak{S} \subset \mathfrak{K}$,

$$\sum_{\mathfrak{S} \subseteq \mathfrak{J} \subseteq \mathfrak{K}} (-1)^{\operatorname{card}(\mathfrak{K} \setminus \mathfrak{J})} = \sum_{\mathfrak{T} \subseteq (\mathfrak{K} \setminus \mathfrak{S})} (-1)^{\operatorname{card} \mathfrak{T}} = 0.$$

Each distribution q of a random subsystem $\mathscr{I}$ of $\mathfrak{L}$ with the property

$$q(\mathscr{I} \supseteqq \mathfrak{S}) = p_{\mathfrak{L}}(\mathscr{I} \supseteqq \mathfrak{S}) \qquad (\mathfrak{S} \subseteqq \mathfrak{L})$$

coincides with $p_{\mathfrak{L}}$. This follows immediately from the uniqueness proposition 1.4.7. when we map $\mathfrak{P}(\mathfrak{L})$ by means of

$$\mathfrak{J} \curvearrowright \chi = \sum_{L \in (\mathfrak{L} \setminus \mathfrak{J})} \delta_L$$

in a one-to-one way onto the set of all simple realizations χ which belong to the finite phase space $\mathfrak{L}$.

Thus we observe that there is exactly one distribution q of a random subsystem $\mathscr{I}$ of $\mathfrak{L}$ which possesses, for all $\mathfrak{S} \subseteqq \mathfrak{L}$, the property

$$q(\mathscr{I} \supseteqq \mathfrak{S}) = f\left(\bigcup_{S \in \mathfrak{S}} S\right),$$

namely the distribution $p_{\mathfrak{L}}$ introduced above.

5. We introduce a semi-ordering relation $\leqq$ in $\mathbf{L}$ by means of the following definition:

$\mathfrak{L}_1 \leqq \mathfrak{L}_2$ if and only if $\mathfrak{L}_1$ is included in the ring $\mathfrak{L}_2^*$ generated by $\mathfrak{L}_2$.

Obviously for two set systems $\mathfrak{L}_1$, $\mathfrak{L}_2$ in $\mathbf{L}$ there exists always an $\mathfrak{L}_3$ in $\mathbf{L}$ with the property $\mathfrak{L}_1, \mathfrak{L}_2 \leqq \mathfrak{L}_3$.

Now suppose that $\mathfrak{L}_1, \mathfrak{L}_2 \in \mathbf{L}$ and $\mathfrak{L}_1 \leqq \mathfrak{L}_2$. To each $\mathfrak{V} \subseteq \mathfrak{L}_2$ we associate an $\mathfrak{J}(\mathfrak{V}) \subseteq \mathfrak{L}_1$ in the following way:

$$L \in \mathfrak{J}(\mathfrak{V}) \quad \text{if and only if} \quad L \in \mathfrak{L}_1 \quad \text{and} \quad L \subseteq \bigcup_{V \in \mathfrak{V}} V.$$

Then this mapping transforms the distribution $p_{\mathfrak{L}_2}$ into the distribution $p_{\mathfrak{L}_1}$.

For if we set q equal to the image of $p_{\mathfrak{L}_2}$ produced by the mapping $\mathfrak{J}(.)$, then in virtue of 4. we obtain, for all $\mathfrak{S} \subseteq \mathfrak{L}_1$:

$$q(\mathscr{I} \supseteq \mathfrak{S}) = p_{\mathfrak{L}_2}(\mathfrak{J}(\mathscr{V}) \supseteq \mathfrak{S}) = p_{\mathfrak{L}_2}\left(\bigcup_{V \in \mathscr{V}} V \supseteq \bigcup_{S \in \mathfrak{S}} S\right) = f\left(\bigcup_{S \in \mathfrak{S}} S\right).$$

As it was stated at the end of 4., this implies that $q = p_{\mathfrak{L}_1}$.

6. Let $H_1, \ldots, H_m$ be a finite sequence of pairwise disjoint sets in $\mathfrak{H}$. Further let $\mathfrak{L}_1, \ldots, \mathfrak{L}_k$ be a given, finite sequence of set systems in $\mathbf{L}$. Let each H_i be included in $\bigcap_{j=1}^{k} \mathfrak{L}_j^*$, i.e. we suppose that

$$H_i = \bigcup_{L \in \mathfrak{J}_{j,i}} L \qquad (1 \leqq i \leqq m,\ 1 \leqq j \leqq k),$$

where $\mathfrak{J}_{j,i} \subseteq \mathfrak{L}_j$.

We select any $\mathfrak{L}$ in $\mathbf{L}$ with the property $\mathfrak{L}_1, \ldots, \mathfrak{L}_k \leqq \mathfrak{L}$. Thus we have

$$L = \bigcup_{V \in \mathfrak{V}_L} V \qquad \left(L \in \bigcup_{j=1}^{k} \mathfrak{L}_j\right)$$

where $\mathfrak{V}_L \subseteq \mathfrak{L}$. To each $\mathfrak{J} \subseteq \mathfrak{L}$ we can associate a k-tuple

$$h = \left[[h_{1,1}, \ldots, h_{1,m}], \ldots, [h_{k,1}, \ldots, h_{k,m}]\right]$$

of m-dimensional vectors having non-negative integer coordinates in the following way:

$h_{j,i}$ = number of those L in $\mathfrak{J}_{j,i}$ for which $(\mathfrak{L} \setminus \mathfrak{J}) \cap \mathfrak{V}_L$ is non-empty.

This mapping transforms $p_{\mathfrak{L}}$ into a distribution $p_{\mathfrak{L}_1,\ldots,\mathfrak{L}_k}^{H_1,\ldots,H_m}$ of a random k-tuple $\xi = [\xi_1, \ldots, \xi_k]$ of m-dimensional vectors with non-negative integer coordinates. In virtue of 5., this distribution does not change if we replace $\mathfrak{L}$ by an $\mathfrak{L}'$ in $\mathbf{L}$ with the property $\mathfrak{L} \leqq \mathfrak{L}'$. Hence $p_{\mathfrak{L}_1,\ldots,\mathfrak{L}_k}^{H_1,\ldots,H_m}$ does not depend on the arbitrary selection of the set system $\mathfrak{L}$.

Let us now state some important properties of $p_{\mathfrak{L}_1,\ldots,\mathfrak{L}_k}^{H_1,\ldots,H_m}$.

6.1. $p_{\mathfrak{L}_1}^{H_1}(\{0\}) = f(H_1)$.

For in this case we can set $\mathfrak{L} = \mathfrak{L}_1$, and obtain

$$p_{\mathfrak{L}_1}^{H_1}(\{0\}) = p_{\mathfrak{L}_1}(\mathscr{I} \supseteq \mathfrak{J}_{1,1}) = f\left(\bigcup_{L \in \mathfrak{J}_{1,1}} L\right) = f(H_1).$$

The following properties result immediately from our introduction of the distributions $p_{\mathfrak{L}_1,\ldots,\mathfrak{L}_k}^{H_1,\ldots,H_m}$.

6.2. For each permutation $i_1, \ldots, i_m$ of $1, \ldots, m$ and for all non-negative integer $l_1, \ldots, l_m$,

$$p_{\mathfrak{L}_1}^{H_1,\ldots,H_m}(\{[l_1, \ldots, l_m]\}) = p_{\mathfrak{L}_1}^{H_{i_1},\ldots,H_{i_m}}(\{[l_{i_1}, \ldots, l_{i_m}]\}).$$

6.3. For all non-negative integer $l_1, \ldots, l_{m-1}$,

$$\begin{aligned} &p_{\mathfrak{L}_1}^{H_1,\ldots,H_m}(\{[v_1, \ldots, v_m] : v_1 = l_1, \ldots, v_{m-1} = l_{m-1}, v_m \geqq 0\}) \\ &= p_{\mathfrak{L}_1}^{H_1,\ldots,H_{m-1}}(\{[l_1, \ldots, l_{m-1}]\}). \end{aligned}$$

6.4. If we denote by H the union of $H_k, \ldots, H_m$, then all non-negative integer $l_1, \ldots, l_k$ satisfy the following condition:

$$\begin{aligned} &p_{\mathfrak{L}_1}^{H_1,\ldots,H_{k-1},H}(\{[l_1, \ldots, l_k]\}) \\ &= p_{\mathfrak{L}_1}^{H_1,\ldots,H_m}(\{[v_1, \ldots, v_m] : v_1 = l_1, \ldots, v_{k-1} = l_{k-1}, v_k + \cdots + v_m = l_k\}). \end{aligned}$$

6.5. If $\mathfrak{L}_1 \leqq \mathfrak{L}_2$, then

$$p_{\mathfrak{L}_1,\mathfrak{L}_2}^{H_1,\ldots,H_m}(\xi_1 \leqq \xi_2) = 1.$$

Here $\leqq$ denotes the usual semi-order relation in R^m.

6.6. For each permutation $i_1, \ldots, i_k$ of $1, \ldots, k$, the following relation holds:

$$p_{\mathfrak{L}_1,\ldots,\mathfrak{L}_k}^{H_1,\ldots,H_m}\Big(\Big\{\big[[h_{1,1},\ldots,h_{1,m}],\ldots,[h_{k,1},\ldots,h_{k,m}]\big]\Big\}\Big)$$
$$= p_{\mathfrak{L}_{i_1},\ldots,\mathfrak{L}_{i_k}}^{H_1,\ldots,H_m}\Big(\Big\{\big[[h_{i_1,1},\ldots,h_{i_1,m}],\ldots,[h_{i_k,1},\ldots,h_{i_k,m}]\big]\Big\}\Big).$$

6.7. For all $h = \big[[h_{1,1},\ldots,h_{1,m}],\ldots,[h_{k-1,1},\ldots,h_{k-1,m}]\big]$,

$$p_{\mathfrak{L}_1,\ldots,\mathfrak{L}_k}^{H_1,\ldots,H_m}(\xi_1 = [h_{1,1},\ldots,h_{1,m}],\ldots,\xi_{k-1} = [h_{k-1,1},\ldots,h_{k-1,m}]) = p_{\mathfrak{L}_1,\ldots,\mathfrak{L}_{k-1}}^{H_1,\ldots,H_m}(\{h\}).$$

7. Now let us suppose in addition that the ring $\mathfrak{H}$ be at most countable. Then also $\mathfrak{L}$ is at most countable. By virtue of 6.6. and 6.7., and in view of Kolmogorov's existence theorem, we can, for all finite sequences $H_1, \ldots, H_m$ of pairwise disjoint sets in $\mathfrak{H}$, regard the family of $p_{\mathfrak{L}_1,\ldots,\mathfrak{L}_k}^{H_1,\ldots,H_m}$ as the family of finite-dimensional distributions of the distribution p of a random family $(\xi_L)_{\mathfrak{L}\in\mathfrak{L},\{H_1,\ldots,H_m\}\subseteq\mathfrak{L}^*}$ of m-dimensional vectors which have non-negative integer coordinates.

In view of 6.5., $(\xi_{\mathfrak{L}})$ is, in the sense of the semi-order relations in $\mathfrak{L}$ and R^m, monotone increasing with probability one. Consequently the limit formed along the directed system $\mathfrak{L}$

$$\xi = \lim_{\mathfrak{L}} \xi_{\mathfrak{L}}$$

exists with probability one. (For, in this context, the set systems $\mathfrak{L}$ with the property $\{H_1, \ldots, H_m\} \subseteq \mathfrak{L}^*$ constitute a subset of $\mathfrak{L}$ which differs only unessentially from $\mathfrak{L}$.)

We shall prove now that all of the coordinates of ξ are finite with probability one.

To begin with, we obtain for all j in $\{1, \ldots, m\}$ and all natural numbers k, using 6.3., 6.6., and 6.7.:

$$p_{\mathfrak{L}_1}^{H_1,\ldots,H_m}(\{[h_{1,1},\ldots,h_{1,m}]: h_{1,j} \leq k\}) = p_{\mathfrak{L}_1}^{H_j}(\{0,\ldots,k\}).$$

From the construction of $p_{\mathfrak{L}_1}^{H_j}$ and from 5. it is dear that $\mathfrak{L}_1$ can be replaced by $\mathfrak{J}_{1,j}$. Accordingly, in what follows we can additionally assume, without restriction of generality, that $\bigcup_{L\in\mathfrak{L}_1} L = H_j$. Now we continue the equation given above as follows:

$$= p_{\mathfrak{L}_1}(L \notin \mathscr{I} \quad \text{for at most } k \text{ sets } L \text{ in } \mathfrak{L}_1).$$

Let n denote the number of elements of $\mathfrak{L}_1$. In view of 4. and the transformations carried out in 2., we can further continue the equation as follows:

$$= p_{\mathfrak{L}_1}(V \in \mathscr{I} \quad \text{for at least } n-k \text{ sets } V \text{ in } \mathfrak{L}_1)$$
$$= \sum_{j=0}^{k} (-1)^j \binom{n-k-1+j}{j} s_{n-k+j}$$

where we set

$$s_r = \sum_{\mathfrak{K} \subseteq \mathfrak{L}_1, \operatorname{card}\mathfrak{K} = r} f\left(\bigcup_{K \in \mathfrak{K}} K\right) \qquad (r = 0, 1, \ldots, n).$$

In view of condition c), for any given $\varepsilon > 0$ the last row of this equation can be made greater than or equal to $1 - \varepsilon$ by selecting a sufficiently large k, this estimation being uniform in $\mathfrak{L}_1$. Hence also

$$\begin{aligned} &p\,(\text{the } j\text{-th coordinate of } \xi \text{ ins} \leqq k) \\ &= \lim_{\mathfrak{L}} p\,(\text{the } j\text{-th coordinate of } \xi_{\mathfrak{L}} \text{ is} \leqq k) \\ &= \lim_{\mathfrak{L}} p_{\mathfrak{L}}^{H_j}(\{0, \ldots, k\}) \geqq 1 - \varepsilon, \end{aligned}$$

and we obtain

$$p\,(\text{the } j\text{-th coordinate of } \xi \text{ is finite}) = 1 \qquad (j = 1, \ldots, m).$$

8. For all finite sequences $H_1, \ldots, H_m$ of pairwise disjoint sets in $\mathfrak{H}$, we set

$$p^{H_1,\ldots,H_m} = p\big(\xi \in (.)\big),$$

thus obtaining a family of distributions of vectors with non-negative integer coordinates.

For all non-negative integer $l_1, \ldots, l_m$,

$$p^{H_1,\ldots,H_m}(\{[l_1, \ldots, l_m]\}) = \lim_{\mathfrak{L}} p_{\mathfrak{L}}{}^{H_1,\ldots,H_m}(\{[l_1, \ldots, l_m]\}).$$

Thus the properties 6.1., 6.2., 6.3., and 6.4. are transferred to the family of the distributions $p^{H_1,\ldots,H_m}$. In view of the validity of d), and by means of Theorem 1.3.5., the existence of a distribution Q in $\boldsymbol{P}$ with the property

$$Q_{H_1,\ldots,H_m} = p^{H_1,\ldots,H_m}$$

can now be inferred. This distribution satisfies the condition

$$Q^*\big(\Phi(H) = 0\big) = Q\big(\Phi(H) = 0\big) = f(H) \qquad (H \in \mathfrak{H}).$$

If we now set $P = Q^*$, then we have found the desired representation of f. In view of 1.4.7., the distribution P is uniquely determined.

9. Now let us drop the additional assumption which requiries that $\mathfrak{H}$ be at most countable.

There is always an at most countable subsystem $\mathfrak{G}$ of $\mathfrak{B}$ which generates $\mathfrak{B}$. For each G in $\mathfrak{G}$, again there exists an at most countable subsystem $\mathfrak{H}_G$ of $\mathfrak{H}$ such that the set G is included in the σ-subring of $\mathfrak{B}$ generated by $\mathfrak{H}_G$. Then the subring $\mathfrak{H}'$ of $\mathfrak{H}$ which is generated by $\bigcup_{G \in \mathfrak{G}} \mathfrak{H}_G$, is at most countable, and generates $\mathfrak{B}$.

Now we substitute $\mathfrak{H}'$ for $\mathfrak{H}$, and restrict the function f to $\mathfrak{H}'$. Then by virtue of 8. there exists exactly one simple distribution P on $\mathfrak{M}$ with the property

$$P\big(\Phi(H) = 0\big) = f(H) \qquad (H \in \mathfrak{H}').$$

If $\mathfrak{H}'$ is different from $\mathfrak{H}$, then we select any H_0 from $\mathfrak{H} \setminus \mathfrak{H}'$, denoting by $\mathfrak{H}''$ the ring generated by $\mathfrak{H}' \cup \{H_0\}$. Also this ring is at most countable, so that again it can be concluded that there exists a simple P' in $\boldsymbol{P}$ with the property

$$P'(\Phi(H) = 0) = f(H) \qquad (H \in \mathfrak{H}'').$$

However, this implies

$$P'(\Phi(H) = 0) = f(H) \qquad (H \in \mathfrak{H}')$$

and consequently $P' = P$.

In view of the arbitrary selection of H_0 in $\mathfrak{H} \setminus \mathfrak{H}'$, it follows that

$$P(\Phi(H) = 0) = f(H) \qquad (H \in \mathfrak{H}).$$

Therefore the distribution P obtained by means of $\mathfrak{H}'$ already satisfies all requirements. ∎

As is shown by the example outlined below, condition c) in Theorem 1.4.10. cannot be omitted, even if $[A, \varrho_A] = R^1$ and $\mathfrak{H}$ coincides with the ring generated by $\{[\alpha, \beta)\}_{-\infty<\alpha\leqq\beta<+\infty}$.

Let ζ be a random real number being uniformly distributed over the interval $[0, 1]$, and let $(\alpha_m)_{m=1,2,\ldots}$ be a sequence of positive real numbers decreasing toward 0 in a strongly monotone way. Without stating all set-theoretical details, we now establish the "random infinite measure" $\omega = \delta_\zeta + \sum_{m=1}^{\infty} \delta_{\zeta+\alpha_m}$, where the random point ζ in $[0, 1]$ initiates a deterministic infinite "cluster" of points. We denote by P the distribution of ω, and set

$$P(\omega(H) = 0) = f(H) \qquad (H \in \mathfrak{H}).$$

Obviously f satisfies condition a) of Theorem 1.4.10. The first step of the proof of 1.4.10. shows that also b) is satisfied. We shall prove now that f satisfies the continuity condition d) as well.

Let $(H_n)_{n=1,2,\ldots}$ be a sequence of sets in $\mathfrak{H}$, which decreases monotonically toward $\emptyset$. If we denote by μ the one-dimensional Lebesgue measure, we have

$$1 - f(H_n) = \mu\left([0, 1] \cap \left(\bigcup_{m=1}^{\infty} (H_n - \alpha_m) \cup H_n\right)\right) \leqq \mu(H_n) + \mu\left(\bigcup_{m=1}^{\infty} (H_n - \alpha_m)\right).$$

Now we select any $\varepsilon > 0$. Due to $\mu(H_n) \xrightarrow[n\to\infty]{} 0$, there exists an n_0 with the property $\mu(H_{n_0}) < \frac{\varepsilon}{2}$.

For sufficiently small positive δ,

$$\mu(H_{n_0} - [0, \delta]) < \mu(H_{n_0}) + \frac{\varepsilon}{2} < \varepsilon$$

and hence also

$$\mu(H_n - [0, \delta]) < \varepsilon \qquad (n \geqq n_0).$$

Now we fix an m_0 such that for all $m > m_0$ the inequality $\alpha_m < \delta$ is satisfied For $n \geqq n_0$ we have:

$$\mu\left(\bigcup_{m=1}^{\infty} (H_n - \alpha_m)\right) \leqq \mu(H_n - [0, \delta]) + \sum_{m=1}^{m_0} \mu(H_n - \alpha_m) \leqq \varepsilon + m_0 \mu(H_n)$$

and hence

$$\overline{\lim_{n\to\infty}}\, \mu\left(\bigcup_{m=1}^{\infty} (H_n - \alpha_m)\right) \leqq \varepsilon .$$

Consequently

$$f(H_n) \xrightarrow[n\to\infty]{} 1 .$$

It remains only to show that f does not satisfy the condition c). For that purpose we set $H = [0, 1 + \alpha_1)$, selecting for $\mathfrak{L}_n$ the decomposition of H into n intervals of equal length, which are closed at their left and open at their right ends. Then for each natural number k we have

$$\begin{aligned}
&\lim_{n\to\infty} \sum_{j=0}^{k} (-1)^j \binom{n-k-1+j}{j} s_{n-k+j} \\
&= \lim_{n\to\infty} P(\omega(L) > 0 \quad \text{for at most } k \text{ sets } L \text{ in } \mathfrak{L}_n) \\
&= P(\omega(H) \leqq k) = 0 .
\end{aligned}$$

The "infinite random measure" ω can be interpreted as a "random closed set". A general theory of random sets without any finiteness conditions was founded by D. G. Kendall A lot of results of the theory of simple stochastics point processes can be generalized in the theory of random sets (cf. Kendall [1] and Matheron [1]).

In the same way as the extension theorem 1.3.5., theorem 1.4.10., too, can be simplified when restricted to probability laws of finite intensity. More specifically the following is true (cf. Kurtz [1]).

1.4.11. Theorem. *Let $\mathfrak{H}$ be a generating subring of $\mathfrak{B}$ and ν a measure in N. Further, let $\mathbf{\mathfrak{L}}$ denote the set of all non-empty, finite systems $\mathfrak{L}$ of pairwise disjoint non-empty sets in $\mathfrak{H}$. If f is a mapping of $\mathfrak{H}$ into $[0, +\infty)$, then there exists a simple distribution P on $\mathfrak{M}$ with the properties*

$$P(\Phi(H) = 0) = f(H) \quad \textit{for all } H \textit{ in } \mathfrak{H},$$

$$\varrho_P = \nu ,$$

if and only if for all non-empty sets H in $\mathfrak{H}$ the condition

e) $$\sup_{\mathfrak{L}\in\mathbf{\mathfrak{L}},\ \bigcup_{X\in\mathfrak{L}} X = H} \ \sum_{X\in\mathfrak{L}} (1 - f(X)) = \nu(H)$$

is fulfilled and the function f satisfies the conditions a) *and* b) *of theorem* 1.4.10.

Proof. 1. Let P be a simple distribution with intensity measure ν. Obviously f satisfies condition a). The first step of the proof of 1.4.10. shows that b) is also satisfied. We shall prove now that f satisfies e).

Because of 1.4.6., for each non-empty H in $\mathfrak{H}$ and each $\varepsilon > 0$ there exists a decomposition $\mathfrak{Z}_\varepsilon''$ of H with the property

$$\frac{\varepsilon}{4} + \sum_{Z\in\mathfrak{Z}_\varepsilon''} (1 - f(Z)) \geqq \nu(H) .$$

Obviously one can select a finite subsystem $\mathfrak{Z}_\varepsilon'$ of $\mathfrak{Z}_\varepsilon''$ such that

$$\frac{\varepsilon}{2} + \sum_{Z \in \mathfrak{Z}_\varepsilon'} \big(1 - f(Z)\big) \geqq \nu(H)$$

If we now replace $\mathfrak{Z}_\varepsilon''$ by $\mathfrak{Z}_\varepsilon' \cup \Big\{ \bigcup_{Z \in (\mathfrak{Z}_\varepsilon'' \setminus \mathfrak{Z}_\varepsilon')} Z \Big\} = \mathfrak{Z}_\varepsilon$ we obtain

$$\frac{\varepsilon}{2} + \sum_{Z \in \mathfrak{Z}_\varepsilon} \big(1 - f(Z)\big) \geqq \nu(H).$$

Applying the technique used in the third step of the proof of 1.4.5. we associate with each Z in $\mathfrak{Z}_\varepsilon$ a set X_Z in $\mathfrak{H}$ such that $\bigcup_{Z \in \mathfrak{Z}_\varepsilon} X_Z = H$ and

$$\frac{\varepsilon}{2} + \sum_{Z \in \mathfrak{Z}_\varepsilon} \big(1 - f(X_Z)\big) \geqq \sum_{Z \in \mathfrak{Z}_\varepsilon} \big(1 - f(Z)\big).$$

If now we put $\mathfrak{L}_\varepsilon = \{X_Z\}_{Z \in \mathfrak{Z}_\varepsilon} \setminus \{\emptyset\}$, we obtain $\mathfrak{L}_\varepsilon \in \mathbf{\mathfrak{L}}$, $\bigcup_{X \in \mathfrak{L}_\varepsilon} X = H$ and

$$\varepsilon + \sum_{X \in \mathfrak{L}_\varepsilon} \big(1 - f(X)\big) \geqq \nu(H).$$

Taking into consideration that for all decompositions $\mathfrak{Z}$ of H the inequality

$$\nu(H) = \sum_{Z \in \mathfrak{Z}} \nu(Z) \geqq \sum_{Z \in \mathfrak{Z}} \big(1 - f(Z)\big)$$

is satisfied we obtain

$$\nu(H) = \sup_{\mathfrak{L} \in \mathbf{\mathfrak{L}},\ \bigcup_{X \in \mathfrak{L}} X = H} \ \sum_{X \in \mathfrak{L}} \big(1 - f(X)\big).$$

This shows that the conditions mentioned above are necessary.

2. Conversely, suppose that the mapping f satisfies the conditions a), b) and e).

The fourth and fifth step of the proof of 1.4.10. can be carried out without changes under the new assumptions. Therefore, under the additional assumption that $\mathfrak{H}$ is at most countable, we conclude from the deductions at the beginning of the sixth step of the proof that for each finite family $H_1, \ldots, H_m$ of pairwise disjoint sets in $\mathfrak{H}$ the random family $(\xi_{\mathfrak{L}})_{\mathfrak{L} \in \mathbf{\mathfrak{L}}, \{H_1, \ldots, H_m\} \subseteq \mathfrak{L}^*}$ is monotone increasing in the sense of the semi-order relations in $\mathbf{\mathfrak{L}}$ and R^m, almost everywhere with respect to p.

We shall prove next, that the limit ξ of $\xi_{\mathfrak{L}}$ formed along the directed system $\mathbf{\mathfrak{L}}$ is finite with probability one. It follows immediately from the definitions that the expected value with respect to p of the i-th coordinate of $\xi_{\mathfrak{L}}$ coincides with

$$\sum_{X \in \mathfrak{L}, X \subseteq H_i} \big(1 - f(X)\big).$$

Because of the condition e) the supremum of all these expectations is equal to $\nu(H_i)$ and, therefore, finite. With the aid of the theorem of B. Levi we deduce that the i-th coordinate of ξ is finite with probability one and has the expectation $\nu(H_i)$.

As in the eighth step of the proof of 1.4.10. we introduce now the distributions $p^{H_1,\ldots,H_m}$ with the aid of the random family $\xi_{\mathfrak{L}}$. We note that these distributions satisfy the conditions a), b) and c) of the extension theorem 1.3.5. Moreover, as we have seen above,

$$\sum_{l=1}^{\infty} l p^H(\{l\}) = \nu(H) \qquad (H \in \mathfrak{H}).$$

In view of the extension theorem 1.3.6., it can be deduced that there exists a well-defined distribution P with the property

$$P_{H_1,\ldots,H_m} = p^{H_1,\ldots,H_m}.$$

This distribution possesses the intensity measure ν. Hence, according to e) for each H in $\mathfrak{H}$ and each $\varepsilon > 0$ there exists a decomposition $\mathfrak{Z}_\varepsilon$ of H with the property

$$\varepsilon + \sum_{Z \in \mathfrak{Z}_\varepsilon} P(\Phi(Z) > 0) \geq \nu(H) = \sum_{Z \in \mathfrak{Z}_\varepsilon} \nu(Z)$$
$$\geq \sum_{Z \in \mathfrak{Z}_\varepsilon} P(\Phi(Z) > 0) + \sum_{Z \in \mathfrak{Z}_\varepsilon} P(\Phi(Z) > 1).$$

Thus we obtain

$$P(\Phi^*(H) < \Phi(H)) \leq \sum_{Z \in \mathfrak{Z}_\varepsilon} P(\Phi(Z) > 1) < \varepsilon,$$

and hence

$$P(\Phi^*(H) \neq \Phi(H)) = 0 \qquad (H \in \mathfrak{H}).$$

Since H is a generating subring of $\mathfrak{B}$, each X in $\mathfrak{B}$ and hence also the whole of A can be covered by a sequence of sets in $\mathfrak{H}$. Thus it follows from 1.4.4. that P is simple.

The additional assumption requiring that $\mathfrak{H}$ be at most countable can now be dropped in the same way as in the last step of the proof of 1.4.10. ∎

A distribution P on $\mathfrak{M}$ is called *orderly*, if for each $\varepsilon > 0$ there exists a decomposition $\mathfrak{Z}$ of A with the property $\sum_{Z \in \mathfrak{Z}} P(\Phi(Z) > 1) < \varepsilon$. Then for all X in $\mathfrak{A}$, $X \cap \mathfrak{Z}$ provides a decomposition $\mathfrak{Z}'$ of X with the property $\sum_{Z \in \mathfrak{Z}'} P(\Phi(Z) > 1) < \varepsilon$. Conversely, if for each $\varepsilon > 0$ and for each X in $\mathfrak{B}$ there exists such a decomposition of X, we can cover A by a sequence (X_n) of pairwise disjoint sets in $\mathfrak{B}$ and associate with each n a decomposition $\mathfrak{Z}_n$ of X_n with the property $\sum_{Z \in \mathfrak{Z}_n} P(\Phi(Z) > 1) < \frac{\varepsilon}{2^n}$. Then the union $\mathfrak{Z} = \bigcup_{n=1}^{\infty} \mathfrak{Z}_n$ represents a decomposition of A with the property $\sum_{Z \in \mathfrak{Z}} P(\Phi(Z) > 1) < \varepsilon$.

For each decomposition $\mathfrak{Z}$ of A, we have

$$P(\Phi \text{ is not simple}) \leq \sum_{Z \in \mathfrak{Z}} P(\Phi^*(Z) \neq \Phi(Z)) \leq \sum_{Z \in \mathfrak{Z}} P(\Phi(Z) > 1)$$

We thus see the validity of

1.4.12. *Each orderly distribution P on $\mathfrak{M}$ is simple.*

In section 1.8., we shall show by means of a counterexample, that the implication of 1.4.12. cannot be reversed. On the other hand we can prove the following proposition (cf. Leadbetter [1]):

1.4.13. Proposition. *For each distribution P on $\mathfrak{M}$ of finite intensity, the following properties are equivalent:*

a) *P is simple.*

b) *For each X in $\mathfrak{B}$ and each distinguished sequence $(\mathfrak{Z}_n)$ of decompositions of X,*

$$\sum_{Z \in \mathfrak{Z}_n} P(\Phi(Z) = 1) \xrightarrow[n \to \infty]{} \varrho_P(X).$$

c) *For each X in $\mathfrak{B}$ and each distinguished sequence $(\mathfrak{Z}_n)$ of decompositions of X,*

$$\sum_{Z \in \mathfrak{Z}_n} P(\Phi(Z) > 0) \xrightarrow[n \to \infty]{} \varrho_P(X).$$

d) *P is orderly.*

Proof. 1. a) implies b). As it is shown by the proof of 1.4.1., for each Φ in M_s and each distinguished sequence $(\mathfrak{Z}_n)$ of decompositions of a set X in $\mathfrak{B}$,

$$\sum_{Z \in \mathfrak{Z}_n} k_{\{1\}}(\Phi(Z)) \xrightarrow[n \to \infty]{} \Phi(X),$$

so that, using Lebesgue's theorem, it can be concluded that

$$\int \Big(\sum_{Z \in \mathfrak{Z}_n} k_{\{1\}}(\Phi(Z))\Big) P(\mathrm{d}\Phi) = \sum_{Z \in \mathfrak{Z}_n} P(\Phi(Z) = 1) \xrightarrow[n \to \infty]{} \varrho_P(X)$$

2. Obviously c) represents a weakened version of b).

3. c) implies d), for we have

$$\begin{aligned}\sum_{Z \in \mathfrak{Z}_n} P(\Phi(Z) > 1) &= \sum_{Z \in \mathfrak{Z}_n} \Big(\sum_{l=2}^{\infty} P(\Phi(Z) = l)\Big) \\ &\leqq \sum_{Z \in \mathfrak{Z}_n} \Big(\sum_{l=1}^{\infty} l P(\Phi(Z) = l) - \sum_{l=1}^{\infty} P(\Phi(Z) = l)\Big) \\ &= \varrho_P(X) - \sum_{Z \in \mathfrak{Z}_n} P(\Phi(Z) > 0).\end{aligned}$$

4. In view of 1.4.12., d) implies a). ∎

In virtue of 1.4.13., each simple P in $\boldsymbol{P}$ of finite intensity is orderly in a stricter sense: For each distinguished sequence $(\mathfrak{Z}_n)$ of decompositions of a set X in $\mathfrak{B}$,

$$\sum_{Z \in \mathfrak{Z}_n} P(\Phi(Z) > 1) \xrightarrow[n \to \infty]{} 0.$$

In section 9.5. we shall construct an orderly distribution which does not have this property.

1.5. The Convolution in $\boldsymbol{E}$

Let $\boldsymbol{E}$ denote the set of all finite, complex-valued measures on $\mathfrak{M}$. With the usual definition

$$(c_1 E_1 + c_2 E_2)(Y) = c_1 E_1(Y) + c_2 E_2(Y)$$

for all c_1, c_2 in the field of complex numbers, all E_1, E_2 in $\boldsymbol{E}$, and all Y in $\mathfrak{M}$, $\boldsymbol{E}$ becomes a vector space over the field of complex numbers.

For all X in $\mathfrak{B}$,

$$[\Phi_1, \Phi_2] \curvearrowright \Phi_1(X) + \Phi_2(X) = (\Phi_1 + \Phi_2)(X)$$

is a measurable mapping of $[M, \mathfrak{M}] \times [M, \mathfrak{M}]$ into $\{0, 1, \ldots\}$. By definition of $\mathfrak{M}$, the correspondence $[\Phi_1, \Phi_2] \curvearrowright \Phi_1 + \Phi_2$ therefore provides a measurable mapping of $[M, \mathfrak{M}] \times [M, \mathfrak{M}]$ into $[M, \mathfrak{M}]$. The sum $\Phi_1 + \Phi_2$ can be interpreted as superposition of Φ_1 and Φ_2.

If E_1, E_2 are any measures in $\boldsymbol{E}$, then we can construct the finite, complex-valued measure $E_1 \times E_2$ on the σ-algebra $\mathfrak{M} \otimes \mathfrak{M}$. The mapping $[\Phi_1, \Phi_2] \curvearrowright \Phi_1 + \Phi_2$ transforms $E_1 \times E_2$ into a finite, complex-valued measure $E_1 * E_2$ on $\mathfrak{M}$. We call $E_1 * E_2$ the *convolution* of E_1 with E_2. Obviously $\delta_\Phi * \delta_\Psi = \delta_{\Phi+\Psi}$ for all Φ, Ψ in M.

Since $\Phi_1 + \Phi_2 = \Phi_2 + \Phi_1$ the convolution in $\boldsymbol{E}$ is commutative. If E_1, E_2, and E_3 are in $\boldsymbol{E}$, then the mapping $[\Phi_1, \Phi_2, \Phi_3] \curvearrowright \Phi_1 + \Phi_2 + \Phi_3$ transforms the product measure $E_1 \times E_2 \times E_3$ into $E_1 * (E_2 * E_3)$, where the latter is regarded as having been built up in two stages:

$$[\Phi_1, \Phi_2, \Phi_3] \curvearrowright [\Phi_1, \Phi_2 + \Phi_3] \curvearrowright \Phi_1 + (\Phi_2 + \Phi_3).$$

However, if we split the mapping as follows:

$$[\Phi_1, \Phi_2, \Phi_3] \curvearrowright [\Phi_1 + \Phi_2, \Phi_3] \curvearrowright (\Phi_1 + \Phi_2) + \Phi_3,$$

we obtain $(E_1 * E_2) * E_3$. Hence the convolution is also associative.

For all Φ in M, $o + \Phi = \Phi$. Consequently $\delta_o * E$ equals E for all E in $\boldsymbol{E}$, i.e. the distribution δ_o is the neutral element of the operation $*$.

For all c_1, c_2 in the field of complex numbers and all E_1, E_2, E in $\boldsymbol{E}$, we have

$$(c_1E_1 + c_2E_2) \times E = c_1(E_1 \times E) + c_2(E_2 \times E)$$

and hence

$$(c_1E_1 + c_2E_2) * E = c_1(E_1 * E) + c_2(E_2 * E).$$

Let us now summarize the above statements:

1.5.1. Proposition. *Endowed with the convolution $*$ the linear space $\boldsymbol{E}$ becomes an associative and commutative algebra over the field of complex numbers which has the unity element δ_o.*

For all $n \geqq 1$, we denote by E^n the n-th *convolution power* of an element E of $\boldsymbol{E}$, setting E^0 equal to the unity element δ_o of $\boldsymbol{E}$.

For all X in $\mathfrak{A}$ and all E in $\boldsymbol{E}$, we define

$$_X E = E\big(_X\Phi \in (.)\big).$$

As an immediate consequence of the definitions we obtain

1.5.2. *For all X in $\mathfrak{A}$, the correspondence $E \curvearrowright {}_XE$ is a homomorphic mapping of the algebra $\boldsymbol{E}$ into itself.*

Obviously $\boldsymbol{P}$ is closed under the operation $*$. The convolution $P_1 * \cdots * P_m$ represents the distribution of the superposition $\Phi_1 + \cdots + \Phi_m$ of independent random Φ_i distributed according to P_i. In the special case $A = \{1, \ldots, m\}$ the convolution of distributions on M corresponds to the conventionally defined convolution of the associated distributions of random m-dimensional vectors.

For all finite sequences $X_1, \ldots, X_m$ of pairwise disjoint sets in $\mathfrak{A}$, the sum

$$_{\bigcup_{i=1}^{m} X_i}\Phi = {}_{X_1}\Phi + \cdots + {}_{X_m}\Phi$$

allows unique conclusions to be drawn for its summands. Consequently, if the summands are random, then the distribution of the sum already determines that of the random vector $[{}_{X_1}\Phi, \ldots, {}_{X_m}\Phi]$. Thus we have:

1.5.3. *Let $X_1, \ldots, X_m$ be a finite sequence of pairwise disjoint sets in $\mathfrak{A}$, and let P be a distribution on $\mathfrak{M}$. Subject to these assumptions, the finite sequence ${}_{X_1}\Phi, \ldots, {}_{X_m}\Phi$ is independent with respect to P if and only if the condition*

$$_{\bigcup_{i=1}^{m} X_i}P = {}_{X_1}P * \cdots * {}_{X_m}P$$

is satisfied.

In the case where P and Q are free from after-effects, for each finite sequence of pairwise disjoint sets in $\mathfrak{A}$, 1.5.2., 1.5.3., and 1.3.4. imply that:

$$_{\bigcup_{i=1}^{m} X_i}(P * Q) = {}_{\bigcup_{i=1}^{m} X_i}P * {}_{\bigcup_{i=1}^{m} X_i}Q = \left(\mathop{\Huge *}_{i=1}^{m} {}_{X_i}P\right) * \left(\mathop{\Huge *}_{i=1}^{m} {}_{X_i}Q\right) = \mathop{\Huge *}_{i=1}^{m} {}_{X_i}(P * Q),$$

i.e. we have

1.5.4. *The convolution $P * Q$ of distributions P, Q on $\mathfrak{M}$ which are free from after-effects, is itself free from after-effects.*

Obviously, the convolution of continuous or of discrete distributions is itself continuous or discrete, respectively.

The convolution of simple distributions is not necessarily simple, as is shown by the example $\delta_{\delta_a} * \delta_{\delta_a} = \delta_{2\delta_a}$; however, the following statement is valid:

1.5.5. Proposition. *The convolution $P * Q$ of two distributions P, Q on $\mathfrak{M}$ is simple if and only if P and Q are simple and the condition*

$$\{a : a \in A,\ P(\Phi(\{a\}) > 0) > 0\} \cap \{a : a \in A,\ Q(\Phi(\{a\}) > 0) > 0\} = \emptyset$$

is satisfied.

Proof. By definition we have:

$$(P * Q)\,(\Phi \text{ is simple}) = (P \times Q)\,(\Phi_1 + \Phi_2 \text{ is simple}).$$

Now, if $P * Q$ is simple, we obtain:

$$\begin{aligned} 1 &= (P \times Q)\,(\Phi_1 + \Phi_2 \text{ is simple}) \leqq (P \times Q)\,(\Phi_1, \Phi_2 \text{ are simple}) \\ &= P\,(\Phi \text{ is simple}) \cdot Q\,(\Phi \text{ is simple}), \end{aligned}$$

and for all a in A

$$\begin{aligned} 0 &= (P \times Q)\big(\Phi_1(\{a\}) > 0,\quad \Phi_2(\{a\}) > 0\big) \\ &= P\big(\Phi(\{a\}) > 0\big)\, Q\big(\Phi(\{a\}) > 0\big). \end{aligned}$$

Hence the stated conditions for the simplicity of $P * Q$ are necessary.

Conversely, suppose that these conditions are satisfied. We obtain

$$(P \times Q)\ (\Phi_1 + \Phi_2 \text{ is simple}) = \int P(\Phi + \Psi \text{ is simple})\, Q(\mathrm{d}\Psi).$$

Almost all Ψ are simple and, moreover, satisfy the condition

$$\big\{a\colon a \in A,\ \Psi(\{a\}) > 0\big\} \cap \big\{a\colon a \in A,\ P\big(\Phi(\{a\}) > 0\big) > 0\big\} = \emptyset.$$

However, for all such Ψ, $P(\Phi + \Psi$ is simple$) = 1$. Consequently

$$\int P(\Phi + \Psi \text{ is simple})\, Q(\mathrm{d}\Psi) = 1. \ ∎$$

Therefore, if two distributions P, Q are both simple and continuous, this is also true for their convolution.

1.5.6. *For all distributions P, Q on $\mathfrak{M}$ and all X in $\mathfrak{A}$ having the property that* $\quad P\big(\Phi(X) = 0\big) > 0, \quad Q\big(\Phi(X) = 0\big) > 0,$

whe have

$$(P * Q)\big((.) \mid \Phi(X) = 0\big) = \Big(P\big((.) \mid \Phi(X) = 0\big)\Big) * \Big(Q\big((.) \mid \Phi(X) = 0\big)\Big).$$

Proof. Due to $(P * Q)\ \big(\Phi(X) = 0\big) = P\big(\Phi(X) = 0\big)\, Q\big(\Phi(X) = 0\big)$, for each Y in $\mathfrak{M}$ we obtain the following relations:

$$\begin{aligned} (P * Q)\big(Y \mid \Phi(X) = 0\big) &= \frac{(P * Q)\big(\Phi \in Y,\ \ \Phi(X) = 0\big)}{P\big(\Phi(X) = 0\big)\, Q\big(\Phi(X) = 0\big)} \\ &= \frac{(P \times Q)\big(\Phi_1 + \Phi_2 \in Y,\ \ \Phi_1(X) = 0,\ \ \Phi_2(X) = 0}{P\big(\Phi(X) = 0\big)\ Q\big(\Phi(X) = 0\big)} \\ &= \Big(P\big((.) \mid \Phi(X) = 0\big) \times Q\big((.) \mid \Phi(X) = 0\big)\Big)\,(\Phi_1 + \Phi_2 \in Y) \\ &= \Big(\Big(P\big((.) \mid \Phi(X) = 0\big)\Big) * \Big(Q\big((.) \mid \Phi(X) = 0\big)\Big)\Big)(Y). \ ∎ \end{aligned}$$

For all finite sequences $X_1, \ldots, X_m$ of sets in $\mathfrak{B}$ and all E in $\boldsymbol{E}$, we denote by $E_{X_1,\ldots,X_m}$ the image of E under the measurable mapping

$$\Phi \curvearrowright \sum_{i=1}^{m} \Phi(X_i)\, \delta_i$$

of $[M, \mathfrak{M}]$ into the measurable space corresponding to the phase space $[A', \varrho_{A'}] = \{1, \ldots, m\}$. On the basis of considerations discussed at the end of section 1.1. the measure $E_{X_1,\ldots,X_m}$ can be interpreted also as a measure on $\mathfrak{P}(\{0, 1, \ldots\}^m)$.

Directly from the definitions we have

1.5.7. *For all finite sequences* $X_1, \ldots, X_m$ *of sets in* $\mathfrak{B}$, *the correspondence*

$$E \rightsquigarrow E_{X_1,\ldots,X_m}$$

is a homomorphic mapping of the algebra **E** *into the algebra* **E**′ *corresponding to the phase space* $\{1, \ldots, m\}$.

Each finite, complex-valued measure V on $\mathfrak{P}(\{0, 1, \ldots\}^m)$ is uniquely determined by the formal power series

$$\sum_{0 \leq n_1,\ldots,n_m < +\infty} V(\{[n_1, \ldots, n_m]\})\, \xi_1^{n_1} \cdots \xi_m^{n_m}.$$

In fact, we obtain in this way exactly those formal power series $\sum_{0 \leq n_1,\ldots,n_m < +\infty} a_{n_1,\ldots,n_m} \xi_1^{n_1} \cdots \xi_m^{n_m}$ which converge absolutely on $[-1, 1]^m$.

Now, the statement of 1.5.7. can be formulated as follows: For all finite sequences $X_1, \ldots, X_m$ of sets in $\mathfrak{B}$ the correspondence

$$E \rightsquigarrow \sum_{0 \leq n_1,\ldots,n_m < +\infty} E\big(\Phi(X_1) = n_1, \ldots, \Phi(X_m) = n_m\big)\, \xi_1^{n_1} \cdots \xi_m^{n_m}$$

is a homomorphic mapping of the algebra **E** into the algebra $C[[\xi_1, \ldots, \xi_m][$ of formal power series in m variables $\xi_1, \ldots, \xi_m$ with complex coefficients.

The algebra $C[[\xi_1, \ldots, \xi_m]]$ has no divisors of zero. Indeed, if

$$\Big(\sum_{0 \leq n_1,\ldots,n_m < +\infty} a_{n_1,\ldots,n_m} \xi_1^{n_1} \cdots \xi_m^{n_m}\Big)\Big(\sum_{0 \leq n_1,\ldots,n_m < +\infty} b_{n_1,\ldots,n_m} \xi_1^{n_1} \cdots \xi_m^{n_m}\Big) = 0$$

and

$$\sum_{0 \leq n_1,\ldots,n_m < +\infty} a_{n_1,\ldots,n_m} \xi_1^{n_1} \cdots \xi_m^{n_m}, \qquad \sum_{0 \leq n_1,\ldots,n_m < +\infty} b_{n_1,\ldots,n_m} \xi_1^{n_1} \cdots \xi_m^{n_m} \neq 0$$

then we select from the sets

$$Z_1 = \{[n_1, \ldots, n_m] : [n_1, \ldots, n_m] \in \{0, 1, \ldots\}^m,\ a_{n_1,\ldots,n_m} \neq 0\},$$
$$Z_2 = \{[n_1, \ldots, n_m] : [n_1, \ldots, n_m] \in \{0, 1, \ldots\}^m,\ b_{n_1,\ldots,n_m} \neq 0\}$$

the minimal (in the sense of lexicographic order) m-tuples, say $[v_1, \ldots, v_m]$ and $[g_1, \ldots, g_m]$ respectively, and we obtain

$$0 = \sum_{\substack{[n_1,\ldots,n_m] \in Z_1, [n_1',\ldots,n_m'] \in Z_2 \\ [n_1+n_1',\ldots,n_m+n_m'] = [v_1+g_1,\ldots,v_m+g_m]}} a_{n_1,\ldots,n_m} b_{n_1',\ldots,n_m'} = a_{v_1,\ldots,v_m} b_{g_1,\ldots,g_m} \neq 0,$$

which is a contradiction.

The proof of the uniqueness proposition 1.3.2. shows that two measures E_1, E_2 in **E** coincide if for all finite sequences $X_1, \ldots, X_m$ of sets in $\mathfrak{B}$ the equation $(E_1)_{X_1,\ldots,X_m} = (E_2)_{X_1,\ldots,X_m}$ holds. Now, if $E_1 * E_2 = 0$ and $E_1, E_2 \neq 0$ is valid, then, first of all, there exists a finite sequence $X_1, \ldots, X_m, X_{m+1}, \ldots, X_{m+k}$ of sets in $\mathfrak{B}$ such that $(E_1)_{X_1,\ldots,X_m}, (E_2)_{X_{m+1},\ldots,X_{m+k}} \neq 0$. We have a fortiori

$$(E_1)_{X_1,\ldots,X_{m+k}}, \qquad (E_2)_{X_1,\ldots,X_{m+k}} \neq 0$$

and

$$(E_1)_{X_1,\ldots,X_{m+k}} * (E_2)_{X_1,\ldots,X_{m+k}} = 0$$

which is a contradiction. Thus we obtain

1.5.8. Proposition. *The algebra* $\boldsymbol{E}$ *has no divisors of zero.*

The convolution quotient of distributions P, Q which are free from after-effects is itself free from after-effects if it exists and belongs to $\boldsymbol{P}$.

1.5.9. *Let P, Q and L be distributions on $\mathfrak{M}$. If $P = Q * L$ and P, Q are free from after-effects, then L, which is uniquely determined by virtue of 1.5.8., is also free from after-effects.*

Proof. Let $X_1, \ldots, X_m$ be any finite sequence of pairwise disjoint sets in $\mathfrak{B}$. We obtain

$$\begin{aligned} P_{X_1} \times \cdots \times P_{X_m} &= P_{X_1,\ldots,X_m} = Q_{X_1,\ldots,X_m} * L_{X_1,\ldots,X_m} \\ &= (Q_{X_1} \times \cdots \times Q_{X_m}) * L_{X_1,\ldots,X_m}. \end{aligned}$$

On the other hand,

$$P_{X_i} = Q_{X_i} * L_{X_i} \qquad (i = 1, \ldots, m)$$

and hence also

$$P_{X_1} \times \cdots \times P_{X_m} = (Q_{X_1} \times \cdots \times Q_{X_m}) * (L_{X_1} \times \cdots \times L_{X_m}).$$

It follows that

$$L_{X_1,\ldots,X_m} = L_{X_1} \times \cdots \times L_{X_m},$$

i.e. L is free from after-effects. ∎

In view of 1.3.4. and 1.5.3. each distribution P which is free from after-effects can be represented as a convolution of the continuous distribution P_c and the discrete distribution P_d, both being free from after-effects.

Now let $P = Q_1 * Q_2$ be an arbitrary representation of P as a convolution of a continuous Q_1 and a discrete Q_2, both being free from after-effects. If we denote by X the countable set $\{a : a \in A,\ P(\Phi(\{a\}) > 0) > 0\}$, then $X = \{a : a \in A,\ Q_2(\Phi(\{a\}) > 0) > 0\}$ and we obtain

$$P_d = {}_X P = {}_X(Q_1) * {}_X(Q_2) = \delta_o * Q_2 = Q_2$$

from which, using 1.5.8., we can conclude that $P_c = Q_1$. Thus we perceive the truth of

1.5.10. Proposition. *Each distribution P on $\mathfrak{M}$ which is free from after-effects can be represented in exactly one way as a convolution $P = P_1 * P_2$ of a continuous distribution P_1 and a discrete distribution P_2, both being free from after-effects.*

If the relation $Q^n = P$ holds for two elements P, Q of $\boldsymbol{P}$ and for a natural number n, we write $Q = \sqrt[n]{P}$ and call Q an n-th *convolution root* of P in $\boldsymbol{P}$.

1.5.11. Proposition. *For all P in $\boldsymbol{P}$ and all natural numbers n, there exists at most one n-th convolution root of P in $\boldsymbol{P}$.*

Proof. By virtue of 1.5.1. and 1.5.8., the ring $\boldsymbol{E}$ can be isomorphically embedded into a field $\boldsymbol{K}$, which in its turn represents an algebra over the field of complex numbers. Now, if Q is any n-th convolution root of P in $\boldsymbol{P}$, and if the sequence $e_1 = 1, e_2, \ldots, e_n$ runs through the n-th roots of unity in the field of complex numbers, then all products $Q_i = e_i Q$ in $\boldsymbol{E}$ satisfy the equation $(Q_i)^n - P = 0$. However, in $\boldsymbol{K}$ an equation of degree n can have at most n different roots. Thus the Q_i provide all of the n-th convolution roots of P in $\boldsymbol{E}$. On the other hand only $Q_1 = Q$ is in $\boldsymbol{P}$. Hence the equation $L^n - P = 0$ has only one solution in $\boldsymbol{P}$, namely Q. ∎

The set of all distributions which are free from after-effects is closed under the formation of roots, i.e. the following is true.

1.5.12. *If for any natural number n the n-th convolution power of a distribution Q on $\mathfrak{M}$ is free from after-effects, then Q itself is free from after-effects.*

Proof. For every finite sequence $X_1, \ldots, X_m$ of pairwise disjoint sets in $\mathfrak{B}$ we obtain

$$(Q_{X_1} \times \cdots \times Q_{X_m})^n = (Q_{X_1})^n \times \cdots \times (Q_{X_m})^n$$
$$= (Q^n)_{X_1} \times \cdots \times (Q^n)_{X_m} = (Q^n)_{X_1,\ldots,X_m} = (Q_{X_1,\ldots,X_m})^n$$

and hence

$$Q_{X_1} \times \cdots \times Q_{X_m} = \sqrt[n]{(Q_{X_1,\ldots,X_m})^n} = Q_{X_1,\ldots,X_m}. \quad ∎$$

The n-th convolution root $\sqrt[n]{P}$ need not exist. Thus for instance the equation $Q^n = \delta_\Phi$ has a solution in $\boldsymbol{P}$ if and only if all multiplicities $\Phi(\{a\})$ of points in Φ are divisible by n.

1.6. The Distributions $\mathcal{U}_E$

A distribution P in $\boldsymbol{P}$ is called *infinitely divisible* if for each $n = 1, 2, \ldots$ there is an n-th convolution root $\sqrt[n]{P}$ of P in $\boldsymbol{P}$. We denote by $\boldsymbol{T}$ the set of all infinitely divisible distributions P on $\mathfrak{M}$.

Using 1.5.2. it is easily shown that

1.6.1. *For all P in $\boldsymbol{T}$ and all X in $\mathfrak{A}$, the distribution ${}_XP$ belongs to $\boldsymbol{T}$.*

Likewise, using 1.5.7. we obtain

1.6.2. *For all P in $\boldsymbol{T}$ and all finite sequences $X_1, \ldots, X_m$ of sets in $\mathfrak{B}$* the *finite-dimensional distribution $P_{X_1,\ldots,X_m}$ is infinitely divisible.*

This assertion can be reversed in a stronger form:

1.6.3. Proposition. *Let $\mathfrak{H}$ be a generating semiring of sets in $\mathfrak{B}$. If each set in $\mathfrak{B}$ can be covered by a finite sequence of sets in $\mathfrak{H}$, then a distribution on $\mathfrak{M}$ belongs to $\boldsymbol{T}$ if for all finite sequences $H_1, \ldots, H_m$ of pairwise disjoint sets in $\mathfrak{H}$ the distribution $P_{H_1,\ldots,H_m}$ is infinitely divisible.*

Proof. Suppose that an arbitrary natural number k is given. For each finite sequence $H_1, \ldots, H_m$ of pairwise disjoint sets in $\mathfrak{H}$, we set

$$p^{H_1,\ldots,H_m} = \sqrt[k]{P_{H_1,\ldots,H_m}}.$$

We shall show that our set-up satisfies the conditions a) through d) of the existence theorem 1.3.5.

Obviously this is true for a), b) and c). Now suppose that some sequence $(H_{n_1} \cup \ldots \cup H_{n_{m_n}})_{n=1,2,\ldots}$ of finite unions of pairwise disjoint sets in $\mathfrak{B}$ decreases monotonically towards $\varnothing$. In view of 1.3.5., we have

$$\lim_{n\to\infty} P_{H_{n_1},\ldots,H_{n_{m_n}}}(\{[0,\ldots,0]\}) = 1.$$

But on the other hand,

$$\left(p^{H_{n_1},\ldots,H_{n_{m_n}}}(\{[0,\ldots,0]\})\right)^k = P_{H_{n_1},\ldots,H_{n_{m_n}}}(\{[0,\ldots,0]\}).$$

Thus we can conclude that

$$\lim_{n\to\infty} p^{H_{n_1},\ldots,H_{n_{m_n}}}(\{[0,\ldots,0]\}) = 1$$

i.e. that the condition d) is also satisfied.

Hence there exists a distribution Q on $\mathfrak{M}$ which satisfies the equation

$$Q_{H_1,\ldots,H_m} = \sqrt[k]{P_{H_1,\ldots,H_m}}$$

for all finite sequences $X_1, \ldots, X_m$ of pairwise disjoint sets in $\mathfrak{H}$. Then by 1.5.7. we always have

$$(Q^k)_{H_1,\ldots,H_m} = P_{H_1,\ldots,H_m},$$

so that, using the uniqueness theorem 1.3.2., it can be concluded that $Q^k = P$. Consequently the distribution P is infinitely divisible. ∎

We can construct infinitely divisible distributions on $\mathfrak{M}$ in the following way: Let $\boldsymbol{E}^+$ denote the set of all finite measures on $\mathfrak{M}$. Now, if we set

$$\mathcal{U}_E = e^{-E(M)} \sum_{n=0}^{\infty} \frac{1}{n!} E^n,$$

then with each E in $\boldsymbol{E}^+$ we have associated a $\mathcal{U}_E$ in $\boldsymbol{P}$. Obviously $\mathcal{U}_0 = \delta_0$.

1.6.4. *For all E_1, E_2 in $\boldsymbol{E}^+$,*

$$\mathcal{U}_{E_1} * \mathcal{U}_{E_2} = \mathcal{U}_{E_1+E_2}.$$

Proof. We observe that

$$\begin{aligned}\mathcal{U}_{E_1} * \mathcal{U}_{E_2} &= \left(e^{-E_1(M)} \sum_{n=0}^{\infty} \frac{1}{n!} {E_1}^n\right)\left(e^{-E_2(M)} \sum_{m=0}^{\infty} \frac{1}{m!} {E_2}^m\right) \\ &= e^{-(E_1(M)+E_2(M))} \sum_{n,m=0}^{\infty} \frac{1}{n!m!} {E_1}^n {E_2}^m \\ &= e^{-(E_1+E_2)(M)} \sum_{k=0}^{\infty} \frac{1}{k!} (E_1 + E_2)^k = \mathcal{U}_{E_1+E_2}. \quad ∎\end{aligned}$$

In particular we obtain

$$(\mathcal{U}_{E/n})^n = \mathcal{U}_E,$$

i.e. all distributions of the form $\mathcal{U}_E$ are infinitely divisible.

Now, if E is any element of $\mathbf{E}^+$, we obtain

$$\mathcal{U}_E = \mathcal{U}_{E((.)\setminus\{o\})+E(\{o\})\delta_o} = \mathcal{U}_{E((.)\setminus\{o\})} * \mathcal{U}_{E(\{o\})\delta_o} = \mathcal{U}_{E((.)\setminus\{o\})},$$

from which we infer the truth of the following proposition.

1.6.5. *For all E in $\mathbf{E}^+$,*

$$\mathcal{U}_E = \mathcal{U}_{E((.)\setminus\{o\})}.$$

Immediate consequences of 1.6.5. and 1.5.2. or 1.5.7., respectively, are

1.6.6. *For all E in $\mathbf{E}^+$ and all X in $\mathfrak{A}$,*

$$_X(\mathcal{U}_E) = \mathcal{U}_{_XE} = \mathcal{U}_{(_XE)((.)\setminus\{o\})}.$$

1.6.7. *For all E in $\mathbf{E}^+$ and all finite sequences $X_1, \ldots, X_m$ of sets in $\mathfrak{B}$,*

$$(\mathcal{U}_E)_{X_1,\ldots,X_m} = \mathcal{U}_{E_{X_1,\ldots,X_m}} = \mathcal{U}_{(E_{X_1,\ldots,X_m})((.)\setminus\{o\})}.$$

The following simple formula will be used frequently.

1.6.8. *For all E in $\mathbf{E}^+$ and all X in $\mathfrak{A}$,*

$$\mathcal{U}_E\big(\Phi(X) = 0\big) = e^{-E(\Psi(X)>0)}.$$

Proof. We have

$$\mathcal{U}_E\big(\Phi(X) = 0\big) = e^{-E(M)} \sum_{n=0}^{\infty} \frac{(E^n)\big(\Phi(X) = 0\big)}{n!}.$$

Bearing in mind that $(E^n)\big(\Phi(X) = 0\big) = \Big(E\big(\Phi(X) = 0\big)\Big)^n$, we obtain an ordinary power series from which it follows that

$$\mathcal{U}_E\big(\Phi(X) = 0\big) = e^{-E(M)+E(\Psi(X)=0)} = e^{-E(\Psi(X)>0)}. \blacksquare$$

The above formula can be generalized.

1.6.9. *For all E in $\mathbf{E}^+$ and all X_1, X_2 in $\mathfrak{A}$,*

$$\mathcal{U}_E\big(\Phi(X_1) = 0 \mid \Phi(X_2) = 0\big) = e^{-E(\Psi(X_1)>0,\Psi(X_2)=0)}.$$

Proof. We have

$$\begin{aligned}\mathcal{U}_E\big(\Phi(X_1) = 0 \mid \Phi(X_2) = 0\big) &= e^{E(\Psi(X_2)>0)}\, \mathcal{U}_E\big(\Phi(X_1 \cup X_2) = 0\big)\\ &= e^{E(\Psi(X_2)>0)}\, e^{-E(\Psi(X_1\cup X_2)>0)} = e^{-\left(E(\Psi(X_1\cup X_2)>0)-E(\Psi(X_2)>0)\right)}\\ &= e^{-E(\Psi(X_1)>0,\Psi(X_2)=0)}. \blacksquare\end{aligned}$$

A number of fundamental properties of the distributions $\mathcal{U}_E$ can be expressed in a simple way in terms of the measures E. We start with

1.6.10. Proposition. *For each measure E in $\mathbf{E}^+$ and each finite sequence $X_1, \ldots, X_m$ of pairwise disjoint sets in $\mathfrak{B}$, the following statements are equivalent:*

a) *The finite sequence ${}_{X_1}\Phi, \ldots, {}_{X_m}\Phi$ is independent with respect to $\mathcal{U}_E$.*
b) *The finite sequence $\Phi(X_1), \ldots, \Phi(X_m)$ is independent with respect to $\mathcal{U}_E$.*
c) *For all i, j satisfying $1 \leqq i < j \leqq m$, $\Phi(X_i)$ is independent of $\Phi(X_j)$ with respect to $\mathcal{U}_E$.*
d) *For all i, j satisfying $1 \leqq i < j \leqq m$, $E\big(\Psi(X_i) > 0, \Psi(X_j) > 0\big) = 0$.*

Proof. Obviously b) is a weakened version of a), and c) is a weakened version of b). Now suppose that $\mathcal{U}_E$ has the property c). For $1 \leqq i < j \leqq m$ we obtain, using 1.6.8. and 1.6.9.,

$$\begin{aligned} &\exp\big(-E\big(\Psi(X_i) > 0, \Psi(X_j) = 0\big) = \mathcal{U}_E\big(\Phi(X_i) = 0 \mid \Phi(X_j) = 0\big) \\ &= \mathcal{U}_E\big(\Phi(X_i) = 0\big) = \exp\big(-E\big(\Psi(X_i) > 0\big) \end{aligned}$$

and hence

$$\begin{aligned} 0 &= E\big(\Psi(X_i) > 0\big) - E\big(\Psi(X_i) > 0, \Psi(X_j) = 0\big) \\ &= E\big(\Psi(X_i) > 0, \Psi(X_j) > 0\big). \end{aligned}$$

Conversely, suppose that E has the property d). If we set $X = X_1 \cup \cdots \cup X_m$, then

$$\begin{aligned} &({}_XE)\big((.) \setminus \{o\}\big) = E\big({}_X\Psi \in (.), \Psi(X) > 0\big) \\ &= \sum_{i=1}^{m} E\big({}_X\Psi \in (.), \Psi(X_i) > 0, \Psi(X_j) = 0 \quad \text{for} \quad j \in \{1, \ldots, m\} \setminus \{i\}\big) \\ &= \sum_{i=1}^{m} E\big({}_{X_i}\Psi \in (.), \Psi(X_i) > 0, \Psi(X_j) = 0 \quad \text{for} \quad j \in \{1, \ldots, m\} \setminus \{i\}\big) \\ &= \sum_{i=1}^{m} ({}_{X_i}E)\big((.) \setminus \{o\}\big). \end{aligned}$$

By virtue of 1.6.4. and 1.6.6. we obtain

$$_X(\mathcal{U}_E) = \mathcal{U}_{({}_XE)((.)\setminus\{o\})} = \mathop{\ast}_{i=1}^{m} \mathcal{U}_{({}_{X_i}E)((.)\setminus\{o\})} = \mathop{\ast}_{i=1}^{m} {}_{X_i}(\mathcal{U}_E).$$

Hence by 1.5.3. we conclude that ${}_{X_1}\Phi, \ldots, {}_{X_m}\Phi$ are independent with respect to $\mathcal{U}_E$. ∎

Using 1.6.10. we can characterize as follows those distributions $\mathcal{U}_E, E \in \mathbf{E}^+$, which are free from after-effects.

1.6.11. Proposition. *For all E in $\mathbf{E}^+$, the distribution $\mathcal{U}_E$ is free from after-effects if and only if*

$$E\,(\Psi \text{ is not of the form } n\delta_a;\ a \in A,\ n = 0, 1, \ldots) = 0.$$

Proof. The set

$$Y = \{\Psi : \Psi \in M, \Psi \text{ has the form } n\delta_a;\ a \in A,\ n = 0, 1, \ldots\}$$

coincides with $\{\Psi : \Psi \in M,\ \Psi^*(A) \leqq 1\}$, and hence belongs to $\mathfrak{M}$ in view of 1.4.2.

Now let $X_1, \ldots, X_m$ be an arbitrary finite sequence of pairwise disjoint sets in $\mathfrak{B}$ and suppose that $E(M \setminus Y) = 0$. Using 1.6.10. we conclude that the finite sequence ${}_{X_1}\Phi, \ldots, {}_{X_m}\Phi$ is independent with respect to P, for we have

$$E\big(\Psi(X_i) > 0,\ \Psi(X_j) > 0\big) = 0 \qquad (1 \leqq i < j \leqq m).$$

Conversely, suppose that $\mathcal{U}_E$ is free from after-effects. For any distinguished sequence $(\mathfrak{Z}_n)$ of decompositions of A, we obtain

$$E\big(\Psi^*(A) > 1\big) = \sup_{n=1,2,\ldots} E\big(\text{there exist sets } Z_1, Z_2 \in \mathfrak{Z}_n \text{ such that } Z_1 \neq Z_2,\ \Phi(Z_1) > 0,\ \Phi(Z_2) > 0\big).$$

Hence, using 1.6.10., we get $E(M \setminus Y) = 0$. ∎

As an immediate consequence of the definition of convolution, we have

1.6.12. *For all E_1, E_2 in $\mathbf{E}^+$,*

$$\varrho_{E_1 * E_2} = E_2(M)\,\varrho_{E_1} + E_1(M)\,\varrho_{E_2}.$$

In particular we obtain $\varrho_{E^n} = n\big(E(M)\big)^{n-1}\varrho_E$ for $n = 1, 2, \ldots$ This in turn leads to

$$\varrho_{\mathcal{U}_E} = e^{-E(M)} \sum_{k=0}^{\infty} (k!)^{-1}\varrho_{E^k} = e^{-E(M)} \sum_{k=1}^{\infty} (k!)^{-1} k\big(E(M)\big)^{k-1}\varrho_E$$
$$= \left(e^{-E(M)} \sum_{k=0}^{\infty} \frac{\big(E(M)\big)^k}{k!}\right)\varrho_E = \varrho_E,$$

from which we deduce the validity of the following proposition:

1.6.13. *For all E in $\mathbf{E}^+$,*

$$\varrho_{\mathcal{U}_E} = \varrho_E.$$

In particular $\mathcal{U}_E$ is continuous if and only if E has this property.

If we consider the second-order moment measure of $\mathcal{U}_E$, the formula corresponding to 1.6.13. becomes somewhat more complicated.

1.6.14. *For all measures E in $\mathbf{E}^+$ of finite intensity,*

$$\varrho_{(\mathcal{U}_E)^{\langle 2\rangle}} = \varrho_E \times \varrho_E + \varrho_{E^{\langle 2\rangle}}.$$

Proof. For all sets Z in $\mathfrak{A} \otimes \mathfrak{A}$, we obtain

$$\varrho_{(\mathscr{U}_E)^{<2>}}(Z) = \int (\Phi \times \Phi)(Z)\, \mathscr{U}_E(\mathrm{d}\Phi)$$

$$= e^{-E(M)} \sum_{n=1}^{\infty} \frac{1}{n!} \int (\Phi \times \Phi)(Z)\, E^n(\mathrm{d}\Phi)$$

$$= e^{-E(M)} \sum_{n=1}^{\infty} \frac{1}{n!} \int \big((\Phi_1 + \cdots + \Phi_n) \times (\Phi_1 + \cdots + \Phi_n)\big)(Z) \cdot (E \times \cdots \times E)(\mathrm{d}[\Phi_1, \ldots, \Phi_n])$$

$$= e^{-E(M)} \sum_{n=1}^{\infty} \frac{1}{n!}\, n\big(E(M)\big)^{n-1} \int (\Phi \times \Phi)(Z)\, E(\mathrm{d}\Phi)$$

$$+ e^{-E(M)} \sum_{n=2}^{\infty} \frac{1}{n!}\, n(n-1)\big(E(M)\big)^{n-2} \int (\Phi_1 \times \Phi_2)(Z)\,(E \times E) \cdot (\mathrm{d}[\Phi_1, \Phi_2])$$

$$= e^{-E(M)} e^{E(M)} \varrho_{E^{<2>}}(Z) + e^{-E(M)} e^{E(M)} \int (\Phi_1 \times \Phi_2)(Z)\,(E \times E)\,\mathrm{d}([\Phi_1, \Phi_2])$$

$$= \varrho_{E^{<2>}}(Z) + \varrho_E \times \varrho_E(Z)$$

because

$$\int (\Phi_1 \times \Phi_2)(Z)\,(E \times E)(\mathrm{d}[\Phi_1, \Phi_2]) = \varrho_E \times \varrho_E(Z)$$

for all measurable rectangles $Z = X_1 \times X_2$ and hence for all sets in $\mathfrak{A} \otimes \mathfrak{A}$. ∎

We conclude this section with

1.6.15. Proposition. *For all E in $\mathbf{E}^+$, the distribution $\mathscr{U}_E$ is simple if and only if E is continuous and simple.*

Proof. Let $P = \mathscr{U}_E$ be simple. For each a in A, $\varrho_P(\{a\}) = 2\varrho_{\sqrt{P}}(\{a\})$. In view of 1.5.5. the simplicity of $P = \sqrt{P} * \sqrt{P}$ implies that $\varrho_{\sqrt{P}}(\{a\}) = 0$. Consequently, $\mathscr{U}_E$ is continuous and hence also E because of 1.6.13. Furthermore,

$$0 = \mathscr{U}_E\,(\Phi \text{ is not simple})$$

$$= e^{-E(M)} \sum_{n=0}^{\infty} \frac{1}{n!} E^n\ (\Phi \text{ is not simple})$$

and hence

$$E\,(\Phi \text{ is not simple}) = 0.$$

Conversely, let us suppose that E is continuous and simple. Then in view of 1.5.5. all convolution powers E^n, $n = 0, 1, \ldots$, are continuous and simple, too. Consequently $\mathscr{U}_E$ is also simple. ∎

1.7. The Poisson Distributions P_ν

For each non-negative real x, let π_x denote the Poisson distribution on $\mathfrak{P}(\{0, 1, \ldots\})$ with expectation x. Now, if ν is any measure in N, then setting

$$p^{H_1, \ldots, H_m} = \pi_{\nu(H_1)} \times \cdots \times \pi_{\nu(H_m)}$$

we obtain a mapping which satisfies the conditions of the existence theorem 1.3.6. with $\mathfrak{H} = \mathfrak{B}$. Consequently, there exists a well-defined distribution P_ν on $\mathfrak{M}$ which is free from after-effects and satisfies the condition $(P_\nu)_X = \pi_{\nu(X)}$ for all X in $\mathfrak{B}$.

We call P_ν the *Poisson distribution* with intensity measure ν. Using 1.3.2., we see the validity of

1.7.1. Proposition. *Let $\mathfrak{H}$ be a generating semiring of sets in $\mathfrak{B}$ and ν a measure in N. Under these assumptions, there exists exactly one distribution P on $\mathfrak{M}$ which is free from after-effects and satisfies the condition $P_H = \pi_{\nu(H)}$ for all H in $\mathfrak{H}$, this distribution being the Poisson distribution P_ν with intensity measure ν.*

Obviously,

1.7.2. *For all ν in N and all X in $\mathfrak{A}$,*

$$_X(P_\nu) = P_{_X\nu}.$$

In the special case $A = \{1, \ldots, m\}$, the Poisson distribution

$$P_\nu = P_{\nu(\{1\})\delta_1 + \cdots + \nu(\{m\})\delta_m}$$

corresponds to the random m-dimensional vector with the distribution

$$\pi_{\nu(\{1\})} \times \cdots \times \pi_{\nu(\{m\})}.$$

Because of 1.5.4. the distribution $P_\nu * P_\varrho$ is always free from after-effects. On the other hand, from 1.5.7. we have for all X in $\mathfrak{B}$

$$(P_\nu * P_\varrho)_X = (P_\nu)_X * (P_\varrho)_X = \pi_{\nu(X)} * \pi_{\varrho(X)} = \pi_{\nu(X)+\varrho(X)} = (P_{\nu+\varrho})_X.$$

Thus we obtain

1.7.3. *For all ν, ϱ in N,*

$$P_\nu * P_\varrho = P_{\nu+\varrho}.$$

In particular

$$(P_{\nu/n})^n = P_\nu$$

from which it follows that all Poisson distributions are infinitely divisible.

For all finite measures ν in N, the measure Q_ν introduced in section 1.2. belongs to $\boldsymbol{E}^+$. In view of 1.6.11., $\mathcal{U}_{Q_\nu}$ is free from after-effects. On the other hand, for all X in $\mathfrak{B}$

$$(\mathcal{U}_{Q_\nu})_X = \mathcal{U}_{(Q_\nu)_X((.)\setminus\{o\})} = \mathcal{U}_{\nu(X)\delta_1}$$

$$= e^{-\nu(X)} \sum_{n=0}^{\infty} \frac{(\nu(X))^n}{n!} (\delta_1)^n = e^{-\nu(X)} \sum_{n=0}^{\infty} \frac{(\nu(X))^n}{n!} \delta_n = \pi_{\nu(X)},$$

i.e. we have the following fundamental formula:

1.7.4. Theorem. *For all finite measures ν on $\mathfrak{A}$,*

$$P_\nu = \mathscr{U}_{Q_\nu}.$$

If now ν is any measure in N and $\nu(X) < +\infty$ for a set X in $\mathfrak{A}$, we obtain

$$(P_\nu)_X = \big(_X(P_\nu)\big)_A = (\mathscr{U}_{Q_{X^\nu}})_A = e^{-\nu(X)} \sum_{n=0}^{\infty} \frac{\big(\nu(X)\big)^n}{n!} \delta_n = \pi_{\nu(X)},$$

and hence

1.7.5. *For all ν in N and all X in $\mathfrak{A}$ satisfying $\nu(X) < +\infty$,*

$$(P_\nu)_X = \pi_{\nu(X)}.$$

In particular, $P_\nu\big(\Phi(X) \text{ is finite}\big) = 1$, if $\nu(X)$ is finite.

In view of 1.6.15. and 1.7.4. a Poisson distribution P_γ with finite intensity measure γ is simple if and only if it is continuous, i.e. if γ is non-atomic. Let now ν be any measure in N and (X_n) a monotone increasing sequence of sets in $\mathfrak{B}$ with union A. If P_ν is simple then the same is true for all $_{X_n}(P_\nu)$, too, and we conclude that all $_{X_n}\nu$ and therefore ν itself are non-atomic. Conversely, if ν is non-atomic, then all $_{X_n}\nu$ are also non-atomic and we conclude that all $_{X_n}(P_\nu)$ are simple. Hence because of 1.4.4. P_ν itself is simple. Summarizing we have

1.7.6. Proposition. *A Poisson distribution P_ν on $\mathfrak{M}$ is simple if and only if its intensity measure ν is diffuse.*

Immediately from 1.7.6. and 1.4.9. we obtain the following characterization of continuous Poisson distributions on $\mathfrak{M}$.

1.7.7. Theorem. *Let $\mathfrak{H}$ be a generating subring of $\mathfrak{B}$ and ν any diffuse measure in N. Then there exists exactly one distribution P on $\mathfrak{M}$ satisfying the condition $P_H = \pi_{\nu(H)}$ for all H in $\mathfrak{H}$, this distribution being the Poisson distribution P_ν.*

In section 1.3. it was illustrated by an example that in the case of the phase space $\{1, 2\}$ there exist distributions $Q \neq P_{\delta_1+\delta_2}$ having all one-dimensional distributions Q_X equal to the one-dimensional distributions $(P_{\delta_1+\delta_2})_X$. Hence the assumption that ν be diffuse cannot be dropped in theorem 1.7.7.

In section 9.5. a distribution Q different from the Poisson distribution will be constructed which for all real a, b with $a \leq b$ satisfies the condition $Q_{[a,b)} = \pi_{b-a}$. Consequently the assumption "$\mathfrak{H}$ is a generating subring of sets in $\mathfrak{B}$" cannot be replaced by the weaker assumption "$\mathfrak{H}$ is a generating semiring of sets in $\mathfrak{B}$" in 1.7.7. and therefore in 1.4.9., neither.

Sometimes the following version of theorem 1.7.7. is handier.

1.7.8. Theorem. *Let $\mathfrak{H}$ be a generating subring of $\mathfrak{B}$ and P a continuous distribution on $\mathfrak{M}$. Then P is Poisson if and only if the one-dimensional distributions P_H, $H \in \mathfrak{H}$, are Poisson.*

Proof. The system of those X in $\mathfrak{B}$, for which P_X is Poisson is monotone closed in $\mathfrak{B}$ and contains the generating subring $\mathfrak{H}$. Therefore all P_X, $X \in \mathfrak{B}$, are Poisson. In particular the intensity measure ν of P belongs to N and is diffuse. With the aid of 1.7.7. we conclude that $P = P_\nu$. ∎

Poisson distributions can be decomposed only into Poisson factors.

1.7.9. Theorem. *If ν is a measure in N, and if $P_\nu = P * Q$ with P, Q in $\mathbf{P}$, then P, Q are Poisson distributions.*

Proof. 1. First we suppose in addition that P_ν is continuous, i.e. that ν is diffuse.

For all X in $\mathfrak{B}$,

$$\pi_{\nu(X)} = (P_\nu)_X = P_X * Q_X.$$

Thus, according to a well-known theorem of Raikov (cf. for instance Loève [1], Chapt. VI, § 19), P_X and Q_X are Poisson distributions. On the other hand, the two distributions P, Q are continuous, since, for all a in A,

$$\varrho_P(\{a\}) + \varrho_Q(\{a\}) = \nu(\{a\}) = 0.$$

Now the assertion is readily deduced from 1.7.8.

2. Suppose next that $[A, \varrho_A] = \{1, \ldots, m\}$. Our assertion then coincides with Teicher's [1] multidimensional version of the Raikov theorem mentioned above.

We set $[A', \varrho_{A'}] = R^1$, and associate with each a in A the uniform distribution p_a on the interval $(a - 1, a)$. If E is an arbitrary measure in $\mathbf{E}^+$, then

$${}^sE = \sum_{0 \leqq k_1, \ldots, k_m < +\infty} E\big(\Phi(\{a\}) = k_a \quad \text{for} \quad 1 \leqq a \leqq m\big) \mathop{\ast}_{a=1}^{m} \big((Q_{p_a})^{k_a}\big)$$

represents a measure ${}^sE$ on the σ-algebra $\mathfrak{M}'$ corresponding to the phase space $[A', \varrho_{A'}]$. The individual points a of the realizations Φ are transformed randomly and independently of each other into points of the new phase space $[A', \varrho_{A'}]$ in accordance with the position-dependent distributions p_a. Obviously the total masses of E and ${}^sE$ coincide, and the following relation holds:

$${}^s(E_1 * E_2) = {}^s(E_1) * {}^s(E_2), \quad {}^s(\mathcal{U}_E) = \mathcal{U}_{({}^sE)}.$$

E can be recovered from ${}^sE$ through

$$E_{\{1\},\ldots,\{m\}} = ({}^sE)_{(0,1),\ldots,(m-1,m)}.$$

Now it follows from $P_\nu = P * Q$ that ${}^s(P_\nu) = {}^sP * {}^sQ$. On the other hand

$${}^s(P_\nu) = {}^s(\mathcal{U}_{Q_\nu}) = \mathcal{U}_{{}^s(Q_\nu)} = \mathcal{U}_{\sum\limits_{a \in A} \nu(\{a\}) Q_{p_a}} = \mathcal{U}_{Q_{\sum\limits_{a \in A} \nu(\{a\}) p_a}} = P_{\sum\limits_{a \in A} \nu(\{a\}) p_a},$$

i.e. ${}^s(P_\nu)$ is a continuous Poisson distribution on $\mathfrak{M}'$ and so 1. can be applied. The two distributions ${}^sP, {}^sQ$ are therefore Poisson. Denoting their intensity measures by γ, ω, we obtain

$$P_{\{1\},\ldots,\{m\}} = ({}^sP)_{(0,1),\ldots,(m-1,m)} = \pi_{\gamma((0,1))} \times \cdots \times \pi_{\gamma((m-1,m))},$$

$$Q_{\{1\},\ldots,\{m\}} = ({}^sQ)_{(0,1),\ldots,(m-1,m)} = \pi_{\omega((0,1))} \times \cdots \times \pi_{\omega((m-1,m))}.$$

Hence the two distributions P, Q are Poisson, too.

3. Now we shall drop all additional assumptions.

Let $X_1, \ldots, X_m$ be an arbitrary sequence of pairwise disjoint sets in $\mathfrak{B}$. We have

$$\pi_{\nu(X_1)} \times \cdots \times \pi_{\nu(X_m)} = (P_\nu)_{X_1,\ldots,X_m} = P_{X_1,\ldots,X_m} * Q_{X_1,\ldots,X_m}.$$

Thus, in view of 2., $P_{X_1,\ldots,X_m}$ and $Q_{X_1,\ldots,X_m}$ are Poisson. Therefore P and Q are free from after-effects and all one-dimensional distributions P_X, $X \in \mathfrak{B}$, and Q_X, $X \in \mathfrak{B}$, are Poisson. From this follows the Poisson character of P and Q. ∎

The following example due to R. Siegmund-Schultze shows that there exist distributions which are free from after-effects and have factors which are not free from after-effects.

The rational functions

$$f(x, y) = \frac{2}{4 - x - y}, \qquad g(x, y) = \frac{1}{2}\left(\frac{1}{2 - x} + \frac{1}{2 - y}\right), \qquad h(x, y) = \frac{1}{2 - x}\frac{1}{2 - y}$$

are generating functions of distributions p, q and r of two-dimensional random vectors with non-negative integer coordinates ζ_1, ζ_2. Obviously ζ_1, ζ_2 are stochastically independent with respect to r. We have $fg = h$, i.e. $p * q = r$. On the other hand, ζ_1, ζ_2 are not stochastically independent with respect to p, q.

1.7.10. *For all σ, ν in N, if P_σ is absolutely continuous with respect to P_ν then σ is absolutely continuous with respect to ν.*

Proof. Let X be a set in A with $\nu(X) = 0$. From this $(P_\nu)_X = \pi_0 = \delta_0$ follows and therefore $P_\nu(\Phi(X) > 0) = 0$. By assumption, this implies $P_\sigma(\Phi(X) > 0) = 0$, i.e. $\delta_0 = (P_\sigma)_X = \pi_{\sigma(X)}$, hence $\sigma(X) = 0$. ∎

Using the structure theorem 1.7.4., one can show (cf. Brown [2]) that for finite σ, ν the converse is also true.

1.7.11. Proposition. *Let σ, ν be finite measures on $\mathfrak{A}$. If σ is absolutely continuous with respect to ν and f is a density function of σ with respect to ν, then P_σ is absolutely continuous with respect to P_ν, and*

$$p_{\sigma,\nu}(\Phi) = e^{\nu(A) - \sigma(A)} \prod_{a \in A, \Phi(\{a\}) > 0} \big(f(a)\big)^{\Phi(\{a\})} \qquad (\Phi \in M, \Phi \text{ finite})$$

is a density function of P_σ with respect to P_ν.

Proof. (cf. Krickeberg [3]). Let h be a bounded real-valued $\mathfrak{M}$-measurable funktion defined on M. Then, by theorem 1.7.4.,

$$\int h(\Phi)\, P_\sigma(\mathrm{d}\Phi$$

$$= e^{-\sigma(A)} \sum_{n=0}^{\infty} \frac{1}{n!} \int h(\delta_{a_1} + \cdots + \delta_{a_n})\, \sigma(\mathrm{d}a_1) \ldots \sigma(\mathrm{d}a_n)$$

$$= e^{-\nu(A)} \sum_{n=0}^{\infty} \frac{1}{n!} \int e^{\nu(A) - \sigma(A)} h(\delta_{a_1} + \cdots + \delta_{a_n})\, f(a_1) \cdot \cdots \cdot f(a_n)\, \nu(\mathrm{d}a_1 \ldots \nu(\mathrm{d}a_n)$$

$$= e^{-\nu(A)} \sum_{n=0}^{\infty} \frac{1}{n!} \int h(\delta_{a_1} + \cdots + \delta_{a_n})\, p_{\sigma,\nu}(\delta_{a_1} + \cdots + \delta_{a_n})\, \nu(\mathrm{d}a_1) \ldots \nu(\mathrm{d}a_n)$$

$$= \int h(\Phi)\, p_{\sigma,\nu}(\Phi)\, P_\nu(\mathrm{d}\Phi)$$

where $p_{\sigma,\nu}$ is the function defined in the proposition. This being true for all such h, the proposition is proved. ∎

In proposition 1.7.11. the assumption of finiteness of the intensity measures σ, ν cannot be dropped. However, the following theorem is true (cf. Brown [2], Liese [2]).

1.7.12. Theorem. *Let σ, ν be measures in N. If σ is absolutely continuous with respect to ν and f is a density function of σ with respect to ν, then*

P_σ is absolutely continuous with respect to P_ν if

$$\int \left(\sqrt{f(a)} - 1\right)^2 \nu(\mathrm{d}a) < +\infty,$$

P_σ is purely singular with respect to P_ν if

$$\int \left(\sqrt{f(a)} - 1\right)^2 \nu(\mathrm{d}a) = +\infty.$$

Proof. 1. If the phase space $[A, \varrho_A]$ is bounded, then we see from 1.7.11. that P_σ is absolutely continuous with respect to P_ν. Furthermore, in this case $\int \left(\sqrt{f(a)} - 1\right)^2 \nu(\mathrm{d}a)$ is finite, for the two measures σ and ν are finite and

$$\sqrt{\int \left(\sqrt{f(a)} - 1\right)^2 \nu(\mathrm{d}a)} \leqq \sqrt{\int f(a)\, \nu(\mathrm{d}a)} + \sqrt{\int 1 \nu(\mathrm{d}a)} = \sqrt{\sigma(A)} + \sqrt{\nu(A)}.$$

Therefore in what follows we can assume without loss of generality that the phase space $[A, \varrho_A]$ is unbounded.

2. Let (X_n) be a sequence of pairwise disjoint non-empty sets in $\mathfrak{B}$ such that every X in $\mathfrak{B}$ can be covered by a finite number of sets X_n.

We set

$$M_n = \{\Phi : \Phi \in M, \Phi(A \setminus X_n) = 0\}, \quad \mathfrak{M}_n = M_n \cap \mathfrak{M}.$$

Then the correspondence

$$\Phi \curvearrowright ({}_{X_n}\Phi)_{n=1,2,\ldots}$$

provides a one-to-one mapping of $[M, \mathfrak{M}]$ into the direct product of the measurable spaces $[M_n, \mathfrak{M}_n]$ which is measurable in both directions. If we now set

$$\sigma_n = {}_{X_n}\sigma, \quad \nu_n = {}_{X_n}\nu,$$

then P_σ and P_ν are mapped into $\mathop{\times}\limits_{n=1}^{\infty} P_{\sigma_n}$ and $\mathop{\times}\limits_{n=1}^{\infty} P_{\nu_n}$, respectively.

Here we note that

$$P_{\sigma_n}(M_n) = 1 = P_{\nu_n}(M_n)$$

is satisfied, so that P_{σ_n}, P_{ν_n} can be considered as distributions on $\mathfrak{M}_n$. Consequently, P_σ is absolutely continuous or purely singular with respect to P_ν according as $\mathop{\times}\limits_{n=1}^{\infty} P_{\sigma_n}$ is absolutely continuous or purely singular with respect to $\mathop{\times}\limits_{n=1}^{\infty} P_{\nu_n}$.

Obviously, for all natural numbers n, the finite measure σ_n is absolutely continuous with respect to the finite measure ν_n. Therefore as the proof of 1.7.11. shows, the distributions P_{σ_n} are absolutely continuous with respect to P_{ν_n} and the functions

$$p_{\sigma_n,\nu_n}(\Phi) = e^{\nu_n(X_n)-\sigma_n(X_n)} \prod_{a\in X_n, \Phi(\{a\})>0} \big(f(a)\big)^{\Phi(\{a\})}$$

are density functions of P_{σ_n} with respect to P_{ν_n}. Hence, by a theorem of Kakutani (cf. for instance Hewitt & Stromberg [1], § 22), $\overset{\infty}{\underset{n=1}{\times}} P_{\sigma_n}$ is either absolutely continuous or purely singular with respect to $\overset{\infty}{\underset{n=1}{\times}} P_{\nu_n}$, where the first case occurs if

$$\prod_{n=1}^{\infty} \left(\int \sqrt{p_{\sigma_n,\nu_n}(\Phi)}\, P_{\nu_n}(\mathrm{d}\Phi)\right) > 0$$

and the second case occurs if this infinite product vanishes. For $n = 1, 2, \ldots$ we obtain

$$\int \sqrt{p_{\sigma_n,\nu_n}(\Phi)}\, P_{\nu_n}(\mathrm{d}\Phi) = \int \sqrt{e^{\nu(X_n)-\sigma(X_n)}} \sqrt{\prod_{a\in X_n, \Phi(\{a\})>0} \big(f(a)\big)^{\Phi(\{a\})}}\, P_{\nu_n}(\mathrm{d}\Phi)$$

$$= \exp\left(\frac{1}{2}\big(\nu(X_n) - \sigma(X_n)\big)\right) \int \left(\prod_{\substack{a\in X_n \\ \Phi(\{a\})>0}} \big(\sqrt{f(a)}\big)^{\Phi(\{a\})}\right) P_{\nu_n}(\mathrm{d}\Phi)$$

$$= \exp\left(\frac{1}{2}\big(\nu(X_n) - \sigma(X_n)\big)\right) \sum_{k=0}^{\infty} \frac{\mathrm{e}^{-\nu(x)}}{k!}$$

$$\times \int \sqrt{f(a_1)\ldots f(a_k)}\, \nu_n(\mathrm{d}a_1)\ldots\nu_n(\mathrm{d}a_k)$$

$$= \exp\left(\frac{1}{2}\big(\nu(X_n) - \sigma(X_n)\big)\right) \exp\left(\int \big(\sqrt{f(a)} - 1\big)\, \nu_n(\mathrm{d}a)\right)$$

$$= \exp\left(-\frac{1}{2}\left(\sigma(X_n) - \nu(X_n) + 2\int_{X_n} \big(1 - \sqrt{f(a)}\big)\, \nu(\mathrm{d}a)\right)\right)$$

$$= \exp\left(-\frac{1}{2}\left(\int_{X_n} \big(f(a) - 2\sqrt{f(a)} + 1\big)\, \nu(\mathrm{d}a)\right)\right)$$

$$= \exp\left(-\frac{1}{2}\left(\int_{X_n} \big(\sqrt{f(a)} - 1\big)^2\, \nu(\mathrm{d}a)\right)\right).$$

Hence

$$\prod_{n=1}^{\infty} \int \sqrt{p_{\sigma_n,\nu_n}(\Phi)}\, P_{\nu_n}(\mathrm{d}\Phi) = \lim_{m\to\infty} \prod_{n=1}^{m} \exp\left(-\frac{1}{2}\int_{X_n} \big(\sqrt{f(a)} - 1\big)^2\, \nu(\mathrm{d}a)\right)$$

$$= \lim_{m\to\infty} \exp\left(-\frac{1}{2}\int_{X_1\cup\cdots\cup X_m} \big(\sqrt{f(a)} - 1\big)^2\, \nu(\mathrm{d}a)\right)$$

$$= \exp\left(-\frac{1}{2}\int \big(\sqrt{f(a)} - 1\big)^2\, \nu(\mathrm{d}a)\right). \blacksquare$$

1.8. The Mixtures $\int P_{l_\nu}(.)\ \sigma(\mathrm{d}l)$

Frequently the following criterion proves useful:

1.8.1. *If* $\mathfrak{H}$ *is a generating semiring of sets in* $\mathfrak{B}$, *then a mapping* $f_{(.)}$ *of a measurable space* $[C, \mathfrak{C}]$ *into* $[\boldsymbol{P}, \mathfrak{P}]$ *is measurable if and only if, for all finite sequences* $H_1, \ldots, H_m$ *of pairwise disjoint sets in* $\mathfrak{H}$ *and for all non-negative integer* $l_1, \ldots, l_m$ *the real function*

$$c \curvearrowright f_{(c)}\big(\Phi(H_1) = l_1, \ldots, \Phi(H_m) = l_m\big)$$

is measurable with respect to $\mathfrak{C}$.

Proof. Let $\mathfrak{B}$ denote the system of those Y in $\mathfrak{M}$ for which $c \curvearrowright f_{(c)}(Y)$ is measurable. Obviously $\mathfrak{B}$ is monotone closed. On the other hand $\mathfrak{B}$ includes all σ-algebras $\mathfrak{M}_{H_1,\ldots,H_m}$ with $H_1, \ldots, H_m$ pairwise disjoint sets in $\mathfrak{H}$, for we have

$$\begin{aligned} &f_{(c)}\big([\Phi(H_1), \ldots, \Phi(H_m)] \in Z\big) \\ &= \sum_{[l_1,\ldots,l_m]\in Z} f_{(c)}\big(\Phi(H_1) = l_1, \ldots, \Phi(H_m) = l_m\big) \end{aligned}$$

for all subsets Z of $\{0, 1, \ldots\}^m$. Therefore $\mathfrak{B}$ includes the union $\mathfrak{M}'$ of all $\mathfrak{M}_{H_1,\ldots,H_m}$. In virtue of 1.3.1. $\mathfrak{M}'$ is a generating subalgebra of $\mathfrak{M}$ and we conclude that $\mathfrak{B} = \mathfrak{M}$. ∎

The convolution is associative with respect to mixtures in the following sense:

1.8.2. *If* $f_{(.)}$ *and* $g_{(.)}$ *are measurable mappings of* $[U, \mathfrak{U}]$ *and* $[V, \mathfrak{B}]$, *respectively, into* $[\boldsymbol{P}, \mathfrak{P}]$, *then the correspondence*

$$[u, v] \curvearrowright f_{(u)} * g_{(v)}$$

provides a measurable mapping of $[U, \mathfrak{U}] \times [V, \mathfrak{B}]$ *into* $[\boldsymbol{P}, \mathfrak{P}]$, *and for all distributions* S *and* T *on* $\mathfrak{U}$ *and* $\mathfrak{B}$, *respectively, the following relation holds*:

$$\begin{aligned} &\Big(\int f_{(u)}(.)\, S(\mathrm{d}u)\Big) * \Big(\int g_{(v)}(.)\, T(\mathrm{d}v)\Big) \\ &= \int \big(f_{(u)} * g_{(v)}\big)(.)\,(S \times T)\,(\mathrm{d}[u, v])\,. \end{aligned}$$

Proof. For all finite sequences $X_1, \ldots, X_m$ of sets in $\mathfrak{B}$ and all non-negative integer $l_1, \ldots, l_m$, we have

$$\begin{aligned} &\big(f_{(u)} * g_{(v)}\big)\big(\Phi(X_i) = l_i \quad \text{for} \quad 1 \leqq i \leqq m\big) \\ &= \sum_{\substack{c_i, d_i \geqq 0 \\ c_i + d_i = l_i \\ \text{for } 1 \leqq i \leqq m}} f_{(u)}\big(\Phi(X_i) = c_i \quad \text{for} \quad 1 \leqq i \leqq m\big)\, g_{(v)}\big(\Phi(X_i) = d_i \quad \text{for} \quad 1 \leqq i \leqq m\big), \end{aligned}$$

so that by means of 1.8.1. we can conclude that $[u, v] \curvearrowright f_{(u)} * g_{(v)}$ is measurable.

We obtain

$$\left(\int f_{(u)}(.)\, S(\mathrm{d}u)\right) * \left(\int g_{(v)}(.)\, T(\mathrm{d}v)\right) \left(\Phi(X_i) = l_i \quad \text{for} \quad 1 \leqq i \leqq m\right)$$

$$= \sum_{\substack{c_i, d_i \geqq 0 \\ c_i + d_i = l_i \\ \text{for } 1 \leqq i \leqq m}} \left(\int f_{(u)}\left(\Phi(X_i) = c_i \quad \text{for} \quad 1 \leqq i \leqq m\right) S(\mathrm{d}u)\right)$$

$$\cdot \left(\int g_{(v)}\left(\Phi(X_i) = d_i \quad \text{for} \quad 1 \leqq i \leqq m\right) T(\mathrm{d}v)\right)$$

$$= \int \Bigg(\sum_{\substack{c_i, d_i \geqq 0 \\ c_i + d_i = l_i \\ \text{for } 1 \leqq i \leqq m}} f_{(u)}\left(\Phi(X_i) = c_i \quad \text{for} \quad 1 \leqq i \leqq m\right)$$

$$\cdot\, g_{(v)}\left(\Phi(X_i) = d_i \quad \text{for} \quad 1 \leqq i \leqq m\right)\Bigg) (S \times T)(\mathrm{d}[u, v])$$

$$= \int (f_{(u)} * g_{(v)})\left(\Phi(X_i) = l_i \quad \text{for} \quad 1 \leqq i \leqq m\right)(S \times T)(\mathrm{d}[u, v]),$$

i.e. the relation to be verified is valid on an algebra generating $\mathfrak{M}$ and hence on the whole of $\mathfrak{M}$. ∎

Using 1.8.1. it can be immediately concluded from the definition of the Poisson distributions on $\mathfrak{M}$ that for all ν in N, the mapping $l \curvearrowright P_{l\nu}$ of $[0, +\infty)$ into $[\boldsymbol{P}, \mathfrak{P}]$ is measurable. Therefore to each distribution σ of a random non-negative number and to each ν in N we can associate the "mixed Poisson distribution" $\int P_{l\nu}(.)\, \sigma(\mathrm{d}l)$.

Using 1.8.2. and 1.7.3., we obtain

$$\left(\int P_{l\nu}(.)\, \sigma(\mathrm{d}l)\right) * \left(\int P_{g\nu}(.)\, \varrho\, (\mathrm{d}g)\right) = \int P_{(l+g)\nu}(.)\, (\sigma \times \varrho)\, (\mathrm{d}[l, g])$$

$$= \int P_{c\nu}(.)\, (\sigma * \varrho)\, (\mathrm{d}c),$$

i.e. we have

1.8.3. *For all ν in N and all distributions σ, ϱ of random non-negative numbers,*

$$\left(\int P_{l\nu}(.)\, \sigma(\mathrm{d}l)\right) * \left(\int P_{l\nu}(.)\, \varrho\, (\mathrm{d}l)\right) = \int P_{l\nu}(.)\, (\sigma * \varrho)\, (\mathrm{d}l).$$

For all X in $\mathfrak{A}$ with $\nu(X) = 0$,

$$_X\left(\int P_{l\nu}(.)\, \sigma(\mathrm{d}l)\right) = \int {}_X(P_{l\nu})\, (\cdot)\, \sigma(\mathrm{d}l) = \delta_0.$$

If, however, $0 < \nu(X) < +\infty$, then by virtue of 1.7.4. we obtain

$$_X\left(\int P_{l\nu}(.)\, \sigma(\mathrm{d}l)\right) = \int {}_X(P_{l\nu})\, (.)\, \sigma(\mathrm{d}l) = \int P_{l\,{}_X\nu}(.)\, \sigma(\mathrm{d}l)$$

$$= \int \left(\sum_{n=0}^{\infty} \frac{e^{-l\nu(X)}}{n!}\, (Q_{l\,{}_X\nu})^n\, (.)\right) \sigma\, (\mathrm{d}l)$$

$$= \sum_{n=0}^{\infty} \left(\int e^{-l\nu(X)}\, \frac{(l\nu(X))^n}{n!}\, \sigma(\mathrm{d}l)\right) (Q_{\nu((.)\mid X)})^n$$

$$= \sum_{n=0}^{\infty} \left(\int (P_{l\nu})_X\, (\{n\})\, \sigma(\mathrm{d}l)\right) (Q_{\nu((.)\mid X)})^n$$

$$= \sum_{n=0}^{\infty} \left(\int P_{l\nu}(.)\, \sigma(\mathrm{d}l)\right)_X\, (\{n\})\, (Q_{\nu((.)\mid X)})^n,$$

where we set

$$\nu\big((.)\mid X\big) = \frac{\nu\big((.)\cap X\big)}{\nu(X)}.$$

Thus we note the validity of

1.8.4. *For all ν in N, all X in $\mathfrak{A}$ satisfying $\nu(X) < +\infty$ and all distributions σ of random non-negative numbers,*

$$x\Big(\int P_{l\nu}(.)\,\sigma(\mathrm{d}l)\Big) = \sum_{n=0}^{\infty}\Big(\int P_{l\nu}(.)\,\sigma(\mathrm{d}l)\Big)_X(\{n\})\,(Q_{\nu((.)\mid X)})^n$$

if $0 < \nu(X) < +\infty$, and

$$x\Big(\int P_{l\nu}(.)\,\sigma(\mathrm{d}l)\Big) = \delta_o$$

if $\nu(X) = 0$.

For all infinite measures ν, the distribution σ describes the "asymptotic density of Φ with respect to ν":

1.8.5. *For each infinite ν in N and each distribution σ of a non-negative random number, there exists a non-negative function $j(.)$ on M which is measurable with respect to $\mathfrak{M}$ and has the property that for each monotone increasing sequence of sets in $\mathfrak{A}$, which satisfies the two conditions*

$$0 < \nu(X_n) < +\infty \quad \text{for} \quad n = 1, 2, \ldots; \quad \bigcup_{n=1}^{\infty} X_n = A$$

the expression $\big((\nu(X_n))^{-1}\,\Phi(X_n)\big)$ tends toward $j(\Phi)$ with probability one with respect to $P = \int P_{l\nu}(.)\,\sigma(\mathrm{d}l)$. The distribution of $j(\Phi)$ with respect to P coincides with σ.

Proof. 1. To start with, let $\sigma = \delta_c$, $c \geqq 0$. We set $\xi_n = \Phi(X_n \setminus X_{n-1})$ with $X_0 = \emptyset$. The sequence (ξ_n) is independent with respect to P, and the n-th term of the sequence has the Poisson distribution $\pi_{c\nu(X_n\setminus X_{n-1})}$.

We have to show that

$$\frac{1}{\nu(X_n)}\sum_{i=1}^{n}\xi_i$$

tends toward c with probability one.

Now let $[A', \varrho_{A'}]$ be the real axis R, μ the Lebesgue measure on the σ-algebra $\mathfrak{R}$ of the Borel sets of R, and $[M', \mathfrak{M}']$ the measurable space of integer-valued measures corresponding to $[A', \varrho_{A'}]$. The sequence $\big(\Phi([k-1, k))\big)_{k=1,2,\ldots}$ is independent and identically distributed with respect to the Poisson distribution $P_{c\mu}$. Therefore, in view of the strong law of large numbers, with increasing n the expression

$$\frac{1}{n}\sum_{k=1}^{n}\Phi\big([k-1, k)\big) = \frac{1}{n}\,\Phi\big([0, n)\big)$$

tends toward the constant c almost everywhere with respect to $P_{c\mu}$.

If we set $t_0 = \nu(X_0)$, $t_1 = \nu(X_1), \ldots$ with $X_0 = \emptyset$, then also the sequence of the measures $\Phi([t_{i-1}, t_i))$ is independent with respect to $P_{c\mu}$, and for $t_m \geqq 1$ we obtain

$$\frac{1}{[t_m]+1} \Phi([0, [t_m])) \leqq \frac{1}{t_m} \sum_{i=1}^{m} \Phi([t_{i-1}, t_i)) \leqq \frac{1}{[t_m]} \Phi([0, [t_m]+1)).$$

Considering now that $\lim_{m\to\infty} t_m = +\infty$ is satisfied which means that, with increasing m,

$$\frac{[t_m]}{[t_m]+1}\left(\frac{1}{[t_m]} \Phi([0, [t_m]))\right)$$

as well as

$$\frac{[t_m]+1}{[t_m]}\left(\frac{1}{[t_m]+1} \Phi([0, [t_m]+1))\right)$$

tend toward c almost everywhere with respect to $P_{c\mu}$, we conclude that

$$\frac{1}{\nu(X_m)} \sum_{k=1}^{m} \Phi([t_{k-1}, t_k))$$

tends toward c with probability one. Since, however, the distribution of $\left(\Phi([t_{k-1}, t_k))\right)_{k=1,2,\ldots}$ under $P_{c\mu}$ coincides with the distribution of $\left(\Phi(X_k \setminus X_{k-1})\right)_{k=1,2,\ldots}$ under $P_{c\nu}$, the expression

$$\frac{1}{\nu(X_m)} \sum_{k=1}^{m} \Phi(X_k \setminus X_{k-1}) = \frac{1}{\nu(X_m)} \Phi(X_m)$$

tends, with increasing m, toward c almost everywhere with respect to $P_{c\nu}$.

2. For arbitrary σ, we have

$$P\left(\left(\frac{\Phi(X_n)}{\nu(X_n)}\right) \text{converges in } R\right)$$
$$= \int P_{l\nu}\left(\left(\frac{\Phi(X_n)}{\nu(X_n)}\right) \text{converges in } R\right) \sigma(\mathrm{d}l) = 1.$$

Consequently $\left((\nu(X_n))^{-1}\Phi(X_n)\right)$ tends toward a non-negative finite value $j(\Phi)$ with probability one as referred to P. For all $x \geqq 0$

$$P(j(\Phi) \leqq x) = \int P_{l\nu}\left(\lim_{n\to 0} \frac{\Phi(X_n)}{\nu(X_n)} \leqq x\right) \sigma(\mathrm{d}l) = \int_{[0,x]} 1\sigma(\mathrm{d}l) = \sigma([0, x]),$$

i.e. $j(\Phi)$ is distributed according to σ.

3. It remains to show that the limits $j(\Phi)$, $j'(\Phi)$ constructed for different sequences (X_n), (X_n') coincide almost everywhere with respect to P.

Using the Chebyshev inequality, we have for all $c > 0$, all $\varepsilon > 0$, and all X in $\mathfrak{A}$ with the property $0 < \nu(X) < +\infty$:

$$P_{c\nu}\Big(\big|\big(\nu(X)\big)^{-1}\,\Phi(X) - c\big| \geqq \varepsilon\Big) \leqq \frac{c\nu(X)}{\big(\varepsilon\nu(X)\big)^2} = \frac{c}{\varepsilon^2\nu(X)}.$$

This gives for all natural numbers k

$$\begin{aligned}
&\overline{\lim_{n\to\infty}}\, P\Big(\big|\big(\nu(X_n')\big)^{-1}\,\Phi(X_n') - j(\Phi)\big| \geqq \varepsilon\Big)\\
&= \overline{\lim_{n\to\infty}} \int P_{l\nu}\left(\left|\big(\nu(X_n')\big)^{-1}\,\Phi(X_n') - \lim_{n\to\infty}\frac{\Phi(X_n)}{\nu(X_n)}\right| \geqq \varepsilon\right)\sigma(\mathrm{d}l)\\
&= \overline{\lim_{n\to\infty}} \int P_{l\nu}\Big(\big|\big(\nu(X_n')\big)^{-1}\,\Phi(X_n') - l\big| \geqq \varepsilon\Big)\,\sigma(\mathrm{d}l)\\
&\leqq \overline{\lim_{n\to\infty}} \int_0^k \frac{l}{\varepsilon^2\nu(X_n')}\,\sigma(\mathrm{d}l) + \sigma\big((k, +\infty)\big)\\
&= \overline{\lim_{n\to\infty}}\, \frac{1}{\varepsilon^2\nu(X_n')} \int_0^k l\sigma(\mathrm{d}l) + \sigma\big((k, +\infty)\big)\\
&= \sigma\big((k, +\infty)\big).
\end{aligned}$$

Hence

$$\overline{\lim_{n\to\infty}}\, P\Big(\big|\big(\nu(X_n')\big)^{-1}\,\Phi(X_n') - j(\Phi)\big| \geqq \varepsilon\Big) = 0,$$

which means the convergence in probability of $\big(\nu(X_n')\big)^{-1}\,\Phi(X_n')$ toward $j(\Phi)$. Consequently $j'(\Phi)$ coincides with $j(\Phi)$ almost everywhere. ∎

From 1.8.5. we immediately have

1.8.6. *If two distributions σ_1, σ_2 of random non-negative numbers satisfy the equation*

$$\int P_{l\nu}(.)\,\sigma_1(\mathrm{d}l) = P_{l\nu}(.)\,\sigma_2(\mathrm{d}l)$$

for any infinite ν in N, then $\sigma_1 = \sigma_2$.

In section 7.1. we shall see that the uniqueness statement 1.8.6. is true even for arbitrary $\nu \neq o$ in N.

Because of 1.8.5., for all infinite measures ν, the relation

$$P_\nu\ (\Phi \text{ is infinite}) = 1$$

holds, while for all finite ν

$$P_\nu\ (\Phi \text{ is finite}) = 1$$

as stated in section 1.7. Consequently, if an infinite ν in N is equivalent to a finite measure σ on $\mathfrak{A}$, then P_ν is nevertheless purely singular with respect to P_σ. Therefore the assumption that ν and σ be finite, cannot be dropped in proposition 1.7.11.

By virtue of 1.2.1., the distribution $P = \int P_{l\nu}(.)\,\sigma(\mathrm{d}l)$ has the intensity measure

$$\varrho_P(X) = \left(\int l\sigma(\mathrm{d}l)\right)\nu(X) \qquad (X \in \mathfrak{A}),$$

where we set $\infty \cdot 0 = 0 \cdot \infty = 0$. Thus for $\nu \neq o$, P is of finite intensity if and only if the expectation $\int l\sigma(\mathrm{d}l)$ is finite.

In view of 1.7.6., the mixture $P = \int P_{l\nu}(.)\,\sigma(\mathrm{d}l)$ is simple for $\sigma \neq \delta_0$ if and only if ν is non-atomic. In this case, however, P is not necessarily also orderly, as it is shown by the following example, which was previously referred to in section 1.4. (cf. Belyaev [2]).

Let $[A, \varrho_A]$ be the real axis R, μ the Lebesgue measure on $\mathfrak{R}$, and

$$P = \int_1^\infty P_{x\mu}(.)\,cx^{-(c+1)}\,\mu(\mathrm{d}x), \quad \text{where} \quad 0 < c < 1.$$

Then P is simple but not orderly.

For all bounded Borel subsets X of $[0, 1]$, we have

$$P(\Phi(X) > 1) = \int_1^\infty \frac{1 - e^{-x\mu(X)}\,(1 + x\mu(X))}{x^{c+1}}\,c\mu(\mathrm{d}x)$$

$$= (\mu(X))^c \int_{\mu(X)}^\infty \frac{1 - e^{-y}\,(1 + y)}{y^{c+1}}\,c\mu(\mathrm{d}y) \geqq \mu(X) \int_1^\infty \frac{1 - e^{-y}\,(1 + y)}{y^{c+1}}\,c\mu(\mathrm{d}y).$$

Now, if $\mathfrak{Z}$ is any decomposition of $[0, 1]$, we obtain

$$\sum_{Z \in \mathfrak{Z}} P(\Phi(Z) > 1) \geqq \Big(\sum_{Z \in \mathfrak{Z}} \mu(Z)\Big) \int_1^\infty \frac{1 - e^{-y}\,(1 + y)}{y^{c+1}}\,c\mu(\mathrm{d}y) = \int_1^\infty \frac{1 - e^{-y}\,(1 + y)}{y^{c+1}}\,c\mu(\mathrm{d}y).$$

Hence the distribution P cannot be orderly, although μ is non-atomic and thus P is simple.

1.9. The Variation Distance in $\boldsymbol{E}_r$

The set $\boldsymbol{E}_r$ of all real-valued E in $\boldsymbol{E}$, i.e. the set of all signed finite measures on $\mathfrak{M}$, forms a real subalgebra of the complex algebra $\boldsymbol{E}$. Each E in $\boldsymbol{E}_r$ can be represented as a difference $E = E_1 - E_2$ of measures in $\boldsymbol{E}^+$. Among these representations there is exactly one, for which $E_1(M) + E_2(M)$ is minimal, namely the Jordan decomposition $E = E^+ - E^-$. This special decomposition of the signed measure E as a difference of two measures can also be characterized by the property that E_1, E_2 are purely singular with respect to each other.

For all E in $\boldsymbol{E}_r$, we denote by $\|E\|$ the total mass $E^+(M) + E^-(M)$. Obviously $\|E\|$ is zero if and only if $E^+ = E^- = 0$ and hence also $E = 0$. For all real c and all E in $\boldsymbol{E}_r$, $\|cE\| = |c|\,\|E\|$. Further, if U, V are any signed measures in $\boldsymbol{E}_r$, then

$$U + V = (U^+ + V^+) - (U^- + V^-)$$

forms a representation of $U + V$ as a difference of measures in $\boldsymbol{E}^+$, and we can conclude that

$$\|U + V\| \leqq (U^+ + V^+)(M) + (U^- + V^-)(M) = \|U\| + \|V\|.$$

Hence $\|.\|$ is a norm on $\boldsymbol{E}_r$.

Obviously $\boldsymbol{E}^+$ and $\boldsymbol{P}$ are closed subsets of $\boldsymbol{E}_r$ with respect to $\|.\|$.

For all E in $\boldsymbol{E}_r$ and all sets X in $\mathfrak{A}$, we obtain

$$\begin{aligned}\|{}_XE\| &= \|{}_X(E^+) - {}_X(E^-)\| \leqq {}_X(E^+)(M) + {}_X(E^-)(M)\\ &= E^+(M) + E^-(M) = \|E\|.\end{aligned}$$

In particular the homomorphism $E \curvearrowright {}_XE$ of $\boldsymbol{E}_r$ into itself is continuous with respect to $\|.\|$. In what follows we write for simplicity ${}_X\|E\|$ instead of $\|{}_XE\|$.

The distance $\|U - V\|$ between signed measures in $\boldsymbol{E}_r$ corresponding to the norm $\|.\|$ is called the *variation distance*. This name is justified by the following formula.

1.9.1. *If $\mathfrak{S}$ is a generating subalgebra of $\mathfrak{M}$, then for all E in $\boldsymbol{E}_r$*

$$\|E\| = \sup_{\mathfrak{Y}} \sum_{Y \in \mathfrak{Y}} |E(Y)|,$$

where $\mathfrak{Y}$ runs through all finite systems of pairwise disjoint sets in $\mathfrak{S}$.

Proof. 1. For each at most countable system $\mathfrak{Y}$ of pairwise disjoint sets in $\mathfrak{M}$ and each E in $\boldsymbol{E}_r$ we have

$$\begin{aligned}\sum_{Y \in \mathfrak{Y}} |E(Y)| &= \sum_{Y \in \mathfrak{Y}} |E^+(Y) - E^-(Y)|\\ &\leqq \sum_{Y \in \mathfrak{Y}} E^+(Y) + \sum_{Y \in \mathfrak{Y}} E^-(Y) \leqq E^+(M) + E^-(M) = \|E\|.\end{aligned}$$

2. Let now $\mathfrak{S}$ be a generating subalgebra of $\mathfrak{M}$ and E a signed measure in $\boldsymbol{E}_r$. Since E^+, E^- are purely singular with respect to each other, there exists a set Y_0 in $\mathfrak{M}$ having the properties

$$E^+ = E((.) \cap Y_0), \qquad -E^- = E((.) \setminus Y_0).$$

We set $|E| = E^+ + E^-$. By the method used in the third step of the proof of 1.4.5. we conclude that for all $\varepsilon > 0$ there exists a set Y_ε in $\mathfrak{S}$ with the property

$$|E|\,((Y_\varepsilon \setminus Y_0) \cup (Y_0 \setminus Y_\varepsilon)) < \varepsilon.$$

Thus we obtain

$$\begin{aligned}&|E(Y_\varepsilon)| + |E(M \setminus Y_\varepsilon)|\\ &= |E^+(Y_\varepsilon) - E^-(Y_\varepsilon)| + |E^+(M \setminus Y_\varepsilon) - E^-(M \setminus Y_\varepsilon)|\\ &\geqq -4\varepsilon + |E^+(Y_0) - E^-(Y_0)| + |E^+(M \setminus Y_0) - E^-(M \setminus Y_0)|\\ &= -4\varepsilon + E^+(M) + E^-(M) = -4\varepsilon + \|E\|. \blacksquare\end{aligned}$$

If a signed measure E in $\boldsymbol{E}_r$ is absolutely continuous with respect to a σ-finite measure L on $\mathfrak{M}$ and f is a Radon-Nikodym-derivative of E with respect to L, then E^+ and E^- are also absolutely continuous with respect to L, for we have $E^+ = E((.) \cap Y_0))$, $-E^- = E((.) \setminus Y_0)$ as was stated in the above proof. Obviously the Radon-Nikodym-derivatives of E^+ and E^- with respect to L are f^+ and f^-, respectively, and we obtain

$$\|E\| = E^+(M) + E^-(M) = \int f^+(\Phi)\, L(\mathrm{d}\Phi) + \int f^-(\Phi)\, L(\mathrm{d}\Phi) = \int |f(\Phi)|\, L(\mathrm{d}\Phi).$$

For each sequence (E_n) of signed measures in $\boldsymbol{E}_r$ there exists a measure L in $\boldsymbol{E}^+$ such that all E_n are absolutely continuous with respect to L, for we can set, for instance

$$L = \sum_{n=1}^{\infty} 2^{-n}(\|E_n\| + 1)^{-1}\,(E_n^+ + E_n^-).$$

Now let (E_n) be a Cauchy sequence with respect to the norm $\|.\|$. For each n we choose a Radon-Nikodym-derivative f_n of E_n with respect to L and obtain

$$\begin{aligned}\|E_n - E_m\| &= \left\| \int_{(.)} f_n(\Phi)\, L(\mathrm{d}\Phi) - \int_{(.)} f_m(\Phi)\, L(\mathrm{d}\Phi) \right\| \\ &= \int |f_n(\Phi) - f_m(\Phi)|\, L(\mathrm{d}\Phi) \qquad (n, m = 1, 2, \ldots).\end{aligned}$$

In virtue of the completeness of the space $\mathscr{L}^1(L)$ there exists a real function f on M, measurable with respect to $\mathfrak{M}$ and such that $\lim\limits_{n\to\infty} \int |f_n(\Phi) - f(\Phi)|\, L(\mathrm{d}\Phi) = 0$. Setting $E = \int\limits_{(.)} f(\Phi)\, L(\mathrm{d}\Phi)$ we obtain a signed measure E in $\boldsymbol{E}_r$ with the property $\|E_n - E\| \xrightarrow[n\to\infty]{} 0$. Hence the normed space $\boldsymbol{E}_r$ is complete.

In the case $A = \{1, \ldots, m\}$ each signed measure E in $\boldsymbol{E}_r$ is uniquely determined by the values

$$E\big(\Phi(\{i\}) = l_i \quad \text{for} \quad 1 \leqq i \leqq m\big) \qquad (l_1, \ldots, l_m \geqq 0),$$

in particular

$$\|E\| = \sum_{l_1,\ldots,l_m \geqq 0} \left|E\big(\Phi(\{i\}) = l_i \quad \text{for} \quad 1 \leqq i \leqq m\big)\right|.$$

Convergence in $\boldsymbol{E}^+$ can easily be expressed in terms of the values $E\big(\Phi(\{i\}) = l_i$ for $1 \leqq i \leqq m\big)$.

1.9.2. *Suppose that $A = \{1, \ldots, m\}$. If (E_n) is a sequence of elements of $\boldsymbol{E}^+$, then (E_n) has a limit E in $\boldsymbol{E}^+$ if and only if the limits*

$$\lim_{n\to\infty} \sum_{l_1,\ldots,l_m \geqq 0} E_n\big(\Phi(\{i\}) = l_i \quad \text{for} \quad 1 \leqq i \leqq m\big) = e,$$

$$\lim_{n\to\infty} E_n\big(\Phi(\{i\}) = l_i \quad \text{for} \quad 1 \leqq i \leqq m\big) = e_{l_1,\ldots,l_m} \qquad (l_1, \ldots, l_m \geqq 0)$$

exist in R and satisfy the condition

$$\sum_{l_1,\ldots,l_m \geqq 0} e_{l_1,\ldots,l_m} = e,$$

in which case

$$E\big(\Phi(\{i\}) = l_i \quad \text{for} \quad 1 \leqq i \leqq m\big) = e_{l_1,\ldots,l_m}.$$

Proof. 1. For $A = \{1, \ldots, m\}$, all V in $\boldsymbol{E}_r$ are absolutely continuous with respect to the σ-finite measure

$$L = \sum_{l_1,\ldots,l_m \geqq 0} \delta_{l_1\delta_1 + \cdots + l_m\delta_m}$$

and the mapping

$$\Phi = \sum_{i=1}^{m} l_i \delta_i \frown V(\Psi(\{i\}) = l_i \quad \text{for} \quad 1 \leqq i \leqq m)$$

can be interpreted as Radon-Nikodym-derivative of V with respect to L.

Let us now assume that all limits $e, e_{l_1,\ldots,l_m}$ exist and are finite and that

$$e = \sum_{l_1,\ldots,l_m \geqq 0} e_{l_1,\ldots l_m}.$$

We set

$$E = \sum_{l_1,\ldots,l_m \geqq 0} e_{l_1,\ldots,l_m} \, \delta_{l_1\delta_1+\cdots+l_m\delta_m}$$

and obtain a measure E in $\mathbf{E}^+$. If we now set

$$f_n\left(\sum_{i=1}^{m} l_i\delta_i\right) = E_n(\Psi(\{i\}) = l_i \quad \text{for} \quad 1 \leqq i \leqq m),$$

$$f\left(\sum_{i=1}^{m} l_i\delta_i\right) = E(\Psi(\{i\}) = l_i \quad \text{for} \quad 1 \leqq i \leqq m),$$

then the functions f_n and f can be interpreted as Radon-Nikodym-derivatives of E_n and E, respectively, with respect to L. By assumption,

$$f_n(\Phi) \xrightarrow[n\to\infty]{} f(\Phi) \quad \text{almost everywhere with respect to } L,$$

$$\int f_n(\Phi)\, L(\mathrm{d}\Phi) \xrightarrow[n\to\infty]{} \int f(\Phi)\, L(\mathrm{d}\Phi).$$

Taking now into consideration the fact that all f_n are non-negative almost everywhere with respect to L, we conclude by a theorem of Scheffé that $\int |f_n(\Phi) - f(\Phi)|\, L(\mathrm{d}\Phi) \xrightarrow[n\to\infty]{} 0$, i.e. $\|E_n - E\| \xrightarrow[n\to\infty]{} 0$. In fact

$$\lim_{n\to\infty} \int |f - f_n|\,(\Phi)\, L(\mathrm{d}\Phi) = \lim_{n\to\infty} \int (f - f_n)^+\,(\Phi)\, L(\mathrm{d}\Phi) + \lim_{n\to\infty} \int (f - f_n)^-\,(\Phi)\, L(\mathrm{d}\Phi)$$

$$= \lim_{n\to\infty} \int (f - f_n)^+\,(\Phi)\, L(\mathrm{d}\Phi) + \lim_{n\to\infty} \int (f - f_n)\,(\Phi)\, L(\mathrm{d}\Phi) + \lim_{n\to\infty} \int (f - f_n)^-\,(\Phi)\, L(\mathrm{d}\Phi)$$

$$= 2 \lim_{n\to\infty} \int (f - f_n)^+\,(\Phi)\, L(\mathrm{d}\Phi) = 2 \lim_{n\to\infty} \int k_{Y_n}(\Phi)\,\big(f(\Phi) - f_n(\Phi)\big)\, L(\mathrm{d}\Phi),$$

where we set

$$Y_n = \{\Phi : \Phi \in M,\ f_n(\Phi) \leqq f(\Phi)\}.$$

The functions $k_{Y_n}(f - f_n)$ converge to zero almost everywhere with respect to L and satisfy the condition

$$|k_{Y_n}(f - f_n)| \leqq f.$$

By the theorem of Lebesgue

$$\lim_{n\to\infty} \int k_{Y_n}(\Phi)\,(f(\Phi) - f_n(\Phi))\, L(\mathrm{d}\Phi) = 0.$$

2. Suppose that $\|E_n - E\| \xrightarrow[n\to\infty]{} 0$ for some E in $\mathbf{E}^+$. Then with increasing n

$$\sum_{l_1,\ldots,l_m \geqq 0} \big|E_n(\Phi(\{i\}) = l_i \quad \text{for} \quad 1 \leqq i \leqq m) - E(\Phi(\{i\}) = l_i \quad \text{for} \quad 1 \leqq i \leqq m)\big|$$

tends toward zero.

Hence all limits $e_{l_1,\ldots,l_m}$ exist and are equal to $E(\Phi(\{i\}) = l_i$ for $1 \leqq i \leqq m)$. In view of $\|E_n - E\| \geqq \big| \|E_n\| - \|E\| \big|$ the limit e of $\|E_n\| = E_n(M)$ also exists and is equal to $E(M)$. ∎

In the special case $A = \{1, \ldots, m\}$, using 1.9.2. we can introduce convergence in $\boldsymbol{E}^+$ in a quite elementary way. Nowhere in the following presentation will it prove necessary to modify this "natural" convergence in any way. Therefore, for finite state spaces we agree to denote convergence in $\boldsymbol{E}^+$ by $\to$.

For all E in $\boldsymbol{E}_r$ and all finite sequences $X_1, \ldots, X_m$ of sets in $\mathfrak{B}$, we obtain

$$\|E_{X_1,\ldots,X_m}\| = \|(E^+)_{X_1,\ldots,X_m} - (E^-)_{X_1,\ldots,X_m}\| \leqq \|(E^+)_{X_1,\ldots,X_m}\| + \|(E^-)_{X_1,\ldots,X_m}\|$$
$$= E^+(M) + E^-(M) = \|E\|.$$

In particular the homomorphism $E \rightsquigarrow E_{X_1,\ldots,X_m}$ of $\boldsymbol{E}_r$ into the algebra $\boldsymbol{E}_r'$ corresponding to the state space $\{1, \ldots, m\}$ is continuous.

Let now $\mathfrak{H}$ be any generating semialgebra of sets in $\mathfrak{B}$. Then, by virtue of 1.3.1., the union of all σ-algebras $\mathfrak{M}_{H_1,\ldots,H_m}$ where $H_1, \ldots, H_m$ runs through the system of all finite sequences of pairwise disjoint sets in $\mathfrak{H}$, is a generating subalgebra of $\mathfrak{M}$. For all finite systems $\mathfrak{Y}$ of pairwise disjoint sets in $\mathfrak{M}_{H_1,\ldots,H_m}$, we have

$$\sum_{Y \in \mathfrak{Y}} |E(Y)| \leqq \|E_{H_1,\ldots,H_m}\|$$

for all E in $\boldsymbol{E}_r$.

However each finite system $\mathfrak{Y}$ of pairwise disjoint sets in the union of all σ-algebras $\mathfrak{M}_{H_1,\ldots,H_m}$ is already contained in one of these σ-algebras. In virtue of 1.9.1., we see the validity of

1.9.3. Proposition. *For each generating semiring $\mathfrak{H}$ of sets in $\mathfrak{B}$ and each signed measure E in $\boldsymbol{E}_r$, $\|E\|$ coincides with the supremum of $\|E_{H_1,\ldots,H_m}\|$ where $H_1, \ldots, H_m$ runs through all finite sequences of pairwise disjoint sets in $\mathfrak{H}$.*

Let now (X_n) be a monotone increasing sequence of sets in $\mathfrak{B}$ with union A. Then, obviously, the system of those X in $\mathfrak{B}$ which are contained in some X_n, forms a generating subring of $\mathfrak{B}$. Hence for all E in $\boldsymbol{E}_r$, $\|E\|$ coincides with the supremum of all $\|E_{H_1,\ldots,H_m}\|$. On the other hand, if $H_1, \ldots, H_m \subseteq X_n$, then

$$\|E_{H_1,\ldots,H_m}\| = \|({}_{X_n}E)_{H_1,\ldots,H_m}\| \leqq \|{}_{X_n}E\| \leqq \|E\|.$$

From this we see that

1.9.4. *For all monotone increasing sequences (X_n) of sets in $\mathfrak{B}$ with union A and all signed measures E in $\boldsymbol{E}_r$,*

$$\|E\| = \sup_{n=1,2,\ldots} \|{}_{X_n}E\|.$$

For all measures E in $\boldsymbol{E}^+$, $\|E\|$ coincides with $E(M)$. On the other hand $\boldsymbol{E}^+$ is closed under convolution. Hence for all C, D in $\boldsymbol{E}^+$, we obtain

$$\|C * D\| = (C * D)(M) = C(M)\, D(M) = \|C\|\, \|D\|.$$

Now, if U, V are any elements of $\boldsymbol{E}_r$, then

$$\begin{aligned}\|U * V\| &= \|(U^+ - U^-) * (V^+ - V^-)\| \\ &\leqq \|U^+ * V^+\| + \|U^+ * V^-\| + \|U^- * V^+\| + \|U^- * V^-\| \\ &= \|U^+\|\,\|V^+\| + \|U^+\|\,\|V^-\| + \|U^-\|\,\|V^+\| + \|U^-\|\,\|V^-\| \\ &= (\|U^+\| + \|U^-\|)(\|V^+\| + \|V^-\|) = \|U\|\,\|V\|.\end{aligned}$$

Summarizing, we can state:

1.9.5. Proposition. *With respect to $*$ and $\|\cdot\|$, the vector space $\boldsymbol{E}_r$ is a commutative real Banach algebra with unit δ_0.*

If $U_1, \ldots, U_m$ and $V_1, \ldots, V_m$ are finite sequences of signed measures in $\boldsymbol{E}_r$, then we have

$$\begin{aligned}&\left\|\mathop{\ast}_{i=1}^{m} U_i - \mathop{\ast}_{i=1}^{m} V_i\right\| \\ &= \left\|\sum_{i=1}^{m} (U_1 * \cdots * U_i * V_{i+1} * \cdots * V_m - U_1 * \cdots * U_{i-1} * V_i * \cdots * V_m)\right\| \\ &= \left\|\sum_{i=1}^{m} U_1 * \cdots * U_{i-1} * V_{i+1} * \cdots * V_m * (U_i - V_i)\right\| \\ &\leqq \sum_{i=1}^{m} \|U_1\| \cdots \|U_{i-1}\|\,\|U_i - V_i\|\,\|V_{i+1}\| \cdots \|V_m\|.\end{aligned}$$

From this follows

1.9.6. *If $P_1, \ldots, P_m$ and $Q_1, \ldots, Q_m$ are finite sequences of distributions on $\mathfrak{M}$, then*

$$\left\|\mathop{\ast}_{i=1}^{m} P_i - \mathop{\ast}_{i=1}^{m} Q_i\right\| \leqq \sum_{i=1}^{m} \|P_i - Q_i\|.$$

As it was stated in the proof of proposition 1.5.11. each E in $\boldsymbol{E}$ different from 0 has within $\boldsymbol{E}$ either no or exactly n different n-th convolution roots. If Q is any n-th convolution root of E, and if $e_1 = 1, e_2, \ldots, e_n$ are the n-th roots of unity in the field of complex numbers, then $Q = e_1Q, e_2Q, \ldots, e_nQ$ exhaust all n-th convolution roots of E. If now E and Q are in $\boldsymbol{E}_r$ and n is odd, then only e_1Q is in $\boldsymbol{E}_r$. Therefore

1.9.7. *For each odd natural number n, each signed measure E in $\boldsymbol{E}_r$ has at most one n-th convolution root in $\boldsymbol{E}_r$.*

For all E in $\boldsymbol{E}_r$, the series $\sum_{k=0}^{\infty} \frac{1}{k!} E^k$ converges absolutely. We denote by $\exp E$ the sum of this series. As in the case of real exponents, for arbitrary E_1, E_2 in $\boldsymbol{E}_r$ we obtain the functional equation

$$\exp(E_1 + E_2) = (\exp E_1) * (\exp E_2).$$

For arbitrary E in $\boldsymbol{E}_r$ it is obvious that

$$_X(\exp E) = \exp\,(_X E)$$

for any X in $\mathfrak{A}$, and

$$(\exp E)_{X_1,\ldots,X_m} = \exp\,(E_{X_1,\ldots,X_m})$$

for arbitrary finite sequences $X_1, \ldots, X_m$ of sets in $\mathfrak{B}$.

1.9.8. *The mapping $E \curvearrowright \exp E$ of $\boldsymbol{E}_r$ into itself is one-to-one.*

Proof. In view of the functional equation for exp, we have only to show that $\exp Z = \delta_o$, $Z \in \boldsymbol{E}_r$, implies $Z = 0$.

For all natural numbers k, $\exp\left(\frac{1}{2k+1} Z\right)$ is a $(2k+1)$-th convolution root of δ_o in $\boldsymbol{E}_r$ and therefore, in view of 1.9.7., coincides with δ_o.

Consequently

$$\sum_{l=1}^{\infty} \frac{1}{l!} \frac{1}{(2k+1)^{l-1}} Z^l = 0 \qquad (k = 1, 2, \ldots)$$

and hence

$$Z + \frac{1}{2k+1} \sum_{l=2}^{\infty} \frac{1}{l!} \frac{1}{(2k+1)^{l-2}} Z^l = 0 \qquad (k = 1, 2, \ldots).$$

From this and

$$\left\| \frac{1}{2k+1} \sum_{l=2}^{\infty} \frac{1}{l!} \frac{1}{(2k+1)^{l-2}} Z^l \right\| \leq \frac{\|Z\|^2}{(2k+1)} e^{\|Z\|} \qquad (k = 1, 2, \ldots)$$

it can be concluded that $Z = 0$. ∎

In place of $\exp E = L$ we write also $E = \ln L$.

1.9.9. *For all signed measures E in $\boldsymbol{E}_r$ satisfying the condition $\|E - \delta_o\| < 1$, $\ln E$ exists and the following relation holds*

$$\ln E = \sum_{k=1}^{\infty} \frac{(-1)^{k+1}}{k} (E - \delta_o)^k.$$

Proof. Because of $\|E - \delta_o\| < 1$ the above infinite series converges absolutely in $\boldsymbol{E}_r$. By standard calculations with absolutely convergent power series we obtain

$$\exp\left(\sum_{k=1}^{\infty} \frac{(-1)^{k+1}}{k} (E - \delta_o)^k\right) = E. \; ∎$$

The domain of definition of ln can be characterized as follows.

1.9.10. *For a signed measure E in $\mathbf{E}_r$,* ln E *exists if and only if the following conditions are fulfilled:*

a) $E(\{o\}) > 0$,

b) *for each odd natural number $2k + 1$, there exists one and therefore exactly one* $(2k+1)$-th *convolution root E_{2k+1} of E in $\mathbf{E}_r$, and*

$$\|E_{2k+1}\| \xrightarrow[k\to\infty]{} 1.$$

Proof. 1. Let $\exp Z = E$, $Z \in \mathbf{E}_r$. Then, for all natural numbers n, $\exp\left(\frac{1}{n} Z\right)$ is an n-th convolution root of E in $\mathbf{E}_r$. Furthermore,

$$E(\{o\}) = \sum_{n=0}^{\infty} \frac{1}{n!} (Z^n)(\{o\}) = \sum_{n=0}^{\infty} \frac{1}{n!} \big(Z(\{o\})\big)^n = e^{Z(\{o\})} > 0$$

and

$$\left| \left\| \exp\left(\frac{1}{n} Z\right) \right\| - 1 \right| = \left| \left\| \delta_o + \sum_{m=1}^{\infty} \frac{1}{m!n^m} Z^m \right\| - \|\delta_o\| \right|$$

$$\leqq \left\| \sum_{m=1}^{\infty} \frac{1}{m!n^m} Z^m \right\| \leqq \sum_{m=1}^{\infty} \frac{1}{m!n^m} \|Z\|^m \xrightarrow[n\to\infty]{} 0.$$

Therefore the conditions stated above are necessary.

2. Conversely, suppose that these conditions are satisfied.

From

$$(E_{2k+1})^{2k+1} = E$$

it follows that

$$\big(E_{2k+1}(\{o\})\big)^{2k+1} = E(\{o\}) > 0$$

and hence

$$E_{2k+1}(\{o\}) \xrightarrow[k\to\infty]{} 1.$$

Consequently

$$\|E_{2k+1} - \delta_o\| = \big\|E_{2k+1}\big((.) \setminus \{o\}\big)\big\| + |E_{2k+1}(\{o\}) - 1|$$

$$= \big(\|E_{2k+1}\| - E_{2k+1}(\{o\})\big) + |E_{2k+1}(\{o\}) - 1| \xrightarrow[k\to\infty]{} 0,$$

so that by 1.9.9. ln E_{2k+1} exists for all sufficiently large k. Consequently ln $E = (2k + 1)$ ln E_{2k+1} exists, too. ∎

Obviously the mapping exp is continuous. Because of 1.9.9. the inverse mapping ln is defined on the open ball $\{E : E \in \mathbf{E}_r, \|E - \delta_o\| < 1\}$ and is continuous. Let now

$$\|\exp Z_n - \exp Z\| \xrightarrow[n\to\infty]{} 0.$$

It follows from this that

$$\|\exp (Z_n - Z) - \delta_o\| = \|\exp (-Z) (\exp Z_n - \exp Z)\|$$

$$\leqq \|\exp (-Z)\| \, \|\exp Z_n - \exp Z\| \xrightarrow[n\to\infty]{} 0.$$

For all sufficiently large n,

$$\|\exp(Z_n - Z) - \delta_o\| < 1,$$

so we conclude

$$\|Z_n - Z\| = \left\|\ln\left(\exp(Z_n - Z)\right) - \ln \delta_o\right\| \xrightarrow[n\to\infty]{} 0.$$

Consequently, the mapping exp is continuous in both directions.

The domain of definition of ln is open. In fact, if $\ln E$ is defined, then we obtain

$$E + V = e^Z + V = e^Z(\delta_o + e^{-Z}V).$$

We have

$$\|e^{-Z}V\| \leqq \|e^{-Z}\| \, \|V\|.$$

Hence $\delta_o + e^{-Z}V$ belongs, according to 1.9.9., to the domain of definition of ln if $\|V\|$ is sufficiently small. Therefore,

$$E + V = e^Z e^{Z'} = e^{Z+Z'},$$

i.e.

$$\ln(E + V) = Z + Z'.$$

Summarizing we can state

1.9.11. Proposition. *The mapping* $E \curvearrowright \exp E$ *is a homeomorphism of the normed space* $\mathbf{E}_r$ *onto an open subset of* $\mathbf{E}_r$.

1.10. A Characterization of the Distributions $\mathcal{U}_E$

For all measures E in $\boldsymbol{E}^+$, we obtain

$$\mathcal{U}_E = e^{-E(M)} \exp E = \exp\left(-E(M)\,\delta_o\right) \exp E = \exp\left(E - E(M)\,\delta_o\right).$$

Based on 1.9.10. we can characterize the distributions of the form $\mathcal{U}_E$, $E \in \boldsymbol{E}^+$, in the following way.

1.10.1. Theorem. *A distribution P on $\mathfrak{M}$ has the form $P = \mathcal{U}_E$, $E \in \boldsymbol{E}^+$, if and only if it is infinitely divisible and satisfies the condition $P(\{o\}) > 0$. Then the measure E is uniquely determined by the additional condition $E(\{o\}) = 0$, the following relation being valid:*

$$\left\| n\left(\sqrt[n]{P}\right)\left((.)\setminus\{o\}\right) - E \right\| \xrightarrow[n\to\infty]{} 0.$$

Proof. 1. It was already stated in section 1.6., that all $\mathcal{U}_E$, $E \in \boldsymbol{E}^+$, belong to $\boldsymbol{T}$. Using 1.6.8. we obtain

$$\mathcal{U}_E(\{o\}) = e^{-E(\Psi \neq o)} > 0.$$

2. From $P = \mathcal{U}_E$, $E(\{o\}) = 0$, it follows for all natural numbers n that

$$n\left(\sqrt[n]{P}\right)\left((.)\setminus\{o\}\right) = n\mathcal{U}_{E/n}\left((.)\setminus\{o\}\right)$$

$$= e^{-\|E\|/n}\sum_{k=1}^{\infty}\frac{1}{k!\,n^{k-1}}E^k = e^{-\|E\|/n}E + e^{-\|E\|/n}\sum_{k=2}^{\infty}\frac{1}{k!\,n^{k-1}}E^k.$$

So we obtain

$$\left\|n\left(\sqrt[n]{P}\right)\left((.)\setminus\{o\}\right) - E\right\| \leqq \|e^{-\|E\|/n}E - E\| + e^{-\|E\|/n}\sum_{k=2}^{\infty}\frac{\|E\|^k}{k!\,n^{k-1}}$$

$$\leqq (1 - e^{-\|E\|/n})\,\|E\| + \frac{1}{n}\,e^{\|E\|},$$

i.e. we have

$$\left\|n\left(\sqrt[n]{P}\right)\left((.)\setminus\{o\}\right) - E\right\| \xrightarrow[n\to\infty]{} 0.$$

3. Now suppose that P is infinitely divisible and $P(\{o\}) > 0$. In virtue of 1.9.10., there exists exactly one representation $P = \exp L$, $L \in \boldsymbol{E}_r$. We set $E = L\left((.)\setminus\{o\}\right)$ and obtain

$$P = \exp\left(E + L(\{o\})\,\delta_o\right) = e^{L(\{o\})}\exp E,$$

with

$$P(\{o\}) = e^{L(\{o\})}.$$

Because of 1.9.7., for all natural numbers k,

$$\exp\left(\frac{1}{2k+1}\,L\right) = \exp\left(\frac{1}{2k+1}\,L(\{o\})\,\delta_o + \frac{1}{2k+1}\,E\right)$$

$$= \sqrt[2k+1]{P(\{o\})}\,\exp\left(\frac{1}{2k+1}\,E\right)$$

is the unique $(2k+1)$-th convolution root of P which belongs to $\boldsymbol{E}_r$, i.e. we have

$$(2k+1)\sqrt[2k+1]{P}\left((.)\setminus\{o\}\right) = \sqrt[2k+1]{P(\{o\})}\sum_{l=1}^{\infty}\frac{1}{l!\,(2k+1)^{l-1}}\,E^l.$$

From this formula we can derive the convergence

$$\left\|(2k+1)\sqrt[2k+1]{P}\left((.)\setminus\{o\}\right) - E\right\| \xrightarrow[k\to\infty]{} 0,$$

proceeding as in 2. As a limit of measures in $\boldsymbol{E}^+$, E also belongs to $\boldsymbol{E}^+$. Consequently $P = P(\{o\})\exp E$ where $E \in \boldsymbol{E}^+$, $E(\{o\}) = 0$, i.e. we have $P = \mathcal{U}_E$, $E \in \boldsymbol{E}^+$. ∎

In virtue of 1.8.5., for all infinite ν in N, we have $P_\nu(\{o\}) = 0$. Hence the Poisson distributions with infinite intensity measures cannot be expressed in the form $P_\nu = \mathcal{U}_E$, $E \in \boldsymbol{E}^+$, although they are infinitely divisible. Using

1.10.1., we obtain a representation of all distributions in $\boldsymbol{T}$ if the phase space $[A, \varrho_A]$ is bounded. This is an immediate consequence of

1.10.2. Proposition. *If an infinitely divisible distribution P on $\mathfrak{M}$ satisfies the condition $P(\Phi(A) < +\infty) = 1$, then it has the form $P = \mathcal{U}_E$, $E \in \boldsymbol{E}^+$.*

Proof. We have only to show that $c = P(\{o\}) > 0$ is satisfied. Suppose that c is equal to zero. Then for all natural numbers n, we obtain

$$P(\Phi(A) \geqq n) = \left(\sqrt[n]{P}\right)^n (\Phi(A) \geqq n) = 1,$$

for in this case we have also $\sqrt[n]{P}\,(\{o\}) = 0$. Hence we conclude that $P(\Phi(A) = +\infty) = 1$, which contradicts our assumptions. ∎

In virtue of 1.6.1., 1.10.1. and 1.10.2. for each distribution P in $\boldsymbol{T}$, the distribution ${}_XP$ has the form

$${}_XP = \mathcal{U}_E, \qquad E \in \boldsymbol{E}^+$$

if the set X is in $\mathfrak{A}$ and satisfies the condition $P(\Phi(X) = 0) > 0$. This condition is certainly satisfied if X is bounded.

Now we shall supplement the representation theorem 1.10.1. by the following continuity theorem.

1.10.3. *For each sequence (E_n) of elements of $\boldsymbol{E}^+$ and each E in $\boldsymbol{E}^+$,*

$$\left\|E_n\left((.)\setminus\{o\}\right) - E\left((.)\setminus\{o\}\right)\right\| \xrightarrow[n\to\infty]{} 0$$

is equivalent to

$$\|\mathcal{U}_{E_n} - \mathcal{U}_E\| \xrightarrow[n\to\infty]{} 0.$$

Proof. In view of 1.6.5., we can assume in addition that $E(\{o\}) = 0 = E_n(\{o\})$, $n = 1, 2, \ldots$, is valid. By 1.9.11. now $\|\mathcal{U}_{E_n} - \mathcal{U}_E\| \to 0$, i.e.

$$\left\|\exp\left(E_n - \|E_n\|\,\delta_o\right) - \exp\left(E - \|E\|\,\delta_o\right)\right\| \xrightarrow[n\to\infty]{} 0$$

is equivalent to

$$\left\|(E_n - \|E_n\|\,\delta_o) - (E - \|E\|\,\delta_o)\right\| \xrightarrow[n\to\infty]{} 0,$$

i.e. to

$$\left(\|E_n - E\| + \left|\|E_n\| - \|E\|\right|\right) \xrightarrow[n\to\infty]{} 0,$$

and hence also to

$$\|E_n - E\| \xrightarrow[n\to\infty]{} 0. \ ∎$$

For each distribution P on $\mathfrak{M}$,

$$\|P - \delta_o\| = 2P(\Phi \neq o).$$

From 1.9.8. and 1.9.9. it follows

1.10.4. *Each distribution P on $\mathfrak{M}$ with the property $P(\{o\}) > \frac{1}{2}$ can be represented in exactly one way in the form $P = \exp L$, $L \in \boldsymbol{E}_r$.*

Because of $1 = P(M) = \exp\big(L(M)\big)$, we have $L(M) = 0$.

Distributions of the form $P = \exp L$, $L \in \boldsymbol{E}_r$, are not necessarily infinitely divisible, i.e. $L\big((.) \setminus \{o\}\big)$ does not necessarily belong to $\boldsymbol{E}^+$.

If, e.g., $A = \{1\}$ and $P = (1-p)\,\delta_0 + p\delta_1$ with $0 < p < \frac{1}{2}$, then the distribution P is not infinitely divisible, however, $P = \exp L$ holds with

$$L = \sum_{n=1}^{\infty} (-1)^{n+1} \frac{1}{n} \left(\frac{p}{1-p}\right)^n \delta_n + \ln(1-p)\,\delta_0.$$

The signed measures L do not tend toward a signed measure in $\boldsymbol{E}_r$ if $p \to \frac{1}{2} - 0$. Therefore, in view of 1.9.11., the distribution $P = \frac{1}{2}(\delta_0 + \delta_1)$ has no logarithm which belongs to $\boldsymbol{E}_r$.

The distributions $P = \exp L$, $L \in \boldsymbol{E}_r$, can be connected with infinitely divisible distributions in the following way. We set $E_1 = L^+$, $E_2 = L^-$ and obtain

$$\begin{aligned}
\mathcal{U}_{E_2} * P &= \exp\big(L^- - L^-(M)\,\delta_o\big) * \exp L \\
&= \exp\Big(\big(L^+ - L^-\big) + \big(L^- - L^-(M)\,\delta_o\big)\Big) = \exp\big(L^+ - L^-(M)\,\delta_o\big) \\
&= \exp\Big(L^+ - \big(L(M) + L^-(M)\big)\,\delta_o\Big) = \exp\big(L^+ - L^+(M)\,\delta_o\big) = \mathcal{U}_{E_1}.
\end{aligned}$$

Conversely, from

$$\mathcal{U}_{E_1} * P = \mathcal{U}_{E_2}, \qquad E_1, E_2 \in \boldsymbol{E}^+; \qquad P \in \boldsymbol{P},$$

it follows that

$$\begin{aligned}
P &= \exp\big(-E_1 + E_1(M)\,\delta_o\big) \exp\big(E_2 - E_2(M)\,\delta_o\big) \\
&= \exp\Big(E_2 - E_1 + \big(E_1(M) - E_2(M)\big)\,\delta_o\Big) = \exp L
\end{aligned}$$

with $L \in \boldsymbol{E}_r$. Thus we see the validity of

1.10.5. Proposition. *A distribution P on $\mathfrak{M}$ has the form $P = \exp L$, $L \in \boldsymbol{E}_r$, if and only if it can be represented as convolution quotient of two infinitely divisible distributions P_1, P_2 with $P_1(\{o\}), P_2(\{o\}) > 0$.*

The question whether a signed measure L in $\boldsymbol{E}_r$ satisfying $L(M) = 0$ is the logarithm of a distribution P is equivalent to the question whether $\mathcal{U}_{L^-}$ is a divisor of $\mathcal{U}_{L^+}$ in the set $\boldsymbol{P}$ of distributions. Therefore, this is quite a complicated question.

Detailed investigations concerning the characterization of the logarithms L of distributions P in the case $A = \{1\}$ were carried out by Lévy [1]. He investigated signed measures L of the form

$$L = \sum_{l=0}^{k} q_k \delta_k \qquad (k = 1, 2, \ldots).$$

Concluding this section we state the following useful inequality (cf. Kerstan & Matthes [5]; the following proof is due to J. Zabczyk).

1.10.6. *For all E_1, E_2 in $\boldsymbol{E}^+$,*

$$\|\mathcal{U}_{E_1} - \mathcal{U}_{E_2}\| \leqq 2\|E_1 - E_2\|.$$

Proof. Setting $H_i = E_i - E_i(M)\,\delta_o$, $i = 1, 2$, we obtain for $n = 1, 2, \ldots$

$$\|\mathcal{U}_{E_1} - \mathcal{U}_{E_2}\| = \|\exp H_1 - \exp H_2\| = \|(\exp n^{-1}H_1)^n - (\exp n^{-1}H_2)^n\|$$

$$= \left\|(\exp n^{-1}H_1 - \exp n^{-1}H_2) \sum_{i=0}^{n-1} (\exp n^{-1}H_1)^{n-1-i} * (\exp n^{-1}H_2)^i\right\|$$

$$\leqq \|\exp n^{-1}H_1 - \exp n^{-1}H_2\| \sum_{i=0}^{n-1} \|(\exp n^{-1}H_1)^{n-1-i} * (\exp n^{-1}H_2)^i\|$$

$$= \left\|n^{-1}H_1 - n^{-1}H_2 + \sum_{k=2}^{\infty} (k!\, n^k)^{-1} (H_1{}^k - H_2{}^k)\right\| n$$

$$\leqq \|H_1 - H_2\| + n^{-1} \sum_{k=2}^{\infty} (k!\, n^{k-2})^{-1} (\|H_1\|^k + \|H_2\|^k)$$

$$\leqq \|E_1 - E_2\| + \|E_2(M)\delta_o - E_1(M)\delta_o\| + n^{-1}(\|H_1\|^2 e^{\|H_1\|} + \|H_2\|^2 e^{\|H_2\|}),$$

and hence

$$\|\mathcal{U}_{E_1} - \mathcal{U}_{E_2}\| \leqq 2\|E_1 - E_2\| + \lim_{n\to\infty} n^{-1}(\|H_1\|^2 e^{\|H_1\|} + \|H_2\|^2 e^{\|H_2\|}). \blacksquare$$

1.11. The Accompanying Distributions

To each P in $\boldsymbol{P}$ we can associate an infinitely divisible distribution in the form of the so-called *accompanying distribution* $\mathcal{U}_P$. If P is close to δ_o, then $\mathcal{U}_P$ represents an approximation to P in the following sense:

1.11.1. *For all P in $\boldsymbol{P}$,*

$$\|P - \mathcal{U}_P\| \leqq 2\big(1 - P(\{o\})\big)^2.$$

Proof. If we set $d = 1 - P(\{o\})$, then

$$\|P - \mathcal{U}_P\| = \left\|P - e^{-d} \sum_{k=0}^{\infty} \frac{1}{k!} \big(P((.)\setminus\{o\})\big)^k\right\|$$

$$= \left\|(1-d)\,\delta_o + P\big((.)\setminus\{o\}\big) - e^{-d}\,\delta_o - e^{-d}P\big((.)\setminus\{o\}\big) - e^{-d}\sum_{k=2}^{\infty}\frac{1}{k!}\big(P((.)\setminus\{o\})\big)^k\right\|$$

$$\leqq e^{-d} - (1-d) + (1 - e^{-d})\,\big\|P\big((.)\setminus\{o\}\big)\big\| + e^{-d}\sum_{k=2}^{\infty}\frac{1}{k!}\,\big\|P\big((.)\setminus\{o\}\big)\big\|^k$$

$$\leqq \frac{d^2}{2} + d(1 - e^{-d}) + e^{-d}\sum_{k=2}^{\infty}\frac{d^k}{k!} \leqq \frac{d^2}{2} + d^2 + \frac{d^2}{2}\left(e^{-d}\sum_{k=2}^{\infty}\frac{2\,d^{k-2}}{k!}\right) < 2d^2. \blacksquare$$

If in the convolution of a finite number of distributions P_i being close to δ_o each factor is replaced by its accompanying distribution, then $\mathop{\ast}\limits_{i=1}^{m} P_i$ is approximated by $\mathop{\ast}\limits_{i=1}^{m} \mathscr{U}_{P_i} = \mathscr{U}_{\sum_{i=1}^{m} P_i}$ in the following way (cf. Le Cam [1]):

1.11.2. *For all finite sequences $P_1, \ldots, P_m$ of distributions on $\mathfrak{M}$,*

$$\left\| \mathop{\ast}\limits_{i=1}^{m} P_i - \mathscr{U}_{\sum_{i=1}^{m} P_i} \right\| \leq 2 \sum_{i=1}^{m} \left(1 - P_i(\{o\})\right)^2 .$$

Proof. The asserted inequality follows immediately from 1.9.6. and 1.11.1. ∎

If all P_i satisfy the condition $P_i\left(\Phi(A) \leq 1\right) = 1$, then all $\mathscr{U}_{P_i}$, and hence also $\mathscr{U}_{\sum_{i=1}^{m} P_i}$, are of the Poisson type.

Immediately from 1.11.2. we have

1.11.3. Let $\nu_1, \ldots, \nu_m$ *be a finite sequence of distributions on $\mathfrak{A}$ and $p_1, \ldots, p_m$ be numbers in $[0, 1]$. Then*

$$\left\| \mathop{\ast}\limits_{i=1}^{m} \left((1 - p_i)\, \delta_o + p_i Q_{\nu_i}\right) - P_{\sum_{i=1}^{m} p_i \nu_i} \right\| \leq 2 \sum_{i=1}^{m} p_i^2 .$$

Combining 1.11.2. with the continuity proposition 1.10.3., we obtain the following convergence theorem:

1.11.4. Theorem. *Suppose that with each natural number n a finite sequence $P_{n,1}, \ldots, P_{n,m_n}$ of distributions on $\mathfrak{M}$ is associated, such that $\min\limits_{1 \leq i \leq m_n} P_{n,i}(\{o\})$ tends toward unity with increasing n. Then for each E in $\mathbf{E}^+$ with the property $E(\{o\}) = 0$,*

$$\left\| \mathop{\ast}\limits_{i=1}^{m_n} P_{n,i} - \mathscr{U}_E \right\| \xrightarrow[n \to \infty]{} 0$$

is equivalent to

$$\left\| \sum_{i=1}^{m_n} P_{n,i}\left((.) \setminus \{o\}\right) - E \right\| \xrightarrow[n \to \infty]{} 0 .$$

Proof. We note that

$$\sum_{i=1}^{m_n} \left(1 - P_{n,i}(\{o\})\right)^2 \leq \max_{1 \leq i \leq m_n} \left(1 - P_{n,i}(\{o\})\right) \sum_{i=1}^{m_n} \left(1 - P_{n,i}(\{o\})\right)$$

$$= \left(1 - \min_{1 \leq i \leq m_n} P_{n,i}(\{o\})\right) \sum_{i=1}^{m_n} P_{n,i}(\Phi \neq o) .$$

If now

$$\left\| \sum_{i=1}^{m_n} P_{n,i}\left((.) \setminus \{o\}\right) - E \right\| \xrightarrow[n \to \infty]{} 0 ,$$

then

$$\sum_{i=1}^{m_n} P_{n,i}(\Phi \neq o) \xrightarrow[n\to\infty]{} E(M)$$

and hence

$$\sum_{i=1}^{m_n} \left(1 - P_{n,i}(\{o\})\right)^2 \xrightarrow[n\to\infty]{} 0,$$

so that by 1.10.3. and 1.11.2. we conclude that

$$\left\| \mathop{\ast}_{i=1}^{m_n} P_{n,i} - \mathcal{U}_E \right\| \xrightarrow[n\to\infty]{} 0.$$

Conversely, suppose that $\mathop{\ast}\limits_{i=1}^{m_n} P_{n,i}$ tends to $\mathcal{U}_E$ in $\boldsymbol{E}^+$. We have

$$\left(\mathop{\ast}_{i=1}^{m_n} P_{n,i}\right)(\{o\}) = \prod_{i=1}^{m_n} P_{n,i}(\{o\}) \xrightarrow[n\to\infty]{} \mathcal{U}_E(\{o\}) = e^{-E(M)},$$

or

$$\sum_{i=1}^{m_n} \Big(-\ln \big(1 - \big(1 - P_{n,i}(\{o\})\big)\big)\Big) \xrightarrow[n\to\infty]{} E(M).$$

In view of

$$-\ln(1 - x) = \sum_{n=1}^{\infty} \frac{x^n}{n} \geqq x \qquad (0 \leqq x < 1)$$

this implies

$$\overline{\lim_{n\to\infty}} \sum_{i=1}^{m_n} \big(1 - P_{n,i}(\{o\})\big) \leqq E(M) < +\infty,$$

so that, using again the chain of inequalities given at the beginning of our proof, we are able to conclude that

$$\sum_{i=1}^{m_n} \big(1 - P_{n,i}(\{o\})\big)^2 \xrightarrow[n\to\infty]{} 0.$$

In view of 1.11.2., this gives

$$\left\| \mathcal{U}_{\sum_{i=1}^{m_n} P_{n,i}} - \mathcal{U}_E \right\| \xrightarrow[n\to\infty]{} 0,$$

so that by 1.10.3. the convergence

$$\left\| \sum_{i=1}^{m_n} P_{n,i}\big((.)\setminus\{o\}\big) - E \right\| \xrightarrow[n\to\infty]{} 0$$

is obtained. ∎

Theorem 1.11.4. does not answer the question whether the limit P of a sequence $\left(\mathop{\ast}\limits_{i=1}^{m_n} P_{n,i}\right)_{n=1,2,\ldots}$ for $\min\limits_{1\leqq i\leqq m_n} P_{n,i}(\{o\}) \xrightarrow[n\to\infty]{} 1$ always has the form $\mathcal{U}_E, E \in \boldsymbol{E}^+$. In view of 1.10.2., the following general statement leads to an affirmative answer to this question in the case of a bounded phase space.

1.11.5. Proposition. *If with each natural number n a finite sequence $P_{n,1},\dots,P_{n,m_n}$ of distributions on $\mathfrak{M}$ is associated such that $\min_{1\leqq i\leqq m_n} P_{n,i}(\{o\}) \xrightarrow[n\to\infty]{} 1$ holds, and if the sequence*

$$\left(\mathop{*}_{i=1}^{m_n} P_{n,i}\right)_{n=1,2,\dots}$$

tends toward P in $\mathbf{E}_r$, then the distribution P is infinitely divisible.

Proof. 1. In view of 1.6.3. and the continuity of the mapping $E \frown E_{X_1,\dots,X_m}$, we can assume in addition without loss of generality that $[A, \varrho_A] = \{1, \dots, m\}$.

2. We have $\overline{\lim}_{n\to\infty} \sum_{i=1}^{m_n} \big(1 - P_{n,i}(\{o\})\big) < +\infty$, and hence

$$\overline{\lim}_{n\to\infty} \sum_{i=1}^{m_n} \big(1 - P_{n,i}(\{o\})\big)^2 = 0.$$

For, suppose that there exists a subsequence $n_1 < n_2 < \cdots$ of the sequence of natural numbers which has the property that $\lim_{s\to\infty} \sum_{1\leqq i\leqq m_{n_s}} \big(1 - P_{n_s,i}(\{o\})\big) = +\infty$. With each natural number s we associate a finite random sequence $\Phi_{s,1}, \dots, \Phi_{s,m_{n_s}}$ of elements of $[M, \mathfrak{M}]$ which is distributed according to $Q_s = P_{n_s,1} \times \cdots \times P_{n_s,m_{n_s}}$. If we now set

$$\xi_{s,i} = \min\big(\Phi_{s,i}(A), 1\big) \qquad (s = 1, 2, \dots; 1 \leqq i \leqq m_{n_s}),$$

then the sequences $\xi_{s,1}, \dots, \xi_{s,m_{n_s}}$ are independent and the distribution of $\xi_{s,i}$ is equal to

$$P_{n_s,i}(\{o\})\delta_0 + \big(1 - P_{n_s,i}(\{o\})\big)\delta_1 .$$

Using Chebyshev's inequality together with our additional assumption, we obtain

$$\overline{\lim}_{s\to\infty} Q_s\left(\sum_{1\leqq i\leqq m_{n_s}} \xi_{s,i} \leqq \frac{1}{2} \sum_{1\leqq i\leqq m_{n_s}} \big(1 - P_{n_s,i}(\{o\})\big)\right)$$

$$\leqq \overline{\lim}_{s\to\infty} Q_s\left(\left|\sum_{1\leqq i\leqq m_{n_s}} \Big(\xi_{s,i} - \big(1 - P_{n_s,i}(\{o\})\big)\Big)\right| \geqq \frac{1}{2} \sum_{1\leqq i\leqq m_{n_s}} \big(1 - P_{n_s,i}(\{o\})\big)\right)$$

$$\leqq \overline{\lim}_{s\to\infty} \frac{4 \sum_{i=1}^{m_{n_s}} P_{n_s,i}(\{o\}) \big(1 - P_{n_s,i}(\{o\})\big)}{\left(\sum_{1\leqq i\leqq m_{n_s}} \big(1 - P_{n_s,i}(\{o\})\big)\right)^2}$$

$$\leqq 4 \overline{\lim}_{s\to\infty} \left(\sum_{1\leqq i\leqq m_{n_s}} \big(1 - P_{n_s,i}(\{o\})\big)\right)^{-1} = 0.$$

Hence for all natural numbers k,

$$\overline{\lim}_{s\to\infty} \left(\mathop{*}_{1\leqq i\leqq m_{n_s}} P_{n_s,i}\right)\big(\Phi(A) \leqq k\big) \leqq \overline{\lim}_{s\to\infty} Q_s\left(\sum_{1\leqq i\leqq m_{n_s}} \xi_{s,i} \leqq k\right) = 0.$$

However, by the hypothesis of the proposition we have

$$\lim_{k\to\infty} \lim_{s\to\infty} \left(\underset{1\leq i\leq m_{n_s}}{\ast} P_{n_s,i} \right) (\Phi(A) \leq k) = \lim_{k\to\infty} P(\Phi(A) \leq k) = 1.$$

Thus, our additional assumption has led to a contradiction. Hence in virtue of 1.11.2. we obtain

$$\left\| \underset{1\leq i\leq m_n}{\ast} P_{n,i} - \mathcal{U}_{\sum\limits_{1\leq i\leq m_n} P_{n,i}} \right\| \xrightarrow[n\to\infty]{} 0,$$

which means that

$$\mathcal{U}_{\sum\limits_{1\leq i\leq m_n} P_{n,i}} \xrightarrow[n\to\infty]{} P.$$

3. If, for a natural number k and for all terms P_n of a sequence of distributions on $\mathfrak{M}$, there exists the k-th root $\sqrt[k]{P_n}$ then the existence of $\sqrt[k]{P}$ can be inferred from $P_n \xrightarrow[n\to\infty]{} P$.

This is so because we can use the diagonal method to select a strongly monotone increasing sequence (n_r) of natural numbers such that all of the limits

$$e_{l_1,\ldots,l_m} = \lim_{r\to\infty} \sqrt[k]{P_{n_r}} \left(\Phi(\{1\}) = l_1, \ldots, \Phi(\{m\}) = l_m\right) \qquad (l_1, \ldots, l_m \geq 0)$$

exist. There is a measure Q in $\boldsymbol{E}^+$ having the property that

$$e_{l_1,\ldots,l_m} = Q\left(\Phi(\{1\}) = l_1, \ldots, \Phi(\{m\}) = l_m\right) \qquad (l_1, \ldots, l_m \geq 0).$$

From $P_{n_r} \xrightarrow[r\to\infty]{} P$ we obtain

$$\lim_{r\to\infty} \sum_{0\leq l_1,\ldots,l_m\leq w} P_{n_r}\left(\Phi(\{1\}) = l_1, \ldots, \Phi(\{m\}) = l_m\right) \xrightarrow[w\to\infty]{} 1,$$

hence also

$$\lim_{r\to\infty} \sum_{0\leq l_1,\ldots,l_m\leq w} \sqrt[k]{P_{n_r}} \left(\Phi(\{1\}) = l_1, \ldots, \Phi(\{m\}) = l_m\right) \xrightarrow[w\to\infty]{} 1,$$

and consequently

$$\sum_{0\leq l_1,\ldots,l_m<+\infty} e_{l_1,\ldots,l_m} = 1 = Q(M),$$

so that by 1.9.2. we can conclude that $\sqrt[k]{P_{n_r}} \xrightarrow[r\to\infty]{} Q$. But this implies $P_{n_r} = \left(\sqrt[k]{P_{n_r}}\right)^k \xrightarrow[r\to\infty]{} Q^k$ which means that $Q^k = P$, and hence that $Q = \sqrt[k]{P}$. ∎

From the third step of the proof of 1.11.5. in connection with 1.6.3. it follows immediately that

1.11.6. *The set* $\boldsymbol{T}$ *is closed in* $\boldsymbol{P}$ *with respect to the variation distance.*

In section 1.5. it was shown that each distribution P on $\mathfrak{M}$ free from after-effects can be represented as the convolution of the discrete part P_d of P with the continuous part P_c of P, P_c and P_d being again free from after-effects.

The structure of discrete distributions free from after-effects was already dealt with in section 1.3. By 1.11.5. we can prove the following assertion about continuous distributions free from after-effects.

1.11.7. Proposition. *Each continuous distribution P on $\mathfrak{M}$ free from after-effects is infinitely divisible.*

Proof. For each a in A and each natural number s let $U_{a,s}$ denote the open ball of radius s^{-1} around a. We have

$$\inf_{s=1,2,\ldots} P\big(\Phi(U_{a,s}) > 0\big) = P\left(\Phi\left(\bigcap_{s=1}^{\infty} U_{a,s}\right) > 0\right) = P\big(\Phi(\{a\}) > 0\big) = 0.$$

Therefore, for each a in A and each natural number n there exists an $s_n(a)$ such that $P\big(\Phi(U_{a,s_n(a)}) > 0\big) < n^{-1}$. We write for simplicity $U_{a,n}$ instead of $U_{a,s_n(a)}$. Since the phase space $[A, \varrho_A]$ satisfies the second axiom of countability, we can associate with each natural number n a sequence $(a_{n,m})_{m=1,2,\ldots}$ of points in A such that $A = \bigcup_{m=1}^{\infty} U_{a_{n,m},n}$.

Let now X be an arbitrary set in $\mathfrak{B}$. We set

$$X_{n,1} = X \cap U_{a_{n,1},n} \quad \text{and} \quad X_{n,k} = (X \cap U_{a_{n,k},n}) \setminus \bigcup_{l=1}^{k-1} X_{n,l}$$

for $k = 2, 3, \ldots$ and obtain, for all natural numbers n, a sequence $(X_{n,m})_{m=1,2,\ldots}$ of pairwise disjoint sets in $\mathfrak{B}$ with the properties

$$\bigcup_{m=1}^{\infty} X_{n,m} = X, \qquad P\big(\Phi(X_{n,m}) > 0\big) < n^{-1} \quad \text{for} \quad n, m = 1, 2, \ldots$$

We set $\bigcup_{m=1}^{k} X_{n,m} = Z_{n,k}$. Obviously, $P\big(\Phi(X \setminus Z_{n,k}) > 0\big)$ tends toward zero with increasing k for all natural numbers n. Hence we can select natural numbers m_n such that

$$P\big(\Phi(X \setminus Z_{n,m_n}) > 0\big) < n^{-1} \quad \text{for} \quad n = 1, 2, \ldots$$

Consequently,

$$P_{Z_{n,m_n}} \xrightarrow[n\to\infty]{} P_X.$$

On the other hand, because P is free from after-effects, we obtain

$$P_{Z_{n,m_n}} = \mathop{\Large *}_{m=1}^{m_n} P_{X_{n,m}},$$

so that, in virtue of 1.11.5., we can conclude that P_X is infinitely divisible.

For each finite sequence $X_1, \ldots, X_m$ of pairwise disjoint sets in $\mathfrak{B}$ and each natural number n,

$$\sqrt[n]{P_{X_1}} \times \cdots \times \sqrt[n]{P_{X_m}} = \sqrt[n]{P_{X_1} \times \cdots \times P_{X_m}} = \sqrt[n]{P_{X_1,\ldots,X_m}}.$$

In virtue of 1.6.3. P is itself infinitely divisible. ∎

Let us conclude this section with the following characterization of the continuous Poisson distributions.

1.11.8. Theorem. *A continuous distribution P on $\mathfrak{M}$ is Poisson if and only if it is both free from after-effects and simple.*

Proof. By virtue of 1.7.6., each continuous Poisson distribution is simple.

Conversely, suppose that P is continuous, free from after-effects and simple. If now X is an arbitrary set in $\mathfrak{B}$, then ${}_XP$ is continuous, free from after-effects and simple, too. Because of 1.11.7. and 1.10.2., ${}_XP$ is infinitely divisible and has the form ${}_XP = \mathscr{U}_E$, $E \in \boldsymbol{E}^+$, $E(\{o\}) = 0$. By virtue of 1.6.11. and 1.6.15. a continuous distribution of this form is free from after-effects and simple if and only if

$$E\big(\Psi^*(A) \neq 1\big) = 0, \qquad E(\Psi \text{ is not simple}) = 0,$$

i.e.

$$E\big(\Psi(A) \neq 1\big) = 0.$$

Hence there exists a finite measure ν on $\mathfrak{A}$ with the property $E = Q_\nu$ and we obtain, according to 1.7.4.

$$P_X = ({}_XP)_X = (\mathscr{U}_{Q_\nu})_X = \pi_{\nu(X)}.$$

Therefore, all one-dimensional distributions P_X, $X \in \mathfrak{B}$, of the distribution P are Poisson, and hence P itself is a Poisson distribution on $\mathfrak{M}$. ∎

In view of 1.4.13., in 1.11.8. the assumption "P is simple" can be replaced by the assumption "P is orderly".

1.12. The Variation Distance of Poisson Distributions

An immediate consequence of 1.10.6. is

1.12.1. Proposition. *For all finite ν, ϱ in N,*

$$\|P_\nu - P_\varrho\| \leqq 2 \operatorname{Var}(\nu - \varrho)$$

Proof. Obviously we have

$$\|P_\nu - P_\varrho\| = \|\mathscr{U}_{Q_\nu} - \mathscr{U}_{Q_\varrho}\| \leqq 2\,\|Q_\nu - Q_\varrho\| = 2 \operatorname{Var}(\nu - \varrho),$$

where the total mass of the not necessarily positive, finite measure $\nu - \varrho$ is denoted by $\operatorname{Var}(\nu - \varrho)$.

The following inequalities (cf. Liese [2]) provide a deeper insight into the dependence of the variation distance $\|P_{\gamma_1} - P_{\gamma_2}\|$ on the intensity measures γ_1, γ_2.

1.12.2. Theorem. *Let γ_1, γ_2 be arbitrary measures in N. If γ is a measure in N such that γ_1, γ_2 are absolutely continuous with respect to γ and g_1, g_2 are density functions of γ_1, γ_2, respectively, with respect to γ, then*

$$2\left(1 - \exp\left(-\frac{1}{2}\int \left(\sqrt{g_1(a)} - \sqrt{g_2(a)}\right)^2 \gamma(\mathrm{d}a)\right)\right) \leqq \|P_{\gamma_1} - P_{\gamma_2}\|$$

$$\leqq \sqrt{8\left(1 - \exp\left(-\frac{1}{2}\int \left(\sqrt{g_1(a)} - \sqrt{g_2(a)}\right)^2 \gamma(\mathrm{d}a)\right)\right)}.$$

Proof. 1. Let γ_1, γ_2 be absolutely continuous with respect to γ. If now (X_n) is a monotone increasing sequence of sets in $\mathfrak{B}$ with union A, then, by virtue of 1.9.4.,

$$\|P_{\gamma_1} - P_{\gamma_2}\| = \lim_{n\to\infty} {}_{X_n}\|P_{\gamma_1} - P_{\gamma_2}\| = \lim_{n\to\infty} \left\|P_{{}_{X_n}(\gamma_1)} - P_{{}_{X_n}(\gamma_2)}\right\|.$$

Obviously $k_{X_n}g_1$ and $k_{X_n}g_2$ can be interpreted as density functions of ${}_{X_n}(\gamma_1)$ and ${}_{X_n}(\gamma_2)$, respectively. Taking additionally into consideration that

$$\lim_{n\to\infty} \int \left(\sqrt{k_{X_n}(a)\, g_1(a)} - \sqrt{k_{X_n}(a)\, g_2(a)}\right)^2 {}_{X_n}\gamma(\mathrm{d}a)$$

$$= \lim_{n\to\infty} \int_{X_n} \left(\sqrt{g_1(a)} - \sqrt{g_2(a)}\right)^2 \gamma(\mathrm{d}a) = \int \left(\sqrt{g_1(a)} - \sqrt{g_2(a)}\right)^2 \gamma(\mathrm{d}a),$$

we see that it will suffice to prove the assertion of 1.12.2. for finite measures ${}_{X_n}(\gamma_1)$, ${}_{X_n}(\gamma_2)$ and ${}_{X_n}(\gamma)$. In what follows, we can therefore assume that γ_1, γ_2 and γ are finite.

2. We set $f(z) = -\sqrt{z}$ and obtain in this way a continuous convex real function f which is defined on $[0, +\infty)$. If now P_1, P_2 are distributions on $\mathfrak{M}$, which are absolutely continuous with respect to a σ-finite measure L on $\mathfrak{M}$ and p_1, p_2 are density functions of P_1, P_2, respectively, we set (cf. Csiszár [1])

$$J(P_1, P_2) = \int f\left(\frac{p_1(\Phi)}{p_2(\Phi)}\right) p_2(\Phi)\, L(\mathrm{d}\Phi) = -\int \sqrt{p_1(\Phi)\, p_2(\Phi)}\, L(\mathrm{d}\Phi).$$

Obviously this definition is independent of the choice of the measure L. We obtain

$$\begin{aligned} 2\left(J(P_1, P_2) + 1\right) &= \int \left(p_1(\Phi) - 2\sqrt{p_1(\Phi)\, p_2(\Phi)} + p_2(\Phi)\right) L(\mathrm{d}\Phi) \\ &= \int \left(\sqrt{p_1(\Phi)} - \sqrt{p_2(\Phi)}\right)^2 L(\mathrm{d}\Phi) \\ &\leqq \int \left(\sqrt{p_1(\Phi)} + \sqrt{p_2(\Phi)}\right) \left|\sqrt{p_1(\Phi)} - \sqrt{p_2(\Phi)}\right| L(\mathrm{d}\Phi) \\ &= \int |p_1(\Phi) - p_2(\Phi)|\, L(\mathrm{d}\Phi), \end{aligned}$$

i.e.

$$2\left(J(P_1, P_2) + 1\right) \leqq \|P_1 - P_2\|.$$

3. For all x, y we have

$$1 - \sin x \sin y - \cos x \cos y = 1 - \cos(x - y)$$
$$\geqq \big(1 - \cos(x - y)\big)\big(\sin(x + y)\cos((x - y)/2)\big)^2$$
$$= 2\big(\sin((x - y)/2)\big)^2 \big(2\sin((x + y)/2)\cos((x + y)/2)\big)^2 \big(\cos((x - y)/2)\big)^2$$
$$= 2^{-1}\big(2\sin((x - y)/2)\cos((x + y)/2)\big)^2 \big(2\sin((x + y)/2)\cos((x - y)/2)\big)^2$$
$$= 2^{-1}(\sin x - \sin y)^2 (\sin x + \sin y)^2 = 2^{-1}\big((\sin x)^2 - (\sin y)^2\big)^2.$$

For all u, v in $[0, 1]$ there exist numbers $x, y \in [0, \pi/2]$ such that $\sqrt{u} = \sin x$, $\sqrt{v} = \sin y$. Thus we obtain

$$1 - \sqrt{uv} - \sqrt{(1 - u)(1 - v)} = 1 - \sin x \sin y - \cos x \cos y$$
$$\geqq 2^{-1}\big((\sin x)^2 - (\sin y)^2\big)^2 = 2^{-1}(u - v)^2.$$

4. We set

$$Y = \{\Phi : \Phi \in M,\ p_2(\Phi) \leqq p_1(\Phi)\}$$

and obtain

$$\|P_1 - P_2\| = (P_1 - P_2)^+(M) + (P_1 - P_2)^-(M)$$
$$= \int_Y \big(p_1(\Phi) - p_2(\Phi)\big) L(\mathrm{d}\Phi) + \int_{M\setminus Y} \big(p_2(\Phi) - p_1(\Phi)\big) L(\mathrm{d}\Phi)$$
$$= P_1(Y) - P_2(Y) + P_2(M \setminus Y) - P_1(M \setminus Y) = 2\big(P_1(Y) - P_2(Y)\big).$$

On the basis of 3. and the Schwarz inequality it follows

$$J(P_1, P_2) + 1 = 1 - \int \sqrt{p_1(\Phi)\, p_2(\Phi)}\, L(\mathrm{d}\Phi)$$
$$= 1 - \int_Y \sqrt{p_1(\Phi)}\sqrt{p_2(\Phi)}\, L(\mathrm{d}\Phi) - \int_{M\setminus Y} \sqrt{p_1(\Phi)}\sqrt{p_2(\Phi)}\, L(\mathrm{d}\Phi)$$
$$\geqq 1 - \sqrt{\int_Y p_1(\Phi)\, L(\mathrm{d}\Phi)}\sqrt{\int_Y p_2(\Phi)\, L(\mathrm{d}\Phi)}$$
$$- \sqrt{\int_{M\setminus Y} p_1(\Phi)\, L(\mathrm{d}\Phi)}\sqrt{\int_{M\setminus Y} p_2(\Phi)\, L(\mathrm{d}\Phi)}$$
$$= 1 - \Big(\sqrt{P_1(Y)\, P_2(Y)} + \sqrt{P_1(M \setminus Y)\, P_2(M \setminus Y)}\Big)$$
$$\geqq \frac{4}{8}\big(P_1(Y) - P_2(Y)\big)^2 = \frac{1}{8}\|P_1 - P_2\|^2,$$

i.e.

$$\|P_1 - P_2\| \leqq \sqrt{8\big(J(P_1, P_2) + 1\big)}.$$

5. Let now γ_1, γ_2 be finite measures in N which are absolutely continuous with respect to a finite measure γ. Further let g_1 and g_2 be density functions of γ_1 and γ_2, respectively, with respect to γ. In virtue of 1.7.11., $P_1 = P_{\gamma_1}$, $P_2 = P_{\gamma_2}$ are absolutely continuous with respect to P_γ so that we can choose $L = P_\gamma$. Applying again 1.7.11. we obtain

$$-J(P_1, P_2) = \int \sqrt{p_{\gamma_1,\gamma}(\Phi)\, p_{\gamma_2,\gamma}(\Phi)}\, P_\gamma(\mathrm{d}\Phi)$$

$$= \exp\left(-\frac{1}{2}\left(\gamma_1(A) + \gamma_2(A) - 2\gamma(A)\right)\right) \int \left(\prod_{\substack{a\in A\\ \Phi(\{a\})>0}} \left(\sqrt{g_1(a)\, g_2(a)}\right)^{\Phi(\{a\})}\right) \mathcal{U}_{Q_\gamma}(\mathrm{d}\Phi)$$

$$= \exp\left(-\frac{1}{2}\left(\gamma_1(A) + \gamma_2(A) - 2\gamma(A)\right)\right) \exp\left(\int \left(\sqrt{g_1(a)\, g_2(a)} - 1\right) \gamma(\mathrm{d}a)\right)$$

$$= \exp\left(-\frac{1}{2}\int \left(g_1(a) - 2\sqrt{g_1(a)\, g_2(a)} + g_2(a)\right) \gamma(\mathrm{d}a)\right)$$

$$= \exp\left(-\frac{1}{2}\int \left(\sqrt{g_1(a)} - \sqrt{g_2(a)}\right)^2 \gamma(\mathrm{d}a)\right).$$

From 1., 2., 4. and 5. we obtain the two asserted inequalities. ∎

The following example is to compare the efficiencies of the estimates 1.12.1. and 1.12.2.

Let $[A, \varrho_A]$ be the real axis R and μ the Lebesgue measure on the σ-algebra $\mathfrak{R}$ of Borel subsets of R. For all natural numbers n, we set

$$\nu_n = \mu((.) \cap [0, n]), \qquad \varrho_n = (1 + n^{-1})\, \nu_n .$$

By 1.12.1. we only obtain the trivial inequality

$$\|P_{\nu_n} - P_{\varrho_n}\| \leqq 2.$$

On the other hand, from 1.12.2. we have the inequality

$$\|P_{\nu_n} - P_{\varrho_n}\| \leqq \sqrt{8\left(1 - \exp\left(-\frac{1}{2}\int\limits_{[0,n]} \left(\sqrt{1} - \sqrt{1 + \frac{1}{n}}\right)^2 \mu(\mathrm{d}x)\right)\right)}$$

$$= \sqrt{8\left(1 - \exp\left(n\left(\sqrt{1 + \frac{1}{n}} - 1\right) - \frac{1}{2}\right)\right)}$$

and, therefore,

$$\|P_{\nu_n} - P_{\varrho_n}\| \xrightarrow[n\to\infty]{} 0.$$

If the assumptions of theorem 1.12.2. are fulfilled, we can conclude from $\int \left(\sqrt{g_1(a)} - \sqrt{g_2(a)}\right)^2 \gamma(\mathrm{d}a) = +\infty$ that

$$2(1 - 0) \leqq \|P_{\gamma_1} - P_{\gamma_2}\|,$$

i.e.

$$\|P_{\gamma_1} - P_{\gamma_2}\| = 2.$$

In this case P_{γ_1} is purely singular with respect to P_{γ_2}.

Conversely, let us suppose that P_{γ_1} is purely singular with respect to P_{γ_2}. We set $L = P_{\gamma_1} + P_{\gamma_2}$, such that P_{γ_1} and P_{γ_2} are absolutely continuous with respect to L. If p_1 and p_2 are density functions of P_{γ_1} and P_{γ_2}, respectively, with respect to L, then the product $p_1(\Phi)\,p_2(\Phi)$ is equal to zero almost everywhere with respect to L. Let now (X_n) be a monotone increasing sequence of sets in $\mathfrak{B}$ with union A. For all natural numbers n we set $\mathfrak{M}_n = {}_{X_n}\mathfrak{M}$ and denote the restrictions of P_{γ_1}, P_{γ_2} and L to $\mathfrak{M}_n$ by $P_{1,n}$, $P_{2,n}$ and L_n, respectively. Obviously $P_{1,n}$ and $P_{2,n}$ are absolutely continuous with respect to L_n. Let $p_{1,n}$ and $p_{2,n}$ be density functions of $P_{1,n}$ and $P_{2,n}$, respectively, with respect to L_n.

Since $\bigcup\limits_{n=1}^{\infty} \mathfrak{M}_n$ is a generating subalgebra of $\mathfrak{M}$, we can deduce (cf. e.g. Hewitt & Stromberg [1], Theorem 20.56) the convergence

$$p_{1,n}(\Phi) \xrightarrow[n\to\infty]{} p_1(\Phi), \qquad p_{2,n}(\Phi) \xrightarrow[n\to\infty]{} p_2(\Phi)$$

almost everywhere with respect to L. The sequences $p_{i,n}(\Phi)$, $i = 1, 2$, are bounded almost everywhere with respect to L.

In fact, the densities $p_1(\Phi)$, $p_2(\Phi)$ are bounded almost everywhere with respect to L by 1. For $Y \in \mathfrak{M}$, we have

$$P_{i,n}(Y) = P_{\gamma_i}(x_n\Phi \in Y) = \int\limits_{\{\Phi:\, x_n\Phi \in Y\}} p_i(\Phi)\, L(\mathrm{d}\Phi).$$

On the other hand,

$$P_{i,n}(Y) = \int\limits_{Y} p_{i,n}(\Phi)\, L_n(\mathrm{d}\Phi) = \int\limits_{\{\Phi:\, x_n\Phi \in Y\}} p_{i,n}(x_n\Phi)\, L(\mathrm{d}\Phi).$$

Therefore for all Y in $\mathfrak{M}$,

$$\int\limits_{\{\Phi:\, x_n\Phi \in Y\}} \big(p_i(\Phi) - p_{i,n}(x_n\Phi)\big)\, L(\mathrm{d}\Phi) = 0.$$

Suppose that for $\varepsilon > 0$ the set $Y_{i,0} = \{\Phi : p_{i,n}(\Phi) > 1 + \varepsilon\}$ has the property $L\,(Y_{i,0}) > 0$. Then we obtain

$$\begin{aligned}
&\int\limits_{\{\Phi:\, x_n\Phi \in Y_{i,0}\}} \big(p_i(\Phi) - p_{i,n}(x_n\Phi)\big)\, L(\mathrm{d}\Phi) \\
&\leqq \int\limits_{\{\Phi:\, x_n\Phi \in Y_{i,0}\}} \big(1 - (1+\varepsilon)\big)\, L(\mathrm{d}\Phi) = -\varepsilon L(Y_{i,0}) < 0,
\end{aligned}$$

which is a contradiction. By Lebesgue's theorem,

$$\int \sqrt{p_{1,n}(\Phi)\, p_{2,n}(\Phi)}\, L(\mathrm{d}\Phi) \xrightarrow[n\to\infty]{} 0.$$

We can identify $P_{1,n}$ and $P_{2,n}$ with

$$x_n(P_{\gamma_1}) = P_{x_n(\gamma_1)} \quad \text{and} \quad x_n(P_{\gamma_2}) = P_{x_n(\gamma_2)},$$

respectively. We have

$$-J\big(P_{x_n(\gamma_1)}, P_{x_n(\gamma_2)}\big) \xrightarrow[n\to\infty]{} 0,$$

and by the aid of the formula derived in the fifth step of the proof of theorem 1.12.2. we conclude that

$$+\infty = \lim_{n\to\infty} \int \left(\sqrt{k_{X_n}(a)\, g_1(a)} - \sqrt{k_{X_n}(a)\, g_2(a)}\right)^2 \gamma(\mathrm{d}a)$$

$$= \lim_{n\to\infty} \int_{X_n} \left(\sqrt{g_1(a)} - \sqrt{g_2(a)}\right)^2 \gamma(\mathrm{d}a) = \int \left(\sqrt{g_1(a)} - \sqrt{g_2(a)}\right)^2 \gamma(\mathrm{d}a).$$

In this way we see the validity of (cf. Brown [2] and Liese [2])

1.12.3. Theorem. *Let γ_1, γ_2 be arbitrary measures in N. If γ is a measure in N such that γ_1, γ_2 are absolutely continuous with respect to γ and g_1, g_2 are density functions of γ_1, γ_2 respectively, with respect to γ, then P_{γ_1} is purely singular with respect to P_{γ_2} if and only if*

$$\int \left(\sqrt{g_1(a)} - \sqrt{g_2(a)}\right)^2 \gamma(\mathrm{d}a) = +\infty.$$

1.13. The Thinning Operator $\mathscr{D}_c$

Let c be a number in $[0, 1]$ and P a distribution on $\mathfrak{M}$. We now find our guidance in the concept of casting dice independently for the individual points a of the realizations Φ to decide whether they survive, i.e. they are preserved, or die, i.e. they are cancelled. Let the "survival probability" be equal to c for all a. As a result of this thinning operation we obtain a new distribution Q on $\mathfrak{M}$ which satisfies the condition

$$Q_{X_1,\ldots,X_m} = \sum_{l_1,\ldots,l_m \geqq 0} P_{X_1,\ldots,X_m}(\{[l_1, \ldots, l_m]\}) \mathop{\times}_{i=1}^{m} \left((1-c)\,\delta_0 + c\delta_1\right)^{l_i}$$

for all finite sequences $X_1, \ldots, X_m$ of pairwise disjoint sets in $\mathfrak{B}$. This set-up is also meaningful for arbitrary finite signed measures on $\mathfrak{M}$, i.e. we have

1.13.1. *For all c in $[0, 1]$ and all E in $\mathbf{E}_r$, there exists exactly one signed measure $\mathscr{D}_c E$ in $\mathbf{E}_r$ which satisfies the condition*

$$(\mathscr{D}_c E)_{X_1,\ldots,X_m} = \sum_{l_1,\ldots,l_m \geqq 0} E\big(\Phi(X_j) = l_j \quad \text{for} \quad 1 \leqq j \leqq m\big) \mathop{\times}_{i=1}^{m} \left((1-c)\,\delta_0 + c\delta_1\right)^{l_i}$$

for all finite sequences $X_1, \ldots, X_m$ of pairwise disjoint sets in $\mathfrak{B}$.

Proof. 1. As it is shown by the proof of the uniqueness proposition 1.3.2., signed finite measures L on $\mathfrak{M}$ are also uniquely determined by the family of the $L_{X_1,\ldots,X_m}$. From this the uniqueness statement concerning $\mathscr{D}_c E$ follows immediately.

2. Let us now suppose in addition that E is a distribution. By the set-up

$$q^{H_1,\ldots,H_m} = \sum_{l_1,\ldots,l_m \geqq 0} E\big(\Phi(H_j) = l_j \quad \text{for} \quad 1 \leqq j \leqq m\big) \mathop{\times}_{i=1}^{m} \left((1-c)\,\delta_0 + c\delta_1\right)^{l_i}$$

we obtain a family of distributions which satisfies the conditions a), b) and c) of the existence theorem 1.3.5. in the case where $\mathfrak{H} = \mathfrak{B}$.

Because of

$$q^{H_1,\ldots,H_m}(\{[0, \ldots, 0]\}) \geqq E_{H_1,\ldots,H_m}(\{[0, \ldots, 0]\}),$$

the condition d) is also satisfied, and we can conclude the existence of $\mathscr{D}_c E$.

3. Each signed measure E on $\mathfrak{M}$ can be represented in the form $E = v_1 P_1 - v_2 P_2$, $v_1 v_2 \geqq 0$; $P_1, P_2 \in \boldsymbol{P}$. Setting now $Q = v_1(\mathscr{D}_c P_1) - v_2(\mathscr{D}_c P_2)$ we obtain a finite signed measure on $\mathfrak{M}$ having the required properties. ∎

From the above-mentioned proof it follows immediately that $\mathscr{D}_c$ transforms the set $\boldsymbol{E}^+$ into itself.

Taking now into consideration the agreements made at the end of section 1.1. we have

1.13.2. *For all E in $\boldsymbol{E}_r$, all c in* $[0, 1]$ *and all finite sequences $X_1, \ldots, X_m$ of pairwise disjoint sets in $\mathfrak{B}$,*

$$(\mathscr{D}_c E)_{X_1,\ldots,X_m} = \mathscr{D}_c(E_{X_1,\ldots,X_m})$$
$$= \sum_{l_1,\ldots,l_m \geqq 0} E\big(\Phi(X_j) = l_j \quad \text{for} \quad 1 \leqq j \leqq m\big) \mathop{\times}_{i=1}^{m} \mathscr{D}_c \delta_{l_i}.$$

For all E in $\boldsymbol{E}_r$, $\mathscr{D}_0 E = E(M)\,\delta_0$. On the other hand all mappings $\mathscr{D}_c$, $0 < c \leqq 1$, are one-to-one.

1.13.3. Proposition. *For all c in* $(0, 1]$, *$\mathscr{D}_c$ is an isomorphism of the real algebra $\boldsymbol{E}_r$ into itself.*

Proof. 1. First we show that all $\mathscr{D}_c$, $0 \leqq c \leqq 1$, are homomorphic mappings.

Obviously $\mathscr{D}_c$ is always linear. Let now E_1, E_2 be signed measures in $\boldsymbol{E}_r$. For all finite sequences $X_1, \ldots, X_m$ of pairwise disjoint sets in $\mathfrak{B}$, we obtain

$$\big(\mathscr{D}_c(E_1 * E_2)\big)_{X_1,\ldots,X_m} = \sum_{l_1,\ldots,l_m \geqq 0} (E_1 * E_2)\big(\Phi(X_j) = l_j \quad \text{for} \quad 1 \leqq j \leqq m\big) \mathop{\times}_{i=1}^{m} \mathscr{D}_c \delta_{l_i}$$
$$= \sum_{l_1,\ldots,l_m \geqq 0} \ \sum_{\substack{v_1,\ldots,v_m,g_1,\ldots,g_m \geqq 0 \\ v_j + g_j = l_j \text{ for } 1 \leqq j \leqq m}} E_1\big(\Phi(X_k) = v_k \quad \text{for} \quad 1 \leqq k \leqq m\big)$$
$$\cdot\, E_2\big(\Phi(X_k) = g_k \quad \text{for} \quad 1 \leqq k \leqq m\big) \mathop{\times}_{i=1}^{m} \mathscr{D}_c \delta_{l_i}.$$

But for $l = g + v$ we have

$$\mathscr{D}_c \delta_l = \big((1 - c)\,\delta_0 + c\delta_1\big)^g * \big((1 - c)\,\delta_0 + c\delta_1\big)^v = (\mathscr{D}_c \delta_g) * (\mathscr{D}_c \delta_v)$$

and we can continue the above sequence of equalities with

$$= \sum_{l_1,\ldots,l_m \geqq 0} \ \sum_{\substack{v_1,\ldots,v_m,g_1,\ldots,g_m \geqq 0 \\ v_j + g_j = l_j \text{ for } 1 \leqq j \leqq m}} E_1\big(\Phi(X_k) = v_k \quad \text{for} \quad 1 \leqq k \leqq m\big)$$
$$\cdot\, E_2\big(\Phi(X_s) = g_s \quad \text{for} \quad 1 \leqq s \leqq m\big) \mathop{\times}_{i=1}^{m} (\mathscr{D}_c \delta_{v_i}) * (\mathscr{D}_c \delta_{g_i})$$

$$= \sum_{l_1,\dots,l_m \geq 0} \left(\sum_{\substack{v_1,\dots,v_m,g_1,\dots,g_m \geq 0 \\ v_j+g_j=l_j \text{ for } 1 \leq j \leq m}} \left(E_1(\Phi(X_k) = v_k \quad \text{for} \quad 1 \leq k \leq m) \mathop{\times}_{i=1}^{m} \mathscr{D}_c \delta_{v_i} \right) \right.$$
$$\left. * \left(E_2(\Phi(X_s) = g_s \quad \text{for} \quad 1 \leq s \leq m) \mathop{\times}_{i=1}^{m} \mathscr{D}_c \delta_{g_i} \right) \right)$$
$$= \left(\sum_{v_1,\dots,v_m \geq 0} E_1(\Phi(X_k) = v_k \quad \text{for} \quad 1 \leq k \leq m) \mathop{\times}_{i=1}^{m} \mathscr{D}_c \delta_{v_i} \right)$$
$$* \left(\sum_{g_1,\dots,g_m \geq 0} E_2(\Phi(X_s) = g_s \quad \text{for} \quad 1 \leq s \leq m) \mathop{\times}_{h=1}^{m} \mathscr{D}_c \delta_{g_h} \right)$$
$$= (\mathscr{D}_c E_1)_{X_1,\dots,X_m} * (\mathscr{D}_c E_2)_{X_1,\dots,X_m} = (\mathscr{D}_c E_1 * \mathscr{D}_c E_2)_{X_1,\dots,X_m}.$$

2. It remains to show that the mapping $\mathscr{D}_c$ is one-to-one for all c in $(0, 1]$. Suppose that for two signed measures E_1, E_2 in $\boldsymbol{E}_r$ the relation $\mathscr{D}_c E_1 = \mathscr{D}_c E_2$ is valid. Let now $X_1, \dots, X_m$ be an arbitrary finite sequence of pairwise disjoint sets in $\mathfrak{B}$. We denote by $f_i(\xi_1, \dots, \xi_m)$ the formal power series

$$\sum_{k_1,\dots,k_m \geq 0} (E_i)_{X_1,\dots,X_m} (\{[k_1, \dots, k_m]\}) \, \xi_1^{k_1} \dots \xi_m^{k_m}$$

which are convergent on $[-1, 1]^m$. By assumption,

$$f_1\big((1-c) + c\eta_1, \dots, (1-c) + c\eta_m\big) = f_2\big((1-c) + c\eta_1, \dots, (1-c) + c\eta_m\big)$$

is valid for all $[\eta_1, \dots, \eta_m]$ in an open neighbourhood of $[1-c, \dots, 1-c]$ which is contained in $(-1, 1)^m$. In view of the identity theorem for power series, $(E_1)_{X_1,\dots,X_m}$ therefore coincides with $(E_2)_{X_1,\dots,X_m}$, i.e. $E_1 = E_2$. ∎

For $c = 1$, $\mathscr{D}_c$ is the identical mapping of $\boldsymbol{E}_r$ onto itself, while $\mathscr{D}_c$ transforms $\boldsymbol{E}_r$ onto a proper subset of $\boldsymbol{E}_r$ for all c in $[0, 1)$.

For all c_1, c_2 in $[0, 1]$, all E in $\boldsymbol{E}_r$ and all finite sequences $X_1, \dots, X_m$ of pairwise disjoint sets in $\mathfrak{B}$,

$$\big(\mathscr{D}_{c_2}(\mathscr{D}_{c_1} E)\big)_{X_1,\dots,X_m} = \mathscr{D}_{c_2}\big(\mathscr{D}_{c_1}(E_{X_1,\dots,X_m})\big)$$
$$= \mathscr{D}_{c_2} \left(\sum_{l_1,\dots,l_m \geq 0} E_{X_1,\dots,X_m}(\{[l_1, \dots, l_m]\}) \mathop{\times}_{i=1}^{m} \mathscr{D}_{c_1}(\delta_1)^{l_i} \right)$$
$$= \sum_{l_1,\dots,l_m \geq 0} E_{X_1,\dots,X_m}(\{[l_1, \dots, l_m]\}) \mathop{\times}_{i=1}^{m} (\mathscr{D}_{c_2} \mathscr{D}_{c_1} \delta_1)^{l_i}$$
$$= \sum_{l_1,\dots,l_m \geq 0} E_{X_1,\dots,X_m}(\{[l_1, \dots, l_m]\}) \mathop{\times}_{i=1}^{m} (\mathscr{D}_{c_1 c_2} \delta_1)^{l_i} = (\mathscr{D}_{c_1 c_2} E)_{X_1,\dots,X_m},$$

i.e. we have

1.13.4. *For all* c_1, c_2 in $[0, 1]$ *and all* E *in* $\boldsymbol{E}_r$,

$$\mathscr{D}_{c_2}(\mathscr{D}_{c_1} E) = \mathscr{D}_{c_1 c_2} E.$$

For all c in $[0, 1]$ and all E in $\boldsymbol{E}_r$,

$$\mathscr{D}_c E = \mathscr{D}_c(E^+) - \mathscr{D}_c(E^-)$$

is a representation of $\mathcal{D}_c E$ as a difference of measures in $\boldsymbol{E}^+$. Taking into consideration that $(\mathcal{D}_c E)(M) = E(M)$ we obtain

$$\|\mathcal{D}_c E\| \leqq (\mathcal{D}_c E^+)(M) + (\mathcal{D}_c E^-)(M) = E^+(M) + E^-(M) = \|E\|,$$

i.e. we have

1.13.5. *For all c in $[0, 1]$ and all E in $\boldsymbol{E}_r$,*

$$\|\mathcal{D}_c E\| \leqq \|E\|.$$

In particular all homomorphic mappings $\mathcal{D}_c$ are continuous. From this

1.13.6. *For all c in $[0, 1]$ and all E in $\boldsymbol{E}_r$,*

$$\mathcal{D}_c(\exp E) = \exp(\mathcal{D}_c E).$$

The mappings $\mathcal{D}_c$ transform $\boldsymbol{P}$ into itself, but $(\mathcal{D}_c)^{-1} P, P \in \boldsymbol{P}, 0 < c < 1$, need not necessarily belong to $\boldsymbol{E}^+$ as it is shown by the following example.

Let $[A, \varrho_A] = \{1\}$. In the form of $V = \exp(-\delta_1 + \delta_2) = \exp(-\delta_1) * \exp \delta_2$ we obtain an element of $\boldsymbol{E}_r$ with the property $V(M) = e^0 = 1$. Because of

$$V(\{1\}) = -1 < 0,$$

V does not belong to $\boldsymbol{P}$. On the other hand for $c = 1/2$

$$\begin{aligned}\mathcal{D}_c V &= \exp \mathcal{D}_c(-\delta_1 + \delta_2) \\ &= \exp\big(-(1 - c)\,\delta_0 - c\delta_1 + ((1 - c)\,\delta_0 + c\delta_1)^2\big) \\ &= \exp(-c^2\delta_0 + c^2\delta_2) = \mathcal{U}_{c^2\delta_2}.\end{aligned}$$

By 1.13.6. we obtain for all non-negative real numbers l and all c in $[0, 1]$

$$\begin{aligned}\mathcal{D}_c\pi_l = \mathcal{D}_c(U_{l\delta_1}) &= \mathcal{D}_c\big(\exp(-l\delta_0 + l\delta_1)\big) \\ &= \exp\big(\mathcal{D}_c(-l\delta_0 + l\delta_1)\big) = \exp\big(-l\delta_0 + l(1 - c)\,\delta_0 + lc\delta_1\big) \\ &= \exp(-lc\delta_0 + lc\delta_1) = U_{lc\delta_1} = \pi_{cl}.\end{aligned}$$

Therefore, if ν is any measure in N and $X_1, \ldots, X_m$ any sequence of pairwise disjoint sets in $\mathfrak{B}$, then

$$\begin{aligned}(\mathcal{D}_c P_\nu)_{X_1,\ldots,X_m} &= \mathcal{D}_c\big((P_\nu)_{X_1,\ldots,X_m}\big) \\ &= \mathcal{D}_c\left(\mathop{\times}_{i=1}^{m} \pi_{\nu(X_i)}\right) = \mathop{\times}_{i=1}^{m} \mathcal{D}_c\pi_{\nu(X_i)} = \mathop{\times}_{i=1}^{m} \pi_{c\nu(X_i)} = (P_{c\nu})_{X_1,\ldots,X_m},\end{aligned}$$

and we see the validity of

1.13.7. Proposition. *For all c in $[0, 1]$ and all ν in N,*

$$\mathcal{D}_c P_\nu = P_{c\nu}.$$

If now $p_1, \ldots, p_m$ is any finite sequence of non-negative numbers with sum equal to one, then by 1.7.3. and 1.13.8. we obtain

$$\mathop{*}_{i=1}^{m} (\mathscr{D}_{p_i} P_\nu) = \mathop{*}_{i=1}^{m} P_{p_i \nu} = P_{\sum\limits_{i=1}^{m} p_i \nu},$$

i.e. for all Poisson distributions P on $\mathfrak{M}$,

$$P = \mathop{*}_{i=1}^{m} \mathscr{D}_{p_i} P.$$

In section 10.2. we shall see that this property is even characteristic for Poisson distributions.

1.14. The Property $g_\nu . I$

If ν is an element of N and X any set in $\mathfrak{B}$, then a signed measure E in $\boldsymbol{E}_r$ is said to have the *property* $g_{\nu,X}$, if

$$_X E = \begin{cases} \sum\limits_{k=0}^{\infty} E\big(\Phi(X) = k\big) \, (Q_{\nu((.)\mid X)})^k & \text{for } \nu(X) > 0 \\ E(M) \, \delta_o & \phantom{\text{for }} \nu(X) = 0 \end{cases}$$

where we set $\nu\big((.) \mid X\big) = \big(\nu(X)\big)^{-1} \nu\big((.) \cap X\big)$. Let $\boldsymbol{E}_{\nu,X}$ denote the set of all E in $\boldsymbol{E}_r$ having the property $g_{\nu,X}$.

1.14.1. *For all ν in N and all Z, X in $\mathfrak{B}$ with $Z \subseteq X$*

$$\boldsymbol{E}_{\nu,X} \subseteq \boldsymbol{E}_{\nu,Z}.$$

Proof. First we suppose that $\nu(X) > 0$. We set $c = \big(\nu(X)\big)^{-1} \nu(Z)$ and obtain

$$_Z E = {}_Z({}_X E) = \sum_{k=0}^{\infty} E\big(\Phi(X) = k\big) \; {}_Z\Big(\big(\big(\nu(X)\big)^{-1} Q_{_X\nu}\big)^k\Big)$$

$$= \sum_{k=0}^{\infty} E\big(\Phi(X) = k\big) \big(\nu(X)\big)^{-k} \big({}_Z(Q_{_X\nu})\big)^k$$

$$= \sum_{k=0}^{\infty} E\big(\Phi(X) = k\big) \big(\nu(X)\big)^{-k} \big(\nu(X \setminus Z) \, \delta_o + Q_{_Z\nu}\big)^k.$$

If now $\nu(Z) > 0$, then we continue with

$$= \sum_{k=0}^{\infty} E_X(\{k\}) \big((1 - c) \, \delta_o + c Q_{\nu((.)\mid Z)}\big)^k$$

$$= \sum_{n=0}^{\infty} (\mathscr{D}_c E_X) \, (\{n\}) \, (Q_{\nu((.)\mid Z)})^n$$

and we conclude that

$$E_Z = \mathscr{D}_c E_X, \qquad {}_Z E = \sum_{k=0}^{\infty} E_Z(\{k\}) \, (Q_{\nu((.)\mid Z)})^k.$$

If however $\nu(Z) = 0$, then

$$ {}_ZE = \sum_{k=0}^{\infty} E\big(\Phi(X) = k\big)\,\delta_o = E(M)\,\delta_o, \qquad E_Z = \mathscr{D}_0 E_X. $$

If now $\nu(X) = 0$, then

$$ {}_ZE = {}_Z({}_XE) = {}_Z\big(E(M)\,\delta_o\big) = E(M)\,\delta_o. \blacksquare $$

From the above mentioned proof it follows

1.14.2. *If ν is a measure in N and X a set in $\mathfrak{B}$ with $\nu(X) > 0$, then for all Borel subsets Z of X and all E in $\boldsymbol{E}_{\nu,X}$,*

$$ E_Z = \mathscr{D}_{(\nu(X))^{-1}\nu(Z)} E_X. $$

If two signed measures E_1, E_2 belong to $\boldsymbol{E}_{\nu,X}$, then obviously the same is true for all real linear combinations $v_1E_1 + v_2E_2$. Furthermore, in the case $\nu(X) > 0$, we obtain

$$ \begin{aligned} {}_X(E_1 * E_2) &= {}_X(E_1) * {}_X(E_2) \\ &= \sum_{n=0}^{\infty} (E_1)_X(\{n\})\big(Q_{\nu((.)|X)}\big)^n * \sum_{m=0}^{\infty} (E_2)_X(\{m\})\big(Q_{\nu((.)|X)}\big)^m \\ &= \sum_{k=0}^{\infty} (E_1)_X * (E_2)_X(\{k\})\big(Q_{\nu((.)|X)}\big)^k \\ &= \sum_{k=0}^{\infty} (E_1 * E_2)_X(\{k\})\big(Q_{\nu((.)|X)}\big)^k. \end{aligned} $$

If $\nu(X) = 0$, then

$$ \begin{aligned} {}_X(E_1 * E_2) &= {}_X(E_1) * {}_X(E_2) = E_1(M)\,\delta_o * E_2(M)\,\delta_o \\ &= E_1(M)\,E_2(M)\,\delta_o = (E_1 * E_2)(M)\,\delta_o, \end{aligned} $$

i.e. $E_1 * E_2$ belongs also to $\boldsymbol{E}_{\nu,X}$.

If a sequence (E_n) of signed measures in $\boldsymbol{E}_{\nu,X}$ converges in variation distance toward an E in $\boldsymbol{E}_r$, then in the case $\nu(X) > 0$ we obtain

$$ \overline{\lim_{n\to\infty}} \left\| \sum_{k=0}^{\infty} (E_n)_X(\{k\})\big(Q_{\nu((.)|X)}\big)^k - \sum_{k=0}^{\infty} E_X(\{k\})\big(Q_{\nu((.)|X)}\big)^k \right\| $$

$$ \leq \overline{\lim_{n\to\infty}} \sum_{k=0}^{\infty} |(E_n)_X(\{k\}) - E_X(\{k\})| = \overline{\lim_{n\to\infty}} \|(E_n)_X - E_X\| = 0. $$

On the other hand,

$$ \overline{\lim_{n\to\infty}} \left\| \sum_{k=0}^{\infty} (E_n)_X(\{k\})\big(Q_{\nu((.)|X)}\big)^k - {}_XE \right\| = \overline{\lim_{n\to\infty}} {}_X\|E_n - E\| = 0, $$

i.e.

$$_XE = \sum_{k=0}^{\infty} E_X(\{k\}) \left(Q_{\nu((.)|X)}\right)^k.$$

If, however, $\nu(X) = 0$, then we conclude from

$$\|_X(E_n) - {}_XE\| \xrightarrow[n\to\infty]{} 0, \qquad E_n(M) \to E(M)$$

that

$$\|_XE - E(M)\,\delta_0\| = \lim_{n\to\infty} \|_X(E_n) - E(M)\,\delta_0\|$$

$$= \lim_{n\to\infty} \|E_n(M)\,\delta_0 - E(M)\,\delta_0\| = 0.$$

Summarizing we see the validity of

1.14.3. *For all ν in N and all X in $\mathfrak{B}$, $\boldsymbol{E}_{\nu,X}$ is a subalgebra of $\boldsymbol{E}_r$ which is closed with respect to $\|.\|$.*

The subalgebras $\boldsymbol{E}_{\nu,X}$ of $\boldsymbol{E}_r$ are closed with respect to the formation of convolution roots in the following sense.

1.14.4. *Let be ν a measure in N, X a set in $\mathfrak{B}$, n a natural number and E a signed measure in $\boldsymbol{E}_{\nu,X}$. If some Q in $\boldsymbol{E}_r$ satisfies the condition $Q^n = E$, then Q belongs to $\boldsymbol{E}_{\nu,X}$, too.*

Proof. If $_XE = 0$, then we conclude from $(_XQ)^n = {}_X(Q^n) = {}_XE = 0$ that $_XQ = 0$ because $\boldsymbol{E}$ has no divisors of 0. Therefore, in what follows we can assume that $_XE \neq 0$.

In the case $\nu(X) = 0$ we have $_XE = E(M)\,\delta_0$, hence also $(_XQ)^n = E(M)\delta_0$. If n is odd, then by 1.9.7. $_XE$ has at most one n-th convolution root in $\boldsymbol{E}_r$. However, $Q^n(M) = \left(Q(M)\right)^n = E(M)$ and therefore $Q(M)\delta_0$ is such a convolution root, i.e. $_XQ = Q(M)\delta_0$. If n is even, then the proof of 1.5.11. shows that in this case a signed measure in $\boldsymbol{E}_r$ which is different from 0 has either no or exactly two n-th convolution roots in $\boldsymbol{E}_r$. Now $\pm\sqrt[n]{E(M)}\,\delta_0$ are two different n-th convolution roots of $_XE$, i.e. we have

$$_XQ = \pm\sqrt[n]{E(M)}\,\delta_0$$

where it is to be noted that $E(M) = \left(Q(M)\right)^n$ is positive.

Now suppose that $\nu(X) > 0$. Because of

$$_XE = \sum_{k=0}^{\infty} E_X(\{k\}) \left(Q_{\nu((.)|X)}\right)^k,$$

$$\left(\sum_{k=0}^{\infty} q(\{k\}) \left(Q_{\nu((.)|X)}\right)^k\right)^n = {}_XE = (_XQ)^n$$

is equivalent to the fact that the finite signed measure q on $\mathfrak{P}(\{0, 1, \ldots\})$ is an n-th convolution root of E_X. On account of $(Q_X)^n = E_X$, Q_X is always such a convolution root. Therefore, if n is odd, then

$$\sum_{k=0}^{\infty} Q_X(\{k\}) \left(Q_{\nu((.)|X)}\right)^k$$

is the uniquely determined n-th convolution root of ${}_XE$ in $\boldsymbol{E}_r$, i.e.

$${}_XQ = \sum_{k=0}^{\infty} Q_X(\{k\}) \left(Q_{\nu((.)|X)}\right)^k.$$

If however n is even, then

$$\pm \sum_{k=0}^{\infty} Q_X(\{k\}) \left(Q_{\nu((.)|X)}\right)^k$$

are the unique n-th convolution roots of this kind, and we can conclude

$${}_XQ = \pm \sum_{k=0}^{\infty} Q_X(\{k\}) \left(Q_{\nu((.)|X)}\right)^k$$

and hence

$${}_XQ = \sum_{k=0}^{\infty} Q_X(\{k\}) \left(Q_{\nu((.)|X)}\right)^k. \blacksquare$$

In virtue of 1.14.3. if E belongs to $\boldsymbol{E}_{\nu,X}$, then $\exp E$ also belongs to $\boldsymbol{E}_{\nu,X}$. Conversely, let us assume that for a signed measure E in $\boldsymbol{E}_r$ the signed measure $\exp E$ belongs to $\boldsymbol{E}_{\nu,X}$. Because of $\left(\exp\left(\frac{1}{n} E\right)\right)^n = \exp E$ and 1.14.4., all measures $\exp\left(\frac{1}{n} E\right)$, $n = 1, 2, \ldots$, belong to $\boldsymbol{E}_{\nu,X}$. For sufficiently large n, $\left\|\delta_o - \exp\left(\frac{1}{n} E\right)\right\| < 1$ so that, according to 1.9.9., the logarithm $\frac{1}{n} E = \ln\left(\exp\left(\frac{1}{n} E\right)\right)$ can be represented as sum of a series whose terms are in $\boldsymbol{E}_{\nu,X}$ which converges absolutely in $\boldsymbol{E}_r$. Hence by 1.14.3. $\frac{1}{n} E$ and therefore E belongs to $\boldsymbol{E}_{\nu,X}$ and we see the validity of

1.14.5. *For all ν in N, all X in $\mathfrak{B}$ and all E in $\boldsymbol{E}_r$,* $\exp E$ *belongs to $\boldsymbol{E}_{\nu,X}$ if and only if E belongs to $\boldsymbol{E}_{\nu,X}$.*

If ν is a measure in N, then a signed measure E in $\boldsymbol{E}_r$ is said to have the *property g_ν* if it belongs to $\boldsymbol{E}_\nu = \bigcap_{X \in \mathfrak{B}} \boldsymbol{E}_{\nu,X}$. Obviously for all $\nu \neq o$ and all positive real l the subalgebras $\boldsymbol{E}_\nu$ and $\boldsymbol{E}_{l\nu}$ of $\boldsymbol{E}_r$ coincide and in the case $0 < \nu(A) < +\infty$ we can go over to the distribution $\nu\big((.) \mid A\big) = \big((\nu A)\big)^{-1} \nu$. In the case $\nu = o$ the subalgebra $\boldsymbol{E}_\nu$ coincides with the set of all real multiples of δ_o.

1.14.6. *Let be ν a measure in N and E_1, E_2 signed measures in $\boldsymbol{E}_\nu$. If $E_2 \neq 0$ and some Q in $\boldsymbol{E}_r$ satisfies the equation $E_2 * Q = E_1$, then Q belongs to $\boldsymbol{E}_\nu$.*

Proof. If $\nu = o$, we obtain $E_1 = E_1(M)\delta_o$, $E_2 = E_2(M)\delta_o$ and hence $E_2(M)\,\delta_o * Q = E_1(M)\,\delta_o$, i.e. $Q = E_1(M)\,\big(E_2(M)\big)^{-1}\delta_o$ and therefore $Q \in \boldsymbol{E}_o$. Now suppose $\nu \neq o$.

Let X be any set in $\mathfrak{B}$ with the property ${}_XE_2\big((.)\setminus\{o\}\big) \neq 0$. Then by assumption $\nu(X) > 0$ and we obtain

$$ {}_XE_1 = \sum_{k=0}^{\infty} (E_1)_X\,(\{k\})\,\big(Q_{\nu((.)|X)}\big)^k, \qquad {}_X(E_2) = \sum_{k=0}^{\infty} (E_2)_X\,(\{k\})\,\big(Q_{\nu((.)|X)}\big)^k. $$

Using

$$ (E_2)_X * Q_X = (E_1)_X $$

it follows

$$ {}_X(E_2) * \sum_{k=0}^{\infty} Q_X(\{k\})\,\big(Q_{\nu((.)|X)}\big)^k = {}_X(E_1). $$

Because of ${}_X(E_2) \neq 0$ and ${}_X(E_2) * {}_XQ = {}_X(E_1)$, by the aid of 1.5.8. the relation ${}_XQ = \sum_{k=0}^{\infty} Q_X(\{k\})\,\big(Q_{\nu((.)|X)}\big)^k$ follows, i.e. Q belongs to $\boldsymbol{E}_{\nu,X}$.

Since $\bigcup_{X\in\mathfrak{B}} {}_X\mathfrak{M}$ is a generating algebra of $\mathfrak{M}$, ${}_X(E_2)$ is different from 0 for sufficiently large X in $\mathfrak{B}$. In virtue of 1.14.1., Q belongs to $\boldsymbol{E}_\nu$. ∎

If E_1, E_2 belong to $\boldsymbol{E}_\nu$ and $(E_1)_Z = (E_2)_Z$ for a Z in $\mathfrak{B}$ having the property $\nu(Z) > 0$, then, with the help of 1.14.2., for all sets X in $\mathfrak{B}$ containing Z and $c = \big(\nu(X)\big)^{-1}\,\nu(Z)$ we obtain

$$ (E_1)_X = (\mathscr{D}_c)^{-1}\,(E_1)_Z = (\mathscr{D}_c)^{-1}\,(E_2)_Z = (E_2)_X $$

and hence

$$ {}_X(E_1) = \sum_{k=0}^{\infty} (E_1)_X\,(\{k\})\,\big(Q_{\nu((.)|X)}\big)^k = \sum_{k=0}^{\infty} (E_2)_X\,(\{k\})\,\big(Q_{\nu((.)|X)}\big)^k = {}_X(E_2). $$

Therefore, E_1, E_2 coincide on the generating subalgebra $\bigcup_{X\in\mathfrak{B},X\supseteq Z} {}_X\mathfrak{M}$ of $\mathfrak{M}$ and we see the validity of (cf. Kallenberg [5])

1.14.7. Proposition. *For all ν in N and all sets X in $\mathfrak{B}$ with $\nu(X) > 0$, from $E_1, E_2 \in \boldsymbol{E}_\nu$ and $(E_1)_X = (E_2)_X$ follows $E_1 = E_2$.*

By virtue of 1.8.4. all mixtures $P = \int P_{l\nu}(.)\,\sigma(\mathrm{d}l)$ have the property g_ν. For infinite measures ν the converse is also true.

1.14.8. Theorem. *For all infinite measures ν in N, a distribution P on $\mathfrak{M}$ has the property g_ν if and only if there is a distribution σ of a random non-negative number which satisfies the equation*

$$ P = \int P_{l\nu}(.)\,\sigma(\mathrm{d}l). $$

Proof. 1. Let X be a set in $\mathfrak{B}$ with the property $\nu(X) > 0$ and P a distribution with the property g_ν. We choose a monotone increasing sequence $X = X_1, X_2, \ldots$ of sets in $\mathfrak{B}$ with union A and denote the distribution of $(\nu(X_n))^{-1}\,\Phi(X_n)$ with respect to P by σ_n, i.e. we set

$$\sigma_n = \sum_{l=0}^{\infty} P(\Phi(X_n) = l)\ \delta_{(\nu(X_n))^{-1}l}\,.$$

For each $\varepsilon > 0$ there exists some $c > 0$ such that the inequality $\sigma_n([c, +\infty)) < \varepsilon$ is satisfied uniformly in n.

Otherwise there would be some $\varepsilon_0 > 0$ for which there would exist an n_s satisfying

$$P(\Phi(X_{n_s}) \geqq s\nu(X_{n_s})) \geqq \varepsilon_0$$

for each natural number s where we can assume in addition that $\nu(X_{n_s}) > s^2$.

For all natural numbers s, h we obtain

$$P(\Phi(X) > h) = \sum_{l=0}^{\infty} P(\Phi(X_{n_s}) = l)\,(Q_{\nu((.)|X_{n_s})})^l\,(\Phi(X) > h).$$

If we denote by u_s the integer part of $s\nu(X_{n_s})$, the equation above can be continued as follows:

$$\begin{aligned}
&\geqq \sum_{l \geqq s\nu(X_{n_s})} P(\Phi(X_{n_s}) = l)\,(Q_{\nu((.)|X_{n_s})})^{u_s}\,(\Phi(X) > h)\\
&= P(\Phi(X_{n_s}) \geqq s\nu(X_{n_s}))\,(\mathscr{D}_{(\nu(X_{n_s}))^{-1}\nu(X)}\delta_{u_s})\,(\{h+1, h+2, \ldots\})\\
&\geqq \varepsilon_0(\mathscr{D}_{(\nu(X_{n_s}))^{-1}\nu(X)}\delta_1)^{u_s}\,(\{h+1, h+2, \ldots\}).
\end{aligned}$$

In view of the inequality 1.11.3. we continue

$$\begin{aligned}
&\geqq \varepsilon_0\left(\pi_{u_s(\nu(X_{n_s}))^{-1}\nu(X)}(\{h+1, h+2, \ldots\}) - 2u_s\left((\nu(X_{n_s}))^{-1}\,\nu(X)\right)^2\right)\\
&\geqq \varepsilon_0\pi_{u_s(\nu(X_{n_s}))^{-1}\nu(X)}(\{h+1, h+2, \ldots\}) - 2\varepsilon_0(\nu(X))^2\,(\nu(X_{n_s}))^{-\frac{1}{2}}.
\end{aligned}$$

Hence, because of $u_s(\nu(X_{n_s}))^{-1}\,\nu(X) \xrightarrow[s\to\infty]{} +\infty$ we obtain for all natural numbers h

$$P(\Phi(X) > h) \geqq \varepsilon_0 \lim_{r\to\infty} \pi_r(\{h+1, h+2, \ldots\}) - 0 = \varepsilon_0$$

and therefore

$$P(\Phi(X) = +\infty) \geqq \varepsilon_0 > 0$$

which contradicts the boundedness of X.

2. In view of 1.11.3., for all natural numbers n and all $s > 0$, we obtain

$$\begin{aligned}
\left\|P_X - \int \pi_{l\nu(X)}(.)\,\sigma_n(dl)\right\| &= \left\|P_X - \sum_{l=0}^{\infty} P(\Phi(X_n) = l)\ \pi_{l(\nu(X_n))^{-1}\nu(X)}\right\|\\
&= \left\|\sum_{l=0}^{\infty} P(\Phi(X_n) = l)\ (\mathscr{D}_{(\nu(X_n))^{-1}\nu(X)}\delta_1)^l - \sum_{l=0}^{\infty} P(\Phi(X_n) = l)\ \pi_{l(\nu(X_n))^{-1}\nu(X)}\right\|
\end{aligned}$$

$$\leq \sum_{0 \leq l < s\nu(X_n)} P(\Phi(X_n) = l)\, 2l\big((\nu(X_n))^{-1}\, \nu(X)\big)^2 + 2P(\Phi(X_n) \geqq s\nu(X_n))$$

$$\leq 2s\nu(X_n)\,\big((\nu(X_n))^{-1}\, \nu(X)\big)^2 + 2P(\Phi(X_n) \geqq s\nu(X_n))$$

$$= 2s(\nu(X))^2\, (\nu(X_n))^{-1} + 2P(\Phi(X_n) \geqq s\nu(X_n)).$$

Thus for all $s > 0$ it follows that

$$\overline{\lim_{n\to\infty}} \left\| P_X - \int \pi_{l\nu(X)}(.)\, \sigma_n(\mathrm{d}l)\right\| \leqq \lim_{n\to\infty} 2P(\Phi(X_n) \geqq s\nu(X_n)).$$

In virtue of 1., with increasing s the right side tends toward zero. Consequently,

$$\int \pi_{l\nu(X)}(.)\, \sigma_n(\mathrm{d}l) \xrightarrow[n\to\infty]{} P_X.$$

3. According to 1., we can select from (σ_n) a subsequence $(\sigma_{n_m})_{m=1,2,\ldots}$ tending weakly towards a distribution σ. For $h = 0, 1, \ldots$, we have

$$\int \pi_{l\nu(X)}(\{h\})\, \sigma_{n_m}(\mathrm{d}l) \xrightarrow[m\to\infty]{} \int \pi_{l\nu(X)}(\{h\})\, \sigma(\mathrm{d}l)$$

and hence, in view of 1.9.2.,

$$\int \pi_{l\nu(X)}(.)\, \sigma_{n_m}(\mathrm{d}l) \xrightarrow[m\to\infty]{} \int \pi_{l\nu(X)}(.)\, \sigma(\mathrm{d}l).$$

In virtue of 2.,

$$P_X = \int \pi_{l\nu(X)}(.)\, \sigma(\mathrm{d}l) = \left(\int P_{l\nu}(.)\, \sigma(\mathrm{d}l)\right)_X.$$

By 1.14.7. from this it follows that

$$P = \int P_{l\nu}(.)\, \sigma(\mathrm{d}l). \blacksquare$$

In the case of infinite measures ν we obtain immediately from 1.14.8. and 1.8.5., the following characterization of the Poisson distributions $P_{l\nu}$, $l \geqq 0$.

1.14.9. Theorem. *Let ν be an infinite measure in N and (X_n) a sequence of sets in $\mathfrak{A}$ with the properties $0 < \nu(X_n) < +\infty$ for $n = 1, 2, \ldots$, and $\bigcup_{n=1}^{\infty} X_n = A$. Subject to these assumptions, a distribution P on $\mathfrak{M}$ has the form $P = P_{l\nu}$, $l \geqq 0$, if and only if this distribution has the property g_ν and, for all $\varepsilon > 0$, the convergence relation*

$$P\left(\left|\frac{\Phi(X_n)}{\nu(X_n)} - l\right| > \varepsilon\right) \xrightarrow[n\to\infty]{} 0$$

is satisfied.

Let now ν be any infinite measure in N. By virtue of 1.8.3. the mixture $P = \int P_{l\nu}(.)\, \sigma(\mathrm{d}l)$ is certainly infinitely divisible if the distribution σ has this property. Conversely, let us suppose that P is infinitely divisible and has the property g_ν. According to 1.14.4., together with P all convolution roots $\sqrt[n]{P}$ belong to $\boldsymbol{E}_\nu$, and, according to 1.14.8., they have the form $\sqrt[n]{P} = \int P_{l\nu}(.)\, \gamma_n(\mathrm{d}l)$. By means of 1.8.3.

$$P = \left(\int P_{l\nu}(.)\, \gamma_n(\mathrm{d}l)\right)^n = \int P_{l\nu}(.)\, (\gamma_n)^n\, (\mathrm{d}l)$$

and by the help of the uniqueness theorem 1.8.6. we conclude that $(\gamma_n)^n = \sigma$ for all natural numbers n. Therefore σ is infinitely divisible and, again applying 1.14.8. and 1.8.3., we see the validity of

1.14.10. Proposition. *For all infinite measures ν in N, the correspondence*

$$\sigma \rightsquigarrow \int P_{l\nu}(.)\, \sigma(\mathrm{d}l)$$

is a one-to-one mapping of the set of all infinitely divisible distributions σ of random non-negative real numbers onto the set of all distributions P in $\boldsymbol{T}$ which have the property g_ν.

The assertion of 1.14.8. cannot be carried over to finite measures ν. Instead of it we have

1.14.11. Proposition. *For any distribution ν on $\mathfrak{A}$, a distribution P on $\mathfrak{M}$ has the property g_ν if and only if*

$$P = \sum_{l=0}^{\infty} P\big(\Phi(A) = l\big)\, (Q_\nu)^l$$

is satisfied.

Proof. 1. If the above representation holds for P, then by the conclusions drawn in the proof of 1.14.1. we obtain for all X in $\mathfrak{B}$

$$_XP = \begin{cases} \sum_{l=0}^{\infty} P\big(\Phi(X) = l\big)\, \big(Q_{\nu((.)|X)}\big)^l & \text{for } \nu(X) > 0 \\ \delta_o & \phantom{\text{for }} \nu(X) = 0. \end{cases}$$

2. Let us conversely assume that the distribution P has the property g_ν. Let (X_n) be a monotone increasing sequence of sets in $\mathfrak{B}$ with the properties $\nu(X_1) > 0$, $\bigcup_{n=1}^{\infty} X_n = A$. For $n > m$ we obtain by the help of 1.14.2.

$$P_{X_n \setminus X_m} = \mathscr{D}_{(\nu(X_n))^{-1}\nu(X_n\setminus X_m)} P_{X_n} = \mathscr{D}_{(\nu(X_n))^{-1}\nu(X_n\setminus X_m)} \big(\mathscr{D}_{(\nu(X_n))^{-1}\nu(X_1)}\big)^{-1} P_{X_1}.$$

If now $\big(\nu(X_1)\big)^{-1} \nu(A \setminus X_m) \leqq 1$ then we obtain

$$\begin{aligned} P_{X_n\setminus X_m}(\{0\}) &= \mathscr{D}_{(\nu(X_1))^{-1}\nu(X_n\setminus X_m)}\, P_{X_1}(\{0\}) \\ &= \sum_{l=0}^{\infty} P_{X_1}(\{l\})\, \Big(1 - \big(\nu(X_1)\big)^{-1} \nu(X_n \setminus X_m)\Big)^l \\ &\geqq \sum_{l=0}^{\infty} P_{X_1}(\{l\})\, \Big(1 - \big(\nu(X_1)\big)^{-1} \nu(A \setminus X_m)\Big)^l, \end{aligned}$$

and hence

$$P_{A\setminus X_m}(\{0\}) \geqq \sum_{l=0}^{\infty} P_{X_1}(\{l\}) \left(1 - \left(\nu(X_1)\right)^{-1} \nu(A \setminus X_m)\right)^l.$$

Therefore,

$$P\left(\Phi(A \setminus X_m) > 0\right) \xrightarrow[m\to\infty]{} 0$$

and, consequently, $P\left(\Phi(A) < +\infty\right) = 1$.

3. Because of

$$\begin{aligned}
\|P - {}_{X_n}P\| &= \left\|P\left(\Phi \in (.)\right) - P\left({}_{X_n}\Phi \in (.)\right)\right\| \\
&= \Big\|P\left(\Phi \in (.), \Phi(A\setminus X_n) = 0\right) + P\left(\Phi \in (.), \Phi(A\setminus X_n) > 0\right) \\
&\quad - P\left({}_{X_n}\Phi \in (.), \Phi(A\setminus X_n) = 0\right) - P\left({}_{X_n}\Phi \in (.), \Phi(A\setminus X_n) > 0\right)\Big\| \\
&= \left\|P\left(\Phi \in (.), \Phi(A\setminus X_n) > 0\right) - P\left({}_{X_n}\Phi \in (.), \Phi(A\setminus X_n) > 0\right)\right\| \\
&\leqq \left\|P\left(\Phi \in (.), \Phi(A\setminus X_n) > 0\right)\right\| + \left\|P\left({}_{X_n}\Phi \in (.), \Phi(A\setminus X_n) > 0\right)\right\| \\
&= 2P\left(\Phi(A\setminus X_n) > 0\right)
\end{aligned}$$

we have, according to 2.,

$$\lim_{n\to\infty} \|P - {}_{X_n}P\| = 0.$$

4. For all natural numbers l_0 it follows by the help of 3.

$$\begin{aligned}
&\left\|P - \sum_{l=0}^{\infty} P\left(\Phi(A) = l\right) (Q_\nu)^l\right\| \\
&\leqq \overline{\lim_{n\to\infty}} \|P - {}_{X_n}P\| + \overline{\lim_{n\to\infty}} \left\|\sum_{l=0}^{\infty} P\left(\Phi(X_n) = l\right) \left(Q_{\nu((.)|X_n)}\right)^l - \sum_{l=0}^{\infty} P\left(\Phi(A) = l\right) (Q_\nu)^l\right\| \\
&\quad \leqq \overline{\lim_{n\to\infty}} \sum_{l=0}^{l_0} \left\|P\left(\Phi(A) = l\right) (Q_\nu)^l - P\left(\Phi(X_n) = l\right) \left(Q_{\nu((.)|X_n)}\right)^l\right\| \\
&\qquad + \overline{\lim_{n\to\infty}} \sum_{l>l_0} \left(P\left(\Phi(A) = l\right) + P\left(\Phi(X_n) = l\right)\right) \\
&\quad \leqq \overline{\lim_{n\to\infty}} \sum_{l=0}^{l_0} P\left(\Phi(A) = l\right) \left\|(Q_\nu)^l - \left(Q_{\nu((.)|X_n)}\right)^l\right\| \\
&\qquad + \lim_{n\to\infty} \sum_{l=0}^{\infty} \left|P\left(\Phi(A) = l\right) - P\left(\Phi(X_n) = l\right)\right| \\
&\qquad + \overline{\lim_{n\to\infty}} \left(P\left(\Phi(A) > l_0\right) + P\left(\Phi(X_n) > l_0\right)\right) \\
&\quad \leqq \overline{\lim_{n\to\infty}} \sum_{l=0}^{l_0} P\left(\Phi(A) = l\right) l \left\|Q_\nu - Q_{\nu((.)|X_n)}\right\| + 0 + 2P\left(\Phi(A) > l_0\right).
\end{aligned}$$

On the other hand,

$$\|Q_\nu - Q_{\nu((.)|X_n)}\| = \mathrm{Var}\left(\nu - \left(\nu(X_n)\right)^{-1} \nu\left((.) \cap X_n\right)\right)$$
$$\leqq \mathrm{Var}\left(\nu - \nu\left((.) \cap X_n\right)\right) + \mathrm{Var}\left(\nu\left((\cdot) \cap X_n\right) - \left(\nu(X_n)\right)^{-1} \nu\left((\cdot) \cap X_n\right)\right)$$
$$= \nu(A \setminus X_n) + 1 - \nu(X_n) \xrightarrow[n\to\infty]{} 0.$$

Consequently, for all natural numbers l_0 we obtain

$$\left\| P - \sum_{l=0}^{\infty} P\left(\Phi(A) = l\right) (Q_\nu)^l \right\| \leqq 2P\left(\Phi(A) > l_0\right)$$

and using 2., we conclude that

$$P = \sum_{l=0}^{\infty} P\left(\Phi(A) = l\right) (Q_\nu)^l . \blacksquare$$

As a counterpart to 1.14.10. we obtain

1.14.12. Proposition. *For all distributions ν in N, the correspondence*

$$(q_n)_{n=1,2,\ldots} \curvearrowright P = \mathcal{U}_{\sum_{n=1}^{\infty} q_n (Q_\nu)^n}$$

is a one-to-one mapping of the set of all sequences (q_n) of non-negative real numbers which satisfy the condition $\sum_{n=1}^{\infty} q_n < +\infty$ onto the set of all distributions P in $\boldsymbol{T}$ which have the property g_ν.

Proof. All distributions of the form $P = \mathcal{U}_{\sum_{n=1}^{\infty} q_n (Q_\nu)^n}$ are infinitely divisible. Because of

$$P = \exp\left(-\left(\sum_{n=1}^{\infty} q_n\right) \delta_o + \sum_{n=1}^{\infty} q_n (Q_\nu)^n\right),$$

by the help of 1.14.11. and 1.14.5., we conclude that P belongs to $\boldsymbol{E}_\nu$.

Let now P be any distribution in $\boldsymbol{T} \cap \boldsymbol{E}_\nu$. By 1.14.11. $P\left(\Phi(A) < +\infty\right) = 1$. by virtue of 1.10.2. there exists a finite measure Q on $\mathfrak{M}$ with the properties $Q(\{o\}) = 0$, $P = \mathcal{U}_Q = \exp\left(Q - Q(M)\, \delta_o\right)$. According to 1.14.5. $Q - Q(M)\,\delta_o$ and hence Q also belongs to $\boldsymbol{E}_\nu$. If now $Q = 0$, then $P = \delta_o = \mathcal{U}_{\sum_{n=1}^{\infty} q_n (Q_\nu)^n}$ with $q_n = 0$, $n = 1, 2, \ldots$. If, however, $Q \neq 0$, then we obtain in the form of $\left(Q(M)\right)^{-1} Q$ a distribution with the property g_ν. Hence by virtue of 1.14.11.

$$\left(Q(M)\right)^{-1} Q = \sum_{l=0}^{\infty} \left(Q(M)\right)^{-1} Q\left(\Phi(A) = l\right) (Q_\nu)^l ,$$

and we obtain

$$P = \mathcal{U}_{\sum_{n=1}^{\infty} q_n (Q_\nu)^n} \quad \text{with} \quad q_n = Q\left(\Phi(A) = n\right), \qquad n = 1, 2, \ldots \blacksquare$$

Given any distribution ν on $\mathfrak{A}$, the distribution $(Q_\nu)^n$ on $\mathfrak{M}$ is generated by successively throwing n independent random points which are distributed according to ν, into the measurable space $[A, \mathfrak{A}]$. Therefore a distribution P on $\mathfrak{M}$ is called *purely random* if it has the following property:

For each X in $\mathfrak{B}$ with $P(\Phi(X) > 0) > 0$, there exists a distribution γ on $\mathfrak{A}$, such that for all $k \geqq 1$ with $P(\Phi(X) = k) > 0$ the following relation holds:

$$P({}_X\Phi \in (.) \mid \Phi(X) = k) = (Q_\gamma)^k.$$

In particular, from this equation we have $\gamma(X) = 1$.

Using 1.14.8. and 1.14.11. we obtain an exhaustive survey of the purely random distributions:

1.14.13.[1]) **Proposition.** *A distribution P on $\mathfrak{M}$ is purely random if and only if it has the property g_ν for some ν in N. If $P \neq \delta_0$, the measure ν is uniquely determined except for a positive factor.*

Proof. Obviously any P in $\boldsymbol{P}$ is certainly purely random if it has the property g_ν for some ν in N.

Conversely, let P be purely random and different from δ_0. To each X in $\mathfrak{B}$ with $P(\Phi(X) > 0) > 0$ we associate a distribution ν_X on $\mathfrak{A}$ by setting $\nu_X = \gamma$. Then in view of the uniqueness of the convolution roots in $\boldsymbol{P}$, Q_γ and hence also γ are uniquely determined. Now let (X_n) be a monotone increasing sequence of sets in $\mathfrak{B}$ having the properties that $P(\Phi(X_1) > 0) > 0$ and that for each X in $\mathfrak{B}$ there exists some n, such that $X \subseteq X_n$.

By assumption, P has the property $g_{\nu_{X_n}, X_n}$ for all n, and hence also the property $g_{\nu_{X_n}((.) \cap X_m), X_m}$ for all $m < n$. Consequently, ν_{X_m} must be a positive multiple of $\nu_{X_n}((.) \cap X_m)$. We set $\nu_n = (\nu_{X_n}(X_1))^{-1} \nu_{X_n}$. Then by the uniqueness of ν_X it follows that for $k < s$ ν_s is an extension of ν_k, i.e. ${}_{X_k}(\nu_s) = \nu_k$. Now, setting

$$\nu = \sum_{k=1}^{\infty} (\nu_{X_k}(X_1))^{-1} \nu_{X_k}((.) \setminus X_{k-1})$$

where $X_0 = \emptyset$, we obtain a ν in N, such that for $n = 1, 2, \ldots$ the measure ν_{X_n} is a positive multiple of $\nu((.) \cap X_n)$. Hence P has the property g_{ν, X_n} for all n. However, since each X in $\mathfrak{B}$ is included in an X_n, we can conclude that P has the property $g_{\nu,X}$. ∎

As we have seen in section 1.8., the mixture $P = \int P_{l\nu}(.)\, \sigma(dl)$ has an intensity measure of the form $\varrho_P(X) = w\nu(X)$, $X \in \mathfrak{B}$, where we set $\infty \cdot 0 = 0$. Obviously, this is also true if P can be represented in the form $P = \sum_{l=0}^{\infty} P(\Phi(A) = l)\, (Q_\nu)^l$ by the help of a distribution ν on $\mathfrak{A}$. In fact, we then have $\varrho_P = \sum_{l=1}^{\infty} P(\Phi(A) = l)\, l\nu$. Hence in virtue of 1.14.8. and 1.14.11., we obtain

1.14.14. *For all ν in N and all distributions P on $\mathfrak{M}$ with the property g_ν there exists a w in $[0, +\infty]$ such that*

$$\varrho_P(X) = w\nu(X), \qquad X \in \mathfrak{A},$$

where we set $\infty \cdot 0 = 0 = 0 \cdot \infty$.

---

[1]) Like 1.14.8. and 1.14.11., theorem 1.14.13. was derived by Kallenberg in [1], independently of the authors of this book.

Closing this section we deal with the structure of the real function $X \rightsquigarrow f(X) = P(\Phi(X) = 0, X \in \mathfrak{B}$ for distributions P on $\mathfrak{M}$ with the property g_ν.

For $\nu = o$ we have $P = \delta_o$, and hence $f(X) = 1$ for all X in $\mathfrak{B}$.

If ν is finite and different from o, we can now consider the distribution $\nu((.) \mid A)$. So we can assume without loss of generality, that $\nu(A) = 1$. According to 1.14.11., P then has the form $P = \sum_{n=0}^{\infty} p_n(Q_\nu)^n$ where $p_n = (P_A)(\{n\})$, $(n = 0, 1, \ldots)$. For all X in $\mathfrak{B}$, we obtain

$$f(X) = \sum_{n=0}^{\infty} p_n(Q_\nu)^n \big(\Phi(X) = 0\big) = \sum_{n=0}^{\infty} p_n\big(1 - \nu(X)\big)^n,$$

i.e. $f(X)$ is equal to the value of the generating function of P_A at the point $1 - \nu(X)$.

Finally, if ν is infinite, then 1.14.8. shows that P has the form $P = \int P_{l\nu}(.)\, \sigma(\mathrm{d}l)$, and hence for all X in $\mathfrak{B}$

$$f(X) = \int P_{l\nu}\big(\Phi(X) = 0\big)\, \sigma(\mathrm{d}l) = \int e^{-l\nu(X)}\, \sigma(\mathrm{d}l),$$

i.e. $f(X)$ is equal to the Laplace transform of the distribution σ at the point $\nu(X)$.

In all cases $f(X)$ depends only on the measure $\nu(X)$ of the set X.

For non-atomic measures ν in N, it was shown by 1.7.6. that all distributions $P_{l\nu}$ $(l \geqq 0)$ are simple. Using 1.5.5., we observe that for non-atomic distributions ν on $\mathfrak{A}$ all convolution powers $(Q_\nu)^n$, $(n = 0, 1, \ldots)$, are simple. Thus we obtain by means of the two structural propositions 1.14.8. and 1.14.11. that for non-atomic measures ν in N all distributions P on $\mathfrak{M}$ having the property g_ν are simple.

For non-atomic measures ν in N, the two properties of distributions on $\mathfrak{M}$ which were derived above, are characteristic for the property g_ν (cf. Kallenberg [1] [2]):

1.14.15. Theorem. *For each diffuse measure ν in N, a distribution P on $\mathfrak{M}$ has the property g_ν if and only if the distribution is simple and $\nu(X_1) = \nu(X_2)$ implies $P\big(\Phi(X_1) = 0\big) = P\big(\Phi(X_2) = 0\big)$ for all X_1, X_2 in $\mathfrak{B}$.*

Proof. 1. Suppose that a distribution P on $\mathfrak{M}$ satisfies the conditions above. In addition we suppose that the phase space $[A, \varrho_A]$ is bounded and that $\nu(A) = 1$.

Let $\mathfrak{R}$ denote the system of all X in $\mathfrak{A}$ which have the property $\nu(X) = 0$, let $\mathfrak{V}$ be the system of all Borel subsets of $[0, 1]$, ω the restriction of the one-dimensional Lebesgue measure to $\mathfrak{V}$, and $\mathfrak{W}$ the system of all Z in $\mathfrak{V}$ which have property $\omega(Z) = 0$. According to a well-known isomorphism theorem of measure theory (cf. Halmos [1], § 41), there exists an isomorphism v of the Boolean algebra $\mathfrak{A}/\mathfrak{R}$ onto $\mathfrak{V}/\mathfrak{W}$ such that for all X in $\mathfrak{A}$ and all Z in $\mathfrak{V}$ belonging to the residue class $v(\dot{X})$ of the image of the residue class $\dot{X}$ of X, the equation $\nu(X) = \omega(Z)$ is satisfied.

For each rational l in $[0, 1]$, we select any Y_l in $\mathfrak{A}$ such that $v(\dot{Y}_l)$ coincides with the residue class of $[0, l)$. Then the set-up

$$X_l = \bigcup_{\substack{0<l'<l,\\ l'\text{rational}}} Y_{l'} \qquad (0 \leqq l \leqq 1)$$

provides a monotone increasing, left-continuous mapping $l \rightsquigarrow X_l$ of $[0, 1]$ into $\mathfrak{A}$ which has the property

$$v(\dot{X}_l) = [0, l) \qquad \left(0 \leqq l \leqq 1\right).$$

For all l in $[0, 1]$, we set

$$h(l) = P\left(\Phi(\overline{X_l}) = 0\right)$$

thus obtaining a left-continuous, monotone increasing real function h on $[0, 1]$ which has the property

$$h(1) = P\left(\Phi(\overline{X_1}) = 0\right) = P\left(\Phi(\emptyset) = 0\right) = 1.$$

For each sequence (l_n) of numbers in $[0, 1]$, which decreases monotonically towards l, we obtain

$$h(l_n) = P\left(\Phi(\overline{X_{l_n}}) = 0\right) \xrightarrow[n\to\infty]{} P\left(\Phi\left(\overline{\bigcap_{n=1}^{\infty} X_{l_n}}\right) = 0\right) = P\left(\Phi(\overline{X_{l+0}}) = 0\right).$$

On the other hand,

$$\nu(X_{l+0} \setminus X_l) = l - l = 0$$

and hence

$$P\left(\Phi(X_{l+0} \setminus X_l) = 0\right) = P\left(\Phi(\emptyset) = 0\right) = 1,$$

so that the equation stated above can be continued as follows:

$$= P\left(\Phi(\overline{X_l}) = 0\right) = h(l).$$

Consequently the function h is continuous on $[0, 1]$.

2. Now let n be an arbitrary natural number. We set

$$\mathfrak{L} = \left\{X_{\frac{1}{n}}, X_{\frac{2}{n}} \setminus X_{\frac{1}{n}}, \ldots, X_1 \setminus X_{\frac{n-1}{n}}\right\},$$

thus obtaining a finite family of pairwise disjoint, non-empty sets in $\mathfrak{B}$. For all natural numbers $r \leqq n$, we obtain

$$\sum_{k=0}^{r} \binom{r}{k} (-1)^{r+k} h\left(\frac{k}{n}\right) = \sum_{k=0}^{r} (-1)^{r+k} \binom{r}{r-k} h\left(\frac{k}{n}\right)$$

$$= \sum_{k=0}^{r} (-1)^{r+k} \sum_{\mathfrak{F} \subseteq \mathfrak{J} \subseteq \mathfrak{L}, \operatorname{card}\mathfrak{J} = n-k} P\left(\Phi\left(\bigcup_{J \in \mathfrak{J}} J\right) = 0\right),$$

where we set

$$\mathfrak{F} = \left\{X_{\frac{r+1}{n}} \setminus X_{\frac{r}{n}}, \ldots, X_1 \setminus X_{\frac{n-1}{n}}\right\}.$$

Now we continue the above equation as follows:

$$= \sum_{k=0}^{r} (-1)^{(n-k)-(n-r)} \sum_{\mathfrak{I} \subseteq \mathfrak{J} \subseteq \mathfrak{L}, \operatorname{card} \mathfrak{J} = n-k} P\left(\Phi\left(\bigcup_{J \in \mathfrak{J}} J\right) = 0\right)$$

$$= \sum_{\mathfrak{I} \subseteq \mathfrak{J} \subseteq \mathfrak{L}} (-1)^{\operatorname{card}(\mathfrak{J} \setminus \mathfrak{I})} P\left(\Phi\left(\bigcup_{J \in \mathfrak{J}} J\right) = 0\right).$$

In view of theorem 1.4.10., the last expression is always non-negative. Hence the function h is also absolutely monotone. Consequently (cf. Feller [2], Chapt. VII, par. 2) there exists a sequence of non-negative numbers $(p_n)_{n=0,1,\ldots}$ having the property $\sum_{n=0}^{\infty} p_n = 1$ such that

$$h(l) = \sum_{n=0}^{\infty} p_n l^n \qquad (0 \leq l \leq 1)$$

is satisfied. Accordingly for all X in $\mathfrak{B}$,

$$P(\Phi(X) = 0) = \sum_{n=0}^{\infty} p_n (1 - \nu(X))^n = \left(\sum_{n=0}^{\infty} p_n (Q_\nu)^n\right) (\Phi(X) = 0),$$

and using the uniqueness theorem 1.4.7., we can conclude that

$$P = \sum_{n=0}^{\infty} p_n (Q_\nu)^n.$$

Consequently the distribution P has the property g_ν.

3. For $\nu = o$ we have $P(\Phi(X) = 0) = P(\Phi(\emptyset) = 0) = 1$ for all X in $\mathfrak{B}$ which means $P = \delta_o$. Therefore in this case it is trivial to conclude that the distribution P has the property g_ν. Now, if ν is different from o, the phase space $[A, \varrho_A]$ can be covered by a monotone increasing sequence (X_n) of bounded, closed subsets such that each X in $\mathfrak{B}$ is included in an X_n and that additionally $\nu(X_n) > 0$, $(n = 1, 2, \ldots)$, is satisfied.

Now, for all n, we pass from P to ${}_{X_n}P$ and subsequently replace the phase space $[A, \varrho_A]$ by the subspace X_n, and ν by the distribution $\nu((.) \mid X_n)$. This makes 2. applicable, and so we observe that ${}_{X_n}P$ has the property $g_{\nu((.)\mid X_n)}$, i.e. that P has the property g_{ν, X_n}. Since each X in $\mathfrak{B}$ is included in an X_n, P has the property $g_{\nu,X}$ for all X in $\mathfrak{B}$, i.e. P has the property g_ν. ∎

1.15. The Metric Space $[M, \varrho_M]$

In introducing a metric ϱ_M in M, we find our guide by the conception that two elements Φ, Ψ in M are close to each other if the points of Φ, Ψ which are contained in an open ball $S_n(z)$ having its centre in z and the radius n are close to each other for a large n where, however, "boundary effects" have to be taken into consideration. We suppose that the point z in A be arbitrary but fixed.

In view of 1.1.2., each element of M can be represented as a sum of an at most countable number of δ-measures where the respective multiplicities are taken into account. Accordingly, let

$$\Phi = \sum_{i \in I} \delta_{a_i}, \qquad \Psi = \sum_{j \in J} \delta_{b_j}.$$

For all real d, we set $I_d = \{i : i \in I, \varrho_A(a_i, z) < d\}$. Let J_d be defined in an analogous manner. For any $\varepsilon > 0$ and any natural n, the measures Φ, Ψ are said to be *ε, n-neighbouring* if the following condition is satisfied: There exists a one-to-one mapping f of a — possible empty — subset D of I into the set J with the properties

a) $I_{n-\varepsilon} \subseteq D, \;\; J_{n-\varepsilon} \subseteq f(D)$,

b) $\varrho_A(a_i, b_{f(i)}) < \varepsilon \quad$ for all i in D.

For all natural n, we set

$$\varrho_n(\Phi, \Psi) = \inf \{\varepsilon : \varepsilon > 0;\; \Phi, \Psi \text{ are } \varepsilon, n\text{-neighbouring}\} \qquad (\Phi, \Psi \in M).$$

1.15.1. *For all natural numbers n, ϱ_n has the following properties:*

α) $0 \leqq \varrho_n(\Phi, \Psi) \leqq n$,

β) $\varrho_n(\Phi, \Psi) = \varrho_n(\Psi, \Phi)$,

γ) $\varrho_n(\Phi, \Psi) \leqq \varrho_n(\Phi, \chi) + \varrho_n(\chi, \Psi)$,

δ) $\varrho_n(\Phi, \Psi) = 0$ *is equivalent to* $\Phi\big((.) \cap S_n(z)\big) = \Psi\big((.) \cap S_n(z)\big)$.

Proof. 1. The two properties α) and β) are immediately deduced from the definition of ϱ_n.

2. We shall now prove the truth of the triangle inequality γ).

Let $\chi = \sum_{v \in V} \delta_{c_v}$. By analogy to the sets I_d, J_d, we define V_d. Suppose now that the two measures Φ, χ be ε, n-neighbouring with respect to f, while χ, Ψ be η, n-neighbouring with respect to g. We set $h = g \circ f$.

Obviously h satisfies the condition b) for $\varepsilon + \eta$. Now let i be an element of $I_{n-(\varepsilon+\eta)}$. Then $f(i) \in V_{n-\eta}$. Consequently h is defined for i. Now we choose any j in $J_{n-(\varepsilon+\eta)}$. The inverse image $g^{-1}(j)$ is contained in $V_{n-\varepsilon}$, and hence in the range of values of f. Consequently the mapping h from I into J likewise satisfies condition a) with respect to $\varepsilon + \eta$.

Thus we have shown that, for all $\varepsilon > \varrho_n(\Phi, \chi)$ and all $\eta > \varrho_n(\chi, \Psi)$, the inequality

$$\varrho_n(\Phi, \Psi) \leqq \varepsilon + \eta$$

is satisfied. From this the property γ) is immediately deduced.

3. Obviously $\Phi\big((.) \cap S_n(z)\big) = \Psi\big((.) \cap S_n(z)\big)$ always implies that $\varrho_n(\Phi, \Psi) = 0$. Conversely, suppose now that $\varrho_n(\Phi, \Psi) = 0$.

Let α denote the minimum distance between any two different points of the finite set $X = \{g : g \in S_n(z), (\Phi + \Psi)(\{g\}) > 0\}$, and β the supremum of all positive γ having the property that $X \subseteq S_{n-2\gamma}(z)$. The measures Φ, Ψ are $\min(\alpha, \beta)$, n-neighbouring with respect to a one-to-one mapping f from I_n into J_n. By virtue of our selection of α and β, f then provides a mapping of I_n onto J_n, and $\varrho_A(a_i, b_{f(i)}) = 0$ for all i in I_n which means that

$$\Phi\big((.) \cap S_n(z)\big) = \Psi\big((.) \cap S_n(z)\big). \ \blacksquare$$

On the basis of 1.15.1., using the set-up

$$\varrho_M(\Phi, \Psi) = \sum_{n=1}^{\infty} \frac{1}{2^n} \varrho_n(\Phi, \Psi)$$

we get a metric ϱ_M on M. We shall write $\Phi_m \overset{M}{\Rightarrow} \Phi$ for the convergence of (Φ_m) toward Φ with respect to ϱ_M. Obviously this convergence is equivalent to the truth of $\varrho_n(\Phi_m, \Phi) \xrightarrow[m\to\infty]{} 0$ for all natural numbers n.

1.15.2. Proposition. *The metric space $[M, \varrho_M]$ is complete and separable.*

Proof. 1. If we choose any at most countable dense subset X in A, then the countable set of all finite sums of measures δ_x, where $x \in X$, is dense in M. Hence $[M, \varrho_M]$ is separable.

2. Now let (Φ_m) be a Cauchy sequence with respect to ϱ_M. The sequence (Φ_m) converges toward some Φ in M if and only if this is the case for an arbitrary subsequence. In proving the convergence of (Φ_m), we therefore may assume without loss of generality that

$$\varrho_M(\Phi_m, \Phi_{m+1}) < 2^{-m} \qquad (m = 1, 2, \ldots)$$

and hence also

$$\varrho_n(\Phi_m, \Phi_{m+1}) < 2^{n-m} \qquad (m, n = 1, 2, \ldots).$$

We put Φ_m into the form $\Phi_m = \sum_{i \in I_m} \delta_{a_{m,i}}$, introducing the sets $I_{m,d}$ by analogy to I_d. For $m > n$, the measures Φ_m, Φ_{m+1} are always 2^{n-m}, n-neighbouring with respect to a mapping f_m from $I_{m,n}$ into $I_{m+1,n}$. We set

$$f_{m,s} = f_{m+s-1} \circ \cdots \circ f_{m+1} \circ f_m \qquad (s = 1, 2, \ldots).$$

As it may be observed from the proof of the triangle inequality for ϱ_n, Φ_m, Φ_{m+s} are then always $\frac{1}{2^{m-n}} + \cdots + \frac{1}{2^{m+s-1-n}}$, n-neighbouring with respect to the mapping $f_{m,s}$ from $I_{m,n}$ into $I_{m+s,n}$.

Now we keep n fixed and choose some $m > n + 1$. All of the mappings $f_{m,s}$ are defined on $I_{m,n-1}$. For all i in $I_{m,n-1}$, we consider the correspondence

$$i \curvearrowright a_{m+s, f_{m,s}(i)}.$$

Obviously $\big(a_{m+s, f_{m,s}(i)}\big)_{s=1,2,\ldots}$ forms a Cauchy sequence with respect to ϱ_A, which tends toward a well-determined element ${}_na_i$ in A. We set

$${}_n\Phi = \sum_{i \in I_{m,n-1}} \delta_{{}_na_i}.$$

For all natural numbers s, let g_s denote the mapping

$$i \curvearrowright f_{m,s}^{-1}(i)$$

of $f_{m,s}(I_{m,n-1})$ onto $I_{m,n-1}$. Then

$$\varrho_A(a_{m+s,i}, {}_n a_{g_s(i)}) \leqq \frac{1}{2^{m+s-n}} + \frac{1}{2^{m+s+1-n}} + \cdots = \frac{1}{2^{m+s-n-1}} \qquad \big(i \in f_{m,s}(I_{m,n-1})\big).$$

The set $I_{m+s,n-2}$ is included in the domain of definition of g_s. Consequently the two measures $\Phi_{m+s}, {}_n\Phi$ are always $\frac{1}{2^{m+s-n-1}}$, $(n-2)$-neighbouring with respect to g_s so that we obtain

$$\varrho_{n-2}(\Phi_{m+s}, {}_n\Phi) \xrightarrow[s\to\infty]{} 0,$$

which means that

$$\varrho_{n-2}(\Phi_m, {}_n\Phi) \xrightarrow[m\to\infty]{} 0.$$

Obviously for all Φ, Ψ in M we have $\varrho_l(\Phi, \Psi) \leqq \varrho_{l+1}(\Phi, \Psi)$ $(l = 1, 2, \ldots)$.
Thus we obtain

$$\varrho_k(\Phi_m, {}_n\Phi) \xrightarrow[m\to\infty]{} 0$$

for all natural numbers $k \leqq n-2$. Hence, if we let n vary, then

$${}_n\Phi\big((.) \cap S_{n-2}(z)\big) = {}_{n+1}\Phi\big((.) \cap S_{n-2}(z)\big) \qquad (n = 3, 4, \ldots)$$

must be valid, i.e. there is some Φ in M having the property that

$$\Phi\big((.) \cap S_{n-2}(z)\big) = {}_n\Phi\big((.) \cap S_{n-2}(z)\big) \qquad (n = 3, 4, \ldots).$$

For all natural numbers k, $\varrho_k(\Phi_m, \Phi)$ tends toward zero as m increases, i.e. we have

$$\Phi_m \overset{M}{\underset{m\to\infty}{\Rightarrow}} \Phi. \ \blacksquare$$

As to its meaning, the construction of the metric ϱ_M is based on a simple conception, but in its details it is affected with elements of arbitrariness. Therefore it is both interesting and of importance for methods of proof that the topology in M generated by ϱ_M can be introduced in a quite different way.

Let F denote the set of those non-negative continuous bounded real functions defined on the metric space $[A, \varrho_A]$ which are identically zero outside of suitable bounded subsets of A. Then we have

1.15.3. Proposition. *The metric ϱ_M generates the coarsest topology in M with respect to which, for all f in F, the real function*

$$\Phi \curvearrowright \int f(a)\, \Phi(\mathrm{d}a) \qquad (\Phi \in M)$$

is continuous.

Proof. 1. Let g be an element of F. We choose some n such that g is identically zero outside of $S_{n-1}(z)$. Now, if (Φ_m) converges toward Φ with respect to ϱ_M, then

$$\varrho_n(\Phi_m, \Phi) \xrightarrow[m\to\infty]{} 0.$$

We set $\Phi_m = \sum_{i\in I_m} \delta_{a_{m,i}}$ and $\Phi = \sum_{i\in I} \delta_{a_i}$. Then, for each ε in $(0, 1)$ and for $m > m_\varepsilon$, the two measures Φ, Φ_m are always ε,n-neighbouring with respect to a suitable mapping f_m from I_m into I. If we now denote by D the subset of I which corresponds to the indices of those points of Φ which are contained in the open ball $S_{n-1}(z)$, then f_m is defined on D for $m > m_\varepsilon$ and we obtain

$$\left|\int g(a)\, \Phi(\mathrm{d}a) - \int g(a)\, \Phi_m(\mathrm{d}a)\right|$$

$$\leqq \sum_{i\in D} |g(a_i) - g(a_{m,f_m(i)})| \leqq \sum_{i\in D} \sup_{b\in S_\varepsilon(a_i)} |g(a_i) - g(b)|.$$

As $\varepsilon \to 0$, the right-hand side tends toward zero. Consequently $\Phi \curvearrowright \int g(a)\, \Phi(\mathrm{d}a)$ is continuous with respect to ϱ_M.

2. Conversely, we have to show that each ϱ_M-neighbourhood of a Φ in M includes a set of the form

$$\left\{\Psi : \Psi \in M, \left|\int g_s(a)\, \Phi(\mathrm{d}a) - \int g_s(a)\, \Psi(\mathrm{d}a)\right| < \eta \quad \text{for} \quad s = 1, \ldots, k\right\}$$

where $g_1, \ldots, g_k \in F$, $\eta > 0$.

Let n be an arbitrary natural number. We choose some real $d > n$ such that $\Phi(\{b\colon b \in A,\ \varrho_A(b, z) = d\}) = 0$. Suppose that $\delta > 0$ be chosen small enough so that, for all a in $S_d(z)$ having the property that $\Phi(\{a\}) > 0$, the distance to z is less than $d - 2\delta$ and the distances between two different a each are greater than 2δ. Now for each a we fix some g_a in F having the properties $0 \leqq g_a \leqq 1$, $g_a(x) = 1$ on $S_{\delta/2}(a)$; $g_a(x) = 0$ outside of $S_\delta(a)$. Moreover suppose that some g in F be choosen such that $0 \leqq g \leqq 1$, $g(x) = 1$ on $S_{d-(\delta/2)}(z)$, $g(x) = 0$ outside of $S_d(z)$ is satisfied.

Now, if $\left|\int g(x)\, \Phi(\mathrm{d}x) - \int g(x)\, \Psi(\mathrm{d}x)\right| < \sigma$ as well as

$$\left|\int g_a(x)\, \Phi(\mathrm{d}x) - \int g_a(x)\, \Psi(\mathrm{d}x)\right| < \sigma \qquad \left(a \in S_d(z),\ \Phi(\{a\}) > 0\right),$$

then, for each given $\varepsilon > 0$ and for $\sigma < \sigma_\varepsilon$ and $\delta < \delta_\varepsilon$

$$\varrho_n(\Phi, \Psi) < \frac{\varepsilon}{2}$$

and hence

$$\varrho_M(\Phi, \Psi) = \sum_{k=1}^{n} \frac{1}{2^k}\, \varrho_k(\Phi, \Psi) + \sum_{k=n+1}^{\infty} \frac{1}{2^k}\, \varrho_k(\Phi, \Psi)$$

$$\leqq \varrho_n(\Phi, \Psi) \sum_{k=1}^{n} \frac{1}{2^k} + \sum_{k=n+1}^{\infty} \frac{k}{2^k} < \frac{\varepsilon}{2} + \sum_{k=n+1}^{\infty} \frac{k}{2^k}.$$

Consequently, by suitable choosing n and ε, we can ensure that $\{\Psi : \Psi \in M, \left|\int g(x)\, \Phi(\mathrm{d}x) - \int g(x)\, \Psi(\mathrm{d}x)\right| < \sigma, \left|\int g_a(x)\, \Phi(\mathrm{d}x) - \int g_a(x)\, \Psi(\mathrm{d}x)\right| < \sigma$ for all a in $S_d(z)$ having the property that $\Phi(\{a\}) > 0\}$ is included in the open ball $S_\varepsilon(\Phi)$. ∎

Thus the convergence relation $\Phi_m \overset{M}{\underset{m\to\infty}{\Rightarrow}} \Phi$ is true if and only if

$$\int f(x)\, \Phi_m(\mathrm{d}x) \xrightarrow[m\to\infty]{} \int f(x)\, \Phi(\mathrm{d}x)$$

is satisfied for all f in F.

An X in $\mathfrak{B}$ is called a *continuity set with respect to a measure* ν in N if $\nu(\partial X) = 0$ is valid. Here ∂X denotes the boundary of X.

1.15.4. Proposition. *For each Φ in M and each sequence (Φ_m) of elemente in M, the following statements are equivalent:*

a) $\Phi_m \overset{M}{\underset{m\to\infty}{\Rightarrow}} \Phi$,

b) $\varlimsup_{m\to\infty} \Phi_m(X') \leqq \Phi(X')$ *for all bounded closed subsets X' of A, and* $\varliminf_{m\to\infty} \Phi_m(X'') \geqq \Phi(X'')$ *for all bounded open subsets X'' of A,*

c) $\Phi_m(X) \xrightarrow[m\to\infty]{} \Phi(X)$ *for all continuity sets X with respect to Φ.*

Proof. 1. a) implies b).

Let X' be a bounded closed subset of A and (X_n') a monotone decreasing sequence of open neighbourhoods of X' having the property that $X' = \bigcap_{n=1}^{\infty} X_n'$. For each n we choose some f_n in F satisfying the condition $k_{X'} \leqq f_n \leqq k_{X_n'}$. Then from $\Phi_m \overset{M}{\underset{m\to\infty}{\Rightarrow}} \Phi$ we obtain, for all n

$$\varlimsup_{m\to\infty} \Phi_m(X') \leqq \lim_{m\to\infty} \int f_n(x)\, \Phi_m(\mathrm{d}x) = \int f_n(x)\, \Phi(\mathrm{d}x) \leqq \Phi(X_n')$$

and hence

$$\varlimsup_{m\to\infty} \Phi_m(X') \leqq \inf_{n=1,2,\ldots} \Phi(X_n') = \Phi(X').$$

Now, if X'' is a bounded open subset of A and (X_n'') a monotone increasing sequence of closed subsets of X'' having the property that $X'' = \bigcup_{n=1}^{\infty} X_n''$, then to each n we choose some g_n in F satisfying the condition $k_{X_n''} \leqq g_n \leqq k_{X''}$. From $\Phi_m \overset{M}{\underset{m\to\infty}{\Rightarrow}} \Phi$ we obtain, for all n,

$$\varliminf_{m\to\infty} \Phi_m(X'') \geqq \lim_{m\to\infty} \int g_n(x)\, \Phi_m(\mathrm{d}x) = \int g_n(x)\, \Phi(\mathrm{d}x) \geqq \Phi(X_n'')$$

and hence

$$\varliminf_{m\to\infty} \Phi_m(X'') \geqq \sup_{n=1,2,\ldots} \Phi(X_n'') = \Phi(X'').$$

2. b) implies c).

Let X be a continuity set with respect to Φ. Denoting the interior of X by X'' and the closure of X by X', we obtain

$$\Phi(X) = \Phi(X') \geqq \overline{\lim_{m\to\infty}}\, \Phi_m(X') \geqq \overline{\lim_{m\to\infty}}\, \Phi_m(X),$$

$$\Phi(X) = \Phi(X'') \leqq \underline{\lim_{m\to\infty}}\, \Phi_m(X'') \leqq \underline{\lim_{m\to\infty}}\, \Phi_m(X),$$

which means that

$$\Phi_m(X) \xrightarrow[m\to\infty]{} \Phi(X).$$

3. c) implies a).

Given any function f in F, the closed sets $f^{-1}(\{c\})$, $c > 0$ are pairwise disjoint. Consequently the inequality $\Phi(f^{-1}(\{c\})) > 0$ can be satisfied for at most countably many c's. Hence to each given $\eta > 0$ there exists a finite monotone increasing sequence

$$0 = c_0 < c_1 < \cdots < c_k$$

having the properties

$$c_k \geqq \sup_{a\in A} f(a), \qquad c_0 \leqq 0 < c_1,$$

$$c_i - c_{i-1} \leqq \eta \quad \text{for} \quad 1 \leqq i \leqq k, \qquad \Phi(f^{-1}(\{c_i\})) = 0 \quad \text{for} \quad 1 \leqq i \leqq k.$$

Since

$$\partial f^{-1}([c_{i-1}, c_i)) \subseteqq f^{-1}(\partial[c_{i-1}, c_i)) = f^{-1}(\{c_{i-1}\}) \cup f^{-1}(\{c_i\})$$

all inverse images $f^{-1}([c_{i-1}, c_i))$, $2 \leqq i \leqq k$, are continuity sets with respect to Φ, and we obtain

$$\int f(x)\, \Phi(\mathrm{d}x) - \eta\Phi(f^{-1}((0, +\infty))) \leqq \sum_{i=2}^{k} c_{i-1}\Phi(f^{-1}([c_{i-1}, c_i)))$$

$$= \lim_{m\to\infty} \sum_{i=2}^{k} c_{i-1}\Phi_m(f^{-1}([c_{i-1}, c_i))) \leqq \underline{\lim_{m\to\infty}} \int f(x)\, \Phi_m(\mathrm{d}x)$$

$$\leqq \overline{\lim_{m\to\infty}} \int f(x)\, \Phi_m(\mathrm{d}x) \leqq \overline{\lim_{m\to\infty}} \left(c_1\Phi_m\left(f^{-1}((0, c_1))\right) + \sum_{i=2}^{k} c_i\Phi_m(f^{-1}([c_{i-1}, c_i)))\right)$$

$$= c_1 \overline{\lim_{m\to\infty}}\, \Phi_m(f^{-1}(0, c_1)) + \sum_{i=2}^{k} c_i\Phi(f^{-1}([c_{i-1}, c_i))).$$

Now, if X is an arbitrary continuity set with respect to Φ which includes $f^{-1}((0, +\infty))$, then we may continue our chain of inequalities by

$$\leqq c_1\Phi(X) + \sum_{i=2}^{k} c_i\Phi(f^{-1}([c_{i-1}, c_i))) \leqq \int f(x)\, \Phi(\mathrm{d}x) + 2\eta\Phi(X).$$

Since $\eta > 0$ can be arbitrarily chosen, it follows that

$$\int f(x)\,\Phi(dx) \leqq \varliminf_{m\to\infty} \int f(x)\,\Phi_m(dx) \leqq \varlimsup_{m\to\infty} \int f(x)\,\Phi_m(dx) = \int f(x)\,\Phi(dx)$$

which means that

$$\int f(x)\,\Phi_m(dx) \xrightarrow[m\to\infty]{} \int f(x)\,\Phi(dx)\,. \ \blacksquare$$

Thus for all non-negative integers k, all closed bounded X', and all open bounded X'', the two sets

$$\{\Phi : \Phi \in M, \Phi(X') \leqq k\}, \qquad \{\Phi : \Phi \in M, \Phi(X'') \geqq k\}$$

are open with respect to ϱ_M.

Consequently, for all X in $\mathfrak{B}$ and all non-negative integers k, the sets $\{\Phi : \Phi \in M,\ \Phi(X) = k,\ \Phi(\partial X) = 0\}$ are open, for we have

$$\{\Phi : \Phi \in M, \Phi(X) = k, \Phi(\partial X) = 0\}$$
$$= \{\Phi : \Phi \in M, \Phi(X') \leqq k\} \cap \{\Phi : \Phi \in M, \Phi(X'') \geqq k\}$$

where X' denotes the closure, X'' the interior of X.

The σ-algebra $\mathfrak{M}$ defined in section 1.1. is consistent with the metric ϱ_M:

1.15.5. *The σ-algebra of the Borel sets of the metric space $[M, \varrho_M]$ coincides with $\mathfrak{M}$.*

Proof. 1. In view of the measurability statement of the Campbell Theorem 1.2.2., for all f in F the real functions

$$\Phi \curvearrowright \int f(a)\,\Phi(da) \qquad (\Phi \in M)$$

are measurable with respect to $\mathfrak{M}$. By virtue of 1.15.3., $\mathfrak{M}$ therefore includes a base of $[M, \varrho_M]$. Hence, since $[M, \varrho_M]$ satisfies the second axiom of countability, $\mathfrak{M}$ includes a countable base; consequently $\mathfrak{M}$ includes all open subsets of $[M, \varrho_M]$, and hence also the σ-algebra of the Borel sets of $[M, \varrho_M]$.

2. Let $\mathfrak{G}$ denote the system of those X in $\mathfrak{B}$ for which $\Phi \curvearrowright \Phi(X)$ is measurable with respect to the σ-algebra of the Borel sets.

To each closed bounded subset K of A there exists a monotone decreasing sequence (g_n) of functions in F having the property that

$$\inf_{n=1,2,\ldots} g_n(x) = k_K(x) \qquad (x \in A)\,.$$

Hence

$$\Phi(K) = \int \Big(\inf_{n=1,2,\ldots} g_n(x)\Big)\,\Phi(dx) = \inf_{n=1,2,\ldots} \int g_n(x)\,\Phi(dx)\,,$$

from which it is concluded that K belongs to $\mathfrak{G}$.

For all closed bounded K_1, K_2, $K_1 \setminus K_2$ belongs to $\mathfrak{G}$, for we have

$$\Phi(K_1 \setminus K_2) = \Phi(K_1) - \Phi(K_1 \cap K_2).$$

If $K_1, \ldots, K_m$ and $K_1', \ldots, K_m'$ are closed bounded subsets of A, and if the $(K_i \setminus K_i')_{1 \leq i \leq m}$ are pairwise disjoint, then

$$\Phi\left(\bigcup_{i=1}^{m} (K_i \setminus K_i')\right) = \sum_{i=1}^{m} \Phi(K_i \setminus K_i')$$

and hence $\left(\bigcup_{i=1}^{m} (K_i \setminus K_i')\right) \in \mathfrak{G}$.

Consequently $\mathfrak{G}$ includes the subring of $\mathfrak{B}$ which is generated by the closed bounded sets. On the other hand, $\mathfrak{G}$ is monotonically closed in $\mathfrak{B}$. Hence the set system $\mathfrak{G}$ coincides with $\mathfrak{B}$, i.e. for all X in $\mathfrak{B}$, the mapping $\Phi \frown \Phi(X)$ is measurable with respect to the σ-algebra of the Borel sets of $[M, \varrho_M]$. ∎

In view of 1.15.3., the convergence $a_n \xrightarrow[n\to\infty]{} a$ with respect to ϱ_A is equivalent with the convergence $\delta_{a_n} \overset{M}{\underset{n}{\Rightarrow}} \delta_a$, i.e. the correspondence $a \frown \delta_a$ provides a homeomorphic mapping of the metric space $[A, \varrho_A]$ into the metric space $[M, \varrho_M]$.

In view of 1.15.5., this leads to a new proof of the measurability proposition 1.1.4.

2. The Canonical Representation of Infinitely Divisible Probability Distributions

2.0. Introduction

Although in the first chapter the concept of an infinitely divisible distribution on $\mathfrak{M}$ was introduced in full generality, the structure of such distributions P could however be disclosed only by means of an additional assumption. A representation of P in the form $P = \mathcal{U}_E$, $E \in \mathbf{E}^+$, $E(\{o\}) = 0$, is possible only in the case where $P(\{o\}) > 0$. It is an essential object of the second chapter to generalize this structural statement. It is possible to extend the above-mentioned one-to-one mapping $E \curvearrowright \mathcal{U}_E$ to a one-to-one mapping $W \curvearrowright \mathcal{E}_W$ of a set $\boldsymbol{W}$ of not necessarily finite measures W on $\mathfrak{M}$ onto the set of all infinitely divisible distributions on $\mathfrak{M}$, thus deriving the canonical representation of arbitrary infinitely divisible distributions on $\mathfrak{M}$.

If P is any infinitely divisible distribution on $\mathfrak{M}$ and X a Borel subset of the phase space $[A, \varrho_A]$ with the property $P(\Phi(X) = 0) > 0$, then the representation theorem 1.10.1. can be applied to the infinitely divisible distribution ${}_XP$ since we have ${}_XP(\{o\}) = P(\Phi(X) = 0) > 0$. Consequently there exists a uniquely determined finite measure W^X on $\mathfrak{M}$ with the properties ${}_XP = \mathcal{U}_{W^X}$, $W^X(\{o\}) = 0$. In view of 1.10.2., each bounded X in $\mathfrak{A}$ satisfies the inequality $P(\Phi(X) = 0) > 0$, so that P is uniquely determined by the family of measures W^X.

For $X_1 \subseteqq X_2$, the relation ${}_{X_1}P = {}_{X_1}({}_{X_2}P)$ permits us to derive ${}_{X_1}P$ from ${}_{X_2}P$, so that we obtain ${}_{X_1}P = {}_{X_1}(\mathcal{U}_{W^{X_2}}) = \mathcal{U}_{{}_{X_1}(W^{X_2})} = \mathcal{U}_{{}_{X_1}(W^{X_2})((.)\setminus\{o\})}$, i.e. $W^{X_1} = {}_{X_1}(W^{X_2})\left((.)\setminus\{o\}\right)$. Using the analogue of Kolmogorov's extension theorem, we observe that there exists exactly one, generally infinite, measure $\tilde{P}$ on $\mathfrak{M}$ such that $\tilde{P}(\{o\}) = 0$ and, for all X in $\mathfrak{A}$ satisfying the condition $P(\Phi(X) = 0) > 0$, the relation $W^X = {}_X(\tilde{P})\left((.)\setminus\{o\}\right)$ holds, i.e. ${}_XP = \mathcal{U}_{{}_X(\tilde{P})((.)\setminus\{o\})}$. The correspondence $P \curvearrowright \tilde{P}$ provides a one-to-one mapping of the set of all infinitely divisible distributions on $\mathfrak{M}$ onto the set $\boldsymbol{W}$ of all measures W on $\mathfrak{M}$ which have the two properties $W(\{o\}) = 0$, $W(\Psi(X) > 0) < +\infty$ for all bounded X in $\mathfrak{A}$ (Theorem 2.1.10.).

The canonical measure $\tilde{P}$ was introduced independently by Kerstan & Matthes [1] and Lee [1]. The canonical representation of the infinitely divisible distributions of finite random measures had been derived already before by Jiřina in [1].

A number of fundamental properties of infinitely divisible distributions P on $\mathfrak{M}$ can be expressed in an elegant way by means of the associated canonical measures $\tilde{P}$. In the following table, statements on P (left column) are compared with their translations written in terms of the measures $\tilde{P}$ (right column).

a) P (Φ is finite) $= 1$	$\tilde{P}$ (Ψ is infinite) $= 0$, $\tilde{P}$ is finite.
b) P is simple	$\tilde{P}$ (Ψ is not simple) $= 0$, $\tilde{P}\big(\Psi(\{a\}) > 0\big) = 0$ for all a in A.
c) P is free from after-effects.	$\tilde{P}$ (Ψ is not of the form $n\delta_a$ $(n = 1, 2, \ldots;\ a \in A)) = 0$.

The equivalences stated above result from the theorems 2.2.7., 2.2.9., and 2.2.13.

Depending on the character of "after-effects", we distinguish two fundamental types of infinitely divisible distributions P on $\mathfrak{M}$: An infinitely divisible distribution P is said to be regular infinitely divisible, if for all bounded X in $\mathfrak{A}$ conditions of the form "$\Phi(X') = 0$" have a uniformly weak influence on the distribution of $_X\Phi$, provided that the distance between the bounded set X' in $\mathfrak{A}$ and the fixed set X is sufficiently large. However if for all bounded X in $\mathfrak{A}$, $P\big(\Phi(X) > 0 \mid \Phi(X') = 0\big)$ can be made arbitrarily small by suitable selection of bounded X' in $\mathfrak{A}$ with the property $X \cap X' = \emptyset$, then P is called a singular infinitely divisible distribution. Each infinitely divisible distribution P on $\mathfrak{M}$ can be represented in exactly one way as a convolution of a regular infinitely divisible distribution P_r and a singular infinitely divisible distribution P_s (Theorem 2.5.4., cf. Matthes [2]). The two theorems 2.5.1., 2.5.3. (cf. Kerstan & Matthes [1], Lee [1]) state that the table begun above can be continued as follows:

d) P is a regular infinitely divisible distribution.	$\tilde{P}$ (Ψ infinite) $= 0$.
e) P is a singular infinitely divisible distribution.	$\tilde{P}$ (Ψ finite) $= 0$.

The transitions from c) to d) as well as from d) to e) show a monotony behaviour on both sides of the table. On the left the "after-effect" with respect to P, and on the right the "density" of "typical" realizations Ψ with respect to $\tilde{P}$, increase with the respective transitions.

Some operations with infinitely divisible distributions P on $\mathfrak{M}$ can be transformed into similar operations with the associated canonical measures: If any X in $\mathfrak{A}$ satisfies the inequality $P\big(\Phi(X) = 0\big) > 0$, then the conditional distribution $P\big((.) \mid \Phi(X) = 0\big)$ can be introduced. This distribution is again infinitely divisible, for we have $\widetilde{P\big((.) \mid \Psi(X) = 0\big)} = \tilde{P}\big(\Psi \in (.),\ \Psi(X) = 0\big)$ (Theorem 2.3.1.). For all X in $\mathfrak{A}$, we have $_X\tilde{P} = \widetilde{_X(P)}\big((.) \setminus \{o\}\big)$ (Theorem 2.2.4.). Finally, for all finite sequences $X_1, \ldots, X_m$ of bounded sets in $\mathfrak{A}$, the finite-dimensional distribution $P_{X_1,\ldots,X_m}$ is infinitely divisible, and we have $\widetilde{P_{X_1,\ldots,X_m}} = (\tilde{P})_{X_1,\ldots,X_m}\big((.) \setminus \{[0, \ldots, 0]\}\big)$ (Theorem 2.2.8.). The expressions for $\widetilde{_XP}$ and $\widetilde{P_{X_1,\ldots,X_m}}$, respectively, are of the same structure. In section 4.2., we shall observe that they can be understood as particular cases of the "clustering theorem".

Instead of $\tilde{P} = W$ we shall also write $P = \mathscr{E}_W$. For all finite measures E in $\boldsymbol{W}$, $\mathscr{E}_E$ coincides with $\mathscr{U}_E$. Note that the mapping $\mathscr{E}$ satisfies the functional equation of the exponential function. Thus if $(P_i)_{i \in I}$ is an at most countable family of infinitely divisible distributions on $\mathfrak{M}$, then the convolution $\underset{i \in I}{\ast} P_i$ exists if and only if the measure $\sum_{i \in I} \widetilde{P_i}$ belongs to $\boldsymbol{W}$; indeed we have $\widetilde{\underset{i \in I}{\ast} P_i} = \sum_{i \in I} \widetilde{P_i}$ (Theorem 2.2.2.). Each measure $\tilde{P}$ can be represented as a sum of an at most countable family $(E_i)_{i \in I}$ of finite measures; thus we obtain $P = \mathscr{E}_{\tilde{P}} = \underset{i \in I}{\ast} \mathscr{U}_{E_i}$. In this way we have established a constructive approach to the parametric representation $W \curvearrowright \mathscr{E}_W$ of the infinitely divisible distributions on $\mathfrak{M}$ and thus gained access to the structure of these distributions which will be conceptually elucidated in section 4.5. Thus the problem set up at the beginning will be solved.

For each infinitely divisible distribution P on $\mathfrak{M}$ and each X in $\mathfrak{A}$ with the property $P(\Phi(X) = 0) > 0$, we have $P = \mathscr{E}_{\tilde{P}} = \mathscr{E}_{\tilde{P}(\Psi \in (.), \Psi(X) = 0) + \tilde{P}(\Psi \in (.), \Psi(X) > 0)}$ $= \big(P((.) \mid \Psi(X) = 0)\big) \ast \mathscr{E}_{\tilde{P}(\Psi \in (.), \Psi(X) > 0)}$. If a distribution P on $\mathfrak{M}$ is continuous, i.e. if it satisfies the condition $P(\Psi(\{a\}) > 0) = 0$ for all a in A, then P is infinitely divisible if and only if the functional equation $P = \big(P((.) \mid \Phi(X) = 0)\big) \ast Q$ has a solution Q in $\boldsymbol{P}$ for all bounded X in $\mathfrak{A}$ with the property $P(\Psi(X) = 0) > 0$ (Theorem 2.3.3., cf. Matthes [4]). From this it is readily concluded that each continuous distribution free from after-effects is infinitely divisible.

2.1. Infinite Convolution Products in *P*

Given any non-empty at most countable set I, let $\mathfrak{M}^I$ denote the smallest σ-algebra of subsets of the power set M^I with respect to which the mappings

$$(\Phi_i)_{i \in I} \curvearrowright \Phi_j \qquad \big((\Phi_i)_{i \in I} \in M^I\big)$$

into $[M, \mathfrak{M}]$ are measurable for all $j \in I$. Then for all X in $\mathfrak{B}$,

$$(\Phi_i)_{i \in I} \curvearrowright \sum_{i \in I} \Phi_i(X) \qquad \big((\Phi_i)_{i \in I} \in M^I\big)$$

is a measurable mapping of $[M^I, \mathfrak{M}^I]$ into $[0, +\infty]$. The measure $\sum_{i \in I} \Phi_i$ is in M if and only if

$$\sum_{i \in I} \Phi_i\big(S_k(z)\big) < +\infty \qquad (k = 1, 2, \ldots)$$

is satisfied. Here z denotes an arbitrary but fixed point in A, and $S_k(z)$ the open ball of radius k around z. Consequently the set L_I of those $(\Phi_i)_{i \in I}$ in M^I for which $\sum_{i \in I} \Phi_i$ is in M, belongs to $\mathfrak{M}^I$, so by going back to the definition of $\mathfrak{M}$, we immediately have

2.1.1. *The correspondence*

$$(\Phi_i)_{i\in I} \curvearrowright \sum_{i\in I} \Phi_i \qquad \left((\Phi_i)_{i\in I} \in L_I\right)$$

is a measurable mapping from $[M^I, \mathfrak{M}^I]$ *into* $[M, \mathfrak{M}]$.

To each family $(P_i)_{i\in I}$ of elements of $\boldsymbol{P}$ we can associate the distribution $\underset{i\in I}{\times} P_i$ on $\mathfrak{M}^I$. Now, if $\left(\underset{i\in I}{\times} P_i\right)(L_I) = 1$, we denote by $\underset{i\in I}{\ast} P_i$ the distribution of $\sum_{i\in I} \Phi_i$ constructed with respect to $\underset{i\in I}{\times} P_i$, calling it the *convolution of the family* $(P_i)_{i\in I}$. Remember that for $I = \{1, \ldots, n\}$ always $\underset{i\in I}{\ast} P_i$ exists and coincides with $P_1 * \cdots * P_n$. The convolution of the empty family $(P_i)_{i\in I}$, $I = \emptyset$ is set equal to δ_o.

The following propositions result immediately from the definition of the convolution of families.

2.1.2. *If* $(P_i)_{i\in I}$ *is an at most countable family of distributions on* $\mathfrak{M}$ *and* $I = \bigcup_{j\in J} I_j$ *is a representation of* I *as a union of an at most countable family of pairwise disjoint sets, then*

$$\underset{i\in I}{\ast} P_i = \underset{j\in J}{\ast}\left(\underset{i\in I_j}{\ast} P_i\right),$$

where the existence of either side entails that of the other.

2.1.3. *If the convolution of an at most countable family* $(P_i)_{i\in I}$ *of distributions on* $\mathfrak{M}$ *exists, then*

a) ${}_X\left(\underset{i\in I}{\ast} P_i\right) = \underset{i\in I}{\ast}\, {}_X(P_i)$ *for all* X *in* $\mathfrak{A}$,

b) $\left(\underset{i\in I}{\ast} P_i\right)_{X_1,\ldots,X_m} = \underset{i\in I}{\ast}\left((P_i)_{X_1,\ldots,X_m}\right)$ *for all finite sequences* $X_1, \ldots, X_m$ *of sets in* $\mathfrak{B}$,

c) $\varrho_{\underset{i\in I}{\ast} P_i} = \sum_{i\in I} \varrho_{P_i}$.

The convolution of a family $(P_i)_{i\in I}$ exists if and only if $\left(\underset{i\in I}{\times} P_i\right)(\Phi_i(X) > 0$ only for finitely many i in $I) = 1$ is satisfied for all X in $\mathfrak{B}$. Using the Borel-Cantelli lemma, we thus obtain

2.1.4. *The convolution of an at most countable family* $(P_i)_{i\in I}$ *of elements of* $\boldsymbol{P}$ *exists if and only if*

$$\sum_{i\in I} P_i\left(\Phi\left(S_k(z)\right) > 0\right) < +\infty \qquad (k = 1, 2, \ldots)$$

is satisfied. Here z *means an arbitrary but fixed point in* A.

Using 2.1.3. c) we obtain

2.1.5. *For each at most countable family* $(P_i)_{i\in I}$ *of distributions on* $\mathfrak{M}$, *the following statements are equivalent:*

a) $\underset{i\in I}{\ast} P_i$ *exists and is of finite intensity,*

b) $\sum_{i\in I} \varrho_{P_i}$ *belongs to* N.

Let $\boldsymbol{W}$ denote the set of those possibly infinite measures W on $\mathfrak{M}$, which satisfy the two conditions

a) $W\big(\Psi(X) > 0\big) < +\infty$ for all X in $\mathfrak{B}$,
b) $W(\{o\}) = 0$.

2.1.6. Proposition. *If* $(E_i)_{i\in I}$ *is an at most countable family of measures in* $\boldsymbol{E}^+$, *then the convolution* $\underset{i\in I}{\ast} \mathcal{U}_{E_i}$ *exists if and only if* $\sum_{i\in I} E_i\big((.) \setminus \{o\}\big)$ *belongs to* $\boldsymbol{W}$.

Proof. In view of 1.6.5., we can additionally suppose without loss of generality that $E_i(\{o\}) = 0$ be true for all i in I.
Then by 1.6.8., for all X in $\mathfrak{B}$ we obtain

$$\sum_{i\in I} (\mathcal{U}_{E_i})\big(\Phi(X) > 0\big) = \sum_{i\in I} \big(1 - e^{-E_i(\Psi(X)>0)}\big) = \sum_{i\in I} E_i\big(\Psi(X) > 0\big)\, e^{-\Theta_i E_i(\Psi(X)>0)}$$

where $0 < \Theta_i < 1$ for $i \in I$.
If the convolution $\underset{i\in I}{\ast} \mathcal{U}_{E_i}$ exists, then according to the Borel-Cantelli lemma the sum $\sum_{i\in I} (\mathcal{U}_{E_i})\big(\Phi(X) > 0\big)$ must converge. From this it follows that $c = \sup_{i\in I} E_i\big(\Psi(X) > 0\big) < +\infty$, for otherwise infinitely many terms $\big(1 - e^{-E_i(\Psi(X)>0)}\big)$ would be greater than 1/2. Consequently, in view of

$$e^{-c} \sum_{i\in I} E_i\big(\Psi(X) > 0\big) \leqq \sum_{i\in I} E_i\big(\Psi(X) > 0\big)\, e^{-\Theta_i E_i(\Psi(X)>0)},$$

the relation

$$\sum_{i\in I} E_i\big(\Psi(X) > 0\big) = \Big(\sum_{i\in I} E_i\Big)\big(\Psi(X) > 0\big) < +\infty$$

must be true for all X in $\mathfrak{B}$. Hence the measure $\sum_{i\in I} E_i$ is in $\boldsymbol{W}$. Conversely, if $\sum_{i\in I} E_i$ is in $\boldsymbol{W}$, then the series $\sum_{i\in I} \mathcal{U}_{E_i}\big(\Phi(X) > 0\big)$ converges for all X in $\mathfrak{B}$ which implies the existence of $\underset{i\in I}{\ast} \mathcal{U}_{E_i}$. ∎

2.1.7. Proposition. *If the sums of two at most countable families* $(D_i)_{i\in I}$, $G_j)_{j\in J}$ *of measures in* $\boldsymbol{E}^+$ *are equal to one and the same measure* W *in* $\boldsymbol{W}$, *then*

$$\underset{i\in I}{\ast} \mathcal{U}_{D_i} = \underset{j\in J}{\ast} \mathcal{U}_{G_j}.$$

Proof. 1. To begin with, let us suppose that the measure W be finite. If I is finite, then by 1.6.4. the convolution $\underset{i \in I}{\ast} \mathcal{U}_{D_i}$ coincides with $\mathcal{U}_W$. Now let $I = \{i_1, i_2, \ldots\}$, where $i_k \neq i_s$ for $k \neq s$. The partial sums $\sum_{k=1}^{n} D_{i_k}$ converge in $\boldsymbol{E}_r$ toward W; for we have

$$\left\| W - \sum_{k=1}^{n} D_{i_k} \right\| = \left\| \sum_{k=n+1}^{\infty} D_{i_k} \right\| = W(M) - \sum_{k=1}^{n} D_{i_k}(M).$$

In view of 1.10.3., $\mathcal{U}_{\sum_{k=1}^{n} D_{i_k}}$ then tends toward $\mathcal{U}_W$ in $\boldsymbol{E}_r$, i.e.

$$\left\| \underset{k=1}{\overset{n}{\ast}} \mathcal{U}_{D_{i_k}} - \mathcal{U}_W \right\| \xrightarrow[n \to \infty]{} 0.$$

On the other hand $\underset{k=1}{\overset{n}{\ast}} \mathcal{U}_{D_{i_k}}$ converges in $\boldsymbol{E}_r$ toward $\underset{k=1}{\overset{\infty}{\ast}} \mathcal{U}_{D_{i_k}}$ as $n \to \infty$, for we have

$$\begin{aligned} \left\| \underset{k=1}{\overset{\infty}{\ast}} \mathcal{U}_{D_{i_k}} - \underset{k=1}{\overset{n}{\ast}} \mathcal{U}_{D_{i_k}} \right\| &= \left\| \left(\underset{k=1}{\overset{n}{\ast}} \mathcal{U}_{D_{i_k}} \right) * \left(\underset{k=n+1}{\overset{\infty}{\ast}} \mathcal{U}_{D_{i_k}} \right) - \underset{k=1}{\overset{n}{\ast}} \mathcal{U}_{D_{i_k}} \right\| \\ &\leqq \left\| \underset{k=n+1}{\overset{\infty}{\ast}} \mathcal{U}_{D_{i_k}} - \delta_o \right\| = 2 \left(\underset{k=n+1}{\overset{\infty}{\ast}} \mathcal{U}_{D_{i_k}} \right) (\Phi \neq o) \\ &\leqq 2 \sum_{k=n+1}^{\infty} \left(\mathcal{U}_{D_{i_k}} \right) (\Phi \neq o) = 2 \sum_{k=n+1}^{\infty} \left(1 - e^{-D_{i_k}(M)} \right). \end{aligned}$$

But in view of $\sum_{k=1}^{\infty} D_{i_k}(M) = W(M) < +\infty$, $\sum_{k=n+1}^{\infty} \left(1 - e^{-D_{i_k}(M)}\right)$ tends to zero as $n \to \infty$. Summarizing, we thus obtain

$$\left\| \underset{i \in I}{\ast} \mathcal{U}_{D_i} - \mathcal{U}_W \right\| \leqq \left\| \underset{k=1}{\overset{n}{\ast}} \mathcal{U}_{D_{i_k}} - \underset{i \in I}{\ast} \mathcal{U}_{D_i} \right\| + \left\| \underset{k=1}{\overset{n}{\ast}} \mathcal{U}_{D_{i_k}} - \mathcal{U}_W \right\| \xrightarrow[n \to \infty]{} 0,$$

i.e. we always have $\underset{i \in I}{\ast} \mathcal{U}_{D_i} = \mathcal{U}_W$. In the same way we can conclude that $\underset{j \in J}{\ast} \mathcal{U}_{G_j} = \mathcal{U}_W$.

2. Now let W be an arbitrary measure in $\boldsymbol{W}$, and (Y_n) a sequence of pairwise disjoint sets in $\mathfrak{M}$ having the properties

$$\bigcup_{n=1}^{\infty} Y_n = M, \qquad W(Y_n) < +\infty \qquad \text{für } n = 1, 2, \ldots.$$

In view of the two previously defined properties a), b) of $\boldsymbol{W}$, such a sequence must always exist. Indeed we can select any countable covering (X_n) of A by sets in $\mathfrak{B}$, and set

$$\begin{aligned} &Y_1 = \{o\},\ Y_2 = \{\Psi : \Psi \in M, \Psi(X_1) > 0\}, \ldots, \\ &Y_n = \{\Psi : \Psi \in M, \Psi(X_{n-1}) > 0, \Psi(X_{n-2}) = \cdots = \Psi(X_1) = 0\}. \end{aligned}$$

By our assumptions, for $n = 1, 2, \ldots$ we have

$$\sum_{i\in I} D_i\big((.) \cap Y_n\big) = W\big((.) \cap Y_n\big) = \sum_{j\in J} G_j\big((.) \cap Y_n\big),$$

and using step 1 of this proof we obtain

$$\mathop{\ast}_{i\in I} \mathscr{U}_{D_i((.)\cap Y_n)} = \mathscr{U}_{W((.)\cap Y_n)} = \mathop{\ast}_{j\in J} \mathscr{U}_{G_j((.)\cap Y_n)},$$

from which in view of 2.1.2. we can conclude that

$$\begin{aligned} \mathop{\ast}_{i\in I} \mathscr{U}_{D_i} &= \mathop{\ast}_{i\in I} \left(\mathop{\ast}_{n=1}^{\infty} \mathscr{U}_{D_i((.)\cap Y_n)}\right) \\ &= \mathop{\ast}_{n=1}^{\infty} \left(\mathop{\ast}_{i\in I} \mathscr{U}_{D_i((.)\cap Y_n)}\right) = \mathop{\ast}_{n=1}^{\infty} \mathscr{U}_{W((.)\cap Y_n)} \\ &= \mathop{\ast}_{n=1}^{\infty} \left(\mathop{\ast}_{j\in J} \mathscr{U}_{G_j((.)\cap Y_n)}\right) = \mathop{\ast}_{j\in J} \left(\mathop{\ast}_{n=1}^{\infty} \mathscr{U}_{G_j((.)\cap Y_n)}\right) = \mathop{\ast}_{j\in J} \mathscr{U}_{G_j}. \blacksquare \end{aligned}$$

Each W in $\mathbf{W}$ can be represented as the sum of a sequence (E_n) of measures in $\mathbf{E}^+$. (As in the second step of the proof of 2.1.7., we can represent the measure W by $\sum_{n=1}^{\infty} W\big((.) \cap Y_n\big)$.)

Now, if we set

$$\mathscr{E}_W = \mathop{\ast}_{n=1}^{\infty} \mathscr{U}_{E_n},$$

then by 2.1.7. we obtain a mapping $\mathscr{E}$ of $\mathbf{W}$ into $\mathbf{P}$ which coincides with $\mathscr{U}$ on $\mathbf{W} \cap \mathbf{E}^+$.

The functional equation $\mathscr{U}_{E_1+E_2} = \mathscr{U}_{E_1} * \mathscr{U}_{E_2}$ also remains true for $\mathscr{E}$. We even have

2.1.8. *If $(W_i)_{i\in I}$ is an at most countable family of measures in* $\mathbf{W}$, *then the convolution $\mathop{\ast}_{i\in I} \mathscr{E}_{W_i}$ exists if and only if $W = \sum_{i\in I} W_i$ is in* $\mathbf{W}$; *in this case*

$$\mathop{\ast}_{i\in I} \mathscr{E}_{W_i} = \mathscr{E}_W.$$

Proof. Each W_i can be represented as the sum of a sequence of finite measures $W_{i,1}, W_{i,2}, \ldots$. Then by 2.1.2. we have

$$\mathop{\ast}_{i\in I} \mathscr{E}_{W_i} = \mathop{\ast}_{i\in I;\, n=1,2,\ldots} \mathscr{U}_{W_{i,n}}.$$

But according to 2.1.6. the right side of this equation exists if and only if $W = \sum_{i\in I, n=1,2,\ldots} W_{i,n}$ is in $\mathbf{W}$. If so, the definition of $\mathscr{E}$ immediately leads to

$$\mathop{\ast}_{i\in I} \mathscr{E}_{W_i} = \mathop{\ast}_{i\in I;\, n=1,2,\ldots} \mathscr{E}_{W_{i,n}} = \mathscr{E}_W. \blacksquare$$

2.1.9. *For all W in $\boldsymbol{W}$ and all X in $\mathfrak{A}$, $({}_XW)\big((.)\setminus\{o\}\big)$ belongs to $\boldsymbol{W}$, and*

$$_X(\mathscr{E}_W) = \mathscr{E}_{({}_XW)((.)\setminus\{o\})}.$$

Proof. Let $W = \sum_{n=1}^{\infty} W_n$ be an arbitrary representation of W as a sum of finite measures. Using 2.1.3. as well as 1.5.2. and 1.6.5., we obtain

$$\begin{aligned} {}_X(\mathscr{E}_W) &= {}_X\left(\mathop{\ast}_{n=1}^{\infty} \mathscr{U}_{W_n}\right) = \mathop{\ast}_{n=1}^{\infty} \mathscr{U}_{{}_X(W_n)} = \mathop{\ast}_{n=1}^{\infty} \mathscr{U}_{({}_X(W_n))((.)\setminus\{o\})} \\ &= \mathscr{E}_{\sum_{n=1}^{\infty}({}_X(W_n))((.)\setminus\{o\})} = \mathscr{E}_{({}_XW)((.)\setminus\{o\})}. \quad \blacksquare \end{aligned}$$

For all W in $\boldsymbol{W}$ and all natural numbers n, $\mathscr{E}_W = (\mathscr{E}_{W/n})^n$. Consequently all distributions of the form $\mathscr{E}_W$, $W \in \boldsymbol{W}$, are infinitely divisible. In this way we cover all of the infinitely divisible distributions on $\mathfrak{M}$, for we have

2.1.10. Theorem. *The correspondence $W \rightsquigarrow \mathscr{E}_W$ provides a one-to-one mapping of $\boldsymbol{W}$ onto $\boldsymbol{T}$.*

Proof. 1. Let $\mathscr{E}_W = \mathscr{E}_S$ where $W, S \in \boldsymbol{W}$. Then for all X in $\mathfrak{B}$, due to the finiteness of the measures $({}_XW)\big((.)\setminus\{o\}\big)$ and $({}_XS)\big((.)\setminus\{o\}\big)$ and in view of 2.1.9. and 1.10.1.,

$$({}_XW)\big((.)\setminus\{o\}\big) = ({}_XS)\big((.)\setminus\{o\}\big),$$

so that the two measures W, S coincide on all sets of the form $Y \cap \{\Psi : \Psi \in M, \Psi(X) > 0\}$, $Y \in {}_X\mathfrak{M}$. Hence, if Y belongs to the algebra $\bigcup_{X \in \mathfrak{B}} {}_X\mathfrak{M}$, then for all X in $\mathfrak{B}$,

$$W\big(\Psi \in Y, \Psi(X) > 0\big) = S\big(\Psi \in Y, \Psi(X) > 0\big).$$

Since $\bigcup_{X \in \mathfrak{B}} {}_X\mathfrak{M}$ generates the σ-algebra $\mathfrak{M}$, this equation holds true for all Y in $\mathfrak{M}$. Now, if we let X run through a monotone increasing sequence (X_n) having the union A, then

$$\begin{aligned} W(Y) &= \sup_{n=1,2,\ldots} W\big(\Psi \in Y, \Psi(X_n) > 0\big) \\ &= \sup_{n=1,2,\ldots} S\big(\Psi \in Y, \Psi(X_n) > 0\big) = S(Y), \end{aligned}$$

i.e. we have $W = S$.

2. Let P be any infinitely divisible distribution on $\mathfrak{M}$. Then according to 1.6.1. ${}_XP$ is also infinitely divisible for all X in $\mathfrak{B}$, and in view of 1.10.2. there exists exactly one finite measure W^X in $\boldsymbol{W}$ with the property

$$_XP = \mathscr{E}_{W^X}.$$

By 2.1.9. we have for all Z in $\mathfrak{B}$ which are included in X:

$$\mathscr{E}_{W^Z} = {}_ZP = {}_Z({}_XP) = {}_Z(\mathscr{E}_{W^X}) = \mathscr{E}_{({}_Z(W^X))((.)\setminus\{o\})},$$

so that we obtain

$$W^Z = \left(z(W^X)\right)\left((.)\setminus\{o\}\right).$$

For an arbitrary but fixed Z in $\mathfrak{B}$, the set-up

$$V^Z(Y) = W^X\left(\Psi \in Y, \Psi(Z) > 0\right) \qquad (X \in \mathfrak{B};\, X \supseteqq Z;\, Y \in {}_X\mathfrak{M})$$

defines a finite, finitely additive measure V^Z on the generating algebra

$$\bigcup_{X\in\mathfrak{B}} {}_X\mathfrak{M} = \bigcup_{X\in\mathfrak{B};\, X\supseteqq Z} {}_X\mathfrak{M},$$

and this measure is totally additive on each σ-algebra ${}_X\mathfrak{M}$, $X \in \mathfrak{B}$.

For extending the measure V^Z to the σ-algebra $\mathfrak{M}$, we can suppose that $V^Z(M) = 1$. For if V^Z is the zero measure, the extension is trivially possible. In the case $V^Z(M) > 0$ however, we obtain the continuation of the finite, finitely additive measure V^Z by multiplying the continuation of the normalized finitely additive measure $\left(V^Z(M)\right)^{-1}V^Z$ by the factor $V^Z(M)$.

Now let $X_1, \ldots, X_m$ be an arbitrary finite sequence of sets in $\mathfrak{B}$ which have the union X. Then for each sequence of non-negative integers $l_1, \ldots, l_m$, the set $Y = \{\Psi : \Psi \in M,\, \Psi(X_1) = l_1, \ldots, \Psi(X_m) = l_m\}$ is included in ${}_X\mathfrak{M}$. V^Z is a totally additive measure on the σ-algebra ${}_X\mathfrak{M}$, and hence on its sub-algebra $\mathfrak{M}_{X_1,\ldots,X_m}$, i.e. we can introduce the distribution $(V^Z)_{X_1,\ldots,X_m}$.

Obviously the finite-dimensional distributions $(V^Z)_{X_1,\ldots,X_m}$ established with respect to V^Z satisfy the conditions a) through d) of the existence theorem 1.3.5., i.e. there exists a distribution H^Z on $\mathfrak{M}$ having the property

$$(H^Z)_{X_1,\ldots,X_m} = (V^Z)_{X_1,\ldots,X_m}$$

for all finite sequences of sets in $\mathfrak{B}$.

Now let X be an arbitrary set in $\mathfrak{B}$. Let $\mathfrak{Y}$ be the union of σ-algebras $\mathfrak{M}_{X_1,\ldots,X_m}$ which extends over all finite sequences $X_1, \ldots, X_m$ of subsets of X which belong to $\mathfrak{B}$. $\mathfrak{Y}$ is an algebra on which V^Z and H^Z coincide. As an immediate consequence of the definition of ${}_X\mathfrak{M}$, it is observed that the σ-algebra ${}_X\mathfrak{M}$ is generated by $\mathfrak{Y}$. Thus V^Z and H^Z coincide on ${}_X\mathfrak{M}$, too. Consequently H^Z is a continuation of the finitely additive measure V^Z defined on $\bigcup_{X\in\mathfrak{B}} {}_X\mathfrak{M}$.

Let z be an arbitrary point in A and Z_k the open ball of radius k around z. For $k_1 < k_2$ we obtain for all $X \supseteqq Z_{k_2}$, $X \in \mathfrak{B}$, and all $Y \in {}_X\mathfrak{M}$:

$$\begin{aligned} &H^{Z_{k_2}}\left(\Psi \in Y, \Psi(Z_{k_1}) > 0\right) \\ &= W^X\left(\Psi \in Y, \Psi(Z_{k_2}) > 0, \Psi(Z_{k_1}) > 0\right) \\ &= W^X\left(\Psi \in Y, \Psi(Z_{k_1}) > 0\right) = H^{Z_{k_1}}(Y). \end{aligned}$$

Since $\bigcup_{X\in\mathfrak{B},\, X\supseteq Z_{k_2}} {}_X\mathfrak{M}$ generates the σ-algebra $\mathfrak{M}$, the following relation holds on $\mathfrak{M}$:

$$H^{Z_{k_2}}\left(\Psi \in (.), \Psi(Z_{k_1}) > 0\right) = H^{Z_{k_1}}.$$

Consequently the sequence of finite measures $(H^{Z_k})_{k=1,2,\ldots}$ is monotone increasing. We denote its supremum by W, so that

$$W(\Psi \in Y, \Psi(Z_k) > 0) = H^{Z_k}(Y) \qquad (Y \in \mathfrak{M}; k = 1, 2, \ldots)$$

is satisfied.

Obviously $W(\{o\}) = \sup_{k=1,2,\ldots} H^{Z_k}(\{o\}) = 0$. Now let X be an arbitrary set in $\mathfrak{B}$. For sufficiently large k, X is included in Z_k, so that we obtain

$$({}_X W)\big((.)\setminus\{o\}\big) = W\big({}_X\Psi \in (.), \Psi(X) > 0\big)$$
$$= H^{Z_k}\big({}_X\Psi \in (.), \Psi(X) > 0\big) = W^{Z_k}\big({}_X\Psi \in (.), \Psi(X) > 0\big) = W^X.$$

Hence $W\big(\Psi(X) > 0\big) < +\infty$; consequently W belongs to $\boldsymbol{W}$, and

$${}_XP = {}_X(\mathscr{E}_W)$$

for all X in $\mathfrak{B}$.

Then in view of the uniqueness proposition 1.3.2., $P = \mathscr{E}_W$ must be true. ∎

2.2. The Mapping $P \rightsquigarrow \tilde{P}$

Instead of $P = \mathscr{E}_W$, $W \in \boldsymbol{W}$, we also write $W = \tilde{P}$. The measure $\tilde{P}$ is called the *canonical measure* of the infinitely divisible distribution P. Obviously $\widetilde{\delta_o}$ is equal to the zero element 0 of $\boldsymbol{E}$. In view of 2.1.10. and 2.1.8., we have

2.2.1. *The correspondence $P \rightsquigarrow \tilde{P}$ provides a one-to-one mapping of $\boldsymbol{T}$ onto $\boldsymbol{W}$.*

2.2.2. *If $(P_i)_{i\in I}$ is an at most countable family of infinitely divisible distributions on $\mathfrak{M}$, then $\mathop{\ast}\limits_{i\in I} P_i$ exists if and only if $\sum_{i\in I} \widetilde{P_i}$ is in $\boldsymbol{W}$; in this case*

$$\widetilde{\mathop{\ast}\limits_{i\in I} P_i} = \sum_{i\in I} \widetilde{P_i}.$$

Thus in particular the equation

$$\widetilde{\sqrt[n]{P}} = \frac{1}{n}\tilde{P}$$

holds for all P in $\boldsymbol{T}$ and all natural numbers n.

As a generalization of 1.6.13., we have

2.2.3. Proposition. *For all infinitely divisible distributions P on $\mathfrak{M}$*

$$\varrho_P = \varrho_{\tilde{P}}.$$

Proof. Let $\tilde{P} = \sum_{n=1}^{\infty} E_n$ be an arbitrary representation of $\tilde{P}$ as a sum of a sequence of measures in $\boldsymbol{E}^+$. Using 2.1.8., we obtain

$$P = \mathscr{E}_{\tilde{P}} = \mathop{\ast}_{n=1}^{\infty} \mathscr{U}_{E_n}$$

from which, using 2.1.3. c) and 1.6.13., it can be concluded that

$$\varrho_P = \sum_{n=1}^{\infty} \varrho_{\mathscr{U}_{E_n}} = \sum_{n=1}^{\infty} \varrho_{E_n} = \varrho_{\sum_{n=1}^{\infty} E_n} = \varrho_{\tilde{P}}. \blacksquare$$

In particular we therefore always have $\varrho_P(\{a\}) = \varrho_{\tilde{P}}(\{a\})$. Therefore we can conclude that a distribution P in $\boldsymbol{T}$ is continuous if and only if $\tilde{P}$ exhibits this property.

As an immediate consequence of 2.1.9., we have

2.2.4. Proposition. *For all infinitely divisible distributions P on $\mathfrak{M}$ and all X in $\mathfrak{A}$, also ${}_XP$ is infinitely divisible, and*

$$\widetilde{{}_XP} = {}_X(\tilde{P})\big((.) \setminus \{o\}\big).$$

Hence, for all X in $\mathfrak{B}$, we obtain by 1.6.8.

$$P\big(\Phi(X) = 0\big) = ({}_XP)\ (\Phi = o) = \mathscr{U}_{{}_X(\tilde{P})((.)\setminus\{o\})}(\Phi = o)$$
$$= e^{-{}_X(\tilde{P})(M\setminus\{o\})} = e^{-\tilde{P}(\Psi(X)>0)}.$$

Each set X in $\mathfrak{A}$ can be represented as a union of a monotone increasing sequence (X_n) of sets in $\mathfrak{B}$, so that we obtain

$$\begin{aligned} P\big(\Phi(X) = 0\big) &= \lim_{n\to\infty} P\big(\Phi(X_n) = 0\big) \\ &= \lim_{n\to\infty} \exp\Big(-\tilde{P}\big(\Psi(X_n) > 0\big)\Big) \\ &= \exp\Big(-\tilde{P}\big(\Psi(X) > 0\big)\Big). \end{aligned}$$

Consequently we have

2.2.5. *Subject to the assumptions of* 2.2.4.,

$$P\big(\Phi(X) = 0\big) = e^{-\tilde{P}(\Psi(X)>0)}.$$

Hence the canonical measure $\tilde{P}$ of P is finite if and only if $P(\Phi = o) > 0$. As a generalization of 1.6.14. we have (cf. Lee [1])

2.2.6. Proposition. *For all infinitely divisible distributions P on $\mathfrak{M}$ of finite intensity,*

$$\varrho_{P^{\langle 2\rangle}} = \varrho_{(\tilde{P})^{\langle 2\rangle}} + \varrho_P \times \varrho_P.$$

Proof. Let (X_n) be an arbitrary monotone increasing sequence of sets in $\mathfrak{B}$ with union A. According to 2.2.4. and 1.6.14., for all Z in $\mathfrak{A} \otimes \mathfrak{A}$, we have

$$\begin{aligned}
\varrho_{P^{\langle 2 \rangle}}(Z) &= \lim_{n\to\infty} \varrho_{P^{\langle 2 \rangle}}\big(Z \cap (X_n \times X_n)\big) \\
&= \lim_{n\to\infty} \int (\Phi \times \Phi) \big(Z \cap (X_n \times X_n)\big)\, P(\mathrm{d}\Phi) \\
&= \lim_{n\to\infty} \int ({}_{X_n}\Phi \times {}_{X_n}\Phi)\, (Z)\, P(\mathrm{d}\Phi) \\
&= \lim_{n\to\infty} \int (\Phi \times \Phi)\, (Z)\, ({}_{X_n}P)\, (\mathrm{d}\Phi) \\
&= \lim_{n\to\infty} \int (\Phi \times \Phi)\, (Z)\, \mathscr{U}_{{}_{X_n}(\tilde{P})((.)\setminus\{o\})}(\mathrm{d}\Phi) \\
&= \lim_{n\to\infty} \Big(\varrho_{({}_{X_n}(\tilde{P})((.)\setminus\{o\}))^{\langle 2 \rangle}}(Z) + \big(\varrho_{{}_{X_n}P} \times \varrho_{{}_{X_n}P}\big)\, (Z)\Big) \\
&= \lim_{n\to\infty} \Big(\int (\Psi \times \Psi)\, (Z)\, {}_{X_n}(\tilde{P})\, (\mathrm{d}\Psi) + \int (\Phi_1 \times \Phi_2)\, (Z) \\
&\qquad\qquad ({}_{X_n}P \times {}_{X_n}P)\, (\mathrm{d}[\Phi_1, \Phi_2])\Big) \\
&= \lim_{n\to\infty} \Big(\int (\Psi \times \Psi) \big(Z \cap (X_n \times X_n)\big)\, \tilde{P}(\mathrm{d}\Psi) \\
&\qquad + \int (\Phi_1 \times \Phi_2) \big(Z \cap (X_n \times X_n)\big)\, (P \times P)\, (\mathrm{d}[\Phi_1, \Phi_2])\Big) \\
&= \int (\Psi \times \Psi)\, (Z)\, \tilde{P}(\mathrm{d}\Psi) + \int (\Phi_1 \times \Phi_2)\, (Z)\, (P \times P)\, (\mathrm{d}[\Phi_1, \Phi_2]) \\
&= \varrho_{(\tilde{P})^{\langle 2 \rangle}}(Z) + (\varrho_P \times \varrho_P)\, (Z). \quad \blacksquare
\end{aligned}$$

Subject to the assumptions of 2.2.6., for all X_1, X_2 in $\mathfrak{B}$, the covariance

$$\int \big(\Phi(X_1) - \varrho_P(X_1)\big) \big(\Phi(X_2) - \varrho_P(X_2)\big)\, P(\mathrm{d}\Phi)$$

is non-negative!

2.2.7. Proposition. *An infinitely divisible distribution P on $\mathfrak{M}$ satisfies the condition $P\big(\Phi(X) < +\infty\big) = 1$ for any X in $\mathfrak{A}$ if and only if $\tilde{P}\big(\Psi(X) > 0\big)$ is finite and $\tilde{P}\big(\Psi(X) = +\infty\big) = 0$.*

Proof. Suppose that $\tilde{P}\big(\Psi(X) > 0\big) < +\infty$ and $\tilde{P}\big(\Psi(X) = +\infty\big) = 0$. Then by 2.2.4. we have

$$\begin{aligned}
P\big(\Phi(X) < +\infty\big) &= ({}_X P)\ \ (\Phi \text{ is finite}) \\
&= \mathscr{U}_{{}_X(\tilde{P})((.)\setminus\{o\})}\ \ (\Phi \text{ is finite}) \\
&= e^{-\tilde{P}(\Psi(X)>0)} \sum_{n=0}^{\infty} \frac{1}{n!} \big({}_X(\tilde{P})\, ((.) \setminus \{o\})\big)^n\ (\Phi \text{ is finite}) \\
&= e^{-\tilde{P}(\Psi(X)>0)} \sum_{n=0}^{\infty} \frac{1}{n!} \Big(\tilde{P}\big(0 < \Psi(X) < +\infty\big)\Big)^n = 1.
\end{aligned}$$

Conversely, let us suppose now that P be infinitely divisible and satisfy the condition $P\big(\Phi(X) < +\infty\big) = 1$. Then by 1.10.2. the infinitely divisible distribution ${}_XP$ has the form ${}_XP = \mathscr{U}_E$ where $E \in \boldsymbol{E}^+$; for $({}_XP)$ (Φ is finite) $= 1$. In other words $\widetilde{{}_XP}$ is finite, i.e. we have

$$_X(\tilde{P})(M \setminus \{o\}) = \tilde{P}(\Psi(X) > 0) < +\infty,$$

and as above we obtain

$$1 = ({}_XP)(\Phi \text{ is finite}) = e^{-\tilde{P}(\Psi(X)>0)} \sum_{n=0}^{\infty} \frac{1}{n!} \left(\tilde{P}\left(0 < \Psi(X) < +\infty\right)\right)^n$$
$$= e^{-\tilde{P}(\Psi(X)>0)} e^{\tilde{P}(0<\Psi(X)<+\infty)}$$

which leads to

$$\tilde{P}\left(\Psi(X) = +\infty\right) = 0. \blacksquare$$

In an analogy to 2.2.4., we have

2.2.8. Proposition. *For all infinitely divisible distributions P on $\mathfrak{M}$ and all finite sequences $X_1, \ldots, X_m$ of sets in $\mathfrak{A}$, the finite-dimensional distribution $P_{X_1,\ldots,X_m}$ exists if and only if $(\tilde{P})_{X_1,\ldots,X_m}$ exists and $(\tilde{P})_{X_1,\ldots,X_m}\left((.) \setminus \{o\}\right)$ is finite. In this case,*

$$\widetilde{P_{X_1,\ldots,X_m}} = (\tilde{P})_{X_1,\ldots,X_m}\left((.) \setminus \{o\}\right).$$

Proof. The existence of $P_{X_1,\ldots,X_m}$ means that $P\left(\Phi\left(\bigcup_{i=1}^m X_i\right) < +\infty\right) = 1$ is satisfied. But in view of 2.2.7. this is equivalent to the validity of $\tilde{P}\left(\Psi\left(\bigcup_{i=1}^m X_i\right) = +\infty\right) = 0$ as well as $\tilde{P}\left(\Psi\left(\bigcup_{i=1}^m X_i\right) > 0\right) < +\infty$, which means that $(\tilde{P})_{X_1,\ldots,X_m}$ exists and $(\tilde{P})_{X_1,\ldots,X_m}\left((.) \setminus \{o\}\right)$ is finite.

Now let $P\left(\Phi(X) < +\infty\right) = 1$ where $X = \bigcup_{i=1}^m X_i$. Using 1.5.7., 1.6.5. and 1.10.2. we obtain

$$P_{X_1,\ldots,X_m} = ({}_XP)_{X_1,\ldots,X_m} = \left(\mathscr{U}_{_X(\tilde{P})((.)\setminus\{o\})}\right)_{X_1,\ldots,X_m}$$
$$= \left(e^{-\tilde{P}(\Psi(X)>0)} \sum_{n=0}^{\infty} \frac{1}{n!} \left({}_X(\tilde{P})\left((.) \setminus \{o\}\right)\right)^n\right)_{X_1,\ldots,X_m}$$
$$= e^{-\tilde{P}(\Psi(X)>0)} \sum_{n=0}^{\infty} \frac{1}{n!} \left(\left({}_X(\tilde{P})\left((.) \setminus \{o\}\right)\right)_{X_1,\ldots,X_m}\right)^n$$
$$= \mathscr{U}_{\left({}_X(\tilde{P})((.)\setminus\{o\})\right)_{X_1,\ldots,X_m}} = \mathscr{U}_{(\tilde{P})_{X_1,\ldots,X_m}((.)\setminus\{o\})},$$

and hence

$$\widetilde{P_{X_1,\ldots,X_m}} = (\tilde{P})_{X_1,\ldots,X_m}\left((.) \setminus \{o\}\right),$$

where in the formulation of 2.2.8. as well as in the proof the random m-dimensional vectors with non-negative integer coordinates have been identified with the corresponding point processes with the phase space $\{1, \ldots, m\}$. $\blacksquare$

2.2.9. Proposition. *An infinitely divisible distribution P on $\mathfrak{M}$ is simple if and only if $\tilde{P}$ is both continuous and simple.*

Proof. Let (X_n) be a monotone increasing sequence of sets in $\mathfrak{B}$ with union A. In virtue of 1.4.4. P and $\tilde{P}$, respectively, are simple if and only if all of the ${}_{X_n}P$ and ${}_{X_n}(\tilde{P})\left((.)\setminus\{o\}\right)$, respectively, are simple. But because of 2.2.4.

$$ {}_{X_n}P = \mathcal{U}_{{}_{X_n}(\tilde{P})((.)\setminus\{o\})}. $$

Hence in virtue of 1.6.15. ${}_{X_n}P$ is simple if and only if ${}_{X_n}(\tilde{P})\left((.)\setminus\{o\}\right)$ is simple and continuous. Obviously ${}_{X_n}(\tilde{P})\left((.)\setminus\{o\}\right)$ is continuous for all natural numbers n if and only if $\tilde{P}$ itself is continuous. ∎

Also the property $g_{\nu,X}$ can readily be expressed by means of the canonical measure. From 1.14.5. and 2.2.4. follows immediately

2.2.10. Proposition. *For all ν in N and all X in $\mathfrak{B}$, an infinitely divisible distribution P on $\mathfrak{M}$ has the property $g_{\nu,X}$ if and only if the same is true for the finite measure ${}_X(\tilde{P})\left((\cdot)\setminus\{o\}\right)$.*

The assertion of 1.14.12. may be read as follows:

2.2.11. Proposition. *If ν is a distribution on $\mathfrak{A}$, then an infinitely divisible distribution P on $\mathfrak{M}$ has the property g_ν if and only if*

$$ \tilde{P} = \sum_{k=1}^{\infty} \tilde{P}\left(\Psi(A) = k\right) (Q_\nu)^k. $$

By the help of 2.2.4. and 2.2.7. from 1.6.10. we obtain

2.2.12. Proposition. *Let be P an infinitely divisible distribution on $\mathfrak{M}$ and $X_1, \ldots, X_m$ a finite sequence of pairwise disjoint sets in $\mathfrak{A}$ which has the property $P\left(\Phi(X_1), \ldots, \Phi(X_m) < +\infty\right) = 1$. Then the following statements are equivalent:*

a) *The finite sequence ${}_{X_1}\Phi, \ldots, {}_{X_m}\Phi$ is independent with respect to P.*
b) *For all i, j satisfyivg $1 \leqq i < j \leqq m$, $\Phi(X_i)$ is independent of $\Phi(X_j)$ with respect to P.*
c) *For all i, j satisfying $1 \leqq i < j \leqq m$,*

$$ \tilde{P}\left(\Psi(X_i) > 0,\ \Psi(X_j) > 0\right) = 0. $$

Within $\boldsymbol{T}$, the distributions which are free from after-effects can be characterized as follows (cf. Lee [3]).

2.2.13. Proposition. *An infinitely divisible distribution P on $\mathfrak{M}$ is free from after-effects if and only if*

$$ \tilde{P}\ (\Psi \text{ is not of the form } n\delta_a;\ a \in A,\ n = 1, 2, \ldots) = 0 $$

is satisfied.

Proof. Let (X_n) be a monotone increasing sequence of sets in $\mathfrak{B}$ such that each X in $\mathfrak{B}$ is contained in some X_n. Then obviously P is free from after-effects if and only if this is true for all ${}_{X_n}P$. By means of 2.2.4. and 1.6.11. this is valid if and only if for all natural numbers n

$$0 = {}_{X_n}(\tilde{P})\left(\Psi^*(A) > 1\right) = \tilde{P}\left(\Psi^*(X_n) > 1\right),$$

i.e.

$$\tilde{P}\left(\Psi^*(A) \neq 1\right) = \tilde{P}\left(\Psi^*(A) > 1\right) = 0. \blacksquare$$

For all ν in N, Q_ν belongs to $\boldsymbol{W}$. In virtue of 2.2.13. $\mathscr{U}_{Q_\nu}$ is always free from after-effects. On the other hand by the help of 2.2.8. for all X in $\mathfrak{B}$ we obtain

$$(\mathscr{E}_{Q_\nu})_X = \mathscr{U}_{(Q_\nu)_X((.)\setminus\{o\})}$$

$$= e^{-\nu(X)} \sum_{n=0}^{\infty} \frac{\left(\nu(X)\right)^n}{n!} (\delta_1)^n = \pi_{\nu(X)},$$

and we see the validity of

2.2.14. Proposition. *For all ν in N*

$$\widetilde{P_\nu} = Q_\nu.$$

If $\nu = \sum_{n=1}^{\infty} \nu_n$ is any representation of some measure ν in N as a sum of a sequence (ν_n) of finite measures in N, then by 2.2.2. and 2.2.14. we obtain

$$\widetilde{\mathop{\ast}_{n=1}^{\infty} P_{\nu_n}} = \sum_{n=1}^{\infty} \widetilde{P_{\nu_n}} = \sum_{n=1}^{\infty} Q_{\nu_n} = Q_\nu = \widetilde{P_\nu}$$

i.e.

$$P_\nu = \mathop{\ast}_{n=1}^{\infty} P_{\nu_n} = \mathop{\ast}_{n=1}^{\infty} \mathscr{U}_{Q_{\nu_n}}.$$

In this way the existence of P_ν for infinite ν in N can be proved by direct construction without using the extension theorem 1.3.5.

According to 1.1.4. each measure W on $\mathfrak{M}$ having the property $W\left(\Psi(A) \neq 1\right) = 0$ is the image of some measure ν on $\mathfrak{A}$ with respect to the mapping $a \frown \delta_a$ of $[A, \mathfrak{A}]$ into $[M, \mathfrak{M}]$. Obviously W belongs to $\boldsymbol{W}$ if and only if ν is in N. Therefore by 2.2.14. we have

2.2.15. Proposition. *An infinitely divisible distribution P on $\mathfrak{M}$ is Poisson if and only if*

$$\tilde{P}\left(\Psi(A) \neq 1\right) = 0$$

is satisfied.

A detailed discussion of infinitely divisible distributions P on $\mathfrak{M}$ having the property that $\tilde{P}(\Psi(A) > 2) = 0$ is found in Milne & Westcott [1].

Sometimes the following theorem will prove useful (cf. Lee [3]):

2.2.16. Proposition. *Let (X_n) be a monotone increasing sequence of sets in $\mathfrak{B}$ having the union A, and P an infinitely divisible distribution on $\mathfrak{M}$. Subject to these assumptions, P is Poisson if and only if this is true for all P_{X_n}.*

Proof. We have only to show that the afore-mentioned condition is sufficient:

For all n, $P_{X_n} = \pi_{c_n}$, $c_n \geqq 0$, and $\widetilde{P_{X_n}} = \tilde{P}\big(\Psi(X_n) \in (.),\, \Psi(X_n) > 0\big)$. If we identify the one-dimensional distribution π_{c_n} with the corresponding point process with phase space $\{1\}$, we can apply 2.2.14. to obtain

$$\widetilde{\pi_{c_n}} = Q_{c_n\delta_1} = c_n\delta_{\delta_1}.$$

Consequently

$$\widetilde{\pi_{c_n}}\big(\Psi(\{1\}) > 1\big) = c_n\delta_{\delta_1}\big(\Psi(\{1\}) > 1\big) = 0$$

and hence

$$\widetilde{P_{X_n}}(\{2, 3, \ldots\}) = \tilde{P}\big(\Psi(X_n) > 1\big) = 0.$$

Thus we have

$$\tilde{P}\big(\Psi(A) \neq 1\big) = \tilde{P}\big(\Psi(A) > 1\big) = \sup_{n=1,2,\ldots} \tilde{P}\big(\Psi(X_n) > 1\big) = 0$$

so that our assertion is immediately deduced from 2.2.15. ∎

We conclude this section with the following consequence of 1.10.6., 2.2.4., and 2.2.5.

2.2.17. Proposition. *For all infinitely divisible distributions P_1, P_2 on $\mathfrak{M}$ and all X in $\mathfrak{A}$ with the property $P_1\big(\Phi(X) = 0\big)$, $P_2\big(\Phi(X) = 0\big) > 0$,*

$$_X\|P_1 - P_2\| \leqq 2\big\|{}_X(\tilde{P}_1)\,\big((.) \setminus \{o\}\big) - {}_X(\tilde{P}_2)\,\big((.) \setminus \{o\}\big)\big\|.$$

2.3. A Characterization of the Continuous Infinitely Divisible Distributions

For all P in $\boldsymbol{T}$, all X in $\mathfrak{A}$ with the property $P\big(\Phi(X) = 0\big) > 0$, and all natural numbers n, the following relation holds as a consequence of 1.5.6.:

$$P\big((.)\,|\,\Phi(X) = 0\big) = \Big(\sqrt[n]{P}\,\big((.)\,|\,\Phi(X) = 0\big)\Big)^n.$$

Thus the conditional distribution $P\big((.) \mid \Phi(X) = 0\big)$ is also infinitely divisible. Let us now determine $\widetilde{P\big((.) \mid \Phi(X) = 0\big)}$. We set

$$W_1 = \tilde{P}\big(\Psi \in (.),\, \Psi(X) = 0\big), \qquad W_2 = \tilde{P}\big(\Psi \in (.),\, \Psi(X) > 0\big).$$

The measures W_1, W_2 are in $\boldsymbol{W}$. Therefore there are distributions P_1, P_2 in $\boldsymbol{T}$ for which the relations $\widetilde{P_1} = W_1$, $\widetilde{P_2} = W_2$ are satisfied. In view of $W_1 + W_2 = \tilde{P}$, we obtain by means of 2.2.2.:

$$P = P_1 * P_2.$$

The definition of W_1 leads to

$$P_1\big(\Phi(X) = 0\big) = e^{-\tilde{P}_1(\Psi(X)>0)} = e^{-W_1(\Psi(X)>0)} = 1;$$

hence

$$P_1\big((.)\,|\,\Phi(X)=0\big)=P_1.$$

On the other hand we obtain

$$P_2\big(\Phi(A)=0\big)=e^{-\widetilde{P_2}(\Psi(A)>0)}=e^{-\tilde{P}(\Psi(A)>0,\,\Psi(X)>0)}=e^{-\tilde{P}(\Psi(X)>0)}$$

$$=e^{-\widetilde{P_2}(\Psi(X)>0)}=P_2\big(\Phi(X)=0\big),$$

which means that

$$P_2\big(\Phi\neq o,\,\Phi(X)=0\big)=0.$$

Using 1.5.6. we conclude that

$$P\big((.)\,|\,\Phi(X)=0\big)=(P_1*P_2)\big((.)\,|\,\Phi(X)=0\big)$$

$$=\Big(P_1\big((.)\,|\,\Phi(X)=0\big)\Big)*\Big(P_2\big((.)\,|\,\Phi(X)=0\big)\Big)=P_1*\delta_o=P_1,$$

and we observe that the following propositions are true.

2.3.1. Proposition. *For all infinitely divisible distributions P on $\mathfrak{M}$ and all sets X in $\mathfrak{A}$ satisfying the condition $P\big(\Phi(X)=0\big)>0$, the conditional distribution $P\big((.)\mid\Phi(X)=0\big)$ is infinitely divisible, and*

$$\widetilde{P\big((.)\,|\,\Phi(X)=0\big)}=\tilde{P}\big(\Psi\in(.),\,\Psi(X)=0\big).$$

2.3.2. *Subject to the assumptions of 2.3.1., there exists exactly one solution Q in $\boldsymbol{P}$ of the equation*

$$P=\Big(P\big((.)\,|\,\Phi(X)=0\big)\Big)*Q.$$

This distribution Q is infinitely divisible, being given by

$$\tilde{Q}=\tilde{P}\big(\Psi\in(.),\,\Psi(X)>0\big).$$

For each P in $\boldsymbol{P}$, let $\mathfrak{S}_P$ denote the system of those X in $\mathfrak{A}$ for which $P\big(\Phi(X)=0\big)>0$ is satisfied and for which there exists one — and hence exactly one — Q in $\boldsymbol{P}$ with the property that $P=\Big(P\big((.)\mid\Phi(X)=0\big)\Big)*Q$.

A continuous P in $\boldsymbol{P}$ is infinitely divisible if $\mathfrak{S}_P$ is sufficiently large:

2.3.3. Theorem. *Let $\mathfrak{H}$ be a subsystem of $\mathfrak{B}$. If each neighbourhood U of a point a in $[A,\varrho_A]$ includes a neighbourhood H of a which belongs to $\mathfrak{H}$, then a continuous distribution P on $\mathfrak{M}$ is infinitely divisible if and only if $\mathfrak{H}$ is included in $\mathfrak{S}_P$.*

Proof. 1. In virtue of 2.2.5. and 2.3.2., the inclusion relation $\mathfrak{B}\subseteqq\mathfrak{S}_P$ is satisfied for all P in $\boldsymbol{T}$.

2. Suppose that $P\in\boldsymbol{P}$ and $X_1,\ldots,X_n\in\mathfrak{S}_P$, and that in particular

$$P=\Big(P\big((.)\,|\,\Phi(X_i)=0\big)\Big)*Q_i\qquad(1\leqq i\leqq n).$$

Then $X_1 \cup \cdots \cup X_n$ also belongs to $\mathfrak{S}_P$, and

$$P = P\big((.)\,|\,\Phi(X_1 \cup \cdots \cup X_n) = 0\big) * \mathop{\ast}_{i=1}^{n} \Big(Q_i\big((.)\,|\,\Phi(X_{i+1} \cup \cdots \cup X_n) = 0\big)\Big).$$

We shall prove this by induction on n. For $n = 1$ the assertion is trivial. On the other hand the relation

$$P = P\big((.)\,|\,\Phi(X_1 \cup \cdots \cup X_{n-1}) = 0\big) * \mathop{\ast}_{i=1}^{n-1} \Big(Q_i\big((.)\,|\,\Phi(X_{i+1} \cup \cdots \cup X_{n-1}) = 0\big)\Big)$$

enables us first to conclude, using

$$\begin{aligned} 0 &< P\big(\Phi(X_n) = 0\big) \\ &= P\big(\Phi(X_n) = 0\,|\,\Phi(X_1 \cup \cdots \cup X_{n-1}) = 0\big) \\ &\quad \cdot \prod_{i=1}^{n-1} Q_i\big(\Phi(X_n) = 0\,|\,\Phi(X_{i+1} \cup \cdots \cup X_{n-1}) = 0\big), \end{aligned}$$

that

$$P\big(\Phi(X_1 \cup \cdots \cup X_n) = 0\big) > 0, \qquad Q_i\big(\Phi(X_{i+1} \cup \cdots \cup X_n) = 0\big) > 0 \qquad (1 \leqq i \leqq n-1),$$

and then, using 1.5.6., that

$$\begin{aligned} &P\big((.)\,|\,\Phi(X_n) = 0\big) \\ &= P\big((.)\,|\,\Phi(X_1 \cup \cdots \cup X_n) = 0\big) * \mathop{\ast}_{i=1}^{n-1} \Big(Q_i\big((.)\,|\,\Phi(X_{i+1} \cup \cdots \cup X_n) = 0\big)\Big). \end{aligned}$$

Consequently

$$\begin{aligned} P &= \Big(P\big((.)\,|\,\Phi(X_n) = 0\big)\Big) * Q_n \\ &= \Big(P\big((.)\,|\,\Phi(X_1 \cup \cdots \cup X_n) = 0\big)\Big) * \mathop{\ast}_{i=1}^{n} \Big(Q_i\big((.)\,|\,\Phi(X_{i+1} \cup \cdots \cup X_n) = 0\big)\Big). \end{aligned}$$

3. We shall suppose in addition that for some ε in $(0, 1)$

$$P\big(\Phi(X_i) > 0\big) < \varepsilon \qquad (1 \leqq i \leqq n).$$

If now any X in $\mathfrak{A}$ satisfies the conditions

$$X_1 \cup \cdots \cup X_n \subseteqq X, \qquad P\big(\Phi(X) = 0\big) > 0,$$

then we obtain

$$Q_i\big(\Phi(X) > 0\,|\,\Phi(X_{i+1} \cup \cdots \cup X_n) = 0\big) < \varepsilon \qquad (1 \leqq i \leqq n).$$

In fact, if we observe that $P\big(\Phi(X_1 \cup \cdots \cup X_n) = 0\big) > 0$ and use 1.5.6., $P = \Big(P\big((.)\mid\Phi(X_i) = 0\big)\Big) * Q_i$ yields the relation

$$\begin{aligned} &P\big((.)\,|\,\Phi(X_{i+1} \cup \cdots \cup X_n) = 0\big) \\ &= P\big((.)\,|\,(X_i \cup \cdots \cup X_n) = 0\big) * Q_i\big((.)\,|\,\Phi(X_{i+1} \cup \cdots \cup X_n) = 0\big) \end{aligned}$$

so that we obtain

$$Q_i\big(\Phi(X) = 0 \mid \Phi(X_{i+1} \cup \cdots \cup X_n) = 0\big)$$
$$= \frac{P\big(\Phi(X)=0, \Phi(X_{i+1} \cup \cdots \cup X_n)=0\big)}{P\big(\Phi(X_{i+1} \cup \cdots \cup X_n) = 0\big)} \cdot \frac{P\big(\Phi(X_i \cup \cdots \cup X_n) = 0\big)}{P\big(\Phi(X) = 0, \Phi(X_i \cup \cdots \cup X_n) = 0\big)}$$
$$= \frac{P\big(\Phi(X_i) = 0, \Phi(X_{i+1} \cup \cdots \cup X_n) = 0\big)}{P\big(\Phi(X_{i+1} \cup \cdots \cup X_n) = 0\big)}$$
$$= P\big(\Phi(X_i) = 0 \mid \Phi(X_{i+1} \cup \cdots \cup X_n) = 0\big).$$

However, since $X_{i+1} \cup \cdots \cup X_n$ belongs to $\mathfrak{S}_P$,

$$P\big(\Phi(X_i) = 0 \mid \Phi(X_{i+1} \cup \cdots \cup X_n) = 0\big) \geqq P\big(\Phi(X_i) = 0\big) > 1 - \varepsilon,$$

which leads to

$$Q_i\big(\Phi(X) = 0 \mid \Phi(X_{i+1} \cup \cdots \cup X_n) = 0\big) > 1 - \varepsilon$$

4. Now suppose that $\mathfrak{H}$ satisfy the afore-mentioned condition and be contained in $\mathfrak{S}_P$.

To each a in A we associate a neighbourhood H_a of a which belongs to $\mathfrak{H}$. Since $[A, \varrho_A]$ satisfies the second axiom of countability, there exists a sequence (a_n) of points in A such that A is covered by the sequence of the interior kernels of the sets H_{a_n}. By virtue of 2., all finite unions $H_{a_1} \cup \cdots \cup H_{a_n}$ are included in $\mathfrak{S}_P$. If we denote by G_n the union of the interiors of $H_{a_1}, \ldots, H_{a_n}$, then $P\big(\Phi(G_n) = 0\big) > 0$ for $n = 1, 2, \ldots$

Now we select an arbitrary m and any $\varepsilon > 0$. To each b in G_m we associate a neighbourhood L_b in $\mathfrak{S}_P$ which is included in G_m and satisfies $P\big(\Phi(L_b) > 0\big) < \varepsilon$. There exists a sequence (b_i) of points in G_m which has the property that $\bigcup_{i=1}^{\infty} L_{b_i} = G_m$. We set $G_m = X$ and $L_{b_i} = X_i$, choosing n so large that

$$P\left(\Phi\left(X \setminus \bigcup_{i=1}^{n} X_i\right) > 0\right) < \varepsilon$$

is satisfied. Since $\bigcup_{i=1}^{n} X_i$ is in $\mathfrak{S}_P$, it follows that

$$P\left(\Phi\left(X \setminus \bigcup_{i=1}^{n} X_i\right) > 0 \mid \Phi(X_1 \cup \cdots \cup X_n) = 0\right) < \varepsilon$$

which means that

$$P\big(\Phi(X) > 0 \mid \Phi(X_1 \cup \cdots \cup X_n) = 0\big) < \varepsilon. \qquad (*)$$

Using 2. we obtain

$$P = P\big((.) \mid \Phi(X_1 \cup \cdots \cup X_n) = 0\big) * \mathop{\Large\ast}_{i=1}^{n} Q_i\big((.) \mid \Phi(X_{i+1} \cup \cdots \cup X_n) = 0\big) \qquad (**)$$

where we set

$$P = \left(P\left((.)\,|\,\Phi(X_i) = 0\right)\right) * Q_i \qquad (1 \leqq i \leqq n).$$

Here we observe that, according to 3.,

$$Q_i\left(\Phi(X) > 0\,|\,\Phi(X_{i+1} \cup \cdots \cup X_n) = 0\right) < \varepsilon \qquad (1 \leqq i \leqq n). \qquad (***)$$

Now let $Z_1, \ldots, Z_k$ be a finite sequence of sets in $\mathfrak{B}$ which are contained in G_m. Using $(*)$, $(**)$, and $(***)$ we conclude that for all $\varepsilon > 0$, $P_{Z_1,\ldots,Z_k}$ can be represented as a finite convolution such that for each convolution factor the probability of the empty realization o is at least equal to $1 - \varepsilon$. Hence, in view of 1.11.5., $P_{Z_1,\ldots,Z_k}$ is infinitely divisible.

For each finite sequence $X_1, \ldots, X_m$ of sets in $\mathfrak{B}$, we obtain

$$P_{X_1 \cap G_n, \ldots, X_m \cap G_n} \xrightarrow[n\to\infty]{} P_{X_1,\ldots,X_m}.$$

Consequently by 1.11.6. all finite-dimensional distributions of P, and hence by 1.6.3. also P itself, are infinitely divisible. ∎

In view of 1.3.4. and 1.5.3. we have for each distribution P on $\mathfrak{M}$ which is free from after-effects, and for all X in $\mathfrak{A}$:

$$P = {}_{\bar{X}}P * {}_XP\,.$$

Therefore if $P\left(\Phi(X) = 0\right) > 0$ we obtain

$$P\left((.)\,|\,\Phi(X) = 0\right) = {}_{\bar{X}}P$$

as well as

$$P = \left(P\left((.)\,|\,\Phi(X) = 0\right)\right) * Q$$

where $Q = {}_XP$. Thus a set X in $\mathfrak{A}$ belongs to $\mathfrak{S}_P$ if and only if $P\left(\Phi(X) = 0\right) > 0$ is satisfied. Now if P is continuous, then all X in $\mathfrak{B}$ which are included in sufficiently small neighbourhoods U_a of the points a in A satisfy this condition. Hence 2.3.3. is a generalization of proposition 1.11.7.

A distribution P of the form $\frac{1}{2}(\delta_o + \delta_{\delta_a})$ is free from after-effects and satisfies the condition $P(\Phi(X) = 0) > 0$ for all A in $\mathfrak{A}$. Nevertheless P is not infinitely divisible. Therefore the assumption of the continuity of P cannot be dropped in either 2.3.3. or 1.11.7.

2.4. Parametrized Semi-groups of Distributions on $\mathfrak{M}$

For each P in $\boldsymbol{T}$, the measures $t\tilde{P}$ belong to $\boldsymbol{W}$ for all $t \geqq 0$, so that, using 2.2.1., we can introduce distributions $P(t)$ in $\boldsymbol{T}$ by means of $\widetilde{P(t)} = t\tilde{P}$. Using 2.2.2. we obtain

$$(*) \qquad P(1) = P, \qquad P(t_1) * P(t_2) = P(t_1 + t_2) \qquad \text{for} \qquad t_1, t_2 \geqq 0\,.$$

Thus we have embedded P in a parametrized semi-group of distributions on $\mathfrak{M}$.

In the special case of Poisson distributions, using 2.2.14. we obtain

$$\widetilde{P_\nu(t)} = tQ_\nu = Q_{t\nu} = \widetilde{P_{t\nu}}$$

for all ν in N, i.e.

$$P_\nu(t) = P_{t\nu} \qquad (t \geqq 0).$$

Conversely, let P be a distribution on $\mathfrak{M}$ and $P(.)$ a mapping of $[0, +\infty)$ into $\boldsymbol{P}$ with the properties $(*)$. In view of

$$\left(P\left(\frac{t}{n}\right)\right)^n = P(t) \qquad (t \geqq 0;\ n = 1, 2, \ldots),$$

all distributions $P(t)$, $t \geqq 0$, are infinitely divisible, and by 2.2.2. it follows that, for all Y in $\mathfrak{M}$, all X in $\mathfrak{B}$, and all $t_1, t_2 \geqq 0$,

$$\widetilde{P(t_1 + t_2)}\big(\Psi \in Y,\ \Psi(X) > 0\big)$$
$$= \widetilde{P(t_1)}\big(\Psi \in Y, \Psi(X) > 0\big) + \widetilde{P(t_2)}\big(\Psi \in Y, \Psi(X) > 0\big).$$

Finally, noting that $\widetilde{P(t)}\big(\Psi \in Y, \Psi(X) > 0\big)$ is non-negative and finite for all $t \geqq 0$, we see that the following relation holds:

$$\widetilde{P(t)}\big(\Psi \in Y, \Psi(X) > 0\big) = t\tilde{P}\big(\Psi \in Y, \Psi(X) > 0\big).$$

Now, letting X run through a monotone increasing sequence (X_n) with the union A, it follows that

$$\widetilde{P(t)} = t\tilde{P} \qquad (t \geqq 0),$$

and we obtain

2.4.1. *A distribution P on $\mathfrak{M}$ is infinitely divisible if and only if there exists a mapping $P(.)$ of $[0, +\infty)$ into $\boldsymbol{P}$ with the properties $(*)$. This mapping is uniquely determined:*

$$P(t) \in \boldsymbol{T} \quad \textit{and} \quad \widetilde{P(t)} = t\tilde{P} \quad \textit{for all } t \geqq 0.$$

For all P in $\boldsymbol{T}$ and all non-negative real t, we denote by P^t the distribution $P(t)$ which is uniquely determined according to 2.4.1. As an immediate consequence of 2.2.8. and 2.2.4., we obtain

$$(P^t)_{X_1,\ldots,X_m} = (P_{X_1,\ldots,X_m})^t \qquad (X_1, \ldots, X_m \in \mathfrak{B})$$

as well as

$$_X(P^t) = (_XP)^t \qquad (X \in \mathfrak{A}).$$

For each P in $\boldsymbol{T}$ and each finite sequence $X_1, \ldots, X_m$ of pairwise disjoint sets in $\mathfrak{B}$, according to 2.2.8. we have, for all non-negative integers $l_1, \ldots, l_m$

and all $t \geqq 0$:

$$
\begin{aligned}
(P^t)_{X_1,\ldots,X_m}(\{[l_1,\ldots,l_m]\}) &= \mathcal{U}_{(\tilde{P}^t)_{X_1,\ldots,X_m}((.)\setminus\{o\})}(\{[l_1,\ldots,l_m]\}) \\
&= U_{t(\tilde{P})_{X_1,\ldots,X_m}((.)\setminus\{o\})}(\{[l_1,\ldots,l_m]\}) \\
&= \exp\Big(-t\tilde{P}\big(\Psi(X_1\cup\cdots\cup X_m)>0\big)\Big) \\
&\quad\cdot\sum_{n=0}^{\infty}\frac{t^n}{n!}\Big(\big((\tilde{P})_{X_1,\ldots,X_m}((.)\setminus\{o\})\big)^n(\{[l_1,\ldots,l_m]\})\Big).
\end{aligned}
$$

Hence, using 1.8.1. we see that the following is true:

2.4.2. *For all P in $\boldsymbol{T}$, the mapping $t \curvearrowright P^t$ of $[0, +\infty)$ into $\boldsymbol{P}$ is measurable with respect to the σ-algebra $\mathfrak{P}$.*

For all distributions P in $\boldsymbol{T}$ and all measures ω which are defined on the σ-Algebra $\mathfrak{R}$ of Borel subsets of the real axis and satisfy the condition that $\omega\big((-\infty, 0)\big) = 0$, we can use 2.4.2. to introduce the measure

$$V = \int P^t(.)\,\omega(\mathrm{d}t)$$

on $\mathfrak{M}$. As a generalization of 1.8.3., we have

2.4.3. Proposition. *If for an at most countable family $(\sigma_i)_{i\in I}$ of distributions of random non-negative numbers the convolution $\sigma = \mathop{\ast}\limits_{i\in I} \sigma_i$ exists, then*

$$\mathop{\ast}_{i\in I}\int P^t(.)\,\sigma_i(\mathrm{d}t) = \int P^t(.)\,\sigma(\mathrm{d}t)$$

for all P in $\boldsymbol{T}$.

Proof. As it was done in the proof of 1.8.3., one shows the validity of 2.4.3. for finite sets of indices I.

Now let I be infinite so that we are allowed to set $I = \{1, 2, \ldots\}$ without loss of generality. Given any set X in $\mathfrak{B}$, we obtain for all natural numbers m:

$$
\begin{aligned}
{}_X\Big(\int P^t(.)\,\sigma(\mathrm{d}t)\Big) &= \int {}_X(P^t)\,(.)\,\sigma(\mathrm{d}t) \\
&= \int {}_X(P^t)\,(.)\Big(\Big(\mathop{\ast}_{n=1}^{m}\sigma_n\Big) * \Big(\mathop{\ast}_{n=m+1}^{\infty}\sigma_n\Big)\Big)(\mathrm{d}t) \\
&= \Big(\mathop{\ast}_{n=1}^{m}\int {}_X(P^t)\,(.)\,\sigma_n(\mathrm{d}t)\Big) * \int {}_X(P^t)\,(.)\Big(\mathop{\ast}_{n=m+1}^{\infty}\sigma_n\Big)(\mathrm{d}t) \\
&= {}_X\Big(\mathop{\ast}_{n=1}^{m}\int P^t(.)\,\sigma_n(\mathrm{d}t)\Big) * \int {}_X(P^t)\,(.)\Big(\mathop{\ast}_{n=m+1}^{\infty}\sigma_n\Big)(\mathrm{d}t).
\end{aligned}
$$

As m increases, $\mathop{\ast}\limits_{n=m+1}^{\infty}\sigma_n$ tends weakly toward δ_0. Hence

$$
\begin{aligned}
&\int {}_X(P^t)\,\big(\Phi(X)=0\big)\Big(\mathop{\ast}_{n=m+1}^{\infty}\sigma_n\Big)(\mathrm{d}t) \\
&= \int \exp\Big(-t\tilde{P}\big(\Psi(X)>0\big)\Big)\Big(\mathop{\ast}_{n=m+1}^{\infty}\sigma_n\Big)(\mathrm{d}t) \xrightarrow[m\to\infty]{} 1,
\end{aligned}
$$

which means that

$$\underset{X}{\left\| \int P^t(.)\,\sigma(\mathrm{d}t) - \mathop{\ast}_{n=1}^{m} \int P^t(.)\,\sigma_n(\mathrm{d}t) \right\|} \xrightarrow[m\to\infty]{} 0.$$

But from this we can conclude that

$$\int P^t(.)\,\sigma(\mathrm{d}t) = \mathop{\ast}_{n=1}^{\infty} \int P^t(.)\,\sigma_n(\mathrm{d}t). \blacksquare$$

In virtue of 2.4.3., for each P in $\boldsymbol{T}$ and each infinitely divisible distribution σ on $\mathfrak{R}$ with the property that $\sigma\big((-\infty, 0)\big) = 0$, the distribution $\int P^t(.)\,\sigma(\mathrm{d}t)$ is infinitely divisible.

For all finite measures ω on $\mathfrak{R}$ satisfying the condition $\omega\big((-\infty, 0)\big) = 0$, the definition

$$\mathcal{U}_\omega = e^{-\omega([0,+\infty))} \sum_{n=0}^{\infty} \frac{\omega^n}{n!}$$

provides an infinitely divisible distribution $\mathcal{U}_\omega$ on $\mathfrak{R}$ with the property $\mathcal{U}_\omega\big((-\infty, 0)\big) = 0$. Obviously $\mathcal{U}_\omega = \mathcal{U}_{\omega((.)\setminus\{0\})}$. Now, if P is a distribution in $\boldsymbol{T}$, then by 2.4.3. we obtain

$$\begin{aligned}\int P^t(.)\,\mathcal{U}_\omega(\mathrm{d}t) &= e^{-\omega([0,+\infty))} \sum_{n=0}^{\infty} \frac{1}{n!} \int P^t(.)\,\omega^n(\mathrm{d}t)\\ &= e^{-\omega([0,+\infty))} \sum_{n=0}^{\infty} \frac{1}{n!} \left(\int P^t(.)\,\omega(\mathrm{d}t)\right)^n\\ &= \mathcal{U}_{\int P^t(.)\omega(\mathrm{d}t)},\end{aligned}$$

i.e. we have (cf. Lee [3]):

2.4.4. *For all P in $\boldsymbol{T}$ and all finite measures ω on $\mathfrak{R}$ which satisfy the condition $\omega\big((-\infty, 0)\big) = 0$, the distribution*

$$V = \int P^t(.)\,\mathcal{U}_\omega(\mathrm{d}t)$$

is infinitely divisible, and

$$\tilde{V} = \int P^t\big((.) \setminus \{o\}\big)\,\omega(\mathrm{d}t).$$

In analogy to the statements made in section 2.1., one can investigate infinite convolution products $\mathop{\ast}_{i\in I} \mathcal{U}_{\omega_i}$, $\omega_i(\{0\}) = 0$. It turns out (see, for instance, Kummer & Matthes [1], [2]) that $\mathop{\ast}_{i\in I} \mathcal{U}_{\omega_i}$ exists if and only if the measure $\omega = \sum_{i\in I} \omega_i$ satisfies the two following conditions:

$$(*) \qquad \omega\big([1, +\infty)\big) < +\infty, \qquad \int_{(0,1)} x\omega(\mathrm{d}x) < +\infty.$$

Also in this case $\mathop{\ast}_{i\in I} \mathcal{U}_{\omega_i}$ depends only on the sum ω but not on the selected representation of ω as a sum of finite measures, so that for all measures ω

on $\mathfrak{R}$ satisfying the conditions

$$(**) \quad \omega\big([1, +\infty)\big) < +\infty, \quad \int\limits_{(0,1)} x\omega(\mathrm{d}x) < +\infty, \quad \omega\big((-\infty, 0]\big) = 0,$$

we can construct the infinitely divisible distribution $\mathscr{E}_\omega$.

It is known (see, for instance, Feller [2], XVII, § 3) that the following proposition is true:

2.4.5. Proposition. *For each infinitely divisible distribution γ on $\mathfrak{R}$ with the property that $\gamma\big((-\infty, 0)\big) = 0$, there exist exactly one non-negative real number x_γ and exactly one measure $\tilde{\gamma}$ on $\mathfrak{R}$ satisfying the conditions (**), such that*

$$\gamma = \delta_{x_\gamma} * \mathscr{E}_{\tilde{\gamma}}$$

is satisfied.

Now, if P is a distribution in $\boldsymbol{T}$, and γ an infinitely divisible distribution on $\mathfrak{R}$ with the property that $\gamma\big((-\infty, 0)\big) = 0$, then by 2.2.2. we obtain for each representation $\tilde{\gamma} = \sum_{n=1}^{\infty} \omega_n$ of $\tilde{\gamma}$ as a sum of finite measures:

$$\widetilde{\int P^t(.)\, \gamma(\mathrm{d}t)} = \widetilde{\int P^t(.)\, (\delta_{x_\gamma} * \mathscr{E}_{\tilde{\gamma}})\, (\mathrm{d}t)}$$

$$= \widetilde{P^{x_\gamma}} + \widetilde{\int P^t(.)\Big(\mathop{\Large *}_{n=1}^{\infty} \mathscr{U}_{\omega_n}\Big)(\mathrm{d}t)}.$$

Using 2.4.3. and 2.4.4., we are able to continue this chain of equations as follows:

$$= x_\gamma \tilde{P} + \widetilde{\mathop{\Large *}_{n=1}^{\infty} \int P^t(.)\, \mathscr{U}_{\omega_n}(\mathrm{d}t)}$$

$$= x_\gamma \tilde{P} + \sum_{n=1}^{\infty} \widetilde{\int P^t(.)\, \mathscr{U}_{\omega_n}(\mathrm{d}t)}$$

$$= x_\gamma \tilde{P} + \sum_{n=1}^{\infty} \int P^t\big((.) \setminus \{o\}\big)\, \omega_n(\mathrm{d}t)$$

$$= x_\gamma \tilde{P} + \int P^t\big((.) \setminus \{o\}\big)\, \tilde{\gamma}(\mathrm{d}t),$$

so that we obtain

2.4.6. Proposition. *For all P in $\boldsymbol{T}$ and all infinitely divisible distributions γ on $\mathfrak{R}$ with the property that $\gamma\big((-\infty, 0)\big) = 0$, the distribution $V = \int P^t(.)\, \gamma(\mathrm{d}t)$ is infinitely divisible, and*

$$\tilde{V} = x_\gamma \tilde{P} + \int P^t\big((.) \setminus \{o\}\big)\, \tilde{\gamma}(\mathrm{d}t).$$

From 2.4.6., 1.14.10., and 2.4.5., we readily obtain

2.4.7. *For all infinite measures ν in N, a distribution P in $\boldsymbol{T}$ has the property g_ν if and only if there exist a non-negative real l as well as a measure ω*

on $\mathfrak{R}$ *satisfying the condition* (**), *such that*

$$\tilde{P} = Q_{l\nu} + \int P_{t\nu}(.)\, \omega(dt)$$

is satisfied, in which case the l, ω *are uniquely determined.*

The following example shows that the set of distributions of the form $\mathcal{U}_E$, $E \in \boldsymbol{E}^+$, is not closed with respect to the variation distance.

In the case $[A, \varrho_A] = R$ we construct a sequence $\mathcal{U}_{E_n}$, $E_n \in \boldsymbol{E}^+$, having the properties

$$\|\mathcal{U}_{E_n} - P\| \xrightarrow[n\to\infty]{} 0, \qquad P(\{o\}) = 0.$$

Let be $P_n = \int P_{l\mu}(.)\, \mathcal{U}_{\sum\limits_{m=1}^{n} \nu_m/m}(dl)$ and $P = \int P_{l\mu}(.)\, \mathcal{E}_{\sum\limits_{m=1}^{\infty} \nu_m/m}(dl)$ where we set $\nu_m = m\mu((.) \cap (0, 1/m))$. Here μ denotes the one-dimensional Lebesgue measure.

The distribution P_n has the form $\mathcal{U}_{E_n}$, $E_n \in \boldsymbol{E}^+$, for we have

$$\int P_{l\mu}(.)\, \mathcal{U}_{\sum\limits_{m=1}^{n} \nu_m/m}(dl) = \mathcal{U}_{\int P_{l\mu}(.)\left(\sum\limits_{m=1}^{n} \nu_m/m\right)(dl)}.$$

On the other hand,

$$P(\{o\}) = \mathcal{E}_{\int P_{l\mu}(.)\left(\sum\limits_{m=1}^{\infty} \nu_m/m\right)(dl)}(\{o\}) = 0.$$

We have to show that the convergence $\|P_n - P\| \xrightarrow[n\to\infty]{} 0$ takes place. Because of

$$\left\|\int P_{l\mu}(.)\, \sigma_1(dl) - \int P_{l\mu}(.)\, \sigma_2(dl)\right\| \leqq \mathrm{Var}\,(\sigma_1 - \sigma_2)$$

it is sufficient to show that

$$\mathrm{Var}\left(\mathcal{U}_{\sum\limits_{m=1}^{n} \nu_m/m} - \mathcal{U}_{\sum\limits_{m=1}^{n} \nu_m/m} * \mathcal{E}_{\sum\limits_{m=n+1}^{\infty} \nu_m/m}\right) \xrightarrow[n\to\infty]{} 0.$$

Let be m_0 fixed and $n > m_0$. We consider the distribution $\mathcal{U}_{\sum\limits_{m=1}^{n} \nu_m/m}$ and split it as follows:

$$\mathcal{U}_{\sum\limits_{m=1}^{n} \nu_m/m} = \mathcal{U}_{\sum\limits_{m=1}^{m_0} \nu_m/m} * \mathcal{U}_{\sum\limits_{m=m_0+1}^{n} \nu_m/m}$$

$$= \left(e^{-\sum\limits_{m=1}^{m_0} 1/m}\, \delta_0 + e^{-\sum\limits_{m=1}^{m_0} 1/m} \sum_{k=1}^{\infty} \frac{1}{k!} \left(\sum_{m=1}^{m_0} \nu_m/m\right)^k\right) * \mathcal{U}_{\sum\limits_{m=m_0+1}^{n} \nu_m/m}.$$

Setting $q_n = e^{-\sum\limits_{m=1}^{m_0} 1/m} \sum\limits_{k=1}^{\infty} \frac{1}{k!} \left(\sum\limits_{m=1}^{m_0} \nu_m/m\right)^k$, $r_n = \mathcal{U}_{\sum\limits_{m=m_0+1}^{n} \nu_m/m}$ and $s_n = \mathcal{E}_{\sum\limits_{m=n+1}^{\infty} \nu_m/m}$, we obtain

$$\mathrm{Var}\left(\mathcal{U}_{\sum\limits_{m=1}^{n} \nu_m/m} - \mathcal{U}_{\sum\limits_{m=1}^{n} \nu_m/m} * s_n\right)$$

$$\leqq 2e^{-\sum\limits_{m=1}^{m_0} 1/m} + \mathrm{Var}\,(q_n * r_n - q_n * r_n * s_n)$$

$$\leqq 2e^{-\sum\limits_{m=1}^{m_0} 1/m} + \mathrm{Var}\,(q_n - q_n * s_n).$$

The measure $q_n - q_n * s_n$ can be written down in the form

$$\sum_{k=1}^{m_0} (\nu_k - \nu_k * s_n) * t_{k,n}$$

where Var $t_{k,n} \leqq 1$. Now we continue the inequality above with

$$\leqq 2e^{-\sum_{m=1}^{m_0} 1/m} + \sum_{k=1}^{m_0} \mathrm{Var}\,(\nu_k - \nu_k * s_n).$$

The mean value of the distribution $\mathscr{E}_{\sum_{m=n+1}^{\infty} \nu_m/m}$ is equal to $\sum_{m=n+1}^{\infty} \int_0^{1/m} x\mu(\mathrm{d}x) = \sum_{m=n+1}^{\infty} \frac{1}{2m^2}$.

Let be $\varepsilon > 0$. We choose an m_0 so that $e^{-\sum_{m=1}^{m_0} 1/m} < \varepsilon/4$. Since the mean value of $\mathscr{E}_{\sum_{m=n+1}^{\infty} \nu_m/m}$ converges towards 0 with increasing n, we can choose an $n_0 > m_0$ such that for all $n \geqq n_0$

$$\sum_{k=1}^{m_0} \mathrm{Var}\,(\nu_k - \nu_k * s_n) < \varepsilon/2,$$

and we obtain

$$\mathrm{Var}\left(\mathscr{U}_{\sum_{m=1}^{n} \nu_m/m} - \mathscr{E}_{\sum_{m=1}^{\infty} \nu_m/m}\right) < \varepsilon \quad \text{for} \quad n \geqq n_0.$$

Therefore, theorem 1.11.5. cannot be strengthened in that direction that the limit distribution P satisfies the condition $P(\{o\}) > 0$.

2.5. Regular and Singular Infinitely Divisible Distributions

Let z be an arbitrary point in A. Any P in $\boldsymbol{T}$ is said to be *regular infinitely divisible* if for all natural numbers k,

$$P\big(\Phi(S_k(z)) = 0 \mid \Phi(S_{n+m}(z) \setminus S_n(z)) = 0\big) \xrightarrow[n,m\to\infty]{} P\big(\Phi(S_k(z)) = 0\big)$$

Here $S_k(z)$ denotes the open ball of radius k around z. Obviously each P in $\boldsymbol{T}$ which is from after-effects is regular infinitely divisible.

The regular infinitely divisible distributions P can be characterized within $\boldsymbol{T}$ as follows:

2.5.1. Proposition. *An infinitely divisible distribution P on $\mathfrak{M}$ is regular infinitely divisible if and only if*

$$\tilde{P}\big(\Psi(A) = +\infty\big) = 0$$

is satisfied.

Proof. First let us suppose that P be regular infinitely divisible. By 2.3.1. and 2.2.5. we have

$$P\big(\Phi(S_k(z)) = 0 \mid \Phi(S_{n+m}(z) \setminus S_n(z)) = 0\big)$$
$$= e^{-\tilde{P}(\Psi(S_k(z))>0,\Psi(S_{n+m}(z)\setminus S_n(z))=0)}$$

as well as

$$P\big(\Phi(S_k(z)) = 0\big) = e^{-\tilde{P}(\Psi(S_k(z))>0)}.$$

Due to

$$\tilde{P}\big(\Psi(S_k(z)) > 0\big) - \tilde{P}\big(\Psi(S_k(z)) > 0,\ \Psi(S_{n+m}(z) \setminus S_n(z)) = 0\big)$$
$$= \tilde{P}\big(\Psi(S_k(z)) > 0,\ \Psi(S_{n+m}(z) \setminus S_n(z)) > 0\big)$$

the following convergence relation results:

$$\tilde{P}\big(\Psi(S_k(z)) > 0,\ \Psi(S_{n+m}(z) \setminus S_n(z)) > 0\big) \xrightarrow[n,m\to\infty]{} 0.$$

In view of

$$\tilde{P}\big(\Psi(S_k(z)) > 0,\ \Psi(S_{n+m}(z) \setminus S_n(z)) > 0\big)$$
$$\xrightarrow[m\to\infty]{} \tilde{P}\big(\Psi(S_k(z)) > 0,\ \Psi(A \setminus S_n(z)) > 0\big),$$

we obtain that

$$\tilde{P}\big(\Psi(S_k(z)) > 0,\ \Psi(A \setminus S_n(z)) > 0\big) \xrightarrow[n\to\infty]{} 0.$$

But we have, for all n

$$\tilde{P}\big(\Psi(S_k(z)) > 0,\ \Psi(A) = +\infty\big) \leqq P\big(\Psi(S_k(z)) > 0,\ \Psi(A \setminus S_n(z)) > 0\big).$$

Consequently

$$\tilde{P}\big(\Psi(S_k(z)) > 0,\ \Psi(A) = +\infty\big) = 0 \qquad (k = 1, 2, \ldots),$$

from which in view of

$$\tilde{P}\big(\Psi(A) = +\infty\big) = \sup_{k=1,2,\ldots} \tilde{P}\big(\Psi(S_k(z)) > 0,\ \Psi(A) = +\infty\big)$$

it can be concluded that

$$\tilde{P}\big(\Psi(A) = +\infty\big) = 0.$$

Conversely, suppose that $\tilde{P}\big(\Psi(A) = +\infty\big) = 0$. Using 2.3.1. and 2.3.2., we obtain for all X in $\mathfrak{B}$:

$$\begin{aligned} a(k, X) &= {}_{S_k(z)}\big\|P\big((.)\,|\,\Phi(X) = 0\big) - P\big\| \\ &= {}_{S_k(z)}\|P_1 - P_1 * P_2\| \\ &= \|{}_{S_k(z)}(P_1) - {}_{S_k(z)}(P_1) * {}_{S_k(z)}(P_2)\| \\ &\leqq \|\delta_o - {}_{S_k(z)}(P_2)\| = 2_{S_k(z)}(P_2)\,(\Phi \neq o) \\ &= 2P_2\big(\Phi(S_k(z)) \neq 0\big) = 2\big(1 - e^{-\tilde{P}(\Psi(S_k(z))>0,\Psi(X)>0)}\big), \end{aligned}$$

where we substituted $P_1 = P\big((.)\mid \Phi(X) = 0\big)$, and where P_2 is determined by $P = P_1 * P_2$.

If we suppose in addition that $X \cap S_n(z) = \emptyset$, then we obtain

$$a(k, X) \leqq 2\big(1 - e^{-\tilde{P}(\Psi(S_k(z))>0,\Psi(A\setminus S_n(z))>0)}\big).$$

Considering that

$$\tilde{P}\big(\Psi(S_k(z)) > 0, \Psi(A \setminus S_n(z)) > 0\big) \xrightarrow[n\to\infty]{} \tilde{P}\big(\Psi(S_k(z)) > 0, \Psi(A) = +\infty\big),$$

we observe that for sufficiently large n,

$$a(k, n) = \sup_{\substack{X \in \mathfrak{B} \\ X \cap S_n(z) = \emptyset}} {}_{S_k(z)}\big\|P\big((.) \mid \Phi(X) = 0\big) - P\big\| < \varepsilon$$

and hence also

$$\sup_{\substack{X \in \mathfrak{B} \\ X \cap S_n(z) = \emptyset}} \big|P\big(\Phi(S_k(z)) = 0 \mid \Phi(X) = 0\big) - P\big(\Phi(S_k(z)) = 0\big)\big| < \varepsilon.$$

Consequently

$$P\big(\Phi(S_k(z)) = 0 \mid \Phi(S_{n+m}(z) \setminus S_n(z)) = 0\big) \xrightarrow[n,m\to\infty]{} P\big(\Phi(S_k(z)) = 0\big). \blacksquare$$

It is immediately observed from 2.5.1. that the arbitrary selection of the point z in A does not play any role in the definition of regular infinitely divisible distributions.

For regular infinitely divisible distributions, conditions of the form $\Phi(X) = 0$ have a uniformly reduced effectiveness on a fixed bounded set Z if it is only ensured that X is outside a sufficiently large, bounded neighbourhood of Z:

2.5.2. *An infinitely divisible distribution P on $\mathfrak{M}$ is regular infinitely divisible if and only if for each Z in $\mathfrak{B}$ and each $\varepsilon > 0$ there exists in $\mathfrak{B}$ some U including Z such that*

$$\sup_{\substack{X \in \mathfrak{B} \\ X \cap U = \emptyset}} {}_Z\big\|P\big((.) \mid \Phi(X) = 0\big) - P\big\| \leqq \varepsilon$$

is satisfied.

Proof. Let P be regular infinitely divisible and $S_k(z) \supseteqq Z$. Then for all X in $\mathfrak{B}$ satisfying $X \cap S_n(z) = \emptyset$,

$${}_Z\big\|P\big((.) \mid \Phi(X) = 0\big) - P\big\| \leqq {}_{S_k(z)}\big\|P\big((.) \mid \Phi(X) = 0\big) - P\big\| \leqq a(k, n)$$

where $a(k, n)$ has the same meaning as in the proof of 2.5.1. But as we observed in the second step of the proof of 2.5.1., $a(k, n)$ tends toward zero as n increases. Consequently the condition stated above is necessary.

Conversely, if this condition is satisfied, we set $Z = S_k(z)$, and first choose n so large that $U \subseteqq S_n(z)$. Then we obtain that, for all natural numbers k, $a(k, n) \xrightarrow[n\to\infty]{} 0$. Consequently P is regular infinitely divisible. $\blacksquare$

By virtue of 2.2.7., any P in $\boldsymbol{T}$ is certainly regular infinitely divisible whenever $P(\Phi$ is finite$) = 1$ is satisfied.

A distribution P in $\boldsymbol{T}$ is called *singular infinitely divisible* if for all natural numbers k,

$$P\big(\Phi(S_k(z)) = 0 \mid \Phi(S_{k+m}(z) \setminus S_k(z)) = 0\big) \xrightarrow[m\to\infty]{} 1.$$

As a counterpart to 2.5.1., we shall now prove

2.5.3. Proposition. *An infinitely divisible distribution P on $\mathfrak{M}$ is singular infinitely divisible if and only if*

$$\tilde{P}(\Psi(A) < +\infty) = 0$$

is satisfied.

Proof. In view of 2.3.1. and 2.2.5. we have, for all P in $\boldsymbol{T}$,

$$P\big(\Phi(S_k(z)) = 0 \mid \Phi(S_{k+m}(z) \setminus S_k(z)) = 0\big) = e^{-\tilde{P}(\Psi(S_k(z))>0,\Psi(S_{k+m}(z)\setminus S_k(z))=0))}.$$

Consequently the distribution P is singular infinitely divisible if and only if the convergence relations

$$\tilde{P}\big(\Psi(S_k(z)) > 0,\ \Psi(S_{k+m}(z) \setminus S_k(z)) = 0\big) \xrightarrow[m\to\infty]{} 0 \qquad (k = 1, 2, \ldots)$$

are valid. But we have

$$\tilde{P}\big(\Psi(S_k(z)) > 0,\ \Psi(S_{k+m}(z) \setminus S_k(z)) = 0\big)$$
$$\xrightarrow[m\to\infty]{} \tilde{P}\big(\Psi(S_k(z)) > 0,\ \Psi(A \setminus S_k(z)) = 0\big).$$

On the other hand, we obtain

$$\tilde{P}\big(\Psi(S_k(z)) > 0, \Psi(A \setminus S_k(z)) = 0\big) \xrightarrow[k\to\infty]{} \tilde{P}(\Psi(A) < +\infty)$$

from which our assertion can be immediately inferred. ∎

For all E in $\boldsymbol{E}^+$ having the property that $E(\Psi(A) < +\infty) = 0$, $\mathcal{U}_E$ is singular infinitely divisible. Obviously δ_o is the only distribution on $\mathfrak{M}$ which is both regular and singular infinitely divisible.

Using 2.2.2., the following proposition can be inferred from 2.5.1. and 2.5.3.:

2.5.4. Proposition. *Each infinitely divisible distribution P on $\mathfrak{M}$ can be represented in exactly one way as a convolution $P = P_r * P_s$ of a regular infinitely divisible P_r and a singular infinitely divisible P_s, and*

$$\widetilde{P_r} = \tilde{P}(\Psi \in (.), \Psi(A) < +\infty),$$
$$\widetilde{P_s} = \tilde{P}(\Psi \in (.), \Psi(A) = +\infty).$$

3. Weak Convergence of Distributions

3.0. Introduction

The convergence of distributions on $\mathfrak{M}$ with respect to the variation distance, which was studied in Section 1.9., is extremely strong. So, for instance, a sequence of the form $(\delta_{\delta_{a_n}})$ converges if and only if the a_n are constant for all indices above a certain n_0. However, weakening the convergence based on the variation distance will enable us to introduce a convenient concept of convergence according to which $(\delta_{\delta_{a_n}})$ tends toward δ_{δ_a} if and only if (a_n) converges toward a with respect to the metric ϱ_A.

A sequence (P_n) of distributions on $\mathfrak{M}$ is said to be weakly convergent toward P, — written as $P_n \underset{n\to\infty}{\Rightarrow} P$ — if the convergence relation $\lim\limits_{n\to\infty} \|(P_n)_{X_1,\ldots,X_m} - P_{X_1,\ldots,X_m}\| = 0$ holds for all finite sequences $X_1, \ldots, X_m$ of bounded sets in $\mathfrak{A}$ which satisfy the condition $P(\Phi$ (boundary of $X_i) > 0) = 0$ for $i = 1, \ldots, m$. The restriction to sets X_i having the property that $P\big(\Phi$ (boundary of $X_i) > 0\big) = 0$ is necessary because otherwise we could not conclude from $a_n \xrightarrow[n\to\infty]{} a$ that $\delta_{\delta_{a_n}} \underset{n\to\infty}{\Rightarrow} \delta_{\delta_a}$. If the phase space $[A, \varrho_A]$ coincides with the real axis R^1, it will suffice to consider bounded intervals $X_i = [\alpha_i, \beta_i)$. The condition $P(\Phi$ (boundary of $X_i) > 0)$ $= 0$ then takes the form $P\big(\Phi(\{\alpha_i\}) > 0\big) = 0 = P\big(\Phi(\{\beta_i\}) > 0\big)$. In this form, the concept of weak convergence has already been used for a longer time (see, for instance, Kerstan & Matthes [1], [4]).

By way of the finite-dimensional distributions, the results of Section 1.9. indirectly form the basis for the limit theorems derived in this chapter. As was already announced in section 1.0., a distribution P on $\mathfrak{M}$ is infinitely divisible if and only if there exists an infinitesimal triangular array $(P_{n,i})_{\substack{n=1,2,\ldots \\ 1\leqq i\leqq m_n}}$ such that $\underset{1\leqq i\leqq m_n}{\ast} P_{n,i} \underset{n\to\infty}{\Rightarrow} P$ (Proposition 3.4.1.) where in view of theorem 1.11.4. this convergence is equivalent to $\underset{1\leqq i\leqq m_n}{\ast} \mathscr{U}_{P_{n,i}} \underset{n\to\infty}{\Rightarrow} P$, i.e. to $\mathscr{E}_{\sum\limits_{1\leqq i\leqq m_n} P_{n,i}((.)\setminus\{o\})} \underset{n\to\infty}{\Rightarrow} \mathscr{E}_{\tilde{P}}$. The continuity Proposition 3.3.6. says that a sequence (P_n) of infinitely divisible distributions on $\mathfrak{M}$ converges weakly toward an infinitely divisible distribution P if and only if, for all finite sequences $X_1, \ldots, X_m$ of bounded sets in $\mathfrak{A}$ satisfying the condition $P(\Phi(\text{boundary of } X_i) > 0) = 0$, the convergence relation

$$\lim_{n\to\infty} \left\|(\tilde{P}_n)_{X_1,\ldots,X_m}\big((.) \setminus \{[0, \ldots, 0]\}\big) - (\tilde{P})_{X_1,\ldots,X_m}\big((.) \setminus \{[0, \ldots, 0]\}\big)\right\| = 0$$

is satisfied (cf. Kerstan & Matthes [1]). We shall denote this by $\tilde{P}_n \underset{n\to\infty}{\overset{W}{\Rightarrow}} \tilde{P}$. For arbitrary infinitesimal triangular arrays $(P_{n,i})$ of distributions on $\mathfrak{M}$, $\underset{1\leqq i\leqq m_n}{\ast} P_{n,i} \underset{n\to\infty}{\Rightarrow} P$ is thus equivalent to $\sum_{1\leqq i\leqq m_n} P_{n,i}\big((.)\setminus\{o\}\big) \underset{n\to\infty}{\overset{W}{\Rightarrow}} \tilde{P}$ (Theorem 3.4.2.). This leads to the well known necessary and sufficient criterion as given by Theorem 3.4.4. for the convergence of convolutions $\underset{1\leqq i\leqq m_n}{\ast} P_{n,i}$ toward a Poisson distribution P_ν (cf. Grigelionis [1], Franken [1]).

In the sets $\boldsymbol{P}$ and $\boldsymbol{W}$, it is possible to introduce distance functions $\varrho_{\boldsymbol{P}}$ and $\varrho_{\boldsymbol{W}}$, respectively, such that the corresponding convergences coincide with $\Rightarrow$ and $\overset{W}{\Rightarrow}$, respectively. To begin with, by 3.1.6. and 3.1.7. $\Rightarrow$ coincides with the weak convergence of distributions on the complete separable metric space $[M, \varrho_M]$. Thus we have established the connection to the general theory of weak convergence (cf. Prohorov [1] as well as Billingsley [1]). In this way we obtain in particular a convenient compactness criterion for $\Rightarrow$ (Proposition 3.2.7.). For $\varrho_{\boldsymbol{P}}$, it is possible to choose the Prohorov metric.

For the topological interpretation of $\overset{W}{\Rightarrow}$, we first provide the set $A^* = M \setminus \{o\}$ with a modification ϱ_{A^*} of ϱ_M so that a complete separable metric space $[A^*, \varrho_{A^*}]$ is obtained. Subsequently, using as a basis the general definition of the Prohorov metric, we introduce a metric ϱ_N in the set N of all measures on $\mathfrak{A}$ which are finite on bounded sets, thus obtaining the complete separable metric space $[N, \varrho_N]$. Now the set $\boldsymbol{W}$ can be identified with the set N^* corresponding to the phase space $[A^*, \varrho_{A^*}]$, and by transforming ϱ_{N^*} we obtain the distance function $\varrho_{\boldsymbol{W}}$; then the convergence with respect to $\varrho_{\boldsymbol{W}}$ coincides with $\overset{W}{\Rightarrow}$ (Proposition 3.3.5.).

3.1. The Metric Space $[\boldsymbol{E}^+, \varrho_{\boldsymbol{E}^+}]$

For all E_1, E_2 in the set $\boldsymbol{E}^+$ of all finite measures on the σ-algebra $\mathfrak{M}$ of the Borel sets of the complete separable metric space $[M, \varrho_M]$, we set

$$\varepsilon_{E_1,E_2} = \inf\left\{\varepsilon : \varepsilon > 0, E_1(Y) < E_2\Big(\bigcup_{\Phi\in Y} S_\varepsilon(\Phi)\Big) + \varepsilon \text{ for all closed subsets } Y \text{ of } M\right\}.$$

The number $\varrho_{\boldsymbol{E}^+}(E_1, E_2) = \max(\varepsilon_{E_1,E_2}, \varepsilon_{E_2,E_1})$ is called the *Prohorov distance* between E_1 and E_2 (cf. Prohorov [1] and Strassen [1]).

By spezialization of general theorems (see, for instance, Prohorov [1], Theorem 1.11.) we obtain the following two statements.

3.1.1. $[\boldsymbol{E}^+, \varrho_{\boldsymbol{E}^+}]$ *is a complete separable metric space.*

3.1.2. *In* $\boldsymbol{E}^+$, *the metric* $\varrho_{\boldsymbol{E}^+}$ *generates the coarsest topology, with respect to which, for all continuous bounded non-negative real functions* h *on* $[M, \varrho_M]$, *the real function*

$$E \curvearrowright \int h(\Phi)\, E(\mathrm{d}\Phi) \qquad (E \in \boldsymbol{E}^+)$$

is continuous.

Consequently the mapping $\Phi \frown \delta_\Phi$ is a homeomorphism of $[M, \varrho_M]$ in $[\boldsymbol{E}^+, \varrho_{\boldsymbol{E}^+}]$. Hence also the mapping $a \frown \delta_{\delta_a}$ of $[A, \varrho_A]$ into $[\boldsymbol{E}^+, \varrho_{\boldsymbol{E}^+}]$ is homeomorphic.

Let $\mathfrak{E}^+$ denote the smallest σ-algebra of subsets of $\boldsymbol{E}^+$ with respect to which the real function

$$E \frown E(Y) \qquad (E \in \boldsymbol{E}^+)$$

is measurable for all Y in $\mathfrak{M}$. By analogy with 1.15.5., we deduce the truth of

3.1.3. *The σ-algebra of the Borel sets of $[\boldsymbol{E}^+, \varrho_{\boldsymbol{E}^+}]$ coincides with $\mathfrak{E}^+$.*

Instead of $\lim_{n\to\infty} \varrho_{\boldsymbol{E}^+}(E_n, E) = 0$, we also write $E_n \underset{n\to\infty}{\Rightarrow} E$, the sequence (E_n) being said to *converge weakly* toward E.

Now, if in 3.1.2. we substitute unity for h, then it is readily observed that $E_n \underset{n\to\infty}{\Rightarrow} E$ implies $E_n(M) \xrightarrow[n\to\infty]{} E(M)$. Hence in particular $\boldsymbol{P}$ is a closed subset of $\boldsymbol{E}^+$. We denote by $\varrho_{\boldsymbol{P}}$ the restriction of the metric $\varrho_{\boldsymbol{E}^+}$ to $\boldsymbol{P}$. In view of 3.1.3., the σ-algebra $\mathfrak{P}$ introduced in section 1.1. coincides with the σ-algebra of the Borel sets of the complete separable metric space $[\boldsymbol{P}, \varrho_{\boldsymbol{P}}]$.

As usual, an Y in $\mathfrak{M}$ is called a *continuity set* with respect to E if $E(\partial Y) = 0$ is satisfied. Here ∂Y denotes the boundary of Y. Obviously the system $\mathfrak{S}_E$ of all continuity sets with respect to E forms a subalgebra of $\mathfrak{M}$.

As is known, the following theorem holds (cf. for instance Prohorov [1], Theorem 1.7., or Billingsley [1], Theorem 2.1.).

3.1.4. *A sequence (E_n) of elements of $\boldsymbol{E}^+$ converges weakly toward E in $\boldsymbol{E}^+$ if and only if the convergence relation*

$$E_n(Y) \to E(Y)$$

is satisfied for all Y in $\mathfrak{S}_E$.

For each measure H on $\mathfrak{M}$, let $\mathfrak{B}_H$ be the system of those X in $\mathfrak{B}$ which satisfy the condition $H\big(\Phi(\partial X) > 0\big) = 0$, i.e. the condition $\varrho_H(\partial X) = 0$. Obviously $\mathfrak{B}_H$ forms a subring of $\mathfrak{B}$.

To each X in $\mathfrak{B}_E$ there corresponds a sequence (Y_n) of sets in $\mathfrak{S}_E$:

3.1.5. *For each E in $\boldsymbol{E}^+$, a set X in $\mathfrak{B}$ belongs to $\mathfrak{B}_E$ if and only if all of the sets*

$$Y_n = \{\Phi : \Phi \in M, \Phi(X) = n\} \qquad (n = 0, 1, \ldots)$$

belong to $\mathfrak{S}_E$.

Proof. 1. Let X' denote the closure and X'' the interior of X. Then

$$\partial Y_n = \{\Phi : \Phi \in M, \Phi(\partial X) > 0, \Phi(X'') \leqq n \leqq \Phi(X')\}.$$

For if Φ belongs to the right side of the equation to be proved, then any point of Φ lying on the boundary of X can be shifted into both X and

$A \setminus X$ by arbitrarily small displacements. Hence in particular it is possible by applying arbitrarily small displacements to make X contain exactly n points, i.e. Φ is a point of contact of Y_n. (In performing the displacements, points of Φ occupying the same places are of course considered to be different.) On the other hand, we can likewise make X not contain exactly n points, i.e. Φ is also a point of contact of $M \setminus Y_n$, and hence a boundary point of Y_n.

Conversely, suppose now that Φ be included in the boundary of Y_n. We choose some sequence (Φ_m) of elements of Y_n which tends toward Φ with respect to ϱ_M. Then in view of 1.15.4. we obtain

$$\Phi(X'') \leqq \varliminf_{m\to\infty} \Phi_m(X'') \leqq \lim_{m\to\infty} \Phi_m(X) = n \leqq \varlimsup_{m\to\infty} \Phi_m(X') \leqq \Phi(X').$$

Now let (Ψ_m) be a sequence of elements of $M \setminus Y_n$ which tends toward Φ with respect to ϱ_M where we may suppose that either $\Psi_m(X) < n$ for all n or $\Psi_m(X) > n$ for all n. If $\Phi(\partial X) = 0$ would be true, then by 1.15.4. we would obtain

$$n \leqq \Phi(X') = \Phi(X'') \leqq \varliminf_{m\to\infty} \Psi_m(X'') \leqq \varliminf_{m\to\infty} \Psi_m(X) < n$$

in the first case, and

$$n \geqq \Phi(X'') = \Phi(X') \geqq \varlimsup_{m\to\infty} \Psi_m(X') \geqq \varlimsup_{m\to\infty} \Psi_m(X) > n,$$

in the second.

2. Using 1. it may be concluded that

$$E\left(\bigcup_{n=0}^{\infty} \partial Y_n\right) = E\left(\bigcup_{n=0}^{\infty} \{\Phi : \Phi \in M,\ \Phi(\partial X) > 0,\ \Phi(X'') \leqq n \leqq \Phi(X')\}\right)$$
$$= E\big(\Phi(\partial X) > 0\big),$$

i.e. $E(\partial Y_n)$ is zero for all n if and only if $\varrho_E(\partial X) = 0$. ∎

We shall now convince ourselves that always "a sufficient number" of sets X in $\mathfrak{B}$ belong to $\mathfrak{B}_E$, especially that $\mathfrak{B}_E$ generates $\mathfrak{B}$.

3.1.6. *To each E in $\mathbf{E}^+$, each closed bounded subset D of A, and each open neighbourhood U of D, there exists a set X in $\mathfrak{B}_E$ which has the property that $D \subseteqq X \subseteqq U$.*

Proof. Let f be a function in the set F defined in section 1.15. which satisfies the condition $k_D \leqq f \leqq k_U$. We shall consider the family $\big(f^{-1}(\{x\})\big)_{0<x<1}$ of pairwise disjoint closed subsets of the bounded Borel set $f^{-1}\big((0, +\infty)\big)$. If E is equal to the zero measure 0, then $E\Big(\Phi\big(f^{-1}(\{x\})\big) > 0\Big) = 0$ for all x in $(0, 1)$. Otherwise we may pass on to $\big(E(M)\big)^{-1} E$ so that, by means of the method used in proving 1.1.5., we see that the inequality $E\Big(\Phi\big(f^{-1}(\{x\})\big) > 0\Big) > 0$ is satisfied by an at most countable number of elements x of $(0, 1)$.

Thus there always exists an x_0 in (0, 1) such that $E\big(\Phi(f^{-1}(\{x_0\})) > 0\big) = 0$. However, we have

$$f^{-1}(\{x_0\}) = f^{-1}\big(\partial(x_0, +\infty)\big) \supseteq \partial f^{-1}\big((x_0, +\infty)\big).$$

Hence $X = f^{-1}\big((x_0, +\infty)\big)$ satisfies all requirements. ∎

Now we are in a position to characterize weak convergence in $\boldsymbol{E}^+$ without using the metric ϱ_M:

3.1.7. Theorem. *Let E be an element of $\boldsymbol{E}^+$ and $\mathfrak{H}$ a semi-ring of sets in $\mathfrak{B}_E$. If to each bounded closed subset D of A and to each open neighbourhood U of D there exists a set X in the ring $\mathfrak{H}^*$ generated by $\mathfrak{H}$ such that $D \subseteq X \subseteq U$, then a sequence (E_n) of elements of $\boldsymbol{E}^+$ tends weakly toward E if and only if $\lim_{n\to\infty} E_n(M) = E(M)$ and the convergence relation*

$$(E_n)_{H_1,\ldots,H_m}(\{[l_1, \ldots, l_m]\}) \xrightarrow[n\to\infty]{} E_{H_1,\ldots,H_m}(\{[l_1, \ldots, l_m]\})$$

holds for each finite sequence $H_1, \ldots, H_m$ of pairwise disjoint sets in $\mathfrak{H}$ and all non-negative integers $l_1, \ldots, l_m$.

Proof. 1. By virtue of 3.1.4. and 3.1.5., the conditions stated above are necessary for the weak convergence of (E_n) toward E.

2. Conversely, suppose that these conditions are satisfied.

For all V in $\boldsymbol{E}^+$ and all finite sequences $f_1, \ldots, f_m$ of functions in F, let $V_{f_1,\ldots,f_m}$ denote the measure on $\mathfrak{R}^m$ derived from V by means of

$$\Phi \curvearrowright \left[\int f_1(a)\, \Phi(\mathrm{d}a), \ldots, \int f_m(a)\, \Phi(\mathrm{d}a)\right].$$

We shall prove now that the sequence $\big((E_n)_{f_1,\ldots,f_m}\big)_{n=1,2,\ldots}$ of finite measures on $\mathfrak{R}^m$ converges weakly toward $E_{f_1,\ldots,f_m}$.

Let X be a set in $\mathfrak{H}^*$ having the property that

$$\left\{a : a \in A,\ \max_{1\le i\le m} f_i(a) > 0\right\} \subseteq X.$$

By our assumption on $\mathfrak{H}$, to each $\eta > 0$ there exists a finite sequence $H_1, \ldots, H_s$ of pairwise disjoint sets in $\mathfrak{H}$ such that

a) $\bigcup_{j=1}^{s} H_j = X$,

b) $\sup_{a,a'\in H_j} |f_i(a) - f_i(a')| \le \eta$ for $i = 1, \ldots, m$ and $j = 1, \ldots, s$.

To show this, observe that to each $i = 1, \ldots, m$ and $k = 1, 2, \ldots$ there is a set $X_{i,k}$ in $\mathfrak{H}^*$ such that

$$f_i^{-1}\left(\left[\frac{k\eta}{2}, \frac{(k+1)\eta}{2}\right]\right) \subseteq X_{i,k} \subseteq f_i^{-1}\left(\left(\frac{k\eta}{2} - \frac{\eta}{4}, \frac{(k+1)\eta}{2} + \frac{\eta}{4}\right)\right).$$

Since f_i is bounded, only finitely many $X_{i,k}$ are different from $\emptyset$. If we choose from $\mathfrak{H}$ a finite sequence $H_1, \ldots, H_s$ of pairwise disjoint sets such that all of the $X_{i,k}$ can be represented as unions of certain sets in this sequence and such that $H_1 \cup \cdots \cup H_s = X$, then the sequence $H_1, \ldots, H_s$ satisfies the conditions a) and b).

Let $\Phi, \Phi' \in M$. If we have $\Phi(H_j) = \Phi'(H_j)$ for $1 \leqq j \leqq s$, we obtain, for $1 \leqq i \leqq m$,

$$\left|\int f_i(a)\, \Phi(\mathrm{d}a) - \int f_i(a)\, \Phi'(\mathrm{d}a)\right| = \left|\int_X f_i(a)\, \Phi(\mathrm{d}a) - \int_X f_i(a)\, \Phi'(\mathrm{d}a)\right|$$
$$\leqq \sum_{j=1}^{s} \eta \Phi(H_j) = \eta \Phi(X).$$

Now let g be an arbitrary bounded continuous real function on R^m. Setting

$$\|g\| = \sup_{x \in R^m} |g(x)|, \qquad \|f_i\| = \sup_{a \in A} f_i(a) \qquad (1 \leqq i \leqq m),$$

we obtain, for all natural numbers k,

$$\overline{\lim_{n\to\infty}} \left| \int g(x)\, (E_n)_{f_1,\ldots,f_m}(\mathrm{d}x) - \int g(x)\, E_{f_1,\ldots,f_m}(\mathrm{d}x) \right|$$
$$= \overline{\lim_{n\to\infty}} \Big| \int g\Big(\int f_1(a)\, \Phi(\mathrm{d}a), \ldots, \int f_m(a)\, \Phi(\mathrm{d}a)\Big)\, E_n(\mathrm{d}\Phi)$$
$$- \int g\Big(\int f_1(a)\, \Phi(\mathrm{d}a), \ldots, \int f_m(a)\, \Phi(\mathrm{d}a)\Big)\, E(\mathrm{d}\Phi) \Big|$$
$$\leqq \overline{\lim_{n\to\infty}} \sum_{\substack{l_1,\ldots,l_s \geqq 0 \\ l_1+\cdots+l_s \leqq k}} \Bigg| \int_{\left\{\substack{\Phi:\Phi\in M, \Phi(H_j)=l_j \\ \text{for } 1\leqq j\leqq s}\right\}} g\Big(\int f_1(a)\, \Phi(\mathrm{d}a), \ldots, \int f_m(a)\, \Phi(\mathrm{d}a)\Big)\, E_n(\mathrm{d}\Phi)$$
$$- \int_{\left\{\substack{\Phi:\Phi\in M, \Phi(H_j)=l_j \\ \text{for } 1\leqq j\leqq s}\right\}} g\Big(\int f_1(a)\, \Phi(\mathrm{d}a), \ldots, \int f_m(a)\, \Phi(\mathrm{d}a)\Big)\, E(\mathrm{d}\Phi) \Bigg|$$
$$+ \overline{\lim_{n\to\infty}} \|g\| \Big(E_n\big(\Phi(X) > k\big) + E\big(\Phi(X) > k\big)\Big)$$
$$\leqq \overline{\lim_{n\to\infty}} \big(E_n(M) + E(M)\big) \cdot \sup_{\substack{|x_i - x_i'| \leqq \eta k \text{ for } 1\leqq i\leqq m \\ 0 \leqq x_i, x_i' \leqq k\|f_i\| \text{ for } 1\leqq i\leqq m}} |g(x_1, \ldots, x_m) - g(x_1', \ldots, x_m')|$$
$$+ \overline{\lim_{n\to\infty}} \sum_{l_1,\ldots,l_s \geqq 0} \|g\| \cdot |(E_n)_{H_1,\ldots,H_s}(\{[l_1, \ldots, l_s]\}) - E_{H_1,\ldots,H_s}(\{[l_1, \ldots, l_s]\})|$$
$$+ 2\|g\| \cdot E\big(\Phi(X) > k\big) + \overline{\lim_{n\to\infty}} \|g\| \cdot \big|E_n\big(\Phi(X) > k\big) - E\big(\Phi(X) > k\big)\big|$$
$$\leqq 2\|g\|\, E\big(\Phi(X) > k\big)$$
$$+ 2E(M) \sup_{\substack{|x_i - x_i'| \leqq \eta k \text{ for } 1\leqq i\leqq m \\ 0 \leqq x_i, x_i' \leqq k\|f_i\| \text{ for } 1\leqq i\leqq m}} |g(x_1, \ldots, x_m) - g(x_1', \ldots, x_m')|$$
$$+ 2\|g\| \overline{\lim_{n\to\infty}} |E_n(M) - E(M)| + \|g\| \overline{\lim_{n\to\infty}} \|(E_n)_{H_1,\ldots,H_s} - E_{H_1,\ldots,H_s}\|$$
$$+ \|g\| \overline{\lim_{n\to\infty}} \big|E_n\big(\Phi(X) > k\big) - E\big(\Phi(X) > k\big)\big|$$

$$\leqq 2\|g\|\, E\bigl(\Phi(X) > k\bigr)$$
$$+\, 2E(M) \sup_{\substack{|x_i - x_i'| \leqq k\eta \text{ for } 1 \leqq i \leqq m \\ 0 \leqq x_i, x_i' \leqq k\|f_i\| \text{ for } 1 \leqq i \leqq m}} |g(x_1, \ldots, x_m) - g(x_1', \ldots, x_m')|$$
$$+\, 4\|g\| \overline{\lim_{n\to\infty}} \|(E_n)_{H_1,\ldots,H_s} - E_{H_1,\ldots,H_s}\|\,.$$

In view of 1.9.2. we have $\overline{\lim\limits_{n\to\infty}} \|(E_n)_{H_1,\ldots,H_m} - E_{H_1,\ldots,H_m}\| = 0$. Since g is uniformly continuous on the bounded closed set $\mathop{\times}\limits_{i=1}^{m} [0, k\|f_i\|]$, the second term on the right side can be made arbitrarily small in dependence on η, so that for all natural numbers k we obtain

$$\overline{\lim_{n\to\infty}} \left|\int g(x) (E_n)_{f_1,\ldots,f_m} (\mathrm{d}x) - \int g(x)\, E_{f_1,\ldots,f_m}(\mathrm{d}x)\right| \leqq 2\|g\|\, E\bigl(\Phi(X) > k\bigr).$$

However, $E\bigl(\Phi(X) > k\bigr)$ tends toward zero as k increases. Hence always

$$\int g(x) (E_n)_{f_1,\ldots,f_m} (\mathrm{d}x) \xrightarrow[n\to\infty]{} \int g(x)\, E_{f_1,\ldots,f_m}(\mathrm{d}x),$$

i.e. the sequence of the $(E_n)_{f_1,\ldots,f_m}$ converges weakly toward $E_{f_1,\ldots,f_m}$.

3. The weak convergence of $(E_n)_{f_1,\ldots,f_m}$ toward $E_{f_1,\ldots,f_m}$ for arbitrary finite sequences $f_1, \ldots, f_m$ of functions in F implies the weak convergence of (E_n) toward E.

Let $f_1, \ldots, f_m$ be a finite sequence of functions in F. For each open continuity set X with respect to $E_{f_1,\ldots,f_m}$, the inverse image Y of X formed with respect to the continuous map

$$\Phi \curvearrowright \left[\int f_1(a)\, \Phi(\mathrm{d}a), \ldots, \int f_m(a)\, \Phi(\mathrm{d}a)\right]$$

is an open continuity set with respect to E. By assumption we have

$$(E_n)_{f_1,\ldots,f_m} (X) \xrightarrow[n\to\infty]{} E_{f_1,\ldots,f_m}(X),$$

and hence

$$E_n(Y) \xrightarrow[n\to\infty]{} E(Y).$$

Let $\mathfrak{S}$ denote the system of all open subsets Y of M which can be represented in this way. Obviously M is contained in $\mathfrak{S}$. In view of 1.15.3., $\mathfrak{S}$ forms a base of the separable metric space $[M, \varrho_M]$. Consequently each open subset of M can be represented as a union of an at most countable subsystem of $\mathfrak{S}$. On the other hand $\mathfrak{S}$ is closed under finite intersection operations. For if Y_1 is the inverse image of X_1 with respect to

$$\Phi \curvearrowright \left[\int f_1(a)\, \Phi(\mathrm{d}a), \ldots, \int f_m(a)\, \Phi(\mathrm{d}a)\right]$$

and Y_2 the inverse image of X_2 with respect to

$$\Phi \curvearrowright \left[\int g_1(a)\, \Phi(\mathrm{d}a), \ldots, \int g_k(a)\, \Phi(\mathrm{d}a)\right],$$

then $Y_1 \cap Y_2$ is the inverse image of

$$X = \{[x_1, \ldots, x_{m+k}] : [x_1, \ldots, x_{m+k}] \in R^{m+k},$$
$$[x_1, \ldots, x_m] \in X_1, [x_{m+1}, \ldots, x_{m+k}] \in X_2\}$$

with respect to

$$\Phi \curvearrowright \left[\int f_1(a)\,\Phi(\mathrm{d}a), \ldots, \int f_m(a)\,\Phi(\mathrm{d}a), \int g_1(a)\,\Phi(\mathrm{d}a), \ldots, \int g_k(a)\,\Phi(\mathrm{d}a)\right]$$

where the following relation holds:

$$E_{f_1,\ldots,f_m,g_1,\ldots,g_k}(\partial X) \leqq E_{f_1,\ldots,f_m,g_1,\ldots,g_k}\Big(\big((\partial X_1) \times R^k\big) \cup \big(R^k \times (\partial X_2)\big)\Big)$$
$$\leqq E_{f_1,\ldots,f_m}(\partial X_1) + E_{g_1,\ldots,g_k}(\partial X_2) = 0\,.$$

For each finite sequence $Y_1, \ldots, Y_s$ of sets in $\mathfrak{S}$, we obtain

$$E_n\left(\bigcup_{i=1}^{s} Y_i\right) = \sum_{i=1}^{s} E_n(Y_i) - \sum_{i,j=1}^{s} E_n(Y_i \cap Y_j) + \cdots$$
$$\xrightarrow[n\to\infty]{} \sum_{i=1}^{s} E(Y_i) - \sum_{i,j=1}^{s} E(Y_i \cap Y_j) + \cdots = E\left(\bigcup_{i=1}^{s} Y_i\right).$$

Now, if Y'' is an arbitrary open subset of the metric space $[M, \varrho_M]$, and (Y_i) a sequence of sets in $\mathfrak{S}$ having the union Y'', then for all s we obtain

$$E\left(\bigcup_{i=1}^{s} Y_i\right) = \lim_{n\to\infty} E_n\left(\bigcup_{i=1}^{s} Y_i\right) \leqq \varliminf_{n\to\infty} E_n(Y'')$$

and hence

$$E(Y'') \leqq \varliminf_{n\to\infty} E_n(Y'')\,.$$

Consequently, if Y' is an arbitrary closed subset of the metric space $[M, \varrho_M]$, then

$$E(Y') = E(M) - E(M \setminus Y') = \lim_{n\to\infty} E_n(M) - E(M \setminus Y')$$
$$\geqq \lim_{n\to\infty} E_n(M) - \varliminf_{n\to\infty} E_n(M \setminus Y') = \varlimsup_{n\to\infty} E_n(Y')\,.$$

If we now choose a continuity set Y with respect to E, denoting by Y' and Y'' its closure and interior, respectively, then it follows that

$$\varlimsup_{n\to\infty} E_n(Y) \leqq \varlimsup_{n\to\infty} E_n(Y') \leqq E(Y') = E(Y) = E(Y'')$$
$$\leqq \varliminf_{n\to\infty} E_n(Y'') \leqq \varliminf_{n\to\infty} E_n(Y)\,,$$

i.e. we have $E_n(Y) \xrightarrow[n\to\infty]{} E(Y)$ so that by 3.1.4. we can conclude that $E_n \underset{n\to\infty}{\Rightarrow} E$. ∎

From the proof of 3.1.7. we see that a sequence (E_n) converges in $\boldsymbol{E}^+$ toward E if and only if, for all $f_1, \ldots, f_m$ in F, the finite measures $(E_n)_{f_1,\ldots,f_m}$ converge weakly toward the finite measure $E_{f_1,\ldots,f_m}$. This characterization of weak convergence in $\boldsymbol{E}^+$ can be simplified as follows (see for instance von Waldenfels [1] and Jagers [1]):

3.1.8. *A sequence (E_n) of measures in $\boldsymbol{E}^+$ converges weakly toward a measure E in $\boldsymbol{E}^+$ if and only if, for all f in F, the sequence of finite measures $(E_n)_f$ converges weakly toward the finite measure E_f.*

Proof. We have only to show that the stated condition implies the weak convergence of $(E_n)_{f_1,\ldots,f_m}$ for all finite sequences $f_1, \ldots, f_m$ of functions in F. In virtue of the continuity theorem for the Laplace transformation, this weak convergence is equivalent to the truth of

$$\lim_{n\to\infty} \int e^{-\sum_{i=1}^{m} x_i y_i} (E_n)_{f_1,\ldots,f_m}(\mathrm{d}[y_1, \ldots, y_m]) = \int e^{-\sum_{i=1}^{m} x_i y_i} E_{f_1,\ldots,f_m}(\mathrm{d}[y_1, \ldots, y_m])$$

for all $x_1, \ldots, x_m \geqq 0$. Now, by assumption we have

$$\begin{aligned}\int e^{-\sum_{i=1}^{m} x_i y_i} (E_n)_{f_1,\ldots,f_m}(\mathrm{d}[y_1, \ldots, y_m]) &= \int e^{-\sum_{i=1}^{m} x_i \int f_i(a)\Phi(\mathrm{d}a)} E_n(\mathrm{d}\Phi) \\ = \int e^{-\int\left(\sum_{i=1}^{m} x_i f_i(a)\right)\Phi(\mathrm{d}a)} E_n(\mathrm{d}\Phi) &\xrightarrow[n\to\infty]{} \int e^{-\int\left(\sum_{i=1}^{m} x_i f_i(a)\right)\Phi(\mathrm{d}a)} E(\mathrm{d}\Phi) \\ = \int e^{-\sum_{i=1}^{m} x_i y_i} E_{f_1,\ldots,f_m}(\mathrm{d}[y_1, \ldots, y_m]),\end{aligned}$$

since $f = \sum_{i=1}^{m} x_i f_i$ belongs to F for all $x_1, \ldots, x_m \geqq 0$. ∎

From 1.9.2., 3.1.6., and 3.1.7. it follows

3.1.9. *A sequence (E_n) of measures in $\boldsymbol{E}^+$ converges weakly toward a measure E in $\boldsymbol{E}^+$ if and only if*

$$(E_n)_{X_1,\ldots,X_m} \xrightarrow[n\to\infty]{} E_{X_1,\ldots,X_m}$$

for all finite sequences $X_1, \ldots, X_m$ of pairwise disjoint sets in $\mathfrak{B}_E$.

Let $E_n \underset{n\to\infty}{\Rightarrow} E$ and $E_n' \underset{n\to\infty}{\Rightarrow} E'$. For all finite sequences $X_1, \ldots, X_m$ of pairwise disjoint sets in $\mathfrak{B}_{E+E'} = \mathfrak{B}_E \cap \mathfrak{B}_{E'}$, we have

$$\begin{aligned}&\overline{\lim_{n\to\infty}}\, \|(E_n * E_n')_{X_1,\ldots,X_m} - (E * E')_{X_1,\ldots,X_m}\| \\ &= \overline{\lim_{n\to\infty}}\, \|(E_n)_{X_1,\ldots,X_m} * (E_n')_{X_1,\ldots,X_m} - (E_{X_1,\ldots,X_m}) * (E'_{X_1,\ldots,X_m})\| \\ &\leqq \overline{\lim_{n\to\infty}}\, \big(\|(E_n)_{X_1,\ldots,X_m} * (E_n')_{X_1,\ldots,X_m} - (E_{X_1,\ldots,X_m}) * (E_n')_{X_1,\ldots,X_m}\| \\ &\quad + \|E_{X_1,\ldots,X_m} * (E_n')_{X_1,\ldots,X_m} - (E_{X_1,\ldots,X_m}) * (E'_{X_1,\ldots,X_m})\|\big) \\ &\leqq \overline{\lim_{n\to\infty}}\, E_n'(M)\, \|(E_n)_{X_1,\ldots,X_m} - E_{X_1,\ldots,X_m}\| \\ &\quad + E(M)\, \overline{\lim_{n\to\infty}}\, \|(E_n')_{X_1,\ldots,X_m} - E'_{X_1,\ldots,X_m}\| = 0.\end{aligned}$$

By 1.6.12., $\mathfrak{B}_{E+E'} \subseteq \mathfrak{B}_{E*E'}$, and by 3.1.6., the ring $\mathfrak{B}_{E+E'}$ fulfils the conditions in 3.1.7., so that we find the

3.1.10. Proposition. *If* $E_n \underset{n\to\infty}{\Rightarrow} E$ *and* $E_n' \underset{n\to\infty}{\Rightarrow} E'$, *then*

$$(E_n * E_n') \underset{n\to\infty}{\Rightarrow} E * E'.$$

For all X in $\mathfrak{B}_E$ and all finite sequences $X_1, \ldots, X_m$ of pairwise disjoint sets in $\mathfrak{B}_E$, from $E_n \underset{n\to\infty}{\Rightarrow} E$ we obtain, using 3.1.9.:

$$({}_XE_n)_{X_1,\ldots,X_m} = (E_n)_{X\cap X_1,\ldots,X\cap X_m} \xrightarrow[n\to\infty]{} E_{X\cap X_1,\ldots,X\cap X_m} = ({}_XE)_{X_1,\ldots,X_m}.$$

Hence, with the use of 3.1.6. and 3.1.7., we get

3.1.11. $E_n \underset{n\to\infty}{\Rightarrow} E$ *implies the convergence* ${}_X(E_n) \underset{n\to\infty}{\Rightarrow} {}_XE$ *for all* X *in* $\mathfrak{B}_E$.

$E_n \underset{n\to\infty}{\Rightarrow} E$ does not imply $\varrho_{E_n}(X) \xrightarrow[n\to\infty]{} \varrho_E(X)$ for X in $\mathfrak{A}$. However, we have

3.1.12. *For all* X *in* $\mathfrak{B}_E$, $E_n \underset{n\to\infty}{\Rightarrow} E$ *implies the truth of the inequality*

$$\varrho_E(X) \leqq \varliminf_{n\to\infty} \varrho_{E_n}(X).$$

Proof. Using 3.1.9. we obtain, for all natural numbers m,

$$\lim_{n\to\infty} \sum_{k=1}^{m} kE_n(\Phi(X) = k) = \sum_{k=1}^{m} kE(\Phi(X) = k)$$

and hence

$$\varliminf_{n\to\infty} \sum_{k=1}^{\infty} kE_n(\Phi(X) = k) \geqq \sum_{k=1}^{m} kE(\Phi(X) = k),$$

consequently

$$\varliminf_{n\to\infty} \varrho_{E_n}(X) \geqq \sum_{k=1}^{\infty} kE(\Phi(X) = k) = \varrho_E(X). \blacksquare$$

The weak convergence toward simple distributions on $\mathfrak{M}$ can be formulated by means of the one-dimensional distributions (cf. Kallenberg [1], [2]).

3.1.13. Theorem. *Let* P *be a simple distribution in* $\mathbf{P}$, *and* $\mathfrak{H}$ *a subring of* $\mathfrak{B}_P$. *If to each bounded closed subset* D *of* A *and to each open neighbourhood* U *of* D *there exists a set* X *in* $\mathfrak{H}$ *such that* $D \subseteq X \subseteq U$, *then a sequence* (P_n) *of distributions in* $\mathbf{P}$ *converges weakly toward* P *if and only if the convergence relation*

$$(P_n)_X \xrightarrow[n\to\infty]{} P_X$$

holds for all X *in* $\mathfrak{H}$.

Proof. By 3.1.7. the afore-mentioned condition for the weak convergence of (P_n) toward P is necessary, and conversely we have only to show that, for all finite sequences $X_1, \ldots, X_m$ of pairwise disjoint sets in $\mathfrak{H}$, the convergence relation

$$(P_n)_{X_1,\ldots,X_m} \xrightarrow[n\to\infty]{} P_{X_1,\ldots,X_m}$$

is true.

We shall prove this in several steps.

1. Using the formula which is proved in the first step of the proof of the proposition 1.4.7., we observe that the following convergence relation is true for each finite family $(V_i)_{i\in I}$ of pairwise disjoint sets in $\mathfrak{H}$ and each subset J of I:

$$P_n\big(\Phi(V_i) > 0 \quad \text{for} \quad i \in J,\, \Phi(V_i) = 0 \quad \text{for} \quad i \in (I \setminus J)\big)$$
$$\xrightarrow[n\to\infty]{} P\big(\Phi(V_i) > 0 \quad \text{for} \quad i \in J,\, \Phi(V_i) = 0 \quad \text{for} \quad i \in (I \setminus J)\big).$$

2. For all finite sequences $X_1, \ldots, X_m$ of pairwise disjoint sets in $\mathfrak{H}$ and all non-negative integers $k_1, \ldots, k_m$,

$$\varlimsup_{n\to\infty} P_n\big(\Phi(X_i) \leqq k_i \quad \text{for} \quad i = 1, \ldots, m\big) \leqq P\big(\Phi(X_i) \leqq k_i \quad \text{for} \quad i = 1, \ldots, m\big).$$

Let ε be an arbitrary positive real number. By 1.4.5. there exists a finite sequence $V_1, \ldots, V_k$ of pairwise disjoint sets in $\mathfrak{H}$ such that each X_i can be represented as a union $X_i = \bigcup_{j\in J_i} V_i$, $J_i \subseteqq \{1, \ldots, k\}$, and that the inequality $P\big(\Phi(V_j) > 1$ for at least one j in $\{1, \ldots, k\}\big) < \varepsilon$ is satisfied.

Using 1. we obtain

$$\begin{aligned} &P_n\big(\Phi(X_i) \leqq k_i \text{ for } i = 1, \ldots, m\big) \\ &\leqq P_n\big(\Phi(V_j) > 0 \text{ for at most } k_i \text{ indices } j \text{ in } J_i,\, i = 1, \ldots, m\big) \\ &\qquad \to P\big(\Phi(V_j) > 0 \text{ for at most } k_i \text{ indices } j \text{ in } J_i,\, i = 1, \ldots, m\big) \\ &\leqq P\big(\Phi(X_i) \leqq k_i \text{ for } i = 1, \ldots, m\big) + \varepsilon. \end{aligned}$$

Consequently, for all $\varepsilon > 0$,

$$\begin{aligned} &\varlimsup_{n\to\infty} P_n\big(\Phi(X_i) \leqq k_i \quad \text{for} \quad i = 1, \ldots, m\big) \\ &\leqq P\big(\Phi(X_i) \leqq k_i \quad \text{for} \quad i = 1, \ldots, m\big) + \varepsilon. \end{aligned}$$

3. For all finite sequences $X_1, \ldots, X_m$ of pairwise disjoint sets in $\mathfrak{H}$ and all non-negative integers $l_1, \ldots, l_m$,

$$P_n\big(\Phi(X_i) = l_i \quad \text{for} \quad i = 1, \ldots, m\big)$$
$$\xrightarrow[n\to\infty]{} P\big(\Phi(X_i) = l_i \quad \text{for} \quad i = 1, \ldots, m\big).$$

We shall prove this by induction on $l = l_1 + \cdots + l_m$. For $l = 0$, our proposition reduces to

$$P_n\left(\Phi\left(\bigcup_{i=1}^{m} X_i\right) = 0\right) \xrightarrow[n\to\infty]{} P\left(\Phi\left(\bigcup_{i=1}^{m} X_i\right) = 0\right)$$

which follows immediately from our assumptions. Now suppose that our assertion be true for a non-negative integer l. If $l_1, \ldots, l_m$ are arbitrary non-negative integers having the sum l, we may use 2. to obtain, for all i in $\{1, \ldots, m\}$,

$$\begin{aligned}
&P\big(\Phi(X_1) = l_1, \ldots, \Phi(X_{i-1}) = l_{i-1}, \Phi(X_i) = l_i + 1, \Phi(X_{i+1}) = l_{i+1}, \ldots, \Phi(X_m) = l_m\big)\\
&\qquad + P\Big(\Phi(X_1) \leqq l_1, \ldots, \Phi(X_{i-1}) \leqq l_{i-1}, \Phi(X_i) \leqq l_i + 1,\\
&\qquad\qquad \Phi(X_{i+1}) \leqq l_{i+1}, \ldots, \Phi(X_m) \leqq l_m, \quad \text{but} \quad \sum_{i=1}^{m} \Phi(X_i) \leqq l\Big)\\
&= P\big(\Phi(X_1) \leqq l_1, \ldots, \Phi(X_{i-1}) \leqq l_{i-1}, \Phi(X_i) \leqq l_i + 1,\\
&\qquad \Phi(X_{i+1}) \leqq l_{i+1}, \ldots, \Phi(X_m) \leqq l_m\big)\\
&\geqq \varlimsup_{n\to\infty} P_n\big(\Phi(X_1) \leqq l_1, \ldots, \Phi(X_{i-1}) \leqq l_{i-1}, \Phi(X_i) \leqq l_i + 1,\\
&\qquad\qquad \Phi(X_{i+1}) \leqq l_{i+1}, \ldots, \Phi(X_m) \leqq l_m\big)\\
&= \varlimsup_{n\to\infty} \Big(P_n\left(\Phi(X_1) = l_1, \ldots, \Phi(X_{i-1}) = l_{i-1}, \Phi(X_i) = l_i + 1,\right.\\
&\qquad\qquad \left.\Phi(X_{i+1}) = l_{i+1}, \ldots, \Phi(X_m) = l_m\right)\\
&\qquad + P_n\Big(\Phi(X_1) \leqq l_1, \ldots, \Phi(X_{i-1}) \leqq l_{i-1}, \Phi(X_i) \leqq l_i + 1,\\
&\qquad\qquad \Phi(X_{i+1}) \leqq l_{i+1}, \ldots, \Phi(X_m) \leqq l_m, \quad \text{but} \quad \sum_{i=1}^{m} \Phi(X_i) \leqq l\Big)\Big).
\end{aligned}$$

Using our induction hypothesis, we may continue as follows:

$$\begin{aligned}
&= \varlimsup_{n\to\infty} P_n\big(\Phi(X_1) = l_1, \ldots, \Phi(X_{i-1}) = l_{i-1}, \Phi(X_i) = l_i + 1,\\
&\qquad\qquad \Phi(X_{i+1}) = l_{i+1}, \ldots, \Phi(X_m) = l_m\big)\\
&\quad + P\Big(\Phi(X_1) \leqq l_1, \ldots, \Phi(X_{i-1}) \leqq l_{i-1}, \Phi(X_i) \leqq l_i + 1,\\
&\qquad\qquad \Phi(X_{i+1}) \leqq l_{i+1}, \ldots, \Phi(X_m) \leqq l_m, \quad \text{but} \quad \sum_{i=1}^{m} \Phi(X_i) \leqq l\Big).
\end{aligned}$$

Thus for all non-negative integers $g_1, \ldots, g_m$ having the sum $l + 1$, we get

$$\varlimsup_{n\to\infty} P_n\big(\Phi(X_i) = g_i \quad \text{for} \quad i = 1, \ldots, m\big) \leqq P\big(\Phi(X_i) = g_i \quad \text{for} \quad i = 1, \ldots, m\big).$$

However, by assumption we have

$$\sum_{\substack{g_1,\ldots,g_m \geqq 0 \\ g_1+\cdots+g_m=l+1}} P_n\big(\Phi(X_i) = g_i \quad \text{for} \quad i = 1, \ldots, m\big) = P_n\left(\Phi\left(\bigcup_{i=1}^{m} X_i\right) = l + 1\right)$$

$$\xrightarrow[n=\infty]{} P\left(\Phi\left(\bigcup_{i=1}^{m} X_i\right) = l + 1\right) = \sum_{\substack{g_1,\ldots,g_m \geqq 0 \\ g_1+\cdots+g_m=l+1}} P\big(\Phi(X_i) = g_i \ \text{for} \ i = 1, \ldots, m\big).$$

Summarizing, we obtain

$$P_n\big(\Phi(X_i) = g_i \quad \text{for} \quad i = 1, \ldots, m\big) \xrightarrow[n\to\infty]{} P\big(\Phi(X_i) = g_i \quad \text{for} \quad i = 1, \ldots, m\big)$$

for all non-negative integers $g_1, \ldots, g_m$ having the sum $l + 1$. ∎

For all E_1, E_2 we obtain

$$\begin{aligned} \varepsilon_{E_1,E_2}, \varepsilon_{E_2,E_1} &\leqq \inf \{\varepsilon : \varepsilon > 0, |E_1(Y) - E_2(Y)| < \varepsilon \quad \text{for all} \quad Y \text{ in } \mathfrak{M}\} \\ &\leqq \|E_1 - E_2\|, \end{aligned}$$

and hence

$$\varrho_{E^+}(E_1, E_2) \leqq \|E_1 - E_2\|.$$

Thus $\|E_n - E\| \xrightarrow[n\to\infty]{} 0$ always implies $E_n \underset{n\to\infty}{\Rightarrow} E$, whereas from $E_n \underset{n\to\infty}{\Rightarrow} E$ it cannot be concluded that $\|E_n - E\| \xrightarrow[n\to\infty]{} 0$. For example, if $\varrho_A(a_n, a) \xrightarrow[n\to\infty]{} 0$ and $a_n \neq a$ for $n = 1, 2, \ldots$, then we obtain $\delta_{\delta_{a_n}} \underset{n\to\infty}{\Rightarrow} \delta_{\delta_a}$, although $\left\|\delta_{\delta_{a_n}} - \delta_{\delta_a}\right\| = 2$, $n = 1, 2, \ldots$, is true.

For finite phase spaces A, the weak convergence in $\boldsymbol{E}^+$ coincides with convergence in variation. For, if A consists of exactly m elements $a_1, \ldots, a_m$, then for all E in $\boldsymbol{E}^+$, the single-element sets $\{a_i\}$ belongs to $\mathfrak{B}_E$ so that by 3.1.9. we can conclude that $E_n \underset{n\to\infty}{\Rightarrow} E$ is equivalent to

$$\|(E_n)_{\{a_1\},\ldots,\{a_m\}} - E_{\{a_1\},\ldots,\{a_m\}}\| \xrightarrow[n\to\infty]{} 0.$$

But the variation distance $\|E_n - E\|$ coincides with

$$\|(E_n)_{\{a_1\},\ldots,\{a_m\}} - E_{\{a_1\},\ldots,\{a_m\}}\|.$$

In the special case $[A, \varrho_A] = R^s$, the weak convergence in $\boldsymbol{E}^+$ can be characterized most easily. For obviously, for an arbitrary E in $\boldsymbol{E}^+$, the system of all bounded half-open axially parallel cuboids $\underset{i=1}{\overset{s}{\times}} [\alpha_i, \beta_i)$ having the property that

$$E\big(\Phi(x_i = \alpha_i) > 0\big) = 0 = E\big(\Phi(x_i = \beta_i) > 0\big) \qquad (1 \leqq i \leqq s)$$

satisfies all conditions of 3.1.7. Observe that, for all i, the set of all those hyperplanes $x_i = \alpha$ for which $E\big(\Phi(x_i = \alpha) > 0\big)$ is positive, is at most countable.

Indeed, for any given natural number n, passing from Φ to

$$\Phi_{n,i} = \Phi\big((R^{i-1} \times (.) \times R^{s-i}) \cap [-n, n]^s\big)$$

yields a measurable mapping of M into the set M' corresponding to the phase space $[A', \varrho_{A'}] = R^1$. By applying 1.1.5. we observe that the set $\{\alpha : E(\Phi_{n,i}(\{\alpha\}) > 0) > 0\}$ is at most countable. Thus also

$$\{\alpha : E(\Phi(x_i = \alpha) > 0) > 0\} = \bigcup_{n=1}^{\infty} \{\alpha : E(\Phi_{n,i}(\{\alpha\}) > 0) > 0\}$$

is at most countable. Hence:

3.1.14. Proposition. *For $[A, \varrho_A] = R^s$, a sequence (E_n) converges weakly in $\mathbf{E}^+$ toward E if and only if $\lim_{n\to\infty} E_n(M) = E(M)$ and, for each finite sequence of pairwise disjoint half-open cuboids*

$$X_j = \mathop{\times}_{i=1}^{s} [\alpha_{i,j}, \beta_{i,j}) \qquad (j = 1, \ldots, m)$$

having the property that

$$E(\Phi(x_i = \alpha_{i,j}) > 0) = 0 = E(\Phi(x_i = \beta_{i,j}) > 0) \qquad (1 \leqq i \leqq s;\ 1 \leqq j \leqq m)$$

and all non-negative integers $l_1, \ldots, l_m$, the convergence relation

$$(E_n)_{X_1,\ldots,X_m}(\{[l_1, \ldots, l_m]\}) \xrightarrow[n\to\infty]{} E_{X_1,\ldots,X_m}(\{[l_1, \ldots, l_m]\})$$

is satisfied.

3.2. Compactness Criteria

The results of Section 1.15. can be transferred to the set N, i.e. they do not depend on the fact that of the considered measures on $\mathfrak{A}$ are integer-valued.

Let z be a point in A and (f_n) a sequence of functions in F having the property that

$$k_{S_n(z)} \leqq f_n \leqq k_{S_{n+1}(z)}.$$

To each ν in N we can associate the sequence $({}^n\nu)$ of the finite measures ${}^n\nu(da) = f_n(a)\,\nu(da)$ in a one-to-one way. Denoting by $\varrho(\alpha, \beta)$ the Prohorov distance of two finite measures α, β on $\mathfrak{A}$, we use the set-up

$$\varrho_N(\nu, \nu') = \sum_{n=1}^{\infty} \frac{\varrho({}^n\nu, {}^n(\nu'))}{2^n\big(1 + \varrho({}^n\nu, {}^n(\nu'))\big)} \qquad (\nu, \nu' \in N)$$

to obtain a non-negative real function ϱ_N on $N \times N$.

3.2.1. Proposition. *$[N, \varrho_N]$ is a complete separable metric space. The metric ϱ_N generates the coarsest topology in N with respect to which the real function*

$$\nu \rightsquigarrow \int f(a)\,\nu(da) \qquad (\nu \in N)$$

is continuous for all f in F.

Proof. 1. Together with ϱ, the function $\frac{\varrho}{1+\varrho}$ is also a metric on the set of all finite measures on $\mathfrak{A}$. Thus ϱ_N is symmetrical and satisfies the triangle inequality. The relation $\varrho_N(\nu, \nu') = 0$ is equivalent to $\varrho\left({}^n\nu, {}^n(\nu')\right) = 0$ $(n = 1, 2, \ldots)$, i.e. to ${}^n\nu = {}^n(\nu')$ $(n = 1, 2, \ldots)$. Hence it is satisfied if and only if ν coincides with ν'.

Consequently $[N, \varrho_N]$ is a metric space.

2. Let f be a function in F and (ν_n) a sequence of elements of N converging toward ν with respect to ϱ_N. By virtue of the definition of F, there exists an m such that f is identically zero outside of $S_m(z)$. We obtain

$$\int f(a)\, \nu_n(\mathrm{d}a) = \int f(a)\, f_m(a)\, \nu_n(\mathrm{d}a) = \int f(a)\, {}^m(\nu_n)\, (\mathrm{d}a)\,.$$

Since $\varrho\left({}^m(\nu_n), {}^m\nu\right) \xrightarrow[n\to\infty]{} 0$ it follows that

$$\int f(a)\, \nu_n(\mathrm{d}a) \xrightarrow[n\to\infty]{} \int f(a)\, {}^m\nu(\mathrm{d}a) = \int f(a)\, f_m(a)\, \nu(\mathrm{d}a) = \int f(a)\, \nu(\mathrm{d}a)\,.$$

Hence, for all f in F, the real function

$$\nu \rightsquigarrow \int f(a)\, \nu(\mathrm{d}a) \qquad (\nu \in N)$$

is continuous with respect to the metric ϱ_N.

Now let U be an open ball with radius $\eta > 0$ formed with respect to ϱ_N around some ν in N. Then there exist a natural number k and an $\varepsilon > 0$, such that the intersection of the open sets

$$U_n = \{\omega : \omega \in N, \varrho({}^n\omega, {}^n\nu) < \varepsilon\} \qquad (n = 1, 2, \ldots, k)$$

is included in U. Note that, in the set of all finite measures γ on $\mathfrak{A}$, ϱ generates the coarsest topology with respect to which, for all non-negative real functions g which are continuous and bounded on A, the real function

$$\gamma \rightsquigarrow \int g(a)\, \gamma(\mathrm{d}a)$$

is continuous. Hence there exist a finite sequence $g_1, \ldots, g_s$ of continuous bounded non-negative real functions on A as well as an $\alpha > 0$ having the property that

$$\begin{aligned} U_n &= \{\omega : \omega \in N, \varrho({}^n\omega, {}^n\nu) < \varepsilon\} \\ &\supseteqq \left\{\omega : \omega \in N, \left|\int g_i(a)\, {}^n\omega(\mathrm{d}a) - \int g_i(a)\, {}^n\nu(\mathrm{d}a)\right| < \alpha \quad \text{for} \quad 1 \leqq i \leqq s\right\} \\ &\qquad\qquad (1 \leqq n \leqq k), \end{aligned}$$

so that we obtain

$$\begin{aligned} U \supseteqq \Big\{\omega : \omega \in N, \Big|\int g_i(a)\, f_n(a)\, \omega(\mathrm{d}a) - \int g_i(a)\, f_n(a)\, \nu(\mathrm{d}a)\Big| < \alpha \\ \text{for} \quad 1 \leqq i \leqq s, 1 \leqq n \leqq k\Big\}. \end{aligned}$$

The functions $g_i f_n$ all belong to F.

Consequently the metric ϱ_N generates the coarsest topology in N with respect to which, for all f in F, the real function

$$\nu \rightsquigarrow \int f(a)\, \nu(\mathrm{d}a) \qquad (\nu \in N)$$

is continuous.

3. Let (ν_m) be a Cauchy sequence with respect to ϱ_N. Then, for all natural numbers n, $\left({}^n(\nu_m)\right)_{m=1,2,\ldots}$ is a Cauchy sequence with respect to ϱ, and hence converges weakly toward a finite measure γ_n on $\mathfrak{A}$, i.e., for all f in F and all n,

$$\int f(a)\, f_n(a)\, \nu_m(\mathrm{d}a) \xrightarrow[m\to\infty]{} \int f(a)\, \gamma_n(\mathrm{d}a).$$

Hence

$$\int f(a)\, \nu_m(\mathrm{d}a) \xrightarrow[m\to\infty]{} \int f(a)\, \gamma_k(\mathrm{d}a)$$

for all $k \geqq n$ and all f in F which are identically zero outside of $S_n(z)$. Consequently

$$\gamma_k\big((.) \cap S_n(z)\big) = \gamma_n\big((.) \cap S_n(z)\big) \qquad (k = n+1, n+2, \ldots),$$

so that we can conclude that there exists a γ in N having the property that

$$\gamma\big((.) \cap S_n(z)\big) = \gamma_n\big((.) \cap S_n(z)\big) \qquad (n = 1, 2, \ldots).$$

Therefore we have

$$\int f(a)\, \nu_m(\mathrm{d}a) \xrightarrow[m\to\infty]{} \int f(a)\, \gamma(\mathrm{d}a)$$

for all f in F which are identically zero outside of $S_n(z)$ $(n = 1, 2, \ldots)$, and accordingly for all f in F. This implies, in view of 2., $\varrho_N(\nu_m, \gamma) \xrightarrow[m\to\infty]{} 0$.

Thus the metric space $[N, \varrho_N]$ is complete.

4. As is known, the metric space of all finite measures on $\mathfrak{A}$ is separable. Now let Z be an at most countable, everywhere dense subset of this space. To each ν in N and each natural number n, there exists a sequence $(\gamma_{n,m})_{m=1,2,\ldots}$ of elements in Z which converges weakly toward ${}^n\nu$, and hence, by virtue of 2., also with respect to ϱ_N. As an immediate consequence of 2. we also obtain the relation $\varrho_N({}^n\nu, \nu) \xrightarrow[n\to\infty]{} 0$. Consequently Z is everywhere dense in N with respect to ϱ_N.

Hence the metric space $[N, \varrho_N]$ is separable. ∎

Instead of $\varrho_N(\nu_n, \nu) \xrightarrow[n\to\infty]{} 0$ we also write $\nu_n \overset{N}{\underset{n\to\infty}{\Rightarrow}} \nu$. An X in $\mathfrak{B}$ is called a *continuity set* with respect to ν if $\nu(\partial X) = 0$ is satisfied.

By analogy with 1.15.4. and 1.15.5. we obtain

3.2.2. *For each ν in N and each sequence (ν_n) of elements of N, the following statements are equivalent:*

a) $\nu_n \overset{N}{\underset{n\to\infty}{\Rightarrow}} \nu$;

b) $\overline{\lim\limits_{n\to\infty}} \nu_n(X') \leqq \nu(X')$ *for all bounded closed subsets X' of A and*

$\underline{\lim\limits_{n\to\infty}} \nu_n(X'') \geqq \nu(X'')$ *for all bounded open subsets X'' of A;*

c) $\nu_n(X) \xrightarrow[n\to\infty]{} \nu(X)$ *for all continuity X with respect to ν.*

3.2.3. *The σ-algebra of the Borel sets of the metric space $[N, \varrho_N]$ coincides with $\mathfrak{N}$.*

The restriction of ϱ_N to M accomplishes the same as the metric ϱ_M introduced in Section 1.15., for we have

3.2.4. *The set M is closed with respect to ϱ_N.*

Proof. Let $\Phi_n \underset{n\to\infty}{\overset{N}{\Rightarrow}} \nu, \quad \Phi_n \in M \quad \text{for} \quad n = 1, 2, \ldots.$

We denote by $\mathfrak{S}$ the ring of those X in $\mathfrak{B}$ for which $\nu(X)$ is integer-valued. In virtue of 3.2.2. c), $\mathfrak{S}$ includes all continuity sets with respect to ν. Obviously $\mathfrak{S}$ is monotone closed in $\mathfrak{B}$, and hence includes the σ-subring of $\mathfrak{B}$ generated by the ring of continuity sets with respect to ν. Similar to the proof of 3.1.6., we can convince ourselves that there is a continuity set with respect to ν between each bounded closed subset X' of A and each open neighbourhood U of X'. The set X' can be represented as an intersection of a decreasing sequence of open neighbourhoods so that we conclude $\mathfrak{S} = \mathfrak{B}$. ∎

The convergence with respect to ϱ_N (and hence also with respect to ϱ_M) can readily be reduced to the weak convergence of finite measures on the σ-algebra $\mathfrak{A}$ of the Borel sets of the complete separable metric space $[A, \varrho_A]$. For by 3.2.1., $\nu_n \underset{n\to\infty}{\overset{N}{\Rightarrow}} \nu$ is equivalent to the fact that, for all f in F, the finite measures $f(a)\,\nu_n(\mathrm{d}a)$ tend weakly toward the finite measure $f(a)\,\nu(\mathrm{d}a)$.

This remark forms the starting-point for showing that

3.2.5. *A subset Y of N is relatively compact with respect to ϱ_N if and only if it satisfies the following conditions for all closed bounded subsets X of A:*

a) $\sup\limits_{\nu\in Y} \nu(X) < +\infty$,

b) *to each $\varepsilon > 0$ there exists a compact subset $B_{X,\varepsilon}$ of X having the property that*

$$\sup_{\nu\in Y} \nu(X \setminus B_{X,\varepsilon}) < \varepsilon .$$

Proof. 1. Let Y be relatively compact with respect to ϱ_N, and X an arbitrary closed bounded subset of A. We choose an f in F having the property that $k_X \leqq f$. The set $\{f(a)\,\nu(\mathrm{d}a)\}_{\nu\in Y}$ is relatively compact in the metric space of all finite measures on $\mathfrak{A}$. Consequently (see for instance Billingsley [1], Theorem 6.2.) we have

$$\sup_{\nu\in Y} \int f(a)\,\nu(\mathrm{d}a) < +\infty,$$

and to each $\varepsilon > 0$ there exists a compact subset B of A having the property that

$$\sup_{\nu\in Y} \int_{A\setminus B} f(a)\, \nu(\mathrm{d}a) < \varepsilon .$$

Using

$$\sup_{\nu\in Y} \nu(X) \leqq \sup_{\nu\in Y} \int f(a)\, \nu(\mathrm{d}a)$$

and

$$\sup_{\nu\in Y} \nu\big(X \setminus (X \cap B)\big) \leqq \sup_{\nu\in Y} \int_{A\setminus B} f(a)\, \nu(\mathrm{d}a) < \varepsilon$$

and setting $B_{X,\varepsilon} = X \cap B$, we obtain

$$\sup_{\nu\in Y} \nu(X) < +\infty, \qquad \sup_{\nu\in Y} \nu(X \setminus B_{X,\varepsilon}) < \varepsilon .$$

Consequently the stated conditions are necessary.

2. Conversely, suppose now that Y satisfies these conditions. Let z denote an arbitrary but fixed point in A. To each natural number n we choose an f_n in F having the property that $k_{S_n(z)} \leqq f_n \leqq k_{S_{n+1}(z)}$, and set

$$Y_n = \{f_n(a)\, \nu(\mathrm{d}a)\}_{\nu\in Y} .$$

Then for all n,

$$\begin{aligned}\sup_{\nu\in Y_n} \nu(A) &= \sup_{\nu\in Y} \int f_n(a)\, \nu(\mathrm{d}a) \\ &\leqq \sup_{\nu\in Y} \nu\big(\{a : a \in A, \varrho_A(a, z) \leqq n + 1\}\big) < +\infty .\end{aligned}$$

To each n and each $\varepsilon > 0$ there exists a compact subset $B_{n,\varepsilon}$ of $\{a : a \in A, \varrho_A(a, z) \leqq n + 1\}$ which has the property that

$$\sup_{\nu\in Y} \nu\big(\{a : a \in A, \varrho_A(a, z) \leqq n + 1\} \setminus B_{n,\varepsilon}\big) < \varepsilon ,$$

so that we obtain

$$\sup_{\nu\in Y_n} \nu(A \setminus B_{n,\varepsilon}) = \sup_{\nu\in Y} \int_{A\setminus B_{n,\varepsilon}} f_n(a)\, \nu(\mathrm{d}a) \leqq \sup_{\nu\in Y} \nu\big(S_{n+1}(z) \setminus B_{n,\varepsilon}\big) < \varepsilon .$$

Thus each Y_n is relatively compact in the metric space of all finite measures on $\mathfrak{A}$.

Now let (ν_m) be a sequence of elements of Y. Since all Y_n are relatively compact, we can use the diagonal method to conclude that there exists a subsequence $(\nu_{m_k})_{k=1,2,\ldots}$, for which all of the sequences $\big(f_n(a)\, \nu_{m_k}(\mathrm{d}a)\big)_{k=1,2,\ldots}$ converge weakly toward finite measures on $\mathfrak{A}$. In the same way as in the third step of the proof of 3.2.1., this enables us to conclude that there exists a ν in N having the property that $\nu_{m_k} \underset{k\to\infty}{\overset{N}{\Rightarrow}} \nu$.

Consequently the set Y is relatively compact with respect to ϱ_N. ∎

From 3.2.5. and 3.2.4. we readily obtain

3.2.6. *A subset Y of M is relatively compact with respect to ϱ_M if and only if it satisfies the following conditions for all closed bounded subsets X of A:*

a) $\sup\limits_{\Phi \in Y} \Phi(X) < +\infty$;

b) *there exists a compact subset B_X of X such that the condition $\Phi(X \setminus B_X) = 0$ is satisfied for all Φ in Y.*

This characterization of the relatively compact subsets of the complete separable metric space $[M, \varrho_M]$ now leads to the following compactness criterion:

3.2.7. Proposition. *A subset $\boldsymbol{Z}$ of $\boldsymbol{E}^+$ is relatively compact with respect to $\varrho_{\boldsymbol{E}^+}$ if and only if $\sup\limits_{Z \in \boldsymbol{Z}} Z(M) < +\infty$ and, for each bounded closed subset X of A and all $\eta > 0$, there exist an $n_{X,\eta}$ having the property that*

$$\sup_{Z \in \boldsymbol{Z}} Z\big(\Phi(X) \geqq n_{X,\eta}\big) < \eta$$

as well as a compact subset $B_{X,\eta}$ of X having the property that

$$\sup_{Z \in \boldsymbol{Z}} Z\big(\Phi(X \setminus B_{X,\eta}) > 0\big) < \eta$$

Proof. 1. As is known (cf. again Billingsley [1], Theorems 6.1., 6.2.), a subset $\boldsymbol{Z}$ of the set $\boldsymbol{E}_+$ of all finite measures on the σ-algebra $\mathfrak{M}$ of the Borel sets of the complete separable metric space $[M, \varrho_M]$ is relatively compact if and only if $\sup\limits_{Z \in \boldsymbol{Z}} Z(M) < +\infty$ and to each $\eta > 0$ there exists a compact subset Y_η of M having the property that $\sup\limits_{Z \in \boldsymbol{Z}} Z(\Phi \notin Y_\eta) < \eta$.

Now let $\boldsymbol{Z}$ be relatively compact, and X an arbitrary bounded closed subset of A. In virtue of 3.2.6., there exist an $n_{X,\eta}$ such that $\sup\limits_{\Phi \in Y_\eta} \Phi(X) < n_{X,\eta}$ as well as a compact subset $B_{X,\eta}$ of X having the property that $\sup\limits_{\Phi \in Y_\eta} \Phi(X \setminus B_{X,\eta}) = 0$, and we obtain

$$\sup_{Z \in \boldsymbol{Z}} Z\big(\Phi(X) \geqq n_{X,\eta}\big) \leqq \sup_{Z \in \boldsymbol{Z}} Z(\Phi \notin Y_\eta) < \eta.$$

as well as

$$\sup_{Z \in \boldsymbol{Z}} Z\big(\Phi(X \setminus B_{X,\eta}) > 0\big) \leqq \sup_{Z \in \boldsymbol{Z}} Z(\Phi \notin Y_\eta) < \eta.$$

Consequently the stated conditions are necessary.

2. Conversely, suppose now that $\boldsymbol{Z}$ satisfies these conditions.

Let z be an arbitrary but fixed point in A, and X_m the closed ball around z which has the radius m. By assumption, to $\eta_m = 2^{-(m+1)}\eta$ there exists an n_{X_m,η_m} as well as a B_{X_m,η_m} with the formulated properties. We set

$$Y_\eta = \{\Phi : \Phi \in M, \Phi(X_m) < n_{X_m,\eta_m}, \\ \Phi(X_m \setminus B_{X_m,\eta_m}) = 0 \quad \text{for} \quad m = 1, 2, \ldots\}.$$

The set Y_η belongs to $\mathfrak{M}$, and by 3.2.6. it is relatively compact. We obtain

$$\sup_{Z\in\mathbf{Z}} Z(\Phi \notin Y_\eta) \leqq \sup_{Z\in\mathbf{Z}} \sum_{m=1}^{\infty} \left(Z\left(\Phi(X_m) \geqq n_{X_m,\eta_m}\right) + Z\left(\Phi(X_m \setminus B_{X_m,\eta_m}) > 0\right)\right)$$

$$< \sum_{m=1}^{\infty} (\eta_m + \eta_m) = \eta .$$

Hence for the compact closure V_η of Y_η it follows that $\sup_{Z\in\mathbf{Z}} Z(\Phi \notin V_\eta) < \eta$, i.e. $\mathbf{Z}$ is relatively compact with respect to $\varrho_{\mathbf{E}^+}$. ∎

Let be $\mathbf{Z}$ a subset of $\mathbf{P}$ and X a set in $\mathfrak{B}$ such that

$$\sup_{P\in\mathbf{Z}} \varrho_P(X) = \sup_{P\in\mathbf{Z}} \int \Phi(X)\, P(\mathrm{d}\Phi) < +\infty .$$

Then to each $\eta > 0$ there exists an $n_{X,\eta}$ which has the property

$$\sup_{P\in\mathbf{Z}} P\left(\Phi(X) \geqq n_{X,\eta}\right) < \eta .$$

Moreover,

$$P\left(\Phi(.) > 0\right) \leqq \varrho_P \qquad (P \in \mathbf{P}).$$

Therefore, by 3.2.5. and 3.2.7.,

3.2.8. Proposition. *Let $\mathbf{Z}$ be a subset of $\mathbf{P}$ such that $\{\varrho_P : P \in \mathbf{Z}\}$ is a subset of N relatively compact with respect to ϱ_N. Then $\mathbf{Z}$ is relatively compact with respect to $\varrho_{\mathbf{P}}$.*

As a counterpart to Proposition 3.1.10. we have

3.2.9. Proposition. *Let (P_n), (V_n), and (Q_n) be sequences of distributions on $\mathfrak{M}$ with the properties*

$$P_n = V_n * Q_n \qquad (n = 1, 2, \ldots)$$

and

$$P_n \underset{n\to\infty}{\Rightarrow} P, \qquad Q_n \underset{n\to\infty}{\Rightarrow} Q .$$

Then the sequence (V_n) also converges weakly toward some V in $\mathbf{P}$ and

$$P = V * Q .$$

Proof. For all natural numbers n and k and all X in $\mathfrak{B}$,

$$V_n\left(\Phi(X) \geqq k\right) \leqq V_n\left((\Phi + \Phi')(X) \geqq k\right) \qquad (\Phi' \in M)$$

and hence

$$V_n\left(\Phi(X) \geqq k\right) \leqq P_n\left(\Phi(X) \geqq k\right).$$

Therefore, by Proposition 3.2.7., as $\{P_n\}_{n=1,2,\ldots}$ is relatively compact with respect to $\varrho_{\mathbf{P}}$ so is $\{V_n\}_{n=1,2,\ldots}$. Thus each subsequence of (V_n) contains some subsequence which converges weakly toward some V in $\mathbf{P}$. By 3.1.10. we obtain $P = V * Q$. Because of Proposition 1.5.8. the distribution V is uniquely determined, i.e. we have $V_n \underset{n\to\infty}{\Rightarrow} V$. ∎

Applying Proposition 3.2.7. and 3.1.10. in the same way as in the above proof one shows that

3.2.10. *If, for each term P_n of a sequence of distributions on $\mathfrak{M}$ which converges weakly toward P, there exists the k-th convolution root $\sqrt[k]{P_n}$, then $\sqrt[k]{P}$ also exists, and*

$$\sqrt[k]{P_n} \underset{n\to\infty}{\Rightarrow} \sqrt[k]{P}.$$

Hence in particular, for each natural number k, the set of those P in $\mathbf{P}$ for which $\sqrt[k]{P}$ exists, is closed with respect to $\varrho_{\mathbf{P}}$. From this follows the statement a) in

3.2.11. *The following subsets of $\mathbf{P}$ are closed with respect to $\varrho_{\mathbf{P}}$:*

a) *The set $\mathbf{T}$ of all infinitely divisible distributions.*
b) *The set of all distributions which are free from after-effects.*
c) *The set of those distributions which have, for a fixed ν in N, the property g_ν.*
d) *The set of all purely random distributions.*

Proof. 1. Let (P_n) be a sequence of distributions on $\mathfrak{M}$ free from after-effects which converges weakly toward P. By 3.1.9. we obtain, for all finite sequences $X_1, \ldots, X_m$ of pairwise disjoint sets in $\mathfrak{B}_P$,

$$(P_n)_{X_1,\ldots,X_m} = \mathop{\times}_{i=1}^{m} (P_n)_{X_i} \underset{n\to\infty}{\longrightarrow} \mathop{\times}_{i=1}^{m} P_{X_i}$$

as well as

$$(P_n)_{X_1,\ldots,X_m} \underset{n\to\infty}{\longrightarrow} P_{X_1,\ldots,X_m}$$

and hence also

$$P_{X_1,\ldots,X_m} = \mathop{\times}_{i=1}^{m} P_{X_i},$$

from which by means of 1.3.3. and 3.1.6. we can conclude that P is free from after-effects.

2. Let ν be a measure in N and (P_n) a sequence of distributions on $\mathfrak{M}$ which have the property g_ν, the sequence converging weakly toward P.

If ν is equal to o, then all P_n, and hence also P, coincide with δ_o. Therefore we can assume without loss of generality that ν be different from o. Then, by 3.1.6., to each X in $\mathfrak{B}$ there exists a V in $\mathfrak{B}_P$ having the properties $X \subseteq V$, $\nu(V) > 0$. Using 3.1.9. we obtain

$$\left\| {}_V(P_n) - \sum_{k=0}^{\infty} P_V(\{k\}) \left(Q_{\nu((.)|V)}\right)^k \right\| = \left\| \sum_{k=0}^{\infty} \left((P_n)_V(\{k\}) - P_V(\{k\})\right) \left(Q_{\nu((.)|V)}\right)^k \right\|$$
$$\leqq \|(P_n)_V - P_V\| \underset{n\to\infty}{\longrightarrow} 0,$$

and hence

$$_V(P_n) \underset{n\to\infty}{\Rightarrow} \sum_{k=0}^{\infty} P_V(\{k\}) \left(Q_{\nu((.)|V)}\right)^k,$$

so that by 3.1.11. we can conclude that

$$_VP = \sum_{k=0}^{\infty} P_V(\{k\}) \left(Q_{\nu((.)|V)}\right)^k.$$

Consequently the distribution P has the property $g_{\nu,V}$, and therefore the property $g_{\nu,X}$ for arbitrary X in $\mathfrak{B}$.

3. Let (P_n) be a sequence of purely random distributions on $\mathfrak{M}$ which converges weakly toward P. Then, by 1.14.13., for all n there exists a ν_n in N such that P_n has the property g_{ν_n}.

In the case $P = \delta_o$ it is trivial that P is purely random, so that without loss of generality we may suppose that P be different from δ_o. To each X in $\mathfrak{B}$ we now choose a set V in $\mathfrak{B}_P$ which includes X and has the property that $P_V \neq \delta_0$. Then for all $n \geqq n_0$, we have $(P_n)_V \neq \delta_o$, and hence

$$_V(P_n) = \sum_{k=0}^{\infty} (P_n)_V (\{k\}) \left(Q_{\nu_n((.)|V)}\right)^k.$$

From $P_n \underset{n\to\infty}{\Rightarrow} P$, $V \in \mathfrak{B}_P$, using 3.1.11. we conclude that

$$_V(P_n) \underset{n\to\infty}{\Rightarrow} {}_VP.$$

From this and by 3.1.5. we obtain, for all natural numbers k having the property that $P_V(\{k\}) > 0$,

$$_V(P_n)\left((\cdot)|\Phi(V) = k\right) \underset{n\to\infty}{\Rightarrow} {}_VP\left((\cdot)|\Phi(V) = k\right)$$

and hence

$$\left(Q_{\nu_n((.)|V)}\right)^k \underset{n\to\infty}{\Rightarrow} {}_VP\left((\cdot)|\Phi(V) = k\right).$$

From this we conclude by means of 3.2.10. that

$$Q_{\nu_n((.)|V)} \underset{n\to\infty}{\Rightarrow} Q = \sqrt[k]{{}_VP((.)|\Phi(V) = k)}.$$

Obviously we have $Q\left(\Phi(A) = \Phi(V) = 1\right) = 1$, i.e. there exists a distribution ν on $\mathfrak{A}$ having the properties

$$\nu(V) = 1, \qquad Q_\nu = \sqrt[k]{{}_VP((\cdot)|\Phi(V) = k)}$$

so that we obtain

$$Q_{\nu_n((.)|V)} \underset{n\to\infty}{\Rightarrow} Q_\nu.$$

Using 3.1.9. and 3.1.10., we now conclude that

$$(P_n)_V \xrightarrow[n\to\infty]{} P_V, \qquad \left(Q_{\nu_n((.)|V)}\right)^l \underset{n\to\infty}{\Rightarrow} (Q_\nu)^l \qquad \text{for} \qquad l = 1, 2, \ldots.$$

Let Y be a continuity set with respect to $\sum_{l=0}^{\infty} P_V(\{l\})\,(Q_\nu)^l$. Then Y is also a continuity set with respect to $P_V(\{l\})\,(Q_\nu)^l$ for $l = 0, 1, \ldots$, and we have

$$\overline{\lim_{n\to\infty}} \left|\nu(P_n)\,(Y) - \sum_{l=0}^{\infty} P_V(\{l\})\,(Q_\nu)^l\,(Y)\right|$$

$$\leqq \overline{\lim_{n\to\infty}} \left|\sum_{l=0}^{m} (P_n)_V\,(\{l\})\,\big(Q_{\nu_n((.)|V)}\big)^l\,(Y) - \sum_{l=0}^{m} P_V(\{l\})\,(Q_\nu)^l\,(Y)\right|$$

$$+ \overline{\lim_{n\to\infty}}\,(P_n)_V\,(\{m+1, m+2, \ldots\}) + P_V(\{m+1, m+2, \ldots\})$$

$$= 0 + 2P_V(\{m+1, m+2, \ldots\})$$

for all natural numbers m. Therefore, we have

$$_VP = \sum_{l=0}^{\infty} (P_V)\,(\{l\})\,(Q_\nu)^l,$$

i.e. that P has the property $g_{\nu,V}$.

Now, if $\nu(X) = 0$, then $P_X = \delta_0$. Otherwise P has the property $g_{\nu((.)|X),X}$. Since X was arbitrarily chosen from $\mathfrak{B}$, it follows that P is purely random. ∎

If (a_n) is a sequence of points in A which converges toward a, all of these points being different from a, then we obtain

$$\delta_{\delta_{a_n}+\delta_a} \underset{n\to\infty}{\Rightarrow} \delta_{2\delta_a}.$$

For non-discrete phase spaces, the set of all simple P in $\boldsymbol{P}$ is therefore not closed with respect to $\varrho_{\boldsymbol{P}}$.

3.3. The Metric Space $[W, \varrho_W]$

Let z be an arbitrary but fixed point in A. By means of

$$r(\Phi) = \sup\left\{d : d \geqq 0,\ \Phi\big(S_d(z)\big) = 0\right\}$$

we obtain a mapping r of M into $[0, +\infty]$. Obviously $r(\Phi)$ is finite if and only if Φ is different from the zero measure o.

3.3.1. *The mapping r is continuous.*

Proof. Let $\Phi_n \overset{M}{\underset{n\to\infty}{\Rightarrow}} \Phi$. For all d in $\big[0, r(\Phi)\big)$ we have $\Phi\big(\partial S_d(z)\big) = 0$, so that by 1.15.4. we can conclude that

$$\lim_{n\to\infty} \Phi_n\big(S_d(z)\big) = \Phi\big(S_d(z)\big) = 0,$$

i.e. that $d \leqq \underline{\lim}_{n\to\infty}\, r(\Phi_n)$. From this, for $r(\Phi) > 0$ we obtain the following inequality which is trivially true in the case $r(\Phi) = 0$:

$$r(\Phi) \leqq \varliminf_{n\to\infty} r(\Phi_n).$$

For all d in $\big(r(\Phi), +\infty\big)$ by 1.15.4. it follows that

$$\varliminf_{n\to\infty} \Phi_n\big(S_d(z)\big) \geqq \Phi\big(S_d(z)\big) \geqq 1$$

and hence $\varlimsup_{n\to\infty} r(\Phi_n) \leqq d$, i.e. for $r(\Phi) < +\infty$ we obtain the following inequality, which is trivially true in the case $r(\Phi) = +\infty$:

$$\varlimsup_{n\to\infty} r(\Phi_n) \leqq r(\Phi).$$

Finally we conclude that $\lim_{n\to\infty} r(\Phi_n) = r(\Phi)$. ∎

For brevity we set $A^* = M \setminus \{o\}$. By

$$\varrho_{A^*}(\Phi, \Psi) = \varrho_M(\Phi, \Psi) + |r(\Phi) - r(\Psi)| \qquad (\Phi, \Psi \in A^*)$$

we obtain a metric ϱ_{A^*} on A^*. In virtue of 3.3.1., ϱ_{A^*} and ϱ_M generate the same topology on A^*. Hence in particular the metric space $[A^*, \varrho_{A^*}]$ is separable. We shall now show that it is complete, too, and thus can be considered a new phase space.

Each Cauchy sequence (Φ_n) with respect to ϱ_{A^*} is also a Cauchy sequence with respect to ϱ_M. Consequently in M there exists a Φ having the property that $\Phi_n \overset{M}{\underset{n\to\infty}{\Rightarrow}} \Phi$. This element Φ is different from the zero measure o. For otherwise we would have, for each natural number n,

$$\varlimsup_{m\to\infty} \varrho_{A^*}(\Phi_n, \Phi_{n+m}) \geqq \varlimsup_{m\to\infty} |r(\Phi_n) - r(\Phi_{n+m})| = +\infty,$$

which would contradict our assumption on (Φ_n). Therefore the sequence (Φ_n) converges toward Φ not only with respect to ϱ_M, but also with respect to ϱ_{A^*}.

We shall distinguish the mathematical structures formed with respect to the phase space $[A^*, \varrho_{A^*}]$ by the superscript *, so that in particular the σ-algebra $\mathfrak{A}^*$, the set F^* as well as the complete separable metric space $[N^*, \varrho_{N^*}]$ are defined in this way.

The bounded subsets of the metric space $[A^*, \varrho_{A^*}]$ can be characterized as follows:

3.3.2. *A subset Z of A^* is bounded with respect to ϱ_{A^*} if and only if there exists an X in $\mathfrak{B}$ having the property that*

$$Z \subseteqq \{\Phi : \Phi \in M, \Phi(X) > 0\}.$$

Proof. 1. Let Z be bounded. For each Φ_0 in A^* we obtain

$$\sup_{\Phi\in Z} |r(\Phi) - r(\Phi_0)| \leqq \sup_{\Phi\in Z} \varrho_{A^*}(\Phi, \Phi_0) < +\infty$$

and hence $\sup_{\Phi \in Z} r(\Phi) < +\infty$. But for each $d > \sup_{\Phi \in Z} r(\Phi)$ we obtain

$$Z \subseteq \{\Phi : \Phi \in M, \Phi(S_d(z)) > 0\},$$

i.e. we can set $X = S_d(z)$.

Thus the stated condition is necessary.

2. Conversely, suppose now that, for some X in $\mathfrak{B}$, Z be included in $\{\Phi : \Phi \in M, \Phi(X) > 0\}$. Then Z is included in $\{\Phi : \Phi \in M, \Phi(S_d(z)) > 0\}$ for some $d > 0$. Hence

$$\sup_{\Phi \in Z} r(\Phi) \leqq \sup \{r(\Phi) : \Phi \in M, \Phi(S_d(z)) > 0\} \leqq d < +\infty.$$

By reason of the boundedness of ϱ_M,

$$\sup_{\Phi \in Z} \varrho_{A^*}(\Phi, \Phi_0) \leqq \sup_{\Phi \in Z} \varrho_M(\Phi, \Phi_0) + \sup_{\Phi \in Z} r(\Phi) + r(\Phi_0) < +\infty$$

for each Φ_0 in A^*.

Consequently Z is bounded. ∎

The σ-algebra $\mathfrak{A}^*$ of the Borel subsets of $[A^*, \varrho_{A^*}]$ consists of those Y in $\mathfrak{M}$ which do not include the zero measure o. Accordingly a measure W in $\boldsymbol{W}$ is uniquely determined by its restriction ν_W to $\mathfrak{A}^*$. In virtue of 3.3.2., the correspondence

$$W \curvearrowright \nu_W \qquad (W \in \boldsymbol{W})$$

provides a one-to-one mapping of $\boldsymbol{W}$ onto N^*. Thus the set-up

$$\varrho_{\boldsymbol{W}}(W_1, W_2) = \varrho_{N^*}(\nu_{W_1}, \nu_{W_2}) \qquad (W_1, W_2 \in \boldsymbol{W})$$

makes $\boldsymbol{W}$ a complete separable metric space $[\boldsymbol{W}, \varrho_{\boldsymbol{W}}]$. For the convergence of (W_n) toward W with respect to the metric $\varrho_{\boldsymbol{W}}$, we shall write $W_n \overset{\boldsymbol{W}}{\underset{n\to\infty}{\Rightarrow}} W$.

In virtue of 3.3.2., we can identify F^* with the set of those bounded non-negative real functions g on M which are continuous with respect to ϱ_M and which, for suitable X in $\mathfrak{B}$, vanish identically outside of the set $\{\Phi : \Phi \in M, \Phi(X) > 0\}$.

This is so because each g in F^* can be continuously extended over the whole M by means of $g(o) = 0$: It follows from $\Phi_n \overset{M}{\underset{n\to\infty}{\Rightarrow}} o$ that $\Phi_n(X) \underset{n\to\infty}{\longrightarrow} 0$ for all X in $\mathfrak{B}$, and hence that $g(\Phi_n) \underset{n\to\infty}{\longrightarrow} 0$.

Thus, by 3.2.1. we readily obtain

3.3.3. *The metric $\varrho_{\boldsymbol{W}}$ generates the coarsest topology in $\boldsymbol{W}$ with respect to which for all X in $\mathfrak{B}$ and all bounded non-negative real functions g on M which are continuous with respect to ϱ_M and vanish identically outside of $\{\Phi : \Phi \in M, \Phi(X) > 0\}$, the real function*

$$W \curvearrowright \int g(\Phi)\, W(\mathrm{d}\Phi) \qquad (W \in \boldsymbol{W})$$

is continuous.

The mapping $a \frown \delta_a$ of A into A^* is continuous with respect to ϱ_A and ϱ_{A^*}. Thus for each g in F^* the real function $a \frown g(\delta_a)$ is belongs to F. Now, if $\nu_n \underset{n\to\infty}{\overset{N}{\Rightarrow}} \nu$, then for all g in F^* it follows that

$$\int g(\Phi)\, Q_{\nu_n}(\mathrm{d}\Phi) = \int g(\delta_a)\, \nu_n(\mathrm{d}a) \xrightarrow[n\to\infty]{} \int g(\delta_a)\, \nu(\mathrm{d}a) = \int g(\Phi)\, Q_\nu(\mathrm{d}\Phi),$$

i.e. we have

$$Q_{\nu_n} \underset{n\to\infty}{\overset{\mathbf{W}}{\Rightarrow}} Q_\nu .$$

For each f in F and each natural number k, the set-up

$$g_k(\Phi) = \min\left(\int f(a)\, \Phi(\mathrm{d}a), k\right)$$

provides a bounded non-negative real function g_k on M which is continuous with respect to ϱ_M. If f is identically zero outside of the set X in $\mathfrak{B}$, then g_k is identically zero outside of $\{\Phi : \Phi \in M, \Phi(X) > 0\}$. In view of 3.3.3. we can therefore conclude from $Q_{\nu_n} \underset{n\to\infty}{\overset{\mathbf{W}}{\Rightarrow}} Q_\nu$ that

$$\int g_k(\Phi)\, Q_{\nu_n}(\mathrm{d}\Phi) \xrightarrow[n\to\infty]{} \int g_k(\Phi)\, Q_\nu(\mathrm{d}\Phi).$$

If k is chosen large enough to satisfy $f \leqq k$, then $\int g_k(\Phi)\, Q_{\nu_n}(\mathrm{d}\Phi)$ coincides with $\int \left(\int f(a)\, \Phi(\mathrm{d}a)\right) Q_{\nu_n}(\mathrm{d}\Phi)$, and hence by 1.2.2. with $\int f(a) \varrho_{Q_{\nu_n}}(\mathrm{d}a) = \int f(a)\, \nu_n(\mathrm{d}a)$. A corresponding statement is true for $\int g_k(\Phi)\, Q_\nu(\mathrm{d}\Phi)$, so that we obtain

$$\int f(a)\, \nu_n(\mathrm{d}a) \xrightarrow[n\to\infty]{} \int f(a)\, \nu(\mathrm{d}a),$$

i.e. we have

$$\nu_n \underset{n\to\infty}{\overset{N}{\Rightarrow}} \nu .$$

Thus we find

3.3.4. *The correspondence $\nu \frown Q_\nu$ provides a homeomorphic mapping of $[N, \varrho_N]$ into $[\mathbf{W}, \varrho_{\mathbf{W}}]$.*

By analogy to 3.1.7., we have

3.3.5. Proposition. *Let W be an element of $\mathbf{W}$ and $\mathfrak{H}$ a semi-ring of sets in $\mathfrak{B}_W$. If to each bounded closed subset D of A and to each open neighbourhood U of D in the ring generated by $\mathfrak{H}$ there exists a set X having the property that $D \subseteqq X \subseteqq U$, then, for a sequence (W_n) of elements of $\mathbf{W}$, $W_n \underset{n\to\infty}{\overset{\mathbf{W}}{\Rightarrow}} W$ if and only if, for each finite sequence $H_1, \ldots, H_m$ of pairwise disjoint sets in $\mathfrak{H}$, the convergence relation*

$$\lim_{n\to\infty} \|(W_n)_{H_1,\ldots,H_m}((.) \setminus \{[0, \ldots, 0]\}) - W_{H_1,\ldots,H_m}((.) \setminus \{[0, \ldots, 0]\})\| = 0$$

is satisfied.

Proof. 1. Suppose that $W_n \underset{n\to\infty}{\overset{\mathbf{W}}{\Rightarrow}} W$. If X is a set in $\mathfrak{B}$, and g a bounded continuous non-negative real function on M which is identically zero outside of $\{\Phi : \Phi \in M, \Phi(X) > 0\}$, then 3.3.3. shows that, for all bounded non-negative real functions h on M which are continuous with respect to ϱ_M, the convergence relation

$$\int h(\Phi)\, g(\Phi)\, W_n(\mathrm{d}\Phi) \xrightarrow[n\to\infty]{} \int h(\Phi)\, g(\Phi)\, W(\mathrm{d}\Phi)$$

is satisfied, i.e. in $\boldsymbol{E}^+$ the sequence of finite measures $g(\Phi)\, W_n(\mathrm{d}\Phi)$ converges weakly toward $g(\Phi)\, W(\mathrm{d}\Phi)$.

Now let $H_1, \ldots, H_m$ be an arbitrary finite sequence of pairwise disjoint sets in $\mathfrak{B}_W$. Then the $H_1, \ldots, H_m$ are also continuity sets with respect to $g(\Phi)\, W(\mathrm{d}\Phi)$. We choose g in such a way that it is equal unity on the set $\{\Phi : \Phi \in M, \Phi(H_1 \cup \cdots \cup H_m) > 0\}$. Then for all n,

$$\big(g(\Phi)\, W_n(\mathrm{d}\Phi)\big)_{H_1,\ldots,H_m}\big((.) \setminus \{[0, \ldots, 0]\}\big) = (W_n)_{H_1,\ldots,H_m}\big((.) \setminus \{[0, \ldots, 0]\}\big),$$

and

$$\big(g(\Phi)\, W(\mathrm{d}\Phi)\big)_{H_1,\ldots,H_m}\big((.) \setminus \{[0, \ldots, 0]\}\big) = W_{H_1,\ldots,H_m}\big((.) \setminus \{[0, \ldots, 0]\}\big),$$

so that by 3.1.9. we can conclude that

$$\lim_{n\to\infty} \big\|(W_n)_{H_1,\ldots,H_m}\big((.) \setminus \{[0, \ldots, 0]\}\big) - W_{H_1,\ldots,H_m}\big((.) \setminus \{[0, \ldots, 0]\}\big)\big\| = 0.$$

Consequently the afore-mentioned condition on the sequence (W_n) is necessary.

2. Conversely, suppose that the sequence (W_n) satisfies the condition stated above. Let X be a set in $\mathfrak{B}$ and g a bounded continuous non-negative real function on M which is identically zero outside of $\{\Phi : \Phi \in M, \Phi(X) > 0\}$. By assumption there exists a finite sequence $V_1, \ldots, V_k$ of pairwise disjoint sets in $\mathfrak{H}$ having the property that $X \subseteq V = V_1 \cup \cdots \cup V_k$. For each finite sequence $H_1, \ldots, H_m$ of pairwise disjoint sets in $\mathfrak{H}$, there exists a finite sequence $X_1, \ldots, X_s$ of pairwise disjoint sets in $\mathfrak{H}$ such that all of the $V_1, \ldots, V_k, H_1, \ldots, H_m$ can be represented as unions of certain X_i. By assumption we have

$$\lim_{n\to\infty} \big\|(W_n)_{X_1,\ldots,X_s}\big((.) \setminus \{[0, \ldots, 0]\}\big) - W_{X_1,\ldots,X_s}\big((.) \setminus \{[0, \ldots, 0]\}\big)\big\| = 0.$$

Thus, if $V = \bigcup_{i\in I} X_i$, we get

$$\begin{aligned} &(W_n)_{X_1,\ldots,X_s}\Big((.) \cap \Big\{[l_1, \ldots, l_s] : \sum_{i\in I} l_i > 0\Big\}\Big) \\ &\xrightarrow[n\to\infty]{} W_{X_1,\ldots,X_s}\Big((.) \cap \Big\{[l_1, \ldots, l_s] : \sum_{i\in I} l_i > 0\Big\}\Big), \end{aligned}$$

so that we obtain

$$\Big(W_n\big(\psi \in (.), \psi(V) > 0\big)\Big)_{H_1,\ldots,H_m} \xrightarrow[n\to\infty]{} \Big(W\big(\psi \in (.), \psi(V) > 0\big)\Big)_{H_1,\ldots,H_m},$$

and hence by 3.1.7.

$$W_n\big(\psi \in (.), \psi(V) > 0\big) \underset{n\to\infty}{\Rightarrow} W\big(\psi \in (.), \psi(V) > 0\big).$$

Thus

$$\int g(\Phi)\, W_n(\mathrm{d}\Phi) \xrightarrow[n\to\infty]{} \int g(\Phi)\, W(\mathrm{d}\Phi),$$

and by 3.3.3. it follows that

$$W_n \underset{n\to\infty}{\overset{\boldsymbol{W}}{\Rightarrow}} W. \quad \blacksquare$$

By virtue of 1.9.2. we may also formulate the convergence condition stated in 3.3.5. as follows. For each finite sequence $H_1, \ldots, H_m$ of pairwise disjoint sets in $\mathfrak{B}_W$ and each finite sequence $l_1, \ldots, l_m$ of non-negative integers satisfying the condition $l_1 + \cdots + l_m > 0$, we have

$$(W_n)_{H_1,\ldots,H_m}(\{[l_1, \ldots, l_m]\}) \xrightarrow[n\to\infty]{} W_{H_1,\ldots,H_m}(\{[l_1, \ldots, l_m]\})$$

as well as

$$W_n\big(\Psi(H_1 \cup \cdots \cup H_m) > 0\big) \xrightarrow[n\to\infty]{} W\big(\Psi(H_1 \cup \cdots \cup H_m) > 0\big).$$

In virtue of 2.2.1., $P \curvearrowright \tilde{P}$ is a one-to-one mapping of $\boldsymbol{T}$ onto $\boldsymbol{W}$, and from 2.2.3. we obtain the relation $\mathfrak{B}_P = \mathfrak{B}_{\tilde{P}}$ for all P in $\boldsymbol{T}$. Thus, in view of 3.1.6., for each W in $\boldsymbol{W}$, each closed set D in $\mathfrak{B}$, and each open neighbourhood U of D, the ring $\mathfrak{B}_W$ includes a set X having the property that $D \subseteq X \subseteq U$. Thus the convergence with respect to $\varrho_{\boldsymbol{W}}$ is completely characterized by 3.3.5.

3.3.6. Proposition. *The one-to-one mapping $P \curvearrowright \tilde{P}$ of $\boldsymbol{T}$ onto $\boldsymbol{W}$ is continuous in both directions with respect to $\varrho_{\boldsymbol{P}}$ and $\varrho_{\boldsymbol{W}}$.*

Proof. Let P be a distribution in $\boldsymbol{T}$ and (P_n) a sequence of distributions in $\boldsymbol{T}$, further let $H_1, \ldots, H_m$ be a finite sequence of pairwise disjoint sets in $\mathfrak{B}_P$.

By virtue of 2.2.8. we have

$$\widetilde{(P_n)_{H_1,\ldots,H_m}} = (\widetilde{P_n})_{H_1,\ldots,H_m}\big((.) \setminus \{[0, \ldots, 0]\}\big),$$

$$\widetilde{P_{H_1,\ldots,H_m}} = (\tilde{P})_{H_1,\ldots,H_m}\big((.) \setminus \{[0, \ldots, 0]\}\big).$$

Hence, in view of 1.10.3.,

$$\|(P_n)_{H_1,\ldots,H_m} - P_{H_1,\ldots,H_m}\| \xrightarrow[n\to\infty]{} 0$$

is equivalent to

$$\lim_{n\to\infty} \Big\|(\widetilde{P_n})_{H_1,\ldots,H_m}\big((.) \setminus \{[0, \ldots, 0]\}\big) - (\tilde{P})_{H_1,\ldots,H_m}\big((.) \setminus \{[0, \ldots, 0]\}\big)\Big\| = 0.$$

By applying 3.1.9. and 3.3.5., we have thus established the equivalence of $P_n \underset{n\to\infty}{\Rightarrow} P$ and $\widetilde{P_n} \underset{n\to\infty}{\overset{\boldsymbol{W}}{\Rightarrow}} \tilde{P}$. $\blacksquare$

As an immediate consequence of 3.3.4. and 3.3.6., using 2.2.14. we obtain

3.3.7. Proposition. *The correspondence $\nu \curvearrowright P_\nu$ provides a homeomorphic mapping of $[N, \varrho_N]$ into $[\boldsymbol{P}, \varrho_{\boldsymbol{P}}]$.*

Note that it is not necessary that $\boldsymbol{P} \cap \boldsymbol{W}$ be closed either in $\boldsymbol{P}$ or in $\boldsymbol{W}$. If for instance we set $[A, \varrho_A] = R^1$, then, in the form of $P_n = \delta_{\delta_n}$ we obtain a sequence of measures in $\boldsymbol{P} \cap \boldsymbol{W}$ which converges toward $\delta_o \notin \boldsymbol{W}$ with respect to ϱ_{E^+}, but toward the zero measure $0 \notin \boldsymbol{P}$ with respect to $\varrho_{\boldsymbol{W}}$.

We have, however,

3.3.8. *The topologies generated by $\varrho_{\boldsymbol{P}}$ and $\varrho_{\boldsymbol{W}}$ coincide on $\boldsymbol{P} \cap \boldsymbol{W}$.*

Proof. By 3.1.2. and 3.3.3., for each sequence (P_n) of distributions in $\boldsymbol{P} \cap \boldsymbol{W}$ and each P in $\boldsymbol{P} \cap \boldsymbol{W}$ we can conclude from $P_n \underset{n\to\infty}{\Rightarrow} P$ that $P_n \overset{\boldsymbol{W}}{\underset{n\to\infty}{\Rightarrow}} P$.

Conversely, suppose now that $P_n \overset{\boldsymbol{W}}{\underset{n\to\infty}{\Rightarrow}} P$ in $\boldsymbol{P} \cap \boldsymbol{W}$. Let $X_1, \ldots, X_m$ be an arbitrary sequence of pairwise disjoint sets in $\mathfrak{B}_P$. By 3.3.5.,

$$\lim_{n\to\infty} \left\|(P_n)_{X_1,\ldots,X_m}\left((\cdot) \setminus \{[0, \ldots, 0]\}\right) - P_{X_1,\ldots,X_m}\left((\cdot) \setminus \{[0, \ldots, 0]\}\right)\right\| = 0.$$

From this it follows

$$\lim_{n\to\infty} (P_n)_{X_1,\ldots,X_m}(\{[0, \ldots, 0]\}) = P_{X_1,\ldots,X_m}(\{[0, \ldots, 0]\}).$$

Summarizing we obtain

$$(P_n)_{X_1,\ldots,X_m} \xrightarrow[n\to\infty]{} P_{X_1,\ldots,X_m}.$$

Now we can use 3.1.9. to conclude that $P_n \underset{n\to\infty}{\Rightarrow} P$. ∎

3.4. Infinitesimal Triangular Arrays

A triangular array $(P_{n,i})_{\substack{n=1,2,\ldots\\ 1\leq i\leq m_n}}$ of distributions on $\mathfrak{M}$ is said to be *infinitesimal*, if for all X in $\mathfrak{B}$ the convergence relation

$$\max_{1\leq i\leq m_n} P_{n,i}\big(\Phi(X) > 0\big) \xrightarrow[n\to\infty]{} 0$$

is satisfied.

3.4.1. Proposition. *A distribution P on $\mathfrak{M}$ is infinitely divisible if and only if there exists an infinitesimal triangular array $(P_{n,i})$ of distributions on $\mathfrak{M}$ such that the relation*

$$\mathop{\Large\ast}_{1\leq i\leq m_n} P_{n,i} \underset{n\to\infty}{\Rightarrow} P$$

is satisfied.

Proof. 1. Given any element P of $\boldsymbol{T}$, we set

$$P_{n,i} = \sqrt[n]{P} \quad \text{for} \quad n = 1, 2, \ldots; \quad 1 \leqq i \leqq m_n = n,$$

so that $\mathop{\ast}_{1\leq i\leq m_n} P_{n,i} \underset{n\to\infty}{\Rightarrow} P$. The triangular array $(P_{n,i})$ is infinitesimal because, for all X in $\mathfrak{B}$,

$$\min_{1\leq i\leq m_n} P_{n,i}\big(\Phi(X) = 0\big) = \sqrt[n]{P}\,\big(\Phi(X) = 0\big) = \sqrt[n]{P(\Phi(X) = 0)} \xrightarrow[n\to\infty]{} 1,$$

ince $P\big(\Phi(X) = 0\big)$ is positive in virtue of 2.2.5.

2. Let $(P_{n,i})$ be an infinitesimal triangular array having the property that $\underset{1 \leq i \leq m_n}{\ast} P_{n,i} \underset{n\to\infty}{\Rightarrow} P$. By 3.1.9. and 1.5.7. we obtain, for all finite sequences $X_1, \ldots, X_m$ of pairwise disjoint sets in $\mathfrak{B}_P$,

$$\underset{1 \leq i \leq m_n}{\ast} (P_{n,i})_{X_1,\ldots,X_m} \underset{n\to\infty}{\longrightarrow} P_{X_1,\ldots,X_m}.$$

By 1.11.5. all $P_{X_1,\ldots,X_m}$ are infinitely divisible from which, by means of 1.6.3. and 3.1.6., we can conclude that P is infinitely divisible. ∎

The weak convergence of $\underset{1 \leq i \leq m_n}{\ast} P_{n,i}$ toward P can be transformed into a ϱ_W-convergence toward $\tilde{P}$:

3.4.2. Theorem. *For each P in $\boldsymbol{T}$ and each infinitesimal triangular array $(P_{n,i})$ of distributions in $\mathbf{P}$, the following statements are equivalent.*

a) $\underset{1 \leq i \leq m_n}{\ast} P_{n,i} \underset{n\to\infty}{\Rightarrow} P$.

b) $\underset{1 \leq i \leq m_n}{\ast} \mathscr{U}_{P_{n,i}} \underset{n\to\infty}{\Rightarrow} P$.

c) $\sum_{1 \leq i \leq m_n} P_{n,i}\big((.) \setminus \{o\}\big) \overset{W}{\underset{n\to\infty}{\Rightarrow}} \tilde{P}$.

Proof. In virtue of 3.1.9., a) is equivalent to the fact that, for all finite sequences $X_1, \ldots, X_m$ of pairwise disjoint sets in $\mathfrak{B}_P$, the convergence relation

$$\underset{1 \leq i \leq m_n}{\ast} (P_{n,i})_{X_1,\ldots,X_m} \underset{n\to\infty}{\longrightarrow} P_{X_1,\ldots,X_m} \qquad (*)$$

is satisfied.

By 2.2.8., $P_{X_1,\ldots,X_m}$ coincides with $\mathscr{U}_{(\tilde{P})_{X_1,\ldots,X_m}((.)\setminus\{[0,\ldots,0]\})}$. Hence, by 1.11.4., $(*)$ is equivalent to

$$\sum_{1 \leq i \leq m_n} (P_{n,i})_{X_1,\ldots,X_m}\big((.) \setminus \{[0,\ldots,0]\}\big) \underset{n\to\infty}{\longrightarrow} (\tilde{P})_{X_1,\ldots,X_m}\big((.) \setminus \{[0,\ldots,0]\}\big).$$

Consequently, in virtue of 3.3.5., and $\mathfrak{B}_P = \mathfrak{B}_{\tilde{P}}$, a) is equivalent to c), since

$$\Big(P_{n,i}\big((.) \setminus \{o\}\big)\Big)_{X_1,\ldots,X_m}\big((.) \setminus \{[0,\ldots,0]\}\big)$$

always coincides with $(P_{n,i})_{X_1,\ldots,X_m}\big((.) \setminus \{[0,\ldots,0]\}\big)$.

Using the continuity proposition 3.3.6., we finally conclude that c) is equivalent to

$$\underset{1 \leq i \leq m_n}{\ast} \mathscr{U}_{P_{n,i}} = \mathscr{U}_{\sum\limits_{1 \leq i \leq m_n} P_{n,i}} = \mathscr{U}_{\sum\limits_{1 \leq i \leq m_n} P_{n,i}((.)\setminus\{o\})} \underset{n\to\infty}{\Rightarrow} \mathscr{E}_{\tilde{P}} = P. \ ∎$$

If we now use, for each P in $\boldsymbol{T}$, the special infinitesimal triangular array constructed in the first step of the proof of 3.4.1., then we readily obtain

3.4.3. Proposition. *For all P in $\boldsymbol{T}$,*

$$n \sqrt[n]{P}\big((.) \setminus \{o\}\big) \overset{W}{\underset{n\to\infty}{\Rightarrow}} \tilde{P}.$$

From 3.4.2. we can derive a convenient criterion for the weak convergence of $\underset{1 \leq i \leq m_n}{*} P_{n,i}$ toward a Poisson distribution P_ν:

3.4.4. Theorem. *Let ν be an element of N, $(P_{n,i})$ an infinitesimal triangular array of distributions on $\mathfrak{M}$, and $\mathfrak{H}$ a semi-ring of continuity sets with respect to ν (which is included in $\mathfrak{B}$). If to each closed set X in $\mathfrak{B}$ and to each open neighbourhood U of X there exists a set H_0 in the ring generated by $\mathfrak{H}$, such that $X \subseteq H_0 \subseteq U$, then*

$$\underset{1 \leq i \leq m_n}{*} P_{n,i} \underset{n\to\infty}{\Rightarrow} P_\nu,$$

is true if and only if the following two conditions are satisfied:

a) *For all finite sequences $H_1, \ldots, H_m$ of pairwise disjoint sets in $\mathfrak{H}$,*

$$\sum_{1 \leq i \leq m_n} P_{n,i}\big(\Phi(H_1 \cup \cdots \cup H_m) > 1\big) \xrightarrow[n\to\infty]{} 0.$$

b) *For all H in $\mathfrak{H}$,*

$$\sum_{1 \leq i \leq m_n} P_{n,i}\big(\Phi(H) = 1\big) \xrightarrow[n\to\infty]{} \nu(H).$$

Proof. In virtue of 3.4.2. and 3.3.5., we have only to show that a) and b) are satisfied if and only if, for all finite sequences $H_1, \ldots, H_m$ of pairwise disjoint sets in $\mathfrak{H}$, the convergence relation

$$\sum_{1 \leq i \leq m_n} (P_{n,i})_{H_1,\ldots,H_m}\big((.) \setminus \{[0, \ldots, 0]\}\big) \xrightarrow[n\to\infty]{} (Q_\nu)_{H_1,\ldots,H_m}\big((.) \setminus \{[0, \ldots, 0]\}\big) \quad (*)$$

is satisfied.

However, we have

$$(Q_\nu)_{H_1,\ldots,H_m}\big((.) \setminus \{[0, \ldots, 0]\}\big) = \nu(H_1)\, \delta_{[1,0,\ldots,0]} + \cdots + \nu(H_m)\, \delta_{[0,\ldots,0,1]}.$$

Consequently $(*)$ is satisfied if and only if

$$\sum_{1 \leq i \leq m_n} (P_{n,i})_{H_1,\ldots,H_m}(\{0, 1, \ldots\}^m \setminus \{[0, \ldots, 0], e_1, \ldots, e_m\}) \xrightarrow[n\to\infty]{} 0 \quad (**)$$

and, for $j = 1, \ldots, m$,

$$\sum_{1 \leq i \leq m_n} (P_{n,i})_{H_1,\ldots,H_m}(\{e_j\}) \xrightarrow[n\to\infty]{} \nu(H_j) \quad (***)$$

are true, where we have set $e_1 = [1, 0, \ldots, 0], \ldots, e_m = [0, \ldots, 0, 1]$.

Obviously $(**)$ is equivalent to the convergence statement

$$\sum_{1 \leq i \leq m_n} (P_{n,i})\big(\Phi(H_1 \cup \cdots \cup H_m) > 1\big) \xrightarrow[n\to\infty]{} 0, \quad (****)$$

and, if $(****)$ is satisfied, $(***)$ is equivalent to

$$\sum_{1 \leq i \leq m_n} P_{n,i}\big(\Phi(H_j) = 1\big) \xrightarrow[n\to\infty]{} \nu(H_j) \qquad (1 \leq j \leq m). \blacksquare$$

4. Cluster Fields

4.0. Introduction

In Chapter 1, the specification of a random mechanism characteristic of infinitely divisible distributions on $\mathfrak{M}$ remained an open question. There we were only able to state a model which was characteristic of infinitely divisible distributions P having the additional property that $P(\{o\}) > 0$. To answer our question, we now utilize the concept of a cluster field. Here we conceive of a cluster field on $[A', \varrho_{A'}]$ with phase space $[A, \varrho_A]$ as a mapping $\varkappa_{(.)}$ of A' into the set $\boldsymbol{P}$ of all distributions on $\mathfrak{M}$, such that, for all Y in $\mathfrak{M}$, the real function $\varkappa_{(.)}(Y)$ is measurable with respect to the σ-algebra $\mathfrak{A}'$ of the Borel sets of $[A', \varrho_{A'}]$. To each point a' in A' there corresponds the distribution $\varkappa_{(a')}$ of a random cluster $\chi_{a'}$ of points in A. The measurability property formulated above is certainly present in cases where $\varkappa_{(.)}$ is continuous with respect to $\varrho_{A'}$ and to the metric $\varrho_{\boldsymbol{P}}$ introduced in Chapter 3. In this case the cluster field $\varkappa$ is said to be continuous.

We may consider $\varkappa_{(a')}$ to be the distribution of the daughter generation of a micro-organism lying at the point a'. Every Φ in M' can be interpreted as a description of a population of such micro-organisms. We now find our guidance in the concept that, independently of each other, the individuals a' in Φ produce daughter generations $\chi_{a'}$ distributed according to $\varkappa_{(a')}$, denoting by $\varkappa_{(\Phi)}$ the distribution $\underset{a' \in A', \Phi(\{a'\})>0}{\ast} \left((\varkappa_{(a')})^{\Phi(\{a'\})}\right)$ of the superposition χ of these daughter generations, i.e. the distribution of the daughter generation generated by Φ. However, $\varkappa_{(\Phi)}$ does not necessarily exist for all Φ in M', since it is possible that with a positive probability there is an infinite number of a' in Φ producing clusters which contain points in a fixed bounded Borel subset X of A. For each measure H on $\mathfrak{M}'$ having the property that H ($\varkappa_{(\Phi)}$ does not exist) $= 0$, we can introduce a measure $H_\varkappa$ on $\mathfrak{M}$ by $H_\varkappa = \int \varkappa_{(\Phi)}(.)\, H(\mathrm{d}\Phi)$. In particular, given any distribution P on $\mathfrak{M}'$, then $P_\varkappa$ can be considered to be the distribution of the daughter generation produced by a random population Φ which is distributed according to P. The distribution $P_\varkappa$ is certainly meaningful if P (Φ finite) $= 1$ is satisfied. In this case one also speaks of population processes (cf. Moyal [1], [2]).

If the convolution $\underset{i \in I}{\ast} P_i$ of an at most countable family $(P_i)_{i \in I}$ of distributions on $\mathfrak{M}'$ exists, then $\left(\underset{i \in I}{\ast} P_i\right)_\varkappa = \underset{i \in I}{\ast} (P_i)_\varkappa$, where the existence of either side of this equation involves that of the other (Proposition 4.3.1.). This leads to the "Clustering theorem" 4.3.3. (cf. Franken, Liemant & Matthes [1]), which states that for all infinitely divisible distributions P

on $\mathfrak{M}'$, the distribution $P_\varkappa$ exists if and only if $(\tilde{P})_\varkappa$ exists and satisfies the condition $(\tilde{P})_\varkappa\big(\Psi(X) > 0\big) < +\infty$ for all bounded X in $\mathfrak{A}$. The distribution $P_\varkappa$ is again infinitely divisible, and $\widetilde{P_\varkappa} = (\tilde{P})_\varkappa\big((.) \setminus \{o\}\big)$. Hence in particular, for a Poisson distribution P_ν on $\mathfrak{M}'$, the distribution $(P_\nu)_\varkappa$ exists if and only if, for all bounded X in $\mathfrak{A}$, the inequality $(Q_\nu)_\varkappa\big(\Psi(X) > 0\big) < +\infty$, i.e. $\int \varkappa_{(a')}\big(\Psi(X) > 0\big)\, \nu(\mathrm{d}a') < +\infty$, is satisfied. $P = (P_\nu)_\varkappa$ is said to be a clustering representation of P on the phase space $[A', \varrho_{A'}]$. A distribution P of this type is always infinitely divisible. Conversely, every infinitely divisible distribution P on $\mathfrak{M}$ has at least one clustering representation on $[A', \varrho_{A'}]$ for each given, unbounded phase space $[A', \varrho_{A'}]$ (Theorem 4.5.1.). Thus we have solved the problem set at the beginning. The infinitely divisible distributions can be characterized by the existence of a clustering representation. This emphasizes the fundamental role played by the Poisson distributions in the general theory of the infinitely divisible distributions of point processes.

Two clustering representations $P = (P_\nu)_\varkappa$, $P = (P_\sigma)_\omega$ of P on one and the same phase space $[A', \varrho_{A'}]$ are called equivalent if $\nu = \sigma$ and $\varkappa_{(a')} = \omega_{(a')}$ almost everywhere with respect to ν. A clustering representation $P = (P_\nu)_\varkappa$ is said to be reduced if the condition $\varkappa_{(a')}(\chi = o) = 0$ is satisfied almost everywhere with respect to ν. To each reduced clustering representation $P = (P_\nu)_\varkappa$, by $H(Y \times X) = \int\limits_X \varkappa_{(a')}(Y)\, \nu(\mathrm{d}a')$, $X \in \mathfrak{A}'$, $Y \in (M \setminus \{o\}) \cap \mathfrak{M}$, we introduce a measure H on the σ-algebra $\big((M \setminus \{o\}) \cap \mathfrak{M}\big) \otimes \mathfrak{A}'$. On the other hand, by $H(Y \times X) = \int\limits_Y c_{(\Psi)}(X)\, \tilde{P}(\mathrm{d}\Psi)$ this measure H can be decomposed with the help of a system $(c_{(\Psi)})_{\Psi \in M \setminus \{o\}}$ of transition probabilities from $[M \setminus \{o\}, (M \setminus \{o\}) \cap \mathfrak{M}]$ into $[A', \mathfrak{A}']$. This system $c_{(.)}$ is uniquely determined except the values on a set having $\tilde{P}$-measure zero. In this way, for any given infinitely divisible P, we get a one-to-one mapping of the set of all equivalence classes of reduced clustering representations of P on $[A', \varrho_{A'}]$ onto the set $\boldsymbol{c}$ of all equivalence classes of transition probabilities $c_{(.)}$ with the property $\int c_{(\Psi)}(X)\, \tilde{P}(\mathrm{d}\Psi) < +\infty$ for all bounded X in $\mathfrak{A}$ (Theorem 4.5.4.). The transition probabilities $c_{(\Psi)}$ have a simple intuitive meaning. To each "realization" Ψ with respect to $\tilde{P}$ there is selected a random "producing point" in A'. With the help of the corresponding systems of transition probabilities, it is possible to find and to describe in a simple way interesting special classes of clustering representations.

A very simple and important class of cluster fields is formed by the random translations. Each point a' in A' is transformed into a point selected at random in A according to some distribution $K\big(a', (.)\big)$. It turns out that the relating formulae and propositions may be substantially generalized by considering substochastic translations. At this it becomes possible that the point to be translated "dies" with a certain probability. Consequently to each point a' in A' there corresponds a measure $K\big(a', (.)\big)$ on $\mathfrak{A}$ with the property $K(a', A) \leqq 1$. If K fulfils the usual measurability condition, we call K a substochastic kernel on $[A', \varrho_{A'}]$ with phase space $[A, \varrho_A]$ and denote the corresponding cluster field $a' \rightsquigarrow \big(1 - K(a', A)\big)\, \delta_o + Q_{K(a', (.))}$ by $\varkappa(K)$. The poisson character of a distribution P on $\mathfrak{M}'$ is preserved in substochastic translations K, if $P_{\varkappa(K)}$ exists (Proposition 4.4.4.).

A distribution P on $\mathfrak{M}$ is said to be cluster-invariant with respect to a cluster field $\varkappa$ on $[A, \varrho_A]$ with phase space $[A, \varrho_A]$, if $P_\varkappa$ exists and coincides with P. If K is a substochastic kernel on $[A, \varrho_A]$ with phase space $[A, \varrho_A]$ and ν is a solution of the equation $\int K(a, (.))\, \nu(\mathrm{d}a) = \nu$ and has finite values on all bounded X in $\mathfrak{A}$, then all distributions P with the property g_ν are cluster-invariant with respect to $\varkappa(K)$. Thus each distribution P with the property g_ν describes a stochastic equilibrium with respect to $\varkappa(K)$ of a random system Φ of particles in A. If ν is a distribution, then $P = \sum_{n=0}^{\infty} P(\Phi(A) = n)\, (Q_\nu)^n$. Then P describes a finite system. However, if ν is an infinite measure, so we have, instead of the distributions $(Q_\nu)^n$, $n = 0, 1, \ldots$, the Poisson distributions $P_{l\nu}$, $l \geqq 0$, describing random infinite systems of particles (except the trivial case $l = 0$). In this way we get a simple probabilistic interpretation of the infinite solutions of the equation $\int K(a, (.))\, \nu(\mathrm{d}a) = \nu$ (cf. Derman [1], Port [1], and Brown [1]).

Conversely, if a distribution P is cluster-invariant with respect to $\varkappa(L)$ for all substochastic kernels L such that $P_{\varkappa(L)}$ exists and $\int L(a, (.))\, \nu(\mathrm{d}a) = \nu$ is satisfied, then P has the property g_ν (Theorem 4.8.1.).

4.1. The Mapping $\Phi \rightsquigarrow \varkappa_{(\Phi)}$

Suppose that, in addition to the phase space $[A, \varrho_A]$, another phase space $[A', \varrho_{A'}]$ be given, which is not necessarily different from $[A, \varrho_A]$. In what follows, we shall always distinguish the mathematical structures formed with respect to $[A', \varrho_{A'}]$ by the superscript $'$.

A *cluster field* on $[A', \varrho_{A'}]$ having the phase space $[A, \varrho_A]$ is conceived of as a mapping $\varkappa_{(.)}$ of A' into $\boldsymbol{P}$ which is measurable with respect to the σ-algebras $\mathfrak{A}'$ and $\mathfrak{P}$.

In the following, several simple types of cluster fields will repeatedly occur:

α) *Deterministic cluster fields.* To each mapping $a' \rightsquigarrow \Phi_{(a')}$ of A' into M there corresponds the mapping $a' \rightsquigarrow \varkappa_{(a')} = \delta_{\Phi_{(a')}}$ of A' into $\boldsymbol{P}$ in a one-to-one way. For all Y in $\mathfrak{M}$, we have

$$\{a' : a' \in A',\ \Phi_{(a')} \in Y\} = \{a' : a' \in A',\ \varkappa_{(a')}(Y) = 1\}.$$

Consequently the mapping $a' \rightsquigarrow \Phi_{(a')}$ is measurable with respect to $\mathfrak{A}'$ and $\mathfrak{M}$ if and only if the mapping $a' \rightsquigarrow \varkappa_{(a')}$ is measurable, i.e. there is a one-to-one correspondence between the measurable mappings of A' into M and the deterministic cluster fields on $[A', \varrho_{A'}]$ having the phase space $[A, \varrho_A]$.

β) *Compoundings.* Let $[A', \varrho_{A'}] = [A, \varrho_A]$. Suppose that to each a in A there corresponds the distribution $\sigma_{(a)}$ of a random non-negative integer. We set

$$\varkappa_{(a)} = \sum_{n=0}^{\infty} \sigma_{(a)}(\{n\})\, \delta_{n\delta_a} \qquad (a \in A).$$

The mapping $\varkappa$ of A into $\boldsymbol{P}$ defined in this way is measurable if and only if, for $n = 0, 1, 2, \ldots$, the real function $\sigma_{(.)}(\{n\})$ defined on A is measurable. Indeed, by 1.1.4. and α) all mappings $a \curvearrowright \delta_{n\delta_a}$ are measurable. Therefore, it follows from the measurability of the functions $\sigma_{(.)}(\{n\})$ that $\varkappa$ is measurable. Conversely, if $\varkappa$ is measurable, then all $\sigma_{(.)}(\{n\}) = \varkappa_{(.)}(\chi(A) = n)$ are measurable.

γ) *Thinnings.* Let $[A,' \varrho_{A'}] = [A, \varrho_A]$ and

$$\varkappa_{(a)} = \big(1 - p(a)\big)\,\delta_o + p(a)\,\delta_{\delta_a} \qquad (a \in A),$$

where p means a real function defined on A which has the property that $0 \leqq p \leqq 1$. Obviously this is a special case of β) where we have to set $\sigma_{(a)} = \big(1 - p(a)\big)\,\delta_0 + p(a)\,\delta_1$. Thus $\varkappa$ is measurable if and only if the "survival probability" p is measurable with respect to $\mathfrak{A}$.

Now let $\varkappa$ be a cluster field on $[A', \varrho_{A'}]$ with phase space $[A, \varrho_A]$, and Φ a counting measure in M'. According to the intuitive concept outlined in 4.0., we set

$$\varkappa_{(\Phi)} = \mathop{\ast}_{a' \in A', \Phi(\{a'\}) > 0} \left((\varkappa_{(a')})^{\Phi(\{a'\})}\right),$$

provided that the convolution on the right side exists. (For cluster fields of the types β) and γ) this is always the case).

Using the associativity of the infinite convolution, we obtain

4.1.1. *For all finite sequences $\Phi_1, \ldots, \Phi_m$ of counting measures in M' and each cluster field $\varkappa$ on $[A', \varrho_{A'}]$ with phase space $[A, \varrho_A]$,*

$$\mathop{\ast}_{i=1}^{n} \varkappa_{(\Phi_i)} = \varkappa_{\left(\sum\limits_{i=1}^{n} \Phi_i\right)},$$

where the existence of either side involves that of the other.

Let $M_\varkappa'$ denote the set of those Φ in M', for which $\varkappa_{(\Phi)}$ exists. By 2.1.4. a counting measure Φ belongs to $M_\varkappa'$ if and only if, for all X in $\mathfrak{B}$,

$$\sum_{a' \in A', \Phi(\{a'\}) > 0} \Phi(\{a'\})\,\varkappa_{(a')}\big(\chi(X) > 0\big) < +\infty.$$

Consequently

4.1.2. *A counting measure Φ in M' belongs to $M_\varkappa'$ if and only if*

$$\int \varkappa_{(a')}\big(\chi(X) > 0\big)\,\Phi(\mathrm{d}a') < +\infty$$

is satisfied for all X in $\mathfrak{B}$.

Obviously it will suffice that X runs through the sequence of the open balls $S_m(z)$ with radius m around a fixed point z in A. According to 1.2.2. each mapping

$$\Phi \curvearrowright \int \varkappa_{(a')}\big(\chi(X) > 0\big)\,\Phi(\mathrm{d}a') \qquad (X \in \mathfrak{A})$$

of M' into $[0, +\infty]$ is measurable with respect to $\mathfrak{M}'$. Consequently

$$M_\varkappa' = \bigcap_{m=1}^{\infty} \left\{\Phi : \Phi \in M', \int \varkappa_{(a')}\big(\chi(S_m(z)) > 0\big)\, \Phi(\mathrm{d}a') < +\infty\right\}$$

belongs to $\mathfrak{M}'$. Moreover

4.1.3. *The mapping*

$$\Phi \rightsquigarrow \varkappa_{(\Phi)} \qquad (\Phi \in M_\varkappa')$$

from M' into $\mathbf{P}$ is measurable with respect to the σ-algebras $\mathfrak{M}'$ and $\mathfrak{P}$.

Proof. We have to prove that, for all Y in $\mathfrak{M}$, the real function $\Phi \rightsquigarrow \varkappa_{(\Phi)}(Y)$ defined on $M_\varkappa'$ is measurable. On account of 1.8.1. we may restrict ourselves to those Y which can be represented in the form

$$Y = \{\Phi : \Phi \in M, \Phi(X_1) = l_1, \ldots, \Phi(X_m) = l_m\}$$

with pairwise disjoint $X_1, \ldots, X_m$ in $\mathfrak{B}$ and non-negative integers $l_1, \ldots, l_m$. For each Y of this type and each Φ in $M_\varkappa'$, we have

$$\varkappa_{(\Phi)}(Y) = \lim_{k\to\infty} \varkappa_{(\Phi((.)\cap S_k(z)))}(Y).$$

Thus it will suffice to study this function restricted to the subsets

$${}_{k,z}M' = \left\{\Phi : \Phi \in M', \quad \Phi\big((.)\setminus S_k(z)\big) = 0\right\}$$

of $M_\varkappa'$.

Since for all L in $\mathfrak{M}$ the real functions $\varkappa_{(.)}(L)$ on A are measurable and bounded, for all natural numbers k and all finite sequences $v_1, \ldots, v_m$ of non-negative integers there exists a sequence $(\mathfrak{Z}_{n,v_1,\ldots,v_m})_{n=1,2,\ldots}$ of finite decompositions of $S_k(z)$ such that

$$\lim_{n\to\infty} \sup_{Z\in\mathfrak{Z}_{n,v_1,\ldots,v_m}} \sup_{a',b'\in Z} \left|(\varkappa_{(a')} - \varkappa_{(b')})\big(\Phi(X_1) = v_1, \ldots, \Phi(X_m) = v_m\big)\right| = 0.$$

For every natural number n let $\mathfrak{B}_n$ denote a finite decomposition of $S_k(z)$ which is a refinement of all $\mathfrak{Z}_{n,v_1,\ldots,v_m}$ with $v_1 \leqq l_1, \ldots, v_m \leqq l_m$. For all Φ in ${}_{k,z}M'$ and all mappings

$$Z \rightsquigarrow a'_{Z,n} \qquad (Z \in \mathfrak{B}_n,\ a'_{Z,n} \in Z),$$

we obtain

$$\begin{aligned}
\varkappa_{(\Phi)}&\big(\chi(X_1) = l_1, \ldots, \chi(X_m) = l_m\big)\\
&= \Big(\mathop{\ast}_{a'\in A',\,\Phi(\{a'\})>0} \big((\varkappa_{(a')})^{\Phi(\{a'\})}\big)\Big)\big(\chi(X_1) = l_1, \ldots, \chi(X_m) = l_m\big)\\
&= \Big(\mathop{\ast}_{a'\in A',\,\Phi(\{a'\})>0} \big((\varkappa_{(a')})_{X_1,\ldots,X_m}\big)^{\Phi(\{a'\})}\Big)(\{[l_1, \ldots, l_m]\})\\
&= \lim_{n\to\infty} \Big(\mathop{\ast}_{Z\in\mathfrak{B}_n} \big((\varkappa_{(a'_{Z,n})})_{X_1,\ldots,X_m}\big)^{\Phi(Z)}\Big)(\{[l_1, \ldots, l_m]\}).
\end{aligned}$$

Hence the mappings

$$\Phi \curvearrowright \varkappa_{(\Phi)}\big(\chi(X_1) = l_1, \ldots, \chi(X_m) = l_m\big) \qquad (\Phi \in M_\varkappa{}')$$

are measurable. ∎

For all X in $\mathfrak{A}$ and all Φ in M', the integral $\int \varkappa_{(a')}\big(\chi(X) > 0\big)\, \Phi(\mathrm{d}a')$ has a simple intuitive meaning: Let $\Phi = \sum_{i \in I} \delta_{c_i}$ and let $(\chi_i)_{i \in I}$ be a random family of counting measures which is distributed according to $\mathop{\times}_{i \in I} \varkappa_{(c_i)}$. Then the expectation of the number of those indices i for which $\chi_i(X) > 0$, is equal to

$$\begin{aligned} &\int \sum_{i \in I} \min\big(\chi_i(X), 1\big) \Big(\mathop{\times}_{i \in I} \varkappa_{(c_i)}\Big) \big(\mathrm{d}(\chi_i)_{i \in I}\big) \\ &= \sum_{i \in I} \int \min\big(\chi(X), 1\big)\, \varkappa_{(c_i)}(\mathrm{d}\chi) \\ &= \sum_{i \in I} \varkappa_{(c_i)}\big(\chi(X) > 0\big) = \int \varkappa_{(a')}\big(\chi(X) > 0\big)\, \Phi(\mathrm{d}a'). \end{aligned}$$

Hence $\int \varkappa_{(a')}\big(\chi(X) > 0\big)\, \Phi(\mathrm{d}a')$ can be interpreted as the expectation of the number of clusters produced by the points in Φ and being "effective" on the set X.

4.2. The Measures $H_\varkappa$

For all measures H on $\mathfrak{M}'$ having the property that $H(\Phi \notin M_\varkappa{}') = 0$, using 4.1.3. we obtain a measure $H_\varkappa$ on $\mathfrak{M}$ by setting

$$H_\varkappa = \int \varkappa_{(\Phi)}(.)\, H(\mathrm{d}\Phi).$$

Obviously $H_\varkappa(M) = H(M')$. Hence in particular the measure $P_\varkappa$, provided that it exists, belongs to $\boldsymbol{P}$ if and only if P belangs to $\boldsymbol{P}'$.

In the special case of a thinning $\varkappa$ with $p(a) = c$, $a \in A$, we have

$$E_\varkappa = \mathscr{D}_c E \qquad (0 \leqq c \leqq 1;\ E \in \boldsymbol{E}^+),$$

thus giving a precise meaning to the considerations at the beginning of section 1.13.

A measure H on $\mathfrak{M}$ is said to be *cluster-invariant* with respect to a cluster field $\varkappa$ on $[A, \varrho_A]$ with phase space $[A, \varrho_A]$, if $H_\varkappa$ exists and coincides with H.

With the use of the intensity measure ϱ_H, we may formulate a simple sufficient criterion for the existence of $H_\varkappa$, i.e. for the validity of $H(\Phi \notin M_\varkappa{}') = 0$.

4.2.1. Proposition. *If, for all X in $\mathfrak{B}$, a measure H on $\mathfrak{M}'$ satisfies the condition*

$$\int \varkappa_{(a')}\big(\chi(X) > 0\big)\, \varrho_H(\mathrm{d}a') < +\infty$$

with respect to a cluster field $\varkappa$ on $[A', \varrho_{A'}]$ with phase space $[A, \varrho_A]$, then $H(\Phi \notin M_\varkappa{}') = 0$.

Proof. By 1.2.2. we obtain, for all natural numbers n and all z in A

$$+\infty > \int \varkappa_{(a')}\big(\chi(S_n(z)) > 0\big)\, \varrho_H(\mathrm{d}a') = \int \Big(\int \varkappa_{(a')}\big(\chi(S_n(z)) > 0\big)\, \Phi(\mathrm{d}a')\Big)\, H(\mathrm{d}\Phi),$$

which gives

$$H\Big(\int \varkappa_{(a')}\big(\chi(S_n(z)) > 0\big)\, \Phi(\mathrm{d}a') = +\infty\Big) = 0 \qquad (n = 1, 2, \ldots).$$

Consequently, in view of 4.1.2. we have

$$H(\Phi \notin M_\varkappa') = H\Big(\int \varkappa_{(a')}\big(\chi(S_n(z)) > 0\big)\, \Phi(\mathrm{d}a') = +\infty \text{ for at least one } n\Big) = 0. \ \blacksquare$$

The following example shows that the sufficient condition stated in 4.2.1. is not necessary.

We set $[A', \varrho_{A'}] = [A, \varrho_A] = R^1$. Furthermore let ζ be a random natural number distributed according to p, the expectation of which is infinite. The distribution of $\Phi = \delta_1 + \cdots + \delta_\zeta$ is denoted by P. Since P (Φ is finite) $= 1$ is satisfied, $P_\varkappa$ exists for all cluster fields $\varkappa$ on R^1. Now we set $\varkappa_{(a)} = \delta_{\delta_0}$ ($a \in R^1$). Then

$$\int \varkappa_{(a)}(\chi(\{0\}) > 0)\, \varrho_P(\mathrm{d}a) = \sum_{n=1}^{\infty} np(\{n\}) = +\infty,$$

i.e. the sufficient condition according to 4.2.1. is violated for $X = \{0\}$.

According to our intuitive interpretation of the integrals

$$\int \varkappa_{(a')}\big(\chi(X) > 0\big)\, \Phi(\mathrm{d}a'),$$

we may use 1.2.2. to interpret $\int \varkappa_{(a')}\big(\chi(X) > 0\big)\, \varrho_H(\mathrm{d}a')$ as the integral

$\int$ (Mean number of clusters produced by the points a' in Φ and being "effective" on X) $H(\mathrm{d}\Phi)$,

whereby we gain a simple intuitive interpretation for the assertion as well as the proof of 4.2.1. If in particular a random counting measure Φ is distributed according to $P \in \boldsymbol{P}'$, then $\int \varkappa_{(a')}\big(\chi(X) > 0\big)\, \varrho_P(\mathrm{d}a')$ can be interpreted as the expectation of the number of clusters produced by the points a' in Φ and being "effective" on the set X.

For all cluster fields $\varkappa$ on $[A', \varrho_{A'}]$ with phase space $[A, \varrho_A]$ and all X in $\mathfrak{A}$, the mapping $a' \rightsquigarrow \varrho_{\varkappa_{(a')}}(X)$ of A' into $[0, +\infty]$ is measurable with respect to $\mathfrak{A}'$, for we have

$$\varrho_{\varkappa_{(a')}}(X) = \sum_{n=1}^{\infty} n \varkappa_{(a')}\big(\Phi(X) = n\big).$$

Hence, if all distributions $\varkappa_{(a')}$, $a' \in A'$, are of finite intensity, then $a' \rightsquigarrow \varrho_{\varkappa_{(a')}}$ provides a measurable mapping of $[A', \varrho_{A'}]$ into $[N, \mathfrak{N}]$.

With the use of the intensity measures ϱ_H and $\varrho_{\varkappa_{(a')}}$, $a' \in A'$, the intensity measure of $H_\varkappa$ can be expressed in the following way:

4.2.2. *For each cluster field $\varkappa$ on $[A', \varrho_{A'}]$ with phase space $[A, \varrho_A]$ and each measure H on $\mathfrak{M}'$ satisfying the condition $H(\Phi \notin M_\varkappa') = 0$,*

$$\varrho_{H_\varkappa} = \int \varrho_{\varkappa_{(a')}}(\cdot)\, \varrho_H(\mathrm{d}a').$$

Proof. For all X in $\mathfrak{A}$, using 1.2.2. and 2.1.3. c) we obtain

$$\begin{aligned}\varrho_{H_\varkappa}(X) &= \int \Psi(X)\, H_\varkappa(\mathrm{d}\Psi) = \int \left(\int \Psi(X)\, \varkappa_{(\Phi)}(\mathrm{d}\Psi)\right) H(\mathrm{d}\Phi) \\ &= \int \varrho_{\varkappa_{(\Phi)}}(X)\, H(\mathrm{d}\Phi) = \int \Bigg(\sum_{a' \in A',\, \Phi(\{a'\})>0} \Phi(\{a'\})\, \varrho_{\varkappa_{(a')}}(X)\Bigg) H(\mathrm{d}\Phi) \\ &= \int \left(\int \varrho_{\varkappa_{(a')}}(X)\, \Phi(\mathrm{d}a')\right) H(\mathrm{d}\Phi) = \int \varrho_{\varkappa_{(a')}}(X)\, \varrho_H(\mathrm{d}a'). \ \blacksquare\end{aligned}$$

From 4.2.1. and 4.2.2. we obtain

4.2.3. Proposition. *For all measures H on $\mathfrak{M}'$ and all cluster fields $\varkappa$ on $[A', \varrho_{A'}]$ with phase space $[A, \varrho_A]$, the following statements are equivalent:*

a) *$H_\varkappa$ exists and is of finite intensity.*
b) *For all X in $\mathfrak{B}$,*

$$\int \varrho_{\varkappa_{(a')}}(X)\, \varrho_H(\mathrm{d}a') < +\infty.$$

Proof. Bearing in mind that $\varrho_{\varkappa_{(a')}}(X) \geqq \varkappa_{(a')}(\chi(X) > 0)$, we may use 4.2.1. and 4.2.2. to infer a) from b). Conversely, using 4.2.2. we may conclude that a) implies b). $\blacksquare$

The set $\{P : P \in \mathbf{P}', P(M_\varkappa') = 1\}$ of those P, for which $P_\varkappa$ exists, belongs to $\mathfrak{P}'$. Moreover

4.2.4. *The mapping*

$$P \curvearrowright P_\varkappa \qquad (P \in \mathbf{P}',\ P_\varkappa \text{ exists})$$

from $\mathbf{P}'$ into $\mathbf{P}$ is measurable with respect to the σ-algebras $\mathfrak{P}'$ and $\mathfrak{P}$.

Proof. For all finite sequences $Y_1, \ldots, Y_m$ of sets in $\mathfrak{M}$ and all non-negative real numbers $v_1, \ldots, v_m$ the real function

$$P \curvearrowright \int \left(\sum_{i=1}^{m} v_i k_{Y_i}(\Phi)\right) P(\mathrm{d}\Phi) \qquad (P \in \mathbf{P}')$$

is measurable with respect to $\mathfrak{P}'$. This property is transferred to the suprema of arbitrary monotone increasing sequences of functions $g_n = \sum_{1 \leqq i \leqq m_n} v_{i,n} k_{Y_{i,n}}$. Consequently, for all Y in $\mathfrak{M}$,

$$P \curvearrowright \int_{M_\varkappa'} \varkappa_{(\Phi)}(Y)\, P(\mathrm{d}\Phi) \qquad (P \in \mathbf{P}',\ P_\varkappa \text{ exists})$$

is measurable with respect to $\mathfrak{P}'$. $\blacksquare$

The "clustering" of distributions and the mixing operation are commutable in the following sense.

4.2.5. *Let $[V, \mathfrak{V}]$ be a measurable space, $P_{(.)}$ a mapping of V into $\mathbf{P}'$ which is measurable with respect to $\mathfrak{V}$, $\mathfrak{P}'$, and $\varkappa$ a cluster field on $[A', \varrho_{A'}]$ with phase space $[A, \varrho_A]$. Subject to these assumptions, $v \curvearrowright (P_{(v)})_\varkappa$ provides a mapping from V into $\mathbf{P}$ which is measurable with respect to $\mathfrak{V}$ and $\mathfrak{P}$. For each measure*

γ on $\mathfrak{V}$, we have

$$\left(\int P_{(v)}(.)\,\gamma(\mathrm{d}v)\right)_\varkappa = \int (P_{(v)})_\varkappa(.)\,\gamma(\mathrm{d}v),$$

where the existence of either side involves that of the other.

Proof. The measurability statement immediately follows from 4.2.4. Because of

$$\left(\int P_{(v)}(.)\,\gamma(\mathrm{d}v)\right)(\Phi \notin M_\varkappa') = \int P_{(v)}(\Phi \notin M_\varkappa')\,\gamma(\mathrm{d}v),$$

the existence of $\left(\int P_{(v)}(.)\,\gamma(\mathrm{d}v)\right)_\varkappa$ is equivalent to the existence of $\int (P_{(v)})_\varkappa(.)\,\gamma(\mathrm{d}v)$. Finally, we obtain, for all Y in $\mathfrak{M}$

$$\begin{aligned}\int (P_{(v)})_\varkappa(Y)\,\gamma(\mathrm{d}v) &= \int\left(\int \varkappa_{(\Phi)}(Y)\,P_{(v)}(\mathrm{d}\Phi)\right)\gamma(\mathrm{d}v)\\ &= \int \varkappa_{(\Phi)}(Y)\left(\int P_{(v)}(.)\,\gamma(\mathrm{d}v)\right)(\mathrm{d}\Phi) = \left(\int P_{(v)}(.)\,\gamma(\mathrm{d}v)\right)_\varkappa(Y). \quad \blacksquare\end{aligned}$$

When running through a system of transition probabilities, the variation distance of two distributions can never become larger. Especially, if both $P_\varkappa$ and $Q_\varkappa$ exist, we have

$$\begin{aligned}\|P_\varkappa - Q_\varkappa\| &= \left\|\int_{M_\varkappa'} \varkappa_{(\Phi)}(.)\,P(\mathrm{d}\Phi) - \int_{M_\varkappa'} \varkappa_{(\Phi)}(.)\,Q(\mathrm{d}\Phi)\right\|\\ &= \left\|\int_{M_\varkappa'} \varkappa_{(\Phi)}(.)\,(P-Q)\,(\mathrm{d}\Phi)\right\|\\ &= \left\|\int_{M_\varkappa'} \varkappa_{(\Phi)}(.)\,(P-Q)^+\,(\mathrm{d}\Phi) - \int_{M_\varkappa'} \varkappa_{(\Phi)}(.)\,(P-Q)^-\,(\mathrm{d}\Phi)\right\|\\ &\leqq \int_{M_\varkappa'} \varkappa_{(\Phi)}(M)\,(P-Q)^+\,(\mathrm{d}\Phi) + \int_{M_\varkappa'} \varkappa_{(\Phi)}(M)\,(P-Q)^-\,(\mathrm{d}\Phi)\\ &= (P-Q)^+\,(M) + (P-Q)^-\,(M) = \|P-Q\|.\end{aligned}$$

In this way we obtain

4.2.6. *Let $\varkappa$ be a cluster field on $[A', \varrho_{A'}]$ with phase space $[A, \varrho_A]$. Then, for all $P, Q \in \{V : V \in \boldsymbol{P}', V_\varkappa \text{ exists}\}$,*

$$\|P_\varkappa - Q_\varkappa\| \leqq \|P - Q\|.$$

We conclude this section with two statements concerning the dependence of the distribution $P_\varkappa$ on the cluster field $\varkappa$.

4.2.7. *Let $\varkappa$, ω be cluster fields on $[A', \varrho_{A'}]$ with phase space $[A, \varrho_A]$. If now, for a distribution P on $\mathfrak{M}'$, the distributions $P_\varkappa$, P_ω exist, then, for all X in $\mathfrak{A}$,*

$${}_X\|P_\varkappa - P_\omega\| \leqq \int {}_X\|\varkappa_{(a')} - \omega_{(a')}\|\,\varrho_P(\mathrm{d}a').$$

Proof 1. Let $\mathfrak{S}$ be any countable generating subalgebra of $\mathfrak{M}$. By 1.9.1. we have for all distributions V_1, V_2 on $\mathfrak{M}$

$$\|V_1 - V_2\| = \sup_{\mathfrak{Y}} \sum_{Y\in\mathfrak{Y}} |V_1(Y) - V_2(Y)|,$$

where $\mathfrak{Y}$ runs through all finite systems of pairwise disjoint sets in $\mathfrak{S}$. Thus, the real functions

$$a' \frown {}_X\|\varkappa_{(a')} - \omega_{(a')}\| = \big\|\varkappa_{(a')}\big(x\chi \in (.)\big) - \omega_{(a')}\big(x\chi \in (.)\big)\big\| \qquad (a' \in A'),$$

$$\Phi \frown {}_X\|\varkappa_{(\Phi)} - \omega_{(\Phi)}\| = \big\|\varkappa_{(\Phi)}\big(x\chi \in (.)\big) - \omega_{(\Phi)}\big(x\chi \in (.)\big)\big\| \qquad (\Phi \in M_\varkappa' \cap M_\omega')$$

are measurable.

2. Using 1. and 1.9.6. we obtain

$$ {}_X\|P_\varkappa - P_\omega\| = \sup_{\mathfrak{Y}} \sum_{Y \in \mathfrak{Y}} \Big| \int \big(x(\varkappa_{(\Phi)})\,(Y) - x(\omega_{(\Phi)})\,(Y)\big)\, P(\mathrm{d}\Phi) \Big|$$

$$\leqq \sup_{\mathfrak{Y}} \int \Big(\sum_{Y \in \mathfrak{Y}} |x(\varkappa_{(\Phi)})\,(Y) - x(\omega_{(\Phi)})\,(Y)|\Big)\, P(\mathrm{d}\Phi)$$

$$\leqq \int \|x(\varkappa_{(\Phi)}) - x(\omega_{(\Phi)})\|\, P(\mathrm{d}\Phi)$$

$$= \int_{M_\varkappa' \cap M_\omega'} \Big\| \mathop{\ast}_{a' \in A',\, \Phi(\{a'\})>0} \Big(\big(x(\varkappa_{(a')})\big)^{\Phi(\{a'\})}\Big) - \mathop{\ast}_{a' \in A',\, \Phi(\{a'\})>0} \Big(\big(x(\omega_{(a')})\big)^{\Phi(\{a'\})}\Big) \Big\|\, P(\mathrm{d}\Phi)$$

$$\leqq \int \Big(\sum_{a' \in A',\, \Phi(\{a'\})>0} \Phi(\{a'\})\; {}_X\|\varkappa_{(a')} - \omega_{(a')}\| \Big)\, P(\mathrm{d}\Phi).$$

In view of 1.2.2., we may continue as follows

$$= \int {}_X\|\varkappa_{(a')} - \omega_{(a')}\|\; \varrho_P(\mathrm{d}a'). \;\blacksquare$$

4.2.8. Proposition. *Let $\varkappa$, ω be cluster fields on $[A', \varrho_{A'}]$ with phase space $[A, \varrho_A]$. Furthermore let γ, α and β be mappings of A' into $\boldsymbol{E}^+$ which are measurable with respect to the σ-algebras $\mathfrak{A}'$, $\mathfrak{E}^+$ and have the properties*

$$\varkappa_{(a')} = \gamma_{(a')} + \alpha_{(a')} \qquad (a' \in A'),$$

$$\omega_{(a')} = \gamma_{(a')} + \beta_{(a')} \qquad (a' \in A').$$

If now, for a distribution P on $\mathfrak{M}'$, the two distributions $P_\varkappa$, P_ω exist, then, for all X in $\mathfrak{A}$,

$$ {}_X\|P_\varkappa - P_\omega\| \leqq 2 \int \Big(\alpha_{(a')}\big(\chi(X) > 0\big) + \beta_{(a')}\big(\chi(X) > 0\big)\Big)\, \varrho_P(\mathrm{d}a').$$

Proof. By 4.2.7.

$$ {}_X\|P_\varkappa - P_\omega\| \leqq \int {}_X\|\varkappa_{(a')} - \omega_{(a')}\|\; \varrho_P(\mathrm{d}a')$$

$$= \int {}_X\|\alpha_{(a')} - \beta_{(a')}\|\; \varrho_P(\mathrm{d}a').$$

By assumption we have

$$x(\alpha_{(a')})\,(M) = x(\beta_{(a')})\,(M),$$

so that we obtain

$$\begin{aligned}\|{}_X(\alpha_{(a')}) - {}_X(\beta_{(a')})\| &= \big\|{}_X(\alpha_{(a')})(\{o\})\,\delta_o - {}_X(\beta_{(a')})(\{o\})\,\delta_o \\ &\quad + {}_X(\alpha_{(a')})\big(\chi \in (.), \chi \neq o\big) - {}_X(\beta_{(a')})\big(\chi \in (.), \chi \neq o\big)\big\| \\ &= \big\|\alpha_{(a')}\big(\chi(X) = 0\big)\,\delta_o - \beta_{(a')}\big(\chi(X) = 0\big)\,\delta_o \\ &\quad + \alpha_{(a')}\big(\chi(X) > 0, {}_X\chi \in (.)\big) - \beta_{(a')}\big(\chi(X) > 0, {}_X\chi \in (.)\big)\big\| \\ &\leqq \big|\alpha_{(a')}\big(\chi(X) = 0\big) - \beta_{(a')}\big(\chi(X) = 0\big)\big| + \alpha_{(a')}\big(\chi(X) > 0\big) + \beta_{(a')}\big(\chi(X) > 0\big) \\ &= \big|\alpha_{(a')}\big(\chi(X) > 0\big) - \beta_{(a')}\big(\chi(X) > 0\big)\big| + \alpha_{(a')}\big(\chi(X) > 0\big) + \beta_{(a')}\big(\chi(X) > 0\big) \\ &\leqq 2\Big(\alpha_{(a')}\big(\chi(X) > 0\big) + \beta_{(a')}\big(\chi(X) > 0\big)\Big).\end{aligned}$$

Consequently

$${}_X\|P_\varkappa - P_\omega\| \leqq 2 \int (\alpha_{(a')} + \beta_{(a')})\big(\chi(X) > 0\big)\,\varrho_P(\mathrm{d}a'). \;\blacksquare$$

4.3. A Commutativity Relation

The results set forth in this section are based on the following commutativity relation.

4.3.1. *If the convolution* $\underset{i \in I}{\ast} P_i$ *of an at most countable family* $(P_i)_{i \in I}$ *of distributions on* $\mathfrak{M}'$ *exists, then for all cluster fields* $\varkappa$ *on* $[A', \varrho_{A'}]$ *with phase space* $[A, \varrho_A]$,

$$\Big(\underset{i \in I}{\ast} P_i\Big)_\varkappa = \underset{i \in I}{\ast} (P_i)_\varkappa,$$

where the existence of either side involves that of the other.

Proof. 1. First suppose that the set of indices I be finite. For empty I, both sides of the formula to be proved are equal to δ_o. Thus we may assume that $I \neq \emptyset$.

In view of 4.1.1.,

$$\Big(\underset{i \in I}{\ast} P_i\Big)(\varkappa_{(\Phi)} \text{ exists}) = 1,$$

i.e.

$$\Big(\underset{i \in I}{\times} P_i\Big)\Big(\varkappa_{\left(\sum\limits_{i \in I} \Phi_i\right)} \text{ exists}\Big) = 1$$

is equivalent to

$$\Big(\underset{i \in I}{\times} P_i\Big)(\varkappa_{(\Phi_i)} \text{ exists for all } i \text{ in } I) = 1,$$

i.e. to

$$\prod_{i \in I} P_i(\varkappa_{(\Phi)} \text{ exists}) = 1,$$

and hence to

$$P_i(\varkappa_{(\Phi)} \text{ exists}) = 1 \quad \text{for all } i \text{ in } I.$$

Consequently the existence of $\Big(\underset{i \in I}{\ast} P_i\Big)_\varkappa$ is equivalent to the existence of $\underset{i \in I}{\ast} (P_i)_\varkappa$.

Now suppose that $(P_i)_\varkappa$ exists for all i in I. Then by 1.8.2. and 4.1.1.,

$$\underset{i\in I}{\ast}(P_i)_\varkappa = \underset{i\in I}{\ast}\left(\int \varkappa_{(\Phi)}(.)\, P_i(\mathrm{d}\Phi)\right)$$

$$= \int \left(\underset{i\in I}{\ast} \varkappa_{(\Phi_i)}\right)(.) \left(\underset{i\in I}{\times} P_i\right)\left(\mathrm{d}(\Phi_i)_{i\in I}\right)$$

$$= \int \varkappa_{\left(\sum\limits_{i\in I}\Phi_i\right)}(.) \left(\underset{i\in I}{\times} P_i\right)\left(\mathrm{d}(\Phi_i)_{i\in I}\right)$$

$$= \int \varkappa_{(\Phi)}(.)\left(\underset{i\in I}{\ast} P_i\right)(\mathrm{d}\Phi) = \left(\underset{i\in I}{\ast} P_i\right)_\varkappa .$$

2. Now suppose that the set of indices I be infinite, so that without loss of generality we may set $I = \{1, 2, \ldots\}$. Using 1., for all natural numbers n we conclude from the existence of $\left(\underset{i\in I}{\ast} P_i\right)_\varkappa$ that

$$\left(\underset{i\in I}{\ast} P_i\right)_\varkappa = \left(\underset{i=1}{\overset{n}{\ast}} P_i\right)_\varkappa \ast \left(\underset{i=n+1}{\overset{\infty}{\ast}} P_i\right)_\varkappa = \underset{i=1}{\overset{n}{\ast}} (P_i)_\varkappa \ast \left(\underset{i=n+1}{\overset{\infty}{\ast}} P_i\right)_\varkappa .$$

Thus for all X in $\mathfrak{B}$ and all k we have

$$\sup_{n=1,2,\ldots} \left(\underset{i=1}{\overset{n}{\ast}} (P_i)_\varkappa\right)(\Phi(X) > k) \leqq \left(\underset{i=1}{\overset{\infty}{\ast}} P_i\right)_\varkappa (\Phi(X) > k)$$

and hence

$$\left(\underset{i=1}{\overset{\infty}{\times}} (P_i)_\varkappa\right)\left(\sum_{i=1}^{\infty} \Phi_i(X) = +\infty\right) = \inf_{k=1,2,\ldots} \left(\underset{i=1}{\overset{\infty}{\times}} (P_i)_\varkappa\right)\left(\sum_{i=1}^{\infty} \Phi_i(X) > k\right)$$

$$= \inf_{k=1,2,\ldots} \sup_{n=1,2,\ldots} \left(\underset{i=1}{\overset{\infty}{\times}} (P_i)_\varkappa\right)\left(\sum_{i=1}^{n} \Phi_i(X) > k\right)$$

$$= \inf_{k=1,2,\ldots} \sup_{n=1,2,\ldots} \left(\underset{i=1}{\overset{n}{\ast}} (P_i)_\varkappa\right)(\Phi(X) > k)$$

$$\leqq \inf_{k=1,2,\ldots} \left(\underset{i=1}{\overset{\infty}{\ast}} (P_i)\right)_\varkappa (\Phi(X) > k) = 0,$$

so that we conclude that $\underset{i\in I}{\ast} (P_i)_\varkappa$ exists.

Conversely, for all natural numbers n it follows from the existence of $\underset{i\in I}{\ast} (P_i)_\varkappa$ that

$$\underset{i\in I}{\ast} (P_i)_\varkappa = \underset{i=1}{\overset{n}{\ast}} (P_i)_\varkappa \ast \underset{i=n+1}{\overset{\infty}{\ast}} (P_i)_\varkappa = \left(\underset{i=1}{\overset{n}{\ast}} P_i\right)_\varkappa \ast \underset{i=n+1}{\overset{\infty}{\ast}} (P_i)_\varkappa$$

$$= \left(\int \varkappa_{\left(\sum\limits_{i=1}^{n}\Phi_i\right)}(.) \left(\underset{i=1}{\overset{n}{\times}} P_i\right)(\mathrm{d}[\Phi_1, \ldots, \Phi_n])\right) \ast \underset{i=n+1}{\overset{\infty}{\ast}} (P_i)_\varkappa$$

$$= \left(\int \varkappa_{\left(\sum\limits_{i=1}^{n}\Phi_i\right)}(.) \left(\underset{i=1}{\overset{\infty}{\times}} P_i\right)(\mathrm{d}(\Phi_i))\right) \ast \underset{i=n+1}{\overset{\infty}{\ast}} (P_i)_\varkappa .$$

Thus for all X in $\mathfrak{B}$ and all k we have

$$\int \sup_{n=1,2,\ldots} \varkappa_{\left(\sum_{i=1}^{n} \Phi_i\right)}(\Psi(X) > k) \left(\mathop{\times}_{i=1}^{\infty} P_i\right) (\mathrm{d}(\Phi_i))$$

$$= \sup_{n=1,2,\ldots} \int \varkappa_{\left(\sum_{i=1}^{n} \Phi_i\right)}(\Psi(X) > k) \left(\mathop{\times}_{i=1}^{\infty} P_i\right) (\mathrm{d}(\Phi_i))$$

$$\leqq \left(\mathop{\ast}_{i=1}^{\infty} (P_i)_\varkappa\right) (\Psi(X) > k).$$

Thus we have

$$\sup_{n=1,2,\ldots} \left(\mathop{\ast}_{i=1}^{n} \varkappa_{(\Phi_i)}\right) (\Phi(X) > k) \xrightarrow[k\to\infty]{} 0$$

almost everywhere with respect to $\mathop{\times}_{i\in I} P_i$, which means that

$$\left(\mathop{\times}_{i=1}^{\infty} P_i\right) \left(\mathop{\ast}_{i=1}^{\infty} \varkappa_{(\Phi_i)} \text{ exists}\right) = 1.$$

In virtue of the associativity of the infinite convolutions and due to the existence of $\mathop{\ast}_{i\in I} P_i$ this statement is equivalent to

$$\left(\mathop{\times}_{i=1}^{\infty} P_i\right) \left(\varkappa_{\left(\sum_{i=1}^{\infty} \Phi_i\right)} \text{ exists}\right) = 1,$$

which proves the existence of $\left(\mathop{\ast}_{i\in I} P_i\right)_\varkappa$.

Consequently the existence of $\left(\mathop{\ast}_{i\in I} P_i\right)_\varkappa$ is equivalent to that of $\mathop{\ast}_{i\in I} (P_i)_\varkappa$.

Now, if the two distributions $\left(\mathop{\ast}_{i\in I} P_i\right)_\varkappa$ and $\mathop{\ast}_{i\in I} (P_i)_\varkappa$ exist, we may infer their equality as follows. For all finite sequences $X_1, \ldots, X_m$ of sets in $\mathfrak{B}$ and all non-negative integers $l_1, \ldots, l_m$, by 2.1.3. b) we have, for $n = 1, 2, \ldots$,

$$\left(\mathop{\ast}_{i\in I} (P_i)_\varkappa\right)_{X_1,\ldots,X_m} (\{[l_1, \ldots, l_m]\}) = \lim_{n\to\infty} \left(\mathop{\ast}_{i=1}^{n} \left((P_i)_\varkappa\right)_{X_1,\ldots,X_m}\right) (\{[l_1, \ldots, l_m]\}),$$

since the existence of $\mathop{\ast}_{i=1}^{\infty} (P_i)_\varkappa$ implies that, as n increases,

$$\left(\mathop{\ast}_{i=n+1}^{\infty} (P_i)_\varkappa\right) (\Phi(X_1 \cup \cdots \cup X_m) > 0)$$

must tend towards zero.

On the other hand, we have

$$\left(\left(\mathop{\ast}_{i\in I} P_i\right)_\varkappa\right)_{X_1,\ldots,X_m} (\{[l_1, \ldots, l_m]\})$$

$$= \int \varkappa_{\left(\sum\limits_{i=1}^{\infty} \Phi_i\right)} (\Phi(X_1) = l_1, \ldots, \Phi(X_m) = l_m) \left(\mathop{\times}_{i=1}^{\infty} P_i\right) (\mathrm{d}(\Phi_i))$$

$$= \int \left(\mathop{\ast}_{i=1}^{\infty} \varkappa_{(\Phi_i)}\right) (\Phi(X_1) = l_1, \ldots, \Phi(X_m) = l_m) \left(\mathop{\times}_{i=1}^{\infty} P_i\right) (\mathrm{d}(\Phi_i))$$

$$= \int \lim_{n\to\infty} \left(\mathop{\ast}_{i=1}^{n} \varkappa_{(\Phi_i)}\right) (\Phi(X_1) = l_1, \ldots, \Phi(X_m) = l_m) \left(\mathop{\times}_{i=1}^{\infty} P_i\right) (\mathrm{d}(\Phi_i))$$

$$= \lim_{n\to\infty} \int \left(\mathop{\ast}_{i=1}^{n} \varkappa_{(\Phi_i)}\right) (\Phi(X_1) = l_1, \ldots, \Phi(X_m) = l_m) \left(\mathop{\times}_{i=1}^{n} P_i\right) (\mathrm{d}[\Phi_1, \ldots, \Phi_n])$$

$$= \lim_{n\to\infty} \left(\mathop{\ast}_{i=1}^{n} P_i\right)_\varkappa (\Phi(X_1) = l_1, \ldots, \Phi(X_m) = l_m)$$

$$= \lim_{n\to\infty} \left(\mathop{\ast}_{i=1}^{n} \left((P_i)_\varkappa\right)\right) (\Phi(X_1) = l_1, \ldots, \Phi(X_m) = l_m)$$

$$= \lim_{n\to\infty} \left(\mathop{\ast}_{i=1}^{n} \left((P_i)_\varkappa\right)_{X_1,\ldots,X_m}\right) (\{[l_1, \ldots, l_m]\}),$$

so that we may conclude that

$$\left(\mathop{\ast}_{i=1}^{\infty} (P_i)_\varkappa\right)_{X_1,\ldots,X_m} = \left(\left(\mathop{\ast}_{i=1}^{\infty} P_i\right)_\varkappa\right)_{X_1,\ldots,X_m} \qquad (X_1, \ldots, X_m \in \mathfrak{B}),$$

i.e. that

$$\mathop{\ast}_{i\in I} (P_i)_\varkappa = \left(\mathop{\ast}_{i\in I} P_i\right)_\varkappa . \quad \blacksquare$$

Now let E be an arbitrary finite measure on $\mathfrak{M}'$, and $\varkappa$ a cluster field on $[A', \varrho_{A'}]$ with phase space $[A, \varrho_A]$. Using the first step in the proof of 4.3.1., for all non-negative integers n we obtain

$$(E_\varkappa)^n = (E^n)_\varkappa ,$$

provided that the measure $E_\varkappa$ exists. We obtain

$$\mathcal{U}_{(E_\varkappa)} = e^{-E(M)} \sum_{n=0}^{\infty} \frac{1}{n!} (E_\varkappa)^n = e^{-E(M)} \sum_{n=0}^{\infty} \frac{1}{n!} (E^n)_\varkappa$$

$$= \left(e^{-E(M)} \sum_{n=0}^{\infty} \frac{1}{n!} E^n\right)_\varkappa = (\mathcal{U}_E)_\varkappa .$$

Conversely, if $(\mathscr{U}_E)_\varkappa$ exists, one gets

$$\mathscr{U}_E(\Phi \notin M_\varkappa') = 0,$$

i.e.

$$e^{-E(M)} \sum_{n=0}^{\infty} \frac{1}{n!} E^n(\Phi \notin M_\varkappa') = 0$$

and hence

$$E(\Phi \notin M_\varkappa') = 0.$$

Thus we observe the truth of

4.3.2. *For all finite measures E on* $\mathfrak{M}'$,

$$(\mathscr{U}_E)_\varkappa = \mathscr{U}_{(E_\varkappa)},$$

where the existence of either side implies that of the other.

For each infinitely divisible distribution P on $\mathfrak{M}'$ the canonical measure $\tilde{P}$ can be represented as the sum of a countable family $(E_i)_{i \in I}$ of finite measures on $\mathfrak{M}'$. Now, if the distribution $P_\varkappa$ exists, then by 1.6.5., 4.3.1. and 4.3.2. we conclude that

$$P_\varkappa = \left(\mathop{*}_{i \in I} \mathscr{U}_{E_i}\right)_\varkappa = \mathop{*}_{i \in I} (\mathscr{U}_{E_i})_\varkappa = \mathop{*}_{i \in I} \mathscr{U}_{(E_i)_\varkappa} = \mathop{*}_{i \in I} \mathscr{U}_{(E_i)_\varkappa((.)\setminus\{o\})} = \mathscr{E}_{\sum_{i \in I} (E_i)_\varkappa((.)\setminus\{o\})}$$

$$= \mathscr{E}_{\left(\sum_{i \in I} E_i\right)_\varkappa ((.)\setminus\{o\})} = \mathscr{E}_{(\tilde{P})_\varkappa((.)\setminus\{o\})},$$

i.e. the measure $(\tilde{P})_\varkappa$ exists, the measure $(\tilde{P})_\varkappa\big((.) \setminus \{o\}\big)$ belongs to $\boldsymbol{W}$, $P_\varkappa$ is infinitely divisible, and

$$\widetilde{P_\varkappa} = (\tilde{P})_\varkappa\big((.) \setminus \{o\}\big).$$

Conversely, suppose that the measure $(\tilde{P})_\varkappa$ exists and $(\tilde{P})_\varkappa\big((.) \setminus \{o\}\big)$ belongs to $\boldsymbol{W}$, then the above chain of equations may be passed in reverse direction, and we may conclude that $P_\varkappa$ exists.

Thus we have now derived the following "*clustering theorem*".

4.3.3. Theorem. *For all infinitely divisible distributions P on* $\mathfrak{M}'$ *and all cluster fields* $\varkappa$ *on* $[A', \varrho_{A'}]$ *with phase space* $[A, \varrho_A]$, *the distribution* $P_\varkappa$ *exists if and only if the measure* $(\tilde{P})_\varkappa$ *exists and* $(\tilde{P})_\varkappa\big((.) \setminus \{o\}\big)$ *belongs to* $\boldsymbol{W}$. *The distribution* $P_\varkappa$ *is again infinitely divisible, and*

$$\widetilde{P_\varkappa} = (\tilde{P})_\varkappa\big((\cdot) \setminus \{o\}\big).$$

The propositions 2.2.4. and 2.2.8. can be considered as special cases of 4.3.3.

To each subset X of A, we may associate a mapping $\varkappa$ of A into $\boldsymbol{P}$ by means of

$$a \curvearrowright k_{A \setminus X}(a)\, \delta_o + k_X(a)\, \delta_{\delta_a} \qquad (a \in A);$$

thus $\varkappa$ is measurable, and hence is a deterministic cluster field on $[A, \varrho_A]$ with phase space $[A, \varrho_A]$, if and only if X belongs to $\mathfrak{A}$. At the same time the cluster field is also of the type γ); the indicator function k_X can be interpreted as a "survival probability".

For all Φ in M, $\varkappa_{(\Phi)}$ exists and coincides with $\delta_{X\Phi}$. Thus for all measures H on $\mathfrak{M}$ we have

$$H_\varkappa = {}_X H,$$

so that 2.1.3. a) and 2.2.4. can be considered as special cases of 4.3.1. and 4.3.3., respectively.

To each finite sequence $X_1, \ldots, X_m$ of sets in $\mathfrak{A}'$, we may associate a measurable mapping $\Phi_{(.)}$ of A' into M by

$$\Phi_{(a')} = \sum_{i=1}^{m} k_{X_i}(a')\, \delta_i,$$

where $[A, \varrho_A]$ has to be set equal to the finite space $\{1, 2, \ldots, m\}$. The associated deterministic cluster field on $[A', \varrho_{A'}]$ with phase space $[A, \varrho_A]$ is denoted by ω.

Obviously the existence of $\omega_{(\Phi)}$ is equivalent to the finiteness of all $\Phi(X_1), \ldots, \Phi(X_m)$. Thus for all measures H on $\mathfrak{M}'$ we have

$$H_\omega = H_{X_1,\ldots,X_m},$$

where the existence of either side involves that of the other; so that 2.1.3. b) and 2.2.8. can be considered as special cases of 4.3.1. and 4.3.3., respectively.

Let ω be a cluster field on $[A, \varrho_A]$ with phase space $[A', \varrho_{A'}]$, and $\varkappa$ a cluster field on $[A', \varrho_{A'}]$ with phase space $[A'', \varrho_{A''}]$. Now, if the distribution $(\omega_{(a)})_\varkappa$ exists for all a in A, then

$$a \curvearrowright (\varkappa \circ \omega)_{(a)} = (\omega_{(a)})_\varkappa \qquad (a \in A)$$

provides a mapping of A into $\boldsymbol{P}''$ in which the two stochastic "development laws" ω and $\varkappa$ are successively applied.

By 4.2.4., $\varkappa \circ \omega$ is measurable with respect to the σ-algebras $\mathfrak{A}$ and $\mathfrak{P}''$, and hence is a cluster field on $[A, \varrho_A]$ with phase space $[A'', \varrho_{A''}]$.

We shall now show the truth of the following associative law:

4.3.4. *Let H be a measure on $\mathfrak{M}$, ω a cluster field on $[A, \varrho_A]$ with phase space $[A', \varrho_{A'}]$, and $\varkappa$ a cluster field on $[A', \varrho_{A'}]$ with phase space $[A'', \varrho_{A''}]$. If the cluster field $\varkappa \circ \omega$ as well as the measure H_ω exist, then the existence of $H_{\varkappa\circ\omega}$ is equivalent to that of the measure $(H_\omega)_\varkappa$, and*

$$(H_\omega)_\varkappa = H_{\varkappa\circ\omega}.$$

Proof. If $\omega_{(\Phi)}$ exists, then by 4.3.1. we obtain

$$(\omega_{(\Phi)})_\varkappa = \Big(\mathop{*}_{a\in A,\, \Phi(\{a\})>0} (\omega_{(a)})^{\Phi(\{a\})}\Big)_\varkappa = \mathop{*}_{a\in A,\, \Phi(\{a\})>0} \big((\omega_{(a)})_\varkappa\big)^{\Phi(\{a\})} = (\varkappa \circ \omega)_{(\Phi)},$$

where the existence of either side involves that of the other.

Consequently, by 4.2.5.

$$\begin{aligned} H_{\varkappa\circ\omega} &= \int (\varkappa \circ \omega)_{(\Phi)}(.)\, H(\mathrm{d}\Phi) = \int (\omega_{(\Phi)})_\varkappa(.)\, H(\mathrm{d}\Phi) \\ &= \Big(\int \omega_{(\Phi)}(.)\, H(\mathrm{d}\Phi)\Big)_\varkappa = (H_\omega)_\varkappa, \end{aligned}$$

where, again, the existence of either side is equivalent to that of the other. ∎

As an immediate consequence of 4.3.4., we have

4.3.5. *Let $\varkappa$ be a cluster field on $[A, \varrho_A]$ with phase space $[A', \varrho_{A'}]$, ω a cluster field on $[A', \varrho_{A'}]$ with phase space $[A'', \varrho_{A''}]$, and γ a cluster field on $[A'', \varrho_{A''}]$ with phase space $[A''', \varrho_{A'''}]$. Now if $\omega \circ \varkappa$ and $\gamma \circ \omega$ exist, then*

$$\gamma \circ (\omega \circ \varkappa) = (\gamma \circ \omega) \circ \varkappa,$$

where the existence of either side involves that of the other.

Let M_e denote the set of finite Φ in M, and let $\mathfrak{M}_e$ denote the σ-algebra of those subsets of M_e which belong to $\mathfrak{M}$. Each cluster field $\varkappa$ on $[A', \varrho_{A'}]$ with phase space $[A, \varrho_A]$ which satisfies the condition

$$\varkappa_{(a')}(M_e) = 1 \qquad \text{for all } a' \text{ in } A'$$

generates a system of transition probabilities from $[M_e', \mathfrak{M}_e']$ into $[M_e, \mathfrak{M}_e]$ in the form of $\Phi \rightsquigarrow \varkappa_{(\Phi)}$, $\Phi \in M_e'$. The operation $\circ$ introduced above then corresponds to the usual successive execution of stochastic transitions.

4.4. Substochastic Translations

A *substochastic kernel* K on $[A', \varrho_{A'}]$ with phase space $[A, \varrho_A]$ is defined to be a mapping K of $A' \times \mathfrak{A}$ into $[0, 1]$ which has the following properties:

a) for all a' in A', $K(a', (.))$ is a measure on $\mathfrak{A}$;
b) for all X in $\mathfrak{A}$, the real function $K((.), X)$ is measurable with respect to $\mathfrak{A}'$.

A substochastic kernel K is said to be *stochastic*, if $K(a', A) = 1$ for all a' in A'.

For all measures ν on $\mathfrak{A}'$, we set

$$(K * \nu)(X) = \int K(a', X)\, \nu(da') \qquad (X \in \mathfrak{A})$$

thus obtaining a measure $K * \nu$ on $\mathfrak{A}$. Now, if K_1 is a substochastic kernel on $[A'', \varrho_{A''}]$ with phase space $[A', \varrho_{A'}]$, and K_2 a substochastic kernel on $[A', \varrho_{A'}]$ with phase space $[A, \varrho_A]$, then

$$(K_2 * K_1)(a'', X) = \int K_2(a', X)\, K_1(a'', da') \qquad (a'' \in A'', X \in \mathfrak{A})$$

provides a substochastic kernel $K_2 * K_1$ on $[A'', \varrho_{A''}]$ with phase space $[A, \varrho_A]$. We call $K * \nu$ and $K_2 * K_1$ the convolutions of K with ν and of K_2 with K_1, respectively, denoting, for all natural numbers n and all substochastic kernels K on $[A, \varrho_A]$ with phase space $[A, \varrho_A]$, the n-th convolution power of K by $K^{[n]}$.

We interprete a substochastic kernel as a description of the following random mechanism. A point a' in A' "dies" with the probability $1 - K(a', A)$; if the "survival probability" $K(a', A)$ is positive, then with this probability a' is transformed, according to the distribution

$K(a', (.))\,(K(a', A))^{-1}$, into a point selected at random from A. We may also say that a' is substochastically translated according to K.

The random mechanism just outlined can be formally described by associating to each a' in A' the distribution

$$\big(\varkappa(K)\big)_{(a')} = \big(1 - K(a', A)\big)\,\delta_o + Q_{K(a',(.))}\,.$$

4.4.1. *The correspondence $K \rightsquigarrow \varkappa(K)$ provides a one-to-one mapping of the set of all substochastic kernels K on $[A', \varrho_{A'}]$ with phase space $[A, \varrho_A]$ onto the set of those cluster fields $\varkappa$ on $[A', \varrho_{A'}]$ with phase space $[A, \varrho_A]$ which satisfy the condition*

$$\varkappa_{(a')}\big(\chi(A) \leqq 1\big) = 1 \qquad (a' \in A').$$

Proof. Let K be a substochastic kernel, $X_1, \ldots, X_m$ a finite sequence of pairwise disjoint sets in $\mathfrak{B}$, $l_1, \ldots, l_m$ a sequence of equal length consisting of non-negative integers, and Y the set

$$\{\Phi : \Phi \in M, [\Phi(X_1), \ldots, \Phi(X_m)] = [l_1, \ldots, l_m]\}.$$

For all a' in A', we obtain

$$\big(\varkappa(K)\big)_{(a')}(Y) = \begin{cases} \big(1 - K(a', A)\big) + K\big(a', A \setminus (X_1 \cup \cdots \cup X_m)\big) & \text{for } l_1 = \ldots = l_m = 0 \\ K(a', X_i) & \text{for } l_i = 1, \sum\limits_{j=1}^{m} l_j = 1; i = 1, \ldots, m \\ 0 & \text{for } \sum\limits_{j=1}^{m} l_j > 1. \end{cases}$$

Consequently, $\big(\varkappa(K)\big)_{(.)}(Y)$ is measurable with respect to $\mathfrak{A}'$ for the Y stated above, and hence for all Y in $\mathfrak{M}$, i.e. $\varkappa(K)$ is a cluster field.

For all a' in A', $K(a', (.)) = \varrho_{(\varkappa(K))_{(a')}}(.)$. Hence in particular K is uniquely determined by $\varkappa(K)$, i.e. the mapping $K \rightsquigarrow \varkappa(K)$ is one-to-one.

Now let $\varkappa$ be an arbitrary cluster field having the property that

$$\varkappa_{(a')}\big(\chi(A) \leqq 1\big) = 1 \qquad (a' \in A').$$

Setting

$$K(a', X) = \varkappa_{(a')}\big(\chi(X) = 1\big) \qquad (a' \in A', X \in \mathfrak{A})$$

we obtain a substochastic kernel K which has the property that $\varkappa = \varkappa(K)$. ∎

A cluster field $\varkappa$ on $[A', \varrho_{A'}]$ with phase space $[A, \varrho_A]$ is said to be of the *type* δ), if it satisfies the condition stated in 4.4.1. Each cluster field $\varkappa$ on $[A, \varrho_A]$ with phase space $[A, \varrho_A]$ which is of the type γ) (cf. 4.1.) is likewise of the type δ); we have

$$\varkappa = \varkappa(K), \qquad K\big(a, (.)\big) = p(a)\,\delta_a \qquad (a \in A),$$

where p denotes the "survival probability" associated with $\varkappa$.

Let $[A', \varrho_{A'}] = [O, \varrho_O]$. Further suppose that a bounded phase space $[K, \varrho_K]$ be given, which, with reference to the statements made at the end of Section 1.0., will be interpreted as a "mark space". We set $[A, \varrho_A] = [O \times K, \varrho_{O\times K}]$ where

$$\varrho_{O\times K}([c_1, k_1], [c_2, k_2]) = \varrho_O(c_1, c_2) + \varrho_K(k_1, k_2).$$

Now, given a mapping $q_{(.)}$ of O into the set of all distributions on the σ-algebra $\mathfrak{K}$ of the Borel subsets of $[K, \varrho_K]$, then the set-up

$$K\big(c, (.)\big) = \delta_c \times q_{(c)} \qquad (c \in O)$$

provides a stochastic kernel on $[O, \varrho_O]$ with phase space $[A, \varrho_A]$ if and only if, for all Z in $\mathfrak{K}$, the real function $q_{(.)}(Z)$ is measurable with respect to the σ-algebra $\mathfrak{O}$ of Borel subsets of O. Cluster fields of the type δ), which are determined by substochastic kernels of the kind described above, are called *independent markings*. To each c in O there is associated at random a mark k in K according to $q_{(c)}$. Subsequently c is transformed into the "marked point" $[c, k]$.

As an immediate consequence of the above definitions, we have

4.4.2. *If K_1 is a substochastic kernel on $[A'', \varrho_{A''}]$ with phase space $[A', \varrho_{A'}]$, and K_2 a substochastic kernel on $[A', \varrho_{A'}]$ with phase space $[A, \varrho_A]$ then $K_2 * K_1$ is a substochastic kernel on $[A'', \varrho_{A''}]$ with phase space $[A, \varrho_A]$, and*

$$\varkappa(K_2 * K_1) = \big(\varkappa(K_2)\big) \circ \big(\varkappa(K_1)\big).$$

4.4.3. *For all ν in N', and all substochastic kernels K on $[A', \varrho_{A'}]$ with phase space $[A, \varrho_A]$ which satisfy the condition $K * \nu \in N$,*

$$(Q_\nu)_{\varkappa(K)} = Q_{K*\nu} + \Big(\int \big(1 - K(a', A)\big)\, \nu(\mathrm{d}a')\Big)\, \delta_o.$$

The truth of the following proposition (cf. Prekopa [1], [2]) is inferred immediately from 4.4.3., 4.3.3., 2.2.1. and 2.2.14.

4.4.4. Proposition. *For all ν in N' and all substochastic kernels K on $[A', \varrho_{A'}]$ with phase space $[A, \varrho_A]$, the distribution $(P_\nu)_{\varkappa(K)}$ exists if and only if $K * \nu$ belongs to N, and*

$$(P_\nu)_{\varkappa(K)} = P_{K*\nu}.$$

Consequently the Poisson character of a distribution is preserved in random translations according to substochastic kernels. Conversely, we have (cf. Szász [1])

4.4.5. Proposition. *Let ν be a measure in N' and $\varkappa$ a cluster field on $[A', \varrho_{A'}]$ with phase space $[A, \varrho_A]$, for which $(P_\nu)_\varkappa$ exists. Subject to these assumptions, the distribution $(P_\nu)_\varkappa$ is Poisson if and only if*

$$\varkappa_{(a')}\big(\chi(A) \leqq 1\big) = 1$$

is satisfied almost everywhere with respect to ν.

Proof. By 4.3.3. $(\tilde{P}_\nu)_\varkappa$ exists, and we obtain

$$\widetilde{(P_\nu)_\varkappa}(\chi(A) > 1) = (\widetilde{P}_\nu)_\varkappa(\chi(A) > 1) = \int \varkappa_{(a')}(\chi(A) > 1)\, \nu(\mathrm{d}a').$$

Now the assertion is readily inferred by means of 2.2.15. ∎

4.5. Clustering Representations

Given a Poisson distribution P_ν on $\mathfrak{M}'$ and a cluster field $\varkappa$ on $[A', \varrho_{A'}]$ with phase space $[A, \varrho_A]$, then by 4.3.3. the distribution $(P_\nu)_\varkappa$ exists if and only if, for all X in $\mathfrak{B}$,

$$(\widetilde{P}_\nu)_\varkappa(\Psi(X) > 0) = (Q_\nu)_\varkappa(\Psi(X) > 0) = \int \varkappa_{(a')}(\Psi(X) > 0)\, \nu(\mathrm{d}a') < +\infty$$

is valid, i.e. if the sufficient condition stated in 4.2.1. is satisfied.

A *clustering representation* of a distribution P in $\boldsymbol{P}$ on the phase space $[A', \varrho_{A'}]$ is conceived of as a representation $P = (P_\nu)_\varkappa$ by means of a Poisson distribution P_ν on $\mathfrak{M}'$ and a cluster field $\varkappa$ on $[A', \varrho_{A'}]$ with phase space $[A, \varrho_A]$. In view of 4.3.3., such a distribution is always infinitely divisible; and we have

$$\tilde{P} = (Q_\nu)_\varkappa((\cdot) \setminus \{o\}) = \int \varkappa_{(a')}((\cdot) \setminus \{o\})\, \nu(\mathrm{d}a').$$

Hence in particular P is a regular or a singular infinitely divisible distribution if the equations $\varkappa_{(a')}(\chi \text{ finite}) = 1$ or $\varkappa_{(a')}(\chi \text{ is zero or infinite}) = 1$, respectively, are satisfied almost everywhere with respect to ν. In view of the statements made in Section 4.2., for all X in $\mathfrak{A}$ the number $\tilde{P}(\Psi(X) > 0)$ can be interpreted as the expectation of the number of clusters "effective" on X, for we have

$$\begin{aligned}\tilde{P}(\Psi(X) > 0) &= \int \varkappa_{(a')}(\Psi(X) > 0)\, \nu(\mathrm{d}a') \\ &= \int \varkappa_{(a')}(\Psi(X) > 0)\, \varrho_{P_\nu}(\mathrm{d}a').\end{aligned}$$

If two cluster fields $\varkappa$, ω on $[A', \varrho_{A'}]$ with the phase space $[A, \varrho_A]$ coincide almost everywhere with respect to a measure ν in N' then we obtain

$$\int \varkappa_{(a')}((\cdot) \setminus \{o\})\, \nu(\mathrm{d}a') = \int \omega_{(a')}((\cdot) \setminus \{o\})\, \nu(\mathrm{d}a'),$$

and it follows that $(P_\nu)_\varkappa = (P_\nu)_\omega$, where the existence of either side implies that of the other. Therefore two clustering representations $P = (P_\nu)_\varkappa$, $P = (P_\sigma)_\omega$ of some P in $\boldsymbol{P}$ on one and the same phase space $[A', \varrho_{A'}]$ are called *equivalent* if $\nu = \sigma$ and $\varkappa_{(a')}$ coincides with $\omega_{(a')}$ almost everywhere with respect to ν.

The existence of a clustering representation is characteristic of infinite divisibility. We even have

4.5.1. Theorem. *Let $[A', \varrho_{A'}]$ be a complete separable metric space. If A' is unbounded with respect to the metric $\varrho_{A'}$, then a distribution P on $\mathfrak{M}$ belongs*

to $\boldsymbol{T}$ *if and only if it has a clustering representation on the phase space* $[A', \varrho_{A'}]$.

Proof. 1. Every P in $\boldsymbol{T}$ has a clustering representation on the discrete phase space $[A'', \varrho_{A''}] = \{1, 2, \ldots\}$.

Let $\tilde{P} = \sum_{n=1}^{\infty} E_n$ be an arbitrary representation of $\tilde{P}$ as a sum of a sequence (E_n) of finite measures on $\mathfrak{M}$. Now, if ν denotes the measure in N'' which is uniquely determined by $\nu(\{n\}) = E_n(M)$, $n = 1, 2, \ldots$, and $\varkappa$ the cluster field

$$\varkappa_{(n)} = \begin{cases} (E_n(M))^{-1} E_n & \\ & \text{for} \\ \delta_o & \end{cases} \begin{matrix} E_n(M) \neq 0 \\ \\ E_n(M) = 0, \end{matrix}$$

then we obtain

$$\begin{aligned} \widetilde{(P_\nu)_\varkappa} &= (\tilde{P}_\nu)_\varkappa\big((.) \setminus \{o\}\big) = \int \varkappa_{(a'')}\big((.) \setminus \{o\}\big)\, \nu(\mathrm{d}a'') \\ &= \sum_{n=1}^{\infty} \nu(\{n\})\, \varkappa_{(n)} = \sum_{n=1}^{\infty} E_n = \tilde{P}, \end{aligned}$$

i.e. we have $P = (P_\nu)_\varkappa$.

2. Now let $[A', \varrho_{A'}]$ be an arbitrary unbounded phase space. We choose a sequence (a_n') of pairwise distinct points in A' such that each bounded subset of A' contains only a finite number of terms of this sequence. Further suppose that an arbitrary clustering representation $P = (P_\nu)_\varkappa$ of P on the phase space $\{1, 2, \ldots\}$ be given. If we now set

$$\sigma = \sum_{n=1}^{\infty} \nu(\{n\})\, \delta_{a_n'}, \qquad \omega_{(a')} = \begin{cases} \varkappa_{(n)} & \\ & \text{for} \\ \delta_o & \end{cases} \begin{matrix} a' = a_n', \quad n = 1, 2, \ldots \\ \\ a' \notin \{a_1', a_2', \ldots\} \end{matrix}$$

then it results that $\sigma \in N'$ and

$$\widetilde{(P_\sigma)_\omega} = \int \omega_{(a')}\big((.) \setminus \{o\}\big)\, \sigma(\mathrm{d}a') = \sum_{n=1}^{\infty} \nu(\{n\})\, \varkappa_{(n)}\big((.) \setminus \{o\}\big) = \tilde{P},$$

i.e. we have $P = (P_\sigma)_\omega$.

Consequently each P in $\boldsymbol{T}$ has a clustering representation on the phase space $[A', \varrho_{A'}]$. On the other hand we had already stated above that a distribution on $\mathfrak{M}$ is infinitely divisible if it permits any clustering representation. ∎

The statement of 4.5.1. becomes false if we drop the assumption that A' be unbounded. Instead we have

4.5.2. Proposition. *Let* $[A', \varrho_{A'}]$ *be a complete separable metric space. If* A' *is bounded with respect to the metric* $\varrho_{A'}$, *then a distribution* P *on* $\mathfrak{M}$ *is of the form* $P = \mathscr{U}_E$, $E \in \boldsymbol{E}^+$, *if and only if it has a clustering representation on the phase space* $[A', \varrho_{A'}]$.

Proof. 1. Let $P = \mathscr{U}_E$, $E \in \boldsymbol{E}^+$, and $P \neq \delta_o$; and hence $E \neq 0$. We choose a point g in A', and set

$$\nu = E(M)\,\delta_g, \qquad \varkappa_{(a')} = \begin{cases} \delta_o & \\ & , \quad \text{if} \\ (E(M))^{-1}\,E & \end{cases} \quad \begin{matrix} a' \neq g \\ \\ a' = g. \end{matrix}$$

Thus we obtain

$$(P_\nu)_\varkappa = \sum_{n=0}^{\infty} e^{-E(M)} \frac{(E(M))^n}{n!} \left((E(M))^{-1}\,E\right)^n = \mathscr{U}_E = P.$$

In the case $P = \delta_o$ we may choose ν arbitrarily from N', setting $\varkappa_{(a')}$ constant and equal to δ_o.

2. Let $P = (P_\nu)_\varkappa$ be a clustering representation of P on the bounded phase space $[A', \varrho_{A'}]$. Then P is infinitely divisible, and

$$\tilde{P}(M) = \int \varkappa_{(a')}(\chi \neq o)\,\nu(\mathrm{d}a') \leqq \nu(A') < +\infty.$$

Hence $\tilde{P}$ is finite and we can conclude that

$$P = \mathscr{E}_{\tilde{P}} = \mathscr{U}_{\tilde{P}}. \quad ▮$$

From the first step in the proof of 4.5.2. we observe that, by virtue of

$$\nu = E(M)\,\delta_1, \qquad \varkappa_{(1)} = \left(E(M)\right)^{-1} E$$

there is a one-to-one correspondence between the representations $P = \mathscr{U}_E$, $E \in \boldsymbol{E}^+$, of a distribution P in $\boldsymbol{T}$ being different from δ_o and satisfying the condition $\tilde{P}(M) < +\infty$ and the clustering representations of P on the single-element phase space $\{1\}$.

If A belongs to $\mathfrak{B}$, then each infinitely divisible distribution P on $\mathfrak{M}$ satisfies the condition $\tilde{P}(M) < +\infty$, and hence it is of the form $P = \mathscr{U}_{\tilde{P}}$. Thus from 4.5.1. and 4.5.2. we may readily infer the truth of

4.5.3. Theorem. *A distribution P on $\mathfrak{M}$ is infinitely divisible if and only if it has a clustering representation on the phase space $[A, \varrho_A]$.*

If $[A, \varrho_A]$ is the direct product $[O \times K, \varrho_{O \times K}]$ of a "position space" $[O, \varrho_O]$ with a bounded "mark space" $[K, \varrho_K]$, then 4.5.3. can be modified by saying that a distribution P on $\mathfrak{M}$ is infinitely divisible if and only if it has a clustering representation on the "position space" $[O, \varrho_O]$.

A clustering representation $P = (P_\nu)_\varkappa$ of some P in $\boldsymbol{T}$ on the phase space $[A', \varrho_{A'}]$ is said to be *reduced*, if the condition $\varkappa_{(a')}(\chi = o) = 0$ is satisfied almost everywhere with respect to ν. As one can see in the first step in the proof of 4.5.1., there is a one-to-one correspondence between the equivalence classes of reduced clustering representations of P on the phase space $\{1, 2, \ldots\}$ and the representations of $\tilde{P}$ as a sum of a sequence of finite measures on $\mathfrak{M}$.

Each clustering representation $P = (P_\nu)_\varkappa$ of P on $[A', \varrho_{A'}]$ can be reduced in the following way. We set

$$\sigma(\mathrm{d}a') = \varkappa_{(a')}(\chi \neq o)\, \nu(\mathrm{d}a'),$$

$$\omega_{(a')} = \begin{cases} \varkappa_{(a')}\big((.) \mid \chi \neq o\big) & \\ & , \quad \text{if} \\ \delta_o & \end{cases} \qquad \begin{matrix} \varkappa_{(a')}(\chi \neq o) > 0 \\ \\ \varkappa_{(a')}(\chi \neq o) = 0. \end{matrix}$$

This yields

$$\int \omega_{(a')}\big((.) \setminus \{o\}\big)\, \sigma(\mathrm{d}a') = \int \varkappa_{(a')}\big((.) \setminus \{o\}\big)\, \nu(\mathrm{d}a'),$$

i.e. we have

$$(P_\nu)_\varkappa = (P_\sigma)_\omega.$$

By reducing the clustering representation on the phase space $\{1\}$ of a P in $\boldsymbol{T}$ being different from δ_o, which is associated to a representation $P = \mathscr{U}_E$, $E \in \boldsymbol{E}^+$, we obtain the clustering representation associated with $P = \mathscr{U}_{E((.)\setminus\{o\})}$.

We shall now give a survey of the equivalence classes of reduced clustering representations of a distribution P in $\boldsymbol{T}$ on a given phase space $[A', \varrho_{A'}]$. We start from the fact that to each reduced clustering representation $P = (P_\nu)_\varkappa$ of P on $[A', \varrho_{A'}]$ there can be associated a well-determined measure H on the σ-algebra $\big((M \setminus \{o\}) \cap \mathfrak{M}\big) \otimes \mathfrak{A}'$ by means of

$$H(Y \times X) = \int_X \varkappa_{(a')}(Y)\, \nu(\mathrm{d}a') \qquad (X \in \mathfrak{A}',\ Y \in \mathfrak{A}^*).$$

Here, for the purpose of abbreviation we set (as was done in Section 3.3.)

$$A^* = M \setminus \{o\}, \qquad \mathfrak{A}^* = A^* \cap \mathfrak{M}.$$

We have

$$\nu = H\big(A^* \times (.)\big), \qquad \tilde{P} = H\big((.) \times A'\big).$$

Obviously two reduced clustering representations of P on $[A', \varrho_{A'}]$ are equivalent if and only if they lead to the same measure H.

Let P be a fixed distribution in $\boldsymbol{T}$. We denote by $\boldsymbol{c}$ the set of those mappings

$$\Psi \curvearrowright c_{(\Psi)}$$

of A^* into the set of all distributions on $\mathfrak{A}'$, which are measurable with respect to the σ-algebras $\mathfrak{A}^*$, $\mathfrak{N}'$ and for which the measure

$$\sigma = \int c_{(\Psi)}(.)\, \tilde{P}(\mathrm{d}\Psi)$$

belongs to N'. Thus the elements of $\boldsymbol{c}$ can be considered to be systems of transition probabilities from $[A^*, \mathfrak{A}^*]$ into $[A', \mathfrak{A}']$. Two elements c, c' of $\boldsymbol{c}$ are said to be *equivalent* if the equation $c_{(\Psi)} = c'_{(\Psi)}$ is satisfied almost everywhere with respect to $\tilde{P}$.

To a given c in $\mathbf{c}$ there exists exactly one measure H on the σ-algebras $\mathfrak{A}^* \otimes \mathfrak{A}'$ having the property that

$$H(Y \times X) = \int_Y c_{(\Psi)}(X)\, \tilde{P}(\mathrm{d}\Psi) \qquad (X \in \mathfrak{A}',\, Y \in \mathfrak{A}^*).$$

So we have

$$\sigma = H(A^* \times (.)), \qquad \tilde{P} = H((.) \times A').$$

Two elements c, c' of $\mathbf{c}$ are equivalent if and only if they lead to the same measure H.

An equivalence class of reduced clustering representations of P on $[A', \varrho_{A'}]$ is said to *correspond* to an equivalence class of elements of $\mathbf{c}$, if the same measure H on $\mathfrak{A}^* \otimes \mathfrak{A}'$ is associated with both equivalence classes.

4.5.4. Theorem. *To each equivalence class of reduced clustering representations of a distribution P on $[A', \varrho_{A'}]$ there corresponds a well-determined equivalence class of elements of $\mathbf{c}$, all of the equivalence classes of elements of $\mathbf{c}$ being covered in this way.*

Proof. 1. For each reduced clustering representation of $P = \delta_o$ we have $\nu = o$. Consequently, in this case there is only one equivalence class of reduced clustering representations. Since $\tilde{P} = 0$, there exists only one equivalence class of elements of $\mathbf{c}$, too. Both equivalence classes lead to the measure $H = 0$, and hence are associated with each other.

From now on we can always suppose in addition that the distribution P be different from δ_o.

2. Given any element c in $\mathbf{c}$, we have to show that there exists a reduced clustering representation $P = (P_\sigma)_\omega$ of P on $[A', \varrho_{A'}]$, such that, for all X in $\mathfrak{A}'$ and all Y in $\mathfrak{A}^*$

$$\int_X \omega_{(a')}(Y)\, \sigma(\mathrm{d}a') = H(Y \times X) = \int_Y c_{(\Psi)}(X)\, \tilde{P}(\mathrm{d}\Psi)$$

is satisfied.

Let $A^* = \bigcup_{i \in I} Y_i$ be a representation of A^* as a union of an at most countable family $(Y_i)_{i \in I}$ of pairwise disjoint sets in $\mathfrak{M}$ having the property that $0 < \tilde{P}(Y_i) < +\infty$. We set

$$H_i = H((.) \cap (Y_i \times A')), \qquad \sigma_i = H_i(A^* \times (.)).$$

Each H_i is a finite measure, being different from 0, on the σ-algebra $\mathfrak{A}^* \otimes \mathfrak{A}'$ of the Borel subsets of the complete separable metric space $[A^* \times A', \varrho_{A^* \times A'}]$, where

$$\varrho_{A^* \times A'}([\Phi_1, a_1'], [\Phi_2, a_2']) = \varrho_{A^*}(\Phi_1, \Phi_2) + \varrho_{A'}(a_1', a_2'),$$

$(\Phi_1, \Phi_2 \in A^*;\, a_1', a_2' \in A')$. Consequently, in virtue of a known theorem of Jiřina (cf. Jiřina [1] or Gikhman & Skorohod [1], Chapt. I, § 3), for each i in I there exists a system $(\omega_i)_{(.)}$, of transition probabilities from $[A', \mathfrak{A}']$

into $[A^*, \mathfrak{A}^*]$, such that

$$H_i(Y \times X) = \int_X (\omega_i)_{(a')}(Y)\, \sigma_i(\mathrm{d}a')$$

is satisfied. We obtain

$$\tilde{P} = \sum_{i \in I} H_i\big((.) \times A'\big) = \sum_{i \in I} \int (\omega_i)_{(a')}(.)\, \sigma_i(\mathrm{d}a')$$

$$= \sum_{i \in I} \int \frac{\mathrm{d}\sigma_i}{\mathrm{d}\sigma}(a')\,(\omega_i)_{(a')}(\cdot)\, \sigma(\mathrm{d}a') = \int \left(\sum_{i \in I} \frac{\mathrm{d}\sigma_i}{\mathrm{d}\sigma}(a')\,(\omega_i)_{(a')}\right)(.)\, \sigma(\mathrm{d}a').$$

Obviously, $\sum_{i \in I} \sigma_i = \sigma$. Therefore, to each i in I, we can associate a non-negative version f_i of $\frac{\mathrm{d}\sigma_i}{\mathrm{d}\sigma}$, such that $\sum_{i \in I} f_i = 1$ is satisfied. The mapping

$$a' \rightsquigarrow \omega_{(a')} = \sum_{i \in I} f_i(a')\,(\omega_i)_{(a')}$$

of A' into the set of all distributions on $\mathfrak{A}^*$ can be interpreted as a reduced cluster field on $[A', \varrho_{A'}]$ with phase space $[A, \varrho_A]$. This yields

$$\tilde{P} = \int \omega_{(a')}(.)\, \sigma(\mathrm{d}a'),$$

i.e. we have $P = (P_\sigma)_\omega$. Moreover, for all X in $\mathfrak{A}'$ and all Y in $\mathfrak{A}^*$ we have

$$\int_Y c_{(\Psi)}(X)\, \tilde{P}(\mathrm{d}\Psi) = H(Y \times X) = \sum_{i \in I} H_i(Y \times X)$$

$$= \sum_{i \in I} \int_X (\omega_i)_{(a')}(Y)\, \sigma_i(\mathrm{d}a') = \int_X \omega_{(a')}(Y)\, \sigma(\mathrm{d}a').$$

3. Given any reduced clustering representation $P = (P_\nu)_\varkappa$ of P on the phase space $[A', \varrho_{A'}]$ we have to show that there exists an element c in $\boldsymbol{c}$, such that, for all X in $\mathfrak{A}'$ and all Y in $\mathfrak{A}^*$,

$$\int_X \varkappa_{(a')}(Y)\, \nu(\mathrm{d}a') = H(Y \times X) = \int_Y c_{(\Psi)}(X)\, \tilde{P}(\mathrm{d}\Psi)$$

is satisfied.

Let $A' = \bigcup_{j \in J} X_j$ be a representation of A' as a union of an at most countable family $(X_j)_{j \in J}$ of pairwise disjoint sets in $\mathfrak{A}'$ having the property that $0 < \nu(X_j) < +\infty$. We set

$$H_j = H\big((.) \cap (A^* \times X_j)\big), \qquad Q_j = H_j\big((.) \times A'\big).$$

Each H_j is a finite measure on $\mathfrak{A}^* \otimes \mathfrak{A}'$, which is different from 0, so that, as it was done in 2., we can conclude that for each j in J there exists a system $(c_j)_{(.)}$ of transition probabilities from $[A^*, \mathfrak{A}^*]$ into $[A', \mathfrak{A}']$, for which the relation

$$H_j(Y \times X) = \int_Y (c_j)_{(\Psi)}(X)\, Q_j(\mathrm{d}\Psi)$$

is satisfied for all X in $\mathfrak{A}'$ and all Y in $\mathfrak{A}^*$. We obtain

$$H(Y \times X) = \sum_{j \in J} H_j(Y \times X) = \sum_{j \in J} \int_Y \frac{\mathrm{d}Q_j}{\mathrm{d}\tilde{P}}(\Psi)\,(c_j)_{(\Psi)}(X)\,\tilde{P}(\mathrm{d}\Psi)$$

$$= \int_Y \sum_{j \in J} \frac{\mathrm{d}Q_j}{\mathrm{d}\tilde{P}}(\Psi)\,(c_j)_{(\Psi)}(X)\,\tilde{P}(\mathrm{d}\Psi).$$

To each j in J we can associate a non-negative version g_j of $\frac{\mathrm{d}Q_j}{\mathrm{d}\tilde{P}}$, such that $\sum_{j \in J} g_j = 1$ is satisfied. Then the measurable mapping

$$\Psi \curvearrowright c_{(\Psi)} = \sum_{j \in J} g_j(\Psi)\,(c_j)_{(\Psi)}$$

of A^* into the set of distributions on $\mathfrak{A}'$ satisfies the condition

$$\int_X \varkappa_{(a')}(Y)\,\nu(\mathrm{d}a') = H(Y \times X) = \int_Y c_{(\Psi)}(X)\,\tilde{P}(\mathrm{d}\Psi) \qquad (X \in \mathfrak{A}',\ Y \in \mathfrak{A}^*).$$

Moreover, the measure $\int c_{(\Psi)}(.)\,\tilde{P}(\mathrm{d}\Psi) = H\big(A^* \times (.)\big) = \nu$ belongs to N', so that c belongs to $\mathbf{c}$. ∎

For some problems it is reasonable further to narrow the concept of the reduced clustering representation on $[A, \varrho_A]$. Thus a reduced clustering representation $P = (P_\sigma)_\omega$ of P on $[A, \varrho_A]$ is called *special* if

$$\omega_{(a)}\big(\chi(\{a\}) > 0\big) = 1$$

is satisfied almost everywhere with respect to σ, i.e. if almost everywhere with respect to σ the point a belongs with probability one to the random cluster produced by it. Obviously this additional property depends only on the equivalence class of the considered clustering representations. We have

4.5.5. *Let P be a distribution in $\mathbf{T}$ which is different from δ_o, $P = (P_\sigma)_\omega$ a reduced clustering representation of P on $[A, \varrho_A]$, and c a corresponding element of $\mathbf{c}$. Subject to these assumptions, $P = (P_\sigma)_\omega$ is a special clustering representation if and only if*

$$c_{(\Psi)}\big(\{a : a \in A,\ \Psi(\{a\}) > 0\}\big) = 1$$

is satisfied almost everywhere with respect to $\tilde{P}$.

Proof. It is obvious that the set

$$Z = \big\{[\Psi, a] : [\Psi, a] \in M \times A,\ \ \Psi(\{a\}) > 0\big\}$$

is closed with respect to the metric $\varrho_{M \times A}$, since by 1.15.4. it follows from $\Psi_n \underset{n \downarrow \infty}{\overset{M}{\Rightarrow}} \Psi$, $a_n \xrightarrow[n \to \infty]{} a$, $\Psi_n(\{a_n\}) > 0$, that the relation $\Psi\big(S_d(a)\big) \geqq 1$ is satisfied for each open ball $S_d(a)$ satisfying the condition $\Psi\big(\partial S_d(a)\big) = 0$.

Consequently Z is a Borel subset of the product space $[M \times A, \varrho_{M\times A}]$. Now, observing that, in view of 1.15.5., $\mathfrak{M}$ coincides with the σ-algebra of the Borel subsets of the complete separable metric space $[M, \varrho_M]$, we conclude that $Z \in \mathfrak{M} \otimes \mathfrak{A}$, and hence that $Z \in \mathfrak{A}^* \otimes \mathfrak{A}$.

For all a in A and all Ψ in A^*, the section $Z\big|_a$ coincides with $\{\Phi : \Phi \in A^*, \Phi(\{a\}) > 0\}$, and the section $Z\big|_\Psi$ with $\{b : b \in A, \Psi(\{b\}) > 0\}$. Therefore all of these sections belong to $\mathfrak{A}^*$ and $\mathfrak{A}$, respectively, and the real functions $a \curvearrowright \omega_{(a)}(\chi(\{a\}) > 0)$ and $\Psi \curvearrowright c_{(\Psi)}(\{a : a \in A, \Psi(\{a\}) > 0\})$ are measurable with respect to $\mathfrak{A}$ and $\mathfrak{A}^*$, respectively. We obtain

$$\int \omega_{(a)}(\chi(\{a\}) = 0)\, \sigma(\mathrm{d}a) = H(\bar{Z}) = \int c_{(\Psi)}(\{a : a \in A, \Psi(\{a\})) = 0\})\, \tilde{P}(\mathrm{d}\Psi). \ \blacksquare$$

Every P in $\boldsymbol{T}$ has a special reduced clustering representation. We even have

4.5.6. *For each infinitely divisible distribution P on $\mathfrak{M}$ and each X in $\mathfrak{B}$, there exists a special reduced clustering representation $P = (P_\sigma)_\omega$ of P on $[A, \varrho_A]$, such that for any corresponding stochastic transition c almost everywhere with respect to $\tilde{P}$ it can be concluded from $\Psi(X) > 0$ that $c_{(\Psi)}(X) = 1$.*

Proof. 1. To each natural number n we associate some covering $(X_{n,m})_{m=1,2,\ldots}$ of X by pairwise disjoint sets in $\mathfrak{B}$, the diameters of which do not exceed $1/n$. For each a in X and each natural number n, we set $f_n(a) = m$ if a belongs to $X_{n,m}$. The correspondence $a \curvearrowright (f_n(a))_{n=1,2,\ldots}$ then provides a one-to-one mapping of X into the power set $\{1, 2, \ldots\}^{\{1,2,\ldots\}}$. Now, if $(f_n(a_1))_{n=1,2,\ldots}$ is less than $(f_n(a_2))_{n=1,2,\ldots}$ in the sense of lexicographic order in $\{1, 2, \ldots\}^{\{1,2,\ldots\}}$, then we write $a_1 < a_2$, thus obtaining a linear order in X. For all Ψ in M having the property that $\Psi(X) > 0$, let $a_1(\Psi)$ denote the smallest point, in the sense of the order relation $<$, in $\{a : a \in X, \Psi(\{a\}) > 0\}$. If we now set

$$X \cap X_{1,i_1} \cap \cdots \cap X_{s,i_s} = V_{i_1,\ldots,i_s},$$

then, for all subsets X' of X which belong to $\mathfrak{B}$, we obtain

$$\begin{aligned} &\{\Psi : \Psi \in M, \Psi(X) > 0, a_1(\Psi) \in X'\} \\ &= \bigcup_{s=1}^{\infty} \bigcup_{i_1,\ldots,i_s=1}^{\infty} \Big\{\Psi : \Psi \in M, \Psi^*(X' \cap V_{i_1,\ldots,i_s}) = \Psi^*(V_{i_1,\ldots,i_s}) = 1, \\ &\qquad \Psi^*\Big(\bigcup_{[l_1,\ldots,l_s] < [i_1,\ldots,i_s]} V_{l_1,\ldots,l_s}\Big) = 0\Big\}, \end{aligned}$$

where Ψ^* denotes the "simplified" counting measure Ψ and $<$ lexicographic order. Consequently the mapping $\Psi \curvearrowright a_1(\Psi)$ from M into A is measurable.

2. We choose a sequence $X_2, X_3, \ldots$ of sets in $\mathfrak{B}$, such that $X = X_1, X_2, \ldots$ forms a sequence of pairwise disjoint sets which covers A and that each set in $\mathfrak{B}$ is covered already by a finite number of terms of this sequence.

The construction method applied in 1. can now be transferred to the sets $X_2, X_3, \ldots$, so that in this way we obtain a sequence of measurable mappings $a_n(.)$ from M into A.

Setting now

$$c_{(\Psi)} = \delta_{a_n(\Psi)}, \quad \text{if} \quad \Psi(X_1 \cup \cdots \cup X_{n-1}) = 0, \qquad \Psi(X_n) > 0,$$

we obtain an element c in $\boldsymbol{c}$ which has, for all Ψ in M satisfying the condition $\Psi(X) > 0$, the property that $c_{(\Psi)}(X) = 1$.

In view of 4.5.4. and 4.5.5., this completes our proof. ∎

Using 4.5.5. and 4.5.6., we obtain

4.5.7. Theorem. *For each infinitely divisible distribution P on $\mathfrak{M}$, the following statements are equivalent:*

a) *P is free from after-effects.*

b) *There exists a clustering representation $P = (P_\nu)_\varkappa$ of P on $[A, \varrho_A]$ such that $\varkappa_{(a)}(\{\delta_a, 2\delta_a, \ldots\}) = 1$ almost everywhere with respect to ν.*

c) *All special reduced clustering representations of P are equivalent.*

Proof. 1. a) implies b) and c). This is immediately clear in the case $P = \delta_o$, because then all of the reduced clustering representations $P = (P_\nu)_\varkappa$ of P on $[A, \varrho_A]$ are equivalent to the representation $P = (P_\sigma)_\omega$, where

$$\sigma = o, \quad \omega_{(a)} = \delta_{\delta_a} \qquad \text{for all } a \text{ in } A.$$

Therefore we can suppose in addition that P be different from δ_o.

In virtue of 2.2.13., we have $\tilde{P}$ (Ψ has mass at more than one point) $= 0$. Thus all c in $\boldsymbol{c}$ satisfying the condition $c_{(\Psi)}\big(\{a : a \in A, \Psi(\{a\}) > 0\}\big) = 1$ almost everywhere with respect to $\tilde{P}$ are equivalent to

$$c_{(n\delta_a)} = \delta_a \qquad (n = 1, 2, \ldots; a \in A).$$

(For $\Psi \neq n\delta_a$ $(n = 1, 2, \ldots; a \in A)$, $c_{(\Psi)}$ need not be defined.) Our set-up really leads to an element of $\boldsymbol{c}$, since for all X in $\mathfrak{B}$ we have

$$\int c_{(\Psi)}(X)\, \tilde{P}(\mathrm{d}\Psi) = \tilde{P}\big(\Psi(X) > 0\big) < +\infty.$$

Consequently, by 4.5.4. and 4.5.5. a) implies c).

Now let $P = (P_\sigma)_\omega$ be a reduced clustering representation of P on $[A, \varrho_A]$ which corresponds to this element c of $\boldsymbol{c}$. We denote by V the set

$$\big\{[\Psi, a] : [\Psi, a] \in A^* \times A, \Psi(A \setminus \{a\}) > 0\big\}.$$

Using for $(A^* \times A) \setminus V$ the method applied in the proof of 4.5.5., we can readily show that V belongs to $\mathfrak{A}^* \otimes \mathfrak{A}$. For the sections of V with a and Ψ we have, respectively,

$$V\big|_a = \big\{\Phi : \Phi \in M, \Phi(A \setminus \{a\}) > 0\big\}, \qquad V\big|_\Psi = \big\{b : b \in A, \Phi(A \setminus \{b\}) > 0\big\}.$$

Consequently the real functions $a \rightsquigarrow \omega_{(a)}\big(\Psi(A \setminus \{a\}) > 0\big)$ and $\Psi \rightsquigarrow c_{(\Psi)}$ $\big(\{a : a \in A,\ \Psi(A \setminus \{a\}) > 0\}\big)$ are measurable with respect to the σ-algebras $\mathfrak{A}$ and $\mathfrak{A}^*$, respectively, and we obtain

$$\int \omega_{(a)}\big(\Psi(A \setminus \{a\}) > 0\big)\, \sigma(\mathrm{d}a) = H(V)$$
$$= \int c_{(\Psi)}\big(\{a : a \in A,\ \Psi(A \setminus \{a\}) > 0\}\big)\, \tilde{P}(\mathrm{d}\Psi) = 0\,.$$

Since the clustering representation $P = (P_\sigma)_\omega$ is reduced, we consequently have

$$\omega_{(a)}(\{\delta_a, 2\delta_a, \ldots\}) = 1\,,$$

almost everywhere with respect to σ, which means that a) implies b).

2. b) implies a). If $P = (P_\nu)_\varkappa$ is a clustering representation of P having the property stated under b), then we obtain

$$\tilde{P}\ (\Psi \text{ has mass at more than one point})$$
$$= \int \varkappa_{(a)}\ (\chi \text{ has mass at more than one point})\ \nu(\mathrm{d}a) = 0\,.$$

Hence, in virtue of 2.2.13., P is free from after-effects.

3. c) implies a). If a distribution P in $\boldsymbol{T}$ is not free from after-effects, and hence also different from δ_o, then, in view of 2.2.12., there exist disjoint sets X_1, X_2 in $\mathfrak{B}$, such that $\tilde{P}\big(\Psi(X_1) > 0,\ \Psi(X_2) > 0\big) > 0$ is satisfied. Thus, by 4.5.6. there exist special reduced clustering representations of P such that, for any associated c_1 and c_2, $\Psi(X_1) > 0$ implies $(c_1)_{(\Psi)}\,(X_1) = 1$ and $\Psi(X_2) > 0$ implies $(c_2)_{(\Psi)}\,(X_2) = 1$, respectively, almost everywhere with respect to $\tilde{P}$. Then in view of our assumptions concerning X_1, X_2, the stochastic transitions c_1, c_2 cannot be equivalent. Consequently the statement c) cannot be valid. ∎

4.6. Some Special Classes of Clustering Representations

The following examples will illustrate the possibilities opened by 4.5.4.

1st Example. Let P be a distribution in $\boldsymbol{T}$ different from δ_o, and $[A', \varrho_{A'}]$ the discrete space $\{1, 2, \ldots\}$. Each c in $\boldsymbol{c}$ is of the form

$$c_{(\Psi)} = \sum_{n=1}^{\infty} f_n(\Psi)\, \delta_n \qquad (\Psi \in M \setminus \{o\}),$$

where the non-negative measurable functions f_n on A^* satisfy the conditions

$$\sum_{n=1}^{\infty} f_n = 1\,, \qquad \sigma(\{n\}) = \int f_n(\Psi)\, \tilde{P}(\mathrm{d}\Psi) < +\infty \qquad (n = 1, 2, \ldots).$$

Then the class of reduced clustering representations of P on $\{1, 2, \ldots\}$ which corresponds to c corresponds to the decomposition

$$\tilde{P} = \sum_{n=1}^{\infty} \int_{(.)} f_n(\Psi)\, \tilde{P}(\mathrm{d}\Psi)$$

(cf. the first step of proof of theorem 4.5.1.).

2nd Example. Let P be a distribution in $\boldsymbol{T}$ different from δ_o and $[A', \varrho_{A'}]$ the complete separable metric space $[A^*, \varrho_{A^*}]$ introduced in section 3.3. Now the following element c in $\boldsymbol{c}$ offers itself:

$$c_{(\Psi)} = \delta_\Psi \qquad (\Psi \in A^*),$$

i.e. the stochastic transition $c_{(.)}$ corresponds to the identity mapping of A^* upon itself. The finiteness condition required for the elements of $\boldsymbol{c}$ is satisfied, since

$$\sigma(Y) = \tilde{P}(Y) < +\infty$$

is valid for all Y in $\mathfrak{B}^*$. We may readily find an associated reduced clustering representation of P on $[A^*, \varrho_{A^*}]$. Let

$$\omega_{(\Phi)} = \delta_\Phi \qquad (\Phi \in A^*).$$

Setting now $\sigma = \tilde{P}$, as was done above, we obtain

$$\int_{Y_1} c_{(\Psi)}(Y_2)\, \tilde{P}(\mathrm{d}\Psi) = H(Y_1 \times Y_2) = \int_{Y_2} \omega_{(\Phi)}(Y_1)\, \sigma(\mathrm{d}\Phi)$$

for all Y_1, Y_2 in $\mathfrak{A}^*$. Consequently, $P = (P_\sigma)_\omega$ is a reduced clustering representation of P on the phase space $[A^*, \varrho_{A^*}]$. It is called the *standard representation* of P (cf. Lee [1], [2]).

By virtue of 4.2.6., for all W_1, W_2 in $\boldsymbol{W} \setminus \{0\}$, we obtain

$$\|\mathscr{E}_{W_1} - \mathscr{E}_{W_2}\| = \|(P_{W_1})_\omega - (P_{W_2})_\omega\| \leqq \|P_{W_1} - P_{W_2}\|,$$

where $(P_{W_1})_\omega$, $(P_{W_2})_\omega$ are the standard representations of $\mathscr{E}_{W_1}$, $\mathscr{E}_{W_2}$. In the case $W_1 = 0$, $W_2 \in \boldsymbol{W}$ we have

$$\begin{aligned}\|\mathscr{E}_{W_1} - \mathscr{E}_{W_2}\| &= \|\delta_o - \mathscr{E}_{W_2}\| \\ &= 2\mathscr{E}_{W_2}(\Phi \neq o) = 2\big(1 - \exp(-W_2(M))\big) \\ &= 2P_{W_2}(\Psi \neq o) = \|\delta_o - P_{W_2}\| = \|P_{W_1} - P_{W_2}\|,\end{aligned}$$

and we observe the truth of

4.6.1. Proposition. *For all* W_1, W_2 *in* $\boldsymbol{W}$,

$$\|\mathscr{E}_{W_1} - \mathscr{E}_{W_2}\| \leqq \|P_{W_1} - P_{W_2}\|.$$

In this way it becomes possible to use results of section 1.12. for arbitrary distributions in $\boldsymbol{T}$ (cf. Liese [2]).

3rd Example. Let P be a regular infinitely divisible distribution on $\mathfrak{M}$ being different from δ_o, and $[A', \varrho_{A'}] = [A, \varrho_A]$. It is natural to employ the following stochastic transition c from $[A^*, \mathfrak{A}^*]$ into $[A, \mathfrak{A}]$:

$$c_{(\Psi)} = \left(\Psi(A)\right)^{-1} \Psi \qquad \left(\Psi \in A^*, \Psi(A) < \infty\right),$$

i.e. for all finite Ψ in M being different from o, we choose any point in Ψ "at random". Since $\tilde{P}(\Psi \text{ is not finite}) = 0$, and as we are interested only in the equivalence class of c, we need not define $c_{(\Psi)}$ for infinite Ψ.

The finiteness condition being characteristic of elements of $\mathbf{c}$ is satisfied, since

$$\begin{aligned}\sigma(X) &= \int c_{(\Psi)}(X)\, \tilde{P}(\mathrm{d}\Psi) = \int \left(\Psi(A)\right)^{-1} \Psi(X)\, \tilde{P}(\mathrm{d}\Psi) \\ &\leqq \tilde{P}\left(\Psi(X) > 0\right) < +\infty\end{aligned}$$

is valid for all X in $\mathfrak{B}$.

By this choice of c, we introduced a well-determined class of special reduced clustering representations $P = (P_\sigma)_\omega$.

4th Example. Let $[A', \varrho_{A'}] = [A, \varrho_A] = R^1$. Further let P be an infinitely divisible distribution on $\mathfrak{M}$, different from δ_o, such that

$$\tilde{P}\left(\Psi\left((-\infty, 0)\right) = +\infty\right) = 0.$$

For each Ψ in A^* having the property that $\Psi\left((-\infty, 0)\right) < +\infty$ we can choose the smallest point $l(\Psi)$ of $\left\{a\colon a \in R^1, \Psi(\{a\}) > 0\right\}$, thus obtaining a measurable mapping l from A^* into R^1. The set-up

$$c_{(\Psi)} = \delta_{l(\Psi)} \qquad \left(\Psi \in A^*, \Psi\left((-\infty, 0)\right) < +\infty\right)$$

defines an equivalence class of elements of $\mathbf{c}$. Now let $P = (P_\sigma)_\omega$ be a corresponding clustering representation of P on R^1. Then

$$\omega_{(a)}\left(\chi\left((-\infty, a)\right) = 0\right) = 1 = \omega_{(a)}\left(\chi(\{a\}) > 0\right)$$

holds almost everywhere with respect to σ, i.e. we obtain a cluster field ω which is "subject to the principle of causality", viz. the cluster produced at the (time-)point a is effective only on the interval $[a, +\infty)$.

The particular importance of this distinguished equivalence class of special reduced clustering representations lies in that the condition $\tilde{P}\left(\Psi\left((-\infty, 0)\right) = +\infty\right) = 0$ is certainly satisfied in cases where P coincides with ${}_{[0,+\infty)}P$. Consequently, given any element P of $\boldsymbol{T}$, then ${}_{[0,+\infty)}P$ can be represented in the way indicated above, if ${}_{[0,+\infty)}P \neq \delta_o$ is satisfied. In this case we can readily state the associated measure σ, viz. for all $t \geqq 0$,

$$\begin{aligned}e^{-\tilde{P}(\Phi([0,t])>0)} &= P\left(\Phi([0, t]) = 0\right) = (P_\sigma)_\omega \left(\Phi([0, t]) = 0\right) \\ &= P_\sigma\left(\Phi([0, t]) = 0\right) = e^{-\sigma([0,t])},\end{aligned}$$

i.e. we have

$$\sigma([0, t]) = -\ln P\left(\Phi([0, t]) = 0\right) = \tilde{P}\left(\Phi([0, t]) > 0\right).$$

4.7. Continuous Cluster Fields

A *continuous cluster field* on $[A', \varrho_{A'}]$ with phase space $[A, \varrho_A]$ is conceived of as a mapping of A' into $\boldsymbol{P}$ which is continuous with respect to $\varrho_{A'}$ and $\varrho_{\boldsymbol{P}}$. In view of 3.1.3., such a mapping is always measurable with respect to the σ-algebras $\mathfrak{A}'$ and $\mathfrak{P}$.

The importance of continuous cluster fields is based on the possibility of exchanging weak convergence in $\boldsymbol{P}'$ and clustering if some additional assumptions are fulfilled.

4.7.1. Proposition. *Let $\varkappa$ be a continuous cluster field on $[A', \varrho_{A'}]$ with phase space $[A, \varrho_A]$, and (P_n) a sequence of distributions in $\boldsymbol{P}'$ which converges weakly toward P in $\boldsymbol{P}'$. If there exists a set X in $\mathfrak{B}'$ with the property*

$$P_n\big(\Phi(A' \setminus X) > 0\big) = 0 \quad \text{for} \quad n = 1, 2, \ldots,$$

then

$$(P_n)_\varkappa \underset{n\to\infty}{\Rightarrow} P_\varkappa .$$

Proof.[1]) 1. Let Z be a closed set in $\mathfrak{B}_P'$ having the property that $X \subseteq Z$. By 3.1.11. the distributions $P_n = {}_Z(P_n)$ converge weakly toward ${}_ZP$, and we obtain $P = {}_ZP$, i.e. $P\big(\Phi(A' \setminus Z) > 0\big) = 0$. Consequently,

$$P(\Phi \text{ finite}) = 1 = P_n(\Phi \text{ finite}) \quad \text{for} \quad n = 1, 2, \ldots,$$

such that $P_\varkappa$, $(P_1)_\varkappa$, $(P_2)_\varkappa$, ... exist.

2. We shall now suppose in addition that there exist a compact subset B of Z having the property that

$$P_n\big(\Phi(A' \setminus B) > 0\big) = 0 \quad \text{for} \quad n = 1, 2, \ldots$$

as well as a natural number k, such that

$$P_n\big(\Phi(A') \leqq k\big) = 1 \quad \text{for} \quad n = 1, 2, \ldots$$

For each natural number $l \leqq k$, in view of 3.1.10., the mapping

$$[x_1, \ldots, x_l] \curvearrowright \mathop{\ast}_{i=1}^{l} \varkappa_{(x_i)}$$

of $(A')^l$ into $\boldsymbol{P}$ is continuous. Hence, in view of 3.1.6., to each $\varepsilon > 0$ there exists a finite covering $X_1, \ldots, X_m$ of B by pairwise disjoint sets in $\mathfrak{B}_P'$ such that, for all Φ, Ψ in M' satisfying the conditions

$$\Phi\big(A' \setminus (X_1 \cup \cdots \cup X_m)\big) = 0 = \Psi\big(A' \setminus (X_1 \cup \cdots \cup X_m)\big),$$
$$\Phi(X_1 \cup \cdots \cup X_m) \leqq k, \quad \Phi(X_i) = \Psi(X_i) \quad \text{for} \quad i = 1, 2, \ldots, m$$

the inequality $\varrho_{\boldsymbol{P}}(\varkappa_{(\Phi)}, \varkappa_{(\Psi)}) \leqq \varepsilon$ is satisfied.

---

[1]) A Short proof is given in Fleischmann & Siegmund-Schultze [1].

Using the definition of $\varrho_{\boldsymbol{P}}$, for all non-negative integral $l_1, \ldots, l_m$ satisfying the condition $\sum_{i=1}^{m} l_i \leqq k$ and all Q_1, Q_2 in $\boldsymbol{P}'$ having the properties

$$Q_j\big(\Phi(A' \setminus (X_1 \cup \cdots \cup X_m)) = 0, \Phi(X_i) = l_i \quad \text{for} \quad i = 1, \ldots, m\big) = 1 \qquad (j = 1, 2)$$

we can now conclude that

$$\varrho_{\boldsymbol{P}}\big((Q_1)_\varkappa, (Q_2)_\varkappa\big) \leqq \varepsilon.$$

If Q_1, Q_2 satisfy only the weaker conditions

$$Q_j\big(\Phi(A' \setminus (X_1 \cup X_2 \cup \cdots \cup X_m)) = 0\big) = 1, \ Q_j\big(\Phi(A') > k\big) = 0 \quad (j = 1, 2),$$

$$Q_1\big(\Phi(X_i) = l_i \quad \text{for} \quad i = 1, \ldots, m\big) = Q_2\big(\Phi(X_i) = l_i \quad \text{for} \quad i = 1, \ldots, m\big) \qquad (l_1, \ldots, l_m \geqq 0),$$

then the kind of inference used above yields

$$\begin{aligned} \varrho_{\boldsymbol{P}}\big((Q_1)_\varkappa, (Q_2)_\varkappa\big) &= \varrho_{\boldsymbol{P}}\Big(\sum_{l_1,\ldots,l_m \geqq 0} Q_1\big(\Phi(X_i) = l_i \quad \text{for} \quad i = 1, \ldots, m\big) \\ &\qquad \cdot \big(Q_1((.) \mid \Phi(X_i) = l_i \quad \text{for} \quad i = 1, \ldots, m)\big)_\varkappa, \\ &\qquad \sum_{l_1,\ldots,l_m \geqq 0} Q_2\big(\Phi(X_i) = l_i \quad \text{for} \quad i = 1, \ldots, m\big) \\ &\qquad \cdot \big(Q_2((.) \mid \Phi(X_i) = l_i \quad \text{for} \quad i = 1, \ldots, m)\big)_\varkappa\Big) \\ &\leqq \max_{\substack{l_1,\ldots,l_m \geqq 0 \\ Q_1(\Phi(X_i) = l_i \text{ for } i=1,\ldots,m) > 0}} \varrho_{\boldsymbol{P}}\Big(\big(Q_1((.) \mid \Phi(X_i) = l_i \quad \text{for} \quad i = 1, \ldots, m)\big)_\varkappa, \\ &\qquad \big(Q_2((.) \mid \Phi(X_i) = l_i \quad \text{for} \quad i = 1, \ldots, m)\big)_\varkappa\Big) \leqq \varepsilon. \end{aligned}$$

In view of 3.1.9., the sequence $\big((P_n)_{X_1,\ldots,X_m}\big)$ converges towards $P_{X_1,\ldots,X_m}$, and we obtain

$$\begin{aligned} \overline{\lim_{n\to\infty}}\, \varrho_{\boldsymbol{P}}\big((P_n)_\varkappa, P_\varkappa\big) &\leqq \overline{\lim_{n\to\infty}}\, \varrho_{\boldsymbol{P}}\Big((P_n)_\varkappa, \sum_{k \geqq l_1,\ldots,l_m \geqq 0} P\big(\Phi(X_i) = l_i \quad \text{for} \quad i = 1, \ldots, m\big) \\ &\qquad \cdot \big(P_n((.) \mid \Phi(X_i) = l_i \quad \text{for} \quad i = 1, \ldots, m)\big)_\varkappa\Big) \\ &+ \overline{\lim_{n\to\infty}}\, \varrho_{\boldsymbol{P}}\Big(\sum_{k \geqq l_1,\ldots,l_m \geqq 0} P\big(\Phi(X_i) = l_i \quad \text{for} \quad i = 1, \ldots, m\big) \\ &\qquad \cdot \big(P_n((.) \mid \Phi(X_i) = l_i \quad \text{for} \quad i = 1, \ldots, m)\big)_\varkappa, \\ &\qquad \sum_{k \geqq l_1,\ldots,l_m \geqq 0} P\big(\Phi(X_i) = l_i \quad \text{for} \quad i = 1, \ldots, m\big) \\ &\qquad \cdot \big(P((.) \mid \Phi(X_i) = l_i \quad \text{for} \quad i = 1, \ldots, m)\big)_\varkappa\Big) \\ &\leqq \overline{\lim_{n\to\infty}} \sum_{l_1,\ldots,l_m \geqq 0} \big| P\big(\Phi(X_i) = l_i \quad \text{for} \quad i = 1, \ldots, m\big) \\ &\qquad - P_n\big(\Phi(X_i) = l_i \quad \text{for} \quad i = 1, \ldots, m\big)\big| + \varepsilon = \varepsilon, \end{aligned}$$

i.e. we have $\lim_{n\to\infty} \varrho_{\boldsymbol{P}}\big((P_n)_\varkappa, P_\varkappa\big) = 0$.

3. Now, first of all, we shall free ourselves from the additional assumption that $P_n(\Phi(A') \leqq k) = 1$, $n = 1, 2, \ldots$, for some natural number k.

In view of 3.1.9.,

$$P_n(\Phi(A') > k) = P_n(\Phi(Z) > k) \xrightarrow[n\to\infty]{} P(\Phi(Z) > k) = P(\Phi(A') > k)$$

for all natural numbers k. In this way, for all sufficiently large k, we obtain

$$\begin{aligned}
&\overline{\lim_{n\to\infty}}\, \varrho_{\mathbf{P}}\big((P_n)_\varkappa, P_\varkappa\big) \\
&\leqq \overline{\lim_{n\to\infty}}\, \varrho_{\mathbf{P}}\big((P_n)_\varkappa, P_n((.)\,|\,\Phi(A') \leqq k)_\varkappa\big) \\
&\quad + \overline{\lim_{n\to\infty}}\, \varrho_{\mathbf{P}}\Big(\big(P_n((.)\,|\,\Phi(A') \leqq k)\big)_\varkappa, \big(P((.)\,|\,\Phi(A') \leqq k)\big)_\varkappa\Big) \\
&\quad + \varrho_{\mathbf{P}}\Big(\big(P((\cdot)\,|\,\Phi(A') \leqq k)\big)_\varkappa, P_\varkappa\Big).
\end{aligned}$$

Since, as n increases, $P_n((.) \mid \Phi(A') \leqq k)$ converges weakly towards $P((.) \mid \Phi(A') \leqq k)$, we may use 2. to conclude that

$$\big(P_n((.)\,|\,\Phi(A') \leqq k)\big)_\varkappa \underset{n\to\infty}{\Rightarrow} \big((P(.)\,|\,\Phi(A') \leqq k)\big)_\varkappa.$$

Therefore the above inequality can be continued as follows:

$$\begin{aligned}
&\leqq \overline{\lim_{n\to\infty}} \Big\| (P_n)_\varkappa - \big(P_n((.)\,|\,\Phi(A') \leqq k)\big)_\varkappa \Big\| + \Big\| P_\varkappa - \big(P((.)\,|\,\Phi(A') \leqq k)\big)_\varkappa \Big\| \\
&= \overline{\lim_{n\to\infty}} \Big\| P_n(\Phi(A') \leqq k)\, \big(P_n((.)\,|\,\Phi(A') \leqq k)\big)_\varkappa \\
&\qquad + P_n(\Phi(A') > k)\, \big(P_n((.)\,|\,\Phi(A') \geqq k)\big)_\varkappa - \big(P_n((.)\,|\,\Phi(A') \leqq k)\big)_\varkappa \Big\| \\
&\quad + \Big\| P(\Phi(A') \leqq k)\, \big(P((.)\,|\,\Phi(A') \leqq k)\big)_\varkappa \\
&\qquad + P(\Phi(A') > k)\, \big(P((.)\,|\,\Phi(A') > k)\big)_\varkappa - \big(P((.)\,|\,\Phi(A') \leqq k)\big)_\varkappa \Big\| \\
&\leqq \overline{\lim_{n\to\infty}}\, 2P_n(\Phi(A') > k) + 2P(\Phi(A') > k) = 4P(\Phi(A') > k),
\end{aligned}$$

i.e. we have

$$\overline{\lim_{n\to\infty}}\, \varrho_{\mathbf{P}}\big((P_n)_\varkappa, P_\varkappa\big) \leqq 4 \inf_{k=1,2,\ldots} P(\Phi(B) > k) = 0$$

and hence

$$(P_n)_\varkappa \underset{n\to\infty}{\Rightarrow} P_\varkappa.$$

4. We now shall completely drop all additional assumptions.

In virtue of the compactness criterion 3.2.7., to each $\varepsilon > 0$ there exists a compact subset B of Z such that

$$P(\Phi(Z \setminus B) > 0) < \varepsilon, \quad P_n(\Phi(Z \setminus B) > 0) < \varepsilon \quad \text{for} \quad n = 1, 2, \ldots$$

is satisfied. Then, for $n = 1, 2, \ldots$, we have

$$\begin{aligned}
&\left\| P_n - P_n\big((.) \mid \Phi(Z \setminus B) = 0\big) \right\| \\
&= \Big\| P_n\big(\Phi(Z \setminus B) = 0,\ \Phi \in (.)\big) + P_n\big(\Phi(Z \setminus B) > 0,\ \Phi \in (.)\big) \\
&\quad - \big(P_n(\Phi(Z \setminus B) = 0)\big)^{-1} P_n\big(\Phi(Z \setminus B) = 0,\ \Phi \in (.)\big) \Big\| \\
&\leqq P_n\big(\Phi(Z \setminus B) > 0\big) + P_n\big(\Phi(Z \setminus B) = 0\big) \Big(\big(P_n(\Phi(Z \setminus B) = 0)\big)^{-1} - 1\Big) \\
&= 2P_n\big(\Phi(Z \setminus B) > 0\big) < 2\varepsilon .
\end{aligned}$$

By 4.2.6., we obtain

$$\left\| (P_n)_\varkappa - \big(P_n((.) \mid \Phi(Z \setminus B) = 0)\big)_\varkappa \right\| < 2\varepsilon \quad \text{for} \quad n = 1, 2, \ldots .$$

By a repeated application of 3.2.7., we observe that to each strictly monotone increasing sequence of natural numbers there exists a subsequence (n_m) such that

$$P_{n_m}\big(\Phi(Z \setminus B) = 0, \Phi \in (.)\big) \underset{m \to \infty}{\Rightarrow} Q$$

for some finite measure Q on $\mathfrak{M}'$. We have

$$P_{n_m}\big(\Phi(Z \setminus B) = 0\big) \xrightarrow[m \to \infty]{} Q(M) \geqq 1 - \varepsilon,$$

so that we can conclude that the conditional distributions

$$P_{n_m}\big((.) \mid \Phi(Z \setminus B) = 0\big)$$

converge weakly towards an H in $\boldsymbol{P}'$, and by 3. it follows that

$$\big(P_{n_m}((.) \mid \Phi(Z \setminus B) = 0)\big)_\varkappa \underset{m \to \infty}{\Rightarrow} H_\varkappa .$$

Immediately from the introduction of Q, we obtain $Q \leqq P$, and hence

$$\begin{aligned}
\|H - P\| &= \left\| (Q(M))^{-1} Q - P \right\| \leqq \big((Q(M))^{-1} - 1\big) + \|Q - P\| \\
&= \big((Q(M))^{-1} - 1\big) + \big(1 - Q(M)\big) = \big(Q(M)\big)^{-1} \big(1 - Q(M)\big) \\
&\quad + \big(1 - Q(M)\big) \leqq \frac{1}{1 - \varepsilon}\, \varepsilon + \varepsilon \leqq \frac{2\varepsilon}{1 - \varepsilon},
\end{aligned}$$

so that by 4.2.6. we can conclude that

$$\|H_\varkappa - P_\varkappa\| \leqq \frac{2\varepsilon}{1 - \varepsilon}.$$

Summarizing, we obtain

$$\begin{aligned}\overline{\lim_{m\to\infty}}\, \varrho_{\boldsymbol{P}}\big((P_{n_m})_\varkappa, P_\varkappa\big) &\leqq \overline{\lim_{n\to\infty}}\, \varrho_{\boldsymbol{P}}\Big((P_n)_\varkappa, \big(P_n((.)\,|\,\Phi(Z\setminus B)>0)\big)_\varkappa\Big)\\ &\quad + \overline{\lim_{m\to\infty}}\, \varrho_{\boldsymbol{P}}\Big(\big(P_{n_m}((.)\,|\,\Phi(Z\setminus B)=0)\big)_\varkappa, H_\varkappa\Big) + \varrho_{\boldsymbol{P}}(H_\varkappa, P_\varkappa)\\ &\leqq \overline{\lim_{n\to\infty}}\, \Big\|(P_n)_\varkappa - \big(P_n((.)\,|\,\Phi(Z\setminus B)=0)\big)_\varkappa\Big\| + 0 + \frac{2\varepsilon}{1-\varepsilon}\\ &\leqq 2\varepsilon\left(1+\frac{1}{1-\varepsilon}\right),\end{aligned}$$

i.e. we have $\overline{\lim_{n\to\infty}}\, \varrho_{\boldsymbol{P}}\big((P_n)_\varkappa, P_\varkappa\big) = 0$ and hence $(P_n)_\varkappa \underset{n\to\infty}{\Rightarrow} P_\varkappa$. ∎

Based on 4.7.1. it is possible to derive the following necessary and sufficient criterion for the possibility of exchanging weak convergence in $\boldsymbol{P}'$ and clustering according to $\varkappa$.

4.7.2. Theorem. *Let $\varkappa$ be a continuous cluster field on $[A', \varrho_{A'}]$ with phase space $[A, \varrho_A]$, and (P_n) a sequence of distributions in $\boldsymbol{P}'$ which converges weakly toward P in $\boldsymbol{P}'$. Under these assumptions the following assertions are equivalent:*

a) $P_\varkappa, (P_1)_\varkappa, (P_2)_\varkappa, \ldots$ *exist, and*

$$(P_n)_\varkappa \underset{n\to\infty}{\Rightarrow} P_\varkappa.$$

b) *For all X_0 in $\mathfrak{B}$ and all $\varepsilon > 0$ we have*

$$\inf_{X\in\mathfrak{B}'}\ \sup_{n=1,2,\ldots} P_n\Big(\int_{A'\setminus X} \varkappa_{(a')}\big(\chi(X_0)>0\big)\,\Phi(\mathrm{d}a') > \varepsilon\Big) = 0.$$

Proof. 1. From b) it follows that for all X_0 in $\mathfrak{B}$ and all $\varepsilon > 0$,

$$\inf_{X\in\mathfrak{B}'} P\Big(\int_{A'\setminus X} \varkappa_{(a')}\big(\chi(X_0)>0\big)\,\Phi(\mathrm{d}a') > \varepsilon\Big) = 0.$$

We select any continuous real function f on $[A, \varrho_A]$ with the property $k_{X_0} \leqq f \leqq 1$. Let X_1 denote the support of f which is assumed to be bounded. For all a' in A' we obtain

$$\varkappa_{(a')}\big(\chi(X_0)>0\big) \leqq \int \min\big(1, \textstyle\int f(a)\,\chi(\mathrm{d}a)\big)\,\varkappa_{(a')}(\mathrm{d}\chi) \leqq \varkappa_{(a')}\big(\chi(X_1)>0\big).$$

In virtue of 1.15.3. the real function $\chi \curvearrowright v(\chi) = \min\big(1, \int f(a)\,\chi(\mathrm{d}a)\big)$ is continuous with respect to ϱ_M. Thus, it is possible with the help of 3.1.2. to deduce the continuity of $a' \curvearrowright \int v(\chi)\,\varkappa_{(a')}(\mathrm{d}\chi)$ with respect to $\varrho_{A'}$. Therefore, for any non-negative bounded real function g on $[A', \varrho_{A'}]$ with bounded support, by 1.15.3. the mapping

$$\Phi \curvearrowright \int g(a') \left(\int v(\chi)\, \varkappa_{(a')}(\mathrm{d}\chi)\right) \Phi(\mathrm{d}a') \qquad (\Phi \in M')$$

is continuous with respect to $\varrho_{M'}$. Consequently, for all $\varepsilon > 0$ we obtain

$$P\left(\int g(a') \left(\int v(\chi)\, \varkappa_{(a')}(\mathrm{d}\chi)\right) \Phi(\mathrm{d}a') > \varepsilon\right)$$
$$\leqq \varliminf_{n\to\infty} P_n\left(\int g(a') \left(\int v(\chi)\, \varkappa_{(a')}(\mathrm{d}\chi)\right) \Phi(\mathrm{d}a') > \varepsilon\right).$$

According to our suppositions, for all $\varepsilon, \eta > 0$ there exists a closed bounded subset $X_{\varepsilon,\eta}$ of A' such that

$$\sup_{n=1,2,\ldots} P_n\left(\int_{A'\setminus X_{\varepsilon,\eta}} \varkappa_{(a')}\left(\chi(X_1) > 0\right) \Phi(\mathrm{d}a') > \varepsilon\right) < \eta$$

holds. Now let (g_m) be any increasing sequence of non-negative continuous real functions on $[A', \varrho_{A'}]$ with bounded supports and the property $\sup_{m=1,2,\ldots} g_m = k_{A'\setminus X_{\varepsilon,\eta}}$. We obtain

$$P\left(\int_{A'\setminus X_{\varepsilon,\eta}} \varkappa_{(a')}\left(\chi(X_0) > 0\right) \Phi(\mathrm{d}a') > \varepsilon\right)$$
$$= P\left(\sup_{m=1,2,\ldots} \int g_m(a')\, \varkappa_{(a')}\left(\chi(X_0) > 0\right) \Phi(\mathrm{d}a') > \varepsilon\right)$$
$$= \sup_{m=1,2,\ldots} P\left(\int g_m(a')\, \varkappa_{(a')}\left(\chi(X_0) > 0\right) \Phi(\mathrm{d}a') > \varepsilon\right)$$
$$\leqq \sup_{m=1,2,\ldots} P\left(\int g_m(a') \left(\int v(\chi)\, \varkappa_{(a')}(\mathrm{d}\chi)\right) \Phi(\mathrm{d}a') > \varepsilon\right)$$
$$\leqq \sup_{m=1,2,\ldots} \varliminf_{n\to\infty} P_n\left(\int g_m(a') \left(\int v(\chi)\, \varkappa_{(a')}(\mathrm{d}\chi)\right) \Phi(\mathrm{d}a') > \varepsilon\right)$$
$$\leqq \sup_{m=1,2,\ldots} \varliminf_{n\to\infty} P_n\left(\int g_m(a')\, \varkappa_{(a')}\left(\chi(X_1) > 0\right) \Phi(\mathrm{d}a') > \varepsilon\right)$$
$$\leqq \sup_{n=1,2,\ldots} P_n\left(\int_{A'\setminus X_{\varepsilon,\eta}} \varkappa_{(a')}\left(\chi(X_1) > 0\right) \Phi(\mathrm{d}a') > \varepsilon\right) < \eta.$$

2. Let $Q \in \boldsymbol{P}'$. If

$$\inf_{X\in\mathfrak{B}'} Q\left(\int_{A'\setminus X} \varkappa_{(a')}\left(\chi(X_0) > 0\right) \Phi(\mathrm{d}a') > 1\right) = 0$$

for all X_0 in $\mathfrak{B}$, then $Q_\varkappa$ exists.

Indeed, under these assumptions it follows that for any fixed z in A', for all $\eta > 0$ and each natural number m there exists a set $X_{\eta,m}$ in $\mathfrak{B}'$ such that

$$Q\left(\int_{A'\setminus X_{\eta,m}} \varkappa_{(a')}\left(\chi\left(S_m(z)\right) > 0\right) \Phi(\mathrm{d}a') \leqq 1\right) \geqq 1 - \eta.$$

Hence

$$Q\left(\int \varkappa_{(a')}\left(\chi(S_m(z)) > 0\right) \Phi(\mathrm{d}a') < +\infty\right) \geqq 1 - \eta$$

for all η and m. Thus, by 4.1.2. $Q(\Phi \in M_\varkappa') = 1$.

3. Now let us assume that b) is fulfilled. From 1. and 2. it follows immediately that $P_\varkappa$, $(P_1)_\varkappa$, $(P_2)_\varkappa$, ... exist.

Let $X_1, \ldots, X_m$ be an arbitrary finite sequence of sets in $\mathfrak{B}_{P_\varkappa}$ and Z a set in $\mathfrak{B}_P'$. Obviously $\mathfrak{B}_{P_\varkappa}$ is a subalgebra of $\mathfrak{B}_{(_ZP)_\varkappa}$. Using 3.1.11. and 4.7.1. we can conclude that

$$\left(_Z(P_n)\right)_\varkappa \underset{n\to\infty}{\Rightarrow} (_ZP)_\varkappa$$

We set $X = X_1 \cup \cdots \cup X_m$ and

$$\omega_{(a')} = \begin{cases} \varkappa_{(a')} & \text{for} \quad a' \in Z \\ \delta_o & \phantom{\text{for}} \quad a' \in A' \setminus Z. \end{cases}$$

Then, for $n = 1, 2, \ldots$

$$\begin{aligned} &\left\|\left(\left(_Z(P_n)\right)_\varkappa\right)_{X_1,\ldots,X_m} - \left((P_n)_\varkappa\right)_{X_1,\ldots,X_m}\right\| \\ &\leqq {}_X\left\|\left(_Z(P_n)\right)_\varkappa - (P_n)_\varkappa\right\| = {}_X\|(P_n)_\omega - (P_n)_\varkappa\|. \end{aligned}$$

Proceeding in the same way as at the beginning of the proof of 4.2.7., for all ε in $(0, 1)$ we may continue as follows

$$\begin{aligned} &\leqq \int {}_X\|\omega_{(\Phi)}(.) - \varkappa_{(\Phi)}(.)\| \, P_n(\mathrm{d}\Phi) \\ &= \int {}_X\|\varkappa_{(_Z\Phi)}(.) - (\varkappa_{(_Z\Phi)} * \varkappa_{(_{A'\setminus Z}\Phi)})(.)\| \, P_n(\mathrm{d}\Phi) \\ &\leqq \int {}_X\|\delta_o - \varkappa_{(_{A'\setminus Z}\Phi)}(.)\| \, P_n(\mathrm{d}\Phi) \\ &= 2 \int \left(\mathop{\ast}_{\substack{a' \in A'\setminus Z \\ \Phi(\{a'\})>0}} \left((\varkappa_{(a')})^{\Phi(\{a'\})}\right)\right) \left(\chi(X) > 0\right) P_n(\mathrm{d}\Phi) \\ &\leqq 2 \int \min\left(1, \sum_{\substack{a' \in A'\setminus Z \\ \Phi(\{a'\})>0}} \Phi(\{a'\}) \, \varkappa_{(a')}\left(\chi(X) > 0\right)\right) P_n(\mathrm{d}\Phi) \\ &= 2 \int \min\left(1, \int_{A'\setminus Z} \varkappa_{(a')}\left(\chi(X) > 0\right) \Phi(\mathrm{d}a')\right) P_n(\mathrm{d}\Phi) \\ &\leqq 2\left(\varepsilon + P_n\left(\int_{A'\setminus Z} \varkappa_{(a')}\left(\chi(X) > 0\right) \Phi(\mathrm{d}a') > \varepsilon\right)\right). \end{aligned}$$

Accordingly we obtain

$$\begin{aligned} &\left\|\left((_ZP)_\varkappa\right)_{X_1,\ldots,X_m} - (P_\varkappa)_{X_1,\ldots,X_m}\right\| \\ &\leqq 2\left(\varepsilon + P\left(\int_{A'\setminus Z} \varkappa_{(a')}\left(\chi(X) > 0\right) \Phi(\mathrm{d}a') > \varepsilon\right)\right), \end{aligned}$$

giving

$$\overline{\lim_{n\to\infty}} \left\| \left((P_n)_\varkappa\right)_{X_1,\dots,X_m} - (P_\varkappa)_{X_1,\dots,X_m} \right\|$$

$$\leqq \overline{\lim_{n\to\infty}} \left\| \left((P_n)_\varkappa\right)_{X_1,\dots,X_m} - \left(\left({}_Z(P_n)\right)_\varkappa\right)_{X_1,\dots,X_m} \right\|$$

$$+ \overline{\lim_{n\to\infty}} \left\| \left(\left({}_Z(P_n)\right)_\varkappa\right)_{X_1,\dots,X_m} - \left(({}_ZP)_\varkappa\right)_{X_1,\dots,X_m} \right\|$$

$$+ \left\| \left(({}_ZP)_\varkappa\right)_{X_1,\dots,X_m} - (P_\varkappa)_{X_1,\dots,X_m} \right\|$$

$$\leqq 2\left(\varepsilon + \sup_{n=1,2,\dots} P_n\left(\int_{A'\setminus Z} \varkappa_{(a')}\left(\chi(X) > 0\right) \Phi(\mathrm{d}a') > \varepsilon\right)\right)$$

$$+ 0 + 2\left(\varepsilon + P\left(\int_{A'\setminus Z} \varkappa_{(a')}\left(\chi(X) > 0\right) \Phi(\mathrm{d}a') > \varepsilon\right)\right).$$

If the set Z is chosen sufficiently large, then by **1.** the right hand side becomes less than 5ε. Hence, for all $X_1, \dots, X_m$ in $\mathfrak{B}_{P_\varkappa}$

$$\left((P_n)_\varkappa\right)_{X_1,\dots,X_m} \xrightarrow[n\to\infty]{} (P_\varkappa)_{X_1,\dots,X_m}.$$

Thus, by 3.1.9. we have

$$(P_n)_\varkappa \underset{n\to\infty}{\Rightarrow} P_\varkappa.$$

4. Let $p_0, p_1, \dots$ be a fixed sequence of non-negative numbers with the property $\sum_{n=0}^{\infty} p_n = 1$. Furthermore, we denote by $\boldsymbol{q}$ the set of all distributions q of random two-dimensional vectors $[\xi_1, \xi_2]$ with non-negative integral coordinates which satisfy the conditions

$$q(\xi_1 \geqq \xi_2) = 1, \qquad q(\xi_1 = n) = p_n \qquad \text{for } n = 0, 1, \dots$$

With the help of 1.9.2. one can easily see that $\boldsymbol{q}$ is compact with respect to the variation distance. Now we will show that to each $\varepsilon > 0$ there exists an $\eta > 0$ such that the inequality

$$q(\xi_2 = 0) > 1 - \varepsilon$$

is satisfied for all $q \in \boldsymbol{q}$ with the property

$$\sum_{n=0}^{\infty} |p_n - q(\xi_1 - \xi_2 = n)| < \eta.$$

Let us suppose this statement is wrong. Then there exists an $\varepsilon_0 > 0$ such that to each natural number k there is a $q_k \in \boldsymbol{q}$ with the properties

$$\sum_{n=0}^{\infty} |p_n - q_k(\xi_1 - \xi_2 = n)| < k^{-1}, \qquad q_k(\xi_2 = 0) \leqq 1 - \varepsilon_0.$$

Because of the compactness of $\boldsymbol{q}$ there exists a subsequence of (q_k), which converges towards a $q \in \boldsymbol{q}$ with respect to the variation distance. Then, besides

$$p_n = q(\xi_1 = n) \qquad \text{for } n = 0, 1, \ldots$$

we have

$$p_n = q(\xi_1 - \xi_2 = n) \qquad \text{for } n = 0, 1, \ldots$$

Consequently, for each natural number m

$$q(\xi_1 < m) = q(\xi_1 - \xi_2 < m) = q(\xi_1 < m) + q(\xi_1 - \xi_2 < m \leqq \xi_1)$$

is satisfied. From this it follows

$$q(\xi_2 > 0) = q(\xi_1 - \xi_2 < \xi_1) \leqq \sum_{m=1}^{\infty} q(\xi_1 - \xi_2 < m \leqq \xi_1) = 0,$$

i.e. $q(\xi_2 = 0) = 1$. But this is a contradiction to $q(\xi_2 = 0) \leqq 1 - \varepsilon_0$.

5. Let be $X_o \in \mathfrak{B}$ and Q a distribution in $\boldsymbol{P}'$ such that $Q_\varkappa$ exists. Then, for each increasing sequence (X_m) of sets in $\mathfrak{B}'$, whose union is A', we have

$$\left\| \left(({}_{X_m}Q)_\varkappa \right)_{X_o} - (Q_\varkappa)_{X_o} \right\| \xrightarrow[m\to\infty]{} 0.$$

In fact,

$$\begin{aligned}
&\left\| \left(({}_{X_m}Q)_\varkappa \right)_{X_o} - (Q_\varkappa)_{X_o} \right\| \\
&\leqq \int \left\| \left(\varkappa_{({}_{X_m}\Phi)} \right)_{X_o} - \left(\varkappa_{(\Phi)} \right)_{X_o} \right\| Q(\mathrm{d}\Phi) \leqq \int \left\| \delta_0 - \left(\varkappa_{({}_{(A' \setminus X_m)}\Phi)} \right)_{X_o} \right\| Q(\mathrm{d}\Phi) \\
&= 2 \int \varkappa_{({}_{(A' \setminus X_m)}\Phi)} \left(\chi(X_o) > 0 \right) Q(\mathrm{d}\Phi) \\
&= 2 \int \Bigl(\mathop{\ast}_{a' \in A' \setminus X_m,\ \Phi(\{a'\}) > 0} \left(\varkappa_{(a')} \right)^{\Phi(\{a'\})} \Bigr) \left(\chi(X_o) > 0 \right) Q(\mathrm{d}\Phi) \xrightarrow[m\to\infty]{} 0
\end{aligned}$$

because the integrand is bounded and converges towards zero almost everywhere with respect to Q as $m \to \infty$.

6. From a) it follows b).

We denote by $[A'', \varrho_{A''}]$ the direct product of $[A, \varrho_A]$ with the finite phase space $\{1, 2\}$. For each X in $\mathfrak{B}_P'$ we get by

$$\omega_{(a')} = \begin{cases} \varkappa_{(a')}\left(\chi \times \delta_1 \in (\cdot) \right) & \text{for} \quad a' \in X \\ \varkappa_{(a')}\left(\chi \times \delta_2 \in (\cdot) \right) & \quad a' \in A' \setminus X \end{cases}$$

a cluster field ω on $[A', \varrho_{A'}]$ with the phase space $[A'', \varrho_{A''}]$. Obviously, for any Q in $\boldsymbol{P}'$ the distribution Q_ω exists if and only if $Q_\varkappa$ exists. We have

$$Q_\omega\left(\Psi\left((\cdot) \times \{1\} \right) \in Y \right) = ({}_X Q)_\varkappa (Y),$$

$$Q_\omega\left(\Psi\left((\cdot) \times \{2\} \right) \in Y \right) = ({}_{(A' \setminus X)} Q)_\varkappa (Y),$$

$$Q_\omega\left(\Psi\left((\cdot) \times \{1, 2\} \right) \in Y \right) = Q_\varkappa(Y)$$

for all Y in $\mathfrak{M}$ (if Q_ω exists).

By 3.2.7., from the relative compactness of $\{(P_n)_\varkappa\}_{n=1,2,\ldots}$ with respect to $\varrho_{\boldsymbol{P}}$ it follows the relative compactness of $\{(P_n)_\omega\}_{n=1,2,\ldots}$ with respect to $\varrho_{\boldsymbol{P}''}$. Let $\big((P_{n_m})_\omega\big)$ be any subsequence of $\big((P_n)_\omega\big)$ which converges towards a distribution ${}^X V$ in $\boldsymbol{P}''$.

In virtue of 3.1.11. and 4.7.1.

$$\big({}_X(P_n)\big)_\varkappa \underset{n\to\infty}{\Rightarrow} ({}_XP)_\varkappa$$

and therefore

$${}^X V\big(\Psi((.)\times\{1\}) \in (.)\big) = ({}_XP)_\varkappa .$$

Besides

$${}^X V\big(\Psi((.)\times\{1,2\}) \in (.)\big) = P_\varkappa$$

because of $(P_n)_\varkappa \underset{n\to\infty}{\Rightarrow} P_\varkappa$.

Let $X_o \in \mathfrak{B}$. The distribution ${}^X V\big(\Psi(X_o\times\{1,2\}) \in (.)\big)$ does not depend on X. By 4., for any $\varepsilon > 0$ there is an $\eta > 0$ such that

$${}^X V\big(\Psi(X_o\times\{2\}) > 0\big) < \varepsilon$$

is always satisfied, if

$$\big\|{}^X V\big(\Psi(X_o\times\{1,2\}) \in (.)\big) - {}^X V\big(\Psi(X_o\times\{1\}) \in (\cdot)\big)\big\| < \eta ,$$

i.e.

$$\big\|(P_\varkappa)_{X_o} - \big(({}_XP)_\varkappa\big)_{X_o}\big\| < \eta .$$

However, the last inequality is fulfilled by 5. for all sufficiently large X. Consequently, uniformly for all convergent $\big((P_{n_m})_\omega\big)$,

$${}^X V\big(\Psi(X_o\times\{2\}) > 0\big) < \varepsilon$$

holds for all sufficiently large X in $\mathfrak{B}_P{}'$.

In addition we suppose that X_o is a closed subset of A. Then the set $\{\Phi : \Phi \in M,\, \Phi(X_o) > 0\}$ is closed with respect to ϱ_M.

Now, from $(P_{n_m})_\omega \underset{m\to\infty}{\Rightarrow} {}^X V$ it follows

$$\big({}_{A\setminus X}(P_{n_m})\big)_\varkappa \underset{m\to\infty}{\Rightarrow} {}^X V\big(\Psi((.)\times\{2\}) \in (.)\big).$$

Therefore

$$\overline{\lim_{m\to\infty}}\, \big({}_{A\setminus X}(P_{n_m})\big)_\varkappa \big(\Phi(X_o) > 0\big) \leqq {}^X V\big(\Psi(X_o\times\{2\}) > 0\big) < \varepsilon$$

and consequently

$$\overline{\lim_{n\to\infty}}\, \big({}_{A\setminus X}(P_n)\big)_\varkappa \big(\Phi(X_o) > 0\big) < \varepsilon .$$

As in 5. we can at least achieve by enlarging X that even

$$\sup_{n=1,2,\ldots} \big({}_{A\setminus X}(P_n)\big)_\varkappa \big(\Phi(X_o) > 0\big) < \varepsilon$$

is satisfied. Therefore

$$\inf_{X\in\mathfrak{B}_P'} \sup_{n=1,2,\dots} \left({}_{A'\setminus X}(P_n)\right)_\varkappa \left(\Phi(X_o) > 0\right) = 0$$

holds for all bounded closed subsets X_o of A, and hence

$$\inf_{X\in\mathfrak{B}'} \sup_{n=1,2,\dots} \left({}_{A'\setminus X}(P_n)\right)_\varkappa \left(\Phi(X_o) > 0\right) = 0$$

for all X_o in $\mathfrak{B}$.

For $\Phi \in M_\varkappa'$ we have

$$\varkappa_{(\Phi)}\left(\chi(X_o) > 0\right) = 1 - \prod_{a'\in A', \Phi(\{a'\})>0} \left(1 - \varkappa_{(a')}\left(\chi(X_o) > 0\right)\right)^{\Phi(\{a'\})}$$

$$\geqq 1 - e^{-\int \varkappa_{(a')}(\chi(X_o)>0)\Phi(\mathrm{d}a')}.$$

Suppose that

$$\inf_{X\in\mathfrak{B}'} \sup_{n=1,2,\dots} P_n\left(\int_{A'\setminus X} \varkappa_{(a')}\left(\chi(X_o) > 0\right) \Phi(\mathrm{d}a') > \delta\right) = c > 0$$

for some $\delta > 0$. Then for each $X \in \mathfrak{B}'$ there is an n_X such that

$$P_{n_X}\left(\int \varkappa_{(a')}\left(\chi(X_o) > 0\right) {}_{(A'\setminus X)}\Phi(\mathrm{d}a') > \delta\right) \geqq c/2$$

and so

$$P_{n_X}\left(\varkappa_{({}_{(A'\setminus X)}\Phi)}\left(\chi(X_o) > 0\right) \geqq 1 - e^{-\delta}\right) \geqq c/2$$

and consequently

$$\left({}_{A'\setminus X}(P_{n_X})\right)_\varkappa \left(\Phi(X_o) > 0\right) \geqq (1 - e^{-\delta})\, c/2$$

is satisfied. From this it follows that

$$\inf_{X\in\mathfrak{B}'} \sup_{n=1,2,\dots} \left({}_{A'\setminus X}(P_n)\right)_\varkappa \left(\Phi(X_o) > 0\right) \geqq (1 - e^{-\delta})\, c/2$$

and we have obtained a contradiction. Finally, we have

$$\inf_{X\in\mathfrak{B}'} \sup_{n=1,2,\dots} P_n\left(\int_{A'\setminus X} \varkappa_{(a')}\left(\chi(X_o) > 0\right) \Phi(\mathrm{d}a') > \delta\right) = 0$$

for all $\delta > 0$. ∎

Sometimes the following sufficient criterion for exchanging weak convergence in $\boldsymbol{P}'$ and clustering is useful (cf. Liemant [1]).

4.7.3. Proposition. *Let $\varkappa$ be a continuous cluster field on $[A', \varrho_{A'}]$ with phase space $[A, \varrho_A]$, and (P_n) a sequence of distributions in $\boldsymbol{P}'$ which converges weakly toward P in $\boldsymbol{P}'$. If there exists a measure ϱ on $\mathfrak{A}'$ having the properties*

a) *for all X in $\mathfrak{B}$, $\int \varkappa_{(a')}\left(\chi(X) > 0\right) \varrho(\mathrm{d}a')$ is finite,*

b) *for all natural numbers n, $\varrho_{P_n} \leqq \varrho$,* *then $P_\varkappa$, $(P_1)_\varkappa$, $(P_2)_\varkappa$, … exist and the sequence $(P_n)_\varkappa$ converges weakly towards $P_\varkappa$.*

Proof. Let $X \in \mathfrak{B}$, $\varepsilon > 0$ and $\eta > 0$. Because of Theorem 4.7.2. we have to show that there is an Z_η in $\mathfrak{B}'$ such that

$$P_n\Big(\int\limits_{A'\setminus Z_\eta} \varkappa_{(a')}\big(\chi(X) > 0\big)\, \Phi(\mathrm{d}a') > \varepsilon\Big) < \eta$$

is satisfied for all n.

Let $Z \in \mathfrak{B}'$. The Markov inequality yields

$$\begin{aligned} &P_n\Big(\int\limits_{A'\setminus Z} \varkappa_{(a')}\big(\chi(X) > 0\big)\, \Phi(\mathrm{d}a') > \varepsilon\Big) \\ &\leqq \varepsilon^{-1} \int \Big(\int\limits_{A'\setminus Z} \varkappa_{(a')}\big(\chi(X) > 0\big)\, \Phi(\mathrm{d}a')\Big)\, P_n(\mathrm{d}\Phi) \\ &= \varepsilon^{-1} \int\limits_{A'\setminus Z} \varkappa_{(a')}\big(\chi(X) > 0\big)\, \varrho_{P_n}(\mathrm{d}a') \\ &\leqq \varepsilon^{-1} \int\limits_{A'\setminus Z} \varkappa_{(a')}\big(\chi(X) > 0\big)\, \varrho(\mathrm{d}a'). \end{aligned}$$

By a) this term is less than η if Z is chosen sufficiently large. ∎

4.8. The Property g_ν. II

Let K be a substochastic kernel on $[A, \varrho_A]$ with phase space $[A, \varrho_A]$. Each distribution P on $\mathfrak{M}$ which is cluster-invariant with respect to $\varkappa(K)$ describes a stochastic equilibrium of a random ensemble of particles in the phase space under the random translation of the individual particles according to K.

In view of 4.4.3., for any distribution ν on $\mathfrak{A}$, the distribution Q_ν is cluster-invariant with respect to $\varkappa(K)$ if and only if $K * \nu = \nu$ is satisfied, i.e. ν describes a stochastic equilibrium for the random position of an individual particle with respect to the random translation according to K. In this case in virtue of 1.14.11., 4.2.5. and 4.3.1. all distributions P on $\mathfrak{M}$ having the property g_ν are cluster-invariant with respect to $\varkappa(K)$.

If an infinite measure ν in N fulfils the equation $K * \nu = \nu$ then by 4.4.4. all Poisson distributions $P_{l\nu}$, $l \geqq 0$, are cluster-invariant with respect to $\varkappa(K)$. In view of 1.14.8. and 4.2.5. all distributions P on $\mathfrak{M}$ with the property g_ν are cluster-invariant with respect to $\varkappa(K)$.

After all, for $\nu = o$ only $P = \delta_o$ has the property g_ν. This distribution is of course cluster-invariant with respect to $\varkappa(K)$.

Consequently, for any ν in N with the property $K * \nu = \nu$, each distribution P on $\mathfrak{M}$ with the property g_ν is cluster-invariant with respect to $\varkappa(K)$.

Conversely, let ν be a measure in N and P be a distribution on $\mathfrak{M}$ which fulfils the following condition: If L is a substochastic kernel on $[A, \varrho_A]$ with phase space $[A, \varrho_A]$ such that $L * \nu = \nu$ holds and $P_{\varkappa(L)}$ exists, then P is cluster-invariant with respect to $\varkappa(L)$. Now we will prove that P has the property g_ν.

For $\nu = o$ we can use $L(a, (.)) = o$, $a \in A$, and we get $P = P_{\varkappa(L)} = \delta_o$. Therefore, we suppose now $\nu \neq o$.

Each X in $\mathfrak{B}$ is contained in an X' in $\mathfrak{B}$ with the property $\nu(X') > 0$. In virtue of

$$L(a, (.)) = \begin{cases} \delta_a & \\ & \text{for} \\ \nu((.) \mid X') & \end{cases} \begin{matrix} a \in A \setminus X' \\ \\ a \in X' \end{matrix}$$

we obtain a stochastic kernel L on $[A, \varrho_A]$ with phase space $[A, \varrho_A]$ such that $P_{\varkappa(L)}$ exists and

$$L * \nu = \nu((.) \cap (A \setminus X')) + \nu(X')\, \nu((.) \mid X') = \nu$$

holds. Hence, by assumption, $P = P_{\varkappa(L)}$ and therefore

$$\begin{aligned} {}_{X'}P &= {}_{X'}(P_{\varkappa(L)}) = \int {}_{X'}\left((\varkappa(L))_{(\Phi)}\right)(.)\, P(\mathrm{d}\Phi) \\ &= \int \left(Q_{\nu((.)\mid X')}\right)^{\Phi(X')}(.)\, P(\mathrm{d}\Phi) \\ &= \sum_{k=0}^{\infty} P(\Phi(X') = k)\left(Q_{\nu((.)\mid X')}\right)^k. \end{aligned}$$

Consequently, P has the property $g_{\nu,X'}$, and by 1.14.1. the property $g_{\nu,X}$, too.

Summarizing we see the truth of

4.8.1. Theorem. *Let ν be a measure in N. If a substochastic kernel K on $[A, \varrho_A]$ with phase space $[A, \varrho_A]$ has the property $K * \nu = \nu$, then every distribution P on $\mathfrak{M}$ with the property g_ν is cluster-invariant with respect to $\varkappa(K)$.*

Conversely, a distribution P on $\mathfrak{M}$ has the property g_ν if it is cluster-invariant with respect to $\varkappa(L)$ for all substochastic kernels L on $[A, \varrho_A]$ with phase space $[A, \varrho_A]$ which fulfil the conditions

$$P_{\varkappa(L)} \quad \textit{exists}, \quad L * \nu = \nu.$$

With the help of the following proposition we can characterize for each distribution ν on $\mathfrak{A}$ those sequences (K_n) of substochastic kernels which have the property: If the "initial distribution" V with V (Φ finite) $= 1$ is clustered according to $\varkappa(K_n)$, then we get as $n \to \infty$ that distribution P with property g_ν which fulfils $P_A = V_A$.

4.8.2. Proposition. *Let ν be a distribution on $\mathfrak{A}$ and (K_n) a sequence of substochastic kernels on $[A', \varrho_{A'}]$ with phase space $[A, \varrho_A]$. Subject to these assumptions, the following statements are equivalent.*

a) *For all a' in A', $\left(K_n(a', (.))\right)$ tends weakly toward the distribution ν, i.e.*

$$K_n(a', (\cdot)) \underset{n\to\infty}{\Rightarrow} \nu \qquad (a' \in A').$$

b) *For all distributions V on $\mathfrak{M}'$ satisfying the condition V (Φ finite) $= 1$,*

$$V_{\varkappa(K_n)} \underset{n\to\infty}{\Rightarrow} \sum_{k=0}^{\infty} V\big(\Phi(A') = k\big)\, (Q_\nu)^k.$$

Proof. 1. A set X in $\mathfrak{B}$ is a continuity set with respect to ν if and only if it belongs to $\mathfrak{B}_{Q_\nu}$. Therefore, by 3.1.13. we have

$$K_n(a', (.)) \underset{n\to\infty}{\Rightarrow} \nu \qquad (a' \in A'),$$

i.e.

$$K_n(a', A) \xrightarrow[n\to\infty]{} \nu(A) = 1, \qquad K_n(a', X) \xrightarrow[n\to\infty]{} \nu(X) \qquad (X \in \mathfrak{B}_{Q_\nu}),$$

if and only if

$$Q_{K_n(a', (.))} \underset{n\to\infty}{\Rightarrow} Q_\nu \qquad (a' \in A').$$

Furthermore, the last convergence assertion is equivalent to

$$\big(\varkappa(K_n)\big)_{(a')} = \big(1 - K_n(a', A)\big)\, \delta_o + Q_{K_n(a', (.))} \underset{n\to\infty}{\Rightarrow} Q_\nu \qquad (a' \in A'),$$

because M and $\{o\}$ are continuity sets with respect to Q_ν.

2. b) implies a). Indeed, for each a' in A' we have

$$(\delta_{\delta_{a'}})_{\varkappa(K_n)} \underset{n\to\infty}{\Rightarrow} Q_\nu,$$

i.e.

$$\big(\varkappa(K_n)\big)_{(a')} \underset{n\to\infty}{\Rightarrow} Q_\nu.$$

3. a) implies b). If $X_1, \ldots, X_m$ is an arbitrary finite sequence of pairwise disjoint sets in $\mathfrak{B}_{Q_\nu}$ then for $n = 1, 2, \ldots$ we obtain

$$\left\| (V_{\varkappa(K_n)})_{X_1,\ldots,X_m} - \left(\sum_{k=0}^{\infty} V\big(\Phi(A') = k\big)\, (Q_\nu)^k\right)_{X_1,\ldots,X_m} \right\|$$

$$= \left\| \int \Big(\big(\varkappa(K_n)\big)_{(\Phi)}\Big)_{X_1,\ldots,X_m}(.)\; V(\mathrm{d}\Phi) - \sum_{k=0}^{\infty} \int_{\{\Phi:\, \Phi(A') = k\}} \big((Q_\nu)^k\big)_{X_1,\ldots,X_m}(.)\; V(\mathrm{d}\Phi) \right\|$$

$$\leqq \sum_{k=0}^{\infty} \int_{\{\Phi:\, \Phi(A') = k\}} \left\| \Big(\big(\varkappa(K_n)\big)_{(\Phi)}\Big)_{X_1,\ldots,X_m} - \big((Q_\nu)^k\big)_{X_1,\ldots,X_m} \right\| V(\mathrm{d}\Phi).$$

Now, by 3.1.10. from

$$\big(\varkappa(K_n)\big)_{(a')} \underset{n\to\infty}{\Rightarrow} Q_\nu \qquad (a' \in A')$$

it follows

$$\big(\varkappa(K_n)\big)_{(\Phi)} \Rightarrow (Q_\nu)^{\Phi(A')} \qquad \big(\Phi \in M',\ \Phi(A') < +\infty\big).$$

Using 3.1.9. and Lebesgue's theorem we may conclude that

$$\lim_{n\to\infty} \left\| (V_{\varkappa(K_n)})_{X_1,\ldots,X_m} - \left(\sum_{k=0}^{\infty} V\big(\Phi(A') = k\big)\, (Q_\nu)^k \right)_{X_1,\ldots,X_m} \right\| = 0,$$

i.e. b) is valid. ∎

If the sequence $(K^{[n]})$ of convolution powers of a substochastic kernel K on $[A, \varrho_A]$ with phase space $[A, \varrho_A]$ fulfils the condition a) in 4.8.2., then we have

$$K^{[n]}(a, A) \underset{n\to\infty}{\longrightarrow} \nu(A) \qquad (a \in A).$$

If we take into consideration that for all a in A the sequence $\big(K^{[n]}(a, A)\big)$ is monotonous decreasing, it follows that K is stochastic.

The following example shows that from the above assumptions it still cannot be concluded that $K * \nu = \nu$.

Let $[A, \varrho_A] = [0, 1]$. We set

$$K(a, (\cdot)) = \begin{cases} \mu((.) \cap [0, 1]) & \text{for} \quad a = 0 \\ \delta_{a/2} & \phantom{\text{for}} \quad a \in (0, 1]. \end{cases}$$

For the stochastic kernel K defined in this way

$$K^{[n]}(a, (.)) \underset{n\to\infty}{\Rightarrow} \delta_0 \qquad (a \in [0, 1]),$$

although $K * \delta_0 = \mu((.) \cap [0, 1])$ is different from δ_0.

A substochastic kernel K on $[A', \varrho_{A'}]$ with phase space $[A, \varrho_A]$ is said to be *continuous* if $a_n' \underset{n\to\infty}{\longrightarrow} a'$ implies $K\big(a_n', (.)\big) \underset{n\to\infty}{\Rightarrow} K\big(a', (.)\big)$.

The continuity of K is stronger than the continuity of the cluster field $\varkappa(K)$ as the following example shows. We choose $[A', \varrho_{A'}] = R = [A, \varrho_A]$ and

$$K(a, (\cdot)) = \begin{cases} \delta_{1/a} & \text{for} \quad a \neq 0 \\ o & \phantom{\text{for}} \quad a = 0. \end{cases}$$

Then $\varkappa(K)$ is continuous but K is not continuous. However, by 3.2.2.c and 3.1.13. we can deduce from the continuity of $\varkappa(K)$ that $a_n' \underset{n\to\infty}{\longrightarrow} a'$ implies $K(a_n', (.)) \overset{N}{\underset{n\to\infty}{\Rightarrow}} K(a', (.))$ and that the converse is also true. Consequently the continuity of $\varkappa(K)$ is equivalent to the continuity of $a' \rightsquigarrow K(a', (.))$ with respect to ϱ_N.

If ν is a distribution on $\mathfrak{A}$ and K a fixed continuous stochastic kernel with the property $K * \nu = \nu$, then each distribution which is cluster-invariant with respect to $\varkappa(K)$ has the property g_ν, if K is ergodic in a certain sense (and therefore each solution of the equation $K * \gamma = \gamma$, $\gamma \in N$, has the form $\gamma = l\nu$, $l \geqq 0$). We have

4.8.3. Theorem. *Let ν be a distribution on $\mathfrak{A}$ and K a continuous stochastic kernel on $[A, \varrho_A]$ with phase space $[A, \varrho_A]$ having for all a in A the property*

$$K^{[n]}\big(a, (.)\big) \underset{n\to\infty}{\Rightarrow} \nu .$$

Under these assumptions a distribution P on $\mathfrak{M}$ is cluster-invariant with respect to $\varkappa(K)$ if and only if it has the property g_ν.

Proof. 1. Since K is continuous, for each bounded real function h on A continuous with respect to ϱ_A

$$g(a) = \int h(x)\, K(a, \mathrm{d}x) \qquad (a \in A)$$

provides a function having the same properties. Hence for all y in A we obtain

$$\int h(x)\, (K * K^{[n]})\, (y, \mathrm{d}x) = \int g(a)\, K^{[n]}(y, \mathrm{d}a)$$

$$\xrightarrow[n\to\infty]{} \int g(a)\, \nu(\mathrm{d}a) = \int h(x)\, (K * \nu)\, (\mathrm{d}x),$$

i.e. $(K * K^{[n]})$ converges pointwise weakly toward $K * \nu$. On the other hand

$$(K * K^{[n]})\, \big(a, (.)\big) = K^{[n+1]}\big(a, (\cdot)\big) \Rightarrow \nu .$$

Thus $K * \nu$ coincides with ν and by 4.8.1. all distributions P on $\mathfrak{M}$ having the property g_ν are cluster-invariant with respect to $\varkappa(K)$.

2. Each distribution P which is cluster-invariant with respect to $\varkappa(K)$ and satisfies the condition P (Φ finite) $= 1$ has the property g_ν. In fact, by 4.8.2. the weak limit of the constant sequence $(P_{\varkappa(K^{[n]})})$ has the property g_ν.

3. Each distribution P cluster-invariant with respect to $\varkappa(K)$ has the property that P (Φ finite) $= 1$.

Let be P (Φ infinite) > 0 and Φ any infinite counting measure in M such that $\big(\varkappa(K^{[n]})\big)_{(\Phi)}$ exists for $n = 1, 2, \ldots$ We choose any X in $\mathfrak{B}$ such that $\nu(X) > 0$ and $\nu(\partial X) = 0$. Then for all natural numbers m we have

$$\big(\varkappa(K^{[n]})\big)_{(\Phi)}\, \big(\chi(X) > m\big) \xrightarrow[n\to\infty]{} 1.$$

Hence, using Lebesgue's theorem

$$P\big(\Psi(X) > m\big) = \lim_{n\to\infty} \int \big(\varkappa(K^{[n]})\big)_{(\Phi)} \big(\Psi(X) > m\big)\, P(\mathrm{d}\Phi)$$

$$\geqq \lim_{n\to\infty} \int_{\{\Phi:\Phi \text{ infinite}\}} \big(\varkappa(\mathfrak{K}^{[n]})\big)_{(\Phi)} \big(\Psi(X) > m\big)\, P(\mathrm{d}\Phi)$$

$$= P\,(\Phi \text{ infinite}) > 0\,,$$

i.e. $P\big(\Psi(X) = +\infty\big) > 0$, which contradicts the definition of M. ∎

We will not discuss here the problem of transferring the above characterization of the distributions with the property g_ν to the case of infinite measures ν in N. Instead, we refer to Kerstan & Debes [1] and Debes, Kerstan, Liemant & Matthes [2].

5. The Campbell Measure

5.0. Introduction

In Section 1.15., the set M was supplemented by a metric ϱ_M to form a bounded complete separable metric space $[M, \varrho_M]$. In the sense of the statements made at the end of Section 1.0., we may conceive of $[M, \varrho_M]$ as a "mark space", and accordingly pass from $[A, \varrho_A]$ to the direct product $[A', \varrho_{A'}]$ of $[A, \varrho_A]$ and $[M, \varrho_M]$. To each Φ in M we now associate a counting measure in the metric space $[M', \varrho_{M'}]$ corresponding to $[A', \varrho_{A'}]$ by

$$\sum_{a \in A, \Phi(\{a\})>0} \Phi(\{a\})\, \delta_{[a,\Phi]} = \Phi \times \delta_\Phi\,,$$

i.e. every point a in Φ is given the mark Φ. The mapping $\Phi \curvearrowright \Phi \times \delta_\Phi$ is continuous, and hence also measurable.

Let C denote the closed subset $\{[a, \Phi] : [a, \Phi] \in A \times M, \Phi(\{a\}) > 0\}$ of $[A', \varrho_{A'}]$, and $\varrho_{C'}$ the restriction of $\varrho_{A'}$ to C'. For all Φ in M, $(\Phi \times \delta_\Phi)\big((A \times M) \setminus C\big) = 0$. Therefore we replace $[A', \varrho_{A'}]$ by the smaller phase space $[C, \varrho_C]$, accordingly applying the restriction $\mathring{\Phi}$ of $\Phi \times \delta_\Phi$ on the σ-algebra $\mathfrak{C}$ of the Borel subsets of $[C, \varrho_C]$. The mapping $\Phi \curvearrowright \mathring{\Phi}$ of M into the metric space $[\mathring{M}, \varrho_{\mathring{M}}]$ belonging to $[C, \varrho_C]$ is continuous, too, and consequently transforms each measure H on $\mathfrak{M}$ into a measure $\mathring{H}$ on the σ-algebra $\mathring{\mathfrak{M}}$ of the Borel subsets of $[\mathring{M}, \varrho_{\mathring{M}}]$. The intensity measure $\varrho_{\mathring{H}}$ of $\mathring{H}$ is called the Campbell measure of H and is denoted by $\mathscr{C}_H$. Observe that, for all non-negative real functions f on C which are measurable with respect to $\mathfrak{C}$,

$$\int \Big(\sum_{a \in A, \Phi(\{a\})>0} f(a, \Phi)\, \Phi(\{a\})\Big)\, H(\mathrm{d}\Phi) = \int f(c)\, \mathscr{C}_H(\mathrm{d}c)$$

(Proposition 5.1.2.), which statement can be considered a refinement of the elementary Campbell theorem 1.2.2. In effect the mapping $H \curvearrowright \mathscr{C}_H$ is one-to-one, since from $\mathscr{C}_{H_1} = \mathscr{C}_{H_2}$ it can be concluded that $H_1\big((.) \setminus \{o\}\big) = H_2\big((.) \setminus \{o\}\big)$ (Proposition 5.1.10.).

As a closed subset of $[A', \varrho_{A'}]$, C belongs to the σ-algebra $\mathfrak{A}' = \mathfrak{A} \otimes \mathfrak{M}$ of the Borel subsets of $[A', \varrho_{A'}]$, and we have $\mathfrak{C} = C \cap (\mathfrak{A} \otimes \mathfrak{M})$. Therefore, in Section 5.1., we introduce the Campbell measure of H without resorting to topological concepts by the definition $\mathscr{C}_H(Z) = \int \mathring{\Phi}(Z)\, H(\mathrm{d}\Phi)$, $Z \in \mathfrak{C}$, considering the above statements as a commentary, which connect our definition with the conceptual system developed in the first chapter.

For all σ-finite measures V and H on $\mathfrak{C}$ and $\mathfrak{M}$, respectively, the mapping $\big[[a, \Phi], \Psi\big] \curvearrowright [a, \Phi + \Psi]$ transforms the measure $V \times H$ into a measure

$V \circledast H$ on $\mathfrak{C}$. This operation is $\circledast$ closely related to the convolution, in that if $(P_i)_{i\in I}$ is an at most countable family of distributions on $\mathfrak{M}$, and if the convolution $\mathop{\ast}\limits_{i\in I} P_i$ exists, then

$$\mathscr{C}_{\mathop{\ast}\limits_{i\in I} P_i} = \sum_{i\in I} \mathscr{C}_{P_i} \circledast \mathop{\ast}\limits_{j\in I\setminus\{i\}} P_j \qquad \text{(Proposition 5.2.3.).}$$

This formula is reminiscent of the rule for the differentiation of products. We may use it to conclude, for all W in $\boldsymbol{W}$, that $\mathscr{C}_{\mathscr{E}_W} = \mathscr{C}_W \circledast \mathscr{E}_W$. This functional equation for $\mathscr{E}_W$ can be considered an analogue of the differential equation of the exponential functions. Hence for all infinitely divisible distributions P on $\mathfrak{M}$, we have $\mathscr{C}_P = \mathscr{C}_{\tilde{P}} \circledast P$. This relation can be developed into a characterization of the infinitely divisible distributions. A distribution P on $\mathfrak{M}$ is infinitely divisible if and only if there exists a σ-finite measure V on $\mathfrak{C}$ having the property that $\mathscr{C}_P = V \circledast P$. This measure V is uniquely determined by P (Theorem 5.2.5., cf. Kummer & Matthes [1]).

For all Poisson distributions P_ν on $\mathfrak{M}$, we have $\widetilde{P_\nu} = Q_\nu$ (Proposition 2.2.14., cf. Kerstan & Matthes [1] and Lee [1]), and hence $\mathscr{C}_{\widetilde{P_\nu}} = \mathscr{C}_{Q_\nu} = \int \delta_{[a,\delta_a]}(.)\, \nu(\mathrm{d}a)$, so that a distribution P on $\mathfrak{M}$ coincides with P_ν if and only if $\mathscr{C}_P = \int \left(\delta_a \times (\delta_{\delta_a} * P)\right)(.)\, \nu(\mathrm{d}a)$ is satisfied (cf. Mecke [1]). This functional equation can be simplified by using, instead of $\mathscr{C}_P$, the reduced Campbell measure

$$\mathscr{C}_P{}^! = \int \Big(\sum_{a\in A,\, \Phi(\{a\})>0} \Phi(\{a\})\, \delta_{[a,\Phi-\delta_a]}(.) \Big)\, H(\mathrm{d}\Phi),$$

so that the equation is transformed into $\mathscr{C}_{P^!} = \nu \times P$ (Theorem 5.4.1.). It turns out that the distributions P having the property g_ν can be characterized by a more general relation: A distribution P on $\mathfrak{M}$ has the property g_ν if and only if there exists a σ-finite measure Q on $\mathfrak{M}$ having the property that $\mathscr{C}_P{}^! = \nu \times Q$ (Theorem 5.4.2., cf. Kallenberg [4]).

With reference to the statements made at the end of Section 1.0., in Sections 5.1., 5.2., and 5.3. it is supposed that the phase space $[A, \varrho_A]$ be the direct product of a "position space" $[A^x, \varrho_{A^x}]$ and a bounded "mark space" $[K, \varrho_K]$. We set

$$C = \left\{[x, \Phi] : [x, \Phi] \in A^x \times M,\ \Phi(\{x\} \times K) > 0\right\},$$

$$\mathring{\Phi}(Z) = \sum_{x\in A^x,\, \Phi(\{x\}\times K)>0} \Phi(\{x\} \times K)\, \delta_{[x,\Phi]}(Z),$$

i.e. the mark k of a marked point $[x, k]$ in Φ is replaced by Φ. This does not lead to any difficulties in the proofs. On the other hand, the reduced Campbell measure $\mathscr{C}_P{}^!$ can be introduced only for one-element mark spaces, i.e. in the non-marked case.

5.1. The Mapping $H \rightsquigarrow \mathscr{C}_H$

Throughout the first three sections of this chapter, we shall assume that the phase space $[A, \varrho_A]$ be the direct product of a complete separable "position space" $[A^x, \varrho_{A^x}]$ and a bounded complete separable "mark space"

$[K, \varrho_K]$, i.e. we let $A = A^x \times K$ as well as

$$\varrho_A([x_1, k_1], [x_2, k_2]) = \varrho_{A^x}(x_1, x_2) + \varrho_K(k_1, k_2) \qquad (x_1, x_2 \in A^x;\, k_1, k_2 \in K).$$

To each Φ in M, by

$$\Phi^x = \Phi\big((.) \times K\big)$$

we associate a Φ^x in the measurable space $[M^x, \mathfrak{M}^x]$ of counting measures, which belongs to the state space $[A^x, \varrho_{A^x}]$; in this way we obtain a measurable mapping of M onto M^x. The transition from Φ to Φ^x means that the marks of the points in Φ are "forgotten". We denote by H^x the image of a measure H on $\mathfrak{M}$ with respect to the mapping $\Phi \rightsquigarrow \Phi^x$. It is obvious that, for all at most countable families $(P_i)_{i\in I}$ of distributions on $\mathfrak{M}$,

$$\Big(\mathop{\ast}_{i\in I} P_i\Big)^x = \mathop{\ast}_{i\in I} (P_i)^x,$$

where the existence of either side involves that of the other. From this it follows that if P is infinitely divisible, so is P^x, where

$$\widetilde{P^x} = (\tilde{P})^x$$

is satisfied, P^x being regular or singular infinitely divisible, respectively, if this is the case with P.

We set

$$C = \big\{[x, \Phi] : [x, \Phi] \in A^x \times M,\, \Phi^x(\{x\}) > 0\big\}.$$

If we now associate to each natural number n a countable covering $(X_{n,m})_{m=1,2,\ldots}$ of A^x by pairwise disjoint sets in the σ-algebra $\mathfrak{A}^x$ of the Borel subsets of $[A^x, \varrho_{A^x}]$, the diameters of which do not exceed $1/n$, then we obtain

$$C = \bigcap_{n=1}^{\infty} \bigcup_{m=1}^{\infty} \big(X_{n,m} \times \{\Phi : \Phi \in M,\, \Phi^x(X_{n,m}) > 0\}\big).$$

Consequently the set C belongs to the σ-algebra $\mathfrak{A}^x \otimes \mathfrak{M}$. Let $\mathfrak{C}$ denote the σ-algebra $C \cap (\mathfrak{A}^x \otimes \mathfrak{M})$ of all sets in $\mathfrak{A}^x \otimes \mathfrak{M}$ which are contained in C.

To each Φ in M we associate a measure $\mathring{\Phi}$ on $\mathfrak{C}$ by means of

$$\mathring{\Phi} = \sum_{x\in A^x,\, \Phi^x(\{x\})>0} \Phi^x(\{x\})\, \delta_{[x,\Phi]} = \Phi^x \times \delta_\Phi .$$

5.1.1. *For all Z in $\mathfrak{C}$, $\Phi \rightsquigarrow \mathring{\Phi}(Z)$ is a measurable mapping of $[M, \mathfrak{M}]$ into $[0, +\infty]$.*

Proof. For all X_0 in the ring $\mathfrak{B}^x$ of all bounded subsets of $[A^x, \varrho_{A^x}]$ let $\mathfrak{B}_{X_0}$ denote the system of those Z in $\mathfrak{A}^x \otimes \mathfrak{M}$, for which

$$\Phi \rightsquigarrow \mathring{\Phi}\big(Z \cap C \cap (X_0 \times M)\big)$$

is measurable. For all X in $\mathfrak{A}^x$ and all Y in $\mathfrak{M}$,

$$\mathring{\Phi}\big((X \times Y) \cap C \cap (X_0 \times M)\big) = \Phi^x(X \cap X_0)\, \delta_\Phi(Y) = \Phi^x(X \cap X_0)\, k_Y(\Phi),$$

and hence $X \times Y \in \mathfrak{B}_{X_0}$. The set of all $X \times Y$ constitutes a semi-algebra. Consequently $\mathfrak{B}_{X_0}$ includes also that sub-algebra of $\mathfrak{A}^x \otimes \mathfrak{M}$ which is generated by the $X \times Y$. On the other hand $\mathfrak{B}_{X_0}$ is monotone closed. Hence $\mathfrak{B}_{X_0} = \mathfrak{A}^x \otimes \mathfrak{M}$ for all X_0 in $\mathfrak{B}$.

Now we choose any monotone increasing sequence (X_n) of sets in $\mathfrak{B}^x$ having the union A^x. For all Z in $\mathfrak{A}^x \otimes \mathfrak{M}$ and all Φ in M, we obtain

$$\mathring{\Phi}(Z \cap C) = \sup_{n=1,2,\dots} \mathring{\Phi}\big(Z \cap C \cap (X_n \times M)\big).$$

Consequently the measurability of the mappings $\Phi \curvearrowright \mathring{\Phi}\big(Z \cap C \cap (X_n \times M)\big)$ is transferred to $\Phi \curvearrowright \mathring{\Phi}(Z \cap C)$. ∎

In virtue of 5.5.1., we can associate to each measure H on $\mathfrak{M}$ a measure $\mathscr{C}_H$ on $\mathfrak{C}$ by means of

$$\mathscr{C}_H(Z) = \int \mathring{\Phi}(Z)\, H(\mathrm{d}\Phi) \qquad (Z \in \mathfrak{C}).$$

Hence

$$\mathscr{C}_H\big((X \times Y) \cap C\big) = \int_Y \Phi^x(X)\, H(\mathrm{d}\Phi) \qquad (X \in \mathfrak{A}^x,\ Y \in \mathfrak{M}).$$

We call $\mathscr{C}_H$ the *Campbell measure* of H. This designation is justified by the following refinement of the Campbell theorem 1.2.2.:

5.1.2. Proposition. *For all measures H on $\mathfrak{M}$ and all non-negative real functions f on $\mathfrak{C}$ which are measurable with respect to $\mathfrak{C}$*

$$\Phi \curvearrowright \sum_{x \in A^x,\, \Phi^x(\{x\})>0} f(x, \Phi)\, \Phi^x(\{x\})$$

is a measurable mapping of $[M, \mathfrak{M}]$ into $[0, +\infty]$, and

$$\int \Bigg(\sum_{x \in A^x,\, \Phi^x(\{x\})>0} f(x, \Phi)\, \Phi^x(\{x\})\Bigg) H(\mathrm{d}\Phi) = \int f(c)\, \mathscr{C}_H(\mathrm{d}c).$$

Proof. For all Φ in M, we obtain

$$\sum_{x \in A^x,\, \Phi^x(\{x\})>0} f(x, \Phi)\, \Phi^x(\{x\}) = \int f(c)\, \mathring{\Phi}(\mathrm{d}c).$$

Thus the assertion is valid for all indicator functions k_Z, $Z \in \mathfrak{C}$, hence also for all non-negative finite linear combinations g of such functions, and consequently also for all limits f of monotone increasing sequences (g_n). ∎

In 5.1.2., the value $f(x, \Phi)$ can be interpreted as a measure of the effect caused by a point x in Φ^x at a fixed point z, where by this effect may also depend on Φ. Then the integrand on the left side represents the superposition of the effects caused at the point z by the individual points of Φ^x, i.e. the total effect at the point z caused by Φ. Now, if H is a distribution on $\mathfrak{M}$, then 5.1.2. provides a representation of the expectation of the total effect caused at the point z.

If f depends only on x, but no longer on Φ, then 5.1.2. becomes the elementary proposition 1.2.2. Indeed we have, as an immediate consequence of the definition of $\mathscr{C}_H$,

5.1.3. *For all measures H on $\mathfrak{M}$ and all X in $\mathfrak{A}^x$,*

$$\varrho_{H^x}(X) = \mathscr{C}_H\big((X \times M) \cap C\big).$$

Consequently the Campbell measure $\mathscr{C}_H$ can be considered a refinement of the intensity measure of H^x. $\mathscr{C}_H$ depends linearly on H in the same way that ϱ_{H^x} does:

5.1.4. *For each at most countable family $(H_i)_{i\in I}$ of measures on $\mathfrak{M}$ and all non-negative v_i, $i \in I$,*

$$\mathscr{C}_{\sum\limits_{i\in I} v_i H_i} = \sum_{i\in I} v_i \mathscr{C}_{H_i}.$$

Obviously the projection

$$[x, \Phi] \sim \Omega(x, \Phi) = \Phi$$

provides a mapping Ω of C onto $M \setminus \{o\}$ which is measurable with respect to $\mathfrak{C}$, $\mathfrak{M}$. This mapping is not one-to-one if A^x contains more than one point since the inverse images of a given Φ in $M \setminus \{o\}$ correspond in a one-to-one way to the points of $\big\{x : x \in A^x, \Phi^x(\{x\}) > 0\big\}$.

We shall gain a substantial simplification of the operations in C if the various at most countable sets

$$\big\{x : x \in A^x, \Phi^x(\{x\}) > 0\big\}, \qquad \Phi \in M \setminus \{o\},$$

are made comparable by indexing, using the set G of integers. Accordingly we introduce the concept of a *measurable indexing relation.* This is understood to mean a mapping

$$[x, \Phi] \sim i(x, \Phi)$$

of C into G for which the correspondence

$$[x, \Phi] \sim [i(x, \Phi), \Phi]$$

effects a one-to-one mapping of C into $G \times M$, which is measurable in both directions with respect to the σ-algebras $\mathfrak{C}, \mathfrak{P}(G) \otimes \mathfrak{M}$. Instead of $i(x,\Phi) = s$, we also write $x = z_s(\Phi)$, so that the inverse mapping of $[x,\Phi] \sim [i(x,\Phi),\Phi]$ takes the form $[s, \Phi] \sim [z_s(\Phi), \Phi]$. For all Φ in $M \setminus \{o\}$, we set

$$I_\Phi = \{n : n \in G, i(x, \Phi) = n \text{ for some } x \text{ in } A^x \text{ with the property that } \Phi^x(\{x\}) > 0\}.$$

5.1.5. *There always exists a measurable indexing relation $[x, \Phi] \sim i(x, \Phi)$.*

Proof. To each natural number n we associate some covering $(X_{n,m})_{m=1,2,\ldots}$ of A^x by pairwise disjoint sets in $\mathfrak{A}^x$, the diameters of which do not exceed $1/n$. For each x in A^x and each natural number n, we set $f_n(x) = m$, if x is in $X_{n,m}$. Then the correspondence $x \sim \big(f_n(x)\big)_{n=1,2,\ldots}$ effects a one-to-one mapping of A^x into the power set $\{1, 2, \ldots\}^{\{1,2,\ldots\}}$. Now, if $\big(f_n(x_1)\big)_{n=1,2,\ldots}$

is less than $\big(f_n(x_2)\big)_{n=1,2,\ldots}$ in the sense of lexicographic order in $\{1, 2, \ldots\}^{\{1,2,\ldots\}}$, we write $x_1 \prec x_2$, thus obtaining a linear order in A^x. For any $[y, \Phi]$ in C, let $(m_1, m_2, \ldots)$ be the sequence $\big(f_n(y)\big)_{n=1,2,\ldots}$ associated to the point y in Φ^x. All points in $\big\{x : x \in A^x, \Phi^x(\{x\}) > 0\big\}$ which precede y in the sense of lexicographic order are contained in the bounded set $\bigcup_{i=1}^{m_1} X_{1,i}$, i.e. the number of such points is finite. Therefore we can consecutively number the points of $\big\{x : x \in A^x, \Phi^x(\{x\}) > 0\big\}$, and hence the inverse images of Φ with respect to Ω, by their order of succession, thus obtaining a sequence $z_1(\Phi), z_2(\Phi), \ldots$ which consists of $(\Phi^x)^*\,(A^x)$ elements, where $(\Phi^x)^*$ denotes the counting measure Φ^x "simplified" according to the definition given in Section 1.4. We now define that $x = z_n(\Phi)$ be equivalent with $i(x, \Phi) = n$.

The one-to-one nature of $[x, \Phi] \curvearrowright [i(x, \Phi), \Phi]$ is immediately derived from the definition of $i(x, \Phi)$.

The range of this mapping is

$$\bigcup_{l=1}^{\infty} \{l\} \times \{\Phi\colon \Phi \in M, (\Phi^x)^*\,(A^x) \geqq l\}$$

which belongs to $\mathfrak{P}(G) \otimes \mathfrak{M}$.

The mapping $[x, \Phi] \curvearrowright i(x, \Phi)$ is measurable with respect to $\mathfrak{C}$. Indeed, if we set

$$X_{1,i_1} \cap \cdots \cap X_{s,i_s} = V_{i_1,\ldots,i_s},$$

then the inverse image of a natural number n is given by

$$C \cap \left(\bigcup_{s=1}^{\infty} \bigcup_{i_1,\ldots,i_s=1}^{\infty} V_{i_1,\ldots,i_s} \times \Bigg\{\Phi\colon \Phi \in M, (\Phi^x)^*\,(V_{i_1,\ldots,i_s}) = 1, \right.$$
$$\left. (\Phi^x)^* \Bigg(\bigcup_{[l_1,\ldots,l_s] \prec [i_1,\ldots,i_s]} V_{l_1,\ldots,l_s}\Bigg) = n - 1\Bigg\}\right).$$

Here $\prec$ denotes the relation "lexicographically less than". Consequently also $[x, \Phi] \curvearrowright [i(x, \Phi), \Phi]$ is measurable.

For all X in $\mathfrak{A}^x$,

$$\{\Phi : \Phi \in M, (\Phi^x)^*\,(A^x) \geqq n, z_n(\Phi) \in X\}$$
$$= \bigcup_{s=1}^{\infty} \bigcup_{i_1,\ldots,i_s=1}^{\infty} \Bigg\{\Phi : \Phi \in M, (\Phi^x)^*\,(X \cap V_{i_1,\ldots,i_s}) = (\Phi^x)^*\,(V_{i_1,\ldots,i_s}) = 1,$$
$$(\Phi^x)^* \Bigg(\bigcup_{[l_1,\ldots,l_s] \prec [i_1,\ldots,i_s]} V_{l_1,\ldots,l_s}\Bigg) = n - 1\Bigg\}.$$

Therefore all mappings $\Phi \curvearrowright z_n(\Phi)$ from M into A^x are measurable, and we are able to infer the measurability of

$$[n, \Phi] \curvearrowright [z_n(\Phi), \Phi] \qquad \big(\Phi \in M, (\Phi^x)^*\,(A^x) \geqq n\big),$$

i.e. the inverse mapping of $[x, \Phi] \curvearrowright [i(x, \Phi), \Phi]$, is measurable. ∎

5.1.6. *For all Z in $\mathfrak{C}$, $\Omega(Z)$ belongs to $\mathfrak{M}$.*

Proof. Indeed, if $[x, \Phi] \rightsquigarrow i(x, \Phi)$ is an arbitrary measurable indexing relation, then $[n, \Phi] \rightsquigarrow \Phi$ gives a one-to-one mapping from $[G \times M, \mathfrak{P}(G) \otimes \mathfrak{M}]$ into $[M, \mathfrak{M}]$ which is measurable in both directions. ∎

The following proposition provides us with the starting points for deriving concrete "inversion formulae" for recovering H from $\mathscr{C}_H$:

5.1.7. Proposition. *Let Y_0 be a set in $\mathfrak{M}$ satisfying the condition $o \notin Y_0$, and h a non-negative real function on $C \cap (A^x \times Y_0)$ which is measurable with respect to $\mathfrak{C}$ and has the property that*

$$\sum_{x \in A^x, \Phi^x(\{x\})>0} h(x, \Phi)\, \Phi^x(\{x\}) = 1 \qquad (\Phi \in Y_0).$$

Subject to these assumptions, for all measures H on $\mathfrak{M}$ and all Y in $\mathfrak{M}$,

$$H(Y \cap Y_0) = \int_{C \cap (A^x \times Y_0)} k_Y(\Phi)\, h(x, \Phi)\, \mathscr{C}_H(\mathrm{d}[x, \Phi]).$$

Proof. By means of

$$h(c) = 0 \quad \text{for} \quad c \in C \setminus (A^x \times Y_0),$$

we continue h over the whole of C; thus by 5.1.2. we obtain

$$\int_{C \cap (A^x \times Y_0)} k_Y(\Phi)\, h(x, \Phi)\, \mathscr{C}_H(\mathrm{d}[x, \Phi]) = \int k_{Y \cap Y_0}(\Phi)\, h(x, \Phi)\, \mathscr{C}_H(\mathrm{d}[x, \Phi])$$

$$= \int \Big(\sum_{x \in A^x, \Phi^x(\{x\})>0} k_{Y \cap Y_0}(\Phi)\, h(x, \Phi)\, \Phi^x(\{x\})\Big)\, H(\mathrm{d}\Phi)$$

$$= \int k_{Y \cap Y_0}(\Phi)\, H(\mathrm{d}\Phi) = H(Y \cap Y_0). \;∎$$

Now, given any set X in $\mathfrak{A}^x$, and if $Y_0 = \{\Phi : \Phi \in M, 0 < \Phi^x(X) < +\infty\}$, we can set

$$h(x, \Phi) = \begin{cases} 0 & \\ & \text{for} \\ \big(\Phi^x(X)\big)^{-1} & \end{cases} \begin{matrix} x \notin X \\ \\ x \in X, \end{matrix}$$

thus obtaining the following inversion formula.

5.1.8. Proposition. *For all X in $\mathfrak{A}^x$, all Y in $\mathfrak{M}$, and all measures H on $\mathfrak{M}$,*

$$H\big(\Phi \in Y, 0 < \Phi^x(X) < +\infty\big) = \sum_{k=1}^{\infty} k^{-1} \mathscr{C}_H\big(x \in X, \Phi \in Y, \Phi^x(X) = k\big).$$

If X even belongs to $\mathfrak{B}^x$, then in 5.1.8. the term $H\big(\Phi \in Y, 0 < \Phi^x(X) < +\infty\big)$ can be replaced by $H\big(\Phi \in Y, \Phi^x(X) > 0\big)$.

Using any measurable indexing relation $[x, \Phi] \rightsquigarrow i(x, \Phi)$ constructed according to the proof of 5.1.5., we can provide a function h satisfying the assumptions of 5.1.7. for $Y_0 = M \setminus \{o\}$. Indeed, if we denote by Z_1 the set of all $[x, \Phi]$ in C having the property that $i(x, \Phi) = 1$, then

$$h(x, \Phi) = \big(\Phi^x(\{x\})\big)^{-1}\, k_{Z_1}(x, \Phi) \qquad ([x, \Phi] \in C)$$

represents a non-negative real function h which is defined on C and has the property

$$\sum_{x \in A^x, \Phi^x(\{x\})>0} h(x, \Phi)\, \Phi^x(\{x\}) = 1 \qquad (\Phi \in M \setminus \{o\}).$$

The measurability of h is immediately derived from 5.1.5. and the following lemma.

5.1.9. *The mapping*

$$[x, \Phi] \rightsquigarrow \Phi^x(\{x\})$$

of C into $\{1, 2, \ldots\}$ is measurable with respect to $\mathfrak{C}$.

Proof. To each natural number n we associate some covering $(X_{n,m})_{m=1,2,\ldots}$ of A^x by pairwise disjoint sets in $\mathfrak{A}^x$, the diameters of which do not exceed $1/n$. Now, if we set

$$f_n(x, \Phi) = \sum_{m=1}^{\infty} k_{X_{n,m}}(x)\, \Phi^x(X_{n,m}),$$

then all f_n are measurable, and

$$f_n(x, \Phi) \xrightarrow[n \to \infty]{} \Phi^x(\{x\}) \qquad ([x, \Phi] \in C). \blacksquare$$

If h satisfies the assumptions of 5.1.7. with respect to $Y_0 = M \setminus \{o\}$, then for all Y in $\mathfrak{M}$ and all measures H on $\mathfrak{M}$,

$$H(Y \setminus \{o\}) = \int k_Y(\Phi)\, h(x, \Phi)\, \mathscr{C}_H(\mathrm{d}[x, \Phi]),$$

so that we obtain

5.1.10. Theorem. *The Campbell measures of two measures H_1, H_2 on $\mathfrak{M}$ coincide if and only if*

$$H_1\big((.) \setminus \{o\}\big) = H_2\big((.) \setminus \{o\}\big)$$

is satisfied.

The Campbell measure of a finite measure is not necessarily finite. However, we have

5.1.11. *The Campbell measure of a measure H on $\mathfrak{M}$ is σ-finite if and only if this is true for $H\big((.) \setminus \{o\}\big)$.*

Proof. 1. First we suppose that $H\big((.) \setminus \{o\}\big)$ be σ-finite. So let (Y_m) be a sequence of sets in $\mathfrak{M}$ having the properties

$$H(Y_m) < +\infty \quad \text{for} \quad m = 1, 2, \ldots, \qquad \bigcup_{m=1}^{\infty} Y_m = M \setminus \{o\}.$$

By virtue of the inversion formula 5.1.8., we obtain, for each monotone increasing sequence (X_n) of sets in $\mathfrak{B}^x$ having the union A^x, and for all natural numbers k, m, n,

$$\mathscr{C}_H\big(x \in X_n, \Phi \in Y_m, \Phi(X_n) = k\big) < +\infty.$$

On the other hand,

$$\bigcup_{k,m,n=1}^{\infty} \{[x, \Phi] : [x, \Phi] \in C, x \in X_n, \Phi \in Y_m, \Phi(X_n) = k\} = C.$$

Consequently the Campbell measure of H is σ-finite.

2. Conversely, suppose that the Campbell measure of $H\big((.) \setminus \{o\}\big)$ be σ-finite. As above, let Z_1 denote the set of all $[x, \Phi]$ in C with the property that $i(x, \Phi) = 1$, where $[x, \Phi] \curvearrowright i(x, \Phi)$ denotes a measurable indexing relation constructed according to the proof of 5.1.5. By assumption there exists a monotone increasing sequence (W_n) of sets in $\mathfrak{C}$ having the properties

$$\mathscr{C}_H(W_n) < +\infty \quad \text{for} \quad n = 1, 2, \ldots, \qquad \bigcup_{n=1}^{\infty} W_n = Z_1.$$

By virtue of 5.1.5., the set Z_1 belongs to $\mathfrak{C}$ and is mapped onto $M \setminus \{o\}$ by Ω, which is a one-to-one mapping and measurable in both directions. Consequently it will suffice to prove the inequalities

$$H\big(\Omega(W_n \cap Z_1)\big) < +\infty \quad \text{for} \quad n = 1, 2, \ldots$$

However, by 5.1.7., for

$$h(x, \Phi) = \big(\Phi^x(\{x\})\big)^{-1}\, k_{Z_1}(x, \Phi)$$

we have the following chain of equations:

$$\begin{aligned} H\big(\Omega(W_n \cap Z_1)\big) &= H\big(\Omega(W_n \cap Z_1) \setminus \{o\}\big) \\ &= \int k_{\Omega(W_n \cap Z_1)}(\Phi)\, h(x, \Phi)\, \mathscr{C}_H(\mathrm{d}[x, \Phi]) \\ &= \sum_{l=1}^{\infty} l^{-1} \int_{\{[x,\Phi] : \Phi^x(\{x\}) = l,\, i(x,\Phi) = 1\}} k_{\Omega(W_n \cap Z_1)}(\Phi)\, \mathscr{C}_H(\mathrm{d}[x, \Phi]) \\ &= \sum_{l=1}^{\infty} l^{-1} \mathscr{C}_H\big([x, \Phi] \in (W_n \cap Z_1), \Phi^x(\{x\}) = l\big) \leqq \mathscr{C}_H(W_n) < +\infty. \blacksquare \end{aligned}$$

We conclude this section with

5.1.12. *For all measures H on $\mathfrak{M}$ and all Y in $\mathfrak{M}$,*

$$\mathscr{C}_H\big((.) \cap (A^x \times Y)\big) = \mathscr{C}_{H((.) \cap Y)}.$$

Proof. Indeed we have

$$\mathscr{C}_H\big((.) \cap (A^x \times Y)\big) = \int \mathring{\Phi}\big((.) \cap (A^x \times Y)\big)\, H(\mathrm{d}\Phi)$$
$$= \int_Y \mathring{\Phi}(.)\, H(\mathrm{d}\Phi) = \int \mathring{\Phi}(.)\, H(Y \cap \mathrm{d}\Phi)\,. \;\blacksquare$$

In particular, $H(\Phi \notin Y, \Phi \neq o) = 0$ is true if and only if $\mathscr{C}_H([x, \Phi] \notin A^x \times Y) = 0$ is satisfied.

5.2. The Functional Equation $\mathscr{C}_P = V \circledast P$

The mapping $\big[[x, \Phi], \Psi\big] \rightsquigarrow [x, \Phi + \Psi]$ of $C \times M$ into C is measurable with respect to the σ-algebras $\mathfrak{C} \otimes \mathfrak{M}, \mathfrak{C}$. It transforms the direct product $V \times H$ of a σ-finite measure V on $\mathfrak{C}$ and a σ-finite measure H on $\mathfrak{M}$ into a measure $V \circledast H$ on $\mathfrak{C}$.

If H_1, H_2 are two σ-finite but not necessarily finite measures on $\mathfrak{M}$, we denote by $H_1 * H_2$ that measure on $\mathfrak{M}$ which results from transforming $H_1 \times H_2$ by $[\Phi, \Psi] \rightsquigarrow \Phi + \Psi$. The operation $\circledast$ is closely related to this generalized convolution:

5.2.1. *Let V be a σ-finite measure on $\mathfrak{C}$. Further let H_1, H_2 be σ-finite measures on $\mathfrak{M}$. If $H_1 * H_2$ and $V \circledast H_1$ are σ-finite, then*

$$(V \circledast H_1) \circledast H_2 = V \circledast (H_1 * H_2)\,.$$

Proof. Let Q denote the measure $V \times H_1 \times H_2$. If we first apply the measurable mapping $\big[[x, \Phi], \Psi_1, \Psi_2\big] \rightsquigarrow \big[[x, \Phi], \Psi_1 + \Psi_2\big]$ and subsequently the measurable mapping $\big[[x, \Phi], \chi\big] \rightsquigarrow [x, \Phi + \chi]$, then Q is transformed into $V \circledast (H_1 * H_2)$. The same measure is also obtained if first $\big[[x, \Phi], \Psi_1, \Psi_2\big]$ is transformed into $\big[[x, \Phi + \Psi_1], \Psi_2\big]$ and subsequently $\big[[x, \chi], \Psi_2\big]$ into $[x, \chi + \Psi_2]$. But in this case Q is transformed into $(V \circledast H_1) \circledast H_2$. $\blacksquare$

The operation $\circledast$ is distributive in the following sense:

5.2.2. *Let (Z_n) be a monotone increasing sequence of sets in $\mathfrak{C}$ which has the union C. Further let $[U, \mathfrak{U}]$, $[L, \mathfrak{L}]$ be measurable spaces. Suppose that to each u in U a measure $V_{(u)}$ on $\mathfrak{C}$ satisfying the condition $V_{(u)}(Z_n) < +\infty$ $(n = 1, 2, \ldots)$ be associated and that to each l in L a measure $E_{(l)}$ in $\mathbf{E}^+$ be associated, the correspondences being defined in such a way that all mappings*

$$u \rightsquigarrow V_{(u)}(Z) \qquad (Z \in \mathfrak{C}),$$
$$l \rightsquigarrow E_{(l)}(Y) \qquad (Y \in \mathfrak{M}),$$

are measurable with respect to $\mathfrak{U}$ and $\mathfrak{L}$, respectively. Then, for all Z in $\mathfrak{C}$,

$$[u, l] \rightsquigarrow (V_{(u)} \circledast E_{(l)})(Z)$$

is measurable with respect to the σ-algebra $\mathfrak{U} \otimes \mathfrak{L}$. *If* α, β *are finite measures on* $\mathfrak{U}, \mathfrak{L}$, *and if* $V = \int V_{(u)}(.)\ \alpha(\mathrm{d}u)$ *is a σ-finite measure and* $E = \int E_{(l)}(.)\ \beta(\mathrm{d}l)$ *a finite measure, then*

$$V \circledast E = \int (V_{(u)} \circledast E_{(l)})(.)\,(\alpha \times \beta)(\mathrm{d}[u, l]).$$

Proof. For all $X, X_1, \ldots, X_m$ in $\mathfrak{B}^x$, all non-negative integers $v_1, \ldots, v_m$, and all natural numbers n, we obtain

$$\begin{aligned}&\Big(V_{(u)}\big((.) \cap Z_n\big) \circledast E_{(l)}\Big)\\&\qquad\Big(\big(X \times \{\Phi : \Phi \in M, [\Phi^x(X_1), \ldots, \Phi^x(X_m)] = [v_1, \ldots, v_m]\}\big) \cap C\Big)\\(*)\qquad &= \sum_{\substack{i_1,\ldots,i_m,j_1,\ldots,j_m \geqq 0\\ i_s + j_s = v_s \text{ for } 1 \leqq s \leqq m}} V_{(u)}\\&\qquad\Big(\big(X \times \{\Phi : \Phi \in M, [\Phi^x(X_1), \ldots, \Phi^x(X_m)] = [i_1, \ldots, i_m]\}\big) \cap Z_n\Big)\\&\qquad\qquad \cdot E_{(l)}\big([\Phi^x(X_1), \ldots, \Phi^x(X_m)] = [j_1, \ldots, j_m]\big).\end{aligned}$$

As was done in the proof of 1.8.2., we can conclude from the above that

$$[u, l] \curvearrowright \Big(V_{(u)}\big((.) \cap Z_n\big) \circledast E_{(l)}\Big)\,(Z \cap C)$$

is measurable for all Z in the σ-algebra $\mathfrak{A}^x \otimes \mathfrak{M}$. On the other hand, we have

$$(V_{(u)} \circledast E_{(l)})\,(Z \cap C) = \sup_{n=1,2,\ldots} \Big(V_{(u)}\big((.) \cap Z_n\big) \circledast E_{(l)}\Big)\,(Z \cap C),$$

so that we obtain the required measurability statement. By integration of $(*)$ we obtain

$$\begin{aligned}&\int \Big(V_{(u)}\big((.) \cap Z_n\big) \circledast E_{(l)}\Big)\\&\qquad\Big(\big(X \times \{\Phi : \Phi \in M, [\Phi^x(X_1), \ldots, \Phi^x(X_m)] = [v_1, \ldots, v_m]\}\big) \cap C\Big)\\&\qquad(\alpha \times \beta)\,(\mathrm{d}[u, l])\\&= \sum_{\substack{i_1,\ldots,i_m,j_1,\ldots,j_m \geqq 0\\ i_s + j_s = v_s \text{ for } 1 \leqq s \leqq m}} \int V_{(u)}\\&\qquad\Big(\big(X \times \{\Phi : \Phi \in M, [\Phi^x(X_1), \ldots, \Phi^x(X_m)] = [i_1, \ldots, i_m]\}\big) \cap Z_n\Big)\\&\qquad\alpha(\mathrm{d}u) \int E_{(l)}\big([\Phi^x(X_1), \ldots, \Phi^x(X_m)] = [j_1, \ldots, j_m]\big)\,\beta(\mathrm{d}l)\\&= \Big(\Big(\int V_{(u)}\big((.) \cap Z_n\big)\,\alpha(\mathrm{d}u)\Big) \circledast \Big(\int E_{(l)}(.)\,\beta(\mathrm{d}l)\Big)\Big)\\&\qquad\Big(\big(X \times \{\Phi : \Phi \in M, [\Phi^x(X_1), \ldots, \Phi^x(X_m)] = [v_1, \ldots, v_m]\}\big) \cap C\Big).\end{aligned}$$

As in the proof of 1.8.2., this gives, for all Z in the σ-algebra $\mathfrak{A}^x \otimes \mathfrak{M}$,

$$\int \left(V_{(u)}((.) \cap Z_n) \circledast E_{(l)}\right) (Z \cap C) (\alpha \times \beta) (\mathrm{d}[u, l])$$
$$= \left(\left(\int V_{(u)}((.) \cap Z_n) \,\alpha(\mathrm{d}u)\right) \circledast E\right) (Z \cap C),$$

and letting $n \to \infty$, we obtain the desired equation. ∎

Using the operation $\circledast$, we are able to derive the following expression for the Campbell measure of a convolution:

5.2.3. Proposition. *If the convolution* $\mathop{\ast}\limits_{i\in I} P_i$ *of an at most countable family* $(P_i)_{i\in I}$ *of distributions on* $\mathfrak{M}$ *exists, then*

$$\mathscr{C}_{\mathop{\ast}\limits_{i\in I} P_i} = \sum_{i\in I} \mathscr{C}_{P_i} \circledast \mathop{\ast}\limits_{j\in I\setminus\{i\}} P_j .$$

Proof. If the set of indices I is empty, then the left-hand side is equal to $\mathscr{C}_{\delta_0}$, and hence equal to the right side, namely to the zero measure on $\mathfrak{C}$. Therefore we are allowed to suppose that I be non-empty.

For each family $(\Phi_i)_{i\in I}$ in M^I with the property that $\Phi = \sum_{i\in I} \Phi_i \in M$, i.e. for each $(\Phi_i)_{i\in I}$ in L_I, we have

$$\mathring{\Phi} = \Phi^x \times \delta_\Phi = \left(\sum_{i\in I} \Phi_i^x\right) \times \delta_\Phi = \sum_{i\in I} \Phi_i^x \times \delta_\Phi$$
$$= \sum_{i\in I} (\Phi_i^x \times \delta_{\Phi_i}) \circledast \delta_{\Phi-\Phi_i} = \sum_{i\in I} \mathscr{C}_{\delta_{\Phi_i}} \circledast \delta_{\sum\limits_{j\in I\setminus\{i\}} \Phi_j} .$$

Using 5.2.2. we obtain

$$\mathscr{C}_{\mathop{\ast}\limits_{i\in I} P_i} = \int \mathring{\Phi}(.) \left(\mathop{\ast}\limits_{i\in I} P_i\right) (\mathrm{d}\Phi) = \int\limits_{L_I} \left(\sum_{i\in I} \mathring{\Phi}_i\right) (.) \left(\mathop{\times}\limits_{i\in I} P_i\right) \left(\mathrm{d}(\Phi_i)_{i\in I}\right)$$
$$= \int\limits_{L_I} \left(\sum_{i\in I} \mathscr{C}_{\delta_{\Phi_i}} \circledast \delta_{\sum\limits_{j\in I\setminus\{i\}} \Phi_j}\right) (.) \left(\mathop{\times}\limits_{i\in I} P_i\right) \left(\mathrm{d}(\Phi_i)_{i\in I}\right)$$
$$= \sum_{i\in I} \int\limits_{M\times L_{I\setminus\{i\}}} \left(\mathscr{C}_{\delta_{\Phi_i}} \circledast \delta_{\sum\limits_{j\in I\setminus\{i\}} \Phi_j}\right) (.) \left(P_i \times \mathop{\times}\limits_{j\in I\setminus\{i\}} P_j\right) \left(\mathrm{d}[\Phi_i, (\Phi_j)_{j\in I\setminus\{i\}}]\right)$$
$$= \sum_{i\in I} \int (\mathscr{C}_{\delta_\Phi} \circledast \delta_\Psi) (.) \left(P_i \times \mathop{\ast}\limits_{j\in I\setminus\{i\}} P_j\right) (\mathrm{d}[\Phi, \Psi])$$
$$= \sum_{i\in I} \left(\int \mathring{\Phi}(.) \, P_i(\mathrm{d}\Phi)\right) \circledast \left(\int \delta_\Psi(.) \left(\mathop{\ast}\limits_{j\in I\setminus\{i\}} P_j\right) (\mathrm{d}\Psi)\right)$$
$$= \sum_{i\in I} \mathscr{C}_{P_i} \circledast \mathop{\ast}\limits_{j\in I\setminus\{i\}} P_j .$$

Indeed, for each increasing sequence (X_n) of sets in $\mathfrak{B}^x$ having the union A^x, we have $\mathring{\Phi}\left((X_n \times M) \cap C\right) < +\infty$ $(n = 1, 2, \ldots)$, so that in applying 5.2.2. we can replace Z_n by the set $(X_n \times M) \cap C$. Moreover, by 5.1.11., the measures $\mathscr{C}_{P_i}$ are all σ-finite. ∎

Each P in $\boldsymbol{T}$ satisfies the following functional equation:

5.2.4. *For all infinitely divisible distributions P on $\mathfrak{M}$,*

$$\mathscr{C}_P = \mathscr{C}_{\tilde{P}} \circledast P.$$

Proof. By 5.1.11., $\mathscr{C}_{\tilde{P}}$ is a σ-finite measure, so that $\mathscr{C}_{\tilde{P}} \circledast P$ is defined. Let $\tilde{P} = \sum_{i \in I} E_i$ be a representation of $\tilde{P}$ as a countable sum of measures in $\boldsymbol{E}^+$. Using 5.2.3. and 2.1.8. we obtain

$$(*) \qquad \mathscr{C}_P = \mathscr{C}_{\underset{i \in I}{*} P_i} = \sum_{i \in I} \mathscr{C}_{P_i} \circledast \underset{j \in I \setminus \{i\}}{*} P_j,$$

where we set $P_i = \mathscr{U}_{E_i}$.

For $E_i = 0$ we have $\mathscr{U}_{E_i} = \delta_o$, and hence $\mathscr{C}_{P_i} = 0 = \mathscr{C}_{\widetilde{P_i}} \circledast P_i$. If $E_i \neq 0$, then by 5.1.4., 5.2.3., and 5.2.2. we obtain

$$\mathscr{C}_{P_i} = \mathscr{C}_{e^{-E_i(M)} \sum_{n=0}^{\infty} \frac{1}{n!} (E_i)^n} = e^{-E_i(M)} \sum_{n=1}^{\infty} \frac{1}{n!} \mathscr{C}_{(E_i)^n}$$

$$= e^{-E_i(M)} \sum_{n=1}^{\infty} \frac{\big(E_i(M)\big)^n}{n!} \mathscr{C}_{(E_i((.)|M))^n} = e^{-E_i(M)} \sum_{n=1}^{\infty} \frac{\big(E_i(M)\big)^n}{n!} \cdot n \mathscr{C}_{E_i((.)|M)} \circledast \big(E_i((.) \mid M)\big)^{n-1}$$

$$= e^{-E_i(M)} \sum_{n=1}^{\infty} \frac{1}{(n-1)!} \mathscr{C}_{E_i} \circledast (E_i)^{n-1}$$

$$= \mathscr{C}_{E_i} \circledast \left(e^{-E_i(M)} \sum_{m=0}^{\infty} \frac{1}{m!} (E_i)^m \right) = \mathscr{C}_{E_i} \circledast P_i.$$

Substituting this formula into $(*)$, we obtain, using 5.2.1., 5.2.2., and 5.1.4.,

$$\mathscr{C}_P = \sum_{i \in I} (\mathscr{C}_{\widetilde{P_i}} \circledast P_i) \circledast \underset{j \in I \setminus \{i\}}{*} P_j = \sum_{i \in I} \mathscr{C}_{\widetilde{P_i}} \circledast \underset{j \in I}{*} P_j$$

$$= \Big(\sum_{i \in I} \mathscr{C}_{\widetilde{P_i}} \Big) \circledast P = \mathscr{C}_{\sum_{i \in I} \widetilde{P_i}} \circledast P = \mathscr{C}_{\tilde{P}} \circledast P. \blacksquare$$

It is characteristic for the infinitely divisible distributions that P can be represented as a solution of a functional equation of the type stated above:

5.2.5. Theorem. *A distribution P on $\mathfrak{M}$ is infinitely divisible if and only if there exists a σ-finite measure V on $\mathfrak{C}$ having the property that*

$$\mathscr{C}_P = V \circledast P.$$

This measure is uniquely determined by P, being given by

$$V = \mathscr{C}_{\tilde{P}}.$$

Proof. 1. Let $\mathscr{C}_P = V \circledast P$. Further let X_0 be a bounded set in $\mathfrak{B}^x$, $X_1 = X_0 \times K$, and $X_2, \ldots, X_k$ a finite sequence of sets in $\mathfrak{B}$. For all

$n_1 > 0$, $n_2, \ldots, n_k \geqq 0$, we obtain, using the inversion formula 5.1.8. as well as 5.2.2.:

$$n_1 P_{X_1,\ldots,X_k}(\{[n_1, \ldots, n_k]\}) = \mathscr{C}_P\big(x \in X_0, \Phi(X_i) = n_i \quad \text{for} \quad 1 \leqq i \leqq k\big)$$
$$= (V \circledast P)\,\big(x \in X_0, \Phi(X_i) = n_i \quad \text{for} \quad 1 \leqq i \leqq k\big)$$
$$= \int (V \circledast \delta_\chi)\,\big(x \in X_0, \Phi(X_i) = n_i \quad \text{for} \quad 1 \leqq i \leqq k\big)\, P(\mathrm{d}\chi)$$
$$= \int V\big(x \in X_0, (\chi + \Phi)\,(X_i) = n_i \quad \text{for} \quad 1 \leqq i \leqq k\big)\, P(\mathrm{d}\chi)$$
$$= \sum_{m_1=0}^{n_1} \cdots \sum_{m_k=0}^{n_k} V\big(x \in X_0, \Phi(X_i) = n_i - m_i \quad \text{for} \quad 1 \leqq i \leqq k\big)$$
$$\cdot\, P_{X_1,\ldots,X_k}(\{[m_1, \ldots, m_k]\}),$$

where we have to set $\infty \cdot 0 = 0$.

2. For $h_1, \ldots, h_k \geqq 0$, we shall abbreviate the expression $V(x \in X_0, \Phi(X_i) = h_i$ for $1 \leqq i \leqq k)$ to

$$V_{X_1,\ldots,X_k}(h_1, \ldots, h_k).$$

If $P_{X_1}(\{0\}) = 1$, then $\mathscr{C}_P(x \in X_0) = 0$, and in view of $\mathscr{C}_P = V \circledast P$ this enables us to conclude that

$$V_{X_1,\ldots,X_k}(h_1, \ldots, h_k) = 0 \qquad (h_1, \ldots, h_k \geqq 0).$$

Now suppose that $P_{X_1}(\{0\}) < 1$. If any of the $V_{X_1,\ldots,X_k}(h_1, \ldots, h_k)$ were infinite, then we might select any positive $P_{X_1,\ldots,X_k}(\{[m_1, \ldots, m_k]\})$ with $m_1 > 0$, and using 1. we would obtain

$$+\infty = V_{X_1,\ldots,X_k}(h_1, \ldots, h_k)\, P_{X_1,\ldots,X_k}(\{[m_1, \ldots, m_k]\})$$
$$\leqq (h_1 + m_1)\, P_{X_1,\ldots,X_k}(\{[h_1 + m_1, \ldots, h_k + m_k]\}) < +\infty.$$

Consequently all $V_{X_1,\ldots,X_k}(h_1, \ldots, h_k)$ are finite.

3. For all $h_2, \ldots, h_k \geqq 0$, we have $V_{X_1,\ldots,X_k}(0, h_2, \ldots, h_k) = 0$ since

$$C \cap \{[x, \Phi] : [x, \Phi] \in X_0 \times M, \Phi(X_0 \times K) = 0\} = \emptyset.$$

If we denote the formal power series

$$\sum_{n_1,\ldots,n_k=0}^{\infty} P_{X_1,\ldots,X_k}(\{[n_1, \ldots, n_k]\})\, \xi_1^{n_1} \ldots \xi_k^{n_k},$$

$$\sum_{n_1,\ldots,n_k=0}^{\infty} V_{X_1,\ldots,X_k}(n_1, \ldots, n_k)\, \xi_1^{n_1} \ldots \xi_k^{n_k}$$

by $U(\xi_1, \ldots, \xi_k)$ and $V(\xi_1, \ldots, \xi_k)$, respectively, then, by virtue of 1., we have

$$\xi_1 \cdot (D_1 U)\,(\xi_1, \ldots, \xi_k) = V(\xi_1, \ldots, \xi_k)\, U(\xi_1, \ldots, \xi_k),$$

where D_i $(1 \leqq i \leqq k)$ denotes the operator of formal differentiation with respect to ξ_i.

Since $V_{X_1,\ldots,X_k}(0, h_2, \ldots, h_k) = 0$, we can divide $V(\xi_1, \ldots, \xi_k)$ by ξ_1, obtaining

$$(D_1 U)(\xi_1, \ldots, \xi_k) = \left(\xi_1^{-1} V(\xi_1, \ldots, \xi_k)\right) U(\xi_1, \ldots, \xi_k).$$

We now select from $\mathfrak{B}^x$ a set X_0', such that $X = X_0' \times K \supseteqq X_1 \cup \cdots \cup X_k$ is satisfied. We obtain

$$D_1 \sum_{n=0}^{\infty} P_X(\{n\}) \xi_1^n = \left(\sum_{n=1}^{\infty} V_X(n) \xi_1^{n-1}\right)\left(\sum_{m=0}^{\infty} P_X(\{m\}) \xi_1^m\right).$$

Now, assuming that $P_X(\{0\}) = 0$, we would successively obtain $P_X(\{1\}) = 0$, $P_X(\{2\}) = 0$, ..., in contradiction to $\sum_{n=0}^{\infty} P_X(\{n\}) = 1$. Hence

$$0 < P_X(\{0\}) \leqq P_{X_1,\ldots,X_k}(\{[0, \ldots, 0]\}),$$

which means that $U(0, \ldots, 0) > 0$.

Consequently the equation derived above can be written as follows:

$$\frac{(D_1 U)(\xi_1, \ldots, \xi_k)}{U(\xi_1, \ldots, \xi_k)} = \xi_1^{-1} V(\xi_1, \ldots, \xi_k). \qquad (*)$$

In what follows, we shall suppose in addition that

$$X_1 \supseteqq X_2, \ldots, X_k.$$

Then

$$P_{X_1,\ldots,X_k}(\{[0, n_2, \ldots, n_k]\}) = 0 \quad \text{for} \quad n_2, \ldots, n_k \geqq 0, \qquad n_2 + \cdots + n_k > 0,$$

i.e. we obtain

$$U(0, \xi_2, \ldots, \xi_k) = U(0, \ldots, 0). \qquad (**)$$

Now we set

$$Q(\xi_1, \ldots, \xi_k) = \sum_{\substack{n_1,\ldots,n_k=0 \\ n_1 \neq 0}}^{\infty} \frac{1}{n_1} V_{X_1,\ldots,X_k}(n_1, \ldots, n_k) \xi_1^{n_1} \ldots \xi_k^{n_k}.$$

Since $Q(0, \ldots, 0) = 0$, we can define the formal power series

$$\exp\left(Q(\xi_1, \ldots, \xi_k)\right) = \sum_{i=0}^{\infty} \frac{\left(Q(\xi_1, \ldots, \xi_k)\right)^i}{i!}$$

where $\left(Q(\xi_1, \ldots, \xi_k)\right)^0 = 1$.

For the formal power series

$$\sum_{n_1,\ldots,n_k=0}^{\infty} h_{n_1,\ldots,n_k} \xi_1^{n_1} \ldots \xi_k^{n_k} = H(\xi_1, \ldots, \xi_k) = U(0, \ldots, 0) \exp\left(Q(\xi_1, \ldots, \xi_k)\right)$$

we obtain

$$\frac{(D_1 H)(\xi_1, \ldots, \xi_k)}{H(\xi_1, \ldots, \xi_k)} = (D_1 Q)(\xi_1, \ldots, \xi_k)$$

$$= \sum_{\substack{n_1,\ldots,n_k=0 \\ n_1 \neq 0}}^{\infty} V_{X_1,\ldots,X_k}(n_1, \ldots, n_k) \xi_1^{n_1 - 1} \xi_2^{n_2} \ldots \xi_k^{n_k} = \xi_1^{-1} V(\xi_1, \ldots, \xi_k).$$

Moreover we have

$$H(0, \xi_2, \ldots, \xi_k) = U(0, \ldots, 0) \exp\big(Q(0, \xi_2, \ldots, \xi_k)\big)$$
$$= U(0, \ldots, 0) \exp\big(Q(0, \ldots, 0)\big) = H(0, \ldots, 0).$$

Consequently the formal power series $H(\xi_1, \ldots, \xi_k)$ satisfies the two conditions $(*)$, $(**)$, just as $U(\xi_1, \ldots, \xi_k)$ does. It remains to be shown that $H(\xi_1, \ldots, \xi_k)$ is the only solution of $(*)$ with the initial condition $(**)$. We write $V(\xi_1, \ldots, \xi_k)$ in the following form:

$$V(\xi_1, \ldots, \xi_k) = v_1(\xi_2, \ldots, \xi_k)\, \xi_1 + v_2(\xi_2, \ldots, \xi_k)\, \xi_1^2 + \cdots = v_1\xi_1 + v_2\xi_1^2 + \cdots,$$

i.e. we have

$$H(\xi_1, \ldots, \xi_k) = U(0, \ldots, 0) \exp\Big(v_1(\xi_2, \ldots, \xi_k)\, \xi_1 + \frac{1}{2} v_2(\xi_2, \ldots, \xi_k)\, \xi_1^2 + \frac{1}{3} v_3(\xi_2, \ldots, \xi_k)\, \xi_1^3 + \cdots\Big).$$

Now we use the following set-up in terms of a power series:

$$W(\xi_1, \ldots, \xi_k) = U(\xi_1, \ldots, \xi_k) - U(0, \ldots, 0) \exp\Big(v_1(\xi_2, \ldots, \xi_k)\, \xi_1 + \frac{1}{2} v_2(\xi_2, \ldots, \xi_k)\, \xi_1^2 + \cdots\Big)$$

and using $(*)$ and $(**)$ we obtain the formal differential equation

$$(D_1 W)(\xi_1, \ldots, \xi_k) = \xi_1^{-1} V(\xi_1, \ldots, \xi_k)\, W(\xi_1, \ldots, \xi_k) \qquad (***)$$

with the initial condition $W(0, \xi_2, \ldots, \xi_k) = 0$. Because of the vanishing initial condition we may represent $W(\xi_1, \ldots, \xi_k)$ in the form

$$W(\xi_1, \ldots, \xi_k) = w_1(\xi_2, \ldots, \xi_k)\, \xi_1 + w_2(\xi_2, \ldots, \xi_k)\, \xi_1^2 + \cdots = w_1\xi_1 + w_2\xi_1^2 + \cdots$$

Thus $(***)$ becomes

$$(w_1 + 2w_2\xi_1 + 3w_3\xi_1^2 + \cdots) = (v_1 + v_2\xi_1 + v_3\xi_1^2 + \cdots)(w_1\xi_1 + w_2\xi_1^2 + \cdots).$$

By comparing the coefficients, we can now conclude that

$$W(\xi_1, \ldots, \xi_k) = U(\xi_1, \ldots, \xi_k) - H(\xi_1, \ldots, \xi_k) \equiv 0.$$

The formal power series $Q(\xi_1, \ldots, \xi_k)$ can be understood as a generating function of a measure q on $\mathfrak{P}(\{0, 1, \ldots\}^k)$; thus for all $[n_1, \ldots, n_k]$ in $\{0, 1, \ldots\}^k$,

$$q(\{[n_1, \ldots, n_k]\}) = \begin{cases} (n_1)^{-1}\, V_{X_1, \ldots, X_k}(n_1, \ldots, n_k), & \text{if } n_1 > 0 \\ 0, & \text{if } n_1 = 0. \end{cases}$$

Hence in particular $q(\{[0, \ldots, 0]\}) = 0$. All convolution powers q^n, $n \geqq 0$, are meaningful, although we have not yet shown, that the measure q is finite.

From

$$U(\xi_1, \ldots, \xi_k) = U(0, \ldots, 0) \exp \big(Q(\xi_1, \ldots, \xi_k)\big)$$

it is concluded that

$$P_{X_1,\ldots,X_k} = P_{X_1,\ldots,X_k}(\{[0, \ldots, 0]\}) \sum_{n=0}^{\infty} \frac{1}{n!} q^n.$$

Therefore the measure q is finite, and we obtain

$$P_{X_1,\ldots,X_k} = e^{-q(\{0,1,\ldots\}^k)} \sum_{n=0}^{\infty} \frac{1}{n!} q^n.$$

Consequently the distribution $P_{X_1,\ldots,X_k}$ is infinitely divisible. Hence also $P_{X_2,\ldots,X_k}$ is infinitely divisible. Since, however, any finite sequence of sets in $\mathfrak{B}$ can be substituted for $X_2, \ldots, X_k$, we can use 1.6.3. to conclude that $P \in \boldsymbol{T}$.

4. Now we shall prove that V is uniquely determined by P.

Let X be a set in $\mathfrak{B}^x$, and n a natural number. We denote by $\mathfrak{B}_{X,n}$ the system of those Z in the σ-algebra $\mathfrak{A}^x \otimes \mathfrak{M}$ for which the value of $V\big(x \in X, \Phi^x(X) = n, [x, \Phi] \in Z\big)$ is uniquely determined by P. By virtue of $(*)$, all of the sets

$$C \cap \big(X_0 \times \{\Phi : \Phi \in M, \Phi(X_i) = n_i \quad \text{for} \quad 1 \leqq i \leqq k\}\big)$$

where $X_0 \in \mathfrak{B}^x$, $X_1 = X_0 \times K$, $X_2, \ldots, X_k \in \mathfrak{B}$, and $n_1, \ldots, n_k \geqq 0$, are included in $\mathfrak{B}_{X,n}$, for we have

$$V\big(x \in X, \Phi(X \times K) = n, x \in X_0, \Phi(X_i) = n_i \quad \text{for} \quad 1 \leqq i \leqq k\big)$$

$$= \sum_{m=0}^{\infty} V_{(X \cap X_0) \times K, X \times K, X_1, \ldots, X_k}(m, n, n_1, \ldots, n_k).$$

The sets of the form

$$X_0 \times Y, \quad (X_0 \in \mathfrak{B}^x, \; Y \in \mathfrak{M}_{X_2,\ldots,X_m}$$

belong to $\mathfrak{B}_{X,n}$ constitute a semi-ring generating $\mathfrak{A}^x \otimes \mathfrak{M}$. Hence $\mathfrak{B}_{X,n}$ includes the generated subring. On the other hand $\mathfrak{B}_{X,n}$ is monotone closed. Hence $\mathfrak{B}_{X,n} = \mathfrak{A}^x \otimes \mathfrak{M}$.

For all Z in the σ-algebra $\mathfrak{A}^x \otimes \mathfrak{M}$ and all X in $\mathfrak{B}^x$,

$$V(x \in X, [x, \Phi] \in Z) = \sum_{n=1}^{\infty} V\big(x \in X, \Phi(X \times K) = n, [x, \Phi] \in Z\big).$$

Hence also $V(x \in X, [x, \Phi] \in Z)$ is uniquely determined by P. Now, if we let X_n run through a monotone increasing sequence of sets in $\mathfrak{B}^x$ having the union A^x, we find that

$$V([x, \Phi] \in Z) = \sup_{n=1,2,\ldots} V(x \in X_n, [x, \Phi] \in Z).$$

Hence all of the values $V([x, \Phi] \in Z)$, $Z \in \mathfrak{A}^x \otimes \mathfrak{M}$, are uniquely determined by P.

5. It remains to show that the functional equation $\mathscr{C}_P = \mathscr{C}_{\tilde{P}} \circledast V$ is satisfied for all infinitely divisible distributions P on $\mathfrak{M}$. But this is just what we stated in 5.2.4. ∎

If V is a σ-finite measure on $\mathfrak{C}$, the functional equation $\mathscr{C}_P = V \circledast P$ can have at most one solution in $\boldsymbol{P}$. Indeed, if we assume that $\mathscr{C}_{P_1} = V \circledast P_1$, $\mathscr{C}_{P_2} = V \circledast P_2$, then using 5.2.5. we find that $\mathscr{C}_{\tilde{P}_1} = V = \mathscr{C}_{\tilde{P}_2}$, from which by 5.1.10. and 2.2.1. we can conclude that $P_1 = P_2$.

In view of 5.1.12., and using 2.5.1., 2.5.3., 2.2.13., and 2.2.15., we observe the truth of

5.2.6. Proposition. *For each σ-finite measure V on $\mathfrak{C}$, the functional equation $\mathscr{C}_P = V \circledast P$ has at most one solution in $\boldsymbol{P}$. This solution is*

regular infinitely divisible		$V\big(\Omega(c)$ *is infinite*$\big) = 0$
singular infinitely divisible	*if and*	$V\big(\Omega(c)$ *is finite*$\big) = 0$
free from after-effects	*only if*	$V\big((\Omega(c))^* (A) \neq 1\big) = 0$
Poisson		$V\big((\Omega(c))(A) \neq 1\big) = 0.$

5.3. Symmetry Properties of Campbell Measures

Let Y be a set in $\mathfrak{M}$, and $[x, \Phi] \curvearrowright \Gamma_{[x,\Phi]}$ a mapping of $C \cap (A^x \times Y)$ into the set of all distributions on $\mathfrak{C}$. We assume that:

(1) For all Φ in $Y \setminus \{o\}$ and all x in A^x having the property that $\Phi^x(\{x\}) > 0$,

$$\Gamma_{[x,\Phi]}\big(C \cap (A^x \times \{\Phi\})\big) = 1.$$

Thus Γ describes a random translation in $C \cap (A^x \times Y)$, which leaves the second coordinate unchanged. In view of (1) we have, for all Φ in $Y \setminus \{o\}$ and all X in A^x with the property that $\Phi^x(\{x\}) > 0$

$$\Gamma_{[x,\Phi]} = \sum_{x' \in A^x, \Phi^x(\{x'\}) > 0} \gamma_{x,x'}(\Phi)\, \delta_{[x',\Phi]},$$

so that we can associate with Φ the stochastic matrix $\big(\gamma_{x,x'}(\Phi)\big)$ with the at most countable phase space $\big\{x : x \in A^x, \Phi^x(\{x\}) > 0\big\}$.

Let us now assume that

$$[x, \Phi] \curvearrowright i(x, \Phi)$$

be a measurable indexing relation. Instead of $\big(\gamma_{x,x'}(\Phi)\big)$ we now have the stochastic matrix

$$\big(\gamma_{i,j}(\Phi)\big) = \big(\gamma_{z_i(\Phi),z_j(\Phi)}(\Phi)\big)$$

with the phase space I_Φ, and for all Φ in $Y \setminus \{o\}$ and all i in I_Φ, we have

$$\Gamma_{[z_i(\Phi),\Phi]} = \sum_{j \in I_\Phi} \gamma_{i,j}(\Phi)\, \delta_{[z_j(\Phi),\Phi]}.$$

In what follows, we shall need the following measurability property:

(2) For all Z in $\mathfrak{C}$, the real function

$$c \rightsquigarrow \Gamma_c(Z) \qquad \big(c \in C \cap (A^x \times Y)\big)$$

is measurable with respect to $\mathfrak{C}$.

This property can be expressed in a quite simple way by means of the matrix elements $\gamma_{i,j}(.)$:

5.3.1. *Given any measurable indexing relation* $[x, \Phi] \rightsquigarrow i(x, \Phi)$, *then* Γ *satisfies the condition* (2) *if and only if the real function* $\gamma_{i,j}(.)$ *defined on* $\{\Phi: \Phi \in Y \setminus \{o\}; i, j \in I_\Phi\}$ *is measurable with respect to* $\mathfrak{M}$ *for all integers* i, j.

Proof. 1. By assumption the range B of the mapping $[x, \Phi] \rightsquigarrow [i(x, \Phi), \Phi]$ belongs to $\mathfrak{P}(G) \otimes \mathfrak{M}$. Hence, for all i, j in G, the sections $\{\Phi : \Phi \in M, [i, \Phi] \in B\}$, $\{\Phi : \Phi \in M, [j, \Phi] \in B\}$ belong to $\mathfrak{M}$, so that we obtain

$$\begin{aligned}&\{\Phi : \Phi \in Y \setminus \{o\}; i, j \in I_\Phi\} \\ &= \{\Phi : \Phi \in M, [i, \Phi] \in B\} \cap \{\Phi : \Phi \in M, [j, \Phi] \in B\} \cap Y \in \mathfrak{M}.\end{aligned}$$

2. Let us first assume that Γ satisfy the condition (2). Since

$$\gamma_{l,s}(\Phi) = \Gamma_{[z_l(\Phi),\Phi]}\big(i(x, \Phi) = s\big) \qquad (\Phi \in Y \setminus \{o\};\ l,\ s \in I_\Phi)$$

the mapping

$$\Phi \rightsquigarrow [l, \Phi] \rightsquigarrow [z_l(\Phi), \Phi] \rightsquigarrow \Gamma_{[z_l(\Phi),\Phi]}\big(i(x, \Phi) = s\big) = \gamma_{l,s}(\Phi)$$

from M into $[0, 1]$, which is defined on $\{\Phi : \Phi \in Y \setminus \{o\};\ l,\ s \in I_\Phi\}$, is measurable with respect to $\mathfrak{M}$ for all l, s in G, because

$$\{[x, \Psi] : [x, \Psi] \in C \cap (A^x \times Y), i(x, \Psi) = s\}$$

is in $\mathfrak{C}$.

3. Conversely, suppose that all of the mappings

$$\Phi \rightsquigarrow \gamma_{i,j}(\Phi) \qquad (\Phi \in Y \setminus \{o\}; i, j \in I_\Phi)$$

from M into $[0, 1]$ be measurable with respect to $\mathfrak{M}$.

For all X in $\mathfrak{A}^x$, all Y' in $\mathfrak{M}$, and $[x, \Phi] \in C \cap (A^x \times Y)$ we obtain

$$\begin{aligned}\Gamma_{[x,\Phi]}\big(C \cap (X \times Y')\big) &= \sum_{j \in I_\Phi} \gamma_{i(x,\Phi),j}(\Phi)\, \delta_{[z_j(\Phi),\Phi]}(X \times Y') \\ &= \Big(\sum_{j \in I_\Phi} \gamma_{i(x,\Phi),j}(\Phi)\, k_X\big(z_j(\Phi)\big)\Big)\, k_{Y'}(\Phi).\end{aligned}$$

By assumption the expression in parentheses on the right-hand side is a measurable function of $[x, \Phi]$. Hence

$$[x, \Phi] \rightsquigarrow \Gamma_{[x,\Phi]}\big(C \cap (X \times Y')\big) \qquad \big([x, \Phi] \in C \cap (A^x \times Y)\big)$$

is measurable with respect to $\mathfrak{C}$, which leads to the conclusion that (2) is satisfied. ∎

In what follows, we shall need only those random translations Γ which satisfy, in addition to (1) and (2), the following condition (3):

(3) For all Φ in $Y \setminus \{o\}$ and all x in A^x having the property that $\Phi^x(\{x\}) > 0$

$$\sum_{x' \in A^x, \Phi^x(\{x'\}) > 0} \Phi^x(\{x'\}) \, \gamma_{x',x}(\Phi) = \Phi^x(\{x\}).$$

Therefore the restriction of Φ^x to $\mathfrak{P}\big(\{x : x \in A^x, \Phi^x(\{x\}) > 0\}\big)$ is a measure invariant with respect to $\big(\gamma_{x,x'}(\Phi)\big)$. If in particular Φ^x is simple, then (3) says that $\big(\gamma_{x,x'}(\Phi)\big)$ is doubly stochastic.

Let $\Gamma(Y)$ denote the set of all Γ having the properties (1), (2), and (3). A Γ in $\Gamma(Y)$ is called *irreducible* if, for all Φ in $Y \setminus \{o\}$, the validity of

$$\sum_{x \in A^x, \Phi^x(\{x\}) > 0} q_x \gamma_{x,x'}(\Phi) = q_{x'} \qquad \big(x' \in A^x, \Phi^x(\{x'\}) > 0\big)$$

as stated for a family $(q_x)_{x \in A^x, \Phi^x(\{x\}) > 0}$ of non-negative numbers implies that there exists a $v \geqq 0$, such that

$$q_x = v\Phi^x(\{x\}) \qquad \big(x \in A^x, \Phi^x(\{x\}) > 0\big).$$

For all measures V on $\mathfrak{C}$ having the property that $V\big(C \setminus (A^x \times Y)\big) = 0$, and all Γ in $\Gamma(Y)$, we may use (2) to introduce a measure ΓV on $\mathfrak{C}$ setting

$$\Gamma V = \int \Gamma_c(.) \, V(\mathrm{d}c).$$

5.3.2. Proposition. *Let Y belong to $\mathfrak{M}$, and Γ be an irreducible element of $\Gamma(Y)$. Then a measure V on $\mathfrak{C}$ is the Campbell measure of a σ-finite measure H on $\mathfrak{M}$ satisfying the condition $H(\Phi \notin Y, \Phi \neq o) = 0$, if and only if V has the properties*

$$V\big(C \setminus (A^x \times Y)\big) = 0, \; V \textit{ is } \sigma\textit{-finite}, \qquad \Gamma V = V.$$

Proof. 1. If V is the Campbell measure of a σ-finite measure H on $\mathfrak{M}$ having the property that $H(\Phi \notin Y, \Phi \neq o) = 0$, then by 5.1.12. we have

$$V\big(C \setminus (A^x \times Y)\big) = 0,$$

and it follows from 5.1.11. that V is a σ-finite measure.

2. It can be seen from (1) and (3) that, for all Φ in $Y \setminus \{o\}$ and all Γ in $\Gamma(Y)$,

$$\begin{aligned}
\Gamma\mathring{\Phi} &= \Gamma\Big(\sum_{x \in A^x, \Phi^x(\{x\}) > 0} \Phi^x(\{x\}) \, \delta_{[x,\Phi]}\Big) = \sum_{x \in A^x, \Phi^x(\{x\}) > 0} \Phi^x(\{x\}) \, \Gamma_{[x,\Phi]} \\
&= \sum_{\substack{x, x' \in A^x \\ \Phi^x(\{x\}), \Phi^x(\{x'\}) > 0}} \Phi^x(\{x\}) \, \gamma_{x,x'}(\Phi) \, \delta_{[x',\Phi]} \\
&= \sum_{x' \in A^x, \Phi^x(\{x'\}) > 0} \Phi^x(\{x'\}) \, \delta_{[x',\Phi]} = \mathring{\Phi}.
\end{aligned}$$

3. For all indicator functions k_Z of sets Z in $\mathfrak{C}$ and all measures H on $\mathfrak{M}$,

$$\int \left(\int k_Z(c)\, \mathring{\Phi}(\mathrm{d}c) \right) H(\mathrm{d}\Phi) = \int k_Z(c)\, \mathscr{C}_H(\mathrm{d}c).$$

Hence this equation holds also for all finite non-negative linear combinations of such functions, and consequently also for all limits f of monotone increasing sequences (g_n), i.e. it holds for all non-negative real functions f on C which are measurable with respect to $\mathfrak{C}$. In particular, subject to the assumptions of 1. we obtain, for all Z in $\mathfrak{C}$ and all Γ in $\Gamma(Y)$,

$$\begin{aligned}(\Gamma V)(Z) &= (\Gamma \mathscr{C}_H)(Z) = \int \Gamma_c(Z)\, \mathscr{C}_H(\mathrm{d}c) \\ &= \int \left(\int \Gamma_c(Z)\, \mathring{\Phi}(\mathrm{d}c) \right) H(\mathrm{d}\Phi) = \int (\Gamma \mathring{\Phi})(Z)\, H(\mathrm{d}\Phi) \\ &= \int \mathring{\Phi}(Z)\, H(\mathrm{d}\Phi) = \mathscr{C}_H(Z).\end{aligned}$$

This proves that the above-mentioned conditions for the representability of V are necessary.

4. Conversely, suppose that V satisfy these conditions. Let $[x, \Phi] \curvearrowright i(x, \Phi)$ be an arbitrary measurable indexing relation. For all j in G, we set

$$Z_j = \{[x, \Phi] : [x, \Phi] \in C \cap (A^x \times Y),\ i(x, \Phi) = j\}.$$

The family $(Z_i)_{i \in G}$ represents a countable covering of $C \cap (A^x \times Y)$ by pairwise disjoint sets in $\mathfrak{C}$, and we get

$$V = \sum_{i \in G} V\big((.) \cap Z_i\big).$$

On each of the Z_i, the mapping $[x, \Phi] \curvearrowright \Omega(x, \Phi) = \Phi$ is one-to-one and measurable in both directions. Using

$$W_i = V\big(\Omega(c) \in (.), c \in Z_i\big) \qquad (i \in G)$$

we obtain

$$V = \sum_{i \in G} \int \delta_{[z_i(\Phi), \Phi]}(.)\, W_i(\mathrm{d}\Phi),$$

where we have to note that

$$W_i\big(M \setminus \Omega(Z_i)\big) = 0 \qquad (i \in G).$$

Now let (S_n) be a countable covering of C by sets in $\mathfrak{C}$ having the property that $V(S_n) < +\infty$ $(n = 1, 2, \ldots)$. For all i in G,

$$W_i\big(\Omega(S_n \cap Z_i)\big) < +\infty \qquad (n = 1, 2, \ldots), \qquad \bigcup_{n=1}^{\infty} \Omega(S_n \cap Z_i) = \Omega(Z_i).$$

Hence all W_i are σ-finite measure, and it follows that there exists a finite measure W on $\mathfrak{M}$, such that $W(\Phi \notin Y \text{ or } \Phi = o) = 0$ is satisfied and that all W_i are absolutely continuous with respect to W. Now let q_i be a density function of W_i with respect to W. Then $q_i(\Phi) < +\infty$ almost everywhere with respect to W. Moreover q_i can be chosen in such a way that $q_i(\Phi)$ vanishes for $\Phi \notin \Omega(Z_i)$. We obtain

$$\begin{aligned} V = \Gamma V &= \Gamma\Big(\sum_{i\in G}\int q_i(\Phi)\,\delta_{[z_i(\Phi),\Phi]}(.)\;W(\mathrm{d}\Phi)\Big)\\ &= \Gamma\int\Big(\sum_{i\in I_\Phi} q_i(\Phi)\,\delta_{[z_i(\Phi),\Phi]}(.)\Big)\;W(\mathrm{d}\Phi)\\ &= \int\Big(\sum_{i\in I_\Phi} q_i(\Phi)\,\Gamma_{[z_i(\Phi),\Phi]}(.)\Big)\;W(\mathrm{d}\Phi)\\ &= \int\Big(\sum_{i\in I_\Phi} q_i(\Phi)\sum_{j\in I_\Phi}\gamma_{i,j}(\Phi)\,\delta_{[z_j(\Phi),\Phi]}(.)\Big)\;W(\mathrm{d}\Phi)\\ &= \int\Big(\sum_{j\in I_\Phi}\Big(\sum_{i\in I_\Phi} q_i(\Phi)\,\gamma_{i,j}(\Phi)\Big)\,\delta_{[z_j(\Phi),\Phi]}(.)\Big)\;W(\mathrm{d}\Phi). \end{aligned}$$

Hence also $\sum_{i\in I_\Phi} q_i(\Phi)\,\gamma_{i,j}(\Phi)$ can be considered a density of W_j with respect to W. Thus for all j in I_Φ,

$$\sum_{i\in I_\Phi} q_i(\Phi)\,\gamma_{i,j}(\Phi) = q_j(\Phi) < +\infty$$

almost everywhere with respect to W, and therefore

$$q_j(\Phi) = v(\Phi)\,\Phi^x\big(\{z_j(\Phi)\}\big).$$

Now, if we set $v(\Phi) = 0$ for $\Phi \notin Y \setminus \{o\}$, then it follows from 5.1.9. that v is measurable, and we obtain

$$V = \int \mathring{\Phi}(.)\,v(\Phi)\,W(\mathrm{d}\Phi).$$

Hence, if we set

$$H = \int_{(.)} v(\Phi)\,W(\mathrm{d}\Phi),$$

then V is the Campbell measure of H. Since $H(\{o\}) = 0$ we may deduce from 5.1.11. that H is σ-finite, and using 5.1.12. we obtain

$$H(\Phi \notin Y, \Phi \neq o) = 0. \blacksquare$$

Now let be $Y = \{\Phi : \Phi \in M, \Phi(A) < \infty\}$. Setting

$$\Lambda_{[x,\Phi]} = \sum_{x'\in A^x,\,\Phi^x(\{x'\})>0} \frac{\Phi^x(\{x'\})}{\Phi^x(A^x)}\,\delta_{[x',\Phi]} \quad \text{for} \quad [x,\Phi]\in C\cap(A^x\times Y),$$

we obtain an element Λ of $\Gamma(Y)$. It is obvious that the two conditions (1) and (3) are satisfied, and observing that, for all X' in $\mathfrak{B}^x$ and all Y' in $\mathfrak{M}$,

$$\Lambda_{[x,\Phi]}\big(C\cap(X'\times Y')\big) = \frac{\Phi^x(X')}{\Phi^x(A^x)}\,k_{Y'}(\Phi), \qquad [x,\Phi]\in C\cap(A^x\times Y),$$

we see that condition (2) is also satisfied. Λ is an irreducible element of $\Gamma(Y)$. As an immediate consequence of 5.3.2., we now have

5.3.3. Theorem. *A measure V on $\mathfrak{C}$ is the Campbell measure of a σ-finite measure H on $\mathfrak{M}$ satisfying the condition $H(\Phi$ infinite$) = 0$, if and only if V has the properties $V\big(\Omega(c)$ infinite$\big) = 0$, V is σ-finite, $\Lambda V = V$.*

In concluding this section, we consider the special case $[A^x, \varrho_{A^x}] = R^1$, where it is possible to derive most simple characteristic symmetry properties of Campbell measures by means of the order relation in R^1.

We number the points $\{x : x \in R^1, \Phi^x(\{x\}) > 0\}$ according to their order, observing the following calibration rule: The first positive point in $\{x : x \in R^1, \Phi^x(\{x\}) > 0\}$ is given the index 0, the last non-positive point in this set gets the index -1. Thus a well-defined integer $k(x, \Phi)$ is associated with each x in R^1 having the property that $\Phi^x(\{x\}) > 0$.

Instead of $k(x, \Phi) = n$ we also write $x = x_n(\Phi)$.

5.3.4. *The correspondence*

$$[x, \Phi] \rightsquigarrow k(x, \Phi)$$

is a measurable indexing relation.

Proof. We shall first show that the mapping $[x, \Phi] \rightsquigarrow k(x, \Phi)$ of C into G is measurable.

For all natural numbers n and all integers k, let $I_{n,k}$ denote the interval $\big(n^{-1}(k-1), n^{-1}k\big]$. We obtain

$$k(x, \Phi) = \lim_{n\to\infty} k_n(x, \Phi) \qquad ([x, \Phi] \in C)$$

where

$$k_n(x, \Phi) = \begin{cases} \text{Number of all } k\text{'s having the properties} \\ \Phi^x(I_{n,k}) > 0, \qquad k \geqq 1, \qquad n^{-1}k < x \\ \text{for } x > 0 \\ -\big(\text{Number of all } k\text{'s having the properties} \\ \Phi^x(I_{n,k}) > 0, \qquad k \leqq 0, \qquad x < n^{-1}k\big) - 1 \\ \text{for } x \leqq 0. \end{cases}$$

Consequently the one-to-one mapping $[x, \Phi] \rightsquigarrow [k(x, \Phi), \Phi]$ of C into $G \times M$ is measurable with respect to $\mathfrak{C}$, $\mathfrak{P}(G) \otimes \mathfrak{M}$.

The inverse mapping of $[x, \Phi] \rightsquigarrow [k(x, \Phi), \Phi]$ has the form $[n, \Phi] \rightsquigarrow [x_n(\Phi), \Phi]$, and is therefore measurable if and only if all of the mappings

$$\Phi \rightsquigarrow x_n(\Phi) \qquad (\Phi \in \{\Psi : \Psi \in M, n \in I_\Psi\})$$

from M into R^1 are measurable with respect to $\mathfrak{M}$.

Now, if $n \geqq 0$, we obtain for all $y > 0$:

$$\{\Phi : \Phi \in M, x_n(\Phi) < y\} = \{\Phi : \Phi \in M, (\Phi^x)^* ((0, y)) > n\}.$$

Moreover we have

$$\{\Phi : \Phi \in M, x_n(\Phi) \leqq 0\} = \emptyset .$$

On the other hand, if n is negative, then for all $y < 0$ we obtain

$$\{\Phi : \Phi \in M, x_n(\Phi) > y\} = \{\Phi : \Phi \in M, (\Phi^x)^* ((y, 0]) \geqq -n\}.$$

Furthermore,

$$\{\Phi : \Phi \in M, x_n(\Phi) > 0\} = \emptyset .$$

Hence all of the mappings $\Phi \curvearrowright x_n(\Phi)$ are measurable. ∎

We can now readily derive a number of special symmetry criteria. Let

$$Y_1 = \{\Phi : \Phi \in M, \Phi^x \text{ simple}, \Phi^x((-\infty, 0)) = +\infty = \Phi^x((0, +\infty))\}.$$

Then, for all Φ in Y_1, I_Φ coincides with the set G of integers. The matrix

$$(\gamma_{i,j}(\Phi)) = \begin{pmatrix} \cdot & \cdot & & & & & \\ \cdot & \cdot & \cdot & & & & \\ & 0 & 0 & 1 & & & \\ & & 0 & 0 & 1 & & \\ & & & 0 & 0 & 1 & \\ & & & & \cdot & \cdot & \cdot \\ & & & & & \cdot & \cdot \end{pmatrix},$$

in which the elements of the right secondary diagonal are unity, while all other elements are zero, is doubly stochastic and, as a constant, depends measurably on Φ; more precisely each point $x_i(\Phi)$ in $\{x : x \in R^1, \Phi^x(\{x\}) > 0\}$ is transformed into its successor point $x_{i+1}(\Phi)$ in this set.

In this way we obtain a Γ_1 in $\Gamma(Y_1)$. Obviously the non-negative multiples of the "equidistribution" $q_i = 1$ $(i \in G)$ are the only families $(q_i)_{i \in G}$ of non-negative numbers which are invariant with respect to the above matrix. Hence Γ_1 is irreducible.

Now we set

$$Y_2 = \{\Phi : \Phi \in M, \Phi^x((-\infty, 0)) = +\infty = \Phi^x((0, +\infty))\}.$$

We can no longer operate with the constant matrix belonging to Γ_1. Indeed, if Φ^x is not simple, condition (3) might be violated. However, if we generate arbitrary Φ's in Y_2 by putting the loci of points together in elements of Y_1, then this heuristic method yields the matrix

$$\left(\gamma_{i,j}(\Phi)\right) = \begin{pmatrix} \ddots & \ddots & & & \\ & \frac{\Phi^x(\{x_{i-1}(\Phi)\}) - 1}{\Phi^x(\{x_{i-1}(\Phi)\})} & \left(\Phi^x(\{x_{i-1}(\Phi)\})\right)^{-1} & & \\ & \ddots & & & \\ & 0 & \frac{\Phi^x(\{x_i(\Phi)\}) - 1}{\Phi^x(\{x_i(\Phi)\})} & \left(\Phi^x(\{x_i(\Phi)\})\right)^{-1} & \\ & & & & \ddots \\ & & 0 & \frac{\Phi^x(\{x_{i+1}(\Phi)\}) - 1}{\Phi^x(\{x_{i+1}(\Phi)\})} & \\ & & \ddots & & \ddots \end{pmatrix},$$

i.e. the right secondary diagonal contains only the elements $\left(\Phi^x(\{x_i(\Phi)\})\right)^{-1}$. The residual probability $1 - \left(\Phi^x(\{x_i(\Phi)\})\right)^{-1}$ has been shifted to the principal diagonal. Thus, the probability of $x_i(\Phi)$ being transformed into $x_{i+1}(\Phi)$ is $\left(\Phi^x(\{x_i(\Phi)\})\right)^{-1}$, while with probability $1 - \left(\Phi^x(\{x_i(\Phi)\})\right)^{-1}$ the transform is again $x_i(\Phi)$. Obviously also the corresponding Γ_2 satisfies the conditions (1) and (3). In view of 5.3.4., 5.3.1., and 5.1.9., condition (2) is also satisfied. Consequently Γ_2 belongs to $\Gamma(Y_2)$.

The non-negative multiples of $\Phi^x(\{x_i(\Phi)\})$ are the only families $(q_i)_{i\in G}$ of non-negative numbers invariant with respect to the stochastic matrix $\left(\gamma_{i,j}(\Phi)\right)$ given above, i.e. Γ_2 is irreducible.

In the case

$$Y_3 = \{\Phi : \Phi \in M,\ \Phi^x \text{ simple and finite},\ \Phi \neq o\}$$

I_Φ is always a finite interval in G. Then the matrix

$$\left(\gamma_{i,j}(\Phi)\right) = \begin{pmatrix} 0 & 1 & & & & \\ & 0 & 1 & & & \\ & & \ddots & \ddots & & \\ & & & 0 & 1 & \\ & & & & 0 & 1 \\ 1 & & & & & 0 \end{pmatrix}$$

leads to an irreducible Γ_3 in $\Gamma(Y_3)$. Each point in $\{x : x \in R^1,\ \Phi^x(\{x\}) > 0\}$ is transformed into its successor point in this set, if such a point exists. The last point is transformed into the first one. If Φ consists of a single point only, then of course $\left(\gamma_{i,j}(\Phi)\right)$ is the unit matrix (1).

Now, setting

$$Y_4 = \{\Phi : \Phi \in M,\ \Phi^x \text{ finite},\ \Phi \neq o\}$$

we use the heuristic method employed in the transition from Γ_1 to Γ_2, thus obtaining the matrix

$$\left(\gamma_{i,j}(\Phi)\right) = \begin{pmatrix} \frac{\Phi^x(\{x_l(\Phi)\}) - 1}{\Phi^x(\{x_l(\Phi)\})} & \left(\Phi^x(\{x_l(\Phi)\})\right)^{-1} & \\ 0 & \frac{\Phi^x(\{x_{l+1}(\Phi)\}) - 1}{\Phi^x(\{x_{l+1}(\Phi)\})} & \left(\Phi^x(\{x_{l+1}(\Phi)\})\right)^{-1} \\ \left(\Phi^x(\{x_k(\Phi)\})\right)^{-1} & 0 & \frac{\Phi^x(\{x_k(\Phi)\}) - 1}{\Phi^x(\{x_k(\Phi)\})} \end{pmatrix},$$

where we set $I_\Phi = \{l, l+1, \ldots, k\}$. In the same way as above, the unit matrix (1) is associated with a Φ in Y_4 in cases where $\{x : x \in R^1, \Phi^x(\{x\}) > 0\}$ has only one element. In this way we obtain an irreducible element Γ_4 of $\Gamma(Y_4)$.

It turns out that calculations with the symmetry operator Γ_4 are very troublesome. Subject to the assumptions used in this context it is better to use the above defined symmetry operator Λ. The corresponding matrix has the form

$$\left(\gamma_{i,j}(\Phi)\right) = \begin{pmatrix} \frac{\Phi^x(\{x_l(\Phi)\})}{\Phi^x(R^1)} & \cdots & \frac{\Phi^x(\{x_k(\Phi)\})}{\Phi^x(R^1)} \\ \frac{\Phi^x(\{x_l(\Phi)\})}{\Phi^x(R^1)} & \cdots & \frac{\Phi^x(\{x_k(\Phi)\})}{\Phi^x(R^1)} \\ \vdots & & \vdots \\ \frac{\Phi^x(\{x_l(\Phi)\})}{\Phi^x(R^1)} & \cdots & \frac{\Phi^x(\{x_k(\Phi)\})}{\Phi^x(R^1)} \end{pmatrix}$$

where we set $I_\Phi = \{l, l+1, \ldots, k\}$.

Now let

$$Y_0 = Y_2 \cup Y_4.$$

Then the set-up

$$\Theta_{[x,\Phi]} = \begin{cases} (\Gamma_2)_{[x,\Phi]} & \text{for} \quad [x, \Phi] \in C \cap (R^1 \times Y_2) \\ \Lambda_{[x,\Phi]} & \phantom{\text{for}} \quad [x, \Phi] \in C \cap (R^1 \times Y_4) \end{cases}$$

yields an irreducible element Θ of $\Gamma(Y_0)$. As an immediate consequence of 5.3.2., we now have

5.3.5. Theorem. *In the case $\lfloor A^x, \varrho_{A^x}\rfloor = R^1$, a measure V on $\mathfrak{C}$ is the Campbell measure of a σ-finite measure H on $\mathfrak{M}$ satisfying the condition $H(\Phi \notin Y_0, \Phi \neq o) = 0$ if and only if V has the properties*

$$V\left(C \setminus (R^1 \times Y_0)\right) = 0, \qquad V \text{ is } \sigma\text{-finite}, \qquad \Theta V = V.$$

5.4. The Property g_ν. III

In the preceding three sections, we always assumed that the phase space $[A, \varrho_A]$ is the direct product of the "position space" $[A^x, \varrho_{A^x}]$ and the "mark space" $[K, \varrho_K]$. The following considerations are valid only for K consisting of a single element. In this case, however, we may identify $[A, \varrho_A]$ with $[A^x, \varrho_{A^x}]$, i.e. we may leave the mark unspecified, and introduce the Campbell measure $\mathscr{C}_H$ of a measure H on $\mathfrak{M}$ by

$$\mathring{\Phi} = \Phi \times \delta_\Phi, \qquad \mathscr{C}_H = \int \mathring{\Phi}(.)\, H(\mathrm{d}\Phi).$$

Then it is trivial that all statements made in the preceding three sections remain valid.

For the sake of simplifying the notation, in Section 5.1. we dispensed with indicating the underlying product representation of the phase space in the symbol C. For the same reasons we shall continue to use C in the case outlined above, where we do not use any product representation at all. In the present section, we shall consider this case alone.

We set

$$C^! = A \times M, \qquad \mathfrak{C}^! = \mathfrak{A} \otimes \mathfrak{M}.$$

By

$$[a, \Phi] \curvearrowright [a, \Phi - \delta_a],$$

C is mapped onto $C^!$ in a one-to-one way. In this way we eliminate a "redundancy" included in the definition of C in the non-marked case. In view of 1.1.4., the mappings

$$[a, \Phi] \curvearrowright [a, \delta_a, \Phi] \curvearrowright [a, \Phi - \delta_a],$$

$$[a, \Phi - \delta_a] \curvearrowright [a, \delta_a, \Phi - \delta_a] \curvearrowright [a, \Phi]$$

are measurable, i.e. $[a, \Phi] \curvearrowright [a, \Phi - \delta_a]$ is measurable in both directions with respect to $\mathfrak{C}, \mathfrak{C}^!$. For each measure V on $\mathfrak{C}$, we denote by $V^!$ the image of V with respect to $[a, \Phi] \curvearrowright [a, \Phi - \delta_a]$, setting

$$(\mathscr{C}_H)^! = \mathscr{C}_H{}^!$$

which means that

$$\mathscr{C}_H{}^! = \int \Big(\sum_{a \in A, \Phi(\{a\}) > 0} \Phi(\{a\})\, \delta_{[a, \Phi - \delta_a]}(.) \Big)\, H(\mathrm{d}\Phi).$$

In virtue of 2.2.14., for all ν in N we have

$$\mathscr{C}_{\tilde{P}_\nu} = \mathscr{C}_{Q_\nu} = \int \mathring{\Phi}(.)\, Q_\nu(\mathrm{d}\Phi) = \int (\mathring{\delta}_a)\,(.)\, \nu(\mathrm{d}a) = \int \delta_{[a, \delta_a]}(.)\, \nu(\mathrm{d}a).$$

Then by 5.2.5. and 5.2.6. a P in $\boldsymbol{P}$ coincides with the Poisson distribution P_ν if and only if it satisfies the functional equation

$$\mathscr{C}_P = \Big(\int \delta_{[a, \delta_a]}(.)\, \nu(\mathrm{d}a) \Big) \circledast P,$$

i.e. if

$$\mathscr{C}_P = \int \left(\delta_a \times (\delta_{\delta_a} * P)\right)(.)\,\nu(\mathrm{d}a)$$

is valid. This relation is, however, equivalent to

$$\mathscr{C}_P{}^! = \int (\delta_a \times P)(.)\,\nu(\mathrm{d}a) = \left(\int \delta_a(.)\,\nu(\mathrm{d}a)\right) \times P,$$

so that we observe the truth of

5.4.1. Theorem. *For all ν in N, a distribution P on $\mathfrak{M}$ coincides with P_ν if and only if it satisfies the functional equation*

$$\mathscr{C}_P{}^! = \nu \times P.$$

For all distributions ν on $\mathfrak{A}$ and all natural numbers k, by means of 5.2.3. we obtain

$$\begin{aligned}\mathscr{C}^!_{(Q_\nu)^k} &= \left(k\mathscr{C}_{Q_\nu} \circledast (Q_\nu)^{k-1}\right)^! \\ &= k\left(\int \left(\delta_a \times \left(\delta_{\delta_a} * (Q_\nu)^{k-1}\right)\right)(.)\,\nu(\mathrm{d}a)\right)^! \\ &= k\int \left(\delta_a \times (Q_\nu)^{k-1}\right)(.)\,\nu(\mathrm{d}a) = k\left(\nu \times (Q_\nu)^{k-1}\right) = \nu \times k(Q_\nu)^{k-1}.\end{aligned}$$

Now, if P is an arbitrary distribution on $\mathfrak{M}$ which has the property g_ν, then by 1.14.11. we obtain

$$\begin{aligned}\mathscr{C}_P{}^! &= \mathscr{C}^!_{\sum\limits_{k=1}^{\infty} P(\Phi(A)=k)(Q_\nu)^k} = \sum_{k=1}^{\infty} P\left(\Phi(A) = k\right)\mathscr{C}^!_{(Q_\nu)^k} \\ &= \sum_{k=1}^{\infty} kP\left(\Phi(A) = k\right)\left(\nu \times (Q_\nu)^{k-1}\right).\end{aligned}$$

Obviously

$$Q = \sum_{k=1}^{\infty} kP\left(\Phi(A) = k\right)(Q_\nu)^{k-1}$$

is a σ-finite measure, so that we may continue the above chain of equations by

$$= \nu \times Q.$$

For $\nu = o$, only $P = \delta_o$ has the property g_ν, and in this case $\mathscr{C}_P{}^! = \nu \times Q$ is valid for all σ-finite measures Q on $\mathfrak{M}$.

Now let ν be an infinite measure in N, and P a distribution on $\mathfrak{M}$ having the property g_ν. In view of 1.14.8., P is of the form $P = \int P_{l\nu}(.)\,\sigma(\mathrm{d}l)$, so that using 5.4.1. we obtain

$$\begin{aligned}\mathscr{C}_P{}^! &= \left(\int \mathring{\Phi}(.)\,P(\mathrm{d}\Phi)\right)^! = \int \left(\int \mathring{\Phi}(.)\,P_{l\nu}(\mathrm{d}\Phi)\right)^! \sigma(\mathrm{d}l) \\ &= \int (l\nu \times P_{l\nu})(.)\,\sigma(\mathrm{d}l) = \int (\nu \times lP_{l\nu})(.)\,\sigma(\mathrm{d}l).\end{aligned}$$

The measure $Q = \int lP_{l\nu}(.)\,\sigma(\mathrm{d}l)$ is σ-finite, since in view of 1.8.5., for each monotone increasing sequence (X_n) of sets in $\mathfrak{A}$ having the properties

$$0 < \nu(X_n) < +\infty \quad \text{for} \quad n = 1, 2, \ldots, \qquad \bigcup_{n=1}^{\infty} X_n = A$$

the following relation holds

$$Q\left(\lim_{n\to\infty} \frac{\Phi(X_n)}{\nu(X_n)} \text{ does not exist in } R\right)$$

$$= \int lP_{l\nu}\left(\lim_{n\to\infty} \frac{\Phi(X_n)}{\nu(X_n)} \text{ does not exist in } R\right) \sigma(\mathrm{d}l) = 0.$$

Moreover we have, for all $x > 0$,

$$Q\left(\lim_{n\to\infty} \frac{\Phi(X_n)}{\nu(X_n)} \leqq x\right) = \int lP_{l\nu}\left(\lim_{n\to\infty} \frac{\Phi(X_n)}{\nu(X_n)} \leqq x\right) \sigma(\mathrm{d}l)$$

$$= \int lk_{[0,x]}(l)\,\sigma(\mathrm{d}l) \leqq x\sigma([0, x]) < +\infty.$$

Consequently

$$\mathscr{C}_P^! = \nu \times Q.$$

Summarizing, we may state that, for all ν in N, to each P in $\boldsymbol{P}$ having the property g_ν there exists a σ-finite measure Q on $\mathfrak{M}$, such that $\mathscr{C}_P^! = \nu \times Q$ is satisfied. Now it turns out that in this way we are able to characterize the distributions having the property g_ν, i.e. we have

5.4.2. Theorem. *If ν is a measure in N, then a distribution P on $\mathfrak{M}$ has the property g_ν if and only if, for some σ-finite measure Q on $\mathfrak{M}$, the equation*

$$\mathscr{C}_P^! = \nu \times Q$$

is satisfied.

Proof. 1. We have only to show that the condition stated above is sufficient for the property g_ν to occur.

For $\nu = o$ we obtain

$$\mathscr{C}_P^! = o \times Q = 0,$$

and hence

$$\mathscr{C}_P = 0,$$

which means that

$$P(\Phi(A) > 0) = 0.$$

Consequently, in this case the distribution P coincides with δ_o, and hence has the property g_ν. Therefore, from now on we may assume that $\nu \neq o$.

Using 5.1.8., for all bounded closed subsets X of A having the property that $\nu(X) > 0$ and all natural numbers k, we obtain

$$kP(\Phi(X) = k) = \mathscr{C}_P(x \in X, \Phi(X) = k)$$

$$= \int (\delta_a \times (\delta_{\delta_a} * Q))\,(x \in X, \Phi(X) = k)\,\nu(\mathrm{d}x)$$

$$= \int_X Q(\Phi(X) = k - 1)\,\nu(\mathrm{d}x) = \nu(X)\,Q(\Phi(X) = k - 1)$$

and hence

$$Q(\Phi(X) = k - 1) < +\infty.$$

Consequently ${}_XQ$ is a σ-finite measure.

As an immediate consequence of the definitions, we have

$$\mathscr{C}^!_{{}_XP} = {}_X\nu \times {}_XQ.$$

Now, if ${}_XP$ has the property $g_{{}_X\nu,X}$ for all of the X considered above, then P itself has the property g_ν. Therefore, from now on, we may additionally suppose that the phase space $[A, \varrho_A]$ be bounded and that ν be different from o. Since, however, $\mathscr{C}_P{}^! = \nu \times Q$ is equivalent to

$$\mathscr{C}_P{}^! = \nu((.) \mid A) \times (\nu(A)Q),$$

we may even assume that ν be a distribution on $\mathfrak{A}$.

2. From

$$\mathscr{C}_P{}^! = \nu \times Q$$

using 5.1.12. we obtain, for all natural numbers k,

$$\begin{aligned}\mathscr{C}^!_{P(\Phi\in(.),\Phi(A)=k)} &= \left(\int (\delta_a \times (\delta_{\delta_a} * Q)) \left((.) \cap \left(A \times \{\Phi : \Phi \in M, \Phi(A) = k\}\right)\right) \nu(\mathrm{d}a)\right)^! \\ &= \int (\delta_a \times Q) \left((.) \cap \left(A \times \{\Phi : \Phi \in M, \Phi(A) = k - 1\}\right)\right) \nu(\mathrm{d}a) \\ &= \nu \times \left(Q(\Phi \in (.), \Phi(A) = k - 1)\right),\end{aligned}$$

from which, for $P(\Phi(A) = k) > 0$, it can be concluded that

$$\mathscr{C}^!_{P((.)\mid\Phi(A)=k)} = \nu \times \left(P(\Phi(A) = k)\right)^{-1} Q(\Phi \in (.), \Phi(A) = k - 1).$$

Thus it will suffice, if for all distributions ν on $\mathfrak{A}$ and all natural numbers k, we conclude from

$$P(\Phi(A) = k) = 1, \qquad \mathscr{C}_P{}^! = \nu \times Q, \qquad Q(\Phi(A) \neq k - 1) = 0$$

that $P = (Q_\nu)^k$.

We shall prove this by complete induction on k.

In the case $k = 1$, we have

$$\mathscr{C}_P{}^! = \nu \times Q, \qquad Q(\Phi \neq o) = 0$$

consequently

$$\begin{aligned}\mathscr{C}_P &= \int \left(\delta_a \times (\delta_{\delta_a} * Q)\right) (.)\, \nu(\mathrm{d}a) \\ &= Q(\{o\}) \int (\delta_a \times \delta_{\delta_a}) (.)\, \nu(\mathrm{d}a) = Q(\{o\})\, \mathscr{C}_{Q_\nu}\end{aligned}$$

and hence

$$P = Q(\{o\})\, Q_\nu$$

which means that

$$Q(\{o\}) = 1, \qquad P = Q_\nu.$$

Now suppose that our assertion be valid for a natural number k. If

$$P(\Phi(A) = k+1) = 1, \qquad \mathscr{C}_P^! = \nu \times Q, \qquad Q(\Phi(A) \neq k) = 0,$$

then by 5.3.3. and 5.1.2. we obtain

$$\begin{aligned}(k+1)\,\mathscr{C}_P &= (k+1) \int \big(\delta_a \times (\delta_{\delta_a} * Q)\big)\,(.)\,\nu(\mathrm{d}a)\\ &= (k+1)\,\Lambda \int \Big(\int \big(\delta_a \times (\delta_{\delta_a} * \delta_\Phi)\big)\,(.)\,Q(\mathrm{d}\Phi)\Big)\,\nu(\mathrm{d}a)\\ &= (k+1) \int \Big(\int \Big(\Lambda\big(\delta_a \times (\delta_{\delta_a} * \delta_\Phi)\big)\Big)\,(.)\,Q(\mathrm{d}\Phi)\Big)\,\nu(\mathrm{d}a)\\ &= \int \Big(\int \Big(\big(\delta_a \times (\delta_{\delta_a} * \delta_\Phi)\big)\,(.) + \int \big(\delta_y \times (\delta_{\delta_a} * \delta_\Phi)\big)\,(.)\,\Phi(\mathrm{d}y)\Big)\,Q(\mathrm{d}\Phi)\Big)\,\nu(\mathrm{d}a)\\ &= \int \Big(\int \big(\delta_a \times (\delta_{\delta_a} * \delta_\Phi)\big)\,(.)\,Q(\mathrm{d}\Phi)\Big)\,\nu(\mathrm{d}a)\\ &\quad + \int \Big(\int \Big(\int \big(\delta_y \times (\delta_{\delta_a} * \delta_\Phi)\big)\,(.)\,\Phi(\mathrm{d}y)\Big)\,Q(\mathrm{d}\Phi)\Big)\,\nu(\mathrm{d}a)\\ &= \mathscr{C}_P + \int \Big(\int \big(\delta_y \times (\delta_{\delta_a} * \delta_\Phi)\big)\,(.)\,\mathscr{C}_Q(\mathrm{d}[y, \Phi])\Big)\,\nu(\mathrm{d}a),\end{aligned}$$

i.e. we have

$$\begin{aligned}k\mathscr{C}_P &= k \int \big(\delta_a \times (\delta_{\delta_a} * Q)\big)\,(.)\,\nu(\mathrm{d}a)\\ &= \int \Big(\int \big(\delta_y \times (\delta_{\delta_a} * \delta_\Phi)\big)\,(.)\,\nu(\mathrm{d}a)\Big)\,\mathscr{C}_Q(\mathrm{d}[y, \Phi])\\ &= \int \big(\delta_y \times (Q_\nu * \delta_\Phi)\big)\,(.)\,\mathscr{C}_Q(\mathrm{d}[y, \Phi]) = \mathscr{C}_Q \circledast Q_\nu.\end{aligned}$$

From this it follows that

$$(*) \qquad \nu \times (kQ) = k(\nu \times Q) = k\mathscr{C}_P^! = (\mathscr{C}_Q \circledast Q_\nu)^! = \mathscr{C}_Q^! \circledast Q_\nu,$$

where also in this case the operation $\circledast$ is introduced by means of the measurable mapping $\big[[x, \Phi], \Psi\big] \curvearrowright [x, \Phi + \Psi]$ of $C^! \times M$ into $C^!$.

By our assumption, we have

$$k + 1 = (k+1)\,P\big(\Phi(A) = k+1\big) = Q\big(\Phi(A) = k\big) = Q(M),$$

and hence

$$Q(M) = (k+1).$$

We shall now show that in $\boldsymbol{E}^+$ there exists a measure S having the property that $Q = S * Q_\nu$.

As an immediate consequence of $(*)$, we have

$$kQ = \big(\mathscr{C}_Q^!(x \in A, \Phi \in (.))\big) * Q_\nu,$$

so that we may set

$$S = k^{-1}\mathscr{C}_Q^!\big(x \in A, \Phi \in (.)\big).$$

Thus from (*) we obtain

$$\nu \times k(S * Q_\nu) = \mathscr{C}_Q{}^! \circledast Q_\nu,$$

which means that

$$(**) \qquad (\nu \times kS) \circledast Q_\nu = \mathscr{C}_Q{}^! \circledast Q_\nu.$$

Now let X be an arbitrary set in $\mathfrak{A}$. If follows immediately from (**) that

$$\big((\nu \times kS)\,\big(x \in X, \Phi \in (.)\big)\big) * Q_\nu = \big(\mathscr{C}_Q{}^!\big(x \in X, \Phi \in (.)\big)\big) * Q_\nu,$$

so that, using 1.5.8., we may conclude that

$$(\nu \times kS)\,\big(x \in X, \Phi \in (.)\big) = \mathscr{C}_Q{}^!\big(x \in X, \Phi \in (.)\big).$$

Consequently the two finite measures $\nu \times (kS)$, $\mathscr{C}_Q{}^!$ coincide, which means that

$$(***) \qquad \mathscr{C}_Q{}^! = \nu \times (kS)$$

where

$$Q\big(\Phi(A) \neq k\big) = 0, \qquad Q(M) = (k+1), \qquad S\big(\Phi(A) \neq k-1\big) = 0.$$

Hence, in view of our induction hypothesis, we have

$$Q = (k+1)\,(Q_\nu)^k,$$

and substitution into the first equation yields

$$\mathscr{C}_P{}^! = \nu \times (k+1)\,(Q_\nu)^k$$

and consequently

$$P = (Q_\nu)^{k+1}. \ \blacksquare$$

The general theory of Gibbs distributions (cf. e.g. Dobrushin [2], Kozlov [1], [2], Preston [1]) leads to another generalization of the relation $\mathscr{C}_P{}^! = \nu \times P$, namely the assumption of the absolute continuity of $\mathscr{C}_P{}^!$ with respect to $\nu \times P$ (cf. Papangelou [3], [4], Georgii [1], Ngyen & Zessin [2], Ledrappier [1], Kallenberg [9], Matthes, Warmuth & Mecke [1]).

6. Stationary Distributions

6.0. Introduction

If the phase space $[A, \varrho_A]$ is of the form R^s, $s \geqq 1$, then each translation $x \curvearrowright x - y$ transforms a counting measure Φ in M into a new element $T_y\Phi$ in M. All of the mappings T_y of M onto itself are measurable, so that for each measure H on $\mathfrak{M}$ we can establish the measures $H\big(T_y\Phi \in (.)\big)$. If all of these measures coincide with H, then H is called stationary. In virtue of 6.3.1., 6.1.4., and 2.2.9., the table begun in section 2.0. can be continued as follows:

f) P is stationary	$\tilde{P}$ is stationary
g) P is stationary and simple	$\tilde{P}$ is stationary and simple.

Using the multidimensional individual ergodic theorem, we show that for each real function h on M which is integrable with respect to a stationary measure H on $\mathfrak{M}$, the mean value $\bar{h}(\Phi) = \lim\limits_{t\to\infty} (2t)^{-s} \int\limits_{[-t,t)^s} h(T_y\Phi)\, \mu(\mathrm{d}y)$ exists almost everywhere with respect to H (Theorem 6.2.1.). Here as well as in the following, μ denotes the Lebesgue measure on the σ-algebra $\mathfrak{R}^s$ of the Borel subsets of R^s. Using now the usual conclusions, in section 6.2. we infer a number of further statements on $\bar{h}$, including in particular the statistical ergodic theorem 6.2.4. In virtue of 6.3.2., 6.3.5., and 6.3.6. (cf. Kerstan & Matthes [5] and Nawrotzki [2], [3]) we may continue the above table as follows:

h) P is stationary and ergodic.	$\tilde{P}$ is stationary and, almost everywhere with respect to $\tilde{P}$, $\overline{k_{\{\Phi:\Phi\in M,\Phi([0,1)^s)>0\}}}(\Psi) = 0$.
i) P is stationary and mixing.	$\tilde{P}$ is stationary and, for all bounded X_1, X_2 in $\mathfrak{R}^s$ $\lim\limits_{\lVert x\rVert\to\infty} \tilde{P}\big(\Psi(X_1) > 0, \Psi(X_2 + x) > 0\big) = 0$.

For the intensity i_H of a stationary measure H on $\mathfrak{M}$, we understand the number $\int \Phi\big([0, 1)^s\big)\, H(\mathrm{d}\Phi)$. In view of Proposition 6.4.2., ϱ_P coincides with $i_P\mu$. Thus the stationarity of P permits it to reduce the intensity measure ϱ_P to a simpler mathematical entity, the intensity i_P.

In view of the statements made in section 6.2., for all stationary distributions P on $\mathfrak{M}$, the sample intensity $s(\Phi) = \lim_{t\to\infty} (2t)^{-s}\, \Phi\big([-t,t)^s\big)$ of Φ exists almost everywhere with respect to P, and $\int s(\Phi)\, P(\mathrm{d}\Phi) = i_P$.

The translation operators T_y can be introduced even if the phase space $[A, \varrho_A]$ is represented as a direct product of R^s, $s \geqq 1$, and a bounded complete separable "mark space" $[K, \varrho_K]$; in this case we set

$$T_y\Phi = \sum_{[x,k]\in A,\, \Phi(\{[x,k]\})>0} \Phi(\{[x,k]\})\, \delta_{[x-y,k]},$$

i.e. we shift the positions of the marked points in Φ but let the marks unchanged. The statements made in the first three sections of this chapter always refer to this general case. Then, for all stationary measures H on $\mathfrak{M}$, i_H is replaced by the measure γ_H defined by $\gamma_H(L) = \int \Phi\big([0,1)^s \times L\big)\, H(\mathrm{d}\Phi)$ on the σ-algebra $\mathfrak{K}$ of the Borel subsets of $[K, \varrho_K]$. If this measure is σ-finite, then ϱ_H coincides with $\mu \times \gamma_H$ (Proposition 6.1.5.).

6.1. The Translation Operator T_y

Throughout the first three sections of this chapter, we shall assume that the phase space $[A, \varrho_A]$ be the direct product of an Euclidean "position space" R^s, $s \geqq 1$, and a bounded complete separable "mark space" $[K, \varrho_K]$. The σ-algebra $\mathfrak{A}$ then coincides with $\mathfrak{R}^s \otimes \mathfrak{K}$, where $\mathfrak{R}^s$ and $\mathfrak{K}$ denote the σ-algebras of the Borel subsets of R^s and K, respectively.

The addition in R^s opens the possibility of defining the translation operator T_y by

$$T_y\Phi = \sum_{[x,k]\in A,\, \Phi(\{[x,k]\})>0} \Phi(\{[x,k]\})\, \delta_{[x-y,k]} \qquad (\Phi \in M).$$

The measurable mapping

$$[x,k] \curvearrowright [x,k] - y = [x-y, k]$$

of $[A, \mathfrak{A}]$ into itself transforms Φ into the measure $T_y\Phi$, i.e. we have

$$(T_y\Phi)(Z) = \Phi(Z+y) \qquad (\Phi \in M;\quad y \in R^s;\quad Z \in \mathfrak{A}).$$

For all y_1, y_2 in R^s, we have

$$T_{y_2}(T_{y_1}\Phi) = T_{y_1+y_2}\Phi.$$

All mappings T_y, $y \in R^s$, of $[M, \mathfrak{M}]$ onto itself are measurable and hence automorphisms of the measurable space $[M, \mathfrak{M}]$. Using the metric ϱ_M introduced in Section 1.15., we even have

6.1.1. *The mapping*

$$[y, \Phi] \curvearrowright T_y\Phi$$

of $R^s \times M$ into M is continuous.

Proof. 1. In Section 1.15., the metric ϱ_M was introduced by means of an arbitrarily selected point z in A. If we now substitute $z-y$ for this point z, then we obtain a topologically equivalent metric ϱ_M'. Obviously the mapping $\Phi \curvearrowright T_y\Phi$ of $[M, \varrho_M]$ onto $[M, \varrho_M']$ is isometric. Hence, for all y in R^s, the mapping $\Phi \curvearrowright T_y\Phi$ is continuous.

2. Let $y_n \xrightarrow[n\to\infty]{} [0, \ldots, 0]$, $\Phi_n \overset{M}{\Rightarrow} \Phi$. Then, for all natural numbers m and all ε in $(0, 1)$, Φ_n, Φ are ε,m-neighbouring above a certain index n_0. If we now choose n_0 large enough so that the inequality $\|y_n\| < \varepsilon$ is satisfied for all $n \geqq n_0$, then $T_{y_n}\Phi_n, \Phi$ are $2\varepsilon, (m-1)$-neighbouring for all $n \geqq n_0$. Consequently $\varrho_{m-1}(T_{y_n}\Phi_n, \Phi) \xrightarrow[n\to\infty]{} 0$ for all $m = 2, 3, \ldots$, i.e. we have $T_{y_n}\Phi_n \underset{n\to\infty}{\overset{M}{\Rightarrow}} \Phi$.

3. In view of 1. and 2., it follows from $y_n \xrightarrow[n\to\infty]{} y$, $\Phi_n \underset{n\to\infty}{\overset{M}{\Rightarrow}} \Phi$, that

$$T_{y_n}\Phi_n = T_y(T_{y_n-y}\Phi_n) \underset{n\to\infty}{\overset{M}{\Rightarrow}} T_y\Phi. \blacksquare$$

A measure H on $\mathfrak{M}$ is called *stationary* if it is transformed into itself by all of the T_y. Using the uniqueness proposition 1.3.2., we readily obtain

6.1.2. *Let $\mathfrak{S}$ be a generating semi-ring of the ring of all bounded Borel subsets of R^s. Then a distribution P on $\mathfrak{M}$ is stationary if and only if, for all y in R^s, all finite sequences $X_1, \ldots, X_m$ of pairwise disjoint sets in $\mathfrak{S}$, and all finite sequences $L_1, \ldots, L_k$ of pairwise disjoint sets in $\mathfrak{K}$, the relation*

$$P\big(\Phi(X_i \times L_j) = m_{i,j} \text{ for } 1 \leqq i \leqq m,\ 1 \leqq j \leqq k\big)$$
$$= P\big(\Phi((X_i + y) \times L_j) = m_{i,j} \text{ for } 1 \leqq i \leqq m,\ 1 \leqq j \leqq k\big)$$

is satisfied for all non-negative integers $m_{i,j}$.

For $\mathfrak{S}$ especially the system of all direct products $\mathop{\times}\limits_{i=1}^{s} [\alpha_i, \beta_i)$ of bounded half-open intervals $[\alpha_i, \beta_i)$ can be selected.

For all $x = [\xi_1, \ldots, \xi_s]$ in R^s, we set

$$G_x = \left\{\Phi : \Phi \in M, \Phi\left(\left(\mathop{\times}_{i=1}^{s} (\xi_i, +\infty)\right) \times K\right) = 0\right\}.$$

Since $T_yG_x = G_{x-y}$, $P(G_x)$ does not depend on x for all stationary P in $\boldsymbol{P}$. We obtain

$$P\big(\Phi((0, +\infty)^s \times K) < +\infty\big) \leqq P\left(\bigcup_{n=1}^{\infty} G_{[n,\ldots,n]}\right)$$

$$= P(G_{[0,\ldots,0]}) = P\left(\bigcap_{n=0}^{\infty} G_{-[n,\ldots,n]}\right) = P(\Phi = o)$$

thus observing that the following proposition is valid:

6.1.3. *For all stationary distributions P on $\mathfrak{M}$,*

$$P\big(\Phi((0, +\infty)^s \times K) < +\infty, \Phi \neq o\big) = 0.$$

Of course, instead of $(0, +\infty)^s$, we could have used also any other "square" of R^s.

For all finite measures γ on $\mathfrak{K}$, the measure $Q_{\mu\times\gamma}$ is stationary. Here as well as in the following, μ denotes the Lebesgue measure on $\mathfrak{R}^s$. Thus the assertion of 6.1.3. cannot be extended to arbitrary stationary measures on $\mathfrak{M}$.

If, for a stationary distribution P on $\mathfrak{M}$, the inequality $P\big(\Phi(\{a\}) > 0\big) > 0$ is satisfied for some a in A, then we obtain $P\big(\Phi(\{a + y\}) > 0\big) > 0$ for all y in R^s. Using 1.1.5., we thus obtain

6.1.4. *Every stationary distribution on $\mathfrak{M}$ is continuous.*

For all stationary measures H on $\mathfrak{M}$, we use

$$\gamma_H(L) = \int \Phi\big([0, 1)^s \times L\big)\, H(\mathrm{d}\Phi) = \varrho_H\big([0, 1)^s \times L\big) \qquad (L \in \mathfrak{K})$$

to define a measure γ_H on $\mathfrak{K}$.

6.1.5. *If, for a stationary measure H on $\mathfrak{M}$, the measure γ_H is σ-finite, then*

$$\varrho_H = \mu \times \gamma_H.$$

Proof. Let (L_n) be an increasing sequence of sets in $\mathfrak{K}$ having the properties $\bigcup_{n=1}^{\infty} L_n = K$, $\gamma_H(L_n) < +\infty$ for $n = 1, 2, \ldots$ We set

$$\varrho_n = \varrho_H\big((.) \cap (R^s \times L_n)\big).$$

For all L in $\mathfrak{K}$, $\varrho_n\big((.) \times L\big)$ is a translation-invariant measure on $\mathfrak{R}^s$ which has the property that $\varrho_n\big([0, 1)^s \times L\big) = \gamma_H(L \cap L_n)$; hence

$$\varrho_n(X \times L) = \mu(X)\, \gamma_H(L \cap L_n) \qquad (X \in \mathfrak{R}^s,\ L \in \mathfrak{K}).$$

Hence, for all natural numbers n,

$$\varrho_n = \mu \times \big(\gamma_H\big((.) \cap L_n\big)\big) = (\mu \times \gamma_H)\,\big((.) \cap (R^s \times L_n)\big).$$

Given now any set Z in $\mathfrak{A}$, then we obtain

$$\varrho_H(Z) = \sup_{n=1,2,\ldots} \varrho_n(Z) = (\mu \times \gamma_H)\,(Z). \ \blacksquare$$

Hence in particular a stationary measure H on $\mathfrak{M}$ is of finite intensity if and only if the measure γ_H is finite.

Obviously, for all ν in N,

$$P_\nu\big(T_x\Phi \in (.)\big) = P_{\nu((.)+x)}.$$

Consequently a Poisson distribution P_ν on $\mathfrak{M}$ is stationary if and only if ν is translation-invariant, i.e. if the equation $\nu = \mu \times \gamma$ is satisfied for a finite measure γ on $\mathfrak{K}$.

Each stationary Poisson distribution $P_{\mu\times\gamma}$ on $\mathfrak{M}$ is simple and free from after-effects. Conversely, each stationary simple distribution on $\mathfrak{M}$ which is free from after-effects is Poisson, by virtue of 1.11.8. and 6.1.4. Thus we have derived the "classical" characterization of the distributions $P_{\mu \times \gamma}$:

6.1.6. Theorem. *For all stationary distributions P on $\mathfrak{M}$, the following statements are equivalent:*

a) *P is simple and free from after-effects.*
b) *There exists a finite measure γ on $\mathfrak{K}$, such that $P = P_{\mu \times \gamma}$ is satisfied.*

6.2. The Ergodic Theorem

A subset Y of M is called *invariant* if, for all y in R^s and all Φ in Y, the counting measure $T_y\Phi$ belongs to Y. Accordingly a mapping h from M into a set Z, which is defined on a subset D of M, is called *invariant* if D is invariant and for all y in R^s,

$$h(T_y\Phi) = h(\Phi) \qquad (\Phi \in D)$$

is satisfied. Obviously a subset Y of M is invariant if and only if its indicator function k_Y has this property.

Let $\mathfrak{J}$ denote the σ-algebra of all invariant sets in $\mathfrak{M}$. A mapping from M into $[-\infty, +\infty]$ which is measurable with respect to $\mathfrak{M}$ is invariant if and only if it is measurable with respect to $\mathfrak{J}$. A stationary distribution P on $\mathfrak{M}$ is called *ergodic* if each Y in $\mathfrak{J}$ has probability zero or one with respect to P.

In view of 6.1.1., for each real function h defined on M and measurable with respect to $\mathfrak{M}$, the real function

$$[y, \Phi] \curvearrowright h(T_y\Phi) \qquad (y \in R^s, \Phi \in M)$$

is measurable with respect to the σ-algebra $\mathfrak{R}^s \otimes \mathfrak{M}$. If we now suppose in addition that the function h be non-negative, then for all X in $\mathfrak{R}^s$ having the property that $0 < \mu(X) < +\infty$ and for all Φ in M there exists the mean value

$$h_X(\Phi) = \big(\mu(X)\big)^{-1} \int_X h(T_y\Phi)\, \mu(dy).$$

Moreover the mapping $\Phi \curvearrowright h_X(\Phi)$ of M into $[0, +\infty]$ is measurable with respect to $\mathfrak{M}$.

For each measurable non-negative real function h defined on M, we define a mapping $\bar{h}$ from M into $[0, +\infty]$ by

$$\bar{h}(\Phi) = \lim_{n\to\infty} h_{[-n,n)^s}(\Phi)$$

where the left-hand side is defined if and only if the right-hand side exists (possibly in an improper sense). In the special case $s = 1$, $\bar{h}$ can be interpreted as the time average of h. Obviously $\bar{h}$ is measurable with respect to $\mathfrak{M}$.

For all Φ in the domain of $\bar{h}$ and all x in R^s, we obtain for $k \geqq \|x\|$, $n > k$

$$\frac{(2n)^{-s}}{\big(2(n-k)\big)^{-s}} \big(2(n-k)\big)^{-s} \int\limits_{[-(n-k),n-k)^s} h(T_y\Phi)\,\mu(\mathrm{d}y)$$
$$\leqq (2n)^{-s} \int\limits_{[-n,n)^s} h(T_{y+x}\Phi)\,\mu(\mathrm{d}y)$$
$$\leqq \frac{(2n)^{-s}}{\big(2(n+k)\big)^{-s}} \big(2(n+k)\big)^{-s} \int\limits_{[-(n+k),n+k)^s} h(T_y\Phi)\,\mu(\mathrm{d}y)$$

and hence, in view of

$$\lim_{n\to\infty} \frac{\big(2(n-k)\big)^s}{(2n)^s} = 1 = \lim_{n\to\infty} \frac{\big(2(n+k)\big)^s}{(2n)^s}$$

we obtain the relation

$$h_{[-n,n)^s}(T_x\Phi) \xrightarrow[n\to\infty]{} \bar{h}(\Phi),$$

i.e. the function $\bar{h}$ is always invariant.

Using the usual notation $[t]$ for the integral part of t, we obtain for $t \geqq 1$:

$$\frac{(2t)^{-s}}{(2[t])^{-s}} (2[t])^{-s} \int\limits_{[-[t],[t])^s} h(T_y\Phi)\,\mu(\mathrm{d}y) \leqq (2t)^{-s} \int\limits_{[-t,t)^s} h(T_y\Phi)\,\mu(\mathrm{d}y)$$
$$\leqq \frac{(2t)^{-s}}{\big(2([t]+1)\big)^{-s}} \big(2([t]+1)\big)^{-s} \int\limits_{[-([t]+1),[t]+1)^s} h(T_y\Phi)\,\mu(\mathrm{d}y),$$

so that, for all Φ in the domain of $\bar{h}$, it can be concluded that

$$h_{[-t,t)^s}(\Phi) \xrightarrow[t\to\infty]{} \bar{h}(\Phi).$$

Using the multidimensional individual ergodic theorem (cf. Dunford & Schwartz [1], Theorem VIII. 6.9.), we obtain the following proposition concerning the domain of definition of $\bar{h}$:

6.2.1. Theorem. *If H is a stationary measure on $\mathfrak{M}$, then for all non-negative real functions h defined on M and measurable with respect to $\mathfrak{M}$, which have the property that $\int h(\Phi)\,H(\mathrm{d}\Phi) < +\infty$ the limit $\bar{h}(\Phi)$ exists almost everywhere with respect to H.*

Subject to the assumption of 6.2.1., we obtain

$$\int h_{[-t,t)^s}(\Phi)\,H(\mathrm{d}\Phi)$$
$$= (2t)^{-s} \int \Big(\int\limits_{[-t,t)^s} h(T_y\Phi)\,\mu(\mathrm{d}y)\Big)\,H(\mathrm{d}\Phi)$$
$$= (2t)^{-s} \int\limits_{[-t,t)^s} \Big(\int h(T_y\Phi)\,H(\mathrm{d}\Phi)\Big)\,\mu(\mathrm{d}y)$$
$$= (2t)^{-s} \int\limits_{[-t,t)^s} \Big(\int h(\Phi)\,H(\mathrm{d}\Phi)\Big)\,\mu(\mathrm{d}y) = \int h(\Phi)\,H(\mathrm{d}\Phi).$$

Using Fatou's lemma we thus observe that the following proposition is valid:

6.2.2. *Subject to the assumptions of* 6.2.1.,

$$\int \bar{h}(\Phi)\, H(\mathrm{d}\Phi) \leqq \int h(\Phi)\, H(\mathrm{d}\Phi).$$

Hence in particular the value $\bar{h}(\Phi)$ is finite almost everywhere with respect to H.

Now let us additionally suppose that H be in $\boldsymbol{P}$. If moreover the function h is bounded from above, then using the theorem of Lebesgue we obtain

$$\int \bar{h}(\Phi)\, H(\mathrm{d}\Phi) = \lim_{t\to\infty} \int h_{[-t,t)^s}(\Phi)\, H(\mathrm{d}\Phi) = \int h(\Phi)\, H(\mathrm{d}\Phi).$$

If h is not bounded from above, we set, for any $c > 0$,

$$h_c = \min(h, c).$$

Then both $\bar{h}(\Phi)$ and $\overline{h_c}(\Phi)$ exist almost everywhere with respect to H, and $\overline{h_c}(\Phi) \leqq \bar{h}(\Phi)$. On the other hand,

$$\int h_c(\Phi)\, H(\mathrm{d}\Phi) = \int \overline{h_c}(\Phi)\, H(\mathrm{d}\Phi) \leqq \int \bar{h}(\Phi)\, H(\mathrm{d}\Phi).$$

As c increases the left-hand side tends toward $\int h(\Phi)\, H(\mathrm{d}\Phi)$. Thus we have

6.2.3. *For each stationary distribution P on $\mathfrak{M}$ and each non-negative real function h defined on M and measurable with respect to $\mathfrak{M}$, which has the property that $\int h(\Phi)\, P(\mathrm{d}\Phi) < +\infty$*

$$\int \bar{h}(\Phi)\, P(\mathrm{d}\Phi) = \int h(\Phi)\, P(\mathrm{d}\Phi).$$

The assertion of 6.2.3. can be given a more stringent form. First we have

$$\lim_{c\to\infty} \int |\bar{h}(\Phi) - \overline{h_c}(\Phi)|\, P(\mathrm{d}\Phi)$$

$$= \lim_{c\to\infty} \left(\int \bar{h}(\Phi)\, P(\mathrm{d}\Phi) - \int \overline{h_c}(\Phi)\, P(\mathrm{d}\Phi)\right) = 0.$$

Moreover, uniformly in t, we obtain

$$\int \left| (2t)^{-s} \int_{[-t,t)^s} h(T_y\Phi)\, \mu(\mathrm{d}y) - (2t)^{-s} \int_{[-t,t)^s} h_c(T_y\Phi)\, \mu(\mathrm{d}y) \right| P(\mathrm{d}\Phi)$$

$$= \int \left((2t)^{-s} \int_{[-t,t)^s} (h - h_c)\,(T_y\Phi)\, \mu(\mathrm{d}y) \right) P(\mathrm{d}\Phi)$$

$$= \int h(\Phi)\, P(\mathrm{d}\Phi) - \int h_c(\Phi)\, P(\mathrm{d}\Phi).$$

Bearing in mind that, in view of Lebesgue's theorem, for all $c > 0$,

$$\int \left| (2t)^{-s} \int_{[-t,t)^s} h_c(T_y\Phi)\, \mu(\mathrm{d}y) - \overline{h_c}(\Phi) \right| P(\mathrm{d}\Phi) \xrightarrow[t\to\infty]{} 0,$$

we are led to the following statistical ergodic theorem:

6.2.4. Proposition. *Subject to the assumptions of* 6.2.3., *the following relation holds:*

$$\int |h_{[-t,t)^s}(\Phi) - \bar{h}(\Phi)|\, P(\mathrm{d}\Phi) \xrightarrow[t\to\infty]{} 0.$$

If we now suppose in addition that the distribution P be ergodic, then the distribution function of the random variable $\bar{h}(\Phi)$, the latter being measurable with respect to $\mathfrak{J}$, takes only the values zero and one. Hence $\bar{h}$ is constant almost everywhere, i.e. by 6.2.3. it is equal to the statistical mean $\int h(\Phi)\, P(\mathrm{d}\Phi)$ almost everywhere.

Conversely, suppose that for all h satisfying the assumptions of 6.2.3. the equation $\bar{h}(\Phi) = \int h(\Phi)\, P(\mathrm{d}\Phi)$ be valid almost everywhere with respect to P.

Now let Y be an arbitrary set in $\mathfrak{J}$. Then k_Y is invariant. Hence for all Φ in M the limit $\overline{k_Y}(\Phi)$ exists and coincides with $k_Y(\Phi)$. On the other hand $k_Y(\Phi)$ is equal to $\int k_Y(\Phi)\, P(\mathrm{d}\Phi) = P(Y)$ almost everywhere with respect to P, i.e. $P(Y) = 0$ or $P(Y) = 1$. Thus we observe the validity of

6.2.5. *A stationary distribution P on $\mathfrak{M}$ is ergodic if and only if, for all non-negative real functions h defined on M and measurable with respect to $\mathfrak{M}$, which have the property that $\int h(\Phi)\, P(\mathrm{d}\Phi) < +\infty$, the relation*

$$\bar{h}(\Phi) = \int h(\Phi)\, P(\mathrm{d}\Phi)$$

is satisfied almost everywhere with respect to P.

Sometimes it proves useful to characterize the ergodicity of distributions in the following way:

6.2.6. *A stationary distribution P on $\mathfrak{M}$ is ergodic if and only if each representation*

$$P = \alpha P_1 + (1 - \alpha)\, P_2 \qquad (0 \leqq \alpha \leqq 1)$$

of P as a mixture of stationary distributions P_1, P_2 is trivial, i.e. fulfills $\alpha = 0$ or $\alpha = 1$ or $P_1 = P_2$.

Proof. If P is not ergodic, we choose any set Y in $\mathfrak{J}$ having the property that $0 < P(Y) < 1$. The two distributions $P\big((.) \mid Y\big), P\big((.) \mid M \setminus Y\big)$ are stationary. Then

$$P = P(Y)\, P\big((.) \mid Y\big) + \big(1 - P(Y)\big)\, P\big((.) \mid M \setminus Y\big)$$

provides a non-trivial mixture representation of P.

Conversely, suppose that a non-trivial mixture representation $P = \alpha P_1 + (1 - \alpha) P_2$ be given. Obviously P_1 is absolutely continuous with respect to P. Let h be a non-negative real function on M, which is measurable with respect to $\mathfrak{M}$ and has the property that

$$P_1(Y) = \int_Y h(\Phi)\, P(\mathrm{d}\Phi) \qquad (Y \in \mathfrak{M}).$$

Since P_1 is stationary, for all y in R^s the translated function $h(T_y(.))$ can also be considered a density of P_1 with respect to P and therefore coincides with h almost everywhere with respect to P. From this we obtain, for all natural numbers n:

$$\begin{aligned}
&\int |h_{[-n,n)^s}(\Phi) - h(\Phi)|\, P(\mathrm{d}\Phi)\\
&= \int \left| (2n)^{-s} \int_{[-n,n)^s} h(T_y\Phi)\, \mu(\mathrm{d}y) - (2n)^{-s} \int_{[-n,n)^s} h(\Phi)\, \mu(\mathrm{d}y) \right| P(\mathrm{d}\Phi)\\
&\leqq \int \left((2n)^{-s} \int_{[-n,n)^s} |h(T_y\Phi) - h(\Phi)|\, \mu(\mathrm{d}y) \right) P(\mathrm{d}\Phi)\\
&= (2n)^{-s} \int_{[-n,n)^s} \left(\int |h(T_y\Phi) - h(\Phi)|\, P(\mathrm{d}\Phi) \right) \mu(\mathrm{d}y) = 0.
\end{aligned}$$

Hence also $\int |\bar{h}(\Phi) - h(\Phi)|\, P(\mathrm{d}\Phi) = 0$, i.e. the invariant function $\bar{h}$, which is defined for almost all Φ, can also be considered a density of P_1 with respect to P.

Now, if $\bar{h}$ were constant almost everywhere and hence, because $\int \bar{h}(\Phi)\, P(\mathrm{d}\Phi) = \int h(\Phi)\, P(\mathrm{d}\Phi)$, equal to unity almost everywhere, then P_1, and hence also P_2, would coincide with P, so that the mixture representation $P = \alpha P_1 + (1 - \alpha)\, P_2$ would be trivial. Consequently the function $\bar{h}$ being measurable with respect to $\mathfrak{J}$ is not constant almost everywhere, and $\mathfrak{J}$ contains a set satisfying the condition $0 < P(Y) < 1$. Hence P cannot be ergodic. ∎

Using 6.2.6. we get

6.2.7. *If P is a stationary distribution on $\mathfrak{M}$, then there is no stationary E in $\mathbf{E}^+$ different from 0 and having the property that $E(\{o\}) = 0$, for which the distribution $Q = \mathcal{U}_E * P$ is ergodic.*

Proof. Observe that

$$\begin{aligned}
Q &= P * \left(e^{-E(M)} \sum_{n=0}^{\infty} \frac{1}{n!} E^n \right) = e^{-E(M)} \sum_{n=0}^{\infty} \frac{1}{n!} (P * E^n)\\
&= e^{-E(M)}\, P + (1 - e^{-E(M)})\, P * \left(\frac{e^{-E(M)}}{1 - e^{-E(M)}} \sum_{n=1}^{\infty} \frac{1}{n!} E^n \right)
\end{aligned}$$

is a non-trivial mixture representation of Q, since by 1.5.8. $P = P * \delta_o$ is different from

$$P * \left(\frac{e^{-E(M)}}{1 - e^{-E(M)}} \sum_{n=1}^{\infty} \frac{1}{n!} E^n \right). \;∎$$

For stationary distributions, the supposition that $\int h(\Phi)\, P(\mathrm{d}\Phi) < +\infty$ can be dropped in 6.2.1. and 6.2.3.:

6.2.8. *If P is a stationary distribution on $\mathfrak{M}$, then for each non-negative real function h defined on M and measurable with respect to $\mathfrak{M}$, the limit $\overline{h}$ exists almost everywhere with respect to P, and*

$$\int \overline{h}(\Phi)\, P(\mathrm{d}\Phi) = \int h(\Phi)\, P(\mathrm{d}\Phi).$$

Proof. As above, we set $h_c = \min(h, c)$. In view of 6.2.1., $\overline{h_n}$ exists almost everywhere with respect to P. Let D_n denote the domain of definition of $\overline{h_n}$. Then for $D = \bigcap_{n=1}^{\infty} D_n$ we have $P(D) = 1$. For all Φ in D, $\overline{h_1}(\Phi) \leqq \overline{h_2}(\Phi) \leqq \ldots$ By

$$g(\Phi) = \sup_{n=1,2,\ldots} \overline{h_n}(\Phi) \qquad (\Phi \in D)$$

we now obtain a mapping g from M into $[0, +\infty]$ which is measurable with respect to $\mathfrak{I}$.

For all Φ in D, we obtain:

$$\overline{h_n}(\Phi) = \lim_{m\to\infty} (h_n)_{[-m,m)^s}(\Phi) \leqq \varliminf_{m\to\infty} h_{[-m,m)^s}(\Phi)$$

and hence

$$g(\Phi) \leqq \varliminf_{m\to\infty} h_{[-m,m)^s}(\Phi).$$

Consequently, for all Φ in D having the property that $g(\Phi) = +\infty$, it follows that

$$g(\Phi) = \lim_{n\to\infty} h_{[-n,n)^s}(\Phi) = \overline{h}(\Phi).$$

Let $b > 0$, and let Y be the set of those Φ in D, for which $g(\Phi) \leqq b$ is satisfied.

Since

$$\int_{[-n,n)^s} k_Y(T_y\Phi)\, h_l(T_y\Phi)\, \mu(\mathrm{d}y) = k_Y(\Phi) \int_{[-n,n)^s} h_l(T_y\Phi)\, \mu(\mathrm{d}y)$$

the limit $\overline{k_Y h_l}\,(\Phi)$ exists for all Φ in Y and all natural numbers l and is equal to $k_Y(\Phi)\, \overline{h_l}(\Phi)$. Thus, using 6.2.3., we obtain:

$$\int (k_Y h)\,(\Phi)\;\; P(\mathrm{d}\Phi) = \sup_{l=1,2,\ldots} \int (k_Y h_l)\,(\Phi)\, P(\mathrm{d}\Phi)$$

$$= \sup_{l=1,2,\ldots} \int_Y \overline{h_l}(\Phi)\, P(\mathrm{d}\Phi) = \int_Y g(\Phi)\, P(\mathrm{d}\Phi) < +\infty.$$

Hence, by virtue of 6.2.1., the limit $\overline{k_Y h}(\Phi)$ exists almost everywhere on M, i.e. the limit $\overline{h}(\Phi)$ exists almost everywhere on the invariant set Y.

In consideration of the arbitrary selection of b, we observe that $\overline{h}$ exists almost everywhere on $\{\Phi : \Phi \in D, g(\Phi) < +\infty\}$. On the other hand $\overline{h}(\Phi)$ exists for all Φ in $\{\Phi : \Phi \in D, g(\Phi) = +\infty\}$. Hence, by $P(D) = 1$, $\overline{h}(\Phi)$ exists almost everywhere with respect to P.

In the same way as was done in the proof of 6.2.3., we may conclude from $\int h(\Phi)\,P(\mathrm{d}\Phi) = +\infty$ that $\int \bar{h}(\Phi)\,P(\mathrm{d}\Phi) = +\infty$. ∎

By 6.1.1. the real function

$$[y, \Phi] \rightsquigarrow k_{Y_2}(T_{-y}\Phi) \qquad (y \in R^s, \Phi \in M)$$

is measurable for all Y_2 in $\mathfrak{M}$. Thus for any stationary distribution P on $\mathfrak{M}$, any Y_1 in $\mathfrak{M}$ and any $t > 0$ we obtain

$$\int_{Y_1} \Big((2t)^{-s} \int_{[-t,t)^s} k_{Y_2}(T_{-y}\Phi)\,\mu(\mathrm{d}y) \Big) P(\mathrm{d}\Phi)$$

$$= (2t)^{-s} \int_{[-t,t)^s} \Big(\int_{Y_1} k_{Y_2}(T_{-y}\Phi)\,P(\mathrm{d}\Phi) \Big) \mu(\mathrm{d}y) = (2t)^{-s} \int_{[-t,t)^s} P(Y_1 \cap T_y Y_2)\,\mu(\mathrm{d}y).$$

Using 6.2.1. and the theorem of Lebesgue it follows that

$$(2t)^{-s} \int_{[-t,t)^s} P(Y_1 \cap T_y Y_2)\,\mu(\mathrm{d}y) \xrightarrow[t\to\infty]{} \int_{Y_1} \overline{k_{Y_2}}(\Phi)\,P(\mathrm{d}\Phi).$$

Thus, if P is ergodic, by 6.2.5. we have

$$(2t)^{-s} \int_{[-t,t)^s} P(Y_1 \cap T_y Y_2)\,\mu(\mathrm{d}y) \xrightarrow[t\to\infty]{} P(Y_1)\,P(Y_2) \qquad (Y_1, Y_2 \in \mathfrak{M}). \quad (*)$$

Each stationary distribution P with the property $(*)$ is ergodic. For, if Y is in $\mathfrak{J}$, it follows that

$$P(Y) = \lim_{t\to\infty} (2t)^{-s} \int_{[-t,t)^s} P(Y \cap T_y Y)\,\mu(\mathrm{d}y) = \big(P(Y)\big)^2,$$

which means that $P(Y)$ is either zero or one.

Thus we observe the truth of

6.2.9. *A stationary distribution P on $\mathfrak{M}$ is ergodic if and only if*

$$(2t)^{-s} \int_{[-t,t)^s} P(Y_1 \cap T_y Y_2)\,\mu(\mathrm{d}y) \xrightarrow[t\to\infty]{} P(Y_1)\,P(Y_2)$$

holds for all Y_1, Y_2 in $\mathfrak{M}$.

A stationary distribution P on $\mathfrak{M}$ is said to be *mixing*, if for all Y_1, Y_2 in $\mathfrak{M}$ and all sequences (y_n) of points in the R^s which have the property that $\|y_n\| \xrightarrow[n\to\infty]{} +\infty$ the convergence relation

$$P(Y_1 \cap T_{y_n} Y_2) \xrightarrow[n\to\infty]{} P(Y_1)\,P(Y_2)$$

is satisfied, so that, if Y_2 is translated towards infinity, then Y_1, $T_{y_n}Y_2$ become stochastically independent.

By 6.2.9. each mixing stationary distribution P is ergodic.

6.3. Infinitely Divisible Stationary Distributions

Let P be an infinitely divisible distribution on $\mathfrak{M}$, and $\tilde{P} = \sum_{n=1}^{\infty} E_n$ a representation of $\tilde{P}$ as a sum of a sequence of finite measures. For all y in R^s and all Y in $\mathfrak{M}$ we now obtain by means of 2.2.2.:

$$P(T_y\Phi \in Y) = \left(\mathop{\ast}_{n=1}^{\infty} \mathscr{U}_{E_n}\right)(T_y\Phi \in Y) = \left(\mathop{\ast}_{n=1}^{\infty} \mathscr{U}_{E_n(T_y\Phi\in(.))}\right)(Y)$$

$$= \left(\mathscr{E}_{\sum_{n=1}^{\infty} E_n(T_y\Phi\in(.))}\right)(Y) = \left(\mathscr{E}_{\tilde{P}(T_y\Phi\in(.))}\right)(Y),$$

i.e. we have

6.3.1. *For all infinitely divisible distributions P on $\mathfrak{M}$ and all y in R^s,*

$$\widetilde{P(T_y\Phi \in (.))} = \tilde{P}(T_y\Psi \in (.)).$$

Hence in particular P is stationary if and only if $\tilde{P}$ has this property. By virtue of 2.2.3. we have $\gamma_P = \gamma_{\tilde{P}}$.

Since $\{\Psi : \Psi \in M, \Psi(A) < +\infty\}$ as well as $\{\Psi : \Psi \in M, \Psi(A) = +\infty\}$ belong to $\mathfrak{I}$, for all stationary P in $\boldsymbol{T}$ the two measures

$$\widetilde{P_r} = \tilde{P}(\Psi \in (.), \Psi(A) < +\infty), \qquad \widetilde{P_s} = \tilde{P}(\Psi \in (.), \Psi(A) = +\infty)$$

are stationary, i.e. both the regular infinitely divisible part P_r and the singular infinitely divisible part P_s of P are stationary.

For all finite stationary measures W in $\boldsymbol{W}$, the stationary distribution $\mathscr{U}_W$ is a singular infinitely divisible one, for by 6.1.3., we have

$$\widetilde{\mathscr{U}_W}(\Psi(A) < +\infty) = W(\Psi(A) < +\infty)$$

$$= W(\Psi \text{ finite}, \Psi \neq o) = 0.$$

If the convolution of an at most countable family $(P_i)_{i\in I}$ of singular infinitely divisible distributions exists, then it is always singular infinitely divisible, for we have

$$\widetilde{\mathop{\ast}_{i\in I} P_i}\,(\Psi(A) < +\infty) = \sum_{i\in I} \widetilde{P_i}(\Psi(A) < +\infty) = 0.$$

A stationary distribution P on $\mathfrak{M}$ is called a *strongly singular infinitely divisible* one if it can be represented as a convolution of an at most countable family of stationary distributions of the form $\mathscr{U}_E$, $E \in \boldsymbol{E}^+$, $E(\{o\}) = 0$.

As it was stated in section 2.4. for each distribution q on $\mathfrak{R}$ and each measure ω on $\mathfrak{R}$ satisfying the conditions (**) the stationary distribution

$$P = \int P_{l(\mu\times q)}(.)\, \mathscr{E}_\omega(\mathrm{d}l)$$

is infinitely divisible and

$$\tilde{P} = \int P_{l(\mu\times q)}(.)\, \omega(\mathrm{d}l)$$

is satisfied.

All distributions P of this form are strongly singular infinitely divisible. In fact, if $\omega = \sum_{n=1}^{\infty} \nu_n$ is any representation of ω as sum of a sequence of finite measures on $\mathfrak{R}$, then we obtain

$$P = \int P_{l(\mu\times q)}(.)\, \mathscr{E}_{\sum_{n=1}^{\infty}\nu_n}(\mathrm{d}l) = \int P_{l(\mu\times q)}(.) \left(\mathop{\ast}_{n=1}^{\infty} \mathscr{U}_{\nu_n}\right)(\mathrm{d}l)$$

$$= \mathop{\ast}_{n=1}^{\infty} \int P_{l(\mu\times q)}(.)\, \mathscr{U}_{\nu_n}(\mathrm{d}l) = \mathop{\ast}_{n=1}^{\infty} \mathscr{U}_{\int P_{l(\mu\times q)}(.)\nu_n(\mathrm{d}l)}.$$

On the other hand,

$$P(\{o\}) = e^{-\tilde{P}(M)} = e^{-\omega(R)}.$$

Consequently P is of the form $\mathscr{U}_E$, $E \in \boldsymbol{E}^+$, $E(\{o\}) = 0$ if and only if the measure ω is finite. In particular we observe that not all strongly singular infinitely divisible distributions P can be put in the form $P = \mathscr{U}_E$, $E \in \boldsymbol{E}^+$.

By analogy with 2.5.1. and 2.5.3., we have

6.3.2. Proposition. *A stationary infinitely divisible distribution P on $\mathfrak{M}$ is a strongly singular infinitely divisible one if and only if the inequality*

$$\overline{k_{\{\Phi:\Phi\in M,\Phi([0,1)^s\times K)>0\}}}(\Psi) > 0$$

is satisfied almost everywhere with respect to $\tilde{P}$.

Proof. Setting $h = k_{\{\Phi:\Phi\in M,\Phi([0,1)^s\times K)>0\}}$, we obtain a non-negative real function defined on M and measurable with respect to $\mathfrak{M}$, which has the property that

$$\int h(\Psi)\, \tilde{P}(\mathrm{d}\Psi) = \tilde{P}\big(\Psi([0,1)^s \times K) > 0\big) < +\infty.$$

In view of 6.2.1., the limit $\bar{h}(\Psi)$ exists almost everywhere with respect to $\tilde{P}$.

Now, if

$$\bar{h}(\Psi) \geqq c > 0$$

almost everywhere with respect to $\tilde{P}$, we obtain

$$c\tilde{P}(M) \leqq \int \bar{h}(\Psi)\, \tilde{P}(\mathrm{d}\Psi) \leqq \int h(\Psi)\, \tilde{P}(\mathrm{d}\Psi) < +\infty$$

and hence $\tilde{P}(M) < +\infty$, i.e. P is of the form

$$\mathscr{U}_E,\ E \in \boldsymbol{E}^+,\ E(\{o\}) = 0.$$

If $\bar{h}(\Psi) > 0$ almost everywhere with respect to $\tilde{P}$, we set

$$Y_1 = \{\Psi : \Psi \in M, \bar{h}(\Psi) \geqq 1\},$$

$$Y_n = \left\{\Psi : \Psi \in M, \frac{1}{n} \leqq \bar{h}(\Psi) < \frac{1}{n-1}\right\} \qquad (n = 2, 3, \ldots).$$

The sets Y_n belong to $\mathfrak{I}$, and $\tilde{P}\left(M \setminus \bigcup_{n=1}^{\infty} Y_n\right) = 0$. Hence

$$\tilde{P} = \sum_{n=1}^{\infty} \tilde{P}\big((.) \cap Y_n\big).$$

For all n, the inequality $\overline{h}(\Psi) \geqq 1/n$ holds almost everywhere with respect to $\tilde{P}((.) \cap Y_n) = \widetilde{\mathscr{E}_{\tilde{P}((.)\cap Y_n)}}$.

Consequently $\tilde{P}((.) \cap Y_n)(M) < +\infty$, and with

$$P = \mathop{\ast}_{n=1}^{\infty} \mathscr{U}_{\tilde{P}((.)\cap Y_n)}$$

we obtain a representation of P as a convolution of a sequence of stationary distributions of the form $\mathscr{U}_E$, $E \in \boldsymbol{E}^+$, $E(\{o\}) = 0$.

Conversely, suppose that P be a strongly singular infinitely divisible distribution, and hence representable as a convolution $P = \mathop{\ast}_{i \in I} P_i$ of an at most countable family of stationary distributions in $\boldsymbol{T}$ having the property that $\widetilde{P}_i(M) < +\infty$ for all $i \in I$. By 6.2.3. we obtain for each i in I:

$$\begin{aligned} 0 &= \int_{\{\Phi : \overline{h}(\Phi) = 0\}} \overline{h}(\Psi)\, \widetilde{P}_i(\mathrm{d}\Psi) \\ &= \int k_{\{\Phi : \Phi \in M, \overline{h}(\Phi) = 0\}}(\Psi)\, \overline{h}(\Psi)\, \widetilde{P}_i(\mathrm{d}\Psi) \\ &= \int \overline{(k_{\{\Phi : \Phi \in M, \overline{h}(\Phi) = 0\}} h)}\,(\Psi)\, \widetilde{P}_i(\mathrm{d}\Psi) \\ &= \int k_{\{\Phi : \Phi \in M, \overline{h}(\Phi) = 0\}}(\Psi)\, h(\Psi)\, \widetilde{P}_i(\mathrm{d}\Psi) \\ &= \widetilde{P}_i\big(\Psi([0, 1)^s \times K) > 0, \overline{h}(\Psi) = 0\big). \end{aligned}$$

Since $\widetilde{P}_i$ is stationary and $\overline{h}$ is invariant, we obtain

$$0 = \widetilde{P}_i\big(\Psi\big(([0, 1)^s + y) \times K\big) > 0, \overline{h}(\Psi) = 0\big) \qquad (y \in R^s)$$

and hence

$$0 = \widetilde{P}_i(\Psi \neq o, \overline{h}(\Psi) = 0) = \widetilde{P}_i(\overline{h}(\Psi) = 0).$$

Consequently

$$0 = \sum_{i \in I} \widetilde{P}_i(\overline{h}(\Psi) = 0) = \tilde{P}(\overline{h}(\Psi) = 0). \blacksquare$$

A stationary singular infinitely divisible distribution P on $\mathfrak{M}$ is called a *weakly singular infinitely divisible* one if it cannot be represented as a convolution $P = P_1 * P_2$ of a stationary infinitely divisible P_1 with a stationary infinitely divisible P_2 satisfying the condition

$$0 < \widetilde{P}_2(M) < +\infty.$$

By analogy with 6.3.2. we obtain

6.3.3. Proposition. *A stationary singular infinitely divisible distribution P on $\mathfrak{M}$ is a weakly singular infinitely divisible one if and only if the limit*

$$\overline{k_{\{\Phi : \Phi \in M, \Phi([0,1)^s \times K) > 0\}}}(\Psi)$$

vanishes almost everywhere with respect to $\tilde{\boldsymbol{P}}$.

Proof. Let P be a weakly singular infinitely divisible distribution. If for

$$h = k_{\{\Phi:\Phi\in M,\Phi([0,1)^s\times K)>0\}}$$

$\bar{h}(\Psi)$ were not zero almost everywhere with respect to $\tilde{P}$, then by

$$\tilde{P} = \tilde{P}\big(\Psi \in (.), \bar{h}(\Psi) < c\big) + \tilde{P}\big(\Psi \in (.), \bar{h}(\Psi) \geqq c\big)$$

for sufficiently small c we would have obtained a decomposition

$$P = \mathscr{E}_{\tilde{P}(\Psi\in(.),\bar{h}(\Psi)<c)} * \mathscr{E}_{\tilde{P}(\Psi\in(.),\bar{h}(\Psi)\geqq c)}$$

for which, as was shown in the proof of theorem 6.3.2.,

$$0 < \widetilde{\mathscr{E}_{\tilde{P}(\Psi\in(.),\bar{h}(\Psi)\geqq c)}}(M) = \tilde{P}\big(\bar{h}(\Psi) \geqq c\big) < +\infty,$$

which leads to a contradiction.

Conversely, suppose now that P is not a weakly singular infinitely divisible distribution. Then there exists a decomposition $P = P_1 * P_2$ having the afore-mentioned properties, so that we obtain

$$\tilde{P}\big(\bar{h}(\Psi) > 0\big) \geqq \widetilde{P_2}\big(\bar{h}(\Psi) > 0\big) > 0,$$

since P_2 is a strongly singular infinitely divisible distribution, and different from δ_o. ∎

For stationary P in $\boldsymbol{T}$, the results derived above permit refining the decomposition theorem 2.5.4. as follows:

6.3.4. Proposition. *Each stationary infinitely divisible distribution P on $\mathfrak{M}$ can be represented in exactly one way as a convolution*

$$P = P_r * P_{s,1} * P_{s,2}$$

of a stationary regular infinitely divisible P_r, a weakly singular infinitely divisible $P_{s,1}$, and a strongly singular infinitely divisible $P_{s,2}$; where

$$\begin{aligned}
\widetilde{P_r} &= \tilde{P}\big(\Psi \in (.), \Psi(A) < +\infty\big),\\
\widetilde{P_{s,1}} &= \tilde{P}\big(\Psi \in (.), \Psi(A) = +\infty, \bar{h}(\Psi) = 0\big),\\
\widetilde{P_{s,2}} &= \tilde{P}\big(\Psi \in (.), \bar{h}(\Psi) > 0\big),
\end{aligned}$$

setting $h = k_{\{\Phi:\Phi\in M,\Phi([0,1)^s\times K)>0\}}$.

Obviously we have

$$\widetilde{P_r * P_{s,1}} = \tilde{P}\big(\Psi \in (.), \bar{h}(\Psi) = 0\big).$$

Now we are able to derive the following characterization of the ergodic stationary distributions P in $\boldsymbol{T}$:

6.3.5. Theorem. *For stationary infinitely divisible distributions P on $\mathfrak{M}$, the following properties are equivalent:*

a) *For all Y_1, Y_2 in $\mathfrak{M}$,*

$$\lim_{t\to\infty} (2t)^{-s} \int_{[-t,t)^s} |P(Y_1 \cap T_y Y_2) - P(Y_1)\,P(Y_2)|\,\mu(\mathrm{d}y) = 0\,.$$

b) *P is ergodic.*

c) $P_{s,2} = \delta_o$.

d) *P cannot be represented as a convolution of a stationary infinitely divisible P_1 with a distribution of the form $\mathscr{U}_E$, $E \in \mathbf{E}^+$, $E(\{o\}) = 0$, which is different from δ_o.*

e) *For all Y in $\mathfrak{F}$, $\tilde{P}(Y) = 0$ or $\tilde{P}(Y) = +\infty$.*

f) *For all bounded Borel subsets X_1, X_2 of R^s,*

$$\lim_{t\to\infty} (2t)^{-s} \int_{[-t,t)^s} \tilde{P}\big(\Psi(X_1 \times K) > 0, \Psi\big((X_2 - y) \times K\big) > 0\big)\,\mu(\mathrm{d}y) = 0\,.$$

Proof. 1. By 6.2.9. a) implies b).

2. d) is equivalent to c).

For if P has a non-trivial stationary factor $\mathscr{U}_E$, then we have

$$\int \bar{h}(\Psi)\,\tilde{P}(\mathrm{d}\Psi) \geqq \int \bar{h}(\Psi)\,E(\mathrm{d}\Psi) = \int h(\Psi)\,E(\mathrm{d}\Psi)$$
$$= E\big(\Psi([0,1)^s \times K) > 0\big) > 0\,,$$

consequently $\tilde{P}\big(\bar{h}(\Psi) > 0\big) > 0$ and hence $P_{s,2} \neq \delta_o$.

On the other hand, if P does not have a non-trivial stationary factor $\mathscr{U}_E$, this is certainly true for the factor P_s of P, i.e. P_s is a weakly singular infinitely divisible distribution. Thus we obtain $P_{s,2} = \delta_o$.

3. b) implies d).

If P has a non-trivial stationary factor $\mathscr{U}_E$, then by 6.2.7. P cannot be ergodic.

4. d) implies e).

Given a set Y in $\mathfrak{F}$ having the property that $0 < \tilde{P}(Y) < +\infty$, then

$$\tilde{P} = \tilde{P}\big((.) \cap Y\big) + \tilde{P}\big((.) \setminus Y\big),$$

i.e. $P = \mathscr{U}_{\tilde{P}((.)\cap Y)} * \mathscr{E}_{\tilde{P}((.)\setminus Y)}$. Consequently the distribution P has a non-trivial stationary factor $\mathscr{U}_E$.

5. e) implies f).

In virtue of 6.1.1., the real function

$$[y, \Psi] \curvearrowright k_{\{\Phi:\Phi\in M, \Phi(X_2\times K)>0\}}(T_{-y}\Psi)$$

is measurable. Hence

$$y \curvearrowright \int_{\{\Phi:\Phi(X_1\times K)>0\}} k_{\{\Phi:\Phi\in M,\Phi(X_2\times K)>0\}}(T_{-y}\Psi)\,\tilde{P}(\mathrm{d}\Psi)$$
$$= \tilde{P}\big(\Psi(X_1 \times K) > 0, \Psi\big((X_2 - y) \times K\big) > 0\big)$$

is a measurable mapping of R^s into $[0, +\infty)$. Consequently the integrand occurring in f) is measurable.

Now we set $f = k_{\{\Phi:\Phi\in M, \Phi(X_2\times K)>0\}}$. The function f is integrable with respect to $\tilde{P}$, so that in view of 6.2.1. the limit $\bar{f}(\Psi)$ exists almost everywhere. Moreover by 6.2.2. we have $\int \bar{f}(\Psi)\, \tilde{P}(\mathrm{d}\Psi) < +\infty$.

If $\bar{f}(\Psi)$ were not zero almost everywhere, then there would be a $c > 0$ having the property that $\tilde{P}\big(\bar{f}(\Psi) \geqq c\big) > 0$. The set $Y = \{\Psi : \Psi \in M, \bar{f}(\Psi) \geqq c\}$ would belong to $\mathfrak{F}$, and, because of $c\tilde{P}(Y) \leqq \int \bar{f}(\Psi)\, \tilde{P}(\mathrm{d}\Psi)$ the inequality $0 < \tilde{P}(Y) < +\infty$ would be satisfied, which contradicts e).

Consequently, $\bar{f}(\Psi) = 0$ almost everywhere with respect to $\tilde{P}$. Noting that

$$(2t)^{-s} \int\limits_{[-t,t)^s} f(T_y\Psi)\, \mu(\mathrm{d}y) \leqq 1$$

is satisfied uniformly with respect to t and Ψ, we obtain by means of Lebesgue's theorem:

$$\begin{aligned}
0 &= \int\limits_{\{\Phi:\Phi(X_1\times K)>0\}} \bar{f}(\Psi)\, \tilde{P}(\mathrm{d}\Psi) \\
&= \lim_{t\to\infty} (2t)^{-s} \int\limits_{\{\Phi:\Phi(X_1\times K)>0\}} \Bigg(\int\limits_{[-t,t)^s} k_{\{\chi:\chi\in M,\chi(X_2\times K)>0\}}(T_{-y}\Psi)\, \mu(\mathrm{d}y)\Bigg)\, \tilde{P}(\mathrm{d}\Psi) \\
&= \lim_{t\to\infty} (2t)^{-s} \int \Bigg(\int\limits_{[-t,t)^s} k_{\{\Phi:\Phi\in M,\Phi(X_1\times K)>0,\Phi((X_2-y)\times K)>0\}}(\Psi)\, \mu(\mathrm{d}y)\Bigg)\, \tilde{P}(\mathrm{d}\Psi) \\
&= \lim_{t\to\infty} (2t)^{-s} \int\limits_{[-t.t)^s} \Big(\int k_{\{\Phi:\Phi\in M,\Phi(X_1\times K)>0,\Phi((X_2-y)\times K)>0\}}(\Psi)\, \tilde{P}(\mathrm{d}\Psi)\Big)\, \mu(\mathrm{d}y) \\
&= \lim_{t\to\infty} (2t)^{-s} \int\limits_{[-t,t)^s} \tilde{P}\Big(\Psi(X_1\times K) > 0,\ \Psi\big((X_2-y)\times K\big) > 0\Big)\, \mu(\mathrm{d}y).
\end{aligned}$$

6. f) implies a).

Given any positive number c, if Y_1, Y_2 belong to the σ-algebra ${}_{(-c,c)^s\times K}\mathfrak{M}$ then T_yY_2 is in ${}_{((-c,c)^s-y)\times K}\mathfrak{M}$ for all y in R^s. Now we select any $v > 0$, such that $\|y\| \geqq v$ implies $(-c,c)^s \cap \big((-c,c)^s - y\big) = \emptyset$. For $\|y\| \geqq v$ we obtain

$$\begin{aligned}
P(Y_1)\,P(Y_2) &= P(Y_1)\,P(T_yY_2) \\
&= P\big({}_{(-c,c)^s\times K}\Phi \in Y_1\big)\, P\big({}_{((-c,c)^s-y)\times K}\Phi \in T_yY_2\big) \\
&= {}_{(-c,c)^s\times K}P(\Phi_1 \in Y_1)\; {}_{((-c,c)^s-y)\times K}P(\Phi_2 \in T_yY_2) \\
&= \big({}_{(-c,c)^s\times K}P \times {}_{((-c,c)^s-y)\times K}P\big)\, (\Phi_1 \in Y_1,\ \Phi_2 \in T_yY_2) \\
&= \big({}_{(-c,c)^s\times K}P \times {}_{((-c,c)^s-y)\times K}P\big) \\
&\qquad\qquad \big({}_{(-c,c)^s\times K}\Phi_1 \in Y_1,\ {}_{((-c,c)^s-y)\times K}\Phi_2 \in T_yY_2\big) \\
&= \big({}_{(-c,c)^s\times K}P * {}_{((-c,c)^s-y)\times K}P\big)\, (Y_1 \cap T_yY_2).
\end{aligned}$$

Using 1.10.6., 2.2.2., and 2.2.4., we thus conclude that

$$
\begin{aligned}
&|P(Y_1 \cap T_y Y_2) - P(Y_1)\,P(Y_2)| \\
&\leqq \left\| {}_{((-c,c)^s \cup ((-c,c)^s - y)) \times K}P - \left({}_{(-c,c)^s \times K}P * {}_{((-c,c)^s - y) \times K}P\right) \right\| \\
&\leqq 2 \left\| \tilde{P}\left({}_{((-c,c)^s \cup ((-c,c)^s - y)) \times K}\Psi \in (.),\right.\right. \\
&\qquad\qquad \left.\Psi\big((-c,c)^s \times K\big) + \Psi\left(\big((-c,c)^s - y\big) \times K\right) > 0\right) \\
&\qquad\qquad - \tilde{P}\left({}_{(-c,c)^s \times K}\Psi \in (.),\ \Psi\big((-c,c)^s \times K\big) > 0\right) \\
&\qquad\qquad \left.- \tilde{P}\left({}_{((-c,c)^s - y) \times K}\Psi \in (.),\ \Psi\left(\big((-c,c)^s - y\big) \times K\right) > 0\right)\right\| \\
&\leqq 2 \left\| \tilde{P}\left({}_{((-c,c)^s \cup ((-c,c)^s - y)) \times K}\Psi \in (.),\ \Psi\big((-c,c)^s \times K\big) > 0,\right.\right. \\
&\qquad\qquad \left.\Psi\left(\big((-c,c)^s - y\big) \times K\right) = 0\right) \\
&\quad \left.- \tilde{P}\left({}_{(-c,c)^s \times K}\Psi \in (.),\ \Psi\big((-c,c)^s \times K\big) > 0,\ \Psi\left(\big((-c,c)^s - y\big) \times K\right) = 0\right)\right\| \\
&\quad + 2 \left\| \tilde{P}\left({}_{((-c,c)^s \cup ((-c,c)^s - y)) \times K}\Psi \in (.),\ \Psi\big((-c,c)^s \times K\big) = 0,\right.\right. \\
&\qquad\qquad \left.\Psi\left(\big((-c,c)^s - y\big) \times K\right) > 0\right) \\
&\qquad\qquad - \tilde{P}\left({}_{((-c,c)^s - y) \times K}\Psi \in (.),\ \Psi\big((-c,c)^s \times K\big) = 0,\right. \\
&\qquad\qquad \left.\left.\Psi\left(\big((-c,c)^s - y\big) \times K\right) > 0\right)\right\| \\
&\qquad\qquad + 2 \cdot 3\tilde{P}\left(\Psi\big((-c,c)^s \times K\big) > 0,\ \Psi\left(\big((-c,c)^s - y\big) \times K\right) > 0\right) \\
&= 0 + 0 + 6\tilde{P}\left(\Psi\big((-c,c)^s \times K\big) > 0,\ \Psi\left(\big((-c,c)^s - y\big) \times K\right) > 0\right),
\end{aligned}
$$

and in view of f), this gives

$$\lim_{t\to\infty} (2t)^{-s} \int_{[-t,t)^s} |P(Y_1 \cap T_y Y_2) - P(Y_1)\,P(Y_2)|\,\mu(\mathrm{d}y) = 0.$$

Hence a) is true for all Y_1, Y_2 in ${}_{(-c,c)^s \times K}\mathfrak{M}$.

We denote by $\mathfrak{G}$ the algebra $\bigcup_{c>0} {}_{(-c,c)^s \times K}\mathfrak{M}$. For each Y in $\mathfrak{M}$, let $\mathfrak{S}_Y$ denote the system of those sets Y' in $\mathfrak{M}$ for which the relation

$$\lim_{t\to\infty} (2t)^{-s} \int_{[-t,t)^s} |P(Y \cap T_y Y') - P(Y)\,P(Y')|\,\mu(\mathrm{d}y) = 0$$

is satisfied.

Now, if Y' is the union of an increasing sequence (Y_n') of sets in $\mathfrak{S}_Y$, then the following relation holds uniformly with respect to y:

$$\begin{aligned}
&|P(Y \cap T_y Y') - P(Y)\,P(Y')| \\
&\leqq |P(Y \cap T_y Y_n') - P(Y)\,P(Y_n')| + |P(Y \cap T_y Y') - P(Y \cap T_y Y_n')| \\
&\quad + |P(Y)\,P(Y') - P(Y)\,P(Y_n')| \\
&\leqq |P(Y \cap T_y Y_n') - P(Y)\,P(Y_n')| + 2P\big((T_y Y') \setminus (T_y Y_n')\big) \\
&= |P(Y \cap T_y Y_n') - P(Y)\,P(Y_n')| + 2P(Y' \setminus Y_n').
\end{aligned}$$

Consequently, for all natural numbers n,

$$\overline{\lim_{t\to\infty}}\,(2t)^{-s} \int_{[-t,t)^s} |P(Y \cap T_y Y') - P(Y)\,P(Y')|\,\mu(\mathrm{d}y) \leqq 2P(Y' \setminus Y_n'),$$

and hence $Y' \in \mathfrak{S}_Y$. Correspondingly it can be concluded that $\mathfrak{S}_Y$ also includes the intersection of each decreasing sequence of sets in $\mathfrak{S}_Y$. Consequently the set systems $\mathfrak{S}_Y$ are monotone closed.

If Y belong to $\mathfrak{G}$, the monotone closed system $\mathfrak{S}_Y$ includes the algebra $\mathfrak{G}$, and hence also the σ-algebra generated by $\mathfrak{G}$, i.e. $\mathfrak{S}_Y = \mathfrak{M}$.

But for any Y_1, Y_2 in $\mathfrak{M}$, the relation $Y_1 \in \mathfrak{S}_{Y_2}$ is equivalent to $Y_2 \in \mathfrak{S}_{Y_1}$ since we have

$$\begin{aligned}
&\int_{[-t,t)^s} |P(Y_1 \cap T_y Y_2) - P(Y_1)\,P(Y_2)|\,\mu(\mathrm{d}y) \\
&= \int_{[-t,t)^s} |P\big((T_{-y} Y_1) \cap Y_2\big) - P(Y_1)\,P(Y_2)|\,\mu(\mathrm{d}y) \\
&= \int_{[-t,t)^s} |P(Y_2 \cap T_y Y_1) - P(Y_2)\,P(Y_1)|\,\mu(\mathrm{d}y).
\end{aligned}$$

Hence the monotone closed system $\mathfrak{S}_Y$ includes the algebra $\mathfrak{G}$ for all Y in $\mathfrak{M}$, i.e. $\mathfrak{S}_Y = \mathfrak{M}$ for all Y in $\mathfrak{M}$. But this is just the assertion of a). ∎

A stationary P in $\boldsymbol{T}$ is *mixing* if and only if it satisfies a stronger version of the condition f) stated in 6.3.5.:

6.3.6. Theorem. *A stationary infinitely divisible distribution P on $\mathfrak{M}$ is mixing if and only if, for all bounded Borel subsets X_1, X_2 of R^s and for each sequence (y_n) of points in R^s having the property that $\|y_n\| \to +\infty$, the following convergence relation holds:*

$$\tilde{P}\left(\Psi(X_1 \times K) > 0,\ \Psi\big((X_2 - y_n) \times K\big) > 0\right) \xrightarrow[n\to\infty]{} 0.$$

Proof. 1. If P is mixing, then for any X_1, X_2 in R^s,

$$\begin{aligned}
&P\left(\Phi(X_1 \times K) = 0,\ \Phi\big((X_2 - y_n) \times K\big) = 0\right) \\
&\xrightarrow[n\to\infty]{} P\big(\Phi(X_1 \times K) = 0\big)\,P\big(\Phi(X_2 \times K) = 0\big)
\end{aligned}$$

i.e.

$$\lim_{n\to\infty} \exp\left(-\tilde{P}\left(\Psi(X_1 \times K) + \Psi\big((X_2 - y_n) \times K\big) > 0\right)\right)$$
$$= \exp\left(-\tilde{P}\big(\Psi(X_1 \times K) > 0\big)\right) \exp\left(-\tilde{P}\big(\Psi(X_2 \times K) > 0\big)\right)$$

from which, in view of

$$\tilde{P}\left(\Psi(X_1 \times K) + \Psi\big((X_2 - y_n) \times K\big) > 0\right)$$
$$= \tilde{P}\big(\Psi(X_1 \times K) > 0\big) + \tilde{P}\left(\Psi\big((X_2 - y_n) \times K\big) > 0\right)$$
$$- \tilde{P}\left(\Psi(X_1 \times K) > 0,\ \Psi\big((X_2 - y_n) \times K\big) > 0\right)$$

the necessity of the condition stated above can be concluded.

2. Conversely, suppose that P satisfy these conditions. Just as in the first part of the 6th step in the proof of 6.3.5., we obtain for all Y_1, Y_2 in the algebra $\mathfrak{G}$:

$$|P(Y_1 \cap T_{y_n} Y_2) - P(Y_1)\, P(Y_2)| \xrightarrow[n\to\infty]{} 0.$$

The conclusions drawn at the end of that 6th step show that the assertion on the mixing behaviour is then valid for all Y_1, Y_2 in $\mathfrak{M}$. ∎

The conclusions drawn in the above proof allow to give a stronger statement. If the assumptions of Theorem 6.3.6. are fulfilled then P is r-mixing for all natural numbers r, i.e. for all sequences $Y_0, \ldots, Y_r$ of sets in $\mathfrak{M}$ and all sequences $\{[t_{1,n}, \ldots, t_{r,n}]\}_{n=1,2,\ldots}$ of r-tuples of real numbers with the properties $\lim_{n\to\infty} \|t_{i,n} - t_{k,n}\| = +\infty$ if $i \neq k$, $i, k = 0, \ldots, r$ and $t_{0,n} = 0$ for $n = 1, 2, \ldots$ the relation

$$\lim_{n\to\infty} P(Y_0 \cap T_{t_{1,n}} Y_1 \cap \cdots \cap T_{t_{r,n}} Y_r) = P(Y_0) \cdots P(Y_r)$$

is satisfied for all $n = 1, 2, \ldots$ (cf. Nawrotzki [2]).

Using 6.3.6. we get

6.3.7. Proposition. *All regular infinitely divisible stationary distributions P on $\mathfrak{M}$ are mixing.*

Proof. We have

$$\tilde{P}\big(\Psi(X_1 \times K) > 0\big) = \tilde{P}\big(\Psi(X_1 \times K) > 0,\ \Psi(A) < +\infty\big)$$
$$= \sup_{m=1,2,\ldots} \tilde{P}\left(\Psi(X_1 \times K) > 0,\ \Psi\Big(\big(R^s \setminus S_m([0, \ldots, 0])\big) \times K\Big) = 0\right),$$

where $S_m([0, \ldots, 0])$ denotes the open ball around $[0, \ldots, 0]$ having the radius m.

For all natural m, we obtain

$$\overline{\lim_{n\to\infty}}\, \tilde{P}\left(\Psi(X_1 \times K) > 0,\ \Psi\big((X_2 - y_n) \times K\big) > 0\right)$$
$$\leqq \tilde{P}\left(\Psi(X_1 \times K) > 0,\ \Psi\Big(\big(R^s \setminus S_m([0, \ldots, 0])\big) \times K\Big) > 0\right)$$
$$+ \overline{\lim_{n\to\infty}}\, \tilde{P}\Big(\Psi(X_1 \times K) > 0,\ \Psi\big((X_2 - y_n) \times K\big) > 0,$$
$$\Psi\Big(\big(R^s \setminus S_m([0, \ldots, 0])\big) \times K\Big) = 0\Big).$$

For sufficiently large n, however,

$$\left\{\Psi : \Psi \in M,\ \Psi\big((X_2 - y_n) \times K\big) > 0,\ \Psi\Big(\big(R^s \setminus S_m([0, \ldots, 0])\big) \times K\Big) = 0\right\} = \emptyset,$$

so that we obtain:

$$\varlimsup_{n\to\infty} \tilde{P}\left(\Psi(X_1 \times K) > 0,\ \Psi\big((X_2 - y_n) \times K\big) > 0\right)$$
$$\leq \inf_{m=1,2,\ldots} \tilde{P}\left(\Psi(X_1 \times K) > 0,\ \Psi\Big(\big(R^s \setminus S_m([0, \ldots, 0])\big) \times K\Big) > 0\right) = 0. \blacksquare$$

In section 9.6. we shall give, on the basis of the renewal theorem, examples of singular infinitely divisible mixing (and hence weakly singular infinitely divisible), stationary distributions on $\mathfrak{M}$.

6.4. The Sample Intensity $s(\Phi)$

The statements made in the preceding three sections can readily be transferred to the "non-marked" case, where the phase space $[A, \varrho_A]$ is of the form R^s, $s \geqq 1$. In this section, we shall consider this case alone.

For all y in R^s and all Φ in M, we now set

$$T_y\Phi = \sum_{x \in R^s, \Phi(\{x\})>0} \Phi(\{x\})\, \delta_{x-y},$$

introducing the concept of the stationary measure just as it was done in section 6.1. Then, in view of 1.4.7., the stationarity criterion 6.1.2. can be modified for simple distributions in the following way:

6.4.1. *Let $\mathfrak{S}$ be a generating subring of $\mathfrak{B}$, and P a simple distribution on $\mathfrak{M}$. Subject to these assumptions, P is stationary if and only if the relation*

$$P\big(\Phi(S) = 0\big) = P\big(\Phi(S + y) = 0\big)$$

is satisfied for all S in $\mathfrak{S}$ and all y in R^s.

For stationary measures H on $\mathfrak{M}$, γ_H is now replaced by the possibly infinite number

$$i_H = \int \Phi\big([0, 1)^s\big)\, H(\mathrm{d}\Phi) = \varrho_H\big([0, 1)^s\big),$$

which is called the *intensity* of H. A stationary measure H on $\mathfrak{M}$ is of finite intensity if and only if i_H is finite, and, in view of 6.1.5., in this case $\varrho_H = i_H\mu$. This statement can be generalized for stationary distributions as follows:

6.4.2. Proposition. *For all stationary distributions P on $\mathfrak{M}$,*

$$\varrho_P(X) = i_P\mu(X) \qquad (X \in \mathfrak{R}^s),$$

where we have to set $\infty \cdot 0 = 0 \cdot \infty = 0$.

Proof. 1. To each Φ in M and to each natural number k we associate the "thinned" counting measure

$$\Phi_k = \sum_{\substack{x \in R^s, \Phi(\{x\})>0 \\ \Phi(x+(-2,2)^s) \leqq k}} \Phi(\{x\})\, \delta_x .$$

Then, for all k, $\Phi \curvearrowright \Phi_k$ provides a measurable mapping of $[M, \mathfrak{M}]$ into itself.

By definition of $\mathfrak{M}$, it will suffice to show the measurability of the real functions $\Phi \curvearrowright \Phi_k(X)$ for all X in $\mathfrak{B}$. To this end first we select, for each natural number m, a countable covering $(X_{m,n})_{n=1,2,\ldots}$ of R^s by pairwise disjoint sets in $\mathfrak{R}^s$, the diameters of which do not exceed $1/m$. For all m, n, let $I_{m,n}$ denote the set of those l, for which all x in $X_{m,n}$ satisfy the condition

$$X_{m,l} \subseteqq x + (-2,2)^s .$$

Now we set

$$f_{m,n}(\Phi) = \begin{cases} \Phi(X_{m,n} \cap X) & \text{if } \Phi\Big(\bigcup\limits_{l \in I_{m,n}} X_{m,l}\Big) \leqq k \\ 0 & \phantom{\text{if }} \Phi\Big(\bigcup\limits_{l \in I_{m,n}} X_{m,l}\Big) > k . \end{cases}$$

All functions $f_{m,n}$, and hence also all sums $f_m = \sum\limits_{n=1}^{\infty} f_{m,n}$, are measurable. Since

$$f_m(\Phi) \xrightarrow[m\to\infty]{} \Phi_k(X) \qquad (\Phi \in M)$$

the mapping $\Phi \curvearrowright \Phi_k(X)$ is also measurable for all k.

2. For all natural numbers k, the measurable mapping $\Phi \curvearrowright \Phi_k$ transforms the distribution P into a stationary distribution P_k on $\mathfrak{M}$. We obtain

$$i_{P_k} = \int \Phi_k\big([0,1)^s\big)\, P(\mathrm{d}\Phi) \leqq kP\Big(\Phi\big([0,1)^s\big) > 0\Big) \leqq k$$

and hence, in view of 6.1.5.,

$$\varrho_{P_k}(X) = i_{P_k}\mu(X) \qquad (X \in \mathfrak{R}^s).$$

As k increases, $\Phi_k(X)$ increases monotonically toward $\Phi(X)$, and we obtain

$$\varrho_P(X) = i_P\mu(X) \qquad (X \in \mathfrak{R}^s),$$

where we have to set $\infty \cdot 0 = 0 \cdot \infty = 0$. ∎

Together with P, the "simplified" distribution P^* is always stationary, too, and we have $i_{P^*} \leqq i_P$. Now, if i_P is finite, then by 6.4.2., we can conclude from $i_{P^*} = i_P$, that $\varrho_{P^*} = \varrho_P \in N$. As it was stated in section 1.4., this implies $P^* = P$. In the case $i_P = +\infty$, in view of 6.4.2. $i_{P^*} = i_P$ still implies $\varrho_P = \varrho_{P^*}$, but P^* may be different from P (cf. Zitek [1]).

As for example, if σ is the distribution of a non-negative random variable with infinite expectation, and Q the stationary distribution $\int P_{l\mu}(.)\, \sigma(\mathrm{d}l)$, then $i_Q = +\infty$. Now let P denote the stationary distribution of 2Φ with respect to Q. Obviously P^* coincides with Q, so that we obtain that $i_{P^*} = +\infty = i_P$, although P is not simple.

Using 6.4.2. we can readily show the truth of the following proposition:

6.4.3. Proposition. *For all stationary distributions P on $\mathfrak{M}$ and all X in R^s having the property that $\mu(X) = 0$, $P(\Phi(X) > 0) = 0$.*

Referring now to the statements made in section 6.2., we set

$$\overline{\Phi([0, 1)^s)} = s(\Phi)$$

calling the, possibly infinite, number $s(\Phi)$ (provided that it exists) the *sample intensity* of Φ. Using the following inequalities, which are valid for $t > 1$,

$$\int_{[-(t-1),t-1)^s} (T_y\Phi)([0, 1)^s)\,\mu(dy) = \int_{[-(t-1),t-1)^s} \Phi([0, 1)^s + y)\,\mu(dy)$$

$$= \sum_{x \in R^s, \Phi(\{x\})>0} \Phi(\{x\}) \int_{[-(t-1),t-1)^s} \delta_x([0, 1)^s + y)\,\mu(dy)$$

$$\leqq \Phi([-t, t)^s) = \sum_{x \in [-t,t)^s, \Phi(\{x\})>0} \Phi(\{x\})$$

$$\leqq \sum_{x \in R^s, \Phi(\{x\})>0} \Phi(\{x\}) \int_{[-(t+1),t+1)^s} \delta_x([0, 1)^s + y)\,\mu(dy)$$

$$= \int_{[-(t+1),t+1)^s} (T_y\Phi)([0, 1)^s)\,\mu(dy) \leqq \Phi\left([-(t + 2), t + 2)^s\right)$$

we obtain

$$s(\Phi) = \lim_{t\to\infty} (2t)^{-s}\, \Phi([-t, t)^s),$$

where the existence of either side implies that of the other. Thus we may interpret $s(\Phi)$ as the asymptotic density of Φ.

By virtue of 6.2.8., for each stationary distribution, and hence also for each stationary finite measure E on $\mathfrak{M}$, the sample intensity $s(\Phi)$ of Φ exists almost everywhere with respect to E, and we have

$$\int s(\Phi)\, E(d\Phi) = i_E,$$

and hence in the case $i_E < +\infty$

$$E(s(\Phi) = +\infty) = 0.$$

For all stationary finite measures E on $\mathfrak{M}$ satisfying the condition $E(s(\Phi) = +\infty) = 0$, we set

$$\sigma_E(.) = E(s(\Phi) \in (.)),$$

thus obtaining a finite measure σ_E on $\mathfrak{R}$ which satisfies the condition $\sigma_E((-\infty, 0)) = 0$.

6.4.4. *If two stationary finite measures E_1, E_2 on $\mathfrak{M}$ satisfy the conditions*

$$E_1(s(\Phi) = +\infty) = 0 = E_2(s(\Phi) = +\infty),$$

then also $(E_1 * E_2)(s(\Phi) = +\infty) = 0$, and $\sigma_{E_1 * E_2} = \sigma_{E_1} * \sigma_{E_2}$.

Proof. If for any counting measures Φ_1, Φ_2 the individual intensities $s(\Phi_1)$, $s(\Phi_2)$ exist, then by definition

$$s(\Phi_1 + \Phi_2) = s(\Phi_1) + s(\Phi_2).$$

Hence

$$\begin{aligned}(E_1 * E_2)\big(s(\Phi) = +\infty\big) &= (E_1 \times E_2)\big(s(\Phi_1) + s(\Phi_2) = +\infty\big)\\ &= (E_1 \times E_2)\big(s(\Phi_1) = +\infty \quad \text{or} \quad s(\Phi_2) = +\infty\big)\\ &\leqq E_1(M)\, E_2\big(s(\Phi) = +\infty\big) + E_2(M)\, E_1\big(s(\Phi) = +\infty\big) = 0.\end{aligned}$$

Moreover we have

$$\begin{aligned}\sigma_{E_1 * E_2} &= (E_1 * E_2)\big(s(\Phi) \in (.)\big)\\ &= (E_1 \times E_2)\Big(\big(s(\Phi_1) + s(\Phi_2)\big) \in (.)\Big) = \sigma_{E_1} * \sigma_{E_2}. \blacksquare\end{aligned}$$

In view of 6.4.4., the distribution σ_P is infinitely divisible for each infinitely divisible stationary distribution P on $\mathfrak{M}$ satisfying the condition $P\big(s(\Phi) < +\infty\big) = 1$; in particular we obtain, using a notation employed in Section 2.4.,

6.4.5. *For all stationary finite measures E on $\mathfrak{M}$ having the property that $E\big(s(\Phi) = +\infty\big) = 0$, the relation $\mathcal{U}_E\big(s(\Phi) < +\infty\big) = 1$ holds, and*

$$\sigma_{\mathcal{U}_E} = \mathcal{U}_{\sigma_E}.$$

If we suppose in addition that $i_P < +\infty$, we may substitute the simpler function $\Phi\big([0, 1)^s\big)$ for the indicator function $k_{\{\Phi : \Phi \in M, \Phi([0,1)^s) > 0\}}$ occurring in the proofs of 6.3.2., 6.3.3., 6.3.4., and 6.3.5. Then $\tilde{P}\big(\Psi([0, 1)^s) > 0\big)$ is replaced by $i_{\tilde{P}}$, i.e. by i_P. In this way we obtain

6.4.6. Proposition. *A stationary infinitely divisible distribution P on $\mathfrak{M}$ of finite intensity i_P is a strongly singular infinitely divisible one if and only if the sample intensity $s(\Phi)$ of Φ is non-vanishing almost everywhere with respect to $\tilde{P}$.*

6.4.7. Proposition. *A stationary singular infinitely divisible distribution P on $\mathfrak{M}$ of finite intensity i_P is a weakly singular infinitely divisible one if and only if the sample intensity $s(\Phi)$ of Φ vanishes almost everywhere with respect to $\tilde{P}$.*

6.4.8. Proposition. *For all stationary infinitely divisible distributions P on $\mathfrak{M}$ of finite intensity i_P,*

$$\widetilde{P_{s,1}} = \tilde{P}\big(\Psi \in (.),\ \Psi(R^s) = +\infty,\ s(\Psi) = 0\big)$$
$$\widetilde{P_{s,2}} = \tilde{P}\big(\Psi \in (.),\ s(\Psi) > 0\big).$$

6.4.9. Theorem. *A stationary infinitely divisible distribution P on $\mathfrak{M}$ of finite intensity i_P is ergodic if and only if the sample intensity $s(\Phi)$ of Φ vanishes almost everywhere with respect to $\tilde{P}$.*

The results obtained hitherto permit an interesting conclusion:

6.4.10. Proposition. *A stationary infinitely divisible distribution P on $\mathfrak{M}$ of finite intensity i_P is ergodic if and only if the sample intensity $s(\Phi)$ of Φ is equal to i_P almost everywhere with respect to P.*

Proof. If P is ergodic, the relation $\sigma_P = \delta_{i_P}$ results immediately from 6.2.5.

Conversely, suppose that P be not ergodic. Then by 6.3.5. P has a stationary factor $\mathscr{U}_E$, $E \in \boldsymbol{E}^+$, $E(\{o\}) = 0$, $E \neq 0$. In view of 6.4.4. and 6.4.5., $P = Q * \mathscr{U}_E$ leads to

$$\sigma_P = \sigma_Q * \sigma_{\mathscr{U}_E} = \sigma_Q * \exp\left(-\sigma_E(R^1)\right) \sum_{n=0}^{\infty} \frac{1}{n!} (\sigma_E)^n$$

$$= \exp\left(-\sigma_E(R^1)\right) \sigma_Q + \left(1 - \exp\left(-\sigma_E(R^1)\right)\right) \gamma$$

where

$$\gamma = \exp\left(-\sigma_E(R^1)\right) \left(1 - \exp\left(-\sigma_E(R^1)\right)\right)^{-1} \sum_{n=1}^{\infty} \frac{1}{n!} (\sigma_E)^n * \sigma_Q.$$

If we now assume that $\sigma_P = \delta_{i_P}$, this would give a representation of δ_{i_P} as a mixture of distributions of non-negative random numbers. From this it follows that $\sigma_Q = \delta_{i_P}$, and hence $i_Q = \int l \sigma_Q(\mathrm{d}l) = i_P$, i.e. that $i_{\mathscr{U}_E} = i_E = 0$, and hence that $E(\Psi \neq o) = 0$, which contradicts our assumptions on E. ∎

Also for non-ergodic stationary infinitely divisible distributions P on $\mathfrak{M}$ of finite intensity i_P, we may, by referring to the statements made at the end of section 2.4., determine the structure of the infinitely divisible distribution σ_P (cf. Nawrotzki [2]).

6.4.11. Proposition. *For all stationary infinitely divisible distributions P on $\mathfrak{M}$ of finite intensity i_P:*

$$\sigma_P = \delta_{(i_{P_r} + i_{P_{s,1}})} * \mathscr{E}_\omega$$

where

$$\omega(.) = \tilde{P}\left(s(\Psi) \in \left((.) \setminus \{0\}\right)\right).$$

Proof. First of all, let us note again that in view of $i_{\tilde{P}} = i_P < +\infty$, it can be concluded by means of 6.2.1. and 6.2.2. that the sample intensity of Ψ exists and is finite almost everywhere with respect to $\tilde{P}$.

By 6.3.4., 6.4.4., and 6.4.9. we have

$$\sigma_P = \sigma_{P_r * P_{s,1}} * \sigma_{P_{s,2}} = \delta_{(i_{P_r} + i_{P_{s,1}})} * \sigma_{P_{s,2}}.$$

It remains only to show that

$$\sigma_{P_{s,2}} = \mathscr{E}_\omega$$

is satisfied.

Let

$$P_{s,2} = \mathop{\Large *}_{n=1}^{\infty} \mathscr{U}_{E_n}$$

be some representation of the strongly singular infinitely divisible distribution $P_{s,2}$ as a convolution of a sequence of stationary distributions $\mathscr{U}_{E_n}$ with $E_n \in \boldsymbol{E}^+$, $E_n(\{o\}) = 0$ for $n = 1, 2, \ldots$

All of the sample intensities $s(\Phi_1)$, $s(\Phi_2)$, ... exist almost everywhere with respect to $\underset{n=1}{\overset{\infty}{\times}} \mathscr{U}_{E_n}$.

Then by definition

$$s\left(\sum_{n=1}^{\infty} \Phi_n\right) \geqq \sum_{n=1}^{\infty} s(\Phi_n),$$

if $\Phi = \sum_{n=1}^{\infty} \Phi_n$ is in M and $s(\Phi)$ exists. On the other hand,

$$\begin{aligned} +\infty > i_{P_{s,2}} &= \int s(\Phi)\, P_{s,2}(\mathrm{d}\Phi) \\ &= \int s\left(\sum_{n=1}^{\infty} \Phi_n\right) \left(\underset{n=1}{\overset{\infty}{\times}} \mathscr{U}_{E_n}\right) \left(\mathrm{d}(\Phi_n)\right) \\ &\geqq \int \left(\sum_{n=1}^{\infty} s(\Phi_n)\right) \left(\underset{n=1}{\overset{\infty}{\times}} \mathscr{U}_{E_n}\right) \left(\mathrm{d}(\Phi_n)\right) \\ &= \sum_{n=1}^{\infty} \int s(\Phi)\, \mathscr{U}_{E_n}(\mathrm{d}\Phi) = \sum_{n=1}^{\infty} i_{\mathscr{U}_{E_n}} = i_{P_{s,2}}, \end{aligned}$$

so that we observe that the relation

$$\sum_{n=1}^{\infty} s(\Phi_n) = s\left(\sum_{n=1}^{\infty} \Phi_n\right)$$

is satisfied almost everywhere with respect to $\underset{n=1}{\overset{\infty}{\times}} \mathscr{U}_{E_n}$. Hence

$$\begin{aligned} \sigma_{P_{s,2}} &= \mathop{\ast}_{n=1}^{\infty} \sigma_{\mathscr{U}_{E_n}} = \mathop{\ast}_{n=1}^{\infty} \mathscr{U}_{\sigma_{E_n}} \\ &= \mathop{\ast}_{n=1}^{\infty} \mathscr{U}_{\sigma_{E_n}((.)\setminus\{0\})} = \mathscr{E}_{\sum_{n=1}^{\infty} \sigma_{E_n}((.)\setminus\{0\})} = \mathscr{E}_{\omega} \end{aligned}$$

where

$$\begin{aligned} \omega &= \sum_{n=1}^{\infty} \sigma_{E_n}\big((.) \setminus \{0\}\big) = \sum_{n=1}^{\infty} E_n \left(s(\Phi) \in \big((.) \setminus \{0\}\big)\right) \\ &= \widetilde{P_{s,2}} \left(s(\Psi) \in \big((.) \setminus \{0\}\big)\right) = \tilde{P} \left(s(\Psi) \in \big((.) \setminus \{0\}\big)\right). \ \blacksquare \end{aligned}$$

In the "non-marked" case, the stationary Poisson distributions on $\mathfrak{M}$ are of the form $P_{l\mu}$, $l \geqq 0$. Now, given any distribution σ of a positive random variable, then for the mixture $P = \int P_{l\mu}(.)\, \sigma(\mathrm{d}l)$ the following statement is obtained for all $x > 0$:

$$P\big(s(\Phi) < x\big) = \int P_{l\mu}\big(s(\Phi) < x\big)\, \sigma(\mathrm{d}l).$$

But in view of 6.4.9. all of the $P_{l\mu}$ are ergodic, so that we may continue this equation as follows:

$$= \int k_{[0,x)}(l)\, \sigma(\mathrm{d}l) = \sigma\big([0, x)\big).$$

Hence

$$P\big(s(\Phi) < +\infty\big) = 1, \qquad \sigma_P = \sigma.$$

If we denote by Q the distribution generated from $P_{\mu/2}$ by applying $\Phi \rightsquigarrow 2\Phi$, then

$$P = \frac{1}{2}\,(P_\mu + Q)$$

represents a non-ergodic stationary distribution P on $\mathfrak{M}$, which satisfies the condition

$$\sigma_P = \delta_1 = \delta_{i_P}.$$

Hence the assumptions on the infinite divisibility of P cannot be dropped in Proposition 6.4.10.

The following theorem (cf. Nawrotzki [1], Westcott [3]) provides a characterization of the mixtures $\int P_{l\mu}(.)\, \sigma(\mathrm{d}l)$.

6.4.12. Theorem. *For each distribution P on $\mathfrak{M}$, the following statements are equivalent:*

a) *P is of the form $\int P_{l\mu}(.)\, \sigma(\mathrm{d}l)$,*
b) *P is stationary and purely random,*
c) *P has the property g_μ.*

Proof. 1. In virtue of 1.14.8., a) and c) are equivalent.

2. a) and c) imply b), since each distribution of the form $\int P_{l\mu}(.)\, \sigma(\mathrm{d}l)$ is stationary.

3. b) implies c).

Let P be stationary and purely random. For $P = \delta_o$, it is trivial that c) is valid. Thus we may additionally suppose that P be different from δ_o.

In view of 1.14.13., P has the property g_ν for some ν in N. If the measure ν were finite, then we might use 1.14.11. to conclude from $P(\Phi \neq o) > 0$ that $P(\Phi \neq o,\ \Phi \text{ finite}) > 0$, which contradicts, however, proposition 6.1.3. Hence the measure ν is infinite. Therefore, in view of 1.14.10., P has the form $\int P_{l\nu}(.)\, \sigma(\mathrm{d}l)$.

Since P is stationary and has the property g_ν, the property $g_{\nu((.)+x)}$ is present for all x in R^s. Hence, in virtue of the uniqueness statement of 1.14.13., there exists a mapping f of R^s into $(0, +\infty)$, such that

$$\nu\big((.) + x\big) = f(x)\, \nu \qquad (x \in R^s).$$

From $P = \int P_{l\nu}(.)\, \sigma(\mathrm{d}l)$ we obtain, translating by x, $P = \int P_{l\nu((.)+x)}(.)\, \sigma(\mathrm{d}l) = \int P_{lf(x)\nu}(.)\, \sigma(\mathrm{d}l)$.

Using the uniqueness proposition 1.8.6., we are now able to conclude that $f(x) = 1$ for all x in R^s. Consequently ν is a translation-invariant measure in N which is different from zero, and hence is a positive multiple of μ. Consequently the distribution P has the property g_μ. ∎

6.5. The Mapping $\Phi \curvearrowright \Phi^x$

Let us now return to the assumption used in the first three sections of this chapter, whereby the phase space $[A, \varrho_A]$ is the direct product of an Euclidean space R^s, $s \geqq 1$, and a bounded complete separable "mark space" $[K, \varrho_K]$. If we denote the "position space" R^s by $[A^x, \varrho_{A^x}]$, then

$$\Phi \curvearrowright \Phi^x = \Phi\big((.) \times K\big)$$

provides a measurable mapping of $[M, \mathfrak{M}]$ onto the measurable space $[M^x, \mathfrak{M}^x]$ of counting measures, which belongs to $[A^x, \varrho_{A^x}]$. We denote by H^x the image of a measure H on $\mathfrak{M}$ with respect to this mapping. Obviously, together with H, H^x is always stationary, too, and

$$i_{H^x} = \gamma_H(K).$$

For each infinitely divisible distribution P on M, P^x is infinitely divisible, too, and

$$\widetilde{P^x} = (\tilde{P})^x.$$

Moreover we obtain

$$\begin{aligned}(\widetilde{P}_r)^x &= \big(\tilde{P}(\Psi \in (.),\ \Psi \text{ finite})\big)^x\\ &= \tilde{P}\big(\Psi^x \in (.),\ \Psi^x \text{ finite}\big)\\ &= (\tilde{P})^x\,\big(\Psi \in (.),\ \Psi \text{ finite}\big) = \big(\widetilde{P^x}\big)_r\end{aligned}$$

and hence

$$(P^x)_r = (P_r)^x.$$

By analogy, we also obtain

$$(P^x)_s = (P_s)^x.$$

A glance at 6.3.4. shows that, for stationary infinitely divisible distributions P on $\mathfrak{M}$, even

$$(P^x)_{s,1} = (P_{s,1})^x, \qquad (P^x)_{s,2} = (P_{s,2})^x$$

is satisfied. Hence $P_{s,2} = \delta_o$ is equivalent to $(P^x)_{s,2} = \delta_o$, so that using 6.3.5. we observe the truth of

6.5.1. Proposition. *A stationary infinitely divisible distribution P on $\mathfrak{M}$ is ergodic if and only if P^x is ergodic.*

From 6.3.6. it is observed that a stationary infinitely divisible P in $\mathbf{P}$ is mixing if and only if P^x has this property.

The following example shows that the assumption of the infinite divisibility of P cannot be dropped in 6.5.1.

Let P_1 and P_2 denote the distributions of $\Phi \times \delta_1$ and $\Phi \times \delta_2$, respectively, with respect to P_μ. Then $P = (P_1 + P_2)/2$ is not ergodic, although $P^x = P_\mu$ is mixing.

7. Cox Distributions

7.0. Introduction

It was as early as in Section 1.8. that we considered special mixtures $\int P_{l\nu}(.)\,\sigma(dl)$ of Poisson distributions on $\mathfrak{M}$. Such a distribution may be interpreted as follows: at first a realization l of a non-negative random variable distributed according to σ is selected as a description of the random "external" conditions, and subsequently a counting measure Φ is realized according to the "internal" stochastic law of the point process considered governed by $P_{l\nu}$. This pattern may be generalized substantially by allowing arbitrary stochastic dependences of the intensity measure on the random "external" conditions to be considered, i.e. mixtures of the form $\int P_\omega(.)\,V(d\omega)$, where now V represents the distribution of the random intensity measure (cf. Cox [1]). In this way we obtain a class of distributions on $\mathfrak{M}$ being of interest both theoretically and for modelling actual processes; these distribution are called Cox (or doubly stochastic Poisson) distributions. The Cox distributions play a fundamental role in stochastic geometry (see, for instance, Davidson [1] and Papangelou [2], [3]). The afore-mentioned special mixtures $\int P_{l\nu}(.)\,\sigma(dl)$ may be characterized within the set of Cox distributions by the fact that they have the property g_ν (Proposition 7.1.4.).

For all "survival probabilities" c in $[0, 1]$ and all distributions P on $\mathfrak{M}$, we may establish the "thinned" distribution $\mathscr{D}_cP$ by casting dice independently for the individual points a of the realizations Φ, to decide whether they are preserved or cancelled, the probabilities being c for the first case and $1 - c$ for the second. For all Cox distributions $P = \int P_\omega(.)\,V(d\omega)$ and all c in $(0, 1]$, we have $P = \mathscr{D}_c \int P_{c^{-1}\omega}(.)\,V(d\omega)$, i.e. P can be represented as a "thinning" to an arbitrary degree c of a suitable distribution P_c on $\mathfrak{M}$. Now it turns out (Theorem 7.2.8.) that this property is characteristic for the Cox distributions.

For all distributions L on $\mathfrak{M}$, $\mathscr{D}_cL$ tends weakly toward δ_o as $c \to 0 + 0$. In the case where $[A, \varrho_A] = R^s$, $s \geqq 1$, the distributions $\mathscr{D}_cL$ may be suitably normalized by means of the operator $\mathscr{K}_c$, which effects a contraction of the realizations and in this way compensates the intensity loss caused by $\mathscr{D}_c$. It turns out that, for all stationary distributions L satisfying the condition $L\big(s(\Phi) < +\infty\big) = 1$, the convergence relation

$$\mathscr{D}_c(\mathscr{K}_cL) = \mathscr{K}_c(\mathscr{D}_cL) \underset{c\to 0+0}{\Rightarrow} \int P_{l\mu}(.)\,\sigma_L(dl)$$

is valid, where μ denotes the Lebesgue measure on $\mathfrak{R}^s$, and σ_L the distribution, formed with respect to L, of the sample intensity $s(\Phi)$ of Φ. The stationarity of L can be weakened in such a way that the existence of a measurable function f_L, defined almost everywhere with respect to L, is assumed, which has the property that uniformly in x the sequence $\big((2n)^{-s}\,\Phi([-n, n)^s + x)\big)$ converges stochastically toward $f_L(\Phi)$ (Theorem 7.3.2., cf. Rényi [1], Belyaev [1] and Nawrotzki [1]. A partial generalization to more general deletion procedures was given in Jagers & Lindvall [1] and Lindvall [1]). This limit theorem for thinnings can be generalized in several directions. It may be concidered (cf. Kallenberg [5]) as a special case of the general limit theorem 7.2.7., which gives necessary and sufficient conditions for the weak convergence of sequences of the form $\mathscr{D}_{c_n}P_n$, where $c_n \to 0 + 0$. This approach to theorem 7.3.2. is presented in Section 7.3. Another approach is based on the fact that thinnings are special cases of substochastic translations. It is possible to derive Theorem 7.3.2. from a general theorem on the convergence of sequences $(L_{\varkappa(K_n)})$, where (K_n) is a sequence of substochastic kernels on $[A', \varrho_{A'}]$ with phase space $[A, \varrho_A]$ (Theorem 7.4.1.).

With the use of Theorem 7.3.2., it is concluded that a stationary distribution P has the property g_μ if and only if, for a given c in $(0, 1)$, the equation $\mathscr{K}_c\mathscr{D}_cP = P$ is satisfied (Theorem 7.3.3.). Hence in particular a stationary distribution P is Poisson if and only if it is ergodic and, for a given c in $(0, 1)$, satisfies the equation $\mathscr{K}_c\mathscr{D}_cP = P$.

The formula $P = \int P_\omega(.)\, V(\mathrm{d}\omega)$ can be conceived not only as a parametric representation of the Cox distribution P. It provides a mapping $V \curvearrowright \mathscr{I}(V) = \int P_\omega(.)\, V(\mathrm{d}\omega)$ of the set of all distributions V of random measures into the set of all distributions of point processes. It is not only linear with respect to mixtures, but transforms also convolutions into convolutions (Theorem 7.1.2.), is of the one-to-one type (Theorem 7.1.2.) and weakly continuous in both directions (Proposition 7.2.1. and 7.2.2.). It can be used to transform statements on distributions of point processes into statements on distributions of random measures (see for example Theorem 7.1.7. and Theorem 7.2.5.). Here it should be considered that the theory of point processes is, in a way, more elementary than that of random measures.

Many concepts and propositions of this book can be transferred to random measures. We shall not discuss this in details; instead we refer to the monographs Kallenberg [6], Grandell [2] and to Jiřina [1], [2], Kingman [1], [2], Kummer & Matthes [1], [2], Lee [2], Mecke [1], [3], Nawrotzki [2] and Tortrat [1], [2].

For much more extended generaliziations of some results to random Schwartz distributions, we refer to Dennler [1] and Nawrotzki [3].

7.1. The Transformation $\mathscr{I}$

Let $\boldsymbol{G}_r$ denote the set of all finite signed measures G on $\mathfrak{N}$. With the usual definitions

$$(c_1G_1 + c_2G_2)\,(.) = c_1G_1(.) + c_2G_2(.), \qquad G_1 * G_2 = (G_1 \times G_2)\,\big(v_1 + v_2 \in (.)\big)$$

for all $c_1, c_2 \in R$ and all $G_1, G_2 \in \boldsymbol{G}_r$, $\boldsymbol{G}_r$ becomes a commutative algebra over R, which has the unity element δ_0. Obviously

$$E \rightsquigarrow E\big((.) \cap M\big)$$

is an isomorphic mapping of the algebra $\boldsymbol{E}_r$ into the algebra $\boldsymbol{G}_r$.

In view of 3.1.3., 3.2.3., and 3.3.7. the mapping $\nu \rightsquigarrow P_\nu$ is measurable with respect to the σ-algebras $\mathfrak{N}$ and $\mathfrak{P}$, such that the set-up

$$\mathcal{J}(G) = \int P_\nu(.)\, G(\mathrm{d}\nu)$$

associates a finite signed measure $\mathcal{J}(G)$ on $\mathfrak{M}$ to each G in $\boldsymbol{G}_r$.

For all non-negative u and all G in $\boldsymbol{G}_r$ we denote by $\omega_u G$ the image of G with respect to the measurable mapping $\nu \rightsquigarrow u\nu$, i.e. we set $\omega_u G = G\big(u\nu \in (.)\big)$ In view of 1.13.7. and 4.2.5. we obtain

7.1.1. *For all u in* $[0, 1]$ *and all G in $\boldsymbol{G}_r$,*

$$\mathcal{D}_u\big(\mathcal{J}(G)\big) = \mathcal{J}(\omega_u G).$$

Consequently, for all $c \in (0, 1]$ and all G in $\boldsymbol{G}_r$,

$$\mathcal{J}(G) = \mathcal{D}_c\big(\mathcal{J}(\omega_{c^{-1}} G)\big).$$

Just as it was done for signed measures in $\boldsymbol{E}_r$, for signed measures G in $\boldsymbol{G}_r$ we may introduce, for each finite sequence $X_1, \ldots, X_m$ of sets in $\mathfrak{B}$, the finite signed measure

$$G_{X_1,\ldots,X_m} = G\big([\nu(X_1), \ldots, \nu(X_m)] \in (.)\big).$$

By means of

$$[x_1, \ldots, x_m] \rightsquigarrow \sum_{i=1}^{m} x_i \delta_i$$

$[0, +\infty)^m$ can be mapped onto the set N' corresponding to the phase space $A' = \{1, \ldots, n\}$, the mapping being one-to-one and measurable in both directions. Thus $G_{X_1,\ldots,X_m}$ can also be conceived of as a signed measure defined on the associated σ-algebra $\mathfrak{N}'$. On the basis of this interpretation, the following relation is immediately concluded from the definition of $\mathcal{J}$:

$$\big(\mathcal{J}(G)\big)_{X_1,\ldots,X_m} = \mathcal{J}(G_{X_1,\ldots,X_m}).$$

The family of all $G_{X_1,\ldots,X_m}$ determines the values of G on a generating subalgebra of $\mathfrak{N}$. Thus, a signed measure G in $\boldsymbol{G}_r$ is uniquely determined by this family.

7.1.2. Theorem. *The mapping $\mathcal{J}$ is an isomorphism of the real algebra $\boldsymbol{G}_r$ into the real algebra $\boldsymbol{E}_r$.*

Proof. 1. In view of 1.7.3. and 1.8.2., for all G_1, G_2 in $\boldsymbol{G}_r$, we have

$$\mathcal{J}(G_1) * \mathcal{J}(G_2) = \int P_{\nu+\varrho}(.)\, (G_1 \times G_2)\,(\mathrm{d}[\nu, \varrho]) = \int P_\sigma(.)\, (G_1 * G_2)\,(\mathrm{d}\sigma)$$
$$= \mathcal{J}(G_1 * G_2).$$

Obviously, for all $c_1, c_2 \in R$ and all $G_1, G_2 \in \boldsymbol{G}_r$,

$$\mathcal{J}(c_1G_1 + c_2G_2) = c_1\mathcal{J}(G_1) + c_2\mathcal{J}(G_2).$$

Consequently, $\mathcal{J}$ is a homomorphic mapping of $\boldsymbol{G}_r$ into $\boldsymbol{E}_r$.

2. For all ν in N, all X in $\mathfrak{B}$, all sequences (c_n) of numbers in $(0, 1]$ which converge to zero, all natural numbers n, and all $\varepsilon > 0$, we obtain by means of the Chebyshev inequality

$$P_{(c_n)^{-1}\nu}\big(|c_n\Phi(X) - \nu(X)| > \varepsilon\big) = P_{(c_n)^{-1}\nu}\big(|\Phi(X) - (c_n)^{-1}\,\nu(X)| > (c_n)^{-1}\,\varepsilon\big)$$
$$\leqq \frac{(c_n)^{-1}\,\nu(X)}{((c_n)^{-1}\,\varepsilon)^2} = c_n\nu(X)\cdot\varepsilon^{-2}.$$

From this it follows that

$$(\omega_{c_n}P_{(c_n)^{-1}\nu})_X \underset{n\to\infty}{\Rightarrow} \delta_{\nu(X)}.$$

Hence, for each finite sequence $X_1, \ldots, X_m$ of sets in $\mathfrak{B}$ and each continuous bounded real function h defined on $[0, +\infty)^m$, we have

$$\int h(x)\,(\omega_{c_n}P_{(c_n)^{-1}\nu})_{X_1,\ldots,X_m}(\mathrm{d}x) \xrightarrow[n\to\infty]{} h\big([\nu(X_1), \ldots, \nu(X_m)]\big).$$

Using Lebesgue's theorem, for all G in $\boldsymbol{G}_r$ we obtain

$$\int h(x)\,\big(\omega_{c_n}(\mathcal{D}_{c_n})^{-1}\,\mathcal{J}(G)\big)_{X_1,\ldots,X_m}(\mathrm{d}x) = \int\Big(\int h(x)\,(\omega_{c_n}P_{(c_n)^{-1}\nu})_{X_1,\ldots,X_m}(\mathrm{d}x)\Big)\,G(\mathrm{d}\nu)$$
$$\xrightarrow[n\to\infty]{} \int h\big([\nu(X_1), \ldots, \nu(X_m)]\big)\,G(\mathrm{d}\nu) = \int h(x)\,G_{X_1,\ldots,X_m}(\mathrm{d}x),$$

i.e. from $\mathcal{J}(G_1) = \mathcal{J}(G_2)$ it follows that

$$(G_1)_{X_1,\ldots,X_m} = (G_2)_{X_1,\ldots,X_m}$$

for all finite sequences $X_1, \ldots, X_m$ of sets in $\mathfrak{B}$. Consequently, the mapping $\mathcal{J}$ is one-to-one. ∎

Let $\boldsymbol{V}$ denote the set of all distributions on the σ-algebra $\mathfrak{N}$, i.e. the set of all distributions of random measures with phase space $[A, \varrho_A]$. Obviously, $\mathcal{J}(V) \in \boldsymbol{P}$ for every V in $\boldsymbol{V}$. A distribution P on $\mathfrak{M}$ is called a *Cox distribution*, if it is of the form $\mathcal{J}(V)$, $V \in \boldsymbol{V}$.

By means of

$$g = \frac{1}{2}\,((3 - e)\,\delta_0 - \delta_1 + e\delta_2)$$

we obtain a finite signed measure on the σ-algebra of the Borel subsets of $[0, +\infty)$. Now

$$\int \pi_l(.)\,g(\mathrm{d}l) = \frac{1}{2}\,((3 - e)\,\delta_0 - \pi_1 + e\pi_2)$$

is a distribution, whereas $g(\{1\}) = -1/2$. Thus, even if $[A, \varrho_A]$ coincides with $\{1\}$, there are distributions of the form $P = \mathcal{J}(G)$, $G \in \boldsymbol{G}_r$, which are not Cox distributions!

For all ν in N, the mapping $l \curvearrowright l\nu$ of $[0, +\infty)$ into $[N, \mathfrak{N}]$ is measurable, and hence transforms a distribution σ of a random non-negative number into a distribution $V_{\nu,\sigma}$ on $\mathfrak{N}$. Obviously,

$$\mathscr{J}(V_{\nu,\sigma}) = \int P_{l\nu}(.)\, \sigma(\mathrm{d}l).$$

For $\nu \neq o$, the mapping $l \curvearrowright l\nu$ is one-to-one and measurable in both directions. Using 7.1.2., we are now able to conclude from

$$\int P_{l\nu}(.)\, \sigma_1(\mathrm{d}l) = \int P_{l\nu}(.)\, \sigma_2(\mathrm{d}l),$$

i.e. from

$$\mathscr{J}(V_{\nu,\sigma_1}) = \mathscr{J}(V_{\nu,\sigma_2})$$

that $V_{\nu,\sigma_1} = V_{\nu,\sigma_2}$ and hence $\sigma_1 = \sigma_2$. In this way we obtain the following generalization of 1.8.6.

7.1.3. Proposition. *If, for any $\nu \neq o$ in N, two distributions σ_1, σ_2 of random non-negative numbers satisfy the equation*

$$\int P_{l\nu}(.)\, \sigma_1(\mathrm{d}l) = \int P_{l\nu}(.)\, \sigma_2(\mathrm{d}l),$$

then $\sigma_1 = \sigma_2$.

By 1.8.4. all distributions of the form $P = \mathscr{J}(V_{\nu,\sigma})$ have the property g_ν. Conversely, if $P = \mathscr{J}(V)$, $V \in \boldsymbol{V}$, has the property g_ν, then, if $\nu = o$, P coincides with δ_o, and $P = \mathscr{J}(V_{\nu,\sigma})$ for arbitrary σ. Now, if $\nu \neq o$, we select any set X in $\mathfrak{B}$ with the property $\nu(X) > 0$. Without loss of generality, we may suppose that $\nu(X) = 1$. Using 1.7.4. we obtain

$$
\begin{aligned}
{}_XP &= \sum_{n=0}^{\infty} P(\Phi(X) = n)\, \big(Q_{\nu((.)\cap X)}\big)^n = \sum_{n=0}^{\infty} \big(\mathscr{J}(V_X)\big)\, (\{n\})\, \big(Q_{\nu((.)\cap X)}\big)^n \\
&= \sum_{n=0}^{\infty} \left(\int_0^{\infty} e^{-l} \frac{l^n}{n!}\, V_X(\mathrm{d}l) \right) \big(Q_{\nu((.)\cap X)}\big)^n \\
&= \int_0^{\infty} \left(e^{-l} \sum_{n=0}^{\infty} \frac{l^n}{n!} \big(Q_{\nu((.)\cap X)}\big)^n \right) V_X(\mathrm{d}l) = \int_0^{\infty} \left(e^{-l} \sum_{n=0}^{\infty} \frac{1}{n!} \big(Q_{l\nu((.)\cap X)}\big)^n \right) V_X(\mathrm{d}l) \\
&= \int_0^{\infty} P_{l\nu((.)\cap X)}(.)\, V_X(\mathrm{d}l) = {}_X\!\left(\int_0^{\infty} P_{l\nu}(.)\, V_X(\mathrm{d}l) \right).
\end{aligned}
$$

In view of 1.14.7. it follows that

$$P = \int P_{l\nu}(.)\, V_X(\mathrm{d}l).$$

Thus, we obtain

7.1.4. Proposition. *For all ν in N, a Cox distribution P on $\mathfrak{M}$ has the property g_ν if and only if P is of the form $P = \int P_{l\nu}(.)\, \sigma(\mathrm{d}l)$.*

A distribution V on $\mathfrak{N}$ is said to be *free from after-effects* if, for all finite sequences $X_1, \ldots, X_m$ of pairwise disjoint sets in $\mathfrak{B}$, the finite sequence $\nu(X_1), \ldots, \nu(X_m)$ is independent with respect to V.

Now let V be free from after-effects, and let $X_1, \ldots, X_m$ be a finite sequence of pairwise disjoint sets in $\mathfrak{B}$. For all non-negative integral $l_1, \ldots, l_m$, we obtain

$$\begin{aligned}
(\mathcal{J}(V))_{X_1,\ldots,X_m}(\{[l_1, \ldots, l_m]\}) &= \mathcal{J}(V_{X_1,\ldots,X_m})(\{[l_1, \ldots, l_m]\}) \\
&= \int \prod_{i=1}^{m} \left(e^{-v_i} \frac{(v_i)^{l_i}}{l_i!}\right) V_{X_1,\ldots,X_m}(\mathrm{d}[v_1, \ldots, v_m]) \\
&= \int \prod_{i=1}^{m} \left(e^{-v_i} \frac{(v_i)^{l_i}}{l_i!}\right) \left(\mathop{\times}_{i=1}^{m} V_{X_i}\right) (\mathrm{d}[v_1, \ldots, v_m]) \\
&= \prod_{i=1}^{m} \left(\int e^{-v_i} \frac{(v_i)^{l_i}}{l_i!} V_{X_i}(\mathrm{d}v_i)\right) = \prod_{i=1}^{m} (\mathcal{J}(V_{X_i})(\{l_i\})) \\
&= \prod_{i=1}^{m} (\mathcal{J}(V))_{X_i}(\{l_i\}) = \left(\mathop{\times}_{i=1}^{m} (\mathcal{J}(V))_{X_i}\right)(\{[l_1, \ldots, l_m]\}).
\end{aligned}$$

Hence, also $\mathcal{J}(V)$ is free from after-effects.

Conversely, suppose that $\mathcal{J}(V)$ be free from after-effects. For each finite sequence $X_1, \ldots, X_m$ of pairwise disjoint sets in $\mathfrak{B}$, we obtain

$$\mathcal{J}(V_{X_1,\ldots,X_m}) = (\mathcal{J}(V))_{X_1,\ldots,X_m} = \mathop{\times}_{i=1}^{m} \left((\mathcal{J}(V))_{X_i}\right) = \mathop{\times}_{i=1}^{m} \mathcal{J}(V_{X_i}).$$

On the basis of the calculations carried out above, we may continue this chain of equations as follows

$$= \mathcal{J}\left(\mathop{\times}_{i=1}^{m} V_{X_i}\right).$$

By 7.1.2. this implies that

$$V_{X_1,\ldots,X_m} = \mathop{\times}_{i=1}^{m} (V_{X_i}).$$

Hence, the distribution V is free from after-effects.

Thus, we have (cf. Krickeberg [2])

7.1.5. Proposition. *A distribution V on $\mathfrak{N}$ is free from after-effects if and only if $\mathcal{J}(V)$ has this property.*

Obviously a Cox distribution $\mathcal{J}(V)$ is infinitely divisible if for all natural numbers n, there exists a n-th convolution root of V in $\mathbf{V}$. It was shown in Kallenberg [6], Example 8.6., and Huff [1], that the converse is not true even in the case of the phase space $\{1\}$!

Let N_d denote the set of all diffuse measures ν in N, i.e. the set $\{\nu : \nu \in N,\ \nu(\{a\}) = 0$ for all a in $A\}$. In view of 1.4.3. and 1.7.6. and the measurability of the mapping $\nu \curvearrowright P_\nu$, we have

$$N_d = \{\nu : \nu \in N,\ P_\nu\,(\Phi \text{ simple}) = 1\} \in \mathfrak{N}.$$

Using anew 1.7.6., for all V in $\mathbf{V}$,

$$(\mathcal{J}(V))\,(\Phi \text{ simple}) = \int P_\nu\,(\Phi \text{ simple})\, V(\mathrm{d}\nu)$$

is equal to unity if and only if the relation $\nu \in N_d$ is satisfied almost everywhere with respect to V. Thus, we observe the truth of (cf. Krickeberg [2], Mönch [2])

7.1.6. Proposition. *A distribution V on $\mathfrak{N}$ satisfies the condition $V(\nu$ diffuse$) = 1$ if and only if $\mathscr{I}(V)$ is simple.*

As an immediate consequence of 1.4.9., and 7.1.6., using the reversibility of $\mathscr{I}$, we obtain (cf. Grandell [1], Kallenberg [3], Mönch [2])

7.1.7. Theorem. *If two distributions V_1, V_2 on $\mathfrak{N}$ satisfy the condition $(V_1)_X = (V_2)_X$ for all sets X in a generating subring $\mathfrak{H}$ of $\mathfrak{B}$, then it can be concluded that $V_1 = V_2$, provided that $V_1(\nu$ diffuse$) = 1$ is satisfied.*

7.2. A Characterization of the Cox Distributions

The set $\boldsymbol{V}$ coincides with the set of all distributions on the σ-algebra of the Borel sets of the complete separable metric space $[N, \varrho_N]$. As it was done in $\boldsymbol{P}$, we may therefore introduce the Prohorov metric $\varrho_{\boldsymbol{V}}$ in $\boldsymbol{V}$, too. A sequence (V_n) of distributions on $\mathfrak{N}$ converges weakly, i.e. with respect to $\varrho_{\boldsymbol{V}}$, toward some V in $\boldsymbol{V}$ if and only if for all bounded real functions h on N which are continuous with respect to ϱ_N, the convergence relation

$$\int h(\nu)\, V_n(\mathrm{d}\nu) \xrightarrow[n\to\infty]{} \int h(\nu)\, V(\mathrm{d}\nu)$$

is satisfied. For this we shall write $V_n \underset{n\to\infty}{\Rightarrow} V$.

In view of 3.3.7., for each bounded real function f on M which is continuous with respect to ϱ_M, the bounded real function

$$\nu \curvearrowright \int f(\Phi)\, P_\nu(\mathrm{d}\Phi) \qquad (\nu \in N)$$

is continuous with respect to ϱ_N, so that from $V_n \underset{n\to\infty}{\Rightarrow} V$ we may conclude that

$$\int \Big(\int f(\Phi)\, P_\nu(\mathrm{d}\Phi)\Big)\, V_n(\mathrm{d}\nu) \xrightarrow[n\to\infty]{} \int \Big(\int f(\Phi)\, P_\nu(\mathrm{d}\Phi)\Big)\, V(\mathrm{d}\nu),$$

i.e. that

$$\int f(\Phi)\, \big(\mathscr{I}(V_n)\big)\, (\mathrm{d}\Phi) \xrightarrow[n\to\infty]{} \int f(\Phi)\, \big(\mathscr{I}(V)\big)\, (\mathrm{d}\Phi).$$

Thus we observe the truth of

7.2.1. Proposition. *The mapping $V \curvearrowright \mathscr{I}(V)$ of $\boldsymbol{V}$ into $\boldsymbol{P}$ is continuous with respect to $\varrho_{\boldsymbol{V}}$ and $\varrho_{\boldsymbol{P}}$.*

Also the inverse of $\mathscr{I}$ is continuous. We even have

7.2.2. Proposition. *If for a sequence (V_n) of distributions on $\mathfrak{N}$ there exists a distribution P on $\mathfrak{M}$, such that $\mathscr{I}(V_n) \underset{n\to\infty}{\Rightarrow} P$, then (V_n) converges weakly toward some V in $\boldsymbol{V}$, and $\mathscr{I}(V) = P$.*

Proof. 1. In view of 3.2.7. we observe that, for each bounded closed subset X of A,

$$\sup_{n=1,2,\dots} \big(\mathscr{I}(V_n)\big)\big(\Phi(X) > k\big) \xrightarrow[k\to\infty]{} 0,$$

i.e. we have

$$\sup_{n=1,2,\dots} \int \left(1 - e^{-\nu(X)} \sum_{i=0}^{k} \frac{(\nu(X))^i}{i!}\right) V_n(\mathrm{d}\nu) \xrightarrow[k\to\infty]{} 0$$

and hence

$$\sup_{n=1,2,\dots} V_n\big(\nu(X) > m\big) \xrightarrow[m\to\infty]{} 0.$$

Consequently, to each $\eta > 0$ there exists an $m_{X,\eta}$ such that

$$\sup_{n=1,2,\dots} V_n\big(\nu(X) > m_{X,\eta}\big) < \eta$$

is satisfied.

By another application of 3.2.7. we observe that to each $\omega > 0$ there exists a compact subset $B_{X,\omega}$ of X having the property that

$$\sup_{n=1,2,\dots} \big(\mathscr{I}(V_n)\big)\big(\Phi(X \setminus B_{X,\omega}) > 0\big) < \omega.$$

Hence

$$\sup_{n=1,2,\dots} \int (1 - e^{-\nu(X\setminus B_{X,\omega})})\, V_n(\mathrm{d}\nu) < \omega.$$

Consequently, to any given $\varepsilon, \eta > 0$ there exists a compact subset $B_{X,\varepsilon,\eta}$ of X having the property that

$$\sup_{n=1,2,\dots} V_n\big(\nu(X \setminus B_{X,\varepsilon,\eta}) > \varepsilon\big) < \eta.$$

2. Using the method of the second step in the proof of 3.2.7., we see that 1. implies that (V_n) is relatively compact with respect to the metric $\varrho_{\mathbf{V}}$. Therefore each subsequence has a subsequence (V_{n_m}) which is weakly convergent in $\mathbf{V}$. Let $V_{n_m} \underset{m\to\infty}{\Rightarrow} V$. Using 7.2.1. we conclude that $\mathscr{I}(V) = P$. Since the inverse of $\mathscr{I}$ exists, the limit V does not depend on the choice of the weakly convergent subsequence of (V_n). Hence each subsequence of (V_n) has a subsequence which converges weakly toward V, i.e. we have $V_n \underset{n\to\infty}{\Rightarrow} V$. ∎

As an immediate consequence of 7.2.2. we obtain

7.2.3. Proposition. *The set of all Cox distributions on $\mathfrak{M}$ is closed with respect to $\varrho_{\mathbf{P}}$.*

Generalizing a notation introduced in Section 3.1., for all V in $\mathbf{V}$ we denote by $\mathfrak{B}_V$ the ring of all sets X in $\mathfrak{B}$ which satisfy the condition $V\big(\nu(\partial X) > 0\big) = 0$. Then obviously $\mathfrak{B}_{\mathscr{I}(V)}$ coincides with $\mathfrak{B}_V$. Using 7.2.1. and 7.2.2., from 3.1.9. we now obtain

7.2.4. *A sequence (V_n) of distributions on $\mathfrak{N}$ converges weakly toward some V in $\mathbf{V}$ if and only if for all finite sequences $X_1, \dots, X_m$ of pairwise disjoint*

sets in $\mathfrak{B}_V$, *the convergence relation*

$$(V_n)_{X_1,\ldots,X_m} \Rightarrow V_{X_1,\ldots,X_m}$$

is satisfied.

By means of $\mathscr{I}$ it is also possible to transform Theorem 3.1.13. into a statement on the distributions of random measures. Using 7.1.6., 7.2.1., and 7.2.2. we obtain (cf. Grandell [1])

7.2.5. Theorem. *Let* V *be a distribution on* $\mathfrak{N}$ *which satisfies the condition* $V(\nu \text{ diffuse}) = 1$ *and* $\mathfrak{H}$ *a subring of* $\mathfrak{B}_V$. *If to each bounded closed subset* D *of* A *and to each open neighbourhood* U *of* D *there exists a set* X *in* $\mathfrak{H}$ *having the property that* $D \subseteqq X \subseteqq U$, *then a sequence* (V_n) *of distributions on* $\mathfrak{N}$ *converges weakly toward* V *if and only if, for all* X *in* $\mathfrak{H}$, *the convergence relation*

$$(V_n)_X \underset{n\to\infty}{\Rightarrow} V_X$$

is satisfied.

Together with 7.2.4., the second step in the proof of Theorem 7.1.2. provides us with the following *inversion formula*:

7.2.6. Proposition. *For all* V *in* $\boldsymbol{V}$ *and each sequence* (c_n) *of numbers in* $(0, 1]$ *which tends toward zero,*

$$\omega_{c_n}\big((\mathscr{D}_{c_n})^{-1}\,\mathscr{I}(V)\big) \underset{n\to\infty}{\Rightarrow} V.$$

If in 7.2.6. we set

$$(\mathscr{D}_{c_n})^{-1}\,\mathscr{I}(V) = P_n, \qquad \mathscr{I}(V) = P,$$

then we obtain

$$\mathscr{D}_{c_n}P_n = P \quad \text{for} \quad n = 1, 2, \ldots, \qquad \omega_{c_n}P_n \underset{n\to\infty}{\Rightarrow} V, \qquad \mathscr{I}(V) = P.$$

In this form, the statement of 7.2.6. can be generalized as follows (cf. Kallenberg [4]).

7.2.7. Theorem. *Let* (P_n) *be a sequence of distributions on* $\mathfrak{M}$, *and* (c_n) *a sequence of numbers in* $[0, 1]$ *which tends toward zero. Subject to these assumptions,* $(\mathscr{D}_{c_n}P_n)$ *tends weakly toward some* P *in* $\boldsymbol{P}$ *if and only if* $(\omega_{c_n}P_n)$ *converges weakly toward some* V *in* $\boldsymbol{V}$; *and* $\mathscr{I}(V) = P$.

Proof. 1. For all c in $[0, 1]$, all X in $\mathfrak{B}$, and all P in $\boldsymbol{P}$, using 1.11.3. we obtain

$$\begin{aligned}
&{}_X\|\mathscr{D}_cP - \mathscr{I}(\omega_cP)\| \\
&= {}_X\Big\|\int (\mathscr{D}_c\delta_\Phi)(.)\,P(\mathrm{d}\Phi) - \int P_{c\Phi}(.)\,P(\mathrm{d}\Phi)\Big\| \\
&\leqq \int {}_X\|\mathscr{D}_c\delta_\Phi - P_{c\Phi}\|\,P(\mathrm{d}\Phi) = \int \Big\| \mathop{\ast}_{a\in X,\,\Phi(\{a\})>0} (\mathscr{D}_c\delta_{\delta_a})^{\Phi(\{a\})} - P_{c_X\Phi}\Big\|\,P(\mathrm{d}\Phi) \\
&\leqq \int 2\min\big(1, c^2\Phi(X)\big)\,P(d\Phi).
\end{aligned}$$

2. Now suppose that $(\mathscr{D}_{c_n} P_n)$ converges weakly toward P. For all X in $\mathfrak{B}_P$ and all natural numbers k, we now obtain with the use of 3.1.9.

$$\begin{aligned} P\big(\Phi(X) > k\big) &= \lim_{n\to\infty} (\mathscr{D}_{c_n} P_n) \big(\Phi(X) > k\big) \\ &= \lim_{n\to\infty} \int (\mathscr{D}_{c_n} \delta_\Psi) \big(\Phi(X) > k\big)\, P_n(\mathrm{d}\Psi) \\ &\geqq \overline{\lim_{n\to\infty}} \int k_{\{\Psi:\Psi(X) > (c_n)^{-3/2}\}}(\omega)\, (\mathscr{D}_{c_n} \delta_\omega) \big(\Phi(X) > k\big)\, P_n(\mathrm{d}\omega) \\ &\geqq \overline{\lim_{n\to\infty}}\, P_n\big(\Psi(X) > (c_n)^{-3/2}\big)\, (\mathscr{D}_{c_n} \delta_{\chi_n}) \big(\Phi(X) > k\big), \end{aligned}$$

where χ_n denotes an arbitrary counting measure in M, for which $\chi_n(X)$ is equal to the integral part $[(c_n)^{-3/2}]$ of $(c_n)^{-3/2}$. For sufficiently large n, k is less than $(c_n)^{-1/2} - (c_n)^{-1/3}$, so that we may continue our above chain of inequalities as follows:

$$\geqq \overline{\lim_{n\to\infty}}\, P_n\big(\Psi(X) > (c_n)^{-3/2}\big)\, (\mathscr{D}_{c_n} \delta_{\chi_n}) \big(|\Phi(X) - [(c_n)^{-3/2}]\, c_n| \leqq (c_n)^{-1/3}\big).$$

Using the Chebyshev inequality, we obtain

$$\begin{aligned} &\geqq \overline{\lim_{n\to\infty}}\, P_n\big(\Psi(X) > (c_n)^{-3/2}\big) \left(1 - \frac{[(c_n)^{-3/2}]\, c_n(1 - c_n)}{(c_n)^{-2/3}}\right) \\ &= \overline{\lim_{n\to\infty}}\, P_n\big(\Psi(X) > (c_n)^{-3/2}\big). \end{aligned}$$

Since the choice of k was arbitrary, we get

$$\lim_{n\to\infty} P_n\big(\Psi(X) > (c_n)^{-3/2}\big) = 0.$$

3. We shall now carry on the considerations of step 1. Let $X_1, \ldots, X_m$ be a finite sequence of sets in $\mathfrak{B}_P$, and $X = \bigcup_{i=1}^{m} X_i$. In view of 1. and 2., we obtain

$$\begin{aligned} &\overline{\lim_{n\to\infty}} \left\| (\mathscr{D}_{c_n} P_n)_{X_1,\ldots,X_m} - \big(\mathscr{J}(\omega_{c_n} P_n)\big)_{X_1,\ldots,X_m} \right\| \\ &\leqq \overline{\lim_{n\to\infty}}\, {}_X\| \mathscr{D}_{c_n} P_n - \mathscr{J}(\omega_{c_n} P_n) \| \\ &\leqq \overline{\lim_{n\to\infty}} \int 2 \min\big(1, (c_n)^2\, \Phi(X)\big)\, P_n(\mathrm{d}\Phi) \\ &\leqq 2\, \overline{\lim_{n\to\infty}} \int_{\{\Psi:\Psi(X) \leqq (c_n)^{-3/2}\}} (c_n)^2\, \Phi(X)\, P_n(\mathrm{d}\Phi) + 2\, \overline{\lim_{n\to\infty}}\, P_n\big(\Phi(X) > (c_n)^{-3/2}\big) \\ &\leqq 2\, \overline{\lim_{n\to\infty}} \big((c_n)^2\, (c_n)^{-3/2}\big) + 0 = 0, \end{aligned}$$

from which by 3.1.9. we conclude that

$$\mathscr{J}(\omega_{c_n}P) \underset{n\to\infty}{\Rightarrow} P.$$

Using 7.2.2., we may conclude from this that $(\omega_{c_n}P)$ converges weakly toward some V in $\boldsymbol{V}$ and that $\mathscr{J}(V) = P$.

4. Now let $\omega_{c_n}P_n \underset{n\to\infty}{\Rightarrow} V$. By 7.2.1. we then observe that $\big(\mathscr{J}(\omega_{c_n}P_n)\big)$ tends toward $P = \mathscr{J}(V)$. We have to show that $\mathscr{D}_{c_n}P_n \underset{n\to\infty}{\Rightarrow} P$ is satisfied.

Let $X_1, \ldots, X_m$ be a finite sequence of sets in $\mathfrak{B}_V$, i.e. in $\mathfrak{B}_P$. Using the estimations performed in 3., we obtain

$$\begin{aligned}
&\overline{\lim_{n\to\infty}} \left\|(\mathscr{D}_{c_n}P_n)_{X_1,\ldots,X_m} - \big(\mathscr{J}(\omega_{c_n}P_n)\big)_{X_1,\ldots,X_m}\right\| \\
&\leqq 2\,\overline{\lim_{n\to\infty}}\, P_n\big(\Phi(X) > (c_n)^{-3/2}\big) \\
&= 2\,\overline{\lim_{n\to\infty}}\, (\omega_{c_n}P_n)\big(\nu(X) > (c_n)^{-1/2}\big) = 0.
\end{aligned}$$

But from this we may conclude that

$$(\mathscr{D}_{c_n}P_n) \underset{n\to\infty}{\Rightarrow} P. \ ■$$

As it was stated in Section 7.1., for all Cox distributions $P = \mathscr{J}(V)$, $V \in \boldsymbol{V}$, and all c in $(0, 1]$, there exists a distribution P_c on $\mathfrak{M}$, such that $P = \mathscr{D}_c P_c$, namely $P_c = \mathscr{J}(\omega_{c^{-1}}V)$.

Now suppose that for a given P in $\boldsymbol{P}$ for all c in $(0, 1]$ there exists a distribution P_c on $\mathfrak{M}$, such that $P = \mathscr{D}_c P_c$. Thus,

$$\mathscr{D}_{1/n}P_{1/n} \underset{n\to\infty}{\Rightarrow} P.$$

Consequently, by 7.2.7. $\omega_{1/n}P_{1/n}$ converges weakly toward some V in $\boldsymbol{V}$; and $\mathscr{J}(V) = P$.

In this way we obtain the following characterization of the Cox distributions (cf. Mecke [2]).

7.2.8. Theorem. *A distribution P on $\mathfrak{M}$ is a Cox distribution if and only if, for all c in* $(0, 1]$ *there exists a distribution P_c on $\mathfrak{M}$ with the property* $\mathscr{D}_c P_c = P$.

7.3. A Limit Theorem for Thinnings

In this section we shall always suppose that the phase space $[A, \varrho_A]$ be of the form R^s, $s \geqq 1$. In the same way as in the procceding chapters, let μ denote the Lebesgue measure on $\mathfrak{R}^s$.

For all c in $[0, 1]$ and all P in $\boldsymbol{P}$, we set

$$\mathscr{K}_c P = P_\omega,$$

where $\omega_{(x)} = \delta_{\delta_{\sqrt[s]{c}\,x}}$ for all x in R^s. The application of the operator $\mathscr{K}_c$ means a compression of the realizations Φ. The operator $\mathscr{K}_c\mathscr{D}_c$ then corresponds to clustering with the cluster field

$$x \rightsquigarrow (1 - c)\,\delta_o + c\delta_{\delta_{\frac{s}{\sqrt{c}}x}} \qquad (x \in R^s),$$

i.e. the random translation by means of the substochastic kernel

$$K_c(x, (.)) = c\delta_{\frac{s}{\sqrt{c}}x} \qquad (x \in R^s).$$

Immediately from 4.4.4., for all $l \geqq 0$ and all c in $(0, 1]$ we now obtain

$$(\mathcal{K}_c \mathcal{D}_c)\, P_{l\mu} = P_{l\mu}.$$

Hence, in view of 4.2.5. and 6.4.12., all stationary purely random distributions are fixed points of the operators $\mathcal{K}_c\mathcal{D}_c$.

Even if P is only stationary and not necessarily purely random, $(\mathcal{K}_c\mathcal{D}_c)P$ can be approximated for $c \to 0 + 0$ by a stationary purely random distribution if the realizations Φ are not too dense:

7.3.1. Theorem. *For all stationary distributions L on $\mathfrak{M}$ having the property that $L(s(\Phi) < +\infty) = 1$, and for all sequences (c_n) of numbers in $(0, 1]$ which tend toward zero,*

$$(\mathcal{K}_{c_n}\mathcal{D}_{c_n})\, L \underset{n\to\infty}{\Rightarrow} \int P_{l\mu}(.)\, \sigma_L(\mathrm{d}l).$$

The assumption on the stationarity of P can be weakened as follows. Let $\boldsymbol{H}$ denote the set of those distributions L on $\mathfrak{M}$ for which there exists a non-negative real function f_L which is defined almost everywhere with respect to L and measurable with respect to $\mathfrak{M}$, such that the stochastic convergence relation

$$(2n)^{-s}\, \Phi([-n, n)^s + x) \underset{L}{\to} f_L(\Phi)$$

is satisfied uniformly in x, i.e. that, for all $\varepsilon > 0$, the condition

$$\lim_{n\to\infty} \sup_{x\in R^s} L(|(2n)^{-s}\, \Phi([-n, n)^s + x) - f_L(\Phi)| > \varepsilon) = 0$$

is satisfied. Obviously the function f_L is uniquely determined except only for the values of a set of probability zero.

For all stationary distributions L on $\mathfrak{M}$ having the property that $L(s(\Phi) < +\infty) = 1$, according to Section 6.4., we have almost everywhere with respect to L

$$(2n)^{-s}\, \Phi([-n, n)^s) \underset{n\to\infty}{\longrightarrow} s(\Phi).$$

Consequently, for all $\varepsilon > 0$ we obtain

$$\lim_{n\to\infty} \sup_{x\in R^s} L(|(2n)^{-s}\, \Phi([-n, n)^s + x) - s(\Phi)| > \varepsilon)$$

$$= \lim_{n\to\infty} L(|(2n)^{-s}\, \Phi([-n, n)^s) - s(\Phi)| > \varepsilon) = 0.$$

Hence the distribution L belongs to $\boldsymbol{H}$, where f_L can be replaced by the restriction of the sample intensity to $\{\Phi : \Phi \in M, s(\Phi) < +\infty\}$.

Generalizing a notation of Section 6.4. we set

$$\sigma_L(.) = L(f_L(\Phi) \in (.)) \qquad (L \in \boldsymbol{H}).$$

The generalization of 7.3.1. previously announced reads as follows:

7.3.2. Theorem. *For all L in $\boldsymbol{H}$ and all sequences (c_n) of numbers in $(0, 1]$ which tend toward zero,*

$$(\mathcal{K}_{c_n}\mathcal{D}_{c_n})\, L \underset{n\to\infty}{\Rightarrow} \int P_{l\mu}(.)\, \sigma_L(\mathrm{d}l).$$

Proof. 1. For all sequences (u_n) of non-negative real numbers which tend toward $+\infty$, letting $n \to \infty$ we obtain

$$(2[u_n])^{-s}\, \Phi([-[u_n], [u_n])^s + x) \underset{L}{\to} f_L(\Phi)$$

which relation holds uniformly in x, and

$$(2[u_n + 1])^{-s}\, \Phi([-[u_n + 1], [u_n + 1])^s + x) \underset{L}{\to} f_L(\Phi).$$

Using the following inequalities, which are valid for $u_n \geqq 1$ and for all Φ in M

$$\left(\frac{[u_n]}{u_n}\right)^s \cdot \frac{\Phi([-[u_n], [u_n])^s + x)}{(2[u_n])^s} \leqq (2u_n)^{-s}\, \Phi([-u_n, u_n)^s + x)$$
$$\leqq \left(\frac{[u_n + 1]}{u_n}\right)^s \frac{\Phi([-[u_n + 1], [u_n + 1])^s + x)}{(2[u_n + 1])^s}$$

we obtain the following relation, which holds uniformly in x:

$$(2u_n)^{-s}\, \Phi([-u_n, u_n)^s + x) \underset{L}{\to} f_L(\Phi).$$

2. Now let X be an arbitrary continuity set with respect to μ, and (v_n) a sequence of non-negative real numbers which tends toward $+\infty$. Then to each $\eta > 0$ there exists a real positive w such that, for suitable unions X_1, X_2 of finite families of half-open cuboids, $[-w, w)^s + 2w[k_1, \ldots, k_s]$, $k_1, \ldots, k_s$ being integers, the two conditions

$$X_1 \subseteqq X \subseteqq X_2, \qquad \mu(X_2 \setminus X_1) < \eta$$

are satisfied.

In view of 1., letting $n \to \infty$ we get uniformly in x

$$(v_n)^{-s}\, \Phi(v_n X_1 + x) \underset{L}{\to} \mu(X_1)\, f_L(\Phi)$$

$$(v_n)^{-s}\, \Phi(v_n X_2 + x) \underset{L}{\to} \mu(X_2)\, f_L(\Phi).$$

Now, if $\varepsilon > 0$, we obtain

$$\begin{aligned}
&\varliminf_{n\to\infty} \inf_{x\in R^s} L\big(|(v_n)^{-s}\, \Phi(v_nX + x) - \mu(X)\, f_L(\Phi)| \leqq \varepsilon\big)\\
&\geqq \varliminf_{n\to\infty} \inf_{x\in R^s} L\Big(f_L(\Phi) \leqq k,\ |(v_n)^{-s}\, \Phi(v_nX_1 + x) - \mu(X_1)\, f_L(\Phi)| \leqq \frac{\varepsilon}{3},\\
&\qquad\qquad |(v_n)^{-s}\, \Phi(v_nX_2 + x) - \mu(X_2)\, f_L(\Phi)| \leqq \frac{\varepsilon}{3}\Big)\\
&= L\big(f_L(\Phi) \leqq k\big)
\end{aligned}$$

for all natural numbers k satisfying the condition $k\eta < \varepsilon/3$. Since η, and hence also k, are arbitrary, it follows that

$$\lim_{n\to\infty} \sup_{x\in R^s} L\big(|(v_n)^{-s}\, \Phi(v_nX + x) - \mu(X)\, f_L(\Phi)| > \varepsilon\big) = 0.$$

3. In virtue of 2., for all continuity sets X with respect to μ and all continuity sets Z with respect to $L\big(\mu(X)\, f_L(\Phi) \in (.)\big)$, we obtain

$$\begin{aligned}
\big(\omega_{c_n}(\mathscr{K}_{c_n}L)\big)_X\,(Z) &= L\big(c_n\Phi\big((c_n)^{-1/s}X\big) \in Z\big)\\
&\xrightarrow[n\to\infty]{} L\big(\mu(X)\, f_L(\Phi) \in Z\big) = \sigma_L\big(\mu(X)\,\xi \in Z\big)\\
&= (V_{\mu,\sigma_L})_X\,(Z),
\end{aligned}$$

i.e. we have

$$\big(\omega_{c_n}(\mathscr{K}_{c_n}L)\big)_X \underset{n\to\infty}{\Rightarrow} (V_{\mu,\sigma_L})_X.$$

In view of 7.2.5., from this it follows that

$$\omega_{c_n}(\mathscr{K}_{c_n}L) \underset{n\to\infty}{\Rightarrow} V_{\mu,\sigma_L}.$$

Now, bearing in mind that

$$(\mathscr{K}_{c_n}\mathscr{D}_{c_n})\, L = \mathscr{D}_{c_n}(\mathscr{K}_{c_n}L)$$

and using 7.2.7. we observe that the following relation holds:

$$(\mathscr{K}_{c_n}\mathscr{D}_{c_n})\, L \underset{n\to\infty}{\Rightarrow} \mathscr{I}(V_{\mu,\sigma_L}) = \int P_{l\mu}(.)\, \sigma_L(\mathrm{d}l).\ \blacksquare$$

Now let P be a stationary distribution on $\mathfrak{M}$ which satisfies the equation $(\mathscr{K}_c\mathscr{D}_c)\, P = P$ for some c in $(0, 1)$. Then, for all natural numbers n, we obtain $(\mathscr{K}_{c^n}\mathscr{D}_{c^n})\, P = P$, so that for $P\big(s(\Phi) < +\infty\big) = 1$, using 7.3.1., we get the representation $P = \int P_{l\mu}(.)\, \sigma_L(\mathrm{d}l)$.

Now, if $P\big(s(\Phi) = +\infty\big) > 0$ were valid, we might use 7.2.7. to show that

$$(\mathscr{K}_{c_n}\mathscr{D}_{c_n})\, P = \mathscr{D}_{c_n}(\mathscr{K}_{c_n}P) \underset{n\to\infty}{\Rightarrow} P$$

would imply that

$$\omega_{c^n}(\mathscr{K}_{c^n}P) \underset{n\to\infty}{\Rightarrow} V, \qquad \mathscr{I}(V) = P.$$

In view of 6.4.2., $X = \left[-\frac{1}{2}, \frac{1}{2}\right)^s$ belongs to in $\mathfrak{B}_P$, and hence also to $\mathfrak{B}_V$, so that we obtain

$$\left(\omega_{c^n}(\mathscr{K}_{c^n} P)\right)_X \underset{n\to\infty}{\Rightarrow} V_X,$$

i.e. we have

$$P\left(c^n \Phi(c^{-n/s} X) \in (.)\right) \underset{n\to\infty}{\Rightarrow} V_X$$

which contradicts the inequality

$$V\left(\nu(X) \geqq k\right) \geqq \overline{\lim_{n\to\infty}}\, P\left(c^n \Phi(c^{-n/s} X) \geqq k\right)$$

$$\geqq P\left(\lim_{n\to\infty} c^n \Phi(c^{-n/s} X) = +\infty\right) = P\left(s(\Phi) = +\infty\right) > 0$$

which is valid for all natural numbers k.

Consequently $P\left(s(\Phi) = +\infty\right) = 0$, and we observe the truth of

7.3.3. Theorem. *For all c in $(0, 1)$, a stationary distribution P on $\mathfrak{M}$ satisfies the equation $(\mathscr{K}_c \mathscr{D}_c) P = P$ if and only if it is purely random.*

We may consider 7.3.3. an analogue of 4.8.3. Here it must however be noted that 4.8.3. characterized the totality of all distributions which are cluster-invariant with respect to $\varkappa(K)$, whereas in 7.3.3. only the stationary distributions which are cluster-invariant with respect to $\varkappa(K_c)$ are considered.

7.4. A Limit Theorem for Sequences of Substochastic Translations

In the proof of Theorem 7.3.2. we used the fact that for all continuity sets H with respect to μ and all sequences (c_n) of numbers in $(0, 1]$ which tend toward zero, the stochastic convergence relation

$$c_n \Phi\left((c_n)^{-1/s} H\right) \underset{L}{\to} f_L(\Phi)\, \mu(H)$$

is satisfied with respect to L, which yields

$$(c_n)^2\, \Phi\left((c_n)^{-1/s} H\right) \underset{L}{\to} 0.$$

If we use the special substochastic kernels K_c, $0 < c \leqq 1$, introduced in the preceding section, then we can transfer the above convergence relations into

$$\int K_{c^n}(x, H)\, \Phi(\mathrm{d}x) \underset{L}{\to} f_L(\Phi)\, \mu(H),$$

$$\int \left(K_{c^n}(x, H)\right)^2 \Phi(\mathrm{d}x) \underset{L}{\to} 0,$$

and the convergence assertion of 7.3.2. gets the form

$$L_{\varkappa(K_{c^n})} \underset{n\to\infty}{\Rightarrow} \int P_{f_L(\Phi)\mu}(.)\, L(\mathrm{d}\Phi).$$

In this form we can generalize Theorem 7.3.2. for arbitrary sequences (K_n) of substochastic kernels on $[A, \varrho_A]$ with phase space $[A, \varrho_A]$. Instead of the Lebesgue measure an arbitrary measure ν in N appears and instead of f_L an arbitrary non-negative measurable real function j, defined almost everywhere with respect to L, which fulfils

$$\int K_n(a, H)\, \Phi(\mathrm{d}a) \underset{L}{\longrightarrow} j(\Phi)\, \nu(H)$$

for all continuity sets H with respect to ν. If all distributions $L_{\varkappa(K_n)}$ exist, then

$$L_{\varkappa(K_n)} \underset{n\to\infty}{\Rightarrow} \int P_{j(\Phi)\nu}(.)\, L(\mathrm{d}\Phi)$$

is equivalent to the fact that for all continuity sets H with respect to ν the deconcentration property

$$\int \big(K_n(a, H)\big)^2\, \Phi(\mathrm{d}a) \underset{L}{\longrightarrow} 0$$

is satisfied.

We will generalize this statement, too. First of all as in Proposition 4.8.2. we investigate substochastic kernels K_n on $[A', \varrho_{A'}]$ with phase space $[A, \varrho_A]$. Then we replace the ring $\mathfrak{B}_{Q_\nu}$ of continuity sets with respect to ν by an arbitrary semiring $\mathfrak{H}$ of sets in $\mathfrak{B}$. In this way we get

7.4.1. Theorem. *Let be ν a measure in N, $\mathfrak{H}$ a semiring of sets in $\mathfrak{B}$, (K_n) a sequence of substochastic kernels on $[A', \varrho_{A'}]$ with phase space $[A, \varrho_A]$, L a distribution on $\mathfrak{M}'$ and j a non-negative measurable real function defined almost everywhere with respect to L. Suppose*

a) *$L_{\varkappa(K_n)}$ exists for $n = 1, 2, \ldots$,*
b) *For all H in $\mathfrak{H}$ and all $\varepsilon > 0$*

$$L\Big(\Big|\int K_n(a', H)\, \Phi(\mathrm{d}a') - j(\Phi)\, \nu(H)\Big| > \varepsilon\Big) \xrightarrow[n\to\infty]{} 0.$$

Then

c) *For all finite sequences $H_1, \ldots, H_m$ of sets in $\mathfrak{H}$,*

$$\big(L_{\varkappa(K_n)}\big)_{H_1,\ldots,H_m} \xrightarrow[n\to\infty]{} \Big(\int P_{j(\Phi)\nu}(.)\, L(\mathrm{d}\Phi)\Big)_{H_1,\ldots,H_m}$$

is equivalent to

d) *For all H in $\mathfrak{H}$ and all $\varepsilon > 0$*

$$L\Big(\int \big(K_n(a', H)\big)^2\, \Phi(\mathrm{d}a') > \varepsilon\Big) \xrightarrow[n\to\infty]{} 0.$$

This theorem can be further generalized, without complicating the proof, if we use a measurable mapping $\Phi \curvearrowright i_{(\Phi)}$ from M' into N instead of $\Phi \curvearrowright j(\Phi)\nu$.

7.4.2. Theorem. *Let be $\mathfrak{H}$ a semiring of sets in $\mathfrak{B}$, (K_n) a sequence of substochastic kernels on $[A', \varrho_{A'}]$, with phase space $[A, \varrho_A]$, L a distribution on $\mathfrak{M}'$*

and $\Phi \rightsquigarrow i_{(\Phi)}$ a measurable mapping from $[M', \mathfrak{M}']$ into $[N, \mathfrak{N}]$, defined almost everywhere with respect to L. Suppose

a) *$L_{\varkappa(K_n)}$ exists for $n = 1, 2, \ldots$,*

b) *For all H in $\mathfrak{H}$ and all $\varepsilon > 0$,*

$$L\left(\left|\int K_n(a', H)\, \Phi(\mathrm{d}a') - i_{(\Phi)}(H)\right| > \varepsilon\right) \xrightarrow[n\to\infty]{} 0.$$

Under these assumptions

c) *For all finite sequences $H_1, \ldots, H_m$ of sets in $\mathfrak{H}$*

$$(L_{\varkappa(K_n)})_{H_1,\ldots,H_m} \xrightarrow[n\to\infty]{} \left(\int P_{i_{(\Phi)}}(.)\, L(\mathrm{d}\Phi)\right)_{H_1,\ldots,H_m}$$

is equivalent to

d) *For all H in $\mathfrak{H}$ and all $\varepsilon > 0$*

$$L\left(\int (K_n(a', H))^2\, \Phi(\mathrm{d}a') > \varepsilon\right) \xrightarrow[n\to\infty]{} 0.$$

Proof. 1. d) implies c).

To each finite sequence $H_1, \ldots, H_m$ of sets in $\mathfrak{H}$ there exists a finite sequence $X_1, \ldots, X_k$ of pairwise disjoint sets in $\mathfrak{H}$ such that each H_i, $1 \leqq i \leqq m$, can be represented as the union of some of the X_j, $1 \leqq j \leqq k$. Now we get

$$\begin{aligned}&\left\|(L_{\varkappa(K_n)})_{H_1,\ldots,H_m} - \left(\int P_{i_{(\Phi)}}(.)\, L(\mathrm{d}\Phi)\right)_{H_1,\ldots,H_m}\right\| \\ &\leqq \left\|(L_{\varkappa(K_n)})_{X_1,\ldots,X_k} - \left(\int P_{i_{(\Phi)}}(.)\, L(\mathrm{d}\Phi)\right)_{X_1,\ldots,X_k}\right\|\end{aligned}$$

for all natural numbers n. Therefore we can now suppose that the sets $H_1, \ldots, H_m$ are pairwise disjoint.

In virtue of 1.11.3. and 1.12.1. we obtain

$$\begin{aligned}
&\overline{\lim_{n\to\infty}} \left\|(L_{\varkappa(K_n)})_{H_1,\ldots,H_m} - \left(\int P_{i_{(\Phi)}}(.)\, L(\mathrm{d}\Phi)\right)_{H_1,\ldots,H_m}\right\| \\
&\leqq \overline{\lim_{n\to\infty}} \int \left\|(\varkappa(K_n)_{(\Phi)})_{H_1,\ldots,H_m} - (P_{i_{(\Phi)}})_{H_1,\ldots,H_m}\right\| L(\mathrm{d}\Phi) \\
&\leqq \overline{\lim_{n\to\infty}} \int \left\|(\varkappa(K_n)_{(\Phi)})_{H_1,\ldots,H_m} - P_{\sum\limits_{1\leqq j\leqq m} (\int K_n(a', H_j)\Phi(\mathrm{d}a'))\delta_j}\right\| L(d\Phi) \\
&\quad + \overline{\lim_{n\to\infty}} \int \left\|P_{\sum\limits_{1\leqq j\leqq m} (\int K_n(a', H_j)\Phi(\mathrm{d}a'))\delta_j} - P_{\sum\limits_{1\leqq j\leqq m} i_{(\Phi)}(H_j)\delta_j}\right\| L(\mathrm{d}\Phi) \\
&\leqq \overline{\lim_{n\to\infty}} \int 2 \min\left(1, \int (K_n(a', H_1 \cup \cdots \cup H_m))^2\, \Phi(\mathrm{d}a')\right) L(\mathrm{d}\Phi) \\
&\quad + \overline{\lim_{n\to\infty}} \int 2 \min\left(1, \sum_{j=1}^{m} \left|\int K_n(a', H_j)\, \Phi(\mathrm{d}a') - i_{(\Phi)}(H_j)\right|\right) L(\mathrm{d}\Phi) \\
&\leqq 2\left(\varepsilon + \overline{\lim_{n\to\infty}} L\left(\sqrt{\int \left(\sum_{1\leqq j\leqq m} K_n(a', H_j)\right)^2 \Phi(\mathrm{d}a')} > \sqrt{\varepsilon}\right)\right) \\
&\quad + 2\left(\varepsilon + \overline{\lim_{n\to\infty}} L\left(\sum_{1\leqq j\leqq m} \left|\int K_n(a', H_j)\, \Phi(\mathrm{d}a') - i_{(\Phi)}(H_j)\right| > \varepsilon\right)\right)
\end{aligned}$$

$$\leqq 4\varepsilon + 2 \varlimsup_{n\to\infty} L\left(\sum_{1\leqq j\leqq m} \sqrt{\int (K_n(a', H_j))^2 \, \Phi(\mathrm{d}a')} > \sqrt{\varepsilon}\right)$$

$$+ 2 \sum_{1\leqq j\leqq m} \varlimsup_{n\to\infty} L\left(\left|\int K_n(a', H_j)\, \Phi(\mathrm{d}a') - i_{(\Phi)}(H_j)\right| > \frac{\varepsilon}{m}\right)$$

$$\leqq 4\varepsilon + 2 \sum_{1\leqq j\leqq m} \varlimsup_{n\to\infty} L\left(\int (K_n(a', H_j))^2 \, \Phi(\mathrm{d}a') > \frac{\varepsilon}{m^2}\right) + 0$$

$$= 4\varepsilon$$

for all $\varepsilon > 0$. Hence

$$(L_{\varkappa(K_n)})_{H_1,\ldots,H_m} \xrightarrow[n\to\infty]{} \left(\int P_{i_{(\Phi)}}(.)\, L(\mathrm{d}\Phi)\right)_{H_1,\ldots,H_m}.$$

2. For each H in $\mathfrak{H}$ and each natural number n we get by

$$l_n(a') = -\ln\left(1 - K_n(a', H)\right) \qquad (a' \in A')$$

a mapping l_n of A' into $[0, +\infty]$ which is measurable with respect to $\mathfrak{A}'$. (We set $-\ln 0 = +\infty$ and $e^{-\infty} = 0$.) Obviously,

$$l_n(a') \geqq K_n(a', H) + \frac{1}{2}\left(K_n(a', H)\right)^2 \qquad (a' \in A').$$

3. c) implies d).

Let H be in $\mathfrak{H}$. Then c) yields

$$(L_{\varkappa(K_n)})_H(\{0\}) \xrightarrow[n\to\infty]{} \left(\int P_{i_{(\Phi)}}(.)\, L(\mathrm{d}\Phi)\right)_H(\{0\}),$$

i.e.

$$\int \left(\varkappa(K_n)\right)_{(\Phi)} \left(\chi(H) = 0\right) L(\mathrm{d}\Phi) \xrightarrow[n\to\infty]{} \int \exp\left(-i_{(\Phi)}(H)\right) L(\mathrm{d}\Phi).$$

For each Φ in $M'_{\varkappa(K_n)}$ we have

$$\varkappa(K_n)_{(\Phi)}\left(\chi(H) = 0\right) = \prod_{a'\in A', \Phi(\{a'\})>0} \left(\left(1 - K_n(a', H)\right)^{\Phi(\{a'\})}\right)$$

$$= \prod_{a'\in A', \Phi(\{a'\})>0} \exp\left(\Phi(\{a'\}) \ln\left(1 - K_n(a', H)\right)\right)$$

$$= \exp\left(-\int l_n(a')\, \Phi(\mathrm{d}a')\right),$$

so that

$$(*) \qquad \int \exp\left(-\int l_n(a')\, \Phi(\mathrm{d}a')\right) L(\mathrm{d}\Phi) \xrightarrow[n\to\infty]{} \int \exp\left(-i_{(\Phi)}(H)\right) L(\mathrm{d}\Phi)$$

holds.

For all $\eta > 0$,

$$\varlimsup_{n\to\infty} \int \left|\exp\left(-\int K_n(a', H)\, \Phi(\mathrm{d}a')\right) - \exp\left(-i_{(\Phi)}(H)\right|\right. L(\mathrm{d}\Phi)$$

$$\leqq \eta + \varlimsup_{n\to\infty} L\left(\left|\int K_n(a', H)\, \Phi(\mathrm{d}a') - i_{(\Phi)}(H)\right| > \eta\right).$$

Consequently, by b) we obtain

$$(**)\quad \lim_{n\to\infty} \int \left|\exp\left(-\int K_n(a', H)\, \Phi(\mathrm{d}a')\right) - \exp\left(-i_{(\Phi)}(H)\right)\right| L(\mathrm{d}\Phi) = 0.$$

With the help of (*) it follows from (**) that

$$\lim_{n\to\infty} \int \left(\exp\left(-\int K_n(a', H)\, \Phi(\mathrm{d}a')\right) - \exp\left(-\int l_n(a')\, \Phi(\mathrm{d}a')\right)\right) L(\mathrm{d}\Phi) = 0.$$

In view of 2., $K_n(a', H) \leqq l_n(a')$ is always satisfied, so that we conclude

$$\lim_{n\to\infty} \int \left|\exp\left(-\int K_n(a', H)\, \Phi(\mathrm{d}a')\right) - \exp\left(-\int l_n(a')\, \Phi(\mathrm{d}a')\right)\right| L(\mathrm{d}\Phi) = 0.$$

By (**) from this we get the following refinement of (*):

$$\lim_{n\to\infty} \int \left|\exp\left(-\int l_n(a')\, \Phi(\mathrm{d}a')\right) - \exp\left(-i_{(\Phi)}(H)\right)\right| L(\mathrm{d}\Phi) = 0.$$

Therefore the sequence $\left(\exp\left(-\int l_n(a')\, \Phi(\mathrm{d}a')\right)\right)$ tends stochastically toward $\exp\left(-i_{(\Phi)}(H)\right)$. Taking into consideration that a stochastic convergence of (ξ_n) towards ξ is equivalent to the fact that each subsequence of (ξ_n) contains a subsequence which converges toward ξ almost everywhere, then we see that

$$-\int l_n(a')\, \Phi(\mathrm{d}a) \underset{L}{\longrightarrow} -i_{(\Phi)}(H),$$

i.e.

$$L\left(\left|\int l_n(a')\, \Phi(\mathrm{d}a') - i_{(\Phi)}(H)\right| > \varepsilon\right) \xrightarrow[n\to\infty]{} 0$$

for all $\varepsilon > 0$. In virtue of this convergence and 2. and b) we obtain

$$\begin{aligned}
&\overline{\lim_{n\to\infty}}\, L\left(\int (K_n(a', H))^2\, \Phi(\mathrm{d}a') > \varepsilon\right) \\
&\leqq \overline{\lim_{n\to\infty}}\, L\left(\int \left(l_n(a') - K_n(a', H)\right) \Phi(\mathrm{d}a') > \frac{\varepsilon}{2}\right) \\
&\leqq \overline{\lim_{n\to\infty}}\, L\left(\left|\int l_n(a')\, \Phi(\mathrm{d}a') - i_{(\Phi)}(H)\right| > \frac{\varepsilon}{4}\right) \\
&\quad + \overline{\lim_{n\to\infty}}\, L\left(\left|\int K_n(a', H)\, \Phi(\mathrm{d}a') - i_{(\Phi)}(H)\right| > \frac{\varepsilon}{4}\right) \\
&= 0 + 0
\end{aligned}$$

for all $\varepsilon > 0$. ∎

An analysis of the first step in the proof of Theorem 7.4.2. shows that the following uniformity assertion is satisfied.

7.4.3. Proposition. *Let be $\mathfrak{H}$ a semiring of sets in $\mathfrak{B}$, (K_n) a sequence of substochastic kernels on $[A', \varrho_{A'}]$ with phase space $[A, \varrho_A]$ and $\mathbf{L}$ a set of distributions on $\mathfrak{M}'$. Suppose that to each L in $\mathbf{L}$ corresponds a measurable map-*

ping $\Phi \rightsquigarrow i_{L,(\Phi)}$ *from* $[M', \mathfrak{M}']$ *into* $[N, \mathfrak{N}]$, *defined almost everywhere with respect to L, such that all distributions* $L_{\varkappa(K_n)}$, $n = 1, 2, \ldots$, *exist and for all* $\varepsilon > 0$ *and all* H *in* $\mathfrak{H}$

$$\lim_{n\to\infty} \sup_{L\in \boldsymbol{L}} L\left(\left|\int K_n(a', H)\, \Phi(\mathrm{d}a') - i_{L,(\Phi)}(H)\right| > \varepsilon\right) = 0,$$

$$\lim_{n\to\infty} \sup_{L\in \boldsymbol{L}} L\left(\int \left(K_n(a', H)\right)^2 \Phi(\mathrm{d}a') > \varepsilon\right) = 0$$

is fulfilled. Under these assumptions for each finite sequence $H_1, \ldots, H_m$ *of sets in* $\mathfrak{H}$, *we have*

$$\lim_{n\to\infty} \sup_{L\in \boldsymbol{L}} \left\|(L_{\varkappa(K_n)})_{H_1,\ldots,H_m} - \left(\int P_{i_{L,(\Phi)}}(.)\, L(\mathrm{d}\Phi)\right)_{H_1,\ldots,H_m}\right\| = 0.$$

Now we return again to Theorem 7.3.2. Let L be in $\boldsymbol{H}$. For all x in R^s we set

$$L_x = L\left(T_x\Phi \in (.)\right), \qquad i_{L_x,(\Phi)} = f_L(T_{-x}\Phi)\, \mu$$

if $f_L(T_{-x}\Phi)$ exists. Therefore $i_{L_x,(.)}$ is defined almost everywhere with respect to L_x. By the second step of the proof of 7.3.2., for each continuity set H with respect to μ and each $\varepsilon > 0$ we have

$$\begin{aligned}
&\sup_{x\in R^s} L_x\left(\left|\int K_{c_n}(a, H)\, \Phi(\mathrm{d}a) - i_{L_x,(\Phi)}(H)\right| > \varepsilon\right) \\
&= \sup_{x\in R^s} L_x\left(\left|c_n\Phi\left((c_n)^{-1/s} H\right) - f_L(T_{-x}\Phi)\, \mu(H)\right| > \varepsilon\right) \\
&= \sup_{x\in R^s} L\left(\left|c_n\Phi\left((c_n)^{-1/s} H + x\right) - f_L(\Phi)\, \mu(H)\right| > \varepsilon\right) \xrightarrow[n\to\infty]{} 0
\end{aligned}$$

and

$$\begin{aligned}
&\sup_{x\in R^s} L_x\left(\int \left(K_{c_n}(a, H)\right)^2 \Phi(\mathrm{d}a) > \varepsilon\right) \\
&= \sup_{x\in R^s} L\left(c_n{}^2\Phi\left((c_n)^{-1/s} H + x\right) > \varepsilon\right) \xrightarrow[n\to\infty]{} 0.
\end{aligned}$$

Because of

$$\begin{aligned}
&\sup_{x\in R^s} \left\|\int \left((L_x)_{\varkappa(K_{c_n})}\right)_{H_1,\ldots,H_m} - \left(\int P_{i_{L_x,(\Phi)}}(.)\, L_x(\mathrm{d}\Phi)\right)_{H_1,\ldots,H_m}\right\| \\
&= \sup_{x\in R^s} \left\|\left((\mathscr{K}_{c_n}\mathscr{D}_{c_n})\, L_x\right)_{H_1,\ldots,H_m} - \left(\int P_{f_L(\Phi)\mu}(.)\, L(\mathrm{d}\Phi)\right)_{H_1,\ldots,H_m}\right\|
\end{aligned}$$

there follows

7.4.4. Theorem. *Under the conditions of 7.3.2., for each finite sequence* $H_1, \ldots, H_m$ *of bounded Borel sets in* R^s *with the property* $\mu(\partial H_i) = 0$, $1 \leqq i \leqq m$,

$$\lim_{n\to\infty} \sup_{x\in R^s} \left\|\left((\mathscr{K}_{c_n}\mathscr{D}_{c_n})\, L\right)_{H_1+x,\ldots,H_m+x} - \left(\int P_{l\mu}(.)\, \sigma_L(\mathrm{d}l)\right)_{H_1,\ldots,H_m}\right\| = 0.$$

Sometimes the following version of 7.4.2. is more suitable.

7.4.5. Theorem. *Let be $\mathfrak{H}$ a semiring of sets in $\mathfrak{B}$, (K_n) a sequence of substochastic kernels on $[A', \varrho_{A'}]$ with phase space $[A, \varrho_A]$, L a distribution on $\mathfrak{M}'$ and $\Phi \rightsquigarrow i_{(\Phi)}$ a measurable mapping from $[M', \mathfrak{M}']$ into $[N, \mathfrak{N}]$ defined almost everywhere with respect to L. Suppose*

a') $K_n * \varrho_L \in N$ *for* $n = 1, 2, \ldots,$

b') *For all H in $\mathfrak{H}$*

$$\int \left| \int K_n(a', H)\, \Phi(\mathrm{d}a') - i_{(\Phi)}(H) \right| L(\mathrm{d}\Phi) \xrightarrow[n\to\infty]{} 0.$$

Under these assumptions

c) *For all finite sequences $H_1, \ldots, H_m$ of sets in $\mathfrak{H}$*

$$(L_{\varkappa(K_n)})_{H_1,\ldots,H_m} \xrightarrow[n\to\infty]{} \left(\int P_{i_{(\Phi)}}(\cdot)\, L(\mathrm{d}\Phi)\right)_{H_1,\ldots,H_m}$$

is equivalent to

d') *For all H in $\mathfrak{H}$*

$$\int \left(K_n(a', H)\right)^2 \varrho_L(\mathrm{d}a') \xrightarrow[n\to\infty]{} 0.$$

Proof. In view of 4.2.3., a') implies a). Obviously, b') implies b), too.

If d') and consequently also d) is satisfied, then Theorem 7.4.2. yields c).

Conversely, let c) be fulfilled. By 7.4.2. we have d), too. Let H be an arbitrary set in $\mathfrak{H}$. For the sake of abbreviation, for all Φ in M' and all natural numbers n, we set

$$f_n(\Phi) = \int K_n(a', H)\, \Phi(\mathrm{d}a').$$

By a') we have $\int f_n(\Phi)\, L(\mathrm{d}\Phi) < +\infty$ and by b')

$$\int i_{(\Phi)}(H)\, L(\mathrm{d}\Phi) \leqq \int |f_n(\Phi) - i_{(\Phi)}(H)|\, L(\mathrm{d}\Phi) + \int f_n(\Phi)\, L(\mathrm{d}\Phi) < +\infty$$

is satisfied for all sufficiently large n.

Now we choose numbers u, v in $(0, +\infty)$ and set $w = u + v$. From d) it follows that

$$\overline{\lim_{n\to\infty}} \int\limits_{\{\Phi : f_n(\Phi) \leqq w\}} \left(\int \left(K_n(a', H)\right)^2 \Phi(\mathrm{d}a')\right) L(\mathrm{d}\Phi)$$

$$\leqq \varepsilon + w\, \overline{\lim_{n\to\infty}}\, L\left(\int \left(K_n(a', H)\right)^2 \Phi(\mathrm{d}a') > \varepsilon\right) = \varepsilon$$

for all $\varepsilon > 0$, i.e.

$$\int\limits_{\{\Phi : f_n(\Phi) \leqq w\}} \left(\int \left(K_n(a', H)\right)^2 \Phi(\mathrm{d}a')\right) L(\mathrm{d}\Phi) \xrightarrow[n\to\infty]{} 0.$$

Furthermore we have

$$\int\limits_{\{\Phi : f_n(\Phi) > w\}} \left(\int \left(K_n(a', H)\right)^2 \Phi(\mathrm{d}a)\right) L(\mathrm{d}\Phi) \leqq \int\limits_{\{\Phi : f_n(\Phi) > w\}} f_n(\Phi)\, L(\mathrm{d}\Phi)$$

$$\leqq \int\limits_{\{\Phi : f_n(\Phi) > w\}} i_{(\Phi)}(H)\, L(\mathrm{d}\Phi) + \int |f_n(\Phi) - i_{(\Phi)}(H)|\, L(\mathrm{d}\Phi).$$

If for a Φ in M' the inequalities

$$f_n(\Phi) > w, \qquad i_{(\Phi)}(H) \leqq u$$

are fulfilled, we obtain $f_n(\Phi) - i_{(\Phi)}(H) > v$. It follows that

$$\int_{\{\Phi: f_n(\Phi) > w\}} i_{(\Phi)}(H)\, L(\mathrm{d}\Phi)$$

$$\leqq \int_{\{\Phi: i_{(\Phi)}(H) > u\}} i_{(\Phi)}(H)\, L(\mathrm{d}\Phi) + uL\big(f_n(\Phi) - i_{(\Phi)}(H) > v\big)$$

$$\leqq \int_{\{\Phi: i_{(\Phi)}(H) > u\}} i_{(\Phi)}(H)\, L(\mathrm{d}\Phi) + v^{-1}u \int |f_n(\Phi) - i_{(\Phi)}(H)|\, L(\mathrm{d}\Phi).$$

Summarizing, with the help of b′) we conclude that

$$\overline{\lim_{n\to\infty}} \int\Big(\int\big(K_n(a', H)\big)^2\, \Phi(\mathrm{d}a')\Big)\, L(\mathrm{d}\Phi)$$

$$\leqq \int_{\{\Phi: i_{(\Phi)}(H) > u\}} i_{(\Phi)}(H)\, L(\mathrm{d}\Phi).$$

Since this inequality holds for all $u > 0$ and, as above mentioned,

$$\int i_{(\Phi)}(H)\, L(\mathrm{d}\Phi) < +\infty$$

holds, too, we have

$$\int\big(K_n(a', H)\big)^2\, \varrho_L(\mathrm{d}a') = \int\Big(\int\big(K_n(a', H)\big)^2\, \Phi(\mathrm{d}a')\Big)\, L(\mathrm{d}\Phi) \xrightarrow[n\to\infty]{} 0. \blacksquare$$

In analogy to 7.4.3. by an analysis of the proof of 7.4.5. we get

7.4.6. Proposition. *Let be $\mathfrak{H}$ a semiring of sets in $\mathfrak{B}$, (K_n) a sequence of substochastic kernels on $[A', \varrho_{A'}]$ with phase space $[A, \varrho_A]$ and $\mathbf{L}$ a set of distributions on $\mathfrak{M}'$. Suppose that to each L in $\mathbf{L}$ corresponds a measurable mapping $\Phi \curvearrowright i_{L,(\Phi)}$ from $[M', \mathfrak{M}']$ into $[N, \mathfrak{N}]$, defined almost everywhere with respect to L, such that all measures $K_n * \varrho_L$, $n = 1, 2, \ldots$,* belong to *N and for all H in $\mathfrak{H}$*

$$\lim_{n\to\infty} \sup_{L\in\mathbf{L}} \int\Big|\int K_n(a', H)\, \Phi(\mathrm{d}a') - i_{L,(\Phi)}(H)\Big|\, L(\mathrm{d}\Phi) = 0,$$

$$\lim_{n\to\infty} \sup_{L\in\mathbf{L}} \int\big(K_n(a', H)\big)^2\, \varrho_L(\mathrm{d}a') = 0$$

is satisfied. Under these assumptions for each finite sequence $H_1, \ldots, H_m$ of sets in $\mathfrak{H}$ we have

$$\lim_{n\to\infty} \sup_{L\in\mathbf{L}} \Big\|\big(L_{\varkappa(K_n)}\big)_{H_1,\ldots,H_m} - \Big(\int P_{i_{L,(\Phi)}}(\cdot)\, L(\mathrm{d}\Phi)\Big)_{H_1,\ldots,H_m}\Big\| = 0.$$

8. The Palm Measure

8.0. Introduction

If the phase space $[A, \varrho_A]$ is of the form R^s, $s \geqq 1$, then the mapping $\Phi \curvearrowright \mathring{\Phi}$ introduced in Section 5.0. can be modified, in a way that not Φ itself but $T_x\Phi$ is considered a mark of a point x in Φ. Accordingly we set

$$\Phi^\wedge = \sum_{x \in R^s, \Phi(\{x\})>0} \Phi(\{x\})\, \delta_{[x, T_x\Phi]}.$$

All marks $T_x\Phi$ occurring in this case belong to in the subset

$$M^0 = \{\Phi : \Phi \in M, \Phi(\{[0, \ldots, 0]\}) > 0\}$$

of M which is closed with respect to ϱ_M. If we now set ϱ_{M^0} equal to the restriction of ϱ_M to M^0, and $[A^\wedge, \varrho_{A^\wedge}]$ equal to the direct product of the "position space" R^s and the "mark space" $[M^0, \varrho_{M^0}]$, then $\Phi \curvearrowright \Phi^\wedge$ provides a measurable mapping of $[M, \mathfrak{M}]$ into the measurable space $[M^\wedge, \mathfrak{M}^\wedge]$ of counting measures, which belongs to $[A^\wedge, \varrho_{A^\wedge}]$. We denote by $H^\wedge$ the image of a measure H on $\mathfrak{M}$ with respect to this mapping. Then the intensity measure $\varrho_{H^\wedge}$ is the image of the Campbell measure $\mathscr{C}_H$ of H with respect to the measurable mapping $[x, \Phi] \curvearrowright [x, T_x\Phi]$ of $[C, \mathfrak{C}]$ onto $[R^s \times M^0, \mathfrak{R}^s \otimes \mathfrak{M}^0]$. We shall denote it by $\mathscr{C}_H{}^T$.

An advantage of the mapping $\Phi \curvearrowright \Phi^\wedge$ as compared to the mapping $\Phi \curvearrowright \mathring{\Phi}$, which is defined for arbitrary phase spaces, consists in the commutativity relation $(T_y\Phi)^\wedge = T_y(\Phi^\wedge)$, which is valid for all y in R^s. Together with H, $H^\wedge$ is always stationary, too. Moreover it can be shown that, for all σ-finite stationary measures H on $\mathfrak{M}$, the measure $\gamma_{H^\wedge} = \varrho_H\big([0,1)^s \times (.)\big)$ is σ-finite. In this case $\gamma_{H^\wedge}$ is called the Palm measure of H and denoted by H^0. Consequently this measure is defined on the σ-algebra $\mathfrak{M}^0 = M^0 \cap \mathfrak{M}$ of the Borel subsets of $[M^0, \varrho_{M^0}]$, but can also be considered to be a measure on $\mathfrak{M}$ satisfying the condition $H^0(\Phi \notin M^0) = 0$. It is related to the Campbell measure of H by the relation $\mathscr{C}_H{}^T = \mu \times H^0$.

The basic properties of the Campbell measures, which were derived in Section 5.1., can readily be translated into corresponding properties of Palm measures. In particular, the measure $H\big((.) \setminus \{o\}\big)$ is uniquely determined by H^0 (Proposition 8.1.3.). If the convolution $H_1 * H_2$ of two σ-finite stationary measures H_1, H_2 on $\mathfrak{M}$ is again σ-finite, then $(H_1 * H_2)^0 = (H_1)^0 * H_2 + H_1 * (H_2)^0$ (Proposition 8.3.7.). Theorem 8.3.8. says that a stationary distribution P on $\mathfrak{M}$ is infinitely divisible if and only if there exists a σ-finite measure Q on $\mathfrak{M}$ having the property that $P^0 = Q * P$. The measure Q is uniquely determined, $Q = (\tilde{P})^0$. This characterization of the stationary infinitely divisible distributions was derived in Kerstan & Matthes [2] for the case $s = 1$ and in Ambartzumian [1] for arbitrary s. For the stationary Poisson distributions $P_{l\mu}, l \geqq 0$, we have $(\widetilde{P_{l\mu}})^0 = l\delta_{\delta_{[0,\ldots,0]}}$ (Proposition 8.3.9.). Consequently a stationary distribution P on $\mathfrak{M}$ coincides with $P_{l\mu}$ if and only if it satisfies the equation $P^0 = l\delta_{\delta_{[0,\ldots,0]}} * P$ (Theorem 8.3.10., cf. Slivnyak [1], [2]).

The transition from $\mathscr{C}_H$ to H^0, which is possible for σ-finite stationary measures H on $\mathfrak{M}$, represents a substantial conceptual simplification. However, it turns out that in the proofs of statements on Palm measures usually Campbell measures are implicitely involved. Even if we are only interested in statements on Palm measures, yet it is reasonable first to derive corresponding, more general statements on Campbell measures, which are shown in a simpler way and permit easier understanding. This is especially true for the characterization of classes of Palm measures by symmetry properties. Here it proves advantageous first to establish characteristic symmetry properties of corresponding classes of Campbell measures (cf. Section 5.3.). In this way not merely does the origin of these symmetry properties clearly stand out, but moreover it is possible not only to verify, but also to find out concrete symmetry properties. The main results in this direction are the characterization of those measures Q on $\mathfrak{M}^0$ for which there exist stationary regular infinitely divisible distributions P on $\mathfrak{M}$ having the property that $(\tilde{P})^0 = Q$, which is derived for arbitrary dimensions s (Theorem 8.4.2., cf. Ambartzumian [1]), the characterization, given by Theorem 8.5.2. for $s = 1$, of all Palm measures of stationary distributions P on $\mathfrak{M}$ having the property that $P(\{o\}) = 0$, and the characterization, given by Theorem 8.5.5. for $s = 1$, of those measures V on $\mathfrak{M}^0$, for which there exist stationary infinitely divisible distributions P on $\mathfrak{M}$ having the property that $(\tilde{P})^0 = V$. The now classical characterization given by Theorem 8.5.1. in the case $s = 1$, of all Palm measures of simple stationary distributions is due to Ryll-Nardzewski [1] and Slivnyak [1], [2], (see also Kerstan & Matthes [2], Neveu [1], Papangelou [1], and Mecke [6]).

Just as in Chapter 5, in this chapter we shall assume as a rule that the phase space $[A, \varrho_A]$ be the direct product of an Euclidean "position space" R^s and a bounded complete separable "mark space" $[K, \varrho_K]$, setting

$$\Phi^\wedge = \sum_{x \in R^s, \Phi(\{x\} \times K) > 0} \Phi(\{x\} \times K)\, \delta_{[x, T_x \Phi]}.$$

The results of this chapter can be generalized in such a way that, instead of Euclidean spaces, arbitrary locally compact Abelian topological groups G satisfying the second axiom of countability can be utilized as "position spaces". Note that for such groups always there exists a suitable metric ϱ_G, so that the connection to the conceptual system used by us can be accomplished. A corresponding generalization of Theorem 8.4.8. was given by Mecke [1] (for the extension to non-Abelian groups, cf. Tortrat [2]).

8.1. The Mapping $H \rightsquigarrow H^0$

In this chapter, we shall as a rule assume that the phase space $[A, \varrho_A]$ be the direct product of an Euclidean space R^s, $s \geqq 1$, and a bounded complete separable "mark space" $[K, \varrho_K]$.

We set

$$M^0 = \{\Phi : \Phi \in M, \Phi^x(\{[0, \ldots, 0]\}) > 0\},$$

denoting by $\mathfrak{M}^0$ the σ-algebra of those sets in $\mathfrak{M}$ which are included in M^0. In virtue of 6.1.1., the one-to-one mapping

$$[x, \Phi] \rightsquigarrow [x, T_x\Phi]$$

of C onto $R^s \times M^0$ is measurable with respect to the σ-algebras $\mathfrak{C}$, $\mathfrak{R}^s \otimes \mathfrak{M}^0$ in both directions. It transforms a measure V on $\mathfrak{C}$ into a measure V^T on $\mathfrak{R}^s \otimes \mathfrak{M}^0$. In particular, for all Φ in M,

$$(\mathring{\Phi})^T = \sum_{x \in R^s, \Phi^x(\{x\})>0} \Phi^x(\{x\})\, \delta_{[x, T_x\Phi]} = \int \delta_{[x, T_x\Phi]}(.)\, \Phi^x(\mathrm{d}x),$$

and for all measures H on $\mathfrak{M}$ we obtain

$$(\mathscr{C}_H)^T = \int \left(\int \delta_{[x, T_x\Phi]}(.)\, \Phi^x(\mathrm{d}x) \right) H(\mathrm{d}\Phi).$$

For the sake of simplicity, in what follows we shall write $\mathscr{C}_H{}^T$ instead of $(\mathscr{C}_H)^T$.

The transition from $\mathscr{C}_H$ to $\mathscr{C}_H{}^T$ is justified by the following proposition:

8.1.1. *Let H be a measure on $\mathfrak{M}$, for which $H\big((.) \setminus \{o\}\big)$ is σ-finite. Subject to this assumption, H is stationary if and only if there exists a σ-finite measure Q on $\mathfrak{M}^0$, such that*

$$\mathscr{C}_H{}^T = \mu \times Q$$

is satisfied.

Proof. 1. For a σ-finite measure Q on $\mathfrak{M}^0$, let $\mathscr{C}_H{}^T = \mu \times Q$. For all bounded X in $\mathfrak{R}^s$, all y in R^s, and all Y in $\mathfrak{M}^0$ satisfying the condition that $Q(Y) < +\infty$,

$$\begin{aligned}
\mathscr{C}_H{}^T(X \times Y) &= \int \Big(\int_X \delta_{T_x\Phi}(Y)\, \Phi^x(\mathrm{d}x) \Big) H(\mathrm{d}\Phi) = \mu(X)\, Q(Y) \\
&= \mu(X + y)\, Q(Y) = \int \Big(\int_{X+y} \delta_{T_x\Phi}(Y)\, \Phi^x(\mathrm{d}x) \Big) H(\mathrm{d}\Phi) \\
&= \int \Big(\int_X \delta_{T_z(T_y\Phi)}(Y)\, (T_y\Phi)^x\, (\mathrm{d}z) \Big) H(\mathrm{d}\Phi) \\
&= \int \Big(\int_X \delta_{T_z\Psi}(Y)\, \Psi^x\, (\mathrm{d}z) \Big) H\big(\mathrm{d}(T_{-y}\Psi)\big) \\
&= \mathscr{C}^T_{H(T_{-y}\Psi \in (.))}(X \times Y).
\end{aligned}$$

Consequently $\mathscr{C}_H{}^T = \mathscr{C}^T_{H(T_{-y}\Psi\in(.))}$ for all y in R^s, i.e. the measure $H\big((.)\setminus\{o\}\big)$ is stationary. Since

$$H = H\big((.)\setminus\{o\}\big) + H(\{o\})\,\delta_o,$$

we can conclude from the above that H is stationary.

2. Conversely, suppose that H be stationary, and $H\big((.)\setminus\{o\}\big)$ σ-finite.

For all non-negative real functions f on $R^s \times R^s \times M^0$ which are measurable with respect to the σ-algebra $\mathfrak{R}^s \otimes \mathfrak{R}^s \otimes \mathfrak{M}^0$, we obtain by means of 5.1.2.

$$\begin{aligned}
&\int f(t, u, \Phi)\, (\mathscr{C}_H{}^T \times \mu)\big(\mathrm{d}[[u, \Phi], t]\big) \\
&= \int\Big(\int f(t, u, \Phi)\, \mathscr{C}_H{}^T(\mathrm{d}[u, \Phi])\Big)\, \mu(\mathrm{d}t) \\
&= \int\Big(\int\Big(\sum_{u\in R^s,\Phi^x(\{u\})>0} f(t, u, T_u\Phi)\, \Phi^x(\{u\})\Big)\, H(\mathrm{d}\Phi)\Big)\, \mu(\mathrm{d}t) \\
&= \int\Big(\int\Big(\sum_{u\in R^s,(T_t\Phi)^x(\{u\})>0} f\big(t, u, T_u(T_t\Phi)\big)\, (T_t\Phi)^x(\{u\})\Big)\, H(\mathrm{d}\Phi)\Big)\, \mu(\mathrm{d}t) \\
&= \int\Big(\int\Big(\sum_{u\in R^s,\Phi^x(\{u+t\})>0} f(t, u, T_{u+t}\Phi)\, \Phi^x(\{u + t\})\Big)\, H(\mathrm{d}\Phi)\Big)\, \mu(\mathrm{d}t) \\
&= \int\Big(\int\Big(\sum_{r\in R^s,\Phi^x(\{r\})>0} f(t, r - t, T_r\Phi)\, \Phi^x(\{r\})\Big)\, H(\mathrm{d}\Phi)\Big)\, \mu(\mathrm{d}t) \\
&= \int\Big(\int f(t, u - t, \Phi)\, \mathscr{C}_H{}^T(\mathrm{d}[u, \Phi])\Big)\, \mu(\mathrm{d}t) \\
&= \int\Big(\int f(t, u - t, \Phi)\, \mu(\mathrm{d}t)\Big)\, \mathscr{C}_H{}^T(\mathrm{d}[u, \Phi]).
\end{aligned}$$

In an analogous way, we are led to

$$\begin{aligned}
&\int f(u, t, \Phi)\, (\mathscr{C}_H{}^T \times \mu)\big(\mathrm{d}[[u, \Phi], t]\big) \\
&= \int\Big(\int f(u - t, t, \Phi)\, \mu(\mathrm{d}t)\Big)\, \mathscr{C}_H{}^T(\mathrm{d}[u, \Phi]).
\end{aligned}$$

On the other hand, we have

$$\int f(t, u - t, \Phi)\, \mu(\mathrm{d}t) = \int f(u - t, t, \Phi)\, \mu(\mathrm{d}t).$$

Summarizing these results, we obtain

$$\begin{aligned}
&\int f(t, u, \Phi)\, (\mathscr{C}_H{}^T \times \mu)\big(\mathrm{d}[[u, \Phi], t]\big) \\
&= \int f(u, t, \Phi)\, (\mathscr{C}_H{}^T \times \mu)\big(\mathrm{d}[[u, \Phi], t]\big),
\end{aligned}$$

from which by rewriting the variables on the right hand side it can be concluded that

$$\begin{aligned}
&\int f(t, u, \Phi)\, (\mathscr{C}_H{}^T \times \mu)\big(\mathrm{d}[[u, \Phi], t]\big) \\
&= \int f(t, u, \Phi)\, (\mathscr{C}_H{}^T \times \mu)\big(\mathrm{d}[[t, \Phi], u]\big). \qquad (*)
\end{aligned}$$

Now, if in particular we choose $f = k_{X_0\times X\times Y}$, where $X_0, X \in \mathfrak{R}^s$, $\mu(X_0) = 1$, $\mu(X) < +\infty$ and $Y \in \mathfrak{M}^0$, then the following relation results:

$$\mathscr{C}_H{}^T(X \times Y) = \mu(X)\, \mathscr{C}_H{}^T(X_0 \times Y).$$

We set

$$Q = \mathscr{C}_H{}^T\big(X_0 \times (.)\big),$$

so that we are now able to write

$$\mathscr{C}_H{}^T(X \times Y) = \mu(X)\, Q(Y). \tag{**}$$

The measure Q is σ-finite. This is because, in view of 5.1.11., firstly $\mathscr{C}_H$, and hence also $\mathscr{C}_H{}^T$, are σ-finite measures. Thus there exists an everywhere positive real function v on $R^s \times M^0$ being measurable with respect to $\mathfrak{R}^s \otimes \mathfrak{M}^0$ and having the property that $\int v(u, \Phi)\, \mathscr{C}_H{}^T(\mathrm{d}[u, \Phi]) < +\infty$. According to (*) we obtain

$$\begin{aligned}
&\int\Big(\int v(u, \Phi)\, \mu(\mathrm{d}u)\Big)\, Q(\mathrm{d}\Phi)\\
&= \int\Big(\int v(u, \Phi)\, \mu(\mathrm{d}u)\Big)\, k_{X_0}(t)\, \mathscr{C}_H{}^T(\mathrm{d}[t, \Phi])\\
&= \int k_{X_0}(t)\, v(u, \Phi)\, (\mathscr{C}_H{}^T \times \mu)\,\big(\mathrm{d}\big[[t, \Phi], u\big]\big)\\
&= \int k_{X_0}(t)\, v(u, \Phi)\, (\mathscr{C}_H{}^T \times \mu)\,\big(\mathrm{d}\big[[u, \Phi], t\big]\big)\\
&= \int v(u, \Phi)\, \mathscr{C}_H{}^T(\mathrm{d}[u, \Phi]) < +\infty.
\end{aligned}$$

Hence $w(\Phi) = \int v(u, \Phi)\, \mu(\mathrm{d}u)$ represents an everywhere positive function on M^0 measurable with respect to $\mathfrak{M}^0$ and having the property that $\int w(\Phi)\, Q(\mathrm{d}\Phi) < +\infty$. From this it follows that Q is σ-finite. Hence, in view of (**),

$$\mathscr{C}_H{}^T = \mu \times Q. \;\blacksquare$$

The σ-finite measure Q on $\mathfrak{M}^0$ which is, subject to the assumptions of 8.1.1., uniquely determined, is called the *Palm measure* of H and denoted by H^0. For all Y in $\mathfrak{M}^0$ and all X in $\mathfrak{R}^s$ satisfying the condition $0 < \mu(X) < +\infty$,

$$H^0(Y) = \big(\mu(X)\big)^{-1} \int\Big(\int_X k_Y(T_x\Phi)\, \Phi^x(\mathrm{d}x)\Big)\, H(\mathrm{d}\Phi).$$

The right side of this formula makes sense also for arbitrary Y in $\mathfrak{M}$, thus enabling us to extend H^0 to a measure on $\mathfrak{M}$. This extension is however trivial, for it is of the form $H^0\big((.) \cap M^0\big)$. In what follows, we shall optionally conceive of H^0 as either a measure on $\mathfrak{M}^0$ or a measure on $\mathfrak{M}$ satisfying the condition $H^0(M \setminus M^0) = 0$.

For stationary measures, the refined Campbell theorem 5.1.2. takes the following form:

8.1.2. Proposition. *Let H be a stationary measure on $\mathfrak{M}$. If $H\big((.) \setminus \{o\}\big)$ is σ-finite, then for all non-negative real functions f on $R^s \times M^0$ which are measurable with respect to the σ-algebra $\mathfrak{R}^s \otimes \mathfrak{M}^0$,*

$$\int\Big(\sum_{x \in R^s, \Phi^x(\{x\}) > 0} f(x, T_x\Phi)\, \Phi^x(\{x\})\Big)\, H(\mathrm{d}\Phi) = \int f(x, \Psi)\, (\mu \times H^0)\, (\mathrm{d}[x, \Psi]).$$

The following propositions are derived immediately from the corresponding statements 5.1.10., 5.1.4., 5.1.3., and 5.1.12. on Campbell measures.

8.1.3. *Let H_1, H_2 be stationary measures on $\mathfrak{M}$, for which $H_1\big((.) \setminus \{o\}\big)$, $H_2\big((.) \setminus \{o\}\big)$ are σ-finite. Subject to these assumptions, $(H_1)^0 = (H_2)^0$ if and only if $H_1\big((.) \setminus \{o\}\big) = H_2\big((.) \setminus \{o\}\big)$ is satisfied.*

8.1.4. *For all at most countable families $(H_i)_{i\in I}$ of stationary measures on $\mathfrak{M}$ and all non-negative v_i, $i \in I$,*

$$\Big(\sum_{i\in I} v_i H_i\Big)^0 = \sum_{i\in I} v_i (H_i)^0,$$

if all $H_i\big((.) \setminus \{o\}\big)$ as well as $\sum_{i\in I} v_i H_i\big((.) \setminus \{o\}\big)$ are σ-finite.

8.1.5. *If H is stationary and $H\big((.) \setminus \{o\}\big)$ is σ-finite, then*

$$i_{H^x} = H^0(M^0).$$

8.1.6. *Subject to the assumptions of* 8.1.5.,

$$\big(H((.) \cap Y)\big)^0 = H^0\big((.) \cap Y\big)$$

for all Y in $\mathfrak{J}$.

Here $\mathfrak{J}$ denotes the σ-algebra of invariant sets in $\mathfrak{M}$ introduced in Section 6.2.

Thus in particular, for all Y in $\mathfrak{J}$, $H\big(\Phi \notin Y, \Phi \neq o\big) = 0$ is equivalent to $H^0(\chi \notin Y) = 0$. Setting especially $Y = \{\Phi : \Phi \in M, \Phi^x \text{ simple}\}$, we observe that H^x is simple if and only if $(H^0)^x$ has this property. We even have

8.1.7. *Subject to the assumptions of* 8.1.5., *the following statements are equivalent:*

a) *H^x is simple.*
b) *$(H^0)^x$ is simple.*
c) $H^0\big(\chi^x(\{[0, \ldots, 0]\}) > 1\big) = 0$.

Proof. It has already been stated above that a) and b) are equivalent. Obviously c) is a consequence of b). Thus it remains only to show that c) implies a).

To this end we use 8.1.2., choosing any set X in $\mathfrak{R}^s$ which satisfies the condition $0 < \mu(X) < +\infty$ and setting

$$f(x, \chi) = \begin{cases} \chi^x(\{[0, \ldots, 0]\}) - 1 & \text{if} \quad x \in X \\ 0 & \phantom{\text{if}} \quad x \notin X. \end{cases}$$

This gives

$$\begin{aligned} &H\big(_X(\Phi^x) \text{ is not simple}\big) \\ &\leqq \int\Big(\sum_{x\in R^s, \Phi^x(\{x\})>0} f(x, T_x\Phi)\, \Phi^x(\{x\})\Big)\, H(\mathrm{d}\Phi) \\ &= \int\big(\int f(x, \chi)\, H^0(\mathrm{d}\chi)\big)\, \mu(\mathrm{d}x) \\ &= \int_X\Big(\int\big(\chi^x(\{[0, \ldots, 0]\}) - 1\big)\, H^0(\mathrm{d}\chi)\Big)\, \mu(\mathrm{d}x) = 0 \cdot \mu(X) = 0. \end{aligned}$$

Bearing in mind that the selection of X was arbitrary, we can conclude from this that H^x is simple. ∎

Like the Campbell measure, the Palm measure can be introduced also in the "non-marked case", i.e. for $[A, \varrho_A] = R^s$, $s \geqq 1$. Then obviously we obtain

8.1.8. *Let H be a stationary measure on $\mathfrak{M}$. If then $H^x\big((.) \setminus \{o\}\big)$ is σ-finite, then*

$$(H^x)^0 = (H^0)^x.$$

8.2. Inversion Formulae

As an immediate consequence of 5.1.7. and 5.1.12., we have

8.2.1. *Let Y be a set in $\mathfrak{J}$ satisfying the condition $o \notin Y$ and h a non-negative real function on $C \cap (R^s \times Y)$ being measurable with respect to $\mathfrak{C}$ and having the property that*

$$\sum_{x \in R^s, \Phi^x(\{x\})>0} h(x, \Phi)\, \Phi^x(\{x\}) = 1 \qquad (\Phi \in Y).$$

Now, if for a stationary measure H on $\mathfrak{M}$ the measure $H\big((.) \setminus \{o\}\big)$ is σ-finite, then, for all Y' in $\mathfrak{M}$,

$$H(Y' \cap Y) = \int_{R^s \times Y} k_{Y'}(T_{-x}\Phi)\, h(x, T_{-x}\Phi)\, (\mu \times H^0)\, (\mathrm{d}[x, \Phi]).$$

We may explicitly state a function h, which satisfies the assumptions of 8.2.1. for $Y = M \setminus \{o\}$:

8.2.2. *The real function h on C, which is defined by*

$$h(x, \Phi) = \begin{cases} \big(\Phi^x(\{y : y \in R^s, \|y\| = \|x\|\})\big)^{-1}, & \textit{if } \Phi^x(\{y : y \in R^s, \|y\| < \|x\|\}) = 0 \\ 0 & \textit{otherwise} \end{cases}$$

satisfies the assumptions of 8.2.1. for $Y = M \setminus \{o\}$.

Proof. It follows immediately from our definition that, for all Φ in $M \setminus \{o\}$,

$$\sum_{x \in R^s, \Phi^x(\{x\})>0} h(x, \Phi)\, \Phi^x(\{x\}) = 1.$$

Thus it only remains to show that the non-negative real function h is measurable with respect to $\mathfrak{C}$.

The measurable mapping $[x, k] \curvearrowright \|x\|$ of $A = R^s \times K$ in R^1 transforms each Φ in M into some $\Phi^!$ in the set $M^!$ of counting measures which corresponds to the phase space $[A^!, \varrho_{A^!}] = R^1$. Obviously the mapping $\Phi \curvearrowright \Phi^!$ is measurable with respect to the σ-algebras $\mathfrak{M}, \mathfrak{M}^!$. If we now use the measurable indexing relation introduced by 5.3.4., we may write:

$$h(x, \Phi) = \begin{cases} \left(\Phi^x(\{[0, \ldots, 0]\})\right)^{-1}, & \text{if} \quad x = [0, \ldots, 0] \\ \left(\Phi^!(\{\|x\|\})\right)^{-1}, & \text{if} \quad k(\|x\|, \Phi^!) = 0 \text{ and } \\ & \qquad \Phi^x(\{[0, \ldots, 0]\}) = 0. \\ 0 & \text{otherwise}. \end{cases}$$

Consequently, in view of 5.3.4. and 5.1.9., h is measurable. ∎

Let V denote the set of those Φ in $M \setminus \{o\}$ for which $h(x, \Phi)$ is positive only at a single point x in R^s which satisfies the condition $\Phi^x(\{x\}) > 0$; in what follows, we shall denote this point x by $y(\Phi)$ for all Φ in V. For all X in $\mathfrak{R}^s$,

$$\{\Phi \cdot \Phi \in V, y(\Phi) \in X\}$$

$$= \bigcup_{\substack{0 \leqq r_1 < r_2 \\ r_1, r_2 \text{rational}}} \Big\{\Phi : \Phi \in M, \Phi^x(\{x : x \in R^s, \|x\| < r_1\}) = 0, \Phi^x(\{x : x \in R^s, r_1 \leqq \|x\| \leqq r_2\})$$

$$= \Phi^x(\{x : x \in R^s, r_1 \leqq \|x\| \leqq r_2\} \cap X) = 1\Big\},$$

so that we observe the validity of:

8.2.3. *The correspondence $\Phi \curvearrowright y(\Phi)$ provides a mapping from M into R^s, which is measurable with respect to the σ-algebras $\mathfrak{M}$, $\mathfrak{R}^s$.*

As referred to each σ-finite stationary measure H on $\mathfrak{M}$, $y(\Phi)$ is defined for almost all Φ. We even have

8.2.4. *If, for a stationary measure H on $\mathfrak{M}$, the measure $H\big((.) \setminus \{o\}\big)$ is σ-finite, then*

H (there exist points x_1, x_2 in R^s, such that

$$x_1 \neq x_2, \|x_1\| = \|x_2\|, \Phi^x(\{x_1\}) > 0, \Phi^x(\{x_2\}) > 0) = 0.$$

Proof. The set

$$Z = \{[x, \Phi] : [x, \Phi] \in C, \text{there exists an } x' \neq x, \text{ such that } [x', \Phi] \in C, \|x\| = \|x'\|\}$$

belongs to $\mathfrak{C}$. For if we use the notation introduced in the proof of 8.2.2., then

$$Z = \big\{[x, \Phi] : [x, \Phi] \in C, \left((\Phi^x)^*\right)^!(\{\|x\|\}) \geqq 2\big\},$$

and it follows from 1.4.2. and 5.1.9. that Z is in $\mathfrak{C}$.

On the basis of the measurability statement of 5.1.2., the correspondence

$$\Phi \curvearrowright f(\Phi) = \sum_{x \in R^s, \Phi^x(\{x\}) > 0} k_Z(x, \Phi) \, \Phi^x(\{x\})$$

provides a measurable mapping of M into $\{0, 1, \ldots, +\infty\}$. Obviously $f(\Phi) > 0$ if and only if there exist points x_1, x_2 in R^s, such that $x_1 \neq x_2$,

$\|x_1\| = \|x_2\|$, $\Phi^x(\{x_1\}) > 0$, $\Phi^x(\{x_2\}) > 0$. Hence the set

$$\{\Phi : \Phi \in M, \text{ there exist points } x_1, x_2 \text{ in } R^s, \text{ such that}$$
$$x_1 \neq x_2, \|x_1\| = \|x_2\|, \Phi^x(\{x_1\}) > 0, \Phi^x(\{x_2\}) > 0\}$$

belongs to $\mathfrak{M}$, and our assertion is equivalent to

$$\int f(\Phi)\, H(\mathrm{d}\Phi) = 0\,.$$

By 8.1.2. we obtain

$$\int f(\Phi)\, H(\mathrm{d}\Phi) = \int \Big(\int k_Z(x, T_{-x}\Phi)\, \mu(\mathrm{d}x) \Big)\, H^0(\mathrm{d}\Phi)$$
$$= \int \mu \Bigg(\bigcup_{\substack{y \in R^s, \Phi^x(\{y\}) > 0 \\ y \neq [0, \ldots, 0]}} \{x : x \in R^s, \|x\| = \|x + y\|\} \Bigg)\, H^0(\mathrm{d}\Phi),$$

since $[x, T_{-x}\Phi] \in Z$ means that there exists an x', such that $x' \neq x$, $\|x'\| = \|x\|$, $(T_{-x}\Phi)^x(\{x'\}) > 0$.

All of the hyperplanes $\{x : x \in R^s, \|x\| = \|x + y\|\}$ have zero Lebesgue measure, and since there are only at most countably many y having the property that $\Phi^x(\{y\}) > 0$, $\mu(\{x : x \in R^s, [x, T_{-x}\Phi] \in Z\})$ is zero for all Φ. Thus we observe the truth of the following relation

$$\int f(\Phi)\, H(\mathrm{d}\Phi) = 0\,. \blacksquare$$

In virtue of 8.2.1. we have, for all Y in $\mathfrak{M}$ and the function h introduced in 8.2.2.,

$$H(Y \setminus \{o\}) = \int \Big(\int k_Y(T_{-x}\Phi)\, h(x, T_{-x}\Phi)\, \mu(\mathrm{d}x) \Big)\, H^0(\mathrm{d}\Phi)\,.$$

By 8.2.4. and 8.1.2. we obtain

$$\int \Big(\int k_{M \setminus V}(T_{-x}\Phi)\, \mu(\mathrm{d}x) \Big)\, H^0(\mathrm{d}\Phi)$$
$$= \int \Bigg(\sum_{x \in R^s, \Phi^x(\{x\}) > 0} k_{M \setminus V}(\Phi)\, \Phi^x(\{x\}) \Bigg)\, H(\mathrm{d}\Phi) = 0\,,$$

which leads to

$$H(Y \setminus \{o\})$$
$$= \int \Bigg(\int_{\{x : x = y(T_{-x}\Phi)\}} k_Y(T_{-x}\Phi) \Big((T_{-x}\Phi)^x \big(\{y(T_{-x}\Phi)\} \big) \Big)^{-1} \mu(\mathrm{d}x) \Bigg)\, H^0(\mathrm{d}\Phi)$$
$$= \int \big(\Phi^x(\{[0, \ldots, 0]\}) \big)^{-1} \Bigg(\int_{\{x : \Phi^x(S_{\|x\|}(-x)) = 0\}} k_Y(T_{-x}\Phi)\, \mu(\mathrm{d}x) \Bigg)\, H^0(\mathrm{d}\Phi)$$
$$= \int \big(\Phi^x(\{[0, \ldots, 0]\}) \big)^{-1} \Bigg(\int_{\{x : \Phi^x(S_{\|x\|}(x)) = 0\}} k_Y(T_x\Phi)\, \mu(\mathrm{d}x) \Bigg)\, H^0(\mathrm{d}\Phi),$$

where $S_d(z)$ means the open ball of radius d around the point z. Thus we obtain the following inversion formula:

8.2.5. Proposition. *Let H be a stationary measure on $\mathfrak{M}$. If the measure $H\big((.) \setminus \{o\}\big)$ is σ-finite, then, for all Y in $\mathfrak{M}$,*

$$H(Y \setminus \{o\}) = \int \big(\Phi^x(\{[0, \ldots, 0]\}) \big)^{-1} \Bigg(\int_{\{x : \Phi^x(S_{\|x\|}(x)) = 0\}} k_Y(T_x\Phi)\, \mu(\mathrm{d}x) \Bigg)\, H^0(\mathrm{d}\Phi)\,.$$

In the special case $s = 1$, the set $\{x : x \in R^1, \Phi^x(S_{\|x\|}(x)) = 0\}$ coincides with the interval $(x_{-2}(\Phi)/2, x_0(\Phi)/2)$ for all Φ in M^0, where we again use the measurable indexing relation introduced in 5.3.4., setting $x_{-2}(\Phi) = -\infty$ and $x_0(\Phi) = +\infty$ if $\Phi^x((-\infty, 0)) = 0$ and $\Phi^x((0, +\infty)) = 0$ respectively. With these conventions, we thus obtain:

8.2.6. *Suppose that $s = 1$ and that H be a stationary measure on $\mathfrak{M}$. If then $H((.) \setminus \{o\})$ is σ-finite, then for all Y in $\mathfrak{M}$,*

$$H(Y \setminus \{o\}) = \int \left(\Phi^x(\{0\})\right)^{-1} \left(\int\limits_{(x_{-2}(\Phi))/2}^{(x_0(\Phi))/2} k_Y(T_x\Phi)\, \mu(\mathrm{d}x) \right) H^0(\mathrm{d}\Phi).$$

More frequently, the following inversion formula is used (cf. Ryll-Nardzewski [1], Slivnjak [1], [2], as well as Port & Stone [2]):

8.2.7. Proposition. *Subject to the assumptions of* 8.2.6.,

$$H\left(\Phi \in Y, \Phi^x((-\infty, 0]) > 0\right)$$
$$= \int \left(\Phi^x(\{0\})\right)^{-1} \left(\int\limits_0^{x_0(\Phi)} k_Y(T_x\Phi)\, \mu(\mathrm{d}x) \right) H^0(\mathrm{d}\Phi),$$

setting $x_0(\Phi) = +\infty$ if $\Phi^x((0, +\infty)) = 0$.

Proof. We set $Y = \{\Phi : \Phi \in M, \Phi^x((-\infty, 0]) > 0\}$, letting

$$h(x, \Phi) = \begin{cases} \left(\Phi^x(\{x\})\right)^{-1} & \text{for} \quad x = x_{-1}(\Phi) \\ 0 & \text{otherwise} \end{cases}$$

define a non-negative real function h on $C \cap (R^1 \times Y)$, which is measurable with respect to $\mathfrak{C}$ and has the property that

$$\sum_{x \in R^1, \Phi^x(\{x\}) > 0} h(x, \Phi)\, \Phi^x(\{x\}) = 1 \qquad (\Phi \in Y).$$

In virtue of 5.1.7. we obtain, for all Y' in $\mathfrak{M}$,

$$H(Y' \cap Y) = \int\limits_{C \cap (R \times Y)} h(x, \Phi)\, k_{Y'}(\Phi)\, \mathscr{C}_H(\mathrm{d}[x, \Phi])$$
$$= \int\limits_{C \cap ((-\infty, 0] \times M)} h(x, \Phi)\, k_{Y'}(\Phi)\, \mathscr{C}_H(\mathrm{d}[x, \Phi])$$
$$= \int\limits_{(-\infty, 0] \times M^0} h(x, T_{-x}\Psi)\, k_{Y'}(T_{-x}\Psi)\, (\mu \times H^0)\, (\mathrm{d}[x, \Psi])$$
$$= \int \left(\int\limits_{(-\infty, 0]} h(x, T_{-x}\Psi)\, k_{Y'}(T_{-x}\Psi)\, \mu(\mathrm{d}x) \right) H^0(\mathrm{d}\Psi).$$

On the other hand, we have

$$h(x, T_{-x}\Psi) = \begin{cases} \left((T_{-x}\Psi)^x\, (\{x\})\right)^{-1}, & \text{if} \quad x = x_{-1}(T_{-x}\Psi) \\ 0 & \text{otherwise,} \end{cases}$$

or

$$h(x, T_{-x}\Psi) = \begin{cases} \left(\Psi^x(\{0\})\right)^{-1}, & \text{if} \quad -x_0(\Psi) < x \leqq 0 \\ 0 & \text{otherwise.} \end{cases}$$

Hence we can continue our above chain of equations as follows:

$$= \int \left(\Psi^x(\{0\})\right)^{-1} \left(\int_0^{x_0(\Psi)} k_{Y'}(T_x\Psi)\ \mu(\mathrm{d}x) \right) H^0(\mathrm{d}\Psi). \ \blacksquare$$

For all $t > 0$, we now obtain

$$\begin{aligned} &H\left(\Phi^x\left((-t, 0]\right) > 0\right) \\ &= \int \left(\Phi^x(\{0\})\right)^{-1} \left(\int_0^{x_0(\Phi)} k_{\{\Psi:\Psi\in M, \Psi^x((-t,0])>0\}}(T_x\Phi)\ \mu(\mathrm{d}x) \right) H^0(\mathrm{d}\Phi) \\ &= \int \left(\Phi^x(\{0\})\right)^{-1} \min\left(t, x_0(\Phi)\right) H^0(\mathrm{d}\Phi), \end{aligned}$$

so that we have

8.2.8. *Subject to the assumptions of* 8.2.6., *for each* $t > 0$,

$$H\left(\Phi^x\left((-t, 0]\right) > 0\right) = \int \left(\Phi^x(\{0\})\right)^{-1} \min\left(t, x_0(\Phi)\right) H^0(\mathrm{d}\Phi),$$

setting $x_0(\Phi) = +\infty$, *if* $\Phi^x\left((0, +\infty)\right) = 0$.

In view of 6.1.2., the term $H\left(\Phi \in Y, \Phi\left((-\infty, 0]\right) > 0\right)$ in Proposition 8.2.7. can be replaced by the simple term $H(\Phi \in Y, \Phi \neq o)$, if H is finite.

By virtue of 8.1.7. the term $\left(\Phi^x(\{0\})\right)^{-1}$ may be dropped in the inversion formulae 8.2.5. through 8.2.8. if H^x is simple.

Changing the order of integration in 8.2.7. we now obtain

$$\begin{aligned} &\int \left(\int_0^{x_0(\Phi)} k_Y(T_x\Phi)\ \mu(\mathrm{d}x) \right) H^0(\mathrm{d}\Phi) \\ &= \int k_{[0,x_0(\Phi))}(x)\ k_Y(T_x\Phi)\ (\mu \times H^0)\ (\mathrm{d}[x, \Phi]) \\ &= \int H^0\left(x_0(\Phi) > x, T_x\Phi \in Y\right) \mu(\mathrm{d}x). \end{aligned}$$

In this way we have

8.2.9. Proposition. *Suppose that* $s = 1$ *and that* H *be a stationary measure on* $\mathfrak{M}$. *If then* $H\left((.) \setminus \{o\}\right)$ *is* σ*-finite and* H^x *is simple, then for all* Y *in* $\mathfrak{M}$,

$$\begin{aligned} &H\left(\Phi \in Y, \Phi^x\left((-\infty, 0]\right) > 0\right) \\ &= \int_0^\infty H^0\left(x_0(\Phi) > x, T_x\Phi \in Y\right) \mu(\mathrm{d}x). \end{aligned}$$

By reflection at the origin every measure H satisfying the assumptions of 8.2.9. is transformed into a measure H^V of the same kind, and we have $(H^V)^0 = (H^0)^V$. If we denote the image of Φ, Y by Φ^V, Y^V, respectively, we obtain

$$\begin{aligned} & H\left(\Phi \in Y, \Phi^x([0, +\infty)) > 0\right) \\ &= H^V\left(\Psi \in Y^V, \Psi^x((-\infty, 0]) > 0\right) \\ &= \int_0^\infty (H^V)^0 \left(x_0(\Psi) > x, T_x\Psi \in Y^V\right) \mu(\mathrm{d}x) \\ &= \int_0^\infty H^0 \left(x_0(\Phi^V) > x, T_x(\Phi)^V \in Y^V\right) \mu(\mathrm{d}x) \\ &= \int_0^\infty H^0 \left(-x_{-2}(\Phi) > x, (T_{-x}\Phi)^V \in Y^V\right) \mu(\mathrm{d}x) \\ &= \int_0^\infty H^0\left(-x_{-2}(\Phi) > x, T_{-x}\Phi \in Y\right) \mu(\mathrm{d}x), \end{aligned}$$

where we set $-x_{-2}(\Phi) = +\infty$ if $\Phi^x((-\infty, 0)) = 0$.

Let t be any positive number and j a non-negative integer. We set

$$Z = \left\{[x, \Phi] : [x, \Phi] \in C,\ 0 < x \leqq t,\ \Phi^x \text{ simple},\ \Phi^x((x, t]) = j\right\}.$$

This set belongs to $\mathfrak{C}$, since, by means of the measurable indexing relation introduced by 5.3.4. it can be put into the following form:

$$\begin{aligned} Z = \bigcup_{0 \leqq l < +\infty} \{[x, \Phi] : [x, \Phi] \in C,\ 0 < x,\ \Phi^x \text{ simple}, \\ & x_l(\Phi) \leqq t < x_{l+1}(\Phi) \quad \text{or} \\ & x_l(\Phi) \leqq t \text{ and } x_{l+1}(\Phi) \quad \text{does not exist}, \\ & k(x, \Phi) = l - j\}. \end{aligned}$$

Obviously

$$\Omega(Z) = \left\{\Phi : \Phi \in M,\ \Phi^x \text{ simple},\ \Phi^x((0, t]) > j\right\}.$$

We denote this set by Y_0.

Setting $h = k_Z$ we obtain

$$\sum_{x \in R^1, \Phi^x(\{x\}) > 0} h(x, \Phi)\, \Phi^x(\{x\}) = 1 \qquad (\Phi \in Y_0),$$

so that, by means of 5.1.7., for all σ-finite measures H on $\mathfrak{M}$,

$$H(Y_0) = \int_{C\cap(R\times Y_0)} h(x, \Phi)\, \mathscr{C}_H(\mathrm{d}[x, \Phi])$$
$$= \mathscr{C}_H(Z)$$
$$= \mathscr{C}_H{}^T\Big(0 < x \leqq t,\ \Phi^x \text{ simple},\ \Phi^x\big((0, t - x]\big) = j\Big).$$

If we now suppose in addition that H is stationary and H^x is simple, then, in virtue of 8.1.1., we have

$$H\Big(\Phi^x\big((0, t]\big) > j\Big) = (\mu \times H^0)\Big(0 < x \leqq t, \Phi^x\big((0, t - x]\big) = j\Big)$$
$$= \int_0^t H^0\Big(\Phi^x\big((0, t - z]\big) = j\Big)\, \mu(\mathrm{d}z).$$

Therefore we obtain the classical Palm-Khinchin formulae (cf. Khinchin [1], § 10).

8.2.10. Proposition. *Subject to the assumptions of 8.2.9., for all $t \geqq 0$ and all non-negative integers j,*

$$H\Big(\Phi^x\big((0, t]\big) > j\Big) = \int_0^t H^0\Big(\Phi^x\big((0, x]\big) = j\Big)\, \mu(\mathrm{d}x).$$

8.3. The Mapping $P \rightsquigarrow (\tilde{P})^0$

In view of 2.2.1. and 8.1.3., we have

8.3.1. Proposition. *The correspondence $P \rightsquigarrow (\tilde{P})^0$ provides a one-to-one mapping of the set of all infinitely divisible stationary distributions on $\mathfrak{M}$ into the set of all σ-finite measures on $\mathfrak{M}^0$.*

Obviously, $\big(\widetilde{\delta_o}\big)^0 = 0$.

For all finite measures γ on $\mathfrak{K}$ and all X in $\mathfrak{R}^s$ having the property that $0 < \mu(X) < +\infty$, we obtain

$$\big(\widetilde{P_{\mu\times\gamma}}\big)^0 = (Q_{\mu\times\gamma})^0 = \big(\mu(X)\big)^{-1} \int \Big(\int_X \delta_{T_x\Phi}(.)\, \Phi^x(\mathrm{d}x)\Big)\, Q_{\mu\times\gamma}(\mathrm{d}\Phi)$$
$$= \big(\mu(X)\big)^{-1} \int \Big(\int_X \delta_{\delta[[0,\ldots,0],k]}(.)\, \delta_y(\mathrm{d}x)\Big)\, (\mu \times \gamma)\, (\mathrm{d}[y, k])$$
$$= \big(\mu(X)\big)^{-1} \int \mu(X)\, \delta_{\delta[[0,\ldots,0],k]}(.)\, \gamma(\mathrm{d}k),$$

which results in

8.3.2. Proposition. *For all finite measures γ on $\mathfrak{K}$*

$$\big(\widetilde{P_{\mu\times\gamma}}\big)^0 = \int \delta_{\delta_{[0,\ldots,0]}\times\delta_k}(.)\, \gamma(\mathrm{d}k).$$

Using 8.1.6., the following proposition is inferred from 2.5.3., 6.3.5., 2.5.1., 2.2.13., 2.2.15., 6.1.4., and 2.2.9.:

8.3.3. Proposition. *A stationary infinitely divisible distribution P on $\mathfrak{M}$ is, respectively,*

singular infinitely divisible		$(\tilde{P})^0\,(\chi \text{ finite}) = 0$
ergodic		$(\tilde{P})^0\left(\overline{k_{\{\Psi:\Psi\in M,\Psi^x([0,1)^s)>0\}}}(\chi) > 0\right) = 0$
regular infinitely divisible	*if and only if*	$(\tilde{P})^0\,(\chi \text{ infinite}) = 0$
free from after-effects		$(\tilde{P})^0\left(\chi^*(A) > 1\right) = 0$
Poisson		$(\tilde{P})^0\left(\chi(A) \neq 1\right) = 0$
simple		$(\tilde{P})^0\,(\chi \text{ not simple}) = 0.$

If the convolution of an at most countable family of stationary infinitely divisible distributions $(P_j)_{j\in J}$ exists, then by 2.2.2. and 8.1.4. it follows that

$$\left(\widetilde{\mathop{\ast}_{j\in J} P_j}\right)^0 = \sum_{j\in J} (\widetilde{P_j})^0 .$$

Hence, using 2.5.4. and 6.3.4., we get

8.3.4. Proposition. *For all stationary infinitely divisible distributions P on $\mathfrak{M}$*

$$(\widetilde{P_r})^0 = (\tilde{P})^0\left(\chi \in (.), \chi \text{ finite}\right), \quad (\widetilde{P_s})^0 = (\tilde{P})^0\left(\chi \in (.), \chi \text{ infinite}\right),$$

$$\left(\widetilde{P_{s,1}}\right)^0 = (\tilde{P})^0\left(\chi \in (.), \chi \text{ infinite}, \overline{k_{\{\Psi:\Psi\in M,\Psi^x([0,1)^s)>0\}}}(\chi) = 0\right),$$

$$\left(\widetilde{P_{s,2}}\right)^0 = (\tilde{P})^0\left(\chi \in (.), \overline{k_{\{\Psi:\Psi\in M,\Psi^x([0,1)^s)>0\}}}(\chi) > 0\right).$$

In view of 2.2.3., the following theorem can be inferred with the use of 8.1.5.

8.3.5. *Subject to the assumptions of* 8.3.4.,

$$i_{P^x} = (\tilde{P})^0\,(M^0).$$

Taking over a notation from Section 5.2., for any σ-finite measures L, H on $\mathfrak{R}^s \otimes \mathfrak{M}^0$, $\mathfrak{M}$ we denote by $L \circledast H$ the image of $L \times H$ produced by the mapping

$$[[x, \Phi], \Psi] \curvearrowright [x, \Phi + \Psi].$$

The following lemma will be required below:

8.3.6. *For all σ-finite measures V on $\mathfrak{C}$ and all σ-finite stationary measures H on $\mathfrak{M}$,*

$$(V \circledast H)^T = V^T \circledast H.$$

Proof. Starting from the measure $V \times H$, we obtain $(V \circledast H)^T$ by first applying the mapping

$$\big[[x, \Phi_1], \Phi_2\big] \rightsquigarrow [x, \Phi_1 + \Phi_2]$$

and subsequently

$$[x, \Psi] \rightsquigarrow [x, T_x\Psi].$$

Now let us look at the transition

$$\big[[x, \Phi_1], \Phi_2\big] \rightsquigarrow \big[[x, \Phi_1], T_x\Phi_2\big].$$

Since H is stationary, $V \times H$ is transformed into $V \times H$ itself. If we apply the mapping

$$\big[[x, \Phi], \Psi\big] \rightsquigarrow \big[[x, T_x\Phi], \Psi\big],$$

followed by

$$\big[[x, \chi], \Psi\big] \rightsquigarrow [x, \chi + \Psi],$$

$V \times H$ is transformed first into $V^T \times H$ and then into $V^T \circledast H$.

Performing these three mappings in succession gives again

$$\big[[x, \Phi_1], \Phi_2\big] \rightsquigarrow [x, T_x\Phi_1 + T_x\Phi_2] = [x, T_x(\Phi_1 + \Phi_2)],$$

so that $V \times H$ is transformed into $(V \circledast H)^T$. Hence

$$(V \circledast H)^T = V^T \circledast H. \blacksquare$$

As is shown by the proof of 5.2.3., the following relation holds for all σ-finite measures H_1, H_2 on $\mathfrak{M}$:

$$\mathscr{C}_{H_1 * H_2} = \mathscr{C}_{H_1} \circledast H_2 + \mathscr{C}_{H_2} \circledast H_1.$$

Now, if H_1, H_2 are stationary, this is also true for $H_1 * H_2$ and in view of 8.3.6. we have:

$$\mathscr{C}^T_{H_1 * H_2} = \mathscr{C}^T_{H_1} \circledast H_2 + \mathscr{C}^T_{H_2} \circledast H_1.$$

If also $H_1 * H_2$ is a σ-finite measure, then using 8.1.1. we may conclude that

$$\begin{aligned}\mu \times (\boldsymbol{H}_1 * \boldsymbol{H}_2)^0 &= \big(\mu \times (\boldsymbol{H}_1)^0\big) \circledast \boldsymbol{H}_2 + \big(\mu \times (\boldsymbol{H}_2)^0\big) \circledast \boldsymbol{H}_1 \\ &= \mu \times \big((\boldsymbol{H}_1)^0 * \boldsymbol{H}_2\big) + \mu \times \big((\boldsymbol{H}_2)^0 * \boldsymbol{H}_1\big) \\ &= \mu \times \big((\boldsymbol{H}_1)^0 * H_2 + (H_2)^0 * H_1\big),\end{aligned}$$

thus observing the validity of

8.3.7. Proposition. *If the convolution $H_1 * H_2$ of two σ-finite stationary measures H_1, H_2 on $\mathfrak{M}$ is again σ-finite, then*

$$(H_1 * H_2)^0 = (H_1)^0 * H_2 + (H_2)^0 * H_1 .$$

For stationary distributions, the assertion of 5.2.5. reads as follows:

8.3.8. Theorem. *A stationary distribution P on $\mathfrak{M}$ is infinitely divisible if and only if there exists a σ-finite measure Q on $\mathfrak{M}$ having the property that*

$$P^0 = Q * P .$$

The measure Q is uniquely determined by

$$Q = (\tilde{P})^0 .$$

Proof. Let P be stationary and infinitely divisible. Then by 8.1.1., 5.2.4., and 8.3.6. we have

$$\mu \times P^0 = \mathscr{C}_P{}^T = (\mathscr{C}_{\tilde{P}} \circledast P)^T = \mathscr{C}_{\tilde{P}}{}^T \circledast P$$
$$= \big(\mu \times (\tilde{P})^0\big) \circledast P = \mu \times \big((\tilde{P})^0 * P\big),$$

and hence

$$P^0 = (\tilde{P})^0 * P .$$

Conversely, suppose that $P^0 = Q * P$. Then

$$Q\big(\Phi^x(\{[0, \ldots, 0]\}) = 0\big) = P^0\big(\Phi^x(\{[0, \ldots, 0]\}) = 0\big) = 0,$$

i.e. $Q(M \setminus M^0) = 0$, and by 8.3.6. we obtain

$$\mathscr{C}_P{}^T = \mu \times P^0 = \mu \times (Q * P) = (\mu \times Q) \circledast P ,$$

and hence

$$\mathscr{C}_P = V \circledast P \qquad \text{where} \qquad V^T = \mu \times Q .$$

Using 5.2.5. we may conclude from this that P is infinitely divisible and that $V = \mathscr{C}_{\tilde{P}}$. Hence

$$\mu \times Q = V^T = \mathscr{C}_{\tilde{P}}{}^T = \mu \times (\tilde{P})^0, \qquad \text{i.e.} \qquad Q = (\tilde{P})^0 . \ \blacksquare$$

From now on, we shall suppose that the phase space be of the form R^s, $s \geqq 1$. As it was stated already in Section 6.4., the stationary Poisson distributions now are of the form $P_{l\mu}$, $l \geqq 0$ and instead of 8.3.2. we have

8.3.9. Proposition. *For all $l \geqq 0$,*

$$\big(\widetilde{P_{l\mu}}\big)^0 = l\delta_{\delta_{[0,\ldots,0]}} .$$

By specialization of Theorem 8.3.8., we now obtain the following characterization of the stationary Poisson distributions

8.3.10. Theorem. *For all $l \geqq 0$, a stationary distribution P on $\mathfrak{M}$ coincides with $P_{l\mu}$ if and only if it satisfies the functional equation*

$$P^0 = l\delta_{\delta_{[0,\ldots,0]}} * P.$$

Now let P be a distribution on $\mathfrak{M}$ having the property g_μ. In view of 1.14.8., P then is of the form $\int P_{l\mu}(.)\,\sigma(dl)$ and hence is stationary.

We choose any X in $\mathfrak{R}^s$ satisfying the condition $0 < \mu(X) < +\infty$ and obtain, for all Y in $\mathfrak{M}^0$,

$$\begin{aligned} P^0(Y) &= \big(\mu(X)\big)^{-1} \int \Big(\int_X k_Y(T_x\Phi)\,\Phi(dx)\Big)\,P(d\Phi) \\ &= \int \Big(\int \Big(\big(\mu(X)\big)^{-1} \int_X k_Y(T_x\Phi)\,\Phi(dx)\Big)\,P_{l\mu}(d\Phi)\Big)\,\sigma(dl) \\ &= \int (P_{l\mu})^0\,(Y)\,\sigma(dl) = \int l\big(\delta_{\delta_{[0,\ldots,0]}} * P_{l\mu}\big)\,(Y)\,\sigma(dl) \\ &= \big(\delta_{\delta_{[0,\ldots,0]}} * \int lP_{l\mu}(.)\,\sigma(dl)\big)\,(Y), \end{aligned}$$

i.e. we have

$$P^0 = \delta_{\delta_{[0,\ldots,0]}} * H,$$

where

$$H = \int lP_{l\mu}(.)\,\sigma(dl).$$

Obviously the measure H is stationary. From the σ-finiteness of P^0 we conclude that H is σ-finite.

It turns out that we have thus derived a characteristic feature of the distributions having the property g_μ, i.e. we have (cf. Mecke [8])

8.3.11. Theorem. *A stationary distribution P on $\mathfrak{M}$ is of the form $\int P_{l\mu}(.)\,\sigma(dl)$ if and only if there exists a σ-finite stationary measure H on $\mathfrak{M}$ such that*

$$P^0 = \delta_{\delta_{[0,\ldots,0]}} * H$$

is satisfied.

Proof. By means of 8.1.1., we conclude from

$$P^0 = \delta_{\delta_{[0,\ldots,0]}} * H$$

that

$$\begin{aligned} \mathscr{C}_P{}^T &= \mu \times P^0 = \mu \times (\delta_{\delta_{[0,\ldots,0]}} * H) = (\mu \times \delta_{\delta_{[0,\ldots,0]}}) \circledast H \\ &= \big(\int (\delta_x \times \delta_{\delta_{[0,\ldots,0]}})\,(.)\,\mu(dx)\big) \circledast H. \end{aligned}$$

In view of 8.3.6., from this we obtain

$$\begin{aligned} \mathscr{C}_P &= \big(\int (\delta_x \times \delta_{\delta_x})\,(.)\,\mu(dx)\big) \circledast H \\ &= \int \big(\delta_x \times (\delta_{\delta_x} * H)\big)\,(.)\,\mu(dx), \end{aligned}$$

i.e. we have

$$\mathscr{C}_P{}^! = \mu \times H,$$

from which, using 5.4.2. and 1.14.8., we conclude that there exists a representation $P = \int P_{l\mu}(.)\, \sigma(\mathrm{d}l)$. ∎

8.4. Symmetry Properties of Palm Measures. I

Let Y be an arbitrary set in $\mathfrak{F}$. A random translation Γ in $\boldsymbol{\Gamma}(Y)$ is called *homogeneous* if, for all x in R^s and all Φ in $Y \setminus \{o\}$,

$$\gamma_{a-x,a'-x}(T_x\Phi) = \gamma_{a,a'}(\Phi) \qquad \big(a, a' \in \{y : y \in R^s, \Phi^x(\{y\}) > 0\}\big).$$

If for instance we take for Y the set of all finite Φ in M, then the random translation Λ is homogeneous.

For each homogeneous Γ in $\boldsymbol{\Gamma}(Y)$, and for each σ-finite measure Q on $\mathfrak{M}^0$ satisfying the condition $Q(M^0 \setminus Y) = 0$, we define a measure $\hat{\Gamma} Q$ on $\mathfrak{M}^0$ by

$$\hat{\Gamma} Q = \int \Big(\sum_{z \in R^s, \Psi^x(\{z\})>0} \gamma_{[0,\ldots,0],z}(\Psi)\, \delta_{T_z\Psi}(.)\Big)\, Q(\mathrm{d}\Psi).$$

This definition is reasonable, since, for all Y' in $\mathfrak{M}^0$,

$$\Psi \curvearrowright \sum_{z \in R^s, \Psi^x(\{z\})>0} \gamma_{[0,\ldots,0],z}(\Psi)\, \delta_{T_z\Psi}(Y') = \big(\Gamma_{[[0,\ldots,0],\Psi]}\big)^T \big(C \cap (R^s \times Y')\big)$$

is a measurable mapping of Y into R^1.

In the case of the symmetry operator Λ we obtain, for all Φ in $Y_4 \cap M^0$,

$$\gamma_{[0,\ldots,0],z}(\Phi) = \frac{\Phi^x(\{z\})}{\Phi^x(R^1)},$$

which means that

$$\sum_{z \in R^1, \Phi^x(\{z\})>0} \gamma_{[0,\ldots,0],z}(\Phi)\, \delta_{T_z\Phi} = \sum_{z \in R^1, \Phi^x(\{z\})>0} \frac{\Phi^x(\{z\})}{\Phi^x(R^1)}\, \delta_{T_z\Phi}.$$

Consequently, a point z in Φ is selected at random and Φ is transformed into $T_z\Phi$.

There exists a well-determined measure V on $\mathfrak{C}$ such that $V^T = \mu \times Q$. This measure is σ-finite and satisfies the condition $V\big(C \setminus (R^s \times Y)\big) = 0$. We obtain

$$\begin{aligned}
(\Gamma V)^T &= \int (\Gamma_{[x,\Phi]})^T(.)\, V(\mathrm{d}[x, \Phi]) \\
&= \int (\Gamma_{[x,T_{-x}\Psi]})^T(.)\, V^T(\mathrm{d}[x, \Psi]) \\
&= \int \Big(\sum_{y \in R^s, (T_{-x}\Psi)^x(\{y\})>0} \gamma_{x,y}(T_{-x}\Psi)\, \delta_{[y, T_y T_{-x}\Psi]}(.)\Big)\, (\mu \times Q)\, (\mathrm{d}[x, \Psi]) \\
&= \int \Big(\int \Big(\sum_{y \in R^s, (T_{-x}\Psi)^x(\{y\})>0} \gamma_{[0,\ldots,0],y-x}(\Psi)\, \delta_{[y, T_{y-x}\Psi]}(.)\Big)\, \mu(\mathrm{d}x)\Big)\, Q(\mathrm{d}\Psi)
\end{aligned}$$

$$= \int \left(\int \left(\sum_{z \in R^s, \Psi^x(\{z\}) > 0} \gamma_{[0,\ldots,0],z}(\Psi)\, \delta_{[z+x, T_z\Psi]}(.) \right) \mu(\mathrm{d}x) \right) Q(\mathrm{d}\Psi)$$

$$= \int \left(\int \left(\sum_{z \in R^s, \Psi^x(\{z\}) > 0} \gamma_{[0,\ldots,0],z}(\Psi)\, \delta_{[y, T_z\Psi]}(.) \right) \mu(\mathrm{d}y) \right) Q(\mathrm{d}\Psi)$$

$$= \int \left(\int \left(\sum_{z \in R^s, \Psi^x(\{z\}) > 0} \gamma_{[0,\ldots,0],z}(\Psi)\, (\delta_y \times \delta_{T_z\Psi})(.) \right) \mu(\mathrm{d}y) \right) Q(\mathrm{d}\Psi)$$

$$= \left(\int \delta_y(.)\, \mu(\mathrm{d}y) \right) \times \left(\int \left(\sum_{z \in R^s, \Psi^x(\{z\}) > 0} \gamma_{[0,\ldots,0],z}(\Psi)\, \delta_{T_z\Psi}(.) \right) Q(\mathrm{d}\Psi) \right)$$

$$= \mu \times \hat{\Gamma} Q .$$

Hence, in view of 5.3.2., 8.1.1., and 8.1.6., we observe the truth of

8.4.1. Proposition. *Let Y be a set in $\mathfrak{F}$, and Γ a homogeneous irreducible element of $\boldsymbol{\Gamma}(Y)$. Subject to these assumptions, a σ-finite measure Q on $\mathfrak{M}^0$ is the Palm measure of a σ-finite stationary measure H on $\mathfrak{M}$ satisfying the condition $H(\Phi \notin Y, \Phi \neq o) = 0$ if and only if $Q(M^0 \setminus Y) = 0$ as well as $\hat{\Gamma} Q = Q$.*

In view of 5.3.3., in the special case $Y = \{\Phi : \Phi \in M, \Phi \text{ finite}\}$ the random translation Λ may be chosen for Γ. This provides a starting point for the proof of

8.4.2. Theorem. *For each measure V on $\mathfrak{M}^0$, the following statements are equivalent:*

a) *There exists a regular infinitely divisible stationary distribution P on $\mathfrak{M}$, such that $(\tilde{P})^0 = V$.*

b) *The measure V satisfies the following conditions*

1. $V(\chi \textit{ infinite}) = 0$.
2. *For all bounded X in $\mathfrak{R}^s$*

$$\int \left(\sum_{k=1}^{\infty} k^{-1} V\big(\chi^x(X - x) > 0,\ \chi(A) = k\big) \right) \mu(\mathrm{d}x) < +\infty .$$

3. $\hat{\Lambda} V = V$.

Proof. 1. Let P be a regular infinitely divisible stationary distribution on $\mathfrak{M}$. Then, in view of 8.3.3., $(\tilde{P})^0$ (χ infinite) $= 0$. Using 5.3.3. and 8.4.1., we conclude that $\hat{\Lambda}(\tilde{P})^0 = (\tilde{P})^0$. Now we set $Y_0 = \{\Phi : \Phi \in M, 0 < \Phi(A) < +\infty\}$, so that, using 5.1.8., we may conclude that, for all bounded X in $\mathfrak{R}^s$

$$\tilde{P}\big(\Psi^x(X) > 0\big) = \tilde{P}\big(\Psi^x(X) > 0,\ \Psi \in Y_0\big)$$
$$= \sum_{k=1}^{\infty} k^{-1} \mathscr{C}_{\tilde{P}}\big(x \in R^s,\ \Psi^x(X) > 0,\ \Psi(A) = k\big).$$

This chain of equations can be continued as follows:

$$= \sum_{k=1}^{\infty} k^{-1} \mathscr{C}_{\tilde{P}}{}^{T}\big(x \in R^s,\ (T_{-x}\Psi)^x\,(X) > 0,\ (T_{-x}\Psi)\,(A) = k\big)$$

$$= \sum_{k=1}^{\infty} k^{-1} \big(\mu \times (\tilde{P})^0\big)\,\big(x \in R^s,\ \Psi^x(X - x) > 0,\ \Psi(A) = k\big).$$

Setting now $L = \sum_{k=1}^{\infty} k^{-1}(\tilde{P})^0 \left(\chi \in (.), \chi(A) = k\right)$, we may continue as follows:

$$= (\mu \times L)\left(x \in R^s, \Psi^x(X - x) > 0\right)$$
$$= \int L\left(\Psi^x(X - x) > 0\right) \mu(\mathrm{d}x).$$

Hence, since $\tilde{P}\left(\Psi^x(X) > 0\right) < +\infty$, we have

$$\int L\left(\Psi^x(X - x) > 0\right) \mu(\mathrm{d}x) < +\infty.$$

Consequently $(\tilde{P})^0$ satisfies all conditions stated under b).

2. Conversely, suppose that a measure V on $\mathfrak{M}^0$ satisfy these conditions. Setting $X = [-1, 1]^s$ and

$$L = \sum_{k=1}^{\infty} k^{-1} V\left(\chi \in (.),\ \chi(A) = k\right),$$

we obtain

$$+\infty > \int L\left(\chi^x([-1, 1]^s - x) > 0\right) \mu(\mathrm{d}x)$$
$$= \sum_{-\infty < m_1, \dots, m_s < +\infty} \int_{\times_{i=1}^{s} [m_i, m_i + 1]} L\left(\chi^x([-1, 1]^s - x) > 0\right) \mu(\mathrm{d}x)$$
$$\geqq \sum_{-\infty < k_1, \dots, k_s < +\infty} L\left(\chi^x\left(\times_{i=1}^{s} [k_i, k_i + 1)\right) > 0\right)$$
$$\geqq L(\chi \neq o) = L(M).$$

Consequently the measure L is finite, from which it follows that V is σ-finite.

In view of 5.3.3., 8.4.1., and 8.1.6., there exists a stationary σ-finite measure H on $\mathfrak{M}$ having the properties

$$H(\{o\}) = 0, \qquad H^0 = V, \qquad H\,(\Psi \text{ infinite}) = 0.$$

For all bounded X in $\mathfrak{R}^s$, the chain of equations derived in the first step of this proof yields:

$$H\left(\Psi^x(X) > 0\right) < +\infty.$$

Hence H belongs to $\boldsymbol{W}$, and by 2.2.1. and 6.3.1. we may conclude that there exists a stationary infinitely divisible distribution P on $\mathfrak{M}$, such that $(\tilde{P})^0 = V$. In view of 8.3.3., this distribution is a regular infinitely divisible one. ∎

8.5. Symmetry Properties of Palm Measures. II

This section is an immediate continuation of the preceding one. Throughout this section, we shall assume that the number of dimensions, s, of the "position space" R^s considered be equal to one, so that we may utilize the special measurable indexing relation $[x, \Phi] \curvearrowright k(x, \Phi)$ introduced by Proposition 5.3.4. Using Proposition 8.4.1., we are now able to derive a number of special symmetry criteria for Palm measures.

Obviously all of the irreducible random translations Γ_i, $1 \leqq i \leqq 4$, which were introduced at the end of Section 5.3., are homogeneous. In the case $\Gamma = \Gamma_1$ we obtain, for all Φ in $Y_1 \cap M^0$

$$\gamma_{0,z}(\Phi) = \begin{cases} 1 & \text{for} \quad z = x_0(\Phi), \\ 0 & \text{otherwise}, \end{cases}$$

which means that

$$\sum_{z \in R^1, \Phi^x(\{z\})>0} \gamma_{0,z}(\Phi)\, \delta_{T_z \Phi} = \delta_{T_{x_0(\Phi)}\Phi}.$$

Thus it is concluded that, for all σ-finite measures Q on $\mathfrak{M}^0$ satisfying the condition $Q(M^0 \setminus Y_1) = 0$, $\hat{\Gamma}_1 Q$ coincides with the image of Q produced by the measurable mapping $\Phi \curvearrowright T_{x_0(\Phi)}\Phi$.

By this mapping Φ is translated to the left until the first point of Φ on the open positive half line arrives at the origin. The mapping $\Phi \curvearrowright T_{x_0(\Phi)}\Phi$ from M^0 into M^0 is measurable with respect to $\mathfrak{M}^0$, because $\Phi \curvearrowright [x_0(\Phi), \Phi]$ is measurable in view of 5.3.4., and $[x, \Phi] \curvearrowright T_x\Phi$ is measurable in view of 6.1.1. The set $Y_1 \cap M^0$ is mapped by $\Phi \curvearrowright T_{x_0(\Phi)}\Phi$ upon $Y_1 \cap M^0$. The inverse mapping is of the form $\Phi \curvearrowright T_{x_{-1}(\Phi)}\Phi$ and hence measurable, too.

By 8.4.1., 8.2.8., and 6.1.3. we obtain

8.5.1. Theorem. *In the case $s = 1$, a measure Q on $\mathfrak{M}^0$ is the Palm measure of a stationary distribution P on $\mathfrak{M}$ satisfying the conditions $P(\{o\}) = 0$, $P(\Phi^x$ simple$) = 1$ if and only if the relations $Q(\Phi \notin Y_1) = 0$, $\hat{\Gamma}_1 Q = Q$ and $\int x_0(\Phi)\, Q(\mathrm{d}\Phi) = 1$ are satisfied.*

In the case $\Gamma = \Gamma_2$ we obtain, for all Φ in $Y_2 \cap M^0$,

$$\gamma_{0,z}(\Phi) = \begin{cases} \left(\Phi^x(\{0\}) - 1\right)\left(\Phi^x(\{0\})\right)^{-1} & \text{for} \quad z = 0 \\ \left(\Phi^x(\{0\})\right)^{-1} & \text{for} \quad z = x_0(\Phi) \\ 0 & \text{otherwise}. \end{cases}$$

Consequently

$$\sum_{z \in R^1, \Phi^x(\{z\})>0} \gamma_{0,z}(\Phi)\, \delta_{T_z\Phi} = \left(\Phi^x(\{0\})\right)^{-1}\left(\left(\Phi^x(\{0\}) - 1\right)\delta_\Phi + \delta_{T_{x_0(\Phi)}\Phi}\right),$$

i.e. the probability of Φ remaining unchanged is $\left(\Phi^x(\{0\})\right)^{-1}\left(\Phi^x(\{0\}) - 1\right)$, whereas the probability of Φ being transformed into $T_{x_0(\Phi)}\Phi$ is $\left(\Phi^x(\{0\})\right)^{-1}$.

In virtue of 6.1.3., each stationary distribution P on $\mathfrak{M}$ satisfying the condition $P(\{o\}) = 0$ has the property that $P(M \setminus Y_2) = 0$; so by 8.1.6. we may conclude that $P^0(M^0 \setminus Y_2) = 0$. From 8.4.1. it follows that $\hat{\Gamma}_2(P^0) = P^0$, and for $Y = M$, the inversion formula 8.2.7. yields the relation

$$\int \left(\Phi^x(\{0\})\right)^{-1} x_0(\Phi)\, P^0(\mathrm{d}\Phi) = 1.$$

Conversely, let Q be a measure on $\mathfrak{M}^0$ satisfying the condition $Q(M^0 \setminus Y_2) = 0$ and having the properties $\hat{\Gamma}_2 Q = Q$, $\int \left(\Phi^x(\{0\})\right)^{-1} x_0(\Phi)\, Q(\mathrm{d}\Phi) = 1$; then

Q is σ-finite and, by virtue of 8.4.1., it is the Palm measure of a well-determined σ-finite stationary measure P on $\mathfrak{M}$ with the property that $P(\{o\}) = 0$. Using 8.2.7. we conclude that $P(M) = 1$, so that we observe the truth of

8.5.2. Theorem. *In the case $s = 1$, the correspondence $P \frown P^0$ provides a one-to-one mapping of the set of all stationary distributions P on $\mathfrak{M}$ satisfying the condition $P(\{o\}) = 0$ onto the set of all measures Q on $\mathfrak{M}^0$ having the properties*

$$Q\left(\Phi^x\big((-\infty, 0)\big) < +\infty \quad \text{or} \quad \Phi^x\big((0, +\infty)\big) < +\infty\right) = 0,$$

$$\int \left(\Phi^x(\{0\})\right)^{-1} x_0(\Phi)\, Q(\mathrm{d}\Phi) = 1, \qquad \hat{\Gamma}_2 Q = Q.$$

In the case $\Gamma = \Gamma_3$ we obtain, for all Φ in $Y_3 \cap M^0$,

$$\gamma_{0,z}(\Phi) = \begin{cases} 1 & \text{for} \quad z = x_0(\Phi) \\ 1 & \text{if} \quad \Phi^x\big((0, +\infty)\big) = 0,\ \Phi^x(\{z\}) > 0, \text{ and } \Phi^x\big((-\infty, z)\big) = 0 \\ 0 & \text{otherwise}, \end{cases}$$

which means that

$$\sum_{z \in R^1, \Phi^x(\{z\}) > 0} \gamma_{0,z}(\Phi)\, \delta_{T_z\Phi}$$

$$= \begin{cases} \delta_{T_{x_0(\Phi)}\Phi}, & \text{if} \quad \Phi^x\big((0, +\infty)\big) > 0 \\ \delta_{T_y\Phi}, & \text{if} \quad \Phi^x\big((0, +\infty)\big) = 0,\ \Phi^x(\{y\}) > 0 \text{ and } \Phi^x\big((-\infty, y)\big) = 0. \end{cases}$$

Hence, if 0 is not the extreme right-hand point in Φ then, as in the case of $\hat{\Gamma}_1$, Φ is transformed into $T_{x_0(\Phi)}\Phi$. Otherwise we choose the extreme left-hand point y in Φ and transform Φ into $T_y\Phi$.

Summarizing, we may use 8.4.1. and 5.3.5. to formulate the following result:

8.5.3. Proposition. *In the case $s = 1$, a σ-finite measure Q on $\mathfrak{M}^0$ is the Palm measure of a σ-finite stationary measure H on $\mathfrak{M}$ having the property*

$$H\left(\Phi^x\big((-\infty, 0)\big) < +\infty,\ \Phi^x\big((0, +\infty)\big) = +\infty\right)$$

$$= 0 = H\left(\Phi^x\big((-\infty, 0)\big) = +\infty,\ \Phi^x\big((0, +\infty)\big) < +\infty\right),$$

if and only if the following relations are satisfied:

$$Q\left(\Phi^x\big((-\infty, 0)\big) < +\infty,\ \Phi^x\big((0, +\infty)\big) = +\infty\right)$$

$$= 0 = Q\left(\Phi^x\big((-\infty, 0)\big) = +\infty,\ \Phi^x\big((0, +\infty)\big) < +\infty\right)$$

and

$$\hat{\Theta} Q = Q.$$

The equation $\hat{\Theta} Q = Q$ means that

$$\hat{\Gamma}_2\Big(Q(\Phi \in (.), \Phi^x((-\infty, 0)) = +\infty = \Phi^x((0, +\infty))\Big)$$
$$= Q\Big((\Phi \in (.), \Phi^x((-\infty, 0)) = +\infty = \Phi^x((0, +\infty))\Big)$$

and

$$\hat{\Lambda}\Big(Q(\Phi \in (.), 0 < \Phi(A) < +\infty)\Big) = Q\big(\Phi \in (.),\ 0 < \Phi(A) < +\infty\big).$$

There are σ-finite stationary measures H on $\mathfrak{M}$ having the property that

$$H\Big(\Phi^x((-\infty, 0)) < +\infty,\ \Phi^x((0, +\infty)) = +\infty\Big) > 0.$$

To consider an example, we choose any element k of K, setting

$$\Phi = \sum_{l=0}^{\infty} \delta_{[l,k]}, \qquad H = \int \delta_{T_x \Phi}(.)\ \mu(\mathrm{d}x).$$

We have, however,

8.5.4. Proposition. *In the case* $s = 1$, *for all stationary measures* W *in* $\mathbf{W}$

$$W\Big(\Psi^x((-\infty, 0)) < +\infty,\ \Psi^x((0, +\infty)) = +\infty\Big)$$
$$= 0 = W\Big(\Psi^x((-\infty, 0)) = +\infty,\ \Psi^x((0, +\infty)) < +\infty\Big).$$

Proof. We set

$$H = W\Big(\Psi \in (.),\ \Psi^x((0, +\infty)) = +\infty,\ \Psi^x((-\infty, 0)) < +\infty\Big)$$

and suppose that H be different from the zero measure 0. Thus we get

$$H = \sum_{-\infty < m < +\infty} H\Big(\Psi \in (.),\ \Psi^x((-\infty, m)) = 0,\ \Psi^x([m, m+1)) > 0\Big).$$

Hence

$$H\Big(\Psi^x([0, 1)) > 0\Big)$$
$$= \sum_{-\infty < m < +\infty} H\Big(\Psi^x([0, 1)) > 0,\ \Psi^x((-\infty, m)) = 0,\ \Psi^x([m, m+1)) > 0\Big)$$
$$= \sum_{m=0}^{\infty} H\Big(\Psi^x([0, 1)) > 0,\ \Psi^x((-\infty, -m)) = 0,\ \Psi^x((-m, -m+1)) > 0\Big).$$

Since the set $\big\{\Psi : \Psi \in M,\ \Psi^x((0, +\infty)) = +\infty,\ \Psi^x((-\infty, 0)) < +\infty\big\}$ belongs to $\mathfrak{J}$, it follows that together with W the measure H is stationary, too, so that we may continue the above chain of equations as follows:

$$= \sum_{m=0}^{\infty} H\Big(\Psi^x([m, m+1)) > 0,\ \Psi^x((-\infty, 0)) = 0,\ \Psi^x([0, 1)) > 0\Big).$$

The measure $H\left(\Psi \in (.), \Psi^x([0, 1)) > 0, \Psi^x((-\infty, 0)) = 0\right)$ is finite. Therefore we may use the Borel-Cantelli lemma to conclude that

$$H\left(\Psi \text{ infinite}, \Psi^x([0, 1)) > 0, \Psi^x((-\infty, 0)) = 0\right) = 0.$$

Hence also, for all integers m,

$$H\left(\Psi \text{ infinite}, \Psi^x([m, m+1)) > 0, \Psi^x((-\infty, m)) = 0\right) = 0,$$

and hence

$$H(\Psi \text{ infinite}) = 0,$$

which contradicts our assumption that $H \neq 0$. Consequently

$$W\left(\Psi^x((-\infty, 0)) < +\infty, \Psi^x((0, +\infty)) = +\infty\right) = 0.$$

In a similar way it is shown that

$$W\left(\Psi^x((-\infty, 0)) = +\infty, \Psi^x((0, +\infty)) < +\infty\right) = 0. \blacksquare$$

In the case $s = 1$ we can now derive a complete symmetry characterization of arbitrary distributions of the form $(\tilde{P})^0$.

8.5.5. Theorem. *In the case $s = 1$, to each measure V on $\mathfrak{M}^0$ there exists a stationary infinitely divisible distribution P on $\mathfrak{M}$ with the property that $(\tilde{P})^0 = V$ if and only if the following conditions are satisfied:*

a) *For all $t > 0$,*

$$\int (\chi^x(\{0\}))^{-1} \min\left(t, x_0(\chi)\right) V(d\chi) < +\infty,$$

where we have to set $x_0(\chi)$ equal to $+\infty$ for $\chi^x((0, +\infty)) = 0$.

b) $$V\left(\chi^x((-\infty, 0)) = +\infty, \chi^x((0, +\infty)) < +\infty\right)$$
$$= 0 = V\left(\chi^x((-\infty, 0)) < +\infty, \chi^x((0, +\infty)) = +\infty\right).$$

c) $\hat{\Theta} V = V$.

Proof. 1. Let P be an infinitely divisible stationary distribution on $\mathfrak{M}$, and $V = (\tilde{P})^0$.

In view of 8.2.8. we have, for all $t > 0$,

$$+\infty > \tilde{P}\left(\Psi^x((-t, 0]) > 0\right) = \int (\chi^x(\{0\}))^{-1} \min\left(t, x_0(\chi)\right) V(d\chi),$$

i.e. V satisfies condition a).

In virtue of propositions 8.5.4. and 8.1.6. this is true also for condition b). By 8.5.3. we may further conclude that condition c) is satisfied.

2. Conversely, suppose that V satisfy all of the conditions stated above. Then by 8.5.3. the σ-finite measure V is the Palm measure of a σ-finite

stationary measure H on $\mathfrak{M}$ satisfying the condition $H(\{o\}) = 0$. It follows from 8.2.8. that, for all $t > 0$,

$$H\left(\Psi^x((-t, 0]) > 0\right) = \int \left(\chi^x(\{0\})\right)^{-1} \min\left(t, x_0(\chi)\right) V(d\chi),$$

i.e. that H belongs to in $\boldsymbol{W}$. Hence

$$V = \left(\widetilde{\mathscr{E}_H}\right)^0. \blacksquare$$

By virtue of 8.3.3. and 8.5.5., we obtain the following counterpart of theorem 8.4.2.:

8.5.6. Theorem. *In the case $s = 1$, to each measure V on $\mathfrak{M}^0$ there exists a stationary singular infinitely divisible distribution P on $\mathfrak{M}$ with the properties $P(\Phi^x \text{ simple}) = 1$, $(\tilde{P})^0 = V$ if and only if the following conditions are satisfied:*

a) *For all $t > 0$,*

$$\int \min\left(t, x_0(\chi)\right) V(d\chi) < +\infty.$$

b) *V^x is simple.*

c)
$$V\left(\chi^x((-\infty, 0)) = +\infty,\ \chi^x((0, +\infty)) < +\infty\right)$$
$$= 0 = V\left(\chi^x((-\infty, 0)) < +\infty,\ \chi^x((0, +\infty)) = +\infty\right).$$

d) $\hat{\Gamma}_1 V = V$.

8.6. A Further Characterization of Palm Measures

The Palm measure H^0 of a σ-finite stationary measure H on $\mathfrak{M}$ is stationary only in the trivial case $H^0 = 0$, i.e. $H(\Phi \neq o) = 0$.

This is because otherwise, in view of the relation $H^0\left(\chi^x(\{[0, \ldots, 0]\}) = 0\right) = 0$, which holds for all Palm measures, the inequality $H^0\left(\chi^x(\{x\}) > 0\right) > 0$ would be satisfied for all x in R^s. Since H^0 is σ-finite, we can now select a sequence (Y_n) of pairwise disjoint sets in $\mathfrak{M}$ having the union M such that all measures $H^0\left((.) \cap Y_n\right)$ are finite. Then by 1.1.5. the set $\left\{x : x \in R^s,\ H^0\left(\chi^x(\{x\}) > 0,\ \chi \in Y_n\right) > 0\right\}$ is at most countable for each natural number n. Hence also $\left\{x : x \in R^s,\ H^0\left(\chi^x(\{x\}) > 0\right) > 0\right\}$ is at most countable, which contradicts the above statement.

In view of 5.1.12., all σ-finite measures Q on $\mathfrak{M}$ having the property that $Q(M \setminus M^0) = 0$ satisfy the equation $\mathscr{C}_Q\left(C \setminus (R^s \times M^0)\right) = 0$, i.e. the total mass of $\mathscr{C}_Q$ is placed on $C \cap (R^s \times M^0)$.

In virtue of 6.1.1., the correspondence

$$[x, \Phi] \curvearrowright [-x, T_x\Phi]$$

provides a one-to-one mapping of $C \cap (R^s \times M^0)$ upon itself which is measurable in both directions. Using this simple mapping, we can now formulate a characteristic symmetry property of the measures $\mathscr{C}_{H^0}$ (cf. Mecke [1]):

8.6.1. Theorem. *A σ-finite measure Q on $\mathfrak{M}$ satisfying the condition $Q(M \setminus M^0) = 0$ is the Palm measure of a σ-finite stationary measure H on $\mathfrak{M}$ if and only if $\mathscr{C}_Q$ is invariant with respect to the mapping*

$$[x, \Phi] \curvearrowright [-x, T_x\Phi]$$

of $C \cap (R^s \times M^0)$ upon itself.

Proof. 1. First suppose that Q be the Palm measure of a σ-finite stationary measure H.

We choose any non-negative real function g on R^s being measurable with respect to $\mathfrak{R}^s$ and having the property that $\int g(y)\,\mu(\mathrm{d}y) = 1$. For each non-negative real function v on $C \cap (R^s \times M^0)$ being measurable with respect to $\mathfrak{C}$, we obtain by means of 5.1.2. and 8.1.2.:

$$\begin{aligned}
&\int v(-x, T_x\Phi)\,\mathscr{C}_Q(\mathrm{d}[x, \Phi]) \\
&= \int \Bigg(\sum_{x\in R^s,\Phi^x(\{x\})>0} v(-x, T_x\Phi)\,\Phi^x(\{x\})\Bigg) H^0(\mathrm{d}\Phi) \\
&= \int \Bigg(\sum_{x\in R^s,\Phi^x(\{x\})>0} g(y)\,v(-x, T_x\Phi)\,\Phi^x(\{x\})\Bigg) (\mu \times H^0)(\mathrm{d}[y, \Phi]) \\
&= \int \Bigg(\sum_{y\in R^s,\Phi^x(\{y\})>0} \Bigg(\sum_{x\in R^s,(T_y\Phi)^x(\{x\})>0} g(y)\,v(-x, T_xT_y\Phi)\,(T_y\Phi)^x(\{x\})\Bigg) \Phi^x(\{y\})\Bigg) H(\mathrm{d}\Phi) \\
&= \int \Bigg(\sum_{y\in R^s,\Phi^x(\{y\})>0} \Bigg(\sum_{z\in R^s,\Phi^x(\{z\})>0} g(y)\,v(y-z, T_z\Phi)\,\Phi^x(\{z\})\Bigg) \Phi^x(\{y\})\Bigg) H(\mathrm{d}\Phi) \\
&= \int \Bigg(\sum_{z\in R^s,\Phi^x(\{z\})>0} \Bigg(\sum_{y\in R^s,\Phi^x(\{y\})>0} g(y)\,v(y-z, T_z\Phi)\,\Phi^x(\{y\})\Bigg) \Phi^x(\{z\})\Bigg) H(\mathrm{d}\Phi) \\
&= \int \Bigg(\sum_{y\in R^s,(T_{-z}\Phi)^x(\{y\})>0} g(y)\,v(y-z, \Phi)\,(T_{-z}\Phi)^x(\{y\})\Bigg) (\mu \times H^0)(\mathrm{d}[z, \Phi]) \\
&= \int \Bigg(\int \Bigg(\sum_{t\in R^s,\Phi^x(\{t\})>0} g(t+z)\,v(t, \Phi)\,\Phi^x(\{t\})\Bigg) \mu(\mathrm{d}z)\Bigg) H^0(\mathrm{d}\Phi) \\
&= \int \Bigg(\sum_{t\in R^s,\Phi^x(\{t\})>0} v(t, \Phi)\,\Phi^x(\{t\})\Bigg) Q(\mathrm{d}\Phi) \\
&= \int v(x, \Phi)\,\mathscr{C}_Q(\mathrm{d}[x, \Phi]),
\end{aligned}$$

i.e. $\mathscr{C}_Q$ possesses the afore-mentioned property of invariance.

2. Conversely, suppose that the Campbell measure $\mathscr{C}_Q$ of a σ-finite measure Q satisfying the condition $Q(M \setminus M^0) = 0$ has the above-stated property of invariance.

Let h be an arbitrary measurable non-negative real function on $[C, \mathfrak{C}]$ having the property that

$$\sum_{x\in R^s,\Phi^x(\{x\})>0} h(x, \Phi)\,\Phi^x(\{x\}) = 1 \qquad (\Phi \in M \setminus \{o\}).$$

By means of

$$H(Y) = \int k_Y(T_{-x}\Phi)\, h(x, T_{-x}\Phi)\, (\mu \times Q)\, (\mathrm{d}[x, \Phi]) \qquad (Y \in \mathfrak{M})$$

we introduce a measure H on $\mathfrak{M}$. Obviously $H(\{o\}) = 0$. For all non-negative real functions g on $R^s \times M^0$ which are measurable with respect to $\mathfrak{R}^s \otimes \mathfrak{M}^0$, we get

$$\int g(y, \Phi)\, \mathscr{C}_H{}^T(\mathrm{d}[y, \Phi])$$

$$= \int \Big(\sum_{y \in R^s, \Phi^x(\{y\}) > 0} g(y, T_y\Phi)\, \Phi^x(\{y\}) \Big)\, H(\mathrm{d}\Phi)$$

$$= \int h(x, T_{-x}\Phi) \Big(\sum_{y \in R^s, (T_{-x}\Phi)^x(\{y\}) > 0} g(y, T_y T_{-x}\Phi)\, (T_{-x}\Phi)^x(\{y\}) \Big)\, (\mu \times Q)\, (\mathrm{d}[x, \Phi])$$

$$= \int \Big(\int \Big(\sum_{z \in R^s, \Phi^x(\{z\}) > 0} h(x, T_{-x}\Phi)\, g(z + x, T_z\Phi)\, \Phi^x(\{z\}) \Big)\, \mu(\mathrm{d}x) \Big)\, Q(\mathrm{d}\Phi)$$

$$= \int \Big(\int \Big(\sum_{z \in R^s, \Phi^x(\{z\}) > 0} h(u - z, T_{-u}T_z\Phi)\, g(u, T_z\Phi)\, \Phi^x(\{z\}) \Big)\, \mu(\mathrm{d}u) \Big)\, Q(\mathrm{d}\Phi)$$

$$= \int \Big(\sum_{z \in R^s, \Phi^x(\{z\}) > 0} \Big(\int h(u - z, T_{-u}T_z\Phi)\, g(u, T_z\Phi)\, \mu(\mathrm{d}u) \Big)\, \Phi^x(\{z\}) \Big)\, Q(\mathrm{d}\Phi)$$

$$= \int \Big(\int h(u - z, T_{-u}T_z\Phi)\, g(u, T_z\Phi)\, \mu(\mathrm{d}u) \Big)\, \mathscr{C}_Q(\mathrm{d}[z, \Phi])$$

$$= \int \Big(\int h(u + z, T_{-u}\Phi)\, g(u, \Phi)\, \mu(\mathrm{d}u) \Big)\, \mathscr{C}_Q(\mathrm{d}[z, \Phi])$$

$$= \int \Big(\sum_{z \in R^s, \Phi^x(\{z\}) > 0} h(u + z, T_{-u}\Phi)\, \Phi^x(\{z\}) \Big)\, g(u, \Phi)\, (\mu \times Q)\, (\mathrm{d}[u, \Phi])$$

$$= \int \Big(\sum_{t \in R^s, (T_{-u}\Phi)^x(\{t\}) > 0} h(t, T_{-u}\Phi)\, (T_{-u}\Phi)^x\, (\{t\}) \Big)\, g(u, \Phi)\, (\mu \times Q)\, (\mathrm{d}[u, \Phi])$$

$$= \int g(y, \Phi)\, (\mu \times Q)\, (\mathrm{d}[y, \Phi]),$$

i.e. we have $\mathscr{C}_H{}^T = \mu \times Q$.

The measure $\mu \times Q$ is σ-finite. By virtue of 5.1.11., this property is transmitted to $H\big((.) \setminus \{o\}\big) = H$. By virtue of 8.1.1., H is stationary and $H^0 = Q$. ∎

9. The Palm Distribution

9.0. Introduction

The introduction of the Palm measure H^0 of a stationary σ-finite measure H on $\mathfrak{M}$, which was outlined at the beginning of Section 8.0., may also be worded as follows: If Φ is a counting measure in M, and Y an arbitrary set in the σ-algebra $\mathfrak{M}^0$, then let Φ^Y denote the counting measure obtained from Φ by cancelling those points x which do not satisfy the condition $T_x\Phi \in Y$. Since $(T_y\Phi)^Y = T_y(\Phi^Y)$, the image H^Y of H under the measurable mapping $\Phi \curvearrowright \Phi^Y$ is stationary. The intensity i_{H^Y} of H^Y then coincides with $H^0(Y)$. Accordingly we may interpret $H^0(Y)$ as the weight contributed to the intensity $i_H = H^0(M^0)$ of H by those points x in Φ which satisfy the condition $T_x\Phi \in Y$. Now, if $0 < i_H < +\infty$ is satisfied, we may pass on to the specific weight, i.e. to $(i_H)^{-1} H^0(Y)$, thus obtaining the Palm distribution $H_0 = (i_H)^{-1} H^0$ of H.

The basic properties of Palm measures, which were derived in the preceding chapter, may be translated into the language of Palm distributions by using additional assumptions on the intensities of the occurring stationary measures. Here it should, however, be noted that, also under the assumption that $H(\{o\}) = 0$, the Palm distribution H_0 determines a stationary measure H only except for a positive factor. In particular this is true for the canonical measure $\tilde{P}$ of a stationary infinitely divisible distribution P on $\mathfrak{M}$ satisfying the condition $0 < i_P < +\infty$: note that, for all $t > 0$, $(\widetilde{P^t})_0 = (t\tilde{P})_0 = (\tilde{P})_0$! On the set of all stationary distributions P satisfying the conditions $P(\{o\}) = 0$, $i_P < +\infty$, the correspondence $P \curvearrowright P_0$ is, however, one-to-one. If the assumption that $P(\{o\}) = 0$ is replaced by the weaker assumption that $0 < i_P$, then P_0 is determined only by the conditional distribution $P\big((.) \mid \Phi \neq o\big)$ (Proposition 9.1.3.).

For all ergodic stationary distributions P on $\mathfrak{M}$ having the property that $0 < i_P < +\infty$, the quotient $\big(\Phi([-n, n)^s)\big)^{-1} \Phi^Y\big([-n, n)^s\big)$ tends toward $P_0(Y)$ almost everywhere with respect to P. Hence, in this case it is possible to interpret $P_0(Y)$ as the relative frequency of the occurrence of the property that $T_x\Phi \in Y$ among the points x in Φ. The Palm distribution P_0 describes the "distribution of $T_x\Phi$ for a point x selected at random from Φ". On the other hand, P describes the "distribution of $T_x\Phi$ for a point x selected at random from the phase space R^s". For ergodic stationary distributions P having the property that $0 < i_P < +\infty$, both ways of viewing one and the same stochastic process are in principle equivalent. The results presented in this chapter allow a changing-over from one point of view to the other.

For all simple stationary measures H on $\mathfrak{M}$ having the property that $0 < i_H < +\infty$, and for all bounded Borel subsets X of R^s with nonempty interior, the conditional distribution $H_0\big((.) \mid \Phi(X) > 0\big)$ can be established. Now it turns out (Proposition 9.3.4.) that $H_0\big((.) \mid \Phi(X_n) > 0\big)$ converges weakly toward H_0 if $\sup_{x \in X_n} \|x\|$ tends toward zero as n increases. Historically, this interpretation was the starting point for the introduction of the Palm distribution. The first steps in this direction were undertaken by Palm (cf. Palm [1]). They were brought into a mathematically exact form and further developed by Khinchin [1], who studied some special Palm probabilities $P_0(Y)$. Independently of each other, Ryll-Nardzewski [1] and Slivnyak [1], [2] presented the explicit introduction of the Palm distribution. In an attempt to reduce unnecessary assumptions and to give consistently designed, purely measure theoretical proofs of fundamental statements on Palm distributions, later on the concepts of the Palm and Campbell measures were developed. However, this did not make the concept of the Palm distribution unnecessary, since it permits the use of a wider spectrum of probabilistic methods in the interpretation, proof, and formulation of propositions. As for example, in the theory of the Palm distribution, the inversion formulae which are already known from the preceding chapter are supplemented, in the case of ergodic distributions, by the simple inversion formula of Proposition 9.1.9.

Palm distributions can be defined also for non-stationary distributions P of finite intensity (cf. Ryll-Nardzewski [1], Papangelou [2], Jagers [2] and Kallenberg [6]). Thus, if $[A, \varrho_A]$ is an arbitrary phase space in the sense of Chapter 1, and P an arbitrary distribution on $\mathfrak{M}$ of finite intensity, then there exists a mapping $a \curvearrowright P_a$ of A into the set of all distributions on $\mathfrak{M}$, such that, for all Y in $\mathfrak{M}$, the real function $a \curvearrowright P_a(Y)$ is measurable and satisfies the equation $\mathscr{C}_P(X \times Y) = \int_X P_a(Y)\, \varrho_P(da)$ for all bounded X in $\mathfrak{A}$. Thus we no longer obtain a single distribution P_0, but rather a family $(P_a)_{a \in A}$. In the case of simple distributions P, we associate with this the intuitive concept that the limiting process associated with the local interpretation of the Palm distribution is performed not only at the origin, but at all points a of the support of the measure ϱ_P. Actually, this limit does of course not necessarily exist at all of these points a. Moreover, it should be noted that the mapping $a \curvearrowright P_a$ is determined only except for the values on a set of the ϱ_P—measure zero. In this book, the authors preferred to operate directly with Campbell measures $\mathscr{C}_P$ rather than with families $(P_a)_{a \in A}$.

As it was done in the preceding chapter, in the theory of the Palm distribution we shall mainly consider the "marked case", as it is constantly encountered especially in the applications of point processes in the queuing theory (see, for example, Franken [2]). In these applications, however, a refinement of the concept of the Palm distribution is usually required in which a Borel subset L of the "mark space" $[K, \varrho_K]$ is fixed. Now, if a stationary distribution P on $\mathfrak{M}$ satisfies the condition that $0 < \gamma_P(L) < +\infty$, we may form the "distribution of $T_x\Phi$ for a point $[x, h]$ selected at random from Φ, the mark h of this point being in L". In this way we obtain the

Palm distribution P_L of P with respect to the subset L. There are no fundamental difficulties in refining already the theory of the Campbell measure so that the Palm distribution with respect to L is incorporated into the general theory. However, the notation becomes much more complicated. Although the corresponding concepts open many more possibilities of applying the Palm distribution (e.g. in the analysis of renewal properties and in the construction of stationary distributions), for the sake of clearness the authors have desisted from presenting this refinement of the general theory; instead we refer to Koenig & Matthes [1], Matthes [3] and Koenig, Matthes & Nawrotzki [1], [2].

9.1. The Mapping $P \rightsquigarrow P_0$

As in the preceding chapter, in the first three sections of this chapter we shall, as a rule, assume that the phase space $[A, \varrho_A]$ be the direct product of an Euclidean "position space" R^s, $s \geqq 1$, with a bounded complete separable "mark space" $[K, \varrho_K]$.

For all stationary measures H on $\mathfrak{M}$ having the property that $0 < i_{H^x} < +\infty$, the measure $H((.) \setminus \{o\})$ belongs to $\boldsymbol{W}$, and hence is σ-finite. Thus the Palm measure H^0 is defined and has the total mass i_{H^x}. The distribution $(i_{H^x})^{-1} H^0$ is called the *Palm distribution* of H, and denoted by H_0. Like H^0, H_0 is optionally considered to be a distribution on $\mathfrak{M}^0$ or a distribution on $\mathfrak{M}$ satisfying the condition $H_0(M \setminus M^0) = 0$.

Now let P be a stationary distribution on $\mathfrak{M}$ which satisfies the condition $0 < i_{P^x} < +\infty$, and Y an arbitrary set in $\mathfrak{I}$ having the property that $o \notin Y$. By 8.1.3. and 8.1.6., $P(Y)$ is positive if and only if this is the case for $P_0(Y)$; so we obtain

$$\begin{aligned}\big(P((.) \mid Y)\big)_0 &= \big(i_{(P((.)\mid Y))^x}\big)^{-1} \big(P((.) \mid Y)\big)^0 \\ &= P(Y) \big(P^0(Y)\big)^{-1} \big((P(Y))^{-1} P((.) \cap Y)\big)^0 \\ &= \big(P^0(Y)\big)^{-1} P^0\big((.) \cap Y\big) = \big(P_0(Y)\big)^{-1} P_0\big((.) \cap Y\big) \\ &= P_0\big((.) \mid Y\big),\end{aligned}$$

so that we observe the truth of

9.1.1. *For all stationary distributions P on $\mathfrak{M}$ satisfying the condition $0 < i_{P^x} < +\infty$ and all Y in $\mathfrak{I}$ having the property that $o \notin Y$, $P(Y)$ is positive if and only if this is the case for $P_0(Y)$, and*

$$\big(P((.) \mid Y)\big)_0 = P_0\big((.) \mid Y\big).$$

Now, if P is ergodic, we first obtain $P(\{o\}) = 0$; for $P(\{o\}) > 0$ would imply $P(\{o\}) = 1$ and hence $i_{P^x} = 0$. For each Y in $\mathfrak{I}$ with the property that $o \notin Y$, $P(Y)$ must be either zero or one. According to 9.1.1., we obtain

$P_0(Y) = 0$ in the first case, and $P_0 = \big(P((.) \mid Y)\big)_0 = P_0((.) \mid Y)$, i.e. $P_0(Y) = 1$, in the second. Since $P(\{o\}) = 0$, P_0 can take only the values zero or one on $\mathfrak{J}$.

Conversely, suppose that $P(\{o\})$ be zero and P_0 takes only the values zero or one on $\mathfrak{J}$. Using 9.1.1. it is then concluded that, for all Y in $\mathfrak{J}$ having the property that $o \notin Y$, P can take only the values zero or one. Hence everywhere on $\mathfrak{J}$ only these two values come into question for P, i.e. P is ergodic. Thus we are led to

9.1.2. *Subject to the assumptions of 9.1.1., P is ergodic if and only if $P(\{o\}) = 0$ and P_0 takes only the values zero or one on $\mathfrak{J}$.*

If in 9.1.1. we substitute the set $M \setminus \{o\}$ for Y, then $P_0 = \big(P((.) \mid \Phi \neq o)\big)_0$. Now, if two stationary distributions P_1, P_2, where $0 < i_{(P_1)^x}, i_{(P_2)^x} < +\infty$, satisfy the equation $(P_1)_0 = (P_2)_0$, then $(P_1)^0$ is a positive multiple of $(P_2)^0$, and hence, according to 8.1.3. and 8.1.4., $P_1((.) \setminus \{o\})$ is a positive multiple of $P_2((.) \setminus \{o\})$. Consequently the distribution $P_1((.) \mid \Phi \neq o)$ is a positive multiple of the distribution $P_2((.) \mid \Phi \neq o)$, and we observe the truth of

9.1.3. Proposition. *The Palm distributions of two distributions P_1, P_2 on $\mathfrak{M}$ satisfying the condition $0 < i_{(P_1)^x}, i_{(P_2)^x} < +\infty$ coincide if and only if*

$$P_1((.) \mid \Phi \neq o) = P_2((.) \mid \Phi \neq o)$$

is satisfied.

For each at most countable family $(P_j)_{j \in J}$ of stationary distributions on $\mathfrak{M}$, and for each family $(p_j)_{j \in J}$ of non-negative numbers satisfying the condition $\sum_{j \in J} p_j = 1$, the intensity i_{P^x} of the mixture $P^x = \sum_{j \in J} p_j (P_j)^x$ is equal to $\sum_{j \in J} p_j i_{(P_j)^x}$. If i_{P^x} is finite, and if all $i_{(P_j)^x}$ are finite positive numbers, then we obtain, using 8.1.4.,

$$P_0 = (i_{P^x})^{-1} \sum_{j \in J} p_j (P_j)^0 = (i_{P^x})^{-1} \sum_{j \in J} p_j i_{(P_j)^x} (P_j)_0,$$

i.e. we have

9.1.4. *For each at most countable family $(P_j)_{j \in J}$ of stationary distributions on $\mathfrak{M}$ having the property that $0 < i_{(P_j)^x} < +\infty$, $j \in J$, and for each family $(p_j)_{j \in J}$ of non-negative numbers satisfying the conditions*

$\sum_{j \in J} p_j = 1, \quad \sum_{j \in J} p_j i_{(P_j)^x} < +\infty$, *we have*

$$\Big(\sum_{j \in J} p_j P_j\Big)_0 = \Big(\sum_{j \in J} p_j i_{(P_j)^x}\Big)^{-1} \sum_{j \in J} p_j i_{(P_j)^x} (P_j)_0 .$$

Given now any set Y in $\mathfrak{M}^0$, then to each Φ in M we may associate the counting measure

$$\Phi^Y = \sum_{x \in R^s,\, T_x \Phi \in Y} \Phi^x(\{x\})\, \delta_x,$$

which corresponds to the phase space R^s, so that we remove those points x from Φ^x for which $T_x\Phi$ is not in Y. For all bounded X in $\mathfrak{R}^s$,

$$\Phi^Y(X) = \int k_X(x)\, k_Y(T_x\Phi)\, \Phi^x(\mathrm{d}x),$$

so that by 6.1.1. and the measurability statement of 5.1.2. it can be concluded that the mapping $\Phi \curvearrowright \Phi^Y(X)$ of M into $\{0, 1, \ldots\}$ is measurable. Consequently the mapping $\Phi \curvearrowright \Phi^Y$ is measurable.

Obviously we have

$$(T_x\Phi)^Y = T_x(\Phi^Y) \qquad (x \in R^s,\ \Phi \in M).$$

Consequently, if a measure H on $\mathfrak{M}$ is stationary the measure H^Y produced from H by $\Phi \curvearrowright \Phi^Y$ is also stationary.

By the help of the mappings $\Phi \curvearrowright \Phi^Y$ the Palm-Khinchin formulae 8.2.10. can be generalized in the following way.

9.1.5. Proposition. *Suppose that $s = 1$ and that H be a stationary measure on $\mathfrak{M}$. If then $H((.) \setminus \{o\})$ is σ-finite and H^x is simple, then for all Y in $\mathfrak{M}^0$, all $t \geq 0$ and all non-negative integers j*

$$H(\Phi^Y((0, t]) > j) = \int_0^t H^0(\Phi \in Y,\ \Phi^Y((0, x]) = j)\, \mu(\mathrm{d}x).$$

Proof. Proceeding in the same way as in the proof of 8.2.10. we have

$$\begin{aligned} Z = \{[x, \Phi] : {} & [x, \Phi] \in C,\quad 0 < x \leq t,\quad \Phi^x \text{ simple},\\ & T_x\Phi \in Y,\quad \Phi^Y((x, t]) = j\}\\ = \bigcup_{0 \leq l < +\infty} \{[x, \Phi] : {} & [x, \Phi] \in C,\quad 0 < x,\quad \Phi^x \text{ simple},\\ & T_x\Phi \in Y,\quad x_l(\Phi^Y) \leq t < x_{l+1}(\Phi^Y)\\ & \text{or } x_l(\Phi^Y) \leq t \text{ and } x_{l+1}(\Phi^Y) \text{ does not exist},\\ & k(x, \Phi^Y) = l - j\}. \end{aligned}$$

We obtain

$$\Omega(Z) = \{\Phi : \Phi \in M,\ \Phi^x \text{ simple},\ \Phi^Y((0, t]) > j\}$$

and by 5.1.7. we conclude that

$$\begin{aligned} H(\Phi^Y((0, t]) > j) &= H(\Phi^x \text{ simple},\ \Phi^Y((0, t]) > j)\\ &= \mathscr{C}_H(Z)\\ &= \mathscr{C}_H^T(0 < x \leq t,\ \Phi^x \text{ simple},\ \Phi \in Y,\ \Phi^Y((0, t - x]) = j)\\ &= \mathscr{C}_H^T(0 < x \leq t,\ \Phi \in Y,\ \Phi^Y((0, t - x]) = j)\\ &= \int_0^t H^0(\Phi \in Y,\ \Phi^Y((0, t - z]) = j)\, \mu(\mathrm{d}z). \quad \blacksquare \end{aligned}$$

Now let P be a stationary distribution on $\mathfrak{M}$ with the property $P(\{o\}) = 0$ and let Y be a set in $\mathfrak{M}^0$. According to Section 6.4., the sample intensity

$s(\Psi)$ of Ψ exists almost everywhere with respect to P^Y, i.e. $s(\Phi^Y)$ exists almost everywhere with respect to P. Using 6.2.3. we obtain

$$\int s(\Phi^Y)\,P(\mathrm{d}\Phi) = \int s(\Psi)\,P^Y(\mathrm{d}\Psi) = i_{P^Y} = P^0(Y).$$

According to Section 6.4., we have

$$(2t)^{-s}\,\Psi\big([-t,t)^s\big) \xrightarrow[t\to\infty]{} s(\Psi)$$

almost everywhere with respect to P^Y, i.e.

$$(2t)^{-s}\,\Phi^Y\big([-t,t)^s\big) \xrightarrow[t\to\infty]{} s(\Phi^Y)$$

almost everywhere with respect to P. The set $\{\Phi : \Phi \in M,\, s(\Phi^x) = 0\}$ belongs to $\mathfrak{F}$. If its probability were positive, we could introduce the stationary distribution $P' = P\big((.) \mid s(\Phi^x) = 0\big)$. Since

$$i_{P'} = \int s(\Psi^x)\,P'(\mathrm{d}\Psi) = 0$$

we would then conclude that $P'(\{o\}) = 1$ and hence also $P(\{o\}) > 0$, which contradicts our assumptions.

Consequently $s(\Phi^x) > 0$ almost everywhere with respect to P, and we obtain, using the additional assumption that $P\big(s(\Phi^x) < +\infty\big) = 1$,

$$\frac{\Phi^Y\big([-t,t)^s\big)}{\Phi^x\big([-t,t)^s\big)} = \left(\frac{\Phi^Y\big([-t,t)^s\big)}{(2t)^s}\right)\left(\frac{(2t)^s}{\Phi^x\big([-t,t)^s\big)}\right) \xrightarrow[t\to\infty]{} \frac{s(\Phi^Y)}{s(\Phi^x)}$$

for almost all Φ. Thus the quotient $\big(s(\Phi^x)\big)^{-1} s(\Phi^Y)$, which is defined almost everywhere, can be considered to be the limit of the relative frequency, within Φ, of those points $[x, k]$ in Φ which have the property that $T_x\Phi \in Y$.

The functions $s(\Phi^Y)$ are invariant, i.e. they are measurable with respect to $\mathfrak{F}$. Hence, if P is ergodic, then $s(\Phi^Y)$ coincides with $\int s(\Phi^Y)\,P(\mathrm{d}\Phi) = P^0(Y)$ almost everywhere, so that we have

9.1.6. Proposition. *For each ergodic stationary distribution P on $\mathfrak{M}$ satisfying the condition that $0 < i_{P^x} < +\infty$, and for each Y in $\mathfrak{M}^0$, the following relation holds for almost all Φ*

$$\Big(\Phi^x\big([-t,t)^s\big)\Big)^{-1}\,\Phi^Y\big([-t,t)^s\big) \xrightarrow[t\to\infty]{} P_0(Y).$$

Consequently, subject to the assumptions of 9.1.6., $P_0(Y)$ can be thought of as a "probability" that a "point $[x, k]$ selected at random in Φ" should have the property that $T_x\Phi \in Y$. If we now drop the assumption that P be ergodic, the expectation

$$P_*(Y) = \int \big(s(\Phi^x)\big)^{-1} s(\Phi^Y)\,P(\mathrm{d}\Phi)$$

should be a possible variant of this "probability":

9.1.7. *For all stationary distributions P on $\mathfrak{M}$ satisfying the condition $P\big(0 < s(\Phi^x) < +\infty\big) = 1$, P_* is a distribution on $\mathfrak{M}^0$.*

Proof. Obviously $P_*(M^0) = 1$ as well as $P_*(Y) \geqq 0$ for all Y in $\mathfrak{M}^0$. Now let (Y_n) be a sequence of pairwise disjoint sets in $\mathfrak{M}^0$, and Y their union. For almost all Φ, all of the $s(\Phi^{Y_n})$ and $s(\Phi^Y)$ exist, and it is readily shown that

$$s(\Phi^Y) \geqq \sum_{n=1}^{m} s(\Phi^{Y_n})$$

for all m, i.e. that

$$s(\Phi^Y) \geqq \sum_{n=1}^{\infty} s(\Phi^{Y_n}).$$

For all sufficiently large m, $P_m = P\big((.) \mid s(\Phi^Y) \leqq m\big)$ exists, and

$$\int s(\Phi^Y)\, P_m(\mathrm{d}\Phi) = i_{(P_m)^Y} = (P_m)^0\,(Y) = \sum_{n=1}^{\infty} (P_m)^0\,(Y_n)$$

$$= \sum_{n=1}^{\infty} i_{(P_m)^{Y_n}} = \sum_{n=1}^{\infty} \int s(\Phi^{Y_n})\, P_m(\mathrm{d}\Phi) = \int \Big(\sum_{n=1}^{\infty} s(\Phi^{Y_n})\Big)\, P_m(\mathrm{d}\Phi).$$

Hence

$$s(\Phi^Y) = \sum_{n=1}^{\infty} s(\Phi^{Y_n})$$

almost everywhere with respect to the P_m, i.e. almost everywhere with respect to P, and consequently

$$P_*(Y) = \int s(\Phi^Y)\, \big(s(\Phi^x)\big)^{-1}\, P(\mathrm{d}\Phi) = \int \Big(\sum_{n=1}^{\infty} s(\Phi^{Y_n})\Big)\, \big(s(\Phi^x)\big)^{-1}\, P(\mathrm{d}\Phi)$$

$$= \sum_{n=1}^{\infty} \int s(\Phi^{Y_n})\, \big(s(\Phi^x)\big)^{-1}\, P(\mathrm{d}\Phi) = \sum_{n=1}^{\infty} P_*(Y_n). \blacksquare$$

Like the Palm distribution of P, P_* is also optionally interpreted either as a distribution on $\mathfrak{M}^0$ or as a distribution on $\mathfrak{M}$ satisfying the condition $P_*(M \setminus M^0) = 0$.

Subject to the assumptions of 9.1.7., P_* coincides with P_0 if and only if $s(\Phi^x)$ is constant almost everywhere with respect to P. More generally, we have (cf. Slivnyak [1], [2]):

9.1.8. Proposition. *Subject to the assumptions of* 9.1.7., *the two measures* P^0, P_* *are absolutely continuous with respect to each other, satisfying the relation*

$$P^0(\mathrm{d}\Phi) = s(\Phi^x)\, P_*(\mathrm{d}\Phi).$$

Proof. Let Y be a set in $\mathfrak{M}^0$. For all natural numbers n, we set

$$Y_n = Y \cap \{\Phi : \Phi \in M,\quad n^{-1} \leqq s(\Phi^x) < n\}.$$

Further let a finite sequence $n^{-1} = t_0 < t_1 < \cdots < t_k = n$ be given. We get

$$\left|P^0(Y_n) - \int_{Y_n} s(\Phi^x)\, P_*(\mathrm{d}\Phi)\right|$$

$$\leqq \sum_{i=1}^{k} \left|P^0\big(Y_n \cap \{\Phi : \Phi \in M, t_{i-1} \leqq s(\Phi^x) < t_i\}\big) - \int_{Y_n \cap \{\Phi : t_{i-1} \leqq s(\Phi^x) < t_i\}} s(\Phi^x)\, P_*(\mathrm{d}\Phi)\right|$$

$$\leqq \sum_{i=1}^{k} \Big|P^0\big(Y_n \cap \{\Phi : \Phi \in M, t_{i-1} \leqq s(\Phi^x) < t_i\}\big)$$

$$- t_{i-1} P_*\big(Y_n \cap \{\Phi : \Phi \in M,\ t_{i-1} \leqq s(\Phi^x) < t_i\}\big)\Big|$$

$$+ \sum_{i=1}^{k} (t_i - t_{i-1})\, P_*\big(Y_n \cap \{\Phi : \Phi \in M, t_{i-1} \leqq s(\Phi^x) < t_i\}\big)$$

$$\leqq \sum_{i=1}^{k} \left| \int_{\{\Phi : t_{i-1} \leqq s(\Phi^x) < t_i\}} s(\Phi^{Y_n})\, P(\mathrm{d}\Phi) - \int_{\{\Phi : t_{i-1} \leqq s(\Phi^x) < t_i\}} t_{i-1} s(\Phi^{Y_n}) \left(s(\Phi^x)\right)^{-1} P(\mathrm{d}\Phi)\right|$$

$$+ \max_{1 \leqq i \leqq k} (t_i - t_{i-1})$$

$$\leqq \sum_{i=1}^{k} \int_{\{\Phi : t_{i-1} \leqq s(\Phi^x) < t_i\}} s(\Phi^{Y_n}) \left(1 - \frac{t_{i-1}}{t_i}\right) P(\mathrm{d}\Phi) + \max_{1 \leqq i \leqq k} (t_i - t_{i-1})$$

$$\leqq \left(\max_{1 \leqq i \leqq k} \left(1 - \frac{t_{i-1}}{t_i}\right)\right) \sum_{i=1}^{k} \int_{\{\Phi : t_{i-1} \leqq s(\Phi^x) < t_i\}} s(\Phi^{Y_n})\, P(\mathrm{d}\Phi) + \max_{1 \leqq i \leqq k} (t_i - t_{i-1})$$

$$\leqq n \max_{1 \leqq i \leqq k} \left(1 - \frac{t_{i-1}}{t_i}\right) + \max_{1 \leqq i \leqq k} (t_i - t_{i-1}).$$

By suitably selecting the finite sequence $t_0, \ldots, t_k$ the right-hand side can be made arbitrarily small, so that we obtain

$$P^0(Y_n) = \int_{Y_n} s(\Phi^x)\, P_*(\mathrm{d}\Phi).$$

Passing to the limit for $n \to \infty$, we find that

$$P^0(Y) = \int_{Y} s(\Phi^x)\, P_*(\mathrm{d}\Phi). \ \blacksquare$$

P can be recovered from P_* by means of the following inversion formula (cf. Slivnyak [1], [2]):

9.1.9. Proposition. *Subject to the assumptions of 9.1.7., the following relation holds for all Y in $\mathfrak{M}$:*

$$P(Y) = \lim_{t \to \infty} (2t)^{-s} \int_{[-t,t)^s} P_*(T_x\Phi \in Y)\, \mu(\mathrm{d}x).$$

Proof. 1. For all Y in $\mathfrak{F}$, $P(Y) = P_*(Y)$. This is because, in this case,

$$s(\Phi^{Y \cap M^0}) = k_Y(\Phi)\, s(\Phi^x)$$

almost everywhere with respect to P, and hence

$$s(\Phi^{Y\cap M^0})\left(s(\Phi^x)\right)^{-1} = k_Y(\Phi).$$

2. Using 6.1.1. we get

$$(2t)^{-s} \int_{[-t,t)^s} P_*(T_x\Phi \in Y)\,\mu(\mathrm{d}x) = (2t)^{-s} \int_{[-t,t)^s\times M} k_Y(T_x\Phi)\,(\mu\times P_*)(\mathrm{d}[x,\Phi])$$

$$= \int\left((2t)^{-s}\int_{[-t,t)^s} k_Y(T_x\Phi)\,\mu(\mathrm{d}x)\right)P_*(\mathrm{d}\Phi).$$

The integrand on the right-hand side converges toward $\overline{k_Y}(\Phi)$ on the whole domain of definition of $\overline{k_Y}$ and hence almost everywhere with respect to P. The domain of definition of $\overline{k_Y}$ belongs to $\mathfrak{J}$ and has the probability one with respect to P. Consequently, in view of 1., $\overline{k_Y}$ is likewise defined almost everywhere with respect to P_*, and using Lebesgue's theorem we get

$$\lim_{t\to\infty} (2t)^{-s} \int_{[-t,t)^s} P_*(T_x\Phi \in Y)\,\mu(\mathrm{d}x) = \int \overline{k_Y}(\Phi)\,P_*(\mathrm{d}\Phi).$$

Since $\overline{k_Y}$ is measurable with respect to $\mathfrak{J}$, we can once more apply 1. to continue the above equations as follows:

$$= \int \overline{k_Y}(\Phi)\,P(\mathrm{d}\Phi) = P(Y).\ \blacksquare$$

As an interesting consequence of Proposition 9.1.9., we have (cf. Matthes [5]):

9.1.10. Proposition. *If a distribution P satisfies the conditions of 9.1.7. and if a stationary distribution Q is a factor of P_*, then Q is also a factor of P.*

Proof. By assumption there exists a distribution S having the property that $P_* = Q * S$. Using 9.1.9. we obtain

$$(2t)^{-s} \int_{[-t,t)^s} (Q * S)\left(T_x\Phi \in (.)\right)\mu(\mathrm{d}x)$$

$$= (2t)^{-s} \int_{[-t,t)^s} \left(Q * (S(T_x\Phi \in (.)))\right)(.)\,\mu(\mathrm{d}x)$$

$$= Q * (2t)^{-s} \int_{[-t,t)^s} S(T_x\Phi \in (\cdot))\,\mu(\mathrm{d}x) \underset{t\to\infty}{\Rightarrow} P.$$

In view of 3.2.9., $\left((2t)^{-s} \int_{[-t,t)^s} S(T_x\Phi \in (.))\,\mu(\mathrm{d}x)\right)$ converges weakly toward some V in $\boldsymbol{P}$ and $P = Q * V$, i.e. Q is a factor of P. $\blacksquare$

With stronger assumptions on P, the inversion formula 9.1.9. becomes much simpler. However, we have now the weak convergence in $\boldsymbol{P}$ instead of the "point-wise convergence" (cf. Franken, Liemant & Matthes [1]):

9.1.11. Proposition. *For all mixing stationary distributions P on $\mathfrak{M}$ which satisfy the condition $0 < i_{P^x} < +\infty$,*

$$P_0\left(T_x\Phi \in (.)\right) \underset{\|x\|\to\infty}{\Rightarrow} P.$$

The proof of this will be postponed to Section 9.3.

We conclude this section by considering the special case $s = 1$. Using 6.1.3. and 8.2.7., we obtain the following inversion formula:

9.1.12. Proposition. *In the case $s = 1$, for all stationary distributions P on $\mathfrak{M}$ satisfying the conditions $P(\{o\}) = 0$, $i_{P^z} < +\infty$,*

$$P(Y) = i_{P^z} \int \left(\Phi^z(\{0\})\right)^{-1} \left(\int_0^{x_0(\Phi)} k_Y(T_x\Phi)\, \mu(\mathrm{d}x) \right) P_0(\mathrm{d}\Phi) \qquad (Y \in \mathfrak{M}).$$

If we now substitute the set M for Y in 9.1.12. we get

9.1.13. *Subject to the assumptions of 9.1.12.,*

$$i_{P^z} = \left(\int \left(\Phi^z(\{0\})\right)^{-1} x_0(\Phi)\, P_0(\mathrm{d}\Phi) \right)^{-1}.$$

Throughout the rest of this section we shall assume that the phase space $[A, \varrho_A]$ coincides with the real axis R^1.

To each simple stationary measure H on $\mathfrak{M}$ having the property $0 < i_H < +\infty$, we associate a distribution function F_H by means of

$$F_H(u) = H_0\left(x_0(\Phi) \leqq u\right) \qquad (u \geqq 0)$$

noting that, here as well as below, events such as "$x_i(\Phi) \in X$" shall be interpreted as "$x_i(\Phi)$ exists and belongs to X". Obviously we always have $F_H(0) = 0$.

Merely by means of the distribution function F_H, the measure, established with respect to H, of several statements on the behaviour of Φ in the neighbourhood of the origin can be expressed in a simple way.

9.1.14. Proposition. *For all $u, v \geqq 0$ and all simple stationary measures H on $\mathfrak{M}$ satisfying the condition $0 < i_H < +\infty$,*

a) $H\left(-x_{-1}(\Phi) \leqq u,\, x_0(\Phi) \leqq v\right)$

$$= i_H \left(\int_0^u \left(1 - F_H(x)\right) \mu(\mathrm{d}x) + \int_0^v \left(1 - F_H(x)\right) \mu(\mathrm{d}x) - \int_0^{u+v} \left(1 - F_H(x)\right) \mu(\mathrm{d}x) \right),$$

b) $H\left(\left(x_0(\Phi) - x_{-1}(\Phi)\right) \leqq u\right) = i_H \int\limits_{(0,u]} x\, \mathrm{d}F_H(x)$,

c) $H\left(-x_{-1}(\Phi) \leqq u\right) = H\left(x_0(\Phi) \leqq u\right) = i_H \int\limits_0^u \left(1 - F_H(x)\right) \mu(\mathrm{d}x)$.

Proof. 1. Using the inversion formula 8.2.9. and setting

$$Y = \{\Phi : \Phi \in M,\, -x_{-1}(\Phi) \leqq u,\, x_0(\Phi) \leqq v\},$$

we obtain

$$H(Y) = i_H \int_0^\infty H_0\big(x_0(\Phi) > x,\, T_x\Phi \in Y\big)\, \mu(\mathrm{d}x)$$

$$= i_H \int_0^u H_0\big(x < x_0(\Phi) \leqq v + x\big)\, \mu(\mathrm{d}x)$$

$$= i_H \int_0^u \big(F_H(v + x) - F_H(x)\big)\, \mu(\mathrm{d}x)$$

$$= i_H \left(\int_0^u \big(1 - F_H(x)\big)\, \mu(\mathrm{d}x) + \int_0^v \big(1 - F_H(x)\big)\, \mu(\mathrm{d}x) - \int_0^{u+v} \big(1 - F_H(x)\big)\, \mu(\mathrm{d}x)\right).$$

2. Setting $Y = \{\Phi : \Phi \in M,\, x_0(\Phi) - x_{-1}(\Phi) \leqq u\}$ we get

$$H(Y) = i_H \int_0^\infty H_0\big(x_0(\Phi) > x,\, x_0(\Phi) \leqq u\big)\, \mu(dx)$$

$$= i_H \int_0^u \big(F_H(u) - F_H(x)\big)\, \mu(\mathrm{d}x)$$

$$= i_H \int_{(0,u]} x\, \mathrm{d}F_H(x).$$

3. We set $Z_1 = \{\Phi : \Phi \in M,\, -x_{-1}(\Phi) \leqq u\}$ and $Z_2 = \{\Phi : \Phi \in M,\, x_0(\Phi) \leqq u\}$, which gives

$$H(Z_1) = i_H \int_0^u H_0\big(x_0(\Phi) > x\big)\, \mu(\mathrm{d}x)$$

$$= i_H \int_0^u \big(1 - F_H(x)\big)\, \mu(\mathrm{d}x).$$

On the other hand we have

$$H(Z_2) = H\big(\Phi((0, u]) > 0\big) = H\big(\Phi((-u, 0]) > 0\big)$$

$$= H\big(\Phi([-u, 0]) > 0\big)$$

$$= i_H \int_0^u \big(1 - F_H(x)\big)\, \mu(\mathrm{d}x). \blacksquare$$

In the case $H = Q_\mu$, we get $F_H(x) = 0$ for all x in $[0, +\infty)$. Consequently the distribution function F_H is not necessarily normalized. We have

9.1.15. *For a simple stationary measure W in $\mathbf{W}$ satisfying the condition $0 < i_W < +\infty$, the relation $W(\Phi((-\infty, 0)) < +\infty$ or $\Phi((0, +\infty)) < +\infty) = 0$ holds if and only if* $\lim\limits_{x\to\infty} F_W(x) = 1$.

Proof. 1. By 8.1.6. the relation $W(\Phi((-\infty,0)) < +\infty$ or $\Phi((0,+\infty)) < +\infty) = 0$ is equivalent to $W^0(\Phi((-\infty, 0)) < +\infty$ or $\Phi((0, +\infty)) < +\infty) = 0$. Consequently, from $W(\Phi((-\infty, 0)) < +\infty$ or $\Phi((0, +\infty)) < +\infty) = 0$ it can be concluded that $W_0(\Phi((-\infty, 0)) < +\infty$ or $\Phi((0, +\infty)) < +\infty) = 0$, hence also that $W_0(x_0(\Phi) < +\infty) = 1$, i.e. that $\lim_{x\to\infty} F_W(x) = 1$.

2. To each finite Φ in M being different from o, we associate the maximum $z(\Phi)$ of the set $\{x : x \in R^1, \Phi(\{x\}) > 0\}$. Since $z(\Phi) < y$ is equivalent to $\Phi([y, +\infty)) = 0$, the correspondence $\Phi \curvearrowright z(\Phi)$ is measurable.

Now, if U is a simple stationary measure in $\boldsymbol{W}$ being different from 0 and having the property that U (Ψ infinite) $= 0$,

$$\nu = U(z(\Phi) \in (.))$$

is a translation-invariant measure in N, which is different from o. Hence $\nu = l\mu$, where $l > 0$.

Setting $Y = \{\Psi : \Psi \in M, x_0(\Psi)$ does not exist$\}$ we obtain

$$\begin{aligned} U^0(x_0(\chi) \text{ does not exist}) &= \int \Big(\int_{(0,1)} k_Y(T_x\Psi)\, \Psi(\mathrm{d}x) \Big)\, U(\mathrm{d}\Psi) \\ &= U(z(\Psi) \in (0, 1)) = l > 0, \end{aligned}$$

and hence also, for $i_U < +\infty$,

$$U_0(x_0(\chi) \text{ does not exist}) > 0$$

i.e.

$$\lim_{x\to\infty} F_U(x) < 1.$$

3. Now let H be an arbitrary simple stationary measure in $\boldsymbol{W}$ having the property that $H(\Phi((-\infty, 0)) < +\infty$ or $\Phi((0, +\infty)) < +\infty) > 0$. Then, by 8.5.4., $H(\Psi$ finite$) > 0$, and hence also H^0 (χ finite) > 0, so that using 8.1.6. and step 2. above, we may conclude that

$$H^0\, (x_0(\chi) \text{ does not exist, } \chi \text{ finite}) > 0$$

and hence, that

$$H^0\, (x_0(\chi) \text{ does not exist}) > 0.$$

If $i_H < +\infty$, it follows that

$$\lim_{x\to\infty} F_H(x) < 1. \blacksquare$$

Like the relation $\lim_{x\to\infty} F_H(x) = 1$, the finiteness of $\int_0^\infty x\, \mathrm{d}F_H(x)$ is also of fundamental importance for the measure H:

9.1.16. *A simple stationary measure W in $\boldsymbol{W}$ having the property that $0 < i_W < +\infty$ is finite if and only if the distribution function F_W satisfies the conditions*

$$\lim_{x\to\infty} F_W(x) = 1, \qquad \int_0^\infty x\, \mathrm{d}F_W(x) < +\infty.$$

Proof. If F_W satisfies the two conditions stated above, then by 9.1.15. we can conclude that $W(\Phi((-\infty, 0)) < +\infty$ or $\Phi((0, +\infty)) < +\infty) = 0$, and then, using proposition b) of 9.1.14. that

$$W(M) = W((x_0(\Phi) - x_{-1}(\Phi)) < +\infty) = \lim_{n\to\infty} W((x_0(\Phi) - x_{-1}(\Phi)) \leqq n)$$

$$= \lim_{n\to\infty} i_W \int_{(0,n]} x \, \mathrm{d}F_W(x) = i_W \int_0^\infty x \, \mathrm{d}F_W(x).$$

Conversely if the measure W is finite, then using 6.1.3. we obtain the relation $W(\Phi((-\infty, 0)) < +\infty \text{ or } \Phi((0, +\infty)) < +\infty) = 0$, so that we can apply 9.1.15. to conclude that $\lim_{x\to\infty} F_W(x) = 1$. The above chain of equations then leads to

$$\int_0^\infty x \, \mathrm{d}F_W(x) < +\infty. \blacksquare$$

9.2. The Mapping $P \rightsquigarrow (\tilde{P})_0$

To each infinitely divisible stationary distribution P on $\mathfrak{M}$ having the property that $0 < i_{P^x} < +\infty$, in view of $i_{(\tilde{P})^x} = i_{P^x}$ we can associate the distribution $(\tilde{P})_0$ on $\mathfrak{M}^0$. The mapping $P \rightsquigarrow (\tilde{P})_0$ established in this way is not one-to-one, because $\tilde{P}$ is determined by $(\tilde{P})_0$ apart from a positive factor. Now, if both i_{P^x} and $(\tilde{P})_0$ are given, then there exists a well-determined W in $\boldsymbol{W}$, such that $W^0 = (\tilde{P})_0$, and we get

$$P = \mathscr{E}_{i_{P^x} W}.$$

Thus the distribution P is uniquely determined by the ordered pair $[i_{P^x}, (\tilde{P})_0]$.

Now we can add two further equivalences to those listed in 8.3.3.:

9.2.1. Proposition. *An infinitely divisible stationary distribution P on $\mathfrak{M}$ having the property that $0 < i_{P^x} < +\infty$ is, respectively,*

$$\left.\begin{array}{l} \textit{ergodic,} \\ \textit{mixing,} \end{array}\right\} \textit{if and only if} \left\{\begin{array}{l} (\tilde{P})_0 \big(s(\chi^x) = 0\big) = 1, \\ \lim_{\|x\|\to\infty} (\tilde{P})_0 \big(\chi^x(X - x) > 0\big) = 0 \textit{ for all bounded } X \textit{ in } \mathfrak{R}^s. \end{array}\right.$$

Proof. 1. In virtue of 6.4.9., and 6.5.1. P is ergodic if and only if the relation $s(\Psi^x) = 0$ holds almost everywhere with respect to $\tilde{P}$. But by 8.1.6. this is equivalent to $(\tilde{P})_0 \big(s(\chi^x) = 0\big) = 1$.

2. For all bounded X in $\mathfrak{R}^s$, let

$$\lim_{\|x\|\to\infty} (\tilde{P})_0 \big(\chi^x(X - x) > 0\big) = 0.$$

Now let X_1, X_2 be arbitrary bounded sets in $\mathfrak{R}^s$. For all y in R^s, we obtain

$$\{\Psi : \Psi \in M, \Psi^x(X_1) > 0, \Psi^x(X_2 - y) > 0\} = \Omega\big(x \in X_1, \Psi^x(X_2 - y) > 0\big)$$

and hence, in view of the inversion formula 5.1.8.,

$$\begin{aligned}
&\tilde{P}\big(\Psi^x(X_1) > 0, \Psi^x(X_2 - y) > 0\big) \leqq \mathscr{C}_{\tilde{P}}\big(x \in X_1, \Psi^x(X_2 - y) > 0\big) \\
&\quad = \mathscr{C}_{\tilde{P}}{}^T\big(x \in X_1, (T_{-x}\Psi)^x\,(X_2 - y) > 0\big) \\
&\quad \leqq \mathscr{C}_{\tilde{P}}{}^T\big(x \in X_1, \Psi^x(X' - y) > 0\big)
\end{aligned}$$

if the bounded set X' in $\Re^s$ includes the difference $X_1 - X_2$.

Our chain of inequalities can now be continued as follows:

$$\begin{aligned}
&= \big(\mu \times (\tilde{P})^0\big)\big(x \in X_1, \chi^x(X' - y) > 0\big) \\
&= \mu(X_1)\,(\tilde{P})^0\,\big(\chi^x(X' - y) > 0\big) \\
&= i_{(\tilde{P})^x}\mu(X_1)\,(\tilde{P})_0\,\big(\chi^x(X' - y) > 0\big) \\
&= \big(i_{P^x}\mu(X_1)\big)\,(\tilde{P})_0\,\big(\chi^x(X' - y) > 0\big).
\end{aligned}$$

By assumption the right-hand side tends toward zero as $\|y\| \to \infty$. Hence, in view of 6.3.6., P is mixing.

3. Conversely, suppose now that P be mixing. Let X be an arbitrary bounded set in $\Re^s$, $X_1 = (-1, 1)^s$ and $X_2 = (-4, 4)^s$. For all y in R^s, and all natural numbers k, we get

$$\begin{aligned}
&2^s(\tilde{P})_0\,\big(\chi^x(X_2) \leqq k, \chi^x(X - y) > 0\big) \\
&\quad = (i_{P^x})^{-1}\,\mathscr{C}_{\tilde{P}}{}^T\big(x \in X_1, \chi^x(X - y) > 0, \chi^x(X_2) \leqq k\big) \\
&\quad = (i_{P^x})^{-1}\,\mathscr{C}_{\tilde{P}}\big(x \in X_1, (T_x\chi)^x\,(X - y) > 0, (T_x\chi)^x\,(X_2) \leqq k\big) \\
&\quad \leqq (i_{P^x})^{-1}\,\mathscr{C}_{\tilde{P}}\big(x \in X_1, \chi^x(X_3 - y) > 0, \chi^x(X_1) \leqq k\big),
\end{aligned}$$

where X_3 means any bounded set in $\Re^s$ including the sum $X + X_1$. Now we are able to continue our chain of inequalities as follows:

$$\begin{aligned}
&\leqq (i_{P^x})^{-1}\,k \sum_{i=1}^{\infty} i^{-1}\mathscr{C}_{\tilde{P}}\big(x \in X_1, \chi^x(X_3 - y) > 0, \chi^x(X_1) = i\big) \\
&= (i_{P^x})^{-1}\,k\tilde{P}\big(\Psi^x(X_1) > 0, \Psi^x(X_3 - y) > 0\big).
\end{aligned}$$

Hence, in view of 6.3.6.

$$\lim_{\|y\| \to \infty} (\tilde{P})_0\,\big(\chi^x(X_2) \leqq k, \chi^x(X - y) > 0\big) = 0$$

for all natural numbers k, giving

$$\begin{aligned}
&\overline{\lim_{\|y\| \to \infty}}\,(\tilde{P})_0\,\big(\chi^x(X - y) > 0\big) \\
&\leqq (\tilde{P})_0\,\big(\chi^x(X_2) > k\big) + \overline{\lim_{\|y\| \to \infty}}\,(\tilde{P})_0\,\big(\chi^x(X_2) \leqq k, \chi^x(X - y) > 0\big) \\
&= (\tilde{P})_0\,\big(\chi^x(X_2) > k\big).
\end{aligned}$$

However, as k increases the right-hand side tends toward zero, i.e.

$$\overline{\lim_{\|y\|\to\infty}} (\tilde{P})_0 \left(\chi^x(X - y) > 0\right) = 0. \blacksquare$$

For Palm distributions, the simple formula 8.3.7. has to be modified as follows:

9.2.2. *For all stationary distributions P_1, P_2 on $\mathfrak{M}$ satisfying the conditions $0 < i_{(P_1)^x}, i_{(P_2)^x} < +\infty$,*

$$(P_1 * P_2)_0 = (i_{(P_1)^x} + i_{(P_2)^x})^{-1} \left(i_{(P_1)^x}(P_1)_0 * P_2 + i_{(P_2)^x}(P_2)_0 * P_1\right).$$

Suppose now that $[A, \varrho_A] \doteq R^1$. Let P be a simple stationary distribution with the property $0 < i_P < +\infty$. According to 9.1.14.,

$$P(\Phi((0, x]) > 0) = i_P \int_0^x (1 - F_P(z))\, \mu(\mathrm{d}z).$$

We set $\overline{F}_P(x) = P(\Phi((0, x]) > 0)$. Then from 9.2.2. we obtain immediately

9.2.3. *For all simple stationary distributions P, Q on $\mathfrak{M}$ satisfying the conditions $0 < i_P, i_Q < +\infty$ and for all $x \geqq 0$,*

$$F_{P*Q}(x) = \frac{i_P}{i_P + i_Q} (1 - (1 - F_P(x))\,(1 - \overline{F}_Q(x)))$$
$$+ \frac{i_Q}{i_P + i_Q} (1 - (1 - \overline{F}_P(x))\,(1 - F_Q(x))).$$

A distribution Q on $\mathfrak{M}$ is called a *factor* of a distribution P on $\mathfrak{M}$ if there exists one (and therefore exactly one) distribution S on $\mathfrak{M}$ with the property $P = Q * S$. Obviously the convolution quotient S is stationary if both P and Q have this property. A stationary distribution P on $\mathfrak{M}$ is called *indecomposable* if it is different from δ_o and has no non-trivial stationary factor.

Remembering that $\overline{F}_P, \overline{F}_Q$ possess the densities $i_P(1 - F_P(x)), i_Q(1 - F_Q(x))$ we get (cf. Matthes [5])

9.2.4. Proposition. *Let P be a simple stationary distribution on $\mathfrak{M}$ with $0 < i_P < +\infty$ and let $(F_P)_c$ be the absolutely continuous component of F_P. Then P is indecomposable if for some $x > 0$ the equation $(F_P)_c\,(x) = 0$ holds.*

Instead of 8.3.8., we now have

9.2.5. Theorem. *A stationary distribution P on $\mathfrak{M}$ satisfying the condition $0 < i_{P^x} < +\infty$ is infinitely divisible if and only if there exists a distribution H on $\mathfrak{M}$ having the property that*

$$P_0 = H * P.$$

The distribution H is uniquely determined by

$$H = (\tilde{P})_0.$$

Consequently a distribution P satisfying the assumptions of 9.2.5. is infinitely divisible if and only if P is a factor of the Palm distribution P_0.

The set of all solutions of the functional equation $P_0 = H * P$ coincides with $\{Q^t\}_{t>0}$, where Q represents the uniquely determined solution having the property that $i_{Q^x} = 1$, i.e. the uniquely determined solution of $P^0 = H * P$.

Now let Q be an arbitrary ergodic stationary infinitely divisible distribution on $\mathfrak{M}$ having the property that $i_{Q^x} = 1$. Further let σ be the distribution of a random non-negative number. Referring to the explanations given in Section 2.4., we set

$$P = \int Q^t(.)\, \sigma(\mathrm{d}t).$$

Since $\left(\widetilde{Q^t}\right)_0 = (t\tilde{Q})_0 = (\tilde{Q})_0$, the distribution Q^t is also ergodic for all $t > 0$ (cf. 9.2.1.). On the other hand, $i_{(Q^t)^x} = i_{(\widetilde{Q^t})^x} = t$, so that we obtain $Q^t\big(s(\Phi^x) = t\big) = 1$. For all Borel subsets Z of $[0, +\infty)$ we obtain

$$\begin{aligned} P\big(s(\Phi^x) \in Z\big) &= \int Q^t\big(s(\Phi^x) \in Z\big)\, \sigma(\mathrm{d}t) \\ &= \int k_Z(t)\, \sigma(\mathrm{d}t) = \sigma(Z), \end{aligned}$$

i.e. σ coincides with the distribution σ_{P^x} formed with respect to P of the sample intensity of Φ^x.

If we now also suppose that $\sigma(\{0\}) = 0$, then in addition to $P\big(s(\Phi^x) < +\infty\big) = 1$, $P(\{o\}) = 0$ is as well satisfied, so that we may form the distribution P_*. For all Y in $\mathfrak{M}^0$, using 9.2.5. we obtain

$$\begin{aligned} P_*(Y) &= \int \Big(\int s(\Phi^Y)\, \big(s(\Phi^x)\big)^{-1}\, Q^t(\mathrm{d}\Phi)\Big)\, \sigma(\mathrm{d}t) \\ &= \int (Q^t)_*\, (Y)\, \sigma(\mathrm{d}t) = \int (Q^t)_0\, (Y)\, \sigma(\mathrm{d}t) \\ &= \int \Big(\big(\widetilde{Q^t}\big)_0 * Q^t\Big)\, (Y)\, \sigma(\mathrm{d}t) \\ &= \int \Big(\big(\tilde{Q}\big)_0 * Q^t\Big)\, (Y)\, \sigma(\mathrm{d}t) \\ &= \Big(\big(\tilde{Q}\big)_0 * \int Q^t(.)\, \sigma(\mathrm{d}t)\Big)\, (Y), \end{aligned}$$

i.e. we have

$$P_* = (\tilde{Q})_0 * P.$$

The following theorem shows that all solutions V of this functional equation are of the form $\int Q^t(.)\, \sigma(\mathrm{d}t)$.

9.2.6. Theorem. *For all distributions* V *on* $\mathfrak{M}$ *satisfying the condition* $V\big(s(\Phi^x) = 0\big) = 1$ *and all stationary distributions* P *on* $\mathfrak{M}$ *having the property* $P\big(0 < s(\Phi^x) < +\infty\big) = 1$, *the relation*

$$P_* = V * P$$

implies the existence of an ergodic stationary infinitely divisible distribution Q having the properties

$$i_{Q^x} = 1, \quad (\tilde{Q})_0 = V,$$

and the following relation holds:

$$P = \int Q^t(.)\, \sigma_{P^x}(\mathrm{d}t).$$

Proof. 1. For all distributions H on $\mathfrak{M}$, the correspondence

$$x \frown [H]_{(x)} = H\big(T_{-x}\chi \in (.)\big)$$

provides a continuous, and hence also measurable, mapping of R^s into $[\boldsymbol{P}, \varrho_{\boldsymbol{P}}]$.

For each sequence (x_n) of elements of R^s which tends toward some x in R^s, and each bounded real function h on M which is continuous with respect to ϱ_M, 6.1.1. showed that, for all Φ in M,

$$h(T_{-x_n}\Phi) \xrightarrow[n\to\infty]{} h(T_{-x}\Phi),$$

so that we may conclude that

$$\int h(\Phi)\, [H]_{(x_n)}\, (\mathrm{d}\Phi) = \int h(T_{-x_n}\Phi)\, H(\mathrm{d}\Phi)$$
$$\xrightarrow[n\to\infty]{} \int h(T_{-x}\Phi)\, H(\mathrm{d}\Phi) = \int h(\Phi)\, [H]_{(x)}\, (\mathrm{d}\Phi).$$

In virtue of 3.1.2., from this it follows that

$$[H]_{(x_n)} \underset{n\to\infty}{\Rightarrow} [H]_{(x)}.$$

2. Now let S be an arbitrary stationary distribution on $\mathfrak{M}$ which satisfies the condition that $S\big(s(\Phi^x) < +\infty\big) = 1$. Since the metric space $[M, \varrho_M]$ is both separable and complete, there exist (cf. Jiřina [1] or Gikhman & Skorochod [1], chap. I, par. 3) conditional distributions of S with respect to the random variable $s(\Phi^x)$. Hence there exists a measurable mapping $l \frown S_{(l)}$ into $[\boldsymbol{P}, \mathfrak{P}]$ such that

$$S = \int S_{(l)}(.)\, \sigma_{S^x}\, (\mathrm{d}l)$$

is satisfied and the equation

$$S_{(l)}\big(s(\Phi^x) = l\big) = 1$$

is satisfied almost everywhere with respect to σ_{S^x}.

Given now an element x of R^s, a set Y in $\mathfrak{M}$, and a Borel subset Z of $[0, +\infty)$. As the function s is invariant, we obtain

$$S\big(\Phi \in Y, s(\Phi^x) \in Z\big) = S\big(T_x\Phi \in Y, s(\Phi^x) \in Z\big) = \int_Z S_{(l)}(T_x\Phi \in Y)\, \sigma_{S^x}(\mathrm{d}l).$$

Thus also the $S_{(l)}\big(T_x\Phi \in (.)\big)$ form a family of conditional distributions of S with respect to $s(\Phi^x)$. Since $[M, \varrho_M]$ is separable, it follows that the two distributions $S_{(l)}, S_{(l)}(T_x\Phi \in (.))$ coincide almost everywhere with respect to σ_{S^x}.

By virtue of 1., it will suffice for $S_{(l)}$ to be stationary, if, for all x in a countable dense subset of R^s, it satisfies the equation

$$S_{(l)} = S_{(l)}(T_x\Phi \in (.)).$$

Thus we observe that the distributions $S_{(l)}$ are stationary almost everywhere with respect to σ_{S^x}.

3. Now suppose that the distributions V, P satisfy all of the assumptions formulated in 9.2.6., According to 2. we set

$$P = \int P_{(l)}(.)\, \sigma_{P^x}(\mathrm{d}l),$$

so that, for all Y in $\mathfrak{M}^0$, we obtain

$$\begin{aligned} &\int \left(\int s(\Phi^Y)\, l^{-1} P_{(l)}(\mathrm{d}\Phi)\right) \sigma_{P^x}(\mathrm{d}l) \\ &= \int \left(\int s(\Phi^Y) \left(s(\Phi^x)\right)^{-1} P_{(l)}(\mathrm{d}\Phi)\right) \sigma_{P^x}(\mathrm{d}l) \\ &= \int s(\Phi^Y) \left(s(\Phi^x)\right)^{-1} P(\mathrm{d}\Phi) = P_*(Y) \\ &= (V * P)(Y) = \left(V * \left(\int P_{(l)}(.)\, \sigma_{P^x}(\mathrm{d}l)\right)\right)(Y) \\ &= \int (V * P_{(l)})(Y)\, \sigma_{P^x}(\mathrm{d}l), \end{aligned}$$

i.e. we have

$$\int (P_{(l)})_0(.)\, \sigma_{P^x}(\mathrm{d}l) = \int (V * P_{(l)})(.)\, \sigma_{P^x}(\mathrm{d}l).$$

By virtue of 2., the relation

$$P_{(l)}\left(s(\Phi^x) = l\right) = 1$$

holds almost everywhere with respect to σ_{P^x}. From this, using 9.1.1. we may conclude that

$$(P_{(l)})_0 \left(s(\Phi^x) = l\right) = 1$$

and, since $V\left(s(\Phi^x) = 0\right) = 1$ we conclude that

$$(V * P_{(l)}) \left(s(\Phi^x) = l\right) = 1$$

almost everywhere with respect to σ_{P^x}. Setting now

$$H = \int (P_{(l)})_0(.)\, \sigma_{P^x}(\mathrm{d}l),$$

we obtain, for all Y in $\mathfrak{M}^0$ and all Borel subsets Z of $[0, +\infty)$

$$\int_Z (P_{(l)})_0(Y)\, \sigma_{P^x}(\mathrm{d}l) = H\left(\Phi \in Y, s(\Phi^x) \in Z\right) = \int_Z (V * P_{(l)})(Y)\, \sigma_{P^x}(\mathrm{d}l),$$

which means that both $((P_{(l)})_0)_{l>0}$ and $(V * P_{(l)})_{l>0}$ can be interpreted as systems of conditional distributions of H with respect to $s(\Phi^x)$. Hence the relation

$$(P_{(l)})_0 = V * P_{(l)}$$

holds almost everywhere with respect to σ_{P^x}. In view of 9.2.5. there exists an infinitely divisible distribution Q on $\mathfrak{M}$ having the properties

$$i_{Q^x} = 1, \quad (\tilde{Q})_0 = V.$$

By 9.2.1. this distribution is ergodic. Now we have

$$(P_{(l)})_0 = (l\tilde{Q})_0 * P_{(l)}, \quad i_{(P_{(l)})^x} = l$$

and hence

$$P_{(l)} = Q^l$$

almost everywhere with respect to σ_{P^x}.

Consequently

$$P = \int Q^l(.)\, \sigma_{P^x}(\mathrm{d}l). \ \blacksquare$$

In the remainder of this section, we shall suppose that the phase space $[A, \varrho_A]$ coincides with the "position space" R^s.

Instead of 8.3.10. we now have

9.2.7. Theorem. *A stationary distribution P on $\mathfrak{M}$ is of the form $P_{l\mu}$, $l > 0$, if and only if $i_P = l$ and*

$$P_0 = \delta_{\delta_{[0,\ldots,0]}} * P$$

is satisfied.

As an immediate consequence of 1.14.8. and 9.2.6., we have (cf. Slivnyak [1], [2]):

9.2.8. Theorem. *A stationary distribution P on $\mathfrak{M}$ satisfying the condition $P(0 < s(\Phi) < +\infty) = 1$ has the property g_μ if and only if*

$$P_* = \delta_{\delta_{[0,\ldots,0]}} * P$$

is satisfied.

9.3. Another Interpretation of the Palm Distribution

A sequence (X_n) of sets in $\mathfrak{R}^s$ is said to satisfy the *condition* (i), if each X_n has a non-empty interior and the convergence relation

$$\sup_{x \in X_n} \|x\| \xrightarrow[n\to\infty]{} 0$$

is satisfied. Hence in particular we can set

$$X_n = S_{n^{-1}}([0, \ldots, 0]) \quad \text{or} \quad X_n = [0, n^{-1})^s.$$

As a generalization of a theorem of Korolyuk (cf. Khinchin [1]), we have

9.3.1. Proposition. *For all stationary measures H on $\mathfrak{M}$, and all sequences of sets in $\mathfrak{R}^s$ satisfying the condition* (i),

$$\lim_{n\to\infty} \left(\mu(X_n)\right)^{-1} H\left(\Phi^x(X_n) > 0\right) = i_{(H^x)^*}.$$

Proof. 1. For $n = 1, 2, \ldots$,

$$H\left(\Phi^x(X_n) > 0\right) = H\left((\Phi^x)^* (X_n) > 0\right) = (H^x)^* \left(\Phi(X_n) > 0\right),$$

so that we may suppose in addition that $[A, \varrho_A]$ coincides with R^s and H be simple.

Now, if $H\left(\Phi(X_n) > 0\right)$ is infinite for $n = 1, 2, \ldots$, then by 6.1.5. the intensity i_H cannot be finite, and

$$\lim_{n\to\infty} \left(\mu(X_n)\right)^{-1} H\left(\Phi(X_n) > 0\right) = i_H.$$

Therefore, in what follows, we are allowed to suppose in addition that $H\left(\Phi(X_{n_0}) > 0\right) < +\infty$ for at least one n_0. Each bounded set X in $\mathfrak{R}^s$ can be covered by a finite family $(X_{n_0} + x_i)_{i=1,\ldots,k}$, so that we obtain

$$\boldsymbol{H}\left(\Phi(X) > 0\right) \leqq \boldsymbol{H}\left(\Phi\left(\bigcup_{i=1}^{k} (X_{n_0} + x_i)\right) > 0\right)$$

$$\leqq \sum_{i=1}^{k} \boldsymbol{H}\left(\Phi(X_{n_0} + x_i) > 0\right) = k\boldsymbol{H}\left(\Phi(X_{n_0}) > 0\right) < +\infty.$$

Consequently $H\left((.) \setminus \{o\}\right)$ belongs to $\boldsymbol{W}$, and hence is a σ-finite measure. Obviously it suffices to derive the convergence proposition of 9.3.1. for $H\left((.) \setminus \{o\}\right)$.

Thus, without loss of generality, we may also suppose that the stationary measure H be simple and in $\boldsymbol{W}$.

2. For $n = 1, 2, \ldots$, the inversion formula 8.2.5. leads to the following relation:

$$\left(\mu(X_n)\right)^{-1} H\left(\Phi(X_n) > 0\right)$$

$$= \int \left(\int_{\{x:\Phi(S_{\|x\|}(x))=0\}} \left(\mu(X_n)\right)^{-1} k_{\{\Psi:\Psi\in M, \Psi(X_n)>0\}}(T_x\Phi)\, \mu(\mathrm{d}x)\right) H^0(\mathrm{d}\Phi)$$

$$= \int \left(\int_{\{x:\Phi(S_{\|x\|}(x))=0\}} \left(\mu(X_n)\right)^{-1} k_{\{y:y\in R^s, \Phi(X_n+y)>0\}}(x)\, \mu(\mathrm{d}x)\right) \boldsymbol{H}^0(\mathrm{d}\Phi)$$

$$= \int \left(\int \left(\mu(X_n)\right)^{-1} k_{\{y:y\in R^s, \Phi(X_n+y)>0, \Phi(S_{\|y\|}(y))=0\}}(x)\, \mu(\mathrm{d}x)\right) \boldsymbol{H}^0(\mathrm{d}\Phi)$$

$$= \int \left(\mu(X_n)\right)^{-1} \mu\left(\{x : x \in R^s, \Phi(X_n + x) > 0, \Phi\left(S_{\|x\|}(x)\right) = 0\}\right) H^0(\mathrm{d}\Phi).$$

Now setting

$$Y_m = \{\Phi : \Phi \in M^0,\ \ \Phi \text{ simple},\ \ \Phi(S_{m^{-1}}([0, \ldots, 0]) \setminus \{[0, \ldots, 0]\}) = 0\},$$

we obtain, for all natural numbers m,

$$\underline{\lim}_{n\to\infty} (\mu(X_n))^{-1} H(\Phi(X_n) > 0)$$

$$\geqq \underline{\lim}_{n\to\infty} \int_{Y_m} (\mu(X_n))^{-1} \mu(\{x : x \in R^s, \Phi(X_n - x) > 0, \Phi(S_{\|x\|}(-x)) = 0\}) H^0(\mathrm{d}\Phi).$$

For $n \geqq n_m$ we have $\sup_{x \in X_n} \|x\| < (2m)^{-1}$.

Then for each Φ in Y_m, it follows from $\Phi(X_n - x) > 0$, $\Phi(S_{\|x\|}(-x)) = 0$, that $\|x\| < (2m)^{-1}$ and hence that $[0, \ldots, 0] \in (X_n - x)$, i.e. $x \in X_n$. Conversely, the two conditions $\Phi(X_n - x) > 0$, $\Phi(S_{\|x\|}(-x)) = 0$ are satisfied for all $x \in X_n$, because $S_{\|x\|}(-x)$ is included in $S_{m^{-1}}([0, \ldots, 0]) \setminus \{[0, \ldots, 0]\}$.

Thus we observe that

$$\underline{\lim}_{n\to\infty} (\mu(X_n))^{-1} H(\Phi(X_n) > 0) \geqq \int_{Y_m} 1 H^0(\mathrm{d}\Phi) = H^0(Y_m)$$

is true for $m = 1, 2, \ldots$, i.e. we have

$$\underline{\lim}_{n\to\infty} (\mu(X_n))^{-1} H(\Phi(X_n) > 0) \geqq H^0(M^0) = i_H. \qquad (*)$$

This completes the proof in the case where $i_H = +\infty$. For finite i_H, we use 6.1.5. to conclude that

$$(\mu(X_n))^{-1} H(\Phi(X_n) > 0) \leqq (\mu(X_n))^{-1} \varrho_H(X_n) = i_H \qquad (n = 1, 2, \ldots).$$

From this it follows that

$$\overline{\lim}_{n\to\infty} (\mu(X_n))^{-1} H(\Phi(X_n) > 0) \leqq i_H. \qquad (**)$$

Combining $(*)$ and $(**)$ yields the required convergence proposition. ∎

The following proposition provides a generalization of a theorem of Dobrushin (cf. Volkonskii [1]).

9.3.2. Proposition. *Let (X_n) be a sequence of sets in $\mathfrak{R}^s$ satisfying the condition* (i). *Then, for all stationary measures H on $\mathfrak{M}$ having the property that $i_{H^x} < +\infty$, the following statements are equivalent:*

a) *H^x is simple,*

b) $\lim_{n\to\infty} (\mu(X_n))^{-1} H(\Phi^x(X_n) = 1) = i_{H^x}$,

c) $\lim_{n\to\infty} (\mu(X_n))^{-1} H(\Phi^x(X_n) > 0) = i_{H^x}$,

d) $\lim_{n\to\infty} (\mu(X_n))^{-1} H(\Phi^x(X_n) > 1) = 0$.

Proof. Obviously we may suppose without loss of generality that $[A, \varrho_A]$ coincides with the "position space" R^s.

1. a) implies b). This is readily shown by substituting $\{\Phi : \Phi \in M, \Phi(X_n) = 1\}$ for the sets $\{\Phi : \Phi \in M, \Phi(X_n) > 0\}$ in the second step of the proof of 9.3.1.

2. Since

$$H(\Phi(X_n) = 1) \leqq H(\Phi(X_n) > 0) \leqq \mu(X_n)\, i_H \qquad (n = 1, 2, \ldots)$$

it is concluded that c) is a weakened version of b).

3. c) implies d). Indeed, for $n = 1, 2, \ldots,$

$$\begin{aligned}
&(\mu(X_n))^{-1} H(\Phi(X_n) > 1) \\
&= (\mu(X_n))^{-1} \sum_{k=2}^{\infty} H(\Phi(X_n) = k) \\
&\leqq (\mu(X_n))^{-1} \left(\sum_{k=1}^{\infty} kH(\Phi(X_n) = k) - \sum_{k=1}^{\infty} H(\Phi(X_n) = k) \right) \\
&= (\mu(X_n))^{-1} \left(\varrho_H(X_n) - H(\Phi(X_n) > 0)\right) \\
&= i_H - (\mu(X_n))^{-1} H(\Phi(X_n) > 0).
\end{aligned}$$

4. d) implies a).
For all Φ in M, we set

$$\Phi^? = \sum_{x \in R^s, \Phi(\{x\}) > 1} \delta_x = (\Phi - \Phi^*)^*.$$

By virtue of 1.4.2., the mapping $\Phi \curvearrowright \Phi^?$ of $[M, \mathfrak{M}]$ into $[M, \mathfrak{M}]$ is measurable. The measure $H(\Phi^? \in (.))$ is denoted by $H^?$. Obviously $H^?$ is simple, stationary, and of finite intensity.
Using 9.3.1. we get

$$\begin{aligned}
&\lim_{n \to \infty} (\mu(X_n))^{-1} H(\Phi(X_n) > 1) \\
&\geqq \lim_{n \to \infty} (\mu(X_n))^{-1} H^?(\Phi(X_n) > 0) = i_{H^?}.
\end{aligned}$$

Hence, in view of our assumption, we can conclude that $i_{H^?} = 0$, i.e. $H^?((.) \setminus \{o\}) = 0$, and consequently $H(\Phi^? \neq o) = 0$, i.e. $H(\Phi$ not simple$) = 0$. ∎

There exist simple, stationary distributions $P = \int P_{l\mu}(\cdot)\, \sigma(dl)$ with infinite intensity satisfying assertion d) (cf. Vasil'ev & Kovalenko [1]).
For a suitable $c > 0$ let

$$\sigma(dx) = \begin{cases} c(x^2 \ln x)^{-1} \mu(dx) & \\ & \text{for} \\ 0 & \end{cases} \quad \begin{matrix} x \geqq 2 \\ \\ x < 2. \end{matrix}$$

We obtain

$$i_P = \int_0^\infty x\sigma(\mathrm{d}x) = c \int_2^\infty \frac{\mu(\mathrm{d}x)}{x \ln x} = +\infty.$$

On the other hand

$$\overline{\lim_{t\to 0+0}} \frac{P(\Phi([0,t)) > 1)}{t}$$

$$= c \overline{\lim_{t\to 0+0}} \frac{1}{t} \left(\int_2^{\left(t \ln\ln\frac{1}{t}\right)^{-1}} \frac{1 - e^{-xt}(1+xt)}{x^2 \ln x} \mu(\mathrm{d}x) + \int_{\left(t\ln\ln\frac{1}{t}\right)^{-1}}^\infty \frac{1 - e^{-xt}(1+xt)}{x^2 \ln x} \mu(\mathrm{d}x) \right).$$

Now we find an upper estimate of the integrand of the first integral on the right-hand side as follows:

$$\frac{1 - e^{-xt}(1+xt)}{x^2 \ln x} = \frac{e^{-xt}(e^{xt} - 1 - xt)}{x^2 \ln x} \leqq \frac{e^{-xt}t^2 e^{xt} x^2}{2x^2 \ln x} \leqq \frac{t^2}{2 \ln 2}.$$

For the second integrand it follows that

$$\frac{1 - e^{-xt}(1+xt)}{x^2 \ln x} \leqq \frac{1}{x^2 \ln\left(\left(t \ln\ln\frac{1}{t}\right)^{-1}\right)},$$

so that we may continue the equation given above as follows:

$$\leqq c \overline{\lim_{t\to 0+0}} \left(\frac{1}{2 \ln 2 \ln\ln\frac{1}{t}} + \frac{\ln\ln\frac{1}{t}}{\ln\frac{1}{t} - \ln\ln\ln\frac{1}{t}} \right) = 0.$$

Hence

$$\lim_{t\to 0+0} \frac{P(\Phi([0,t)) > 1)}{t} = 0.$$

Let (X_n) be any sequence of sets in $\mathfrak{R}^1$ satisfying the condition (i). We obtain

$$\lim_{n\to\infty} (\mu(X_n))^{-1} P(\Phi(X_n) > 1) = \lim_{n\to\infty} (\mu(X_n))^{-1} P(\Phi([0, \mu(X_n))) > 1) = 0.$$

This distribution P provides also an example of an orderly distribution with infinite intensity.

For all X in $\mathfrak{R}^s$, we set, using the notation adapted in 8.2.3.,

$$y(\Phi, X) = y({}_{X\times K}\Phi),$$

provided that the right-hand side proves meaningful. Thus $y(\Phi, X)$ is defined if the set $\{a : a \in X, \Phi^x(\{a\}) > 0\}$ is non-empty and has exactly one point x with minimum distance to $[0, \ldots, 0]$; in this case $y(\Phi, X) = x$. In view of 8.2.3., and since $\Phi \curvearrowright {}_{X\times K}\Phi$ is measurable, $\Phi \curvearrowright y(\Phi, X)$ is a mapping from M into X which is measurable with respect to the σ-algebras $\mathfrak{M}$, $\mathfrak{R}^s$.

Now let H be a stationary measure on $\mathfrak{M}$ having the properties $0 < i_{H^x} < +\infty$, H^x simple. Then by 8.2.4. the value $y(\Phi, X)$ is defined for almost all Φ in $\{\Phi : \Phi \in M, \Phi^x(X) > 0\}$. Now, if $\mu(X)$ is positive and finite, it can be concluded by means of 6.1.5., that $0 < H(\Phi^x(X) > 0) < +\infty$, so that the distribution $H((.) \mid \Phi^x(X) > 0)$ is found meaningful. Later on we shall see that, for all sequences (X_n) of sets in $\mathfrak{R}^s$ satisfying the condition (i), the sequence of distributions $H((.) \mid \Phi^x(X_n) > 0)$ converges weakly toward the Palm distribution H_0. This convergence can be strengthened by using the distribution $H(T_{y(\Phi,X_n)}\Phi \in (.) \mid \Phi^x(X_n) > 0)$ instead of $H((.) \mid \Phi^x(X_n) > 0)$. (Remember that, in view of 6.1.1., the correspondence

$$\Phi \curvearrowright [\Phi, y(\Phi, X)] \curvearrowright T_{y(\Phi,X)}\Phi$$

provides a measurable mapping from M into M). We have (cf. Koenig & Matthes [1], Matthes [3]):

9.3.3. Proposition. *For all sequences (X_n) of sets in $\mathfrak{R}^s$ satisfying the condition* (i), *and all stationary measures H on $\mathfrak{M}$ having the properties $0 < i_{H^x} < +\infty$, H^x simple,*

$$\left\| H\left(T_{y(\Phi,X_n)}\Phi \in (.) \mid \Phi^x(X_n) > 0\right) - H_0 \right\| \xrightarrow[n\to\infty]{} 0.$$

Proof. Using 9.3.2. we get

$$\overline{\lim_{n\to\infty}} \left\| H_0 - H\left(T_{y(\Phi,X_n)}\Phi \in (.) \mid \Phi^x(X_n) > 0\right) \right\|$$

$$= \overline{\lim_{n\to\infty}} \left\| H_0 - \left(H(\Phi^x(X_n) > 0)\right)^{-1} \int_{\{\Phi : \Phi^x(X_n) > 0\}} k_{(.)}(T_{y(\Phi,X_n)}\Phi)\, H(\mathrm{d}\Phi) \right\|$$

$$\leqq \overline{\lim_{n\to\infty}} \left\| H_0 - \left(i_{H^x}\mu(X_n)\right)^{-1} \int_{\{\Phi : \Phi^x(X_n) > 0\}} k_{(.)}(T_{y(\Phi,X_n)}\Phi)\, H(\mathrm{d}\Phi) \right\|$$

$$+ \overline{\lim_{n\to\infty}} \left\| \left((i_{H^x}\mu(X_n))^{-1} H(\Phi^x(X_n) > 0) - 1 \right) H\left(T_{y(\Phi,X_n)}\Phi \in (.) \mid \Phi^x(X_n) > 0\right) \right\|$$

$$= (i_{H^x})^{-1} \overline{\lim_{n\to\infty}} \left\| H^0 - \left(\mu(X_n)\right)^{-1} \int_{\{\Phi : \Phi^x(X_n) > 0\}} k_{(.)}(T_{y(\Phi,X_n)}\Phi)\, H(\mathrm{d}\Phi) \right\|$$

$$\leqq 2(i_{H^x})^{-1} \overline{\lim_{n\to\infty}} \sup_{Y \in \mathfrak{M}^0} \left| H^0(Y) - \left(\mu(X_n)\right)^{-1} \int_{\{\Phi : \Phi^x(X_n) > 0\}} k_Y(T_{y(\Phi,X_n)}\Phi)\, H(\mathrm{d}\Phi) \right|$$

$$= 2(i_{H^x})^{-1} \overline{\lim_{n\to\infty}} \sup_{Y \in \mathfrak{M}^0} \left(\mu(X_n)\right)^{-1} \left| \int \left(\sum_{x \in X_n, \Phi^x(\{x\}) > 0} k_Y(T_x\Phi) \right) H(\mathrm{d}\Phi) - \int_{\{\Phi : \Phi^x(X_n) > 0\}} k_Y(T_{y(\Phi,X_n)}\Phi)\, H(\mathrm{d}\Phi) \right|.$$

For each Φ in $\{\Phi : \Phi \in M, \Phi^x(X_n) > 0\}$ the point $y(\Phi, X_n)$ belongs to the set $\{x : x \in X_n, \Phi^x(\{x\}) > 0\}$, and we may continue the above chain of inequalities as follows:

$$\leq 2(i_{H^x})^{-1} \varlimsup_{n\to\infty} \left(\mu(X_n)\right)^{-1} \int_{\{\Phi:\Phi^x(X_n)>0\}} \left(\Phi^x(X_n) - 1\right) H(\mathrm{d}\Phi)$$

$$= 2(i_{H^x})^{-1} \varlimsup_{n\to\infty} \left(\mu(X_n)\right)^{-1} \left(\varrho_{H^x}(X_n) - H\left(\Phi^x(X_n) > 0\right)\right)$$

$$= 2(i_{H^x})^{-1} \varlimsup_{n\to\infty} \left(i_{H^x} - \left(\mu(X_n)\right)^{-1} H\left(\Phi^x(X_n) > 0\right)\right) = 0. \blacksquare$$

In view of 9.3.2., the convergence proposition of 9.3.3. remains valid, if $H\left(T_{y(\Phi,X_n)}\Phi \in (.) \mid \Phi^x(X_n) > 0\right)$ is replaced by the distribution $H\left(T_{y(\Phi,X_n)}\Phi \in (.) \mid \Phi^x(X_n) = 1\right)$, which is meaningful for sufficiently large n.

We shall now present the proposition which was announced above (cf. Ryll-Nardzewski [1]):

9.3.4. Proposition. *Subject to the assumptions of 9.3.3., the following relation holds:*

$$H\left((.) \mid \Phi^x(X_n) > 0\right) \underset{n\to\infty}{\Rightarrow} H_0.$$

Proof. Let h be an arbitrary bounded non-negative real function on M which is continuous with respect to ϱ_M. By 9.3.3. we have

$$\lim_{n\to\infty} \int h(T_{y(\Phi,X_n)}\Phi)\, H\left(\mathrm{d}\Phi \mid \Phi^x(X_n) > 0\right) = \int h(\Phi)\, H_0(\mathrm{d}\Phi)$$

so that we obtain

$$\varlimsup_{n\to\infty} \left| \int h(\Phi)\, H\left(\mathrm{d}\Phi \mid \Phi^x(X_n) > 0\right) - \int h(\Phi)\, H_0(\mathrm{d}\Phi) \right|$$

$$= \varlimsup_{n\to\infty} \left| \int h(\Phi)\, H\left(\mathrm{d}\Phi \mid \Phi^x(X_n) > 0\right) - \int h(T_{y(\Phi,X_n)}\Phi)\, H\left(\mathrm{d}\Phi \mid \Phi^x(X_n) > 0\right) \right|$$

$$\leq \varlimsup_{n\to\infty} \int |h(\Phi) - h(T_{y(\Phi,X_n)}\Phi)|\, H\left(\mathrm{d}\Phi \mid \Phi^x(X_n) > 0\right).$$

Setting now $d_n = \sup_{x\in X_n} \|x\|$ as well as $H_{(X_n)} = H\left(T_{y(\Phi,X_n)}\Phi \in (.) \mid \Phi^x(X_n) > 0\right)$, we may once more use 9.3.3. to continue the above chain of inequalities as follows

$$\leq \varlimsup_{n\to\infty} \int \sup_{\|x\|\leq d_n} |h(T_x\Psi) - h(\Psi)|\, H_{(X_n)}(\mathrm{d}\Psi)$$

$$= \varlimsup_{n\to\infty} \int \sup_{\|x\|\leq d_n} |h(T_x\Psi) - h(\Psi)|\, H_0(\mathrm{d}\Psi).$$

In view of 6.1.1., for all Ψ in M, $\sup_{\|x\|\leq d_n} |h(T_x\Psi) - h(\Psi)|$ tends toward zero as n increases. Hence, in view of Lebesgue's theorem, from this it follows that

$$\overline{\lim_{n\to\infty}} \int \sup_{\|x\|\leq d_n} |h(T_x\Psi) - h(\Psi)|\, H_0(\mathrm{d}\Psi) = 0,$$

which means that

$$\lim_{n\to\infty} \int h(\Phi)\, H\big(\mathrm{d}\Phi \mid \Phi^x(X_n) > 0\big) = \int h(\Phi)\, H_0(\mathrm{d}\Phi). \blacksquare$$

The assumption that $H^x = (H^x)^*$ cannot be dropped in 9.3.3. Let us now outline how the definition of the Palm distribution has to be modified in order that 9.3.3. be valid also in the case $H^x \neq (H^x)^*$.

To each stationary measure H on $\mathfrak{M}$ satisfying the condition $0 < i_{(H^x)^*} < +\infty$, we can associate a distribution $H_\square$ on $\mathfrak{M}^0$ by

$$H_\square(Y) = \big(i_{(H^x)^*}\mu(X)\big)^{-1} \int \Big(\sum_{x\in X,\, \Phi^x(\{x\})>0} k_Y(T_x\Phi)\Big)\, H(\mathrm{d}\Phi) \qquad (Y \in \mathfrak{M}^0),$$

where the selection of the set X in $\mathfrak{R}^s$ having the property that $0 < \mu(X) < +\infty$ is without. We interpret the multiplicity of each point in $\{x : x \in R^s,\ \Phi^x(\{x\}) > 0\}$ as a mark, i.e. we use the modified Campbell measure $\int \big((\Phi^x)^* \times \delta_\Phi\big)\, (.)\, H(\mathrm{d}\Phi)$.

As an analogue to 9.3.3., we now have

9.3.5. Proposition. *For all sequences (X_n) of sets in $\mathfrak{R}^s$ satisfying the condition* (i), *and all stationary measures H on $\mathfrak{M}$ having the property $0 < i_{H^x} < +\infty$,*

$$\big\| H\big(T_{y(\Phi,X_n)}\Phi \in (.) \mid \Phi^x(X_n) > 0\big) - H_\square \big\| \xrightarrow[n\to\infty]{} 0.$$

From the definitions of H_0 and $H_\square$, it is immediately concluded that, for all Y in $\mathfrak{M}^0$ and all natural numbers k,

$$\begin{aligned} &k i_{(H^x)^*} H_\square\big(\Phi \in Y,\ \Phi^x(\{[0, \ldots, 0]\}) = k\big) \\ &= i_{H^x} H_0\big(\Phi \in Y,\ \Phi^x(\{[0, \ldots, 0]\}) = k\big), \end{aligned}$$

i.e. we have

9.3.6. Proposition. *Subject to the assumptions of* 9.3.5.,

$$H_\square = \sum_{k=1}^{\infty} (k i_{(H^x)^*})^{-1}\, i_{H^x} H_0\big(\Phi \in (.),\ \Phi^x(\{[0, \ldots, 0]\}) = k\big).$$

Hence in particular

$$i_{(H^x)^*} = i_{H^x} \sum_{k=1}^{\infty} k^{-1} H_0\big(\Phi^x(\{[0, \ldots, 0]\}) = k\big).$$

In view of 9.3.6., the two distributions $H_0, H_\square$ coincide if and only if the equation

$$H_0\big(\Phi^x(\{[0, \ldots, 0]\}) = k_0\big) = 1$$

i.e. if

$$H\big(\Phi^x \neq k_0(\Phi^x)^*\big) = 0$$

is satisfied for some k_0.

Applying the above statements on $H_\square$, we obtain

9.3.7. Proposition. *Subject to the assumptions of proposition* 9.3.5., *the following relation holds for all natural numbers k satisfying the condition* $H_0\big(\Phi^x(\{[0, \ldots, 0]\}) = k\big) > 0$:

$$\Big\| \qquad {}_{y(\Phi,X_n)}\Phi \in (.)\,|\,\Phi^x(X_n) = k\big) - H_0\big((.)\,|\,\Phi^x(\{[0, \ldots, 0]\}) = k\big)\Big\| \xrightarrow[n\to\infty]{} 0.$$

Proof. Using 9.3.2., 9.3.5. and 9.3.6. we get

$$\begin{aligned}
&\lim_{n\to\infty} \big(H\big(\Phi^x(X_n) > 0\big)\big)^{-1} H\big(\Phi^x(X_n) = k\big)\\
&= \lim_{n\to\infty} H\big(\Phi^x(X_n) = k \mid \Phi^x(X_n) > 0\big)\\
&= \lim_{n\to\infty} H\big((T_{y(\Phi,x_n)}\Phi)^x\ (\{[0, \ldots, 0]\}) = k \,|\, \Phi^x(X_n) > 0\big)\\
&= H_\square\big(\Phi^x(\{[0, \ldots, 0]\}) = k\big) = (k i_{(H^x)^*})^{-1}\, i_{H^x} H_0\big(\Phi^x(\{[0, \ldots, 0]\}) = k\big).
\end{aligned}$$

Hence

$$\begin{aligned}
&\overline{\lim_{n\to\infty}} \Big\|H\big(T_{y(\Phi,X_n)}\Phi \in (.) \mid \Phi^x(X_n) = k\big) - H_0\big((.) \mid \Phi^x(\{[0, \ldots, 0]\}) = k\big)\Big\|\\
&= \overline{\lim_{n\to\infty}} \left\| H\big(T_{y(\Phi,X_n)}\Phi \in (.),\, \Phi^x(X_n) = k \mid \Phi^x(X_n) > 0\big) \frac{H\big(\Phi^x(X_n) > 0\big)}{H\big(\Phi^x(X_n) = k\big)}\right.\\
&\qquad \left. - \frac{H_0\big(\Phi \in (.),\, \Phi^x(\{[0, \ldots, 0]\}) = k\big)}{H_0\big(\Phi^x(\{[0, \ldots, 0]\}) = k\big)}\right\|\\
&= \overline{\lim_{n\to\infty}} \Big\|H\big(T_{y(\Phi,X_n)}\Phi \in (.),\, (T_{y(\Phi,X_n)}\Phi)^x\ (\{[0, \ldots, 0]\}) = k \mid \Phi^x(X_n) > 0\big)\\
&\qquad\qquad \cdot \big(i_{H^x} H_0\big(\Phi^x(\{[0, \ldots, 0]\}) = k\big)\big)^{-1} k i_{(H^x)^*}\\
&\qquad - \big(i_{H^x} H_0\big(\Phi^x(\{[0, \ldots, 0]\}) = k\big)\big)^{-1} k i_{(H^x)^*} H_\square\big(\Phi \in (.), \Phi^x(\{[0, \ldots, 0]\}) = k\big)\Big\|\\
&\leqq \big(i_{H^x} H_0\big(\Phi^x(\{[0, \ldots, 0]\}) = k\big)\big)^{-1} k i_{(H^x)^*}\\
&\qquad \cdot \overline{\lim_{n\to\infty}} \Big\|H\big(T_{y(\Phi,X_n)}\Phi \in (.) \mid \Phi^x(X_n) > 0\big) - H_\square\Big\| = 0. \quad \blacksquare
\end{aligned}$$

We conclude this section with the

Proof of Proposition 9.1.11. which has not yet been supplied. 1. Let P be a mixing stationary distribution on $\mathfrak{M}$ having the property that $0 < i_{P^x} < +\infty$. For each sequence (X_n) of sets in $\mathfrak{R}^s$ which satisfies the condition (i), and for each natural number k having the property that

$P_0(\Phi^x(\{[0, \ldots, 0]\}) = k) > 0$, 9.3.7. shows that, for all sufficiently large n, the distribution $P_{X_n,k} = P(T_{y(\Phi, X_n)}\Phi \in (.) \mid \Phi^x(X_n) = k)$ is meaningful. For all bounded real functions f on M which are uniformly continuous with respect to ϱ_M, and for all x in R^s, we obtain

$$\begin{aligned}
&\left|\int f(T_x\Phi)\, P_0(\mathrm{d}\Phi \mid \Phi^x(\{[0, \ldots, 0]\}) = k) - \int f(T_x\Phi)\, P(\mathrm{d}\Phi \mid \Phi^x(X_n) = k)\right| \\
&\leqq \left|\int f(T_x\Phi)\, P_0(\mathrm{d}\Phi \mid \Phi^x(\{[0, \ldots, 0]\}) = k)\right. \\
&\quad \left. - \int f(T_x T_{y(\Phi, X_n)}\Phi)\, P(\mathrm{d}\Phi \mid \Phi^x(X_n) = k)\right| \\
&\quad + \left|\int \big(f(T_x\Phi) - f(T_x T_{y(\Phi, X_n)}\Phi)\big)\, P(\mathrm{d}\Phi \mid \Phi^x(X_n) = k)\right| \\
&\leqq \Big(\sup_{\Psi \in M} |f(\Psi)|\Big)\, \big\|P_0((.) \mid \Phi^x(\{[0, \ldots, 0]\}) = k) - P_{X_n,k}\big\| \\
&\quad + \sup_{x \in R^s, \chi} |f(T_x\chi) - f(T_x T_{y(\chi, X_n)}\chi)|,
\end{aligned}$$

where χ runs through those counting measures in M for which $y(\chi, X_n)$ exists.

It is immediately observed from the definition of the metric ϱ_M that $\sup_{\|y\| \leqq \eta} \varrho_M(T_y\Psi, \Psi)$ for $\eta \to 0 + 0$ tends toward zero, the convergence being uniform in Ψ. Hence to each $\varepsilon > 0$ there exists a natural number l_ε such that, for all $n \geqq l_\varepsilon$

$$\sup_{x \in R^s, \chi} |f(T_x\chi) - f(T_x T_{y(\chi, X_n)}\chi)| < \varepsilon$$

is satisfied.

In view of 9.3.7., there exists a natural number $r_{\varepsilon,k}$ such that, for all $n \geqq r_{\varepsilon,k}$

$$\Big(\sup_{\Psi \in M} |f(\Psi)|\Big)\, \big\|P_0((.) \mid \Phi^x(\{[0, \ldots, 0]\}) = k) - P_{X_n,k}\big\| < \varepsilon.$$

Thus, for all $n \geqq \max(l_\varepsilon, r_{\varepsilon,k})$ we obtain uniformly in x

$$\begin{aligned}
&\left|\int f(T_x\Phi)\, P_0(\mathrm{d}\Phi \mid \Phi^x(\{[0, \ldots, 0]\}) = k)\right. \\
&\quad \left. - \int f(T_x\Phi)\, P(\mathrm{d}\Phi \mid \Phi^x(X_n) = k)\right| < 2\varepsilon,
\end{aligned}$$

which gives

$$\begin{aligned}
&\overline{\lim_{\|x\| \to \infty}} \left|\int f(T_x\Phi)\, P_0(\mathrm{d}\Phi \mid \Phi^x(\{[0, \ldots, 0]\}) = k) - \int f(\Phi)\, P(\mathrm{d}\Phi)\right| \\
&\leqq 2\varepsilon + \overline{\lim_{\|x\| \to \infty}} \left|\int f(T_x\Phi)\, P(\mathrm{d}\Phi \mid \Phi^x(X_n) = k) - \int f(\Phi)\, P(\mathrm{d}\Phi)\right| \\
&= 2\varepsilon + \overline{\lim_{\|x\| \to \infty}} \left|\big(P(\Phi^x(X_n) = k)\big)^{-1} \int f(T_x\Phi)\, k_{\{\Psi : \Psi \in M, \Psi^x(X_n) = k\}}(\Phi)\, P(\mathrm{d}\Phi)\right. \\
&\qquad \left. - \int f(\Phi)\, P(\mathrm{d}\Phi)\right|.
\end{aligned}$$

Bearing in mind that P is mixing, we may continue as follows

$$= 2\varepsilon + \overline{\lim_{\|x\|\to\infty}} \Big| \big(P(\Phi^x(X_n) = k)\big)^{-1} \int f(\Phi)\, P(\mathrm{d}\Phi)$$

$$\cdot \int k_{\{\Psi : \Psi \in M, \Psi^x(X_n) = k\}}(\Phi)\, P(\mathrm{d}\Phi) - \int f(\Phi)\, P(\mathrm{d}\Phi) \Big| = 2\varepsilon .$$

By suitably selecting n, we can make ε arbitrarily small, so that we obtain

$$\int f(T_x\Phi)\, P_0\big(\mathrm{d}\Phi \mid \Phi^x(\{[0, \ldots, 0]\}) = k\big) \xrightarrow[\|x\|\to\infty]{} \int f(\Phi)\, P(\mathrm{d}\Phi).$$

Thus for all natural numbers m we have

$$\overline{\lim_{\|x\|\to\infty}} \left| \int f(T_x\Phi)\, P_0(\mathrm{d}\Phi) - \int f(\Phi)\, P(\mathrm{d}\Phi) \right|$$

$$= \overline{\lim_{\|x\|\to\infty}} \Big| \sum_k P_0\big(\Phi^x(\{[0, \ldots, 0]\}) = k\big)$$

$$\cdot \Big(\int f(T_x\Phi)\, P_0\big(\mathrm{d}\Phi \mid \Phi^x(\{[0, \ldots, 0]\}) = k\big) - \int f(\Phi)\, P(\mathrm{d}\Phi) \Big) \Big|$$

$$\leqq \Big(2 \sup_{\Psi \in M} |f(\Psi)| \Big)\, P_0\big(\Phi^x(\{[0, \ldots, 0]\}) > m\big) + \sum_{k \leqq m} P_0\big(\Phi^x(\{[0, \ldots, 0]\}) = k\big)$$

$$\cdot \overline{\lim_{\|x\|\to\infty}} \left| \int f(T_x\Phi)\, P_0\big(\mathrm{d}\Phi \mid \Phi^x(\{[0, \ldots, 0]\}) = k\big) - \int f(\Phi)\, P(\mathrm{d}\Phi) \right|$$

$$= 2 \sup_{\Psi \in M} |f(\Psi)|\, P_0\big(\Phi^x(\{[0, \ldots, 0]\}) > m\big),$$

where the summation always extends only over those k which satisfy the condition that $P_0\big(\Phi^x(\{[0, \ldots, 0]\}) = k\big) > 0$.

Now, letting m tend to infinity, we obtain

$$\int f(T_x\Phi)\, P_0(\mathrm{d}\Phi) \xrightarrow[\|x\|\to\infty]{} \int f(\Phi)\, P(\mathrm{d}\Phi).$$

It remains only to apply the known fact that (cf. Billingsley [1], Theorem 2.1.) the weak convergence $P_n \underset{n\to\infty}{\Rightarrow} P$ is valid if, for all bounded real functions f on M which are uniformly continuous with respect to ϱ_M, the convergence relation

$$\int f(\Phi)\, P_n(\mathrm{d}\Phi) \xrightarrow[n\to\infty]{} \int f(\Phi)\, P(\mathrm{d}\Phi)$$

is satisfied. ∎

9.4. The Mapping $P \rightsquigarrow (P_0)^{\#}$

In this section we shall always suppose that the dimension s of the "position space" R^s be equal to one.

As was done in Section 5.3. and 8.5., we set

$$Y_1 = \big\{\Phi : \Phi \in M, \Phi^x \text{ simple}, \Phi^x\big((-\infty, 0)\big) = +\infty = \Phi^x\big((0, +\infty)\big)\big\}.$$

Using the special measurable indexing relation introduced according to 5.3.4., we are able to represent every Φ in Y_1 by

$$\Phi = \sum_{-\infty<n<+\infty} \delta_{[x_n(\Phi),k_n(\Phi)]}.$$

Setting now

$$y_n(\Phi) = x_n(\Phi) - x_{n-1}(\Phi) \qquad (\Phi \in Y_1, -\infty < n < +\infty),$$

we associate to each Φ in Y_1 the doubly infinite sequence

$$\Phi^{\#} = \big([y_n(\Phi), k_{n-1}(\Phi)]\big)_{-\infty<n<+\infty}$$

of elements of $(0, +\infty) \times K$. According to 5.3.4. and 6.1.1., the mapping

$$\Phi \rightsquigarrow [x_{n-1}(\Phi), x_n(\Phi), \Phi] \rightsquigarrow [y_n(\Phi), T_{x_{n-1}(\Phi)}\Phi]$$

is measurable for all n. Noting that, for all Borel subsets L of the "mark space" $[K, \varrho_K]$ and all Φ in Y_1, the relation $k_{n-1}(\Phi) \in L$ is equivalent to $(T_{x_{n-1}(\Phi)}\Phi)(\{0\} \times L) > 0$, we obtain

9.4.1. *The correspondence* $\Phi \rightsquigarrow \Phi^{\#}$ *provides a mapping of* Y_1 *into* $\big((0, +\infty) \times K\big)^G$, *which is measurable with respect to the σ-algebras* $\mathfrak{M}$, $\Big(\big((0, +\infty) \cap \mathfrak{R}^1\big) \otimes \mathfrak{K}\Big)^G$.

(Here G denotes the set of all integers.)

Obviously an element $([h_n, l_n])$ of $\big((0, +\infty) \times K\big)^G$ belongs to the range V of the mapping $\Phi \rightsquigarrow \Phi^{\#}$ if and only if

$$\sum_{n=1}^{\infty} h_{-n} = +\infty = \sum_{n=1}^{\infty} h_n$$

is satisfied. Hence the set V is measurable.

The mapping $\Phi \rightsquigarrow \Phi^{\#}$ is not one-to-one, since in order to recover Φ from $\Phi^{\#}$ the "phase" $-x_{-1}(\Phi)$ of Φ is required in addition. However, if Φ belongs to $Y_1 \cap M^0$, then $x_{-1}(\Phi) = 0$, and for $\Phi^{\#} = ([h_n, l_n])$ we obtain

$$\Phi = \sum_{n=1}^{\infty} \delta_{[-(h_{-1}+\cdots+h_{-n}),l_{-n}]} + \delta_{[0,l_0]} + \sum_{n=0}^{\infty} \delta_{[(h_0+\cdots+h_n),l_{n+1}]}.$$

Now let X be an arbitrary set in $\big((0, +\infty) \cap \mathfrak{R}^1\big) \otimes \mathfrak{K}$. Then, for all $m \geqq 0$ the real function

$$([h_n, l_n]) \rightsquigarrow \delta_{[(h_0+\cdots+h_m),l_{m+1}]}(X) = k_X\big([(h_0 + \cdots + h_m), l_{m+1}]\big)$$

is measurable with respect to $\Big(\big((0, +\infty) \cap \mathfrak{R}^1\big) \otimes \mathfrak{K}\Big)^G$. The same holds for

$$([h_n, l_n]) \rightsquigarrow \delta_{[0,l_0]}(X)$$

and for

$$([h_n, l_n]) \rightsquigarrow \delta_{[-(h_{-1}+\cdots+h_{-m}),l_{-m}]}(X) \quad (m = 1, 2, \ldots).$$

Consequently the real function

$$([h_n, l_n]) \curvearrowright \Phi(X)$$

which is defined on V, is measurable, too. Thus it follows from the definition of $\mathfrak{M}$ that $([h_n, l_n]) \curvearrowright \Phi$, i.e. the inverse mapping of $\Phi \curvearrowright \Phi^\#$ is measurable, and we observe the truth of

9.4.2. *The one-to-one mapping $\Phi \curvearrowright \Phi^\#$ of $Y_1 \cap M^0$ into $\big((0, +\infty) \times K\big)^G$ is measurable in both directions with respect to the σ-algebras $\mathfrak{M}^0$, $\Big(\big((0, +\infty) \cap \mathfrak{R}^1\big) \otimes \mathfrak{K}\Big)^G$.*

We denote the image of a measure H on $\mathfrak{M}^0$ with respect to the mapping $\Phi \curvearrowright \Phi^\#$ by $H^\#$. Then the correspondence $H \curvearrowright H^\#$ provides a one-to-one mapping of the set of all finite measures H on $\mathfrak{M}^0$ satisfying the condition $H(M^0 \setminus Y_1) = 0$ upon the set of all finite measures q on $\Big(\big((0, +\infty) \cap \mathfrak{R}^1\big) \otimes \mathfrak{K}\Big)^G$, which satisfy the condition $q\Big(\big((0, +\infty) \times K\big)^G \setminus V\Big) = 0$. Obviously $\hat{\Gamma}_1 H = H$ is equivalent to the fact that $H^\#$ is stationary with respect to the "time shift" $([h_n, l_n]) \curvearrowright ([h_n{}^*, l_n{}^*])$, where $h_n{}^* = h_{n+1}$, $l_n{}^* = l_{n+1}$ for all integer n. Bearing in mind that, in view of the individual ergodic theorem, every stationary finite measure q automatically satisfies the condition

$$q(V) = 1$$

we may bring Theorem 8.5.1. into the following convenient form, where we use the additional assumption that $i_{P^x} < +\infty$ (cf. Ryll—Nardzewski [1] and Slivnyak [1], [2]):

9.4.3. Theorem. *The correspondence $P \curvearrowright (P_0)^\#$ provides a one-to-one mapping of the set of all stationary distributions P on $\mathfrak{M}$ satisfying the conditions $P(\{o\}) = 0$, P^x is simple, $i_{P^x} < +\infty$ upon the set of those distributions q on $\Big(\big((0, +\infty) \cap \mathfrak{R}^1\big) \otimes \mathfrak{K}\Big)^G$, stationary with respect to time shift, for which the expectation $\mathrm{E}_q h_0$ is finite.*

Subject to the assumptions of 9.4.3., the relation

$$i_{P^x} = \Big(\int x_0(\Phi)\, P_0(\mathrm{d}\Phi)\Big)^{-1} = (\mathrm{E}_q h_0)^{-1}$$

can be interpreted as showing that the "mean distance between a point x in Φ^x selected at random and the successor point in Φ^x" equals the reciprocal of the "mean asymptotic density of Φ^x". The property of invariance of P_0, i.e. the stationarity of $(P_0)^\#$, corresponds to the picture that the "distribution of $T_x\Phi$ for a point x in Φ^x selected at random" coincides with the corresponding "distribution" as referred to the successor point in Φ^x.

If P can be represented as a non-trivial mixture

$$P = \alpha P_1 + (1 - \alpha)\, P_2$$

of stationary distributions, then using 9.1.4. we get

$$P_0 = (i_{P^x})^{-1} \left(\alpha i_{(P_1)^x}(P_1)_0 + (1-\alpha)\, i_{(P_2)^x}(P_2)_0\right),$$

thus obtaining a non-trivial representation

$$(P_0)^\# = (i_{P^x})^{-1} \left(\alpha i_{(P_1)^x}\big((P_1)_0\big)^\# + (1-\alpha)\, i_{(P_2)^x}\big((P_2)_0\big)^\#\right)$$

of $(P_0)^\#$ as a mixture of stationary distributions. Hence, if P is not ergodic, this is true also for $(P_0)^\#$.

Conversely, let

$$(P_0)^\# = \gamma L_1 + (1-\gamma)\, L_2$$

be a non-trivial representation of $(P_0)^\#$ as a mixture of stationary distributions. Together with $(P_0)^\#$, L_1 and L_2 also satisfy the condition of 9.4.3., and hence they are of the form $L_1 = \big((P_1)_0\big)^\#$, $L_2 = \big((P_2)_0\big)^\#$. We get

$$P_0 = \gamma (P_1)_0 + (1-\gamma)\, (P_2)_0,$$

i.e.

$$P = i_{P^x}\big((i_{(P_1)^x})^{-1} \gamma P_1 + (i_{(P_2)^x})^{-1} (1-\gamma)\, P_2\big).$$

Hence, if $(P_0)^\#$ is not ergodic, neither is P.

Thus we observe the truth of (cf. Ryll—Nardzewski [1])

9.4.4. Proposition. *Subject to the assumptions of* 9.4.3., P *is ergodic if and only if* $(P_0)^\#$ *is ergodic with respect to time shift.*

If $(P_0)^\#$ is mixing with respect to time shift, this is not necessarily true for P.

To give an example of this, we choose an arbitrary $[h, l]$ from $(0, +\infty) \times K$, and set

$$(P_0)^\# \left([h_n, l_n] = [h, l] \quad \text{for all integer } n\right) = 1.$$

It is, however, possible to derive the following counterpart to 9.1.11. (cf. Franken, Liemant Matthes [1]):

9.4.5. Proposition. *If, for a stationary distribution* P *on* $\mathfrak{M}$ *satisfying the conditions* $P(\{o\}) = 0$, P^x *is simple,* $i_{P^x} < +\infty$, *the distribution* $(P_0)^\#$ *is mixing with respect to time shift, then*

$$P(T_{x_n(\Phi)}\Phi \in Y) \xrightarrow[n\to\infty]{} P_0(Y)$$

for all Y *in* $\mathfrak{M}^0$.

Proof. In view of 9.1.12., for all integers n we have

$$\begin{aligned} P(T_{x_n(\Phi)}\Phi \in Y) &= i_{P^x} \int \left(\int_0^{x_0(\Phi)} k_Y(T_{x_n(\Phi)}\Phi)\, \mu(\mathrm{d}x) \right) P_0(\mathrm{d}\Phi) \\ &= i_{P^x} \int y_0(\Phi)\, k_Y(T_{x_n(\Phi)}\Phi)\, P_0(\mathrm{d}\Phi). \end{aligned}$$

Since $\hat{\Gamma}_1 P_0 = P_0$, we may now continue this as follows

$$= i_{P^x} \int y_{-(n+1)}(\Phi)\, k_Y(\Phi)\, P_0(\mathrm{d}\Phi)$$

$$= i_{P^x} \int h_{-(n+1)} k_Z\big(([h_n, l_n])\big)\, (P_0)^{\#}\, \big(\mathrm{d}([h_n, l_n])\big),$$

where Z denotes the image of Y produced by the mapping

$$\Phi \curvearrowright \Phi^{\#} \text{ of } Y_1 \cap M^0 \text{ into } \big((0, +\infty) \times K\big)^G.$$

As $(P_0)^{\#}$ is mixing with respect to time shift, we obtain by means of 9.1.13.

$$\lim_{n\to\infty} P(T_{x_n(\Phi)}\Phi \in Y) = i_{P^x} \int h_0 (P_0)^{\#}\, \big(\mathrm{d}([h_n, l_n])\big) \int_Z 1 (P_0)^{\#}\, \big(\mathrm{d}([h_n, l_n])\big)$$
$$= P_0(Y). \;\blacksquare$$

As we observed in the proof of 9.4.5., for all stationary distributions P on $\mathfrak{M}$ satisfying the conditions $P(\{o\}) = 0$, P^x is simple, $i_{P^x} < +\infty$, and for all Y in $\mathfrak{M}^0$, we have

$$P(T_{x_{-1}(\Phi)}\Phi \in Y) = i_{P^x} \int_Y x_0(\Phi)\, P_0(\mathrm{d}\Phi),$$

i.e. the distribution $P\big(T_{x_{-1}(\Phi)}\Phi \in (.)\big)$ has the density $i_{P^x} x_0(\Phi)$ with respect to P_0. (Under the stronger assumptions as they are used here, this can be considered a generalization of proposition b) in 9.1.14.) Subject to the above assumptions, we may now interpret the inversion formula 9.1.12. as follows. To pass from P_0 (or the equivalent distribution $(P_0)^{\#}$) to P, we first derive from P_0 the distribution $i_{P^x} \int_{(.)} x_0(\Phi)\, P_0(\mathrm{d}\Phi)$, thus obtaining the distribution $P\big(T_{x_{-1}(\Phi)}\Phi \in (.)\big)$. Subsequently, for each Φ in Y_1, we toss a die to find the "phase" x "on the off chance", i.e. according to $\big(x_0(\Phi)\big)^{-1} \mu\big((.) \cap [0, x_0(\Phi))\big)$, and then pass to $T_x\Phi$. Thus we finally obtain the distribution

$$i_{P^x} \int \left(\int_0^{x_0(\Phi)} k_{(.)}(T_x\Phi)\, \big(x_0(\Phi)\big)^{-1} \mu(\mathrm{d}x) \right) x_0(\Phi)\, P_0(\mathrm{d}\Phi),$$

i.e. the distribution P.

9.5. Simple Stationary Recurrent Distributions

Throughout this section we shall assume that the phase space $[A, \varrho_A]$ coincides with the real axis R^1.

Let $\boldsymbol{F}$ denote the set of those right-continuous, monotone increasing real functions F on the interval $[0, +\infty)$ which satisfy the condition

$$0 \leqq F(x) \leqq 1 \qquad (0 \leqq x < +\infty),$$

i.e. the set of all distribution functions of random non-negative, possibly infinite numbers.

To each F in $\boldsymbol{F}$ having the two properties $F(0) = 0$, $\lim_{x\to\infty} F(x) = 1$ we can associate the distribution q_F of an independent, doubly infinite sequence of random positive numbers distributed according to F. Then by 9.4.2. there exists a well-determined H_F on $\mathfrak{M}^0$ having the property that $(H_F)^{\#} = q_F$, which satisfies the condition $H_F(Y_1) = 1$. (Note that the statements made in the preceding section can trivially be extended to the "non-marked case".) Due to the stationarity of q_F with respect to time shift, H_F satisfies the condition $\hat{\Gamma}_1 H_F = H_F$.

Every Φ in Y_1 can be considered to describe a renewal process where the $x_n(\Phi)$ are the points of renewal, and the $y_n(\Phi)$ the ages of the objects failing at the times $x_n(\Phi)$. If we use this interpretation in the case of the distributions H_F, these are characterized by the fact that the lifetimes $y_n(\Phi)$ are mutually independent and that a renewal takes place at the origin. Therefore H_F is also called the distribution of the *recurrent point process* associated with the distribution function F.

In view of 9.4.3., H_F is the Palm distribution of a well-determined stationary simple distribution P_F satisfying the condition $P(\{o\}) = 0$ and having a finite intensity, if and only if $\Delta_F = \int_0^\infty x \, \mathrm{d}F(x) < +\infty$; then obviously we have

$$F_{(P_F)} = F, \quad (i_{P_F})^{-1} = \Delta_F.$$

In view of 9.4.4., all distributions P_F are ergodic.

It is known that (cf., for instance, Matthes [1]) a distribution P_F is mixing if and only if there is no real $d > 0$, for which the relation

$$\sum_{n=0}^{\infty} (F(nd) - F(nd - 0)) = 1$$

is satisfied. Hence in particular, for

$$F(x) = 2^{-1}(k_{[1,+\infty)}(x) + k_{[\pi,+\infty)}(x)) \qquad (x \geqq 0),$$

$P = P_F$ satisfies all assumptions of 9.1.11. Consequently, $P_0(T_x\Phi \in (.)) \underset{\|x\|\to\infty}{\Rightarrow} P$, i.e. for all continuity sets $P_0(T_x\Phi \in Y) \xrightarrow[\|x\|\to\infty]{} P(Y)$. However, $P_0(T_x\Phi \in Y) \xrightarrow[\|x\|\to\infty]{} P(Y)$ does not hold for all Y in $\mathfrak{M}$! For, given any sequence (x_n) of real numbers which tends toward $+\infty$, there exists a countable subset Z of R^1, such that

$$H_F((T_{x_n}\Phi)(R^1 \setminus Z) = 0 \quad \text{for} \quad n = 1, 2, \ldots) = 1$$

is satisfied, and for $Y = \{\Phi : \Phi \in M, \; \Phi(R^1 \setminus Z) = 0\}$ we obtain

$$\lim_{n\to\infty} P_0(T_{x_n}\Phi \in Y) = 1, \quad P(Y) = 0.$$

Setting now $P = P_{k_{[1,\infty)}}$, we get

$$P_0(T_n\Phi \in (.)) \underset{n\to\infty}{\Rightarrow} P_0,$$

from which we observe that the assumption that "P be mixing" in 9.1.11. cannot be replaced by the weaker assumption that "P be ergodic".

From the statements made at the end of the preceding section, the following exhaustive structural statement on P_F is readily derived:

9.5.1. Proposition. *For all F in $\mathbf{F}$ satisfying the conditions* $F(0) = 0$, $\lim_{x\to\infty} F(x) = 1$, $\Delta_F = \int_0^\infty x \,\mathrm{d}F(x) < +\infty$, *the doubly infinite sequence of the* $y_i(\Phi) = \big(x_i(\Phi) - x_{i-1}(\Phi)\big)$ *is independent with respect to P_F. In particular* $y_0(\Phi)$ *has the distribution function* $(\Delta_F)^{-1} \int_0^x u \,\mathrm{d}F(u) \quad (x \geqq 0)$, *and the* $y_i(\Phi)$, $i \neq 0$, *have the distribution function F. The conditional distribution of* $-x_{-1}(\Phi)$, *given* $\big(y_i(\Phi)\big)_{-\infty<i<+\infty}$, *is the uniform distribution on the interval* $[0, y_0(\Phi)]$. *Thus the "phase"* $-x_{-1}(\Phi)$ *is independent of* $\big(y_i(\Phi)\big)_{i\neq 0}$.

In view of 9.5.1., P_F is called the distribution of the *stationary recurrent point process* associated with the distribution function F.

For all $l > 0$, we denote by E_l the distribution function $1 - e^{-lx}$ $(x \geqq 0)$ of the negative exponential distribution. Since

$$\int_0^\infty x \,\mathrm{d}E_l(x) = \int_0^\infty \big(1 - E_l(x)\big)\, \mu(\mathrm{d}x) = \int_0^\infty e^{-lx} \mu(\mathrm{d}x) = l^{-1},$$

we can form the distribution P_{E_l}.

The Palm distribution of P_{E_l} is given by the recurrent distribution H_{E_l}. Using 9.1.14. a) we obtain, for all $u, v \geqq 0$

$$\begin{aligned} &P_{E_l}\big(-x_{-1}(\Phi) \leqq u, x_0(\Phi) \leqq v\big) \\ &= l\left(\int_0^u e^{-lx}\mu(\mathrm{d}x) + \int_0^v e^{-lx}\mu(\mathrm{d}x) - \int_0^{u+v} e^{-lx}\mu(\mathrm{d}x)\right) \\ &= l\big(l^{-1}(1 - e^{-lu}) + l^{-1}(1 - e^{-lv}) - l^{-1}(1 - e^{-l(u+v)})\big) \\ &= (1 - e^{-lu})\,(1 - e^{-lv}) = E_l(u)\, E_l(v). \end{aligned}$$

Consequently, for $F = E_l$, the distribution of the vector $[-x_{-2}(\Phi), x_0(\Phi)]$, which is formed with respect to H_F, coincides with the distribution of $[-x_{-1}(\Phi), x_0(\Phi)]$, formed with respect to P_F. Hence, in view of 9.5.1.,

$$(P_F)_0 = H_F = \delta_{\delta_0} * P_F.$$

Consequently, in virtue of 9.2.7., we obtain the following classical structural proposition:

9.5.2. Proposition. *For all* $l > 0$, $P_{E_l} = P_{l\mu}$.

Of course we could also have verified by direct calculation that the simple stationary distribution P_{E_l} is free from after-effects, since

$$(1 - E_l(x))^{-1}\,(1 - E_l(x + y)) = 1 - E_l(y) \qquad (x, y \geqq 0).$$

In Haberland [1] it was proved that an infinitely divisible distribution of the form P_F is Poisson, i.e. $F = E_l$, $l > 0$, holds, if F possesses a continuous derivative on the half-line $[0, +\infty)$ (at 0 the right-hand derivative). In Daley [2] a non-Poisson distribution P_F was constructed which possesses a non-trivial stationary Poisson factor, i.e. can be brought into the form

$$P_F = P_{l\mu} * Q \qquad (l > 0,\ Q \text{ stationary}).$$

In this case the second factor Q cannot be recurrent because (cf. Mecke [4]) no non-Poisson distribution of the form P_F can be represented as convolution $P_{F_1} * P_{F_2}$.

In view of 9.2.4. P_F is surely indecomposable, i.e. there exists only a trivial representation $Q_1 * Q_2$ as convolution of stationary distributions, if for some $t > 0$ the absolutely continuous component F_c of F satisfies the condition $F_c(t) = 0$.

For simple stationary distributions P on $\mathfrak{M}$ having the properties $P(\{o\}) = 0$, $i_P < +\infty$, it is observed from 9.1.14. b) that usually the distribution of $y_0(\Phi) = x_0(\Phi) - x_{-1}(\Phi)$ does not coincide with the distribution of this length as referred to P_0; indeed, a point x on the real axis, which is "selected at random", falls in general into intervals $[x_{i-1}(\Phi), x_i(\Phi)]$ of greater lengths! The two distributions coincide if and only if, for all $x \geqq 0$

$$F_P(x) = i_P \int_{(0,x]} u \, \mathrm{d}F_P(u)$$

is satisfied, where by 9.1.13. $(i_P)^{-1}$ coincides with $\Delta_{(F_P)}$. The only solutions F of the equation

$$F(x)\,\Delta_F = xF(x) - \int_0^x F(z)\,\mu(\mathrm{d}z),$$

i.e. of

$$F(x)\,(x - \Delta_F) = \int_0^x F(z)\,\mu(\mathrm{d}z),$$

which belong to $\boldsymbol{F}$ and have the properties $F(0) = 0$, $\lim\limits_{x\to\infty} F(x) = 1$, $\Delta_F < +\infty$ are the step functions $F = k_{[z,\infty)}$, $z > 0$. Hence in cases where the two distributions considered are equal, $x_0(\Phi)$ is, as referred to P_0, almost certainly equal to a positive constant z. Then, in view of the stationarity of $(P_0)^{\#}$ with respect to time shift, all $y_i(\Phi)$ are, as referred to P_0, almost certainly equal to z, i.e. P_0 coincides with the recurrent distribution $H_{k_{[z,\infty)}}$, so that it can be concluded that

$$P = P_{k_{[z,\infty)}}.$$

In view of the statements made at the end of the preceding section, the distribution $P\big(T_{x_{-1}(\Phi)}\Phi \in (.)\big)$ likewise coincides with P_0 if and only if $x_0(\Phi)$ is constant almost everywhere with respect to P_0, so that as a matter of fact, the last non-positive point in Φ just is not "selected at random" from Φ!

Summarizing, we observe the truth of

9.5.3. Proposition. *For all simple stationary distributions P on $\mathfrak{M}$ satisfying the conditions $P(\{o\}) = 0$, $i_P < +\infty$, the following propositions are equivalent:*

a) $P_0 = P\big(T_{x_{-1}(\Phi)}\Phi \in (.)\big)$,

b) $P = P_{k[z,\infty)}$ *for some* $z > 0$,

c) $\big(x_0(\Phi) - x_{-1}(\Phi)\big)$ *has the distribution function* F_P *with respect to* P.

In view of 6.1.1. for all Y in $\mathfrak{M}$, the real function

$$z \curvearrowright P_{k[z,\infty)}(Y) = z^{-1} \int \left(\int_0^z k_Y(T_x\Phi)\, \mu(\mathrm{d}x) \right) H_{k[z,\infty)}(\mathrm{d}\Phi)$$

$$= z^{-1} \int_0^z k_Y\Big(T_x \sum_{-\infty<i<+\infty} \delta_{iz}\Big)\, \mu(\mathrm{d}x)$$

is measurable on $(0, +\infty)$, i.e.

$$z \curvearrowright P_{k[z,\infty)}$$

is a measurable mapping of $(0, +\infty)$ into $[\boldsymbol{P}, \mathfrak{P}]$, so that we can introduce the mixture

$$D_\sigma = \int_0^\infty P_{k[z,\infty)}(.)\, \sigma(\mathrm{d}z)$$

for all distributions σ of random positive numbers. All D_σ are simple, stationary, satisfy the condition $D_\sigma(\{o\}) = 0$, and have the intensity

$$i_{D_\sigma} = \int_0^\infty z^{-1}\sigma(\mathrm{d}z).$$

We have (cf. Koenig, Matthes & Nawrotzki [1])

9.5.4. Proposition. *A stationary distribution P on $\mathfrak{M}$ has the form D_σ if and only if it is simple, satisfies $P(\{o\}) = 0$ and has the property that the doubly infinite sequence of the $y_i(\Phi) = x_i(\Phi) - x_{i-1}(\Phi)$ is stationary with respect to P.*

Proof. Obviously each of the D_σ satisfies the conditions stated above.

Conversely, suppose that the conditions stated above are satisfied for a stationary P.

Since $P(\{o\}) = 0$, the limit

$$s(\Phi) = \lim_{t\to\infty} (2t)^{-1}\, \Phi\big([-t, t)\big)$$

which exists, according to Section 6.4., for almost all Φ, is positive almost everywhere. Obviously

$$\big(s(\Phi)\big)^{-1} = \lim_{n\to\infty} (2n + 1)^{-1} \sum_{i=-n}^{n} y_i(\Phi).$$

Now, if $s(\Phi)$ were not finite almost everywhere, then applying the individual ergodic theorem to the stationary, doubly infinite sequence $(y_i(\Phi))$ would lead us to the conclusion that the probability of $y_0(\Phi) = 0$ is positive, which would contradict the definition of the $y_i(\Phi)$.

Consequently, $0 < s(\Phi) < +\infty$ is valid almost everywhere, so that we can represent P as a mixture in the following way:

$$P = \sum_i P\big((i-1) < s(\Phi) \leqq i\big)\, P\big((.) \mid (i-1) < s(\Phi) \leqq i\big),$$

where only those summands are taken into account for which $P\big((i-1) < s(\Phi) \leqq i\big)$ is positive. All of the $P_i = P\big((.) \mid (i-1) < s(\Phi) \leqq i\big)$ are stationary, because $s(\Phi)$ is known to be invariant. It will suffice to show that each P_i is of the form D_{σ_i}, for this would imply that

$$P = \sum_i P\big((i-1) < s(\Phi) \leqq i\big)\, D_{\sigma_i} = D_{\sum\limits_i P((i-1)<s(\Phi)\leqq i)\sigma_i}.$$

All P_i have finite intensities and satisfy the conditions stated above, so that, from now on, we can suppose in addition that the intensity of P be finite.

As was stated in the proof of 9.4.5., the following relation holds for all integers i and all Y in $\mathfrak{M}^0$:

$$P(T_{x_i(\Phi)}\Phi \in Y) = i_P \int_Y y_{-(i+1)}(\Phi)\, P_0(\mathrm{d}\Phi).$$

Due to our assumptions on $(y_i(\Phi))$, the $T_{x_i(\Phi)}\Phi$ are identically distributed with respect to P, so that

$$x_0(\Phi) = x_i(\Phi) - x_{i-1}(\Phi) \qquad (-\infty < i < +\infty)$$

must be valid almost everywhere with respect to P_0. But it was just this we had to show. ∎

Now we present an example (cf. Daley [1]) – which was announced earlier in section 1.4. – of a stationary distribution P, for which, though being orderly, not every distinguished sequence $(\mathfrak{Z}_n)$ of decompositions of $(0, 1]$ satisfies the condition

$$\sum_{Z \in \mathfrak{Z}_n} P(\Phi(Z) > 1) \xrightarrow[n\to\infty]{} 0.$$

Let $(t_r)_{r=1,2,\ldots}$ be a sequence of real numbers taken from the interval $(0, 1)$, which tends monotonically toward zero. We define a distribution function F of a random positive real number by means of

$$F(x) = \begin{cases} \zeta_r & \text{for} \quad \zeta_r \leqq x < \zeta_{r-1} \quad (r = 1, 2, \ldots) \\ 1 & \phantom{\text{for}} \quad 1 \leqq x, \end{cases}$$

setting $\zeta_r = t_r \zeta_{r-1}$, $\zeta_0 = 1$. The corresponding distribution is denoted by σ. For a given ε in $(0, 1)$, the relation $t_r < 1 - \varepsilon$ is valid for sufficiently large r. The sequence of the ζ_r likewise tends toward zero, so that for sufficiently large r there exists a natural number n_r, such that the inequality

$$(\zeta_{r-1})^{-1} < n_r < (1-\varepsilon)^{-1}(\zeta_{r-1})^{-1}$$

is satisfied. Hence

$$\zeta_r < (1-\varepsilon)\,\zeta_{r-1} < h_r < \zeta_{r-1},$$

where $h_r = (n_r)^{-1}$.

Using these inequalities, we show that D_σ is orderly. To this end we calculate

$$D_\sigma(\Phi((0,h]) > 1) = \int_0^\infty P_{k[z,+\infty)}(\Phi((0,h]) > 1)\,\mathrm{d}F(z)$$

$$= \int_0^h P_{k[z,+\infty)}(\Phi((0,h]) > 1)\,\mathrm{d}F(z) = F\left(\frac{h}{2}\right) + \int_{(h/2,h]} P_{k[z,+\infty)}(\Phi((0,h]) > 1)\,\mathrm{d}F(z)$$

$$= F\left(\frac{h}{2}\right) + \int_{(h/2,h]} z^{-1}(h-z)\,\mathrm{d}F(z) = h\int_{h/2}^h z^{-2}F(z)\,\mu(\mathrm{d}z).$$

Now let $\mathfrak{Z}_r$ be the equidistant decomposition of the interval $(0, 1]$ into intervals of the length h_r. Then we obtain

$$\sum_{Z\in\mathfrak{Z}_r} D_\sigma(\Phi(Z) > 1) = n_r D_\sigma(\Phi((0,h_r]) > 1)$$

$$= n_r\left(h_r\int_{h_r/2}^{h_r} z^{-2}F(z)\,\mu(\mathrm{d}z)\right) \leqq 2n_r\zeta_r \leqq \frac{2\zeta_r}{(1-\varepsilon)\,\zeta_{r-1}} = \frac{2}{1-\varepsilon}\,t_r,$$

which proves the orderliness of the stationary distribution D_σ.

On the other hand, for each $\varepsilon > 0$ there is a natural number n_r' being sufficiently large, such that

$$((1+\varepsilon)\,2\zeta_r)^{-1} < n_r' < (2\zeta_r)^{-1}$$

as well as $(\zeta_{r-1})^{-1} < n_r'$ are valid. This means however that

$$\zeta_r < 2^{-1}h_r' < (1+\varepsilon)\,\zeta_r \quad \text{and} \quad h_r' < \zeta_{r-1},$$

where again $h_r' = (n_r')^{-1}$. Let $(\mathfrak{Z}_r')$ be the sequence of decompositions of the interval $(0, 1]$ into intervals of the length h_r', i.e. a distinguished sequence of decompositions. Here we get

$$\sum_{Z\in\mathfrak{Z}_r'} D_\sigma(\Phi(Z) > 1) = (h_r')^{-1}\,D_\sigma(\Phi((0,h_r']) > 1)$$

$$= \int_{h_r'/2}^{h_r'} z^{-2}F(z)\,\mu(\mathrm{d}z) > \frac{\zeta_r}{2h_r'} > \frac{1}{4(1+\varepsilon)},$$

which means that

$$\sum_{Z\in\mathfrak{Z}_n} D_\sigma(\Phi(Z) > 1) \xrightarrow[n\to\infty]{} 0$$

is not valid for each distinguished sequence $(\mathfrak{Z}_n)$ of decompositions of $(0, 1]$.

To conclude this section, we shall keep a promise made in section 1.4. We show (cf. Moran [2], Goldman [2], Lee [3], as well as Szász [2], where an even stronger counter-example is constructed):

9.5.5. Proposition. *In the case* $s = 1$, *there exists a simple stationary distribution* P *on* $\mathfrak{M}$, *which is different from* P_μ, *but satisfies the condition* $P_I = (P_\mu)_I$ *for all bounded intervals* I *of the real axis* R^1.

Proof. 1. First we construct a probability density function f of a random vector $[\zeta_1, \zeta_2]$ with positive co-ordinates such that

a) ζ_1, ζ_2 are distributed according to E_1, i.e. they have the probability density function e^{-x} $(x \geqq 0)$;
b) $\zeta_1 + \zeta_2$ is distributed according to $E_1 * E_2$, i.e. it has the probability density function xe^{-x} $(x \geqq 0)$;
c) ζ_1, ζ_2 are not independent.

For that purpose we select a number ε in $(0, e^{-6})$, setting

$$f(x_1, x_2) = e^{-(x_1+x_2)} + g(x_1, x_2) \qquad (x_1, x_2 \geqq 0)$$

where

$$g(x_1, x_2) = \begin{cases} \varepsilon & \text{on } [0,1)\times[2,3) \cup [1,2)\times[3,4) \cup [2,3)\times[1,2) \cup [3,4)\times[0,1) \\ -\varepsilon & \text{on } [0,1)\times[3,4) \cup [1,2)\times[2,3) \cup [2,3)\times[0,1) \cup [3,4)\times[1,2) \\ 0 & \text{otherwise.} \end{cases}$$

Then $f(x_1, x_2) \geqq 0$ for $x_1, x_2 \geqq 0$, and

$$\int_0^\infty g(x_1, x_2)\, \mu(\mathrm{d}x_1) = \int_0^\infty g(x_1, x_2)\, \mu(\mathrm{d}x_2) = \int_0^a g(x, a - x)\, \mu(\mathrm{d}x) = 0$$

for all $x_1, x_2, a \geqq 0$. Hence f is a probability density function having the following properties:

a′) $\int_0^\infty f(x_1, x_2)\, \mu(\mathrm{d}x_1) = e^{-x_2}$,
$\int_0^\infty f(x_1, x_2)\, \mu(\mathrm{d}x_2) = e^{-x_1}$ $\quad (x_1, x_2 \geqq 0)$,

b′) $\int_0^a f(x, a - x)\, \mu(\mathrm{d}x) = ae^{-a} \qquad (a \geqq 0)$,

c′) $f(x_1, x_2) \neq e^{-(x_1+x_2)}$ on a Borel subset X of $[0, +\infty)^2$, the Lebesgue measure of which is positive.

These properties of f do however, just correspond to the conditions a) b) and c).

2. Now we construct two simple distributions Q_1, Q_2 on $\mathfrak{M}^0$ by means of the following prescription: With respect to Q_1 and Q_2,

$$\big([y_{2k}(\Phi), y_{2k+1}(\Phi)]\big)_{-\infty<k<+\infty} \quad \text{or} \quad \big([y_{2k-1}(\Phi), y_{2k}(\Phi)]\big)_{-\infty<k<+\infty},$$

respectively, represent an independent sequence of random vectors distributed according to f. Now, if we set $Q = 2^{-1}(Q_1 + Q_2)$, then obviously

$$Q\big(\Phi \text{ simple}, \Phi\big((-\infty, 0)\big) = +\infty = \Phi\big((0, +\infty)\big)\big) = 1,$$

$$\hat{\Gamma}_1 Q = Q, \quad \text{and} \quad \int x_0(\Phi)\, Q(\mathrm{d}\Phi) = 1.$$

Hence, in view of 8.5.1., Q is the Palm measure of a well-determined simple stationary distribution P on $\mathfrak{M}$ having the properties $P(\{o\}) = 0$, $i_P = 1$. This distribution is different from P_μ, since the distribution of $[-x_{-2}(\Phi), x_0(\Phi)]$ with respect to P_0 has the density $2^{-1}\big(f(x_1, x_2) + e^{-(x_1+x_2)}\big)$, $(x_1, x_2 \geqq 0)$, and hence is not almost everywhere equal to the density $e^{-(x_1+x_2)}$, $(x_1, x_2 \geqq 0)$ of the distribution of this vector as referred to $(P_\mu)_0 = H_{E_1}$.

3. For all $t > 0$ and all natural numbers n,

$$\begin{aligned} Q_i\big(\Phi((0, t]) \geqq n\big) &= Q_i\big(x_0(\Phi) + y_1(\Phi) + \cdots + y_{n-1}(\Phi) \leqq t\big) \\ &= (E_1)^n (t) = P_\mu\big(x_0(\Phi) + y_1(\Phi) + \cdots + y_{n-1}(\Phi) \leqq t\big) \\ &= P_\mu\big(\Phi((0, t]) \geqq n\big), \end{aligned}$$

for $i = 1, 2$, i.e.

$$(Q_i)_{(0,t]} = (P_\mu)_{(0,t]} = \pi_t$$

and hence also

$$Q_{(0,t]} = \pi_t$$

for all positive t.

4. In order to prove that $P_I = (P_\mu)_I$, we can restrict ourselves to intervals of the form $(0, t]$, $t > 0$, since P and P_μ are stationary. Using 8.2.10. we obtain, for all non-negative t and all non-negative integers j,

$$\begin{aligned} P_{(0,t]}(\{j+1, j+2, \ldots\}) &= i_P \int_0^t P_0\big(\Phi((0, z]) = j\big)\, \mu(\mathrm{d}z) \\ &= i_{P_\mu} \int_0^t (P_\mu)_0 \big(\Phi((0, z]) = j\big)\, \mu(\mathrm{d}z) = (P_\mu)_{(0,t]} (\{j+1, j+2, \ldots\}), \end{aligned}$$

which means that

$$P_{(0,t]} = \pi_t. \quad \blacksquare$$

9.6. A Class of Singular Infinitely Divisible Stationary Distributions

We shall assume that the dimension s of the "position space" R^s be equal to one. Now, noting that $(\tilde{P})^0 (M) = i_{(\tilde{P})^x} = i_{P^x}$, which is valid for all stationary infinitely divisible distributions P on $\mathfrak{M}$, then from 8.5.6. we obtain, as a counterpart to Theorem 9.4.3.

9.6.1. Theorem. *The correspondence* $P \rightsquigarrow \big((\tilde{P})^0\big)^\#$ *provides a one-to-one mapping of the set of all singular infinitely divisible stationary distributions* P *on* $\mathfrak{M}$ *which satisfy the conditions* $P(\Phi^x \text{ simple}) = 1$, $i_{P^x} < +\infty$ *upon the set of all finite measures* q *on* $\big(((0, +\infty) \cap \mathfrak{R}^1) \otimes \mathfrak{K}\big)^G$, *which are stationary with respect to time shift.*

We shall now proceed to the non-marked case, i.e. we set $[A, \varrho_A]$ equal to R^1.

In view of 9.6.1. for each $l \geqq 0$ and each F in $\boldsymbol{F}$ satisfying the conditions $F(0) = 0$, $\lim_{x\to\infty} F(x) = 1$, there exists a well-determined singular infinitely divisible stationary simple distribution P on $\mathfrak{M}$ having the property that

$$(\tilde{P})^0 = lH_F.$$

If in addition $\Delta_F < +\infty$, then, as it was stated in the preceding section,

$$(P_F)^0 = i_{P_F}(P_F)_0 = (\Delta_F)^{-1} H_F,$$

and we obtain

$$\tilde{P} = l\Delta_F P_F.$$

Now suppose that $\Delta_F = +\infty$. If F is non-periodic, i.e. if there exists no positive real d such that

$$\sum_{n=0}^{\infty} \big(F(nd) - F(nd - 0)\big) = 1,$$

then, in view of Blackwell's renewal theorem (see, for instance, Feller [2], chapt. XI, par. 1), for all α, β satisfying $0 < \alpha < \beta < +\infty$,

$$\varrho_{H_F}\big((\alpha + t, \beta + t]\big) = \int \chi\big((\alpha + t, \beta + t]\big)\, H_F(\mathrm{d}\chi) \xrightarrow[t\to\infty]{} 0,$$

and subject to the above assumptions we get, for $l > 0$,

$$\begin{aligned}&(\tilde{P})_0 \big(\chi\big((\alpha + t, \beta + t]\big) > 0\big)\\ &= H_F\big(\chi\big((\alpha + t, \beta + t]\big) > 0\big) \leqq \varrho_{H_F}\big((\alpha + t, \beta + t]\big) \xrightarrow[t\to\infty]{} 0.\end{aligned}$$

By reflection in the origin, H_F is transformed into itself. Therefore the corresponding convergence proposition remains valid also for $t \to -\infty$. (The restriction to intervals on the right hand side of the origin is inessential, because each bounded interval, if translated by sufficiently large positive t, can be brought to the right hand side of the origin.)

Consequently, in view of theorem 9.2.1., P is mixing. (Note that $P = \delta_o$ for $l = 0$). If F is periodic, we can draw the same conclusions if we use, instead of Blackwell's renewal theorem, its discrete analogue (see, for instance, Feller [1], chapt. XIII, par. 3).

Summarizing, we observe the truth of

9.6.2. Proposition. *For each function F in $\boldsymbol{F}$ satisfying the conditions $F(0) = 0$, $\lim_{x\to\infty} F(x) = 1$, and for each non-negative real l, there exists exactly one stationary infinitely divisible distribution P having the property*

$$(\tilde{P})^0 = lH_F.$$

This distribution is always simple. If $\Delta_F < +\infty$, it has the form

$$P = \mathcal{U}_{l\Delta_F P_F}$$

and hence is a strongly singular infinitely divisible distribution. On the other hand, if $\Delta_F = +\infty$, it is a mixing, and hence a weakly singular infinitely divisible distribution.

The structural statement 9.5.1., together with its proof, can be transferred to the infinite measures $\tilde{P}$ in $\boldsymbol{W}$ which are associated through $(\tilde{P})^0 = lH_F$, $l > 0$, with distribution functions F in $\boldsymbol{F}$ having the properties

$$F(0) = 0, \qquad \lim_{x\to\infty} F(x) = 1, \qquad \Delta_F = +\infty.$$

We proceed as follows: We observe that $F_{\tilde{P}}$ coincides with F. The "distribution" of the length of the interval around the origin, $[x_{-1}(\Phi), x_0(\Phi)]$, has the "distribution function"

$$l \int_0^x u \,\mathrm{d}F(u) \qquad (x \geqq 0).$$

The "distribution", established with respect to $\tilde{P}$, of the doubly infinite sequence $\big(y_n(\Phi)\big)_{-\infty<n<+\infty}$ is obtained as

$$\left(\mathop{\times}_{n=-\infty}^{-1} p_n\right) \times p_0 \times \left(\mathop{\times}_{n=1}^{\infty} p_n\right)$$

where, for $n \neq 0$, p_n denotes the distribution associated with F, and p_0 coincides with the infinite measure associated with the "distribution function" $l \int_0^x u \,\mathrm{d}F(u)$. The transition to the "distribution" of Φ is now carried out along the same lines as in the case $\Delta_F < +\infty$.

We see that the infinite stationary measures $\tilde{P}$ considered here are natural generalizations of the stationary recurrent distributions to the case $\Delta_F = +\infty$.

10. The Generalized Palm-Khinchin Theorem

10.0. Introduction

If the phase space $[A, \varrho_A]$ is of the form R^s, $s \geqq 1$, then it is possible to associate, in a one-to-one way, a finite measure $(\tilde{P})^0$ with each infinitely divisible stationary distribution P on $\mathfrak{M}$ which satisfies the condition that $i_P < +\infty$. The continuity proposition 10.3.8. says that, for each sequence (P_n) of infinitely divisible stationary distributions having the properties $i_{P_n} < +\infty$ for $n = 1, 2, \ldots$; $i_{P_n} \xrightarrow[n\to\infty]{} l$, $l < +\infty$, the relations $P_n \underset{n\to\infty}{\Rightarrow} P$, $i_P = l$ are satisfied if and only if $(\widetilde{P_n})^0 \underset{n\to\infty}{\Rightarrow} (\tilde{P})^0$. Hence, if $(P_{n,j})_{\substack{n=1,2,\ldots \\ 1\leqq j\leqq m_n}}$ is an infinitesimal triangular array of stationary distributions on $\mathfrak{M}$ having the properties $i_{P_{n,j}} < +\infty$ for $n = 1, 2, \ldots$; $1 \leqq j \leqq m_n$; $\sum_{1\leqq j\leqq m_n} i_{P_{n,j}} \xrightarrow[n\to\infty]{} l$, $l < +\infty$, then $\underset{1\leqq j\leqq m_n}{\ast} P_{n,j} \underset{n\to\infty}{\Rightarrow} P$, $i_P = l$, or $\mathscr{U}_{\sum_{1\leqq j\leqq m_n} P_{n,j}} \Rightarrow P$, $i_P = l$, is equivalent to the validity of $\sum_{1\leqq j\leqq m_n} (P_{n,j})^0 \underset{n\to\infty}{\Rightarrow} (\tilde{P})^0$ (Theorem 10.3.9., cf. Kerstan & Matthes [4] and Port & Stone [2]). In Section 10.5. we shall apply this limit theorem to infinitesimal triangular arrays of stationary recurrent distributions $P_{F_{i,j}}$ (cf. Franken, Liemant & Matthes [1] and Kovalenko [1]).

Theorem 10.3.9. generalizes a classical theorem concerning the convergence of convolutions $\underset{1\leqq j\leqq m_n}{\ast} P_{n,j}$ toward stationary Poisson distributions. This theorem has been set up by Palm [1], exactly formulated and proved by Khinchin [1], and transformed into the version currently used by Ososkov [1] (Proposition 10.3.13.).

All of the statements made in Section 10.3. on stationary distributions are specializations of more general results for arbitrary phase spaces $[A, \varrho_A]$, where the corresponding Campbell measures have to be substituted for the Palm measures. The considerations presented in Section 10.1. are devoted to these general results (cf. Kallenberg [1], [6]).

10.1. Continuity Properties of the Mapping $W \rightsquigarrow \mathscr{C}_W$

If a sequence (W_n) of measures in $\boldsymbol{W}$ converges toward W with respect to the metric ϱ_W, then, using 3.3.5. for all X in $\mathfrak{B}_W$ and all natural numbers k, we obtain

$$\sum_{l=1}^{k} l W_n\big(\Psi(X) = l\big) \xrightarrow[n\to\infty]{} \sum_{l=1}^{k} l W\big(\Psi(X) = l\big).$$

Hence, by analogy with 3.1.12., we get

$$\begin{aligned}\varrho_W(X) &= \sup_{k=1,2,\dots} \sum_{l=1}^{k} l W(\Psi(X) = l) \\ &= \sup_{k=1,2,\dots} \lim_{n\to\infty} \sum_{l=1}^{k} l W_n(\Psi(X) = l) \\ &\leqq \varliminf_{n\to\infty} \sum_{l=1}^{\infty} l W_n(\Psi(X) = l) = \varliminf_{n\to\infty} \varrho_{W_n}(X).\end{aligned}$$

Now let us suppose in addition that $\varrho_{W_n}(X) < +\infty$ for $n = 1, 2, \dots$ as well as $\varrho_W(X) < +\infty$.

If $\lim_{n\to\infty} \varrho_{W_n}(X) = \varrho_W(X)$, we first choose, for any $\varepsilon > 0$, some k_ε having the property that

$$\sum_{l>k_\varepsilon} l W(\Psi(X) = l) < \varepsilon.$$

Since

$$\begin{aligned}\sum_{l>k_\varepsilon} l W_n(\Psi(X) = l) &= \varrho_{W_n}(X) - \sum_{l=1}^{k_\varepsilon} l W_n(\Psi(X) = l) \\ &\xrightarrow[n\to\infty]{} \varrho_W(X) - \sum_{l=1}^{k_\varepsilon} l W(\Psi(X) = l) = \sum_{l>k_\varepsilon} l W(\Psi(X) = l),\end{aligned}$$

there exists an n_ε, such that

$$\sum_{l>k_\varepsilon} l W_n(\Psi(X) = l) < \varepsilon$$

is satisfied for all $n \geqq n_\varepsilon$. By increasing k_ε, this inequality can be made valid also for $n = 1, \dots, n_\varepsilon - 1$.

Conversely, if $\sum_{l>k} l W_n(\Psi(X) = l)$ tends uniformly in n toward zero as k increases, then for all natural numbers k we obtain

$$\begin{aligned}&\varlimsup_{n\to\infty} |\varrho_{W_n}(X) - \varrho_W(X)| \\ &\leqq \varlimsup_{n\to\infty} \sum_{l>k} l W_n(\Psi(X) = l) + \sum_{l>k} l W(\Psi(X) = l) \\ &\quad + \varlimsup_{n\to\infty} \left| \sum_{l\leqq k} l W_n(\Psi(X) = l) - \sum_{l\leqq k} l W(\Psi(X) = l) \right| \\ &= \varlimsup_{n\to\infty} \sum_{l>k} l W_n(\Psi(X) = l) + \sum_{l>k} l W(\Psi(X) = l).\end{aligned}$$

For $k \to \infty$, the right hand side becomes arbitrarily small, i.e. we have

$$\varrho_{W_n}(X) \xrightarrow[n\to\infty]{} \varrho_W(X).$$

Bearing in mind that each X in $\mathfrak{B}$ is included in a set in $\mathfrak{B}_W$, we may use 3.2.2. to observe the truth of

10.1.1. *If a sequence (W_n) of measures in $\boldsymbol{W}$ converges toward a measure W in $\boldsymbol{W}$ with respect to the metric $\varrho_{\boldsymbol{W}}$, then, for all X in $\mathfrak{B}_W$, the inequality*

$$\varrho_W(X) \leqq \underline{\lim}_{n\to\infty} \varrho_{W_n}(X)$$

is satisfied.

If, moreover, all ϱ_{W_n} as well as ϱ_W are in N, then

$$\varrho_{W_n} \overset{N}{\underset{n\to\infty}{\Rightarrow}} \varrho_W$$

is true if and only if, for all X in $\mathfrak{B}$, the convergence relation

$$\sum_{l>k} l W_n\big(\Psi(X) = l\big) \xrightarrow[k\to\infty]{} 0$$

is satisfied uniformly in n.

We shall now return to a consideration outlined earlier in Section 5.0. Let $[A^?, \varrho_{A^?}]$ denote the direct product of the phase space $[A, \varrho_A]$ with $[M, \varrho_M]$. Then the correspondence $\Phi \rightsquigarrow \Phi \times \delta_\Phi$ provides a mapping of $[M, \varrho_M]$ into the space $[M^?, \varrho_{M^?}]$ of counting measures associated with $[A^?, \varrho_{A^?}]$.

10.1.2. *The mapping $\Phi \rightsquigarrow \Phi \times \delta_\Phi$ is continuous.*

Proof. Let Φ be a counting measure in M, and f a bounded non-negative real function on $A^?$ which is continuous with respect to $\varrho_{A^?}$ and identically zero outside of a bounded subset. There exists a bounded open subset X of $[A, \varrho_A]$ such that f is identically zero outside of $X \times M$ and $\Phi(\partial X) = 0$ is satisfied. For all sufficiently small $\varepsilon > 0$, the finitely many open balls $S_\varepsilon(x)$, $x \in X$, $\Phi(\{x\}) > 0$ are then pairwise disjoint and included in X.

Now suppose that $\Phi_n \overset{M}{\Rightarrow} \Phi$. Then, in virtue of 1.15.4., there exists an n_ε, such that the conditions

$$\Phi_n(X) = \Phi(X), \quad \Phi_n\big(S_\varepsilon(x)\big) = \Phi\big(S_\varepsilon(x)\big) = \Phi(\{x\})$$

are satisfied for all $x \in X$ having the property that $\Phi(\{x\}) > 0$, and for all $n \geqq n_\varepsilon$. Thus we obtain

$$\overline{\lim_{n\to\infty}} \left|\int f(x, \Psi)\,(\Phi \times \delta_\Phi)\,(\mathrm{d}[x, \Psi]) - \int f(x, \Psi)\,(\Phi_n \times \delta_{\Phi_n})\,(\mathrm{d}[x, \Psi])\right|$$

$$= \overline{\lim_{n\to\infty}} \left|\int f(x, \Phi)\,\Phi(\mathrm{d}x) - \int f(x, \Phi_n)\,\Phi_n(\mathrm{d}x)\right|$$

$$\leqq \sum_{x\in X, \Phi(\{x\})>0} \Phi(\{x\})\,\overline{\lim_{n\to\infty}} \sup_{y\in S_\varepsilon(x)} |f(x, \Phi) - f(y, \Phi_n)|\,.$$

In virtue of the continuity of f, all summands on the right hand side tend toward zero as $\varepsilon \to 0 + 0$, and we get

$$\int f(x, \Psi)\, (\Phi_n \times \delta_{\Phi_n})\, (\mathrm{d}[x, \Psi]) \xrightarrow[n\to\infty]{} \int f(x, \Psi)\, (\Phi \times \delta_\Phi)\, (\mathrm{d}[x, \Psi])$$

so that, using 1.15.3., we may conclude that

$$\Phi_n \times \delta_{\Phi_n} \underset{n\to\infty}{\overset{M^?}{\Rightarrow}} \Phi \times \delta_\Phi . \blacksquare$$

Obviously $C = \{[x, \Psi] : [x, \Psi] \in A^?,\ \Psi(\{x\}) > 0\}$ is a subset of $A^?$ closed with respect to $\varrho_{A^?}$. Denoting by ϱ_C the restriction of $\varrho_{A^?}$ to C, we obtain a complete separable metric space $[C, \varrho_C]$. The σ-algebra of the Borel subsets of this space coincides with $C \cap \mathfrak{A}^? = C \cap (\mathfrak{A} \otimes \mathfrak{M})$, i.e. with the σ-algebra $\mathfrak{C}$, introduced in Section 5.1., in the non-marked case. Now, if to each Φ in M we associate the restriction $\mathring{\Phi}$ of $\Phi \times \delta_\Phi$ to $\mathfrak{C}$, we obtain a mapping of $[M, \varrho_M]$ into the space $[\mathring{M}, \varrho_{\mathring{M}}]$ of counting measures, which is associated with the phase space $[C, \varrho_C]$.

10.1.3. *The mapping $\Phi \rightsquigarrow \mathring{\Phi}$ is continuous.*

Proof. Given any bounded non-negative real function h on C which is continuous on C and identically zero outside of a bounded subset Z of C, there exists an extension h' of h to a bounded non-negative real function on $A^?$ which is continuous with respect to $\varrho_{A^?}$. Let Z_1 denote the closure of Z, and Z_2 an arbitrary bounded open subset of $A^?$ which includes Z_1. There exists a real function z on $A^?$ continuous with respect to $\varrho_{A^?}$ and satisfying the condition $k_{Z_1} \leqq z \leqq 1$ which is identically zero outside of Z_2. Now $g = zh'$ provides an extension of h to a bounded non-negative real function on $A^?$ which is continuous with respect to $\varrho_{A^?}$ and identically zero outside of Z_2.

In view of 10.1.2. and 1.15.3., it follows from $\Phi_n \overset{M}{\Rightarrow} \Phi$ that

$$\int h(c)\, (\mathring{\Phi}_n)\, (dc) = \int g(x, \Psi)\, (\Phi_n \times \delta_{\Phi_n})\, (\mathrm{d}[x, \Psi])$$
$$\xrightarrow[n\to\infty]{} \int g(x, \Psi)\, (\Phi \times \delta_\Phi)\, (\mathrm{d}[x, \Psi]) = \int h(c)\, \mathring{\Phi}(\mathrm{d}c),$$

so that, using 1.15.3., we may conclude that

$$(\mathring{\Phi}_n) \underset{n\to\infty}{\overset{\mathring{M}}{\Rightarrow}} \mathring{\Phi} . \blacksquare$$

We denote by $\mathring{H}$ the image of a measure H on $\mathfrak{M}$ subjected to the mapping $\Phi \rightsquigarrow \mathring{\Phi}$. Obviously we have $\mathring{H}(\{o\}) = H(\mathring{\Phi} = o) = H(\{o\})$. Further, if Z is an arbitrary bounded set in $\mathfrak{C}$, we choose a set X in $\mathfrak{B}$ which satisfies the condition $Z \subseteqq X \times M$, thus obtaining

$$\mathring{H}(\chi(Z) > 0) \leqq H(\mathring{\Phi}(C \cap (X \times M)) > 0) = H(\Phi(X) > 0).$$

Hence, if H is in $\boldsymbol{W}$, then $\mathring{H}$ belongs to the set $\mathring{\boldsymbol{W}}$ corresponding to the phase space $[C, \varrho_C]$.

Because of the boundedness of M, for all measures H on $\mathfrak{M}$, obviously the relation $\varrho_H \in N$ is equivalent to $\varrho_{\mathring{H}} \in \mathring{N}$. Here $\mathring{N}$ denotes the set of those measures on $\mathfrak{C}$ which are finite on all bounded sets in $\mathfrak{C}$.

10.1.4. $W_n \underset{n\to\infty}{\overset{\boldsymbol{W}}{\Rightarrow}} W$ *implies* $(\mathring{W}_n) \underset{n\to\infty}{\overset{\mathring{\boldsymbol{W}}}{\Rightarrow}} \mathring{W}$.

Proof. Let Z be an arbitrary bounded set in $\mathfrak{C}$, and f a bounded non-negative real function on $\mathring{M}$ continuous with respect to $\varrho_{\mathring{M}}$ and identically zero outside of $\{\chi : \chi \in \mathring{M}, \chi(Z) > 0\}$. In view of 10.1.3., the bounded real function $\Phi \frown g(\Phi) = f(\mathring{\Phi})$ defined on M is continuous with respect to ϱ_M. If we now choose any set X in $\mathfrak{B}$ which satisfies the condition $Z \subseteqq X \times M$, this results in

$$\{\Phi : \Phi \in M, \mathring{\Phi}(Z) > 0\} \subseteqq \{\Phi : \Phi \in M, \Phi(X) > 0\}.$$

Thus g is identically zero outside of $\{\Phi : \Phi \in M, \Phi(X) > 0\}$. Using 3.3.3. we now obtain

$$\int f(\chi)\,(\mathring{W}_n)\,(\mathrm{d}\chi) = \int g(\Phi)\,W_n(\mathrm{d}\Phi) \underset{n\to\infty}{\longrightarrow} \int g(\Phi)\,W(\mathrm{d}\Phi) = \int f(\chi)\,\mathring{W}(\mathrm{d}\chi),$$

so that, once more applying 3.3.3., we may conclude that

$$(\mathring{W}_n) \underset{n\to\infty}{\overset{\mathring{\boldsymbol{W}}}{\Rightarrow}} \mathring{W}. \blacksquare$$

Now suppose that $W_n \underset{n\to\infty}{\overset{\boldsymbol{W}}{\Rightarrow}} W$, $\varrho_{W_n} \in N$ for $n = 1, 2, \ldots$, $\varrho_W \in N$ as well as $\varrho_{W_n} \underset{n\to\infty}{\overset{N}{\Rightarrow}} \varrho_W$. According to 10.1.4. we obtain $(\mathring{W}_n) \underset{n\to\infty}{\overset{\mathring{\boldsymbol{W}}}{\Rightarrow}} \mathring{W}$. Given any bounded set Z in $\mathfrak{C}$, we choose a set X in $\mathfrak{B}$, such that $Z \subseteqq X \times M$. In view of 10.1.1., $\sum_{l>k} l W_n(\Psi(X) = l)$ tends toward zero as k increases, the convergence being uniform in n. However, we always have

$$\sum_{l>k} l(\mathring{W}_n)\,(\chi(Z) = l) = \int_{\{\chi : \chi(Z) > k\}} \chi(Z)\,(\mathring{W}_n)\,(\mathrm{d}\chi)$$

$$\leqq \int_{\{\chi : \chi(C \cap (X \times M)) > k\}} \chi(C \cap (X \times M))\,(\mathring{W}_n)\,(\mathrm{d}\chi) = \int_{\{\Psi : \Psi(X) > k\}} \Psi(X)\,W_n(\mathrm{d}\Psi).$$

Hence also $\sum_{l>k} l(\mathring{W}_n)\,(\chi(Z) = l)$ tends toward zero as k increases, the convergence being uniform in n. Consequently, using 10.1.1. we obtain

$$\varrho_{(\mathring{W}_n)} \underset{n\to\infty}{\overset{\mathring{N}}{\Rightarrow}} \varrho_{\mathring{W}}.$$

Instead of $\overset{\mathring{N}}{\Rightarrow}$, we shall write $\overset{\circ}{\Rightarrow}$ as an abbreviation. Now, noting that, for all measures H on $\mathfrak{M}$, the intensity measure $\varrho_{\mathring{H}}$ coincides with the Campbell measure $\mathscr{C}_H$, we observe the truth of

10.1.5. Proposition. *For each sequence* (W_n) *of measures in* $\mathbf{W}$ *having the properties*

$$\varrho_{W_n} \in N, \quad \textit{for} \quad n = 1, 2, \ldots, \quad W_n \underset{n\to\infty}{\overset{\mathbf{W}}{\Rightarrow}} W, \quad \varrho_W \in N, \quad \varrho_{W_n} \underset{n\to\infty}{\overset{N}{\Rightarrow}} \varrho_W,$$

the convergence relation $\mathscr{C}_{W_n} \underset{n\to\infty}{\overset{\circ}{\Rightarrow}} \mathscr{C}_W$ *is satisfied.*

The following converse of the continuity proposition 10.1.5. can be shown:

10.1.6. Proposition. *Let* (W_n) *be a sequence of measures in* $\mathbf{W}$ *having the property* $\varrho_{W_n} \in N$ *for* $n = 1, 2, \ldots$. *If there exists a measure* V *in* $\mathring{N}$, *such that* $\mathscr{C}_{W_n} \underset{n\to\infty}{\overset{\circ}{\Rightarrow}} V$, *then* V *is the Campbell measure of a measure* W *in* $\mathbf{W}$ *satisfying the condition* $\varrho_W \in N$; *and*

$$W_n \underset{n\to\infty}{\overset{\mathbf{W}}{\Rightarrow}} W, \qquad \varrho_{W_n} \underset{n\to\infty}{\overset{N}{\Rightarrow}} \varrho_W.$$

Proof. 1. Given any set X in $\mathfrak{B}$, we choose a real function f on A which is continuous with respect to ϱ_A, satisfies the condition $k_X \leqq f \leqq 1$, and is identically zero outside of a set X' in $\mathfrak{B}$. Then, in view of 1.15.3., the real function

$$[x, \Psi] \curvearrowright g(x, \Psi) = f(x) \int f(y)\, \Psi(\mathrm{d}y)$$

which is defined on C, is continuous with respect to ϱ_C. Moreover, g is identically zero outside of the bounded subset $Z = C \cap (X' \times M)$ of C.

The closed sets $g^{-1}(\{v\})$, $v > 0$, are pairwise disjoint and included in the bounded set Z. Hence the inequality $V\big(g^{-1}(\{v\})\big) > 0$ holds only for at most countably many $v > 0$. For all other $v > 0$, and hence for all v in a set being dense in $(0, +\infty)$, we then have

$$V\Big(\partial g^{-1}\big((v, +\infty)\big)\Big) \leqq V\Big(g^{-1}\big(\partial(v, +\infty)\big)\Big) = V\big(g^{-1}(\{v\})\big) = 0.$$

According to 3.2.2., for all those $v > 0$ we may therefore conclude from $\mathscr{C}_{W_n} \underset{n\to\infty}{\overset{\circ}{\Rightarrow}} V$ that

$$\mathscr{C}_{W_n}\Big(g^{-1}\big((v, +\infty)\big)\Big) \xrightarrow[n\to\infty]{} V\Big(g^{-1}\big((v, +\infty)\big)\Big).$$

Given now any $\varepsilon > 0$, since $g^{-1}\big((v, +\infty)\big)$ is monotonically depending on v, and since $\bigcap\limits_{v>0} g^{-1}\big((v, +\infty)\big) = \emptyset$, there exists a $v_\varepsilon > 0$ having the properties

$$V\Big(g^{-1}\big((v_\varepsilon, +\infty)\big)\Big) < \varepsilon, \qquad V\big(g^{-1}(\{v_\varepsilon\})\big) = 0.$$

Then, for all $n \geqq n_\varepsilon$,

$$\mathscr{C}_{W_n}\Big(g^{-1}\big((v_\varepsilon, +\infty)\big)\Big) < \varepsilon.$$

By increasing v_ε, this inequality can be made valid also for $n = 1, \ldots, n_\varepsilon - 1$. We thus observe that the convergence relation

$$\mathscr{C}_{W_n}\Big(g^{-1}\big((v, +\infty)\big)\Big) \xrightarrow[v \to +\infty]{} 0$$

is satisfied uniformly in n.

For all natural numbers k, n, the inversion formula 5.1.8. yields

$$\sum_{l>k} l W_n\big(\Psi(X) = l\big) = \sum_{l>k} l\Big(l^{-1}\mathscr{C}_{W_n}\big(x \in X, \Psi(X) = l\big)\Big)$$

$$= \mathscr{C}_{W_n}\big(x \in X, \Psi(X) > k\big) \leqq \mathscr{C}_{W_n}\Big(g^{-1}\big((k, +\infty)\big)\Big).$$

Hence, for all X in $\mathfrak{B}$, the relation

$$\sum_{l>k} l W_n\big(\Psi(X) = l\big) \xrightarrow[k \to \infty]{} 0$$

is satisfied uniformly in n.

2. By

$$\nu(X) = V\big(C \cap (X \times M)\big), \qquad (X \in \mathfrak{A})$$

we obtain a measure ν in N.

Now let f be a bounded non-negative real function on A continuous with respect to ϱ_A and identically zero outside of a bounded subset X of A. Then with

$$g(x, \Psi) = f(x) \qquad ([x, \Psi] \in C)$$

we have a bounded real function g on C continuous with respect to ϱ_C and identically zero outside of the bounded set $\{[x, \Psi] : [x, \Psi] \in C, x \in X\}$. Now, in view of 5.1.3. and 3.2.1., we are able to conclude from $\mathscr{C}_{W_n} \overset{\circ}{\underset{n\to\infty}{\Rightarrow}} V$ that

$$\int f(x)\, \varrho_{W_n}(\mathrm{d}x) = \int g(c)\, \mathscr{C}_{W_n}(\mathrm{d}c) \xrightarrow[n \to \infty]{} \int g(c)\, V(\mathrm{d}c) = \int f(x)\, \nu(\mathrm{d}x).$$

Hence by 3.2.1. we have

$$\varrho_{W_n} \overset{N}{\underset{n\to\infty}{\Rightarrow}} \nu.$$

3. According to 2.2.1., for all natural numbers n, in $\boldsymbol{T}$ there exists a well-determined distribution P_n, such that $\widetilde{P_n} = W_n$. Since $\varrho_{P_n} = \varrho_{\tilde{P}_n} = \varrho_{W_n}$, we may use step 2 to conclude that

$$\varrho_{P_n} \overset{N}{\underset{n\to\infty}{\Rightarrow}} \nu, \qquad \nu \in N.$$

By 3.2.8. from this it follows that $\{P_n\}_{n=1,2,\ldots}$ is relatively compact with respect to $\varrho_{\boldsymbol{P}}$. Therefore in each subsequence of (P_n) we may select a subsequence (P_{n_m}) which is convergent with respect to $\varrho_{\boldsymbol{P}}$ to some P in $\boldsymbol{P}$.

By 3.2.11. a) we observe that P is in $\boldsymbol{T}$, so that we may use 3.3.6. to conclude that

$$\tilde{P}_{n_m} \underset{m\to\infty}{\overset{\boldsymbol{W}}{\Rightarrow}} \tilde{P}.$$

Every X in $\mathfrak{B}$ is included in an X_1 in $\mathfrak{B}_{\tilde{P}}$. By 10.1.1. we obtain

$$\varrho_{\tilde{P}}(X) \leqq \varrho_{\tilde{P}}(X_1) \leqq \varliminf_{m\to\infty} \varrho_{W_{n_m}}(X_1).$$

If we now choose an arbitrary continuity set X_2 with respect to ν which includes X_1, then by means of 2. and 3.2.2. we obtain

$$\varliminf_{m\to\infty} \varrho_{W_{n_m}}(X_1) \leqq \lim_{n\to\infty} \varrho_{W_n}(X_2) = \nu(X_2).$$

Hence

$$\varrho_{\tilde{P}}(X) \leqq \nu(X_2) < +\infty,$$

i.e. $\varrho_{\tilde{P}}$ belongs to N.

Taking into account 1., we may now use 10.1.1. to conclude that

$$\varrho_{W_{n_m}} \underset{m\to\infty}{\overset{N}{\Rightarrow}} \varrho_{\tilde{P}}.$$

In view of 10.1.5., we thus have

$$\mathscr{C}_{W_{n_m}} \underset{m\to\infty}{\overset{\circ}{\Rightarrow}} \mathscr{C}_{\tilde{P}},$$

and hence

$$\mathscr{C}_{\tilde{P}} = V.$$

Consequently, in view of the uniqueness proposition 5.1.10., $\tilde{P}$ does not depend on the choice of the subsequence, i.e. we have

$$W_n \underset{n\to\infty}{\overset{\boldsymbol{W}}{\Rightarrow}} \tilde{P}, \qquad \varrho_{W_n} \underset{n\to\infty}{\overset{N}{\Rightarrow}} \varrho_{\tilde{P}}. \blacksquare$$

The following two propositions result immediately from 3.3.6., 10.1.5., and 10.1.6.

10.1.7. Proposition. *The set* $\{\mathscr{C}_{\tilde{P}}\}_{P\in\boldsymbol{T},\,\varrho_P\in N}$ *is closed with respect to* $\varrho_{\mathring{N}}$.

10.1.8. Proposition. *If* $P, P_1, P_2, \ldots$ *are infinitely divisible distributions on* $\mathfrak{M}$, *the intensity measures of which belong to* N, *then*

$$P_n \underset{n\to\infty}{\Rightarrow} P, \qquad \varrho_{P_n} \underset{n\to\infty}{\overset{N}{\Rightarrow}} \varrho_P$$

is equivalent to

$$\mathscr{C}_{\tilde{P}_n} \underset{n\to\infty}{\overset{\circ}{\Rightarrow}} \mathscr{C}_{\tilde{P}}.$$

In virtue of 3.4.2., for each infinitesimal triangular array $(P_{n,j})_{\substack{n=1,2,\ldots\\1\leqq j\leqq m_n}}$ of distributions on $\mathfrak{M}$ and each infinitely divisible distribution P on $\mathfrak{M}$, the weak convergence of $\underset{1\leqq j\leqq m_n}{\ast} P_{n,j}$ toward P is equivalent to the weak convergence of $\mathscr{U}_{\sum\limits_{1\leqq j\leqq m_n} P_{n,j}}$ toward P, so that, using 10.1.8., we observe the truth of

10.1.9. Theorem. *If* $(P_{n,j})_{\substack{n=1,2,\ldots\\1\leqq j\leqq m_n}}$ *is an infinitesimal triangular array of distributions on* $\mathfrak{M}$, P *an infinitely divisible distribution on* $\mathfrak{M}$, *and if the following relations hold*

$$\varrho_{P_{n,j}} \in N \quad \textit{for} \quad n = 1, 2, \ldots,\ 1 \leqq j \leqq m_n, \quad \textit{and} \quad \varrho_P \in N,$$

then

$$\underset{1\leqq j\leqq m_n}{\ast} P_{n,j} \underset{n\to\infty}{\Rightarrow} P, \qquad \sum_{1\leqq j\leqq m_n} \varrho_{P_{n,j}} \underset{n\to\infty}{\overset{N}{\Rightarrow}} \varrho_P$$

is equivalent to

$$\sum_{1\leqq j\leqq m_n} \mathscr{C}_{P_{n,j}} \underset{n\to\infty}{\overset{\circ}{\Rightarrow}} \mathscr{C}_{\tilde{P}}.$$

With the use of 10.1.6. we obtain

10.1.10. Theorem. *Let* $(P_{n,j})_{\substack{n=1,2,\ldots\\1\leqq j\leqq m_n}}$ *be an infinitesimal triangular array of distributions on* $\mathfrak{M}$, *the intensity measures* $\varrho_{P_{n,j}}$ *of which belong to* N. *If there exists a measure* V *in* $\mathring{N}$, *such that* $\sum\limits_{1\leqq j\leqq m_n} \mathscr{C}_{\tilde{P}_{n,j}} \underset{n\to\infty}{\overset{\circ}{\Rightarrow}} V$ *is satisfied, then* V *is of the form* $\mathscr{C}_{\tilde{P}}$, *where* $P \in \mathbf{T}$, $\varrho_P \in N$; *and*

$$\underset{1\leqq j\leqq m_n}{\ast} P_{n,j} \underset{n\to\infty}{\Rightarrow} P, \qquad \sum_{1\leqq j\leqq m_n} \varrho_{P_{n,j}} \underset{n\to\infty}{\overset{N}{\Rightarrow}} \varrho_P.$$

Subject to the assumptions of 10.1.10., $(P_{n,j})$ is infinitesimal in a sharpened sense:

10.1.11. *If a triangular array* $(P_{n,j})_{\substack{n=1,2,\ldots\\1\leqq j\leqq m_n}}$ *satisfies the assumptions of Theorem* 10.1.10., *then, for all* X *in* $\mathfrak{B}$,

$$\max_{1\leqq j\leqq m_n} \varrho_{P_{n,j}}(X) \xrightarrow[n\to\infty]{} 0.$$

Proof. Obviously we are allowed to confine ourselves to considering sets X in $\mathfrak{B}_P$. Now suppose that $\lim\limits_{n\to\infty} \max\limits_{1\leqq j\leqq m_n} \varrho_{P_{n,j}}(X) = 0$ be not valid. By omitting rows and rearranging the terms in the remaining rows, it will then be possible to make all $\varrho_{P_{n,1}}(X)$ greater than a fixed $c > 0$. Now we have

$$(P_{n,2} * \cdots * P_{n,m_n}) \underset{n\to\infty}{\Rightarrow} P$$

and hence, in view of 10.1.1.,

$$\varliminf_{n\to\infty} \sum_{2\leq j\leq m_n} \varrho_{P_{n,j}}(X) \geq \varrho_P(X).$$

On the other hand, by assumption we have

$$\varrho_{P_{n,1}}(X) + \sum_{2\leq j\leq m_n} \varrho_{P_{n,j}}(X) \xrightarrow[n\to\infty]{} \varrho_P(X)$$

which leads to a contradiction set up by

$$\varrho_P(X) \leq \varliminf_{n\to\infty} \sum_{2\leq j\leq m_n} \varrho_{P_{n,j}}(X) \leq \varrho_P(X) - c. \blacksquare$$

We shall conclude this section by showing the following special limit theorem:

10.1.12. Theorem. *If an infinitesimal triangular array* $(P_{n,j})_{\substack{n=1,2,\ldots\\ 1\leq j\leq m_n}}$ *of distributions on* $\mathfrak{M}$ *as well as some* ν *in* N *satisfy the conditions*

$$\varrho_{P_{n,j}} \in N \quad \textit{for} \quad n = 1, 2, \ldots, \; 1 \leq j \leq m_n; \qquad \sum_{1\leq j\leq m_n} \varrho_{P_{n,j}} \underset{n\to\infty}{\Rightarrow} \nu,$$

then $\underset{1\leq j\leq m_n}{\ast} P_{n,j} \underset{n\to\infty}{\Rightarrow} P_\nu$ *is equivalent to the validity of the convergence relation*

$$\sum_{1\leq j\leq m_n} \mathscr{C}_{P_{n,j}}\big(x \in X, \Psi(X) > 1\big) \xrightarrow[n\to\infty]{} 0$$

for all X *in* $\mathfrak{B}$.

Proof. 1. Suppose that, for all X in $\mathfrak{B}$,

$$\sum_{1\leq j\leq m_n} \mathscr{C}_{P_{n,j}}\big(x \in X, \Psi(X) > 1\big) \xrightarrow[n\to\infty]{} 0.$$

Hence, in view of the inversion formula 5.1.8., it follows that

$$\sum_{1\leq j\leq m_n} \sum_{l=2}^{\infty} l P_{n,j}\big(\Phi(X) = l\big) = \sum_{1\leq j\leq m_n} \sum_{l=2}^{\infty} l\Big(l^{-1}\mathscr{C}_{P_{n,j}}\big(x \in X, \Psi(X) = l\big)\Big)$$

$$= \sum_{1\leq j\leq m_n} \mathscr{C}_{P_{n,j}}\big(x \in X, \Psi(X) > 1\big) \xrightarrow[n\to\infty]{} 0.$$

Thus in particular we have, for all X in $\mathfrak{B}$,

$$(*) \qquad \sum_{1\leq j\leq m_n} P_{n,j}\big(\Phi(X) > 1\big) \xrightarrow[n\to\infty]{} 0.$$

If we now suppose in addition that X be a continuity set with respect to ν, then using 3.2.2. we obtain

$$(**)\qquad \sum_{1\leqq j\leqq m_n} P_{n,j}\big(\Phi(X)=1\big)$$

$$= \sum_{1\leqq j\leqq m_n} \varrho_{P_{n,j}}(X) - \sum_{1\leqq j\leqq m_n} \mathscr{C}_{P_{n,j}}\big(x\in X,\, \Psi(X)>1\big) \xrightarrow[n\to\infty]{} \nu(X) - 0.$$

In view of 3.4.4., from (*), (**) it may be concluded that $\underset{1\leqq j\leqq m_n}{*} P_{n,j} \underset{n\to\infty}{\Rightarrow} P_\nu$, i.e. that the condition stated above is sufficient.

2. Conversely, suppose that

$$\underset{1\leqq j\leqq m_n}{*} P_{n,j} \underset{n\to\infty}{\Rightarrow} P_\nu .$$

In view of 10.1.9., we may then conclude that

$$\sum_{1\leqq j\leqq m_n} \mathscr{C}_{P_{n,j}} \overset{\circ}{\underset{n\to\infty}{\Rightarrow}} \mathscr{C}_{Q_\nu}.$$

Given now any bounded closed subset X of A, as was already stated in Section 1.15., the set $\{\Phi : \Phi \in M,\, \Phi(X) \leqq 1\}$ is open with respect to ϱ_M, hence $\{\Phi : \Phi \in M,\, \Phi(X) > 1\}$ is closed with respect to ϱ_M. Consequently, the bounded set $C \cap X \times \{\Phi : \Phi \in M,\, \Phi(X) > 1\}$ is closed with respect to ϱ_C. Thus, according to 3.2.2., from

$$\sum_{1\leqq j\leqq m_n} \mathscr{C}_{P_{n,j}} \overset{\circ}{\underset{n\to\infty}{\Rightarrow}} \mathscr{C}_{Q_\nu}$$

it may be concluded that

$$\overline{\lim_{n\to\infty}} \sum_{1\leqq j\leqq m_n} \mathscr{C}_{P_{n,j}}\big(x\in X,\, \Psi(X)>1\big) \leqq \mathscr{C}_{Q_\nu}\big(x\in X,\, \Psi(X)>1\big).$$

Consequently

$$\sum_{1\leqq j\leqq m_n} \mathscr{C}_{P_{n,j}}\big(x\in X,\, \Psi(X)>1\big) \xrightarrow[n\to\infty]{} 0$$

holds for all bounded closed subsets X of A, and hence also for all X in $\mathfrak{B}$. Thus we observe that the condition stated above is necessary, too. ∎

10.2. A Characterization of the Poisson Distributions

Let $p_1, \ldots, p_m$ be an arbitrary finite sequence of non-negative numbers having the property that $\sum_{i=1}^{m} p_i = 1$. If we set $K = \{1, \ldots, m\}$ and denote by $[A', \varrho_{A'}]$ the direct product of the phase spaces $[A, \varrho_A]$ and K, then

$$\varkappa_{(a)} = \sum_{i=1}^{m} p_i \delta_{\delta_{[a,i]}} \qquad (a \in A)$$

represents a cluster field on $[A, \varrho_A]$ with phase space $[A', \varrho_{A'}]$, by which marks k selected at random according to the distribution $\sum_{i=1}^{m} p_i \delta_i$ are independently associated with the points a in A. Hence we have a special independent marking in the sense of Section 4.4.

Obviously the cluster field $\varkappa$ corresponds to the stochastic kernel

$$K(a, (.)) = \sum_{i=1}^{m} p_i \delta_{[a,i]} = \delta_a \times \left(\sum_{i=1}^{m} p_i \delta_i\right) \qquad (a \in A).$$

Consequently, if P is a Poisson distribution on $\mathfrak{M}$, then by 4.4.4. the distribution $P_\varkappa$ is Poisson, too, and hence free from after-effects. Thus, in this case, the finite sequence ${}_{A\times\{1\}}\Psi, \ldots, {}_{A\times\{m\}}\Psi$, and hence also the finite sequence $\Psi((.) \times \{1\}), \ldots, \Psi((.) \times \{m\})$, are stochastically independent with respect to $P_\varkappa$. This "property of randomness" is characteristic of Poisson distributions in the following sense (cf. Moran [1], Rényi [2], and Srivastava [1]):

10.2.1. Theorem. *Let $p_1, \ldots, p_m$ be a finite sequence of numbers in $[0, 1)$ with sum unity, and $\varkappa$ the cluster field associated with the stochastic kernel*

$$K(a, (.)) = \sum_{i=1}^{m} p_i \delta_{[a,i]} \qquad (a \in A).$$

Subject to these assumptions, a distribution P on $\mathfrak{M}$ is Poisson if and only if the finite sequence $\Psi((.) \times \{1\}), \ldots, \Psi((.) \times \{m\})$ is stochastically independent with respect to $P_\varkappa$.

As it was done in Section 4.2., for all c in $[0, 1]$ and all measures H on $\mathfrak{M}$ we define $\mathcal{D}_c H = H_\gamma$ with $\gamma_{(a)} = (1 - c)\,\delta_o + c\delta_{\delta_a}$ for all a in A. Subject to the above assumptions, we obviously have

$$P_\varkappa\big(\Psi((.) \times \{i\}) \in (.)\big) = \mathcal{D}_{p_i} P \qquad (i \in K),$$

i.e. the distribution of $\Psi((.) \times \{i\})$ formed with respect to $P_\varkappa$ is derived from P by applying the thinning operator $\mathcal{D}_{p_i}$. If we observe that

$$\varkappa_{(\Phi)}\left(\left(\sum_{i=1}^{m} \Psi((.) \times \{i\})\right) \in (.)\right) = \delta_\Phi$$

is satisfied for all Φ in M, and hence

$$P_\varkappa\left(\left(\sum_{i=1}^{m} \Psi((.) \times \{i\})\right) \in (.)\right) = P$$

for all P in $\boldsymbol{P}$, we see that the independence of $\Psi((.) \times \{1\}), \ldots, \Psi((.) \times \{m\})$ with respect to $P_\varkappa$ implies the relation $P = \mathop{\Large *}_{i=1}^{m} \mathcal{D}_{p_i} P$. Now, it turns out that the following sharpened version of Theorem 10.2.1. can be shown (cf. Fichtner [2]):

10.2.2. Theorem. *Let $p_1, \ldots, p_m$ be a finite sequence of numbers in $[0, 1)$ with sum unity. Then a distribution P on $\mathfrak{M}$ is Poisson if and only if it satisfies the condition*

$$P = \mathop{\ast}_{i=1}^{m} (\mathcal{D}_{p_i} P).$$

The proof of Theorem 10.2.2. is based on the following limit theorem (cf. Fichtner [2], [3]):

10.2.3. Proposition. *Let P be a distribution on $\mathfrak{M}$, the intensity measure ν of which belongs to N. If to each natural number n a finite sequence $p_{n,1}, \ldots, p_{n,m_n}$ of non-negative real numbers is associated such that*

$$\sum_{1 \leq j \leq m_n} p_{n,j} = 1 \quad \textit{for} \quad n = 1, 2, \ldots, \qquad \max_{1 \leq j \leq m_n} p_{n,j} \xrightarrow[n \to \infty]{} 0$$

is satisfied, then

$$\mathop{\ast}_{1 \leq j \leq m_n} (\mathcal{D}_{p_{n,j}} P) \underset{n \to \infty}{\Rightarrow} P_\nu .$$

Proof. Obviously the triangular array $(\mathcal{D}_{p_{n,j}} P)_{\substack{n=1,2,\ldots \\ 1 \leq j \leq m_n}}$ is infinitesimal. Further, for $n = 1, 2, \ldots$, according to 4.2.2. we have

$$\sum_{1 \leq j \leq m_n} \varrho_{\mathcal{D}_{p_{n,j}} P} = \sum_{1 \leq j \leq m_n} p_{n,j} \varrho_P = \nu .$$

In view of 10.1.12., it only remains to show that the convergence relation

$$\sum_{1 \leq j \leq m_n} \int_{\{\Phi : \Phi(X) > 1\}} \Phi(X) \, (\mathcal{D}_{p_{n,j}} P) \, (\mathrm{d}\Phi)$$

$$= \sum_{1 \leq j \leq m_n} \mathcal{C}_{\mathcal{D}_{p_{n,j}} P} \big(x \in X, \Psi(X) > 1\big) \xrightarrow[n \to \infty]{} 0$$

is satisfied for all X in $\mathfrak{B}$.

So let X be an arbitrary set in $\mathfrak{B}$. First let us suppose in addition that, for a natural number k,

$$P\big(\Phi(X) \leq k\big) = 1 .$$

In view of the relation $(\mathcal{D}_c P)_X = \mathcal{D}_c(P_X)$ which is valid for all c in $[0, 1]$, we obtain for $n = 1, 2, \ldots$

$$\sum_{1 \leq j \leq m_n} \int_{\{\Phi : \Phi(X) > 1\}} \Phi(X) \, (\mathcal{D}_{p_{n,j}} P) \, (\mathrm{d}\Phi)$$

$$\leq \sum_{1 \leq j \leq m_n} \sum_{l=2}^{\infty} l (\mathcal{D}_{p_{n,j}} \delta_k) \, (\{l\})$$

$$= \sum_{1 \leq j \leq m_n} \sum_{l=2}^{\infty} l (\mathcal{D}_{p_{n,j}} \delta_1)^k \, (\{l\}) .$$

By 1.11.3. we have, for $n = 1, 2, \ldots,$

$$\left\| \mathop{*}_{1 \leqq j \leqq m_n} (\mathscr{D}_{p_{n,j}} \delta_1) - \pi_1 \right\| \leqq 2 \sum_{1 \leqq j \leqq m_n} (p_{n,j})^2 .$$

Hence, as $n \to \infty$, $\mathop{*}_{1 \leqq j \leqq m_n} (\mathscr{D}_{p_{n,j}} \delta_1)$ tends toward the Poisson distribution π_1 with expectation one. Consequently

$$\Big(\mathop{*}_{1 \leqq j \leqq m_n} (\mathscr{D}_{p_{n,j}} \delta_1) \Big)^k = \mathop{*}_{1 \leqq j \leqq m_n} (\mathscr{D}_{p_{n,j}} \delta_1)^k = \mathop{*}_{1 \leqq j \leqq m_n} (\mathscr{D}_{p_{n,j}} \delta_k) \xrightarrow[n \to \infty]{} \pi_k .$$

The triangular array $(\mathscr{D}_{p_{n,j}} \delta_k)_{\substack{n=1,2,\ldots \\ 1 \leqq j \leqq m_n}}$ of distributions of random non-negative integers is infinitesimal. Noting that $\sum_{1 \leqq j \leqq m_n} \sum_{l=1}^{\infty} l (\mathscr{D}_{p_{n,j}} \delta_k) (\{l\})$ is constantly equal to k for $n = 1, 2, \ldots,$ we can use 10.1.12. to conclude that

$$\sum_{1 \leqq j \leqq m_n} \sum_{l=2}^{\infty} l (\mathscr{D}_{p_{n,j}} \delta_k) (\{l\}) \xrightarrow[n \to \infty]{} 0 .$$

This implies that

$$\sum_{1 \leqq j \leqq m_n} \int_{\{\Phi : \Phi(X) > 1\}} \Phi(X) (\mathscr{D}_{p_{n,j}} P) (\mathrm{d}\Phi) \xrightarrow[n \to \infty]{} 0$$

is valid.

Now we shall free ourselves from the additional assumption that $P\big(\Phi(X) \leqq k\big) = 1$ for some k. For all sufficiently large u, $P\big(\Phi(X) \leqq u\big)$ is positive, so that we obtain

$$\begin{aligned}
&\overline{\lim_{n \to \infty}} \sum_{1 \leqq j \leqq m_n} \int_{\{\Phi : \Phi(X) > 1\}} \Phi(X) (\mathscr{D}_{p_{n,j}} P) (\mathrm{d}\Phi) \\
&\leqq \overline{\lim_{n \to \infty}} \sum_{1 \leqq j \leqq m_n} \int_{\{\Phi : \Phi(X) > 1\}} \Phi(X) \Big(\mathscr{D}_{p_{n,j}} P\big((.) \cap \{\chi : \chi(X) > u\} \big) \Big) (\mathrm{d}\Phi) \\
&\quad + \overline{\lim_{n \to \infty}} \sum_{1 \leqq j \leqq m_n} \int_{\{\Phi : \Phi(X) > 1\}} \Phi(X) \Big(\mathscr{D}_{p_{n,j}} P\big((.) \cap \{\chi : \chi(X) \leqq u\} \big) \Big) (\mathrm{d}\Phi) \\
&\leqq \overline{\lim_{n \to \infty}} \sum_{1 \leqq j \leqq m_n} \int \Phi(X) \Big(\mathscr{D}_{p_{n,j}} P\big((.) \cap \{\chi : \chi(X) > u\} \big) \Big) (\mathrm{d}\Phi) \\
&\quad + \overline{\lim_{n \to \infty}} \sum_{1 \leqq j \leqq m_n} \int_{\{\Phi : \Phi(X) > 1\}} \Phi(X) \Big(\mathscr{D}_{p_{n,j}} P\big((.) \mid \chi(X) \leqq u \big) \Big) (\mathrm{d}\Phi) \\
&\leqq \overline{\lim_{n \to \infty}} \sum_{1 \leqq j \leqq m_n} \varrho_{\mathscr{D}_{p_{n,j}} (P((.) \cap \{\chi : \chi(X) > u\}))} (X) + 0 \\
&= \overline{\lim_{n \to \infty}} \sum_{1 \leqq j \leqq m_n} p_{n,j} \varrho_{P((.) \cap \{\chi : \chi(X) > u\})} (X) \\
&= \varrho_{P((.) \cap \{\chi : \chi(X) > u\})} (X) \\
&= \int_{\{\Phi : \Phi(X) > u\}} \Phi(X) \, P(\mathrm{d}\Phi) \xrightarrow[u \to \infty]{} 0 .
\end{aligned}$$

Therefore we obtain

$$\sum_{1 \leq j \leq m_n} \int_{\{\Phi : \Phi(X) > 1\}} \Phi(X) \, (\mathscr{D}_{p_{n,j}} P) \, (\mathrm{d}\Phi) \xrightarrow[n\to\infty]{} 0. \blacksquare$$

Now let P be a distribution on $\mathfrak{M}$, and $p_1, \ldots, p_m$ a finite sequence of numbers in $[0, 1)$ which have the properties

$$\sum_{i=1}^{m} p_i = 1, \qquad P = \mathop{\ast}_{i=1}^{m} (\mathscr{D}_{p_i} P).$$

By successive insertion we obtain, for all natural numbers n,

$$P = \mathop{\ast}_{i_1,\ldots,i_n=1}^{m} \left(\mathscr{D}_{p_{i_1}} \cdots \mathscr{D}_{p_{i_n}} P\right) = \mathop{\ast}_{i_1,\ldots,i_n=1}^{m} \left(\mathscr{D}_{p_{i_1} \cdots p_{i_n}} P\right).$$

If the intensity measure ν of P is in N, we may use Proposition 10.2.3., to conclude that

$$\mathop{\ast}_{i_1,\ldots,i_n=1}^{m} \left(\mathscr{D}_{p_{i_1} \cdots p_{i_n}} P\right) \xrightarrow[n\to\infty]{} P_\nu,$$

and hence that $P = P_\nu$.

If the intensity measure of P is not in N, we choose some set X in $\mathfrak{B}$, such that $\varrho_P(X) = +\infty$. For all natural numbers l, we set

$$q_l = P_X\big((.) \cap \{0, 1, \ldots, l\}\big) + P_X(\{l+1, \ldots\}) \, \delta_0.$$

In view of Proposition 10.2.3., we have

$$\mathop{\ast}_{i_1,\ldots,i_n=1}^{m} \left(\mathscr{D}_{p_{i_1} \cdots p_{i_n}} q_l\right) \xrightarrow[n\to\infty]{} \pi_c, \qquad c = \int_{\{\Phi : \Phi(X) \leq l\}} \Phi(X) \, P(\mathrm{d}\Phi).$$

On the other hand, for all non-negative integers k,

$$q_l(\{k, k+1, \ldots\}) \leq P_X(\{k, k+1, \ldots\}),$$

and hence

$$\left(\mathop{\ast}_{i_1,\ldots,i_n=1}^{m} \left(\mathscr{D}_{p_{i_1} \cdots p_{i_n}} q_l\right)\right) (\{k, k+1, \ldots\})$$

$$\leq \left(\mathop{\ast}_{i_1,\ldots,i_n=1}^{m} \left(\mathscr{D}_{p_{i_1} \cdots p_{i_n}} P_X\right)\right) (\{k, k+1, \ldots\}) = P_X(\{k, k+1, \ldots\}).$$

From this it follows that

$$\pi_c(\{k, k+1, \ldots\}) = \lim_{n\to\infty} \left(\mathop{\ast}_{i_1,\ldots,i_n=1}^{m} \left(\mathscr{D}_{p_{i_1} \cdots p_{i_n}} q_l\right)\right) (\{k, k+1, \ldots\})$$

$$\leq P_X(\{k, k+1, \ldots\}).$$

But c can be made arbitrarily large by a suitable choice of l so that, for $k = 0, 1, \ldots$, we obtain

$$1 \leqq P_X(\{k, k+1, \ldots\}),$$

which contradicts the boundedness of X.

This completes the proof of Theorem 10.2.2.

Now let ν be a measure in N and P be a distribution on $\mathfrak{M}$ with the property g_ν. Let us suppose that

$$X_1 \cap X_2 = \emptyset, \qquad \nu(X_1)\,\nu(X_2) > 0, \qquad P_{X_1, X_2} = P_{X_1} \times P_{X_2}$$

holds for two sets X_1, X_2 in $\mathfrak{B}$. We set $X = X_1 \cup X_2$. Then by 1.14.2. we have

$$P_X = P_{X_1} * P_{X_2} = (\mathscr{D}_{\nu(X_1|X)} P_X) * (\mathscr{D}_{\nu(X_2|X)} P_X).$$

But in virtue of Theorem 10.2.2. this equality is only fulfilled if P_X is Poisson. We choose some $l > 0$ such that $P_X = (P_{l\nu})_X$ is satisfied. Then 1.14.7. even yields $P = P_{l\nu}$.

In particular we obtain

10.2.4. Proposition. *If a measure ν in N has not the form $\nu = c\delta_a$, $c \geqq 0$, $a \in A$, then a distribution P on $\mathfrak{M}$ has the form $P = P_{l\nu}$, $l \geqq 0$, if and only if it is free from after-effects and has the property g_ν.*

10.3. Continuity Properties of the Mapping $W \rightsquigarrow W^0$

Throughout this section, we shall assume that the phase space $[A, \varrho_A]$ be of the form R^s, $s \geqq 1$.

Taking up an idea outlined in Section 8.0., we now denote by ϱ_{M^0} the restriction of ϱ_M to the closed subset

$$M^0 = \{\Phi : \Phi \in M, \Phi(\{[0, \ldots, 0]\}) > 0\}.$$

Then the correspondence

$$[x, \Phi] \rightsquigarrow [x, T_x\Phi] \qquad ([x, \Phi] \in C)$$

provides a one-to-one mapping of $[C, \varrho_C]$ onto the direct product $[A^\wedge, \varrho_{A^\wedge}]$ of R^s with $[M^0, \varrho_{M^0}]$, the mapping being continuous in both directions in view of 6.1.1. Obviously a subset Z of C is bounded if and only if its image with respect to $[x, \Phi] \rightsquigarrow [x, T_x\Phi]$ has this property. If we denote by $N^\wedge$ the set of those measures on the σ-algebra $\mathfrak{A}^\wedge$ of the Borel subsets of $[A^\wedge, \varrho_{A^\wedge}]$ which are finite on bounded sets, where for brevity we write $\overset{\wedge}{\Rightarrow}$ instead of $\overset{N^\wedge}{\Rightarrow}$, then obviously we have

10.3.1. *Let $V, V_1, V_2, \ldots$ be measures in $\mathring{N}$. Then $V_n \underset{n\to\infty}{\overset{\circ}{\Rightarrow}} V$ is equivalent to $(V_n)^T \underset{n\to\infty}{\overset{\wedge}{\Rightarrow}} V^T$.*

Now we prove

10.3.2. *If $Q, Q_1, Q_2, \ldots$ are finite measures on $\mathfrak{M}$ with the property $Q(M \setminus M^0) = Q_1(M \setminus M^0) = \cdots = 0$, then $Q_n \underset{n\to\infty}{\Rightarrow} Q$ is equivalent to $\mu \times Q_n \underset{n\to\infty}{\overset{\wedge}{\Rightarrow}} \mu \times Q$.*

Proof. From $\mu \times Q_n \underset{n\to\infty}{\overset{\wedge}{\Rightarrow}} \mu \times Q$ it follows that $Q_n \underset{n\to\infty}{\Rightarrow} Q$.

Let f be an arbitrary non-negative continuous real function on R^s which is identically zero outside a bounded set, but different from the constant 0. For each bounded real function g on M which is continuous with respect to ϱ_M, we now have, in view of 3.2.1.,

$$\int f(x)\, g(\Phi)\, (\mu \times Q_n)\, (\mathrm{d}[x, \Phi]) \xrightarrow[n\to\infty]{} \int f(x)\, g(\Phi)\, (\mu \times Q)\, (\mathrm{d}[x, \Phi]),$$

which means that

$$\left(\int f(x)\, \mu(\mathrm{d}x)\right) \int g(\Phi)\, Q_n(\mathrm{d}\Phi) \xrightarrow[n\to\infty]{} \left(\int f(x)\, \mu(\mathrm{d}x)\right) \int g(\Phi)\, Q(d\Phi),$$

and hence

$$\int g(\Phi)\, Q_n(\mathrm{d}\Phi) \xrightarrow[n\to\infty]{} \int g(\Phi)\, Q(\mathrm{d}\Phi)$$

so that by 3.1.2. we are able to conclude that $Q_n \underset{n\to\infty}{\Rightarrow} Q$.

Conversely, suppose that, as $n \to \infty$, Q_n tends toward Q with respect to ϱ_{E^+}. Now, given any bounded non-negative real function h on C which is continuous with respect to ϱ_C and identically zero outside of a bounded subset of C, then there exists a bounded subset X of R^s, such that f is identically zero outside of $C \cap (X \times M)$. Thus the real function $\Phi \curvearrowright \int h(x, \Phi)\, \mu(\mathrm{d}x)$, which is defined on M, is bounded. Using Lebesgue's theorem, we observe that this function is continuous with respect to ϱ_M, too, and by means of 3.1.2. we conclude that

$$\int h(x, \Phi)\, (\mu \times Q_n)\, (\mathrm{d}[x, \Phi]) = \int \left(\int h(x, \Phi)\, \mu(\mathrm{d}x)\right) Q_n(\mathrm{d}\Phi)$$

$$\xrightarrow[n\to\infty]{} \int \left(\int h(x, \Phi)\, \mu(\mathrm{d}x)\right) Q(\mathrm{d}\Phi) = \int h(x, \Phi)\, (\mu \times Q)\, (\mathrm{d}[x, \Phi]).$$

In view of 3.2.1., this implies that

$$\mu \times Q_n \underset{n\to\infty}{\overset{\wedge}{\Rightarrow}} \mu \times Q. \blacksquare$$

As was stated in sections 6.4. and 8.1., for all stationary measures W in $\boldsymbol{W}$ we have

$$\varrho_W = \varrho_{\mathscr{E}_W} = i_{\mathscr{E}_W}\mu = i_W\mu, \qquad \mathscr{C}_W^T = \mu \times W^0.$$

Now, bearing in mind what was stated by 10.3.1. and 10.3.2., we observe the truth of

10.3.3. *If a sequence (W_n) of stationary measures in $\boldsymbol{W}$ as well as a stationary measure W in $\boldsymbol{W}$ satisy the conditions*

$$i_{W_n} < +\infty \quad \textit{for} \quad n = 1, 2, \ldots, \quad i_W < +\infty,$$

then $\mathscr{C}_{W_n} \overset{\circ}{\underset{n\to\infty}{\Rightarrow}} \mathscr{C}_W$ *is equivalent to* $(W_n)^0 \underset{n\to\infty}{\Rightarrow} W^0$.

By virtue of the first step in the proof of Theorem 9.2.6. all of the mappings $P \rightsquigarrow P(T_y\Phi \in (.))$, $y \in R^s$, are continuous with respect to $\varrho_{\boldsymbol{P}}$. From this it follows that the set $\bigcap_{y \in R^s} \{P : P \in \boldsymbol{P},\, P(T_y\Phi \in (.)) = P\}$, i.e. the set of all stationary distributions on $\mathfrak{M}$, is closed with respect to $\varrho_{\boldsymbol{P}}$. Noting that a measure W in $\boldsymbol{W}$ is stationary if and only if $\mathscr{E}_W$ has this property, the corresponding closure proposition is then obtained for the set of all stationary measures in $\boldsymbol{W}$. Thus we observe the truth of

10.3.4. *The sets of all stationary distributions on $\mathfrak{M}$ and of all stationary measures in $\boldsymbol{W}$ are closed with respect to $\varrho_{\boldsymbol{P}}$ and $\varrho_{\boldsymbol{W}}$, respectively.*

The sequence of the $W_n = n^{-1}P_{n\mu}$ shows that from $W_n \overset{\boldsymbol{W}}{\underset{n\to\infty}{\Rightarrow}} W$, $i_{W_n} < +\infty$ for $n = 1, 2, \ldots$, and $i_W < +\infty$, it cannot yet be concluded that $i_{W_n} \xrightarrow[n\to\infty]{} i_W$, and hence even less that $(W_n)^0 \underset{n\to\infty}{\Rightarrow} W^0$. However, using 10.1.5., 10.3.3., and 10.3.4., we get

10.3.5. Proposition. *For every sequence (W_n) of stationary measures in $\boldsymbol{W}$ having the properties*

$$i_{W_n} < +\infty \quad \textit{for} \quad n = 1, 2, \ldots, \qquad W_n \overset{\boldsymbol{W}}{\underset{n\to\infty}{\Rightarrow}} W, \qquad i_W < +\infty, \qquad \lim_{n\to\infty} i_{W_n} = i_W,$$

the convergence relation $(W_n)^0 \underset{n\to\infty}{\Rightarrow} W^0$ *is satisfied.*

Using 10.3.2., 10.3.1., 10.1.6., and 10.3.4., the continuity proposition 10.3.5. can be inverted as follows:

10.3.6. Proposition. *Let (W_n) be a sequence of stationary measures in $\boldsymbol{W}$ having finite intensities i_{W_n}. If the sequence of the $(W_n)^0$ converges weakly in $\boldsymbol{E}^+$ toward a measure Q, then Q is the Palm measure of a stationary measure W in $\boldsymbol{W}$ with finite intensity i_W, and*

$$W_n \overset{\boldsymbol{W}}{\underset{n\to\infty}{\Rightarrow}} W, \qquad i_{W_n} \xrightarrow[n\to\infty]{} i_W.$$

The following two propositions are immediately derived from 3.3.6., 10.3.5., and 10.3.6.:

10.3.7. Proposition. *The set* $\{(\tilde{P})^0\}_{P\in \boldsymbol{T},\, P \text{ stationary},\, i_P<+\infty}$ *is closed with respect to* $\varrho_{\boldsymbol{E}^+}$.

10.3.8. Proposition. *If $P, P_1, P_2, \ldots$ are stationary infinitely divisible distributions on $\mathfrak{M}$ having finite intensities $i_P, i_{P_1}, i_{P_2}, \ldots$, then*

$$P_n \underset{n\to\infty}{\Rightarrow} P, \qquad i_{P_n} \xrightarrow[n\to\infty]{} i_P$$

is equivalent to

$$(\widetilde{P_n})^0 \underset{n\to\infty}{\Rightarrow} (\tilde{P})^0.$$

As a consequence of 10.1.9. and 10.3.3., we have

10.3.9. Theorem. *For each infinitesimal triangular array $(P_{n,j})_{\substack{n=1,2,\ldots\\ 1\leq j\leq m_n}}$ of stationary distributions $P_{n,j}$ on $\mathfrak{M}$ having finite intensities $i_{P_{n,j}}$, and for each stationary P in $\boldsymbol{T}$ with finite intensity i_P,*

$$\mathop{\ast}_{1\leq j\leq m_n} P_{n,j} \underset{n\to\infty}{\Rightarrow} P, \qquad \sum_{1\leq j\leq m_n} i_{P_{n,j}} \xrightarrow[n\to\infty]{} i_P$$

is equivalent to

$$\sum_{1\leq j\leq m_n} (P_{n,j})^0 \underset{n\to\infty}{\Rightarrow} (\tilde{P})^0.$$

In the case $i_{P_{n,j}} > 0$ for $n = 1, 2, \ldots, 1 \leq j \leq m_n$; and $i_P > 0$, we may also write

$$\sum_{1\leq j\leq m_n} i_{P_{n,j}}(P_{n,j})_0 \underset{n\to\infty}{\Rightarrow} i_P(\tilde{P})_0$$

instead of $\sum_{1\leq j\leq m_n} (P_{n,j})^0 \underset{n\to\infty}{\Rightarrow} (\tilde{P})^0$. Since $\sum_{1\leq j\leq m_n} i_{P_{n,j}} \xrightarrow[n\to\infty]{} i_P$, this convergence statement is equivalent to

$$\Big(\sum_{1\leq j\leq m_n} i_{P_{n,j}}\Big)^{-1} \sum_{1\leq j\leq m_n} i_{P_{n,j}}(P_{n,j})_0 \underset{n\to\infty}{\Rightarrow} (\tilde{P})_0.$$

Using 10.3.7. we obtain

10.3.10. Theorem. *Let $(P_{n,j})_{\substack{n=1,2,\ldots\\ 1\leq j\leq m_n}}$ be an infinitesimal triangular array of stationary distributions on $\mathfrak{M}$ having finite intensities $i_{P_{n,j}}$. If, as n increases, $\sum_{1\leq j\leq m_n} (P_{n,j})^0$ converges weakly toward a finite measure Q, then there exists a stationary infinitely divisible distribution P on $\mathfrak{M}$ with finite intensity i_P for which $(\tilde{P})^0 = Q$, and*

$$\mathop{\ast}_{1\leq j\leq m_n} P_{n,j} \underset{n\to\infty}{\Rightarrow} P, \qquad \sum_{1\leq j\leq m_n} i_{P_{n,j}} \xrightarrow[n\to\infty]{} i_P.$$

As an immediate consequence of 10.3.9., 10.3.10., and 3.4.1. (cf. Ambartzumian [2]), we have the following convergence proposition for special infinitesimal triangular arrays:

10.3.11. *Let (P_n) be a sequence of stationary distributions on $\mathfrak{M}$ having the properties*

$$0 < i_{P_n} < +\infty \quad \text{for} \quad n = 1, 2, \ldots, \quad \lim_{n\to\infty} i_{P_n} = 0.$$

If a sequence (m_n) of natural numbers satisfies the condition

$$m_n i_{P_n} \xrightarrow[n\to\infty]{} i, \qquad 0 < i < +\infty,$$

then the sequence $(P_n)^{m_n}$ converges weakly toward a stationary distribution P on $\mathfrak{M}$ with intensity i if and only if the sequence $(P_n)_0$ converges weakly toward a distribution Q. The distribution P is infinitely divisible, and $(\tilde{P})_0 = Q$.

Hence, in particular we have, for all stationary P in $\boldsymbol{T}$ having the property that $0 < i_P < +\infty$,

$$\left(\sqrt[n]{P}\right)_0 \underset{n\to\infty}{\Rightarrow} (\tilde{P})_0.$$

As an immediate consequence of 10.1.11., we have

10.3.12. *If a triangular array $(P_{n,j})_{\substack{n=1,2,\ldots\\ 1\leqq j\leqq m_n}}$ satisfies the assumptions of Theorem* 10.3.10., *then*

$$\max_{1\leqq j\leqq m_n} i_{P_{n,j}} \xrightarrow[n\to\infty]{} 0.$$

If a triangular array $(P_{n,j})_{\substack{n=1,2,\ldots\\ 1\leqq j\leqq m_n}}$ satisfies the assumptions of 10.3.10., then the triangular array produced by adding $P_{n,m_n+1} = \mathscr{U}_{P_{n\mu}/n}$ is infinitesimal, too, and

$$\mathop{\ast}_{1\leqq j\leqq m_n+1} P_{n,j} \underset{n\to\infty}{\Rightarrow} P, \qquad \sum_{1\leqq j\leqq m_n+1} i_{P_{n,j}} \xrightarrow[n\to\infty]{} i_P + 1.$$

By analogy with 10.1.12., we obtain

10.3.13. Theorem. *If an infinitesimal triangular array $(P_{n,j})_{\substack{n=1,2,\ldots\\ 1\leqq j\leqq m_n}}$ of stationary distributions on $\mathfrak{M}$ having finite intensities $i_{P_{n,j}}$ satisfies the condition*

$$\sum_{1\leqq j\leqq m_n} i_{P_{n,j}} \xrightarrow[n\to\infty]{} l,$$

for some $l \geqq 0$, then $\mathop{\ast}\limits_{1\leqq j\leqq m_n} P_{n,j} \underset{n\to\infty}{\Rightarrow} P_{l\mu}$ is equivalent to the validity of the convergence relation

$$\sum_{1\leqq j\leqq m_n} (P_{n,j})^0 \left(\Phi\left(S_k([0,\ldots,0])\right) > 1\right) \xrightarrow[n\to\infty]{} 0$$

for all natural numbers k.

Proof. For all $l \geqq 0$ we have, by 8.3.9.,

$$(\widetilde{P_{l\mu}})^0 = l\delta_{\delta_{[0,\ldots,0]}}.$$

Hence, according to 10.3.9., we have to prove that, subject to the above assumptions, the relation

$$(*) \qquad \sum_{1\leqq j\leqq m_n} (P_{n,j})^0 \underset{n\to\infty}{\Rightarrow} l\delta_{\delta_{[0,\ldots,0]}}$$

holds if and only if, for all k

$$(**) \qquad \sum_{1 \leq j \leq m_n} (P_{n,j})^0 \left(\Phi\left(S_k([0, \ldots, 0])\right) > 1\right) \xrightarrow[n\to\infty]{} 0$$

is satisfied.

By definition of the Palm measure, every measure $(P_{n,j})^0$ can be given the form

$$(P_{n,j})^0 = \delta_{\delta_{[0,\ldots,0]}} * Q_{n,j}, \qquad Q_{n,j} \in \boldsymbol{E}^+,$$

so that we have

$$Q_{n,j}(M) = (P_{n,j})^0\,(M) = i_{P_{n,j}}.$$

Relation $(*)$ is equivalent to

$$(\mathrm{o}) \qquad \sum_{1 \leq j \leq m_n} Q_{n,j} \underset{n\to\infty}{\Rightarrow} l\delta_o,$$

and relation $(**)$ is equivalent to the fact that

$$(\mathrm{oo}) \qquad \sum_{1 \leq j \leq m_n} Q_{n,j}\left(\Phi\left(S_k([0, \ldots, 0])\right) > 0\right) \xrightarrow[n\to\infty]{} 0$$

is satisfied for all k. But by 3.1.9., (o) is equivalent to (oo). ∎

In the case $s = 1$, this convergence proposition can be given the following form:

10.3.14. Theorem. *If an infinitesimal triangular array* $(P_{n,j})_{\substack{n=1,2,\ldots \\ 1 \leq j \leq m_n}}$ *of stationary simple distributions on* $\mathfrak{M}$ *having positive finite intensities* $i_{P_{n,j}}$ *satisfies the condition*

$$\sum_{1 \leq j \leq m_n} i_{P_{n,j}} \xrightarrow[n\to\infty]{} l$$

for some $l \geqq 0$*, then, in the case* $s = 1$*,*

$$\mathop{\Large *}_{1 \leq j \leq m_n} P_{n,j} \underset{n\to\infty}{\Rightarrow} P_{l\mu}$$

is equivalent to the fact that, for all $t > 0$*, the convergence relation*

$$\sum_{1 \leq j \leq m_n} i_{P_{n,j}} F_{P_{n,j}}(t) \xrightarrow[n\to\infty]{} 0$$

is satisfied.

Proof. Let $\mathop{\Large *}_{1 \leq j \leq m_n} P_{n,j} \xrightarrow[n\to\infty]{} P_{l\mu}$. By 10.3.13., for each $t > 0$ we are able to conclude that

$$\sum_{1 \leq j \leq m_n} i_{P_{n,j}} (P_{n,j})_0 \left(\Phi\left((0, t)\right) > 0\right) \xrightarrow[n\to\infty]{} 0$$

and hence that

$$\sum_{1 \leq j \leq m_n} i_{P_{n,j}} F_{P_{n,j}}(t) \xrightarrow[n\to\infty]{} 0.$$

Conversely, suppose that this convergence relation holds for all $t > 0$. For all k we obtain, using 9.1.14.,

$$\sum_{1 \leq j \leq m_n} (P_{n,j})^0 \left(\Phi\big((-k, k)\big) > 1\right)$$

$$\leq \sum_{1 \leq j \leq m_n} \left((P_{n,j})^0 \left(\Phi\big((-k, 0)\big) > 0\right) + (P_{n,j})^0 \left(\Phi\big((0, k)\big) > 0\right)\right)$$

$$= 2 \sum_{1 \leq j \leq m_n} (P_{n,j})^0 \left(\Phi\big((0, k)\big) > 0\right) \leq 2 \sum_{1 \leq j \leq m_n} i_{P_{n,j}} F_{P_{n,j}}(k),$$

so that by 10.3.13. we are able to conclude that

$$\mathop{\ast}_{1 \leq j \leq m_n} P_{n,j} \underset{n \to \infty}{\Rightarrow} P_{l\mu}. \blacksquare$$

Sometimes the following lemma is useful.

10.3.15. *Let U be any bounded open neighbourhood of the origin. Then for any sequence (P_n) of stationary infinitely divisible distributions on $\mathfrak{M}$ of finite intensity, the set $\left\{\left(\widetilde{P_n}\right)^0\right\}_{n=1,2,\ldots}$ is relatively compact with respect to ϱ_{E^+} if and only if the set $\left\{\left(\left(\widetilde{P_n}\right)^0\right)_U\right\}_{n=1,2,\ldots}$ of one-dimensional distributions is relatively compact.*

Proof. By 3.2.7. the set $\left\{\left(\widetilde{P_n}\right)^0\right\}_{n=1,2,\ldots}$ is relatively compact if and only if $\left\{\left(\left(\widetilde{P_n}\right)^0\right)_W\right\}_{n=1,2,\ldots}$ is relatively compact for all bounded neighbourhoods W of $[0, \ldots, 0]$.

We select a symmetric open neighbourhood G of the origin satisfying $G + G \subseteq U$, and get for $k, n = 1, 2, \ldots$

$$\mu(U) \left(\widetilde{P_n}\right)^0 \left(\Phi(U) \geq k\right) = \mathscr{C}^T_{\widetilde{P_n}}\left(x \in U, \Phi(U) \geq k\right)$$

$$\geq \mathscr{C}_{\widetilde{P_n}}\left(x \in G, \Phi(G) \geq k\right) = \int_{\{\Phi : \Phi(G) \geq k\}} \Phi(G)\, \widetilde{P_n}(\mathrm{d}\Phi).$$

Hence, if $\left\{\left(\left(\widetilde{P_n}\right)^0\right)_U\right\}_{n=1,2,\ldots}$ is relatively compact, it follows that

$$\int_{\{\Phi : \Phi(G) \geq k\}} \Phi(G)\, \widetilde{P_n}(\mathrm{d}\Phi) \xrightarrow[k \to \infty]{} 0$$

is satisfied uniformly in n. Taking into consideration the stationarity of the canonical measures $\widetilde{P_n}$, we then obtain the same assertion for arbitrary bounded sets Z in $\mathfrak{R}^s$. Thus, for any bounded open neighbourhood W of $[0, \ldots, 0]$, we have for all sufficiently large Z

$$\mu(W) \left(\widetilde{P_n}\right)^0 \left(\Phi(W) \geq k\right) = \mathscr{C}^T_{\widetilde{P_n}}\left(x \in W, \Phi(W) \geq k\right)$$

$$\leq \mathscr{C}_{\widetilde{P_n}}\left(x \in Z, \Phi(Z) \geq k\right) = \int_{\{\Phi : \Phi(Z) \geq k\}} \Phi(Z)\, \widetilde{P_n}(\mathrm{d}\Phi).$$

Consequently,

$$\inf_{k=1,2,\ldots} \sup_{n=1,2,\ldots} \left(\widetilde{P_n}\right)^0 \left(\Phi(W) \geq k\right) = 0.$$

On the other hand

$$\sup_{n=1,2,\ldots} \left\|\left((\widetilde{P}_n)^0\right)_W\right\| = \sup_{n=1,2,\ldots} i_{P_n} = \sup_{n=1,2,\ldots} \left\|\left((\widetilde{P}_n)^0\right)_U\right\|.$$

Hence $\left\{\left((\widetilde{P}_n)^0\right)_W\right\}_{n=1,2,\ldots}$ is relatively compact. ∎

10.4. The Distributions H_F. I

Throughout this section, we shall assume that the phase space $[A, \varrho_A]$ coincides with the real axis R^1.

Referring to the statements made in Section 9.5., an element Φ of M satisfying the condition $\Phi\big((-\infty, 0)\big) = +\infty = \Phi\big((0, +\infty)\big)$ can still be considered as a description of a realization of a renewal process, even if Φ is not simple. If $\Phi(\{x\}) > 1$, then we consider this an expression of the fact that at the time x first of all $\Phi(\{x\}) - 1$ "unsuccessful" renewals, i.e. renewals with zero lifetimes occurred before a "successful" renewal with positive lifetime occurred.

Now let F be a distribution function in $\boldsymbol{F}$ such that $F(0) < 1$, $\lim_{x\to+\infty} F(x) = 1$. The construction of the distribution H_F of the recurrent point process associated with F, which was carried out in Section 9.5. with the additional assumption that $F(0) = 0$, can be generalized in the following way. Let q_F be the distribution of a doubly infinite sequence of independent random non-negative numbers identically distributed according to F. If we associate the measure

$$\Phi = \sum_{n=1}^{\infty} \delta_{-(a_{-1}+\cdots+a_{-n})} + \delta_0 + \sum_{n=0}^{\infty} \delta_{a_0+\cdots+a_n}$$

to each doubly infinite sequence $(a_n)_{-\infty<n<+\infty}$ of non-negative real numbers, then Φ is in M if and only if

$$\sum_{n=1}^{\infty} a_{-n} = +\infty = \sum_{n=1}^{\infty} a_n$$

is satisfied, which is the case almost everywhere with respect to q_F. The afore-mentioned mapping $(a_n) \rightsquigarrow \Phi$ is measurable and transforms q_F into a distribution H_F on $\mathfrak{M}^0$. Obviously

$$H_F(Y_2) = H_F\Big(\Phi\big((-\infty, 0)\big) = +\infty = \Phi\big((0, +\infty)\big)\Big) = 1$$

holds always true. Furthermore, H_F is simple if and only if $F(0) = 0$. In this case our definition coincides with that given in Section 9.5.

Now let F be a distribution function in $\boldsymbol{F}$ such that $\lim_{x\to+\infty} F(x) < 1$. Similarly, let q_F denote the distribution of a doubly infinite sequence of independent random elements of $[0, +\infty]$, identically distributed according to F. Then there exist, for almost all realizations (a_n), with respect to q_F both a lowest non-negative index $k_1 + 1$ having the property that

$a_{k_1+1} = +\infty$ and a highest negative index $-(k_2+1)$ having the property that $a_{-(k_2+1)} = +\infty$. The values a_{k_1+1} and $a_{-(k_2+1)}$ can be interpreted as infinite lifetimes. Setting now

$$\Phi = \sum_{k=1}^{k_2} \delta_{-(a_{-1}+\cdots+a_{-k})} + \delta_0 + \sum_{k=0}^{k_1} \delta_{a_0+\cdots+a_k}$$

we obtain a measurable mapping into M^0 defined almost everywhere with respect to q_F. Here q_F is transformed into a distribution H_F on $\mathfrak{M}^0$. Obviously we have

$$H_F(Y_4) = H_F(\Phi \text{ finite}) = 1.$$

In the case $F(x) = 0$ for all $x \geqq 0$, the distribution H_F coincides with δ_{δ_0}

For all F in $\boldsymbol{F}$ having the property that $F(0) < 1$ we set

$$F^*(x) = \big(1 - F(0)\big)^{-1} \big(F(x) - F(0)\big),$$

thus obtaining a distribution function F^* in $\boldsymbol{F}$ having the property that $F^*(0) = 0$. Obviously, we have

10.4.1. *For all F in $\boldsymbol{F} \setminus \{k_{[0,+\infty)}\}$,*

$$(H_F)^* = H_{F^*}.$$

In particular, H_F is simple if and only if $F = F^*$.

The distributions H_F can be represented with the help of H_{F^*} in the following way.

10.4.2. *Let F be a distribution function in $\boldsymbol{F} \setminus \{k_{[0,+\infty)}\}$ and $\varkappa$ be the cluster field on R^1 defined by*

$$\varkappa_{(x)} = \sum_{n=1}^{\infty} \big(1 - F(0)\big) \big(F(0)\big)^{n-1} \delta_{n\delta_x} \qquad (x \in R^1).$$

Then we have

$$H_F = \sum_{n=1}^{\infty} n\big(1 - F(0)\big)^2 \big(F(0)\big)^{n-1} (H_{F^*})_\varkappa \big(\Phi \in (.) \mid \Phi(\{0\}) = n\big).$$

Proof. The continuous cluster field $\varkappa$ is of type β), hence $(H_{F^*})_\varkappa$ exists. Then, for all natural numbers n, we can compute

$$\begin{aligned} &H_F\big(\Phi(\{0\}) = n\big) \\ &= \sum_{k_1,k_2 \geqq 0,\, k_1+k_2=n-1} \big(1 - F(0)\big) \big(F(0)\big)^{k_1} \big(1 - F(0)\big) \big(F(0)\big)^{k_2} \\ &= n\big(1 - F(0)\big)^2 \big(F(0)\big)^{n-1} \end{aligned}$$

and

$$H_F\big(\Phi \in (.) \mid \Phi(\{0\}) = n\big) = (H_{F^*})_\varkappa \big(\Phi \in (.) \mid \Phi(\{0\}) = n\big).$$

(if these expressions are meaningful). ∎

10.4.3. *The correspondence*

$$F \curvearrowright H_F$$

provides a one-to-one mapping of $\boldsymbol{F} \setminus \{k_{[0,+\infty)}\}$ *into the set of all distributions on* $\mathfrak{M}^0$.

Proof. First the relation $H_F(\Phi(\{0\}) = 1) = (1 - F(0))^2$ uniquely determines $F(0)$ by H_F. For each $t > 0$ we further have $H_F(\Phi((0, t]) = 0) = 1 - F^*(t)$. Thus $F(t) = F^*(t)(1 - F(0)) + F(0)$ is determined by H_F for all t. ∎

To each distribution function F in $\boldsymbol{F}$ there corresponds, in a one-to-one way, a distribution p_F on the σ-algebra of the Borel sets of the compact space $[0, +\infty]$. Therefore, we can introduce a metric $\varrho_{\boldsymbol{F}}$ in $\boldsymbol{F}$ such that the convergence with respect to $\varrho_{\boldsymbol{F}}$ coincides with the weak convergence of the corresponding distributions. Consequently, $\lim_{n\to\infty} \varrho_{\boldsymbol{F}}(F_n, F) = 0$ if and only if the convergence relation $F_n(t) \xrightarrow[n\to\infty]{} F(t)$ is valid for all non-negative t satisfying the condition $F(t - 0) = F(t)$. For this we may also write $F_n \underset{n\to\infty}{\Rightarrow} F$. In virtue of the compactness of $[0, +\infty]$ the space $[\boldsymbol{F}, \varrho_{\boldsymbol{F}}]$ is compact, too.

10.4.4. Proposition. *The one-to-one mapping* $F \curvearrowright H_F$ *of* $\boldsymbol{F} \setminus \{k_{[0,+\infty)}\}$ *into* $\boldsymbol{P}$ *is continuous in both directions with respect to* $\varrho_{\boldsymbol{F}}$ *and* $\varrho_{\boldsymbol{P}}$. *Moreover, the set* $\{H_F\}_{F \in \boldsymbol{F} \setminus \{k_{[0,+\infty)}\}}$ *is closed with respect to* $\varrho_{\boldsymbol{P}}$.

Proof. 1. From $F_n \underset{n\to\infty}{\Rightarrow} F$, $\{F, F_1, F_2, \ldots\} \subseteqq \boldsymbol{F} \setminus \{k_{[0,+\infty)}\}$ it follows that $H_{F_n} \underset{n\to\infty}{\Rightarrow} H_F$.

Let $t_{-s} < \cdots < t_{-1} < 0 < t_1 < \cdots < t_s$ as well as $H_F(\Phi(\{t_j\}) > 0) = 0$ for $j = \pm 1, \ldots, \pm s$. Further let l_j, $j = 0, \pm 1, \ldots, \pm(s-1)$ be arbitrary non-negative integers. We set $l = \sum_{-(s-1) \leqq j \leqq s-1} l_j$. Denoting the distributions of an independent family $(\xi_i)_{i=\pm 1, \ldots, \pm(l+1)}$ of random elements of $[0, +\infty]$ with distribution functions F_n and F, by p_n and p, respectively, we obtain, for $n = 1, 2, \ldots$,

$$H_{F_n}\big(\Phi([t_{-j}, t_{-j+1})) = l_{-j+1} \ \text{ for } \ j = 2, \ldots, s; \ \ \Phi([t_{-1}, t_1)) = l_0 + 1;$$
$$\Phi([t_{j-1}, t_j)) = l_{j-1} \ \text{ for } \ j = 2, \ldots, s\big)$$

$$= \sum_{0 \leqq k \leqq l_0} H_{F_n}\big(\Phi([t_{-j}, t_{-j+1})) = l_{-j+1} \ \text{ for } \ j = 2, \ldots, s;$$
$$\Phi([t_{-1}, 0)) = k, \ \ \Phi([0, t_1)) = l_0 + 1 - k;$$
$$\Phi([t_{j-1}, t_j)) = l_{j-1} \ \text{ for } \ j = 2, \ldots, s\big)$$

$$= \sum_{0 \leqq k \leqq l_0} p_n(Z_k)$$

where

$$Z_k = \Big\{(x_i)_{i=\pm 1,\ldots,\pm(l+1)} : x_i \in [0, +\infty],\quad i = \pm 1, \ldots, \pm(l+1);$$

$$-\sum_{1 \le i \le l_{-j+1}+\cdots+l_{-1}+k+1} x_{-i} < t_{-j} \le -\sum_{1 \le i \le l_{-j+1}+\cdots+l_{-1}+k} x_{-i} \quad \text{for} \quad j = 2, \ldots, s;$$

$$-\sum_{1 \le i \le k+1} x_i < t_{-1} \le -\sum_{1 \le i \le k} x_i;$$

$$\sum_{1 \le i \le l_0+\cdots+l_{j-1}-k} x_i < t_j \le \sum_{1 \le i \le l_0+\cdots+l_{j-1}-k+1} x_i \quad \text{for} \quad j = 1, \ldots, s\Big\}.$$

We have

$$p(\partial Z_k) \le p\Big(\text{there is an } m \in \{1, \ldots, (l+1)\} \text{ and a } j \in \{1, \ldots, s\}$$

$$\text{with} \quad -\sum_{1 \le i \le m} \xi_{-i} = t_{-j} \quad \text{or} \quad \sum_{1 \le i \le m} \xi_i = t_j\Big)$$

$$\le H_F\big(\Phi(\{t_i\}) > 0 \text{ for at least one } j = \pm 1, \ldots, \pm s\big) = 0.$$

On the other hand, our assumptions imply that, as n increases, p_n converges weakly toward p; thus we may conclude that

$$\lim_{n\to\infty} H_{F_n}\Big(\Phi\big([t_{-j}, t_{-j+1})\big) = l_{-j+1} \quad \text{for} \quad j = 2, \ldots, s;$$

$$\Phi\big([t_{-1}, t_1)\big) = l_0 + 1; \quad \Phi\big([t_{j-1}, t_j)\big) = l_{j-1} \quad \text{for} \quad j = 2, \ldots, s\Big)$$

$$= \lim_{n\to\infty} \sum_{0 \le k \le l_0} p_n(Z_k) = \sum_{0 \le k \le l_0} p(Z_k)$$

$$= H_F\Big(\Phi\big([t_{-j}, t_{-j+1})\big) = l_{-j+1} \quad \text{for} \quad j = 2, \ldots, s;$$

$$\Phi\big([t_{-1}, t_1)\big) = l_0 + 1; \quad \Phi\big([t_{j-1}, t_j)\big) = l_{j-1} \quad \text{for} \quad j = 2, \ldots, s\Big).$$

Since the non-negative integers l_j may be chosen arbitrarily, and since $H_{F_n}\Big(\Phi\big([t_{-1}, t_1)\big) = 0\Big) = 0$ $(n = 1, 2, \ldots)$ as well as $H_F\Big(\Phi\big([t_{-1}, t_1)\big) = 0\Big) = 0$ are satisfied, we may now use 1.9.2. to conclude that

$$\lim_{n\to\infty} \big\| (H_{F_n})_{[t_{-s},t_{-s+1}),\ldots,[t_{-2},t_{-1}),[t_{-1},t_1),[t_1,t_2),\ldots,[t_{s-1},t_s)}$$

$$- (H_F)_{[t_{-s},t_{-s+1}),\ldots,[t_{-2},t_{-1}),[t_{-1},t_1),[t_1,t_2),\ldots,[t_{s-1},t_s)} \big\| = 0.$$

Now, given any finite sequence $I_1, \ldots, I_m$ of pairwise disjoint non-empty bounded intervals $[\alpha_k, \beta_k)$ which have the property that

$$H_F\big(\Phi(\{\alpha_k\}) > 0\big) = 0 = H_F\big(\Phi(\{\beta_k\}) > 0\big) \quad \text{for} \quad k = 1, \ldots, m,$$

then always there exists a finite sequence $t_{-s} < \cdots < t_{-1} < 0 < t_1 < \cdots < t_s$, such that all of the I_l can be represented as unions of certain intervals $[t_{-s}, t_{-s+1}), \ldots, [t_{-2}, t_{-1}), [t_{-1}, t_1), [t_1, t_2), \ldots [t_{s-1}, t_s)$ and all probabilities

$H_F(\Phi(\{t_j\}) > 0)$, $j = \pm 1, \ldots, \pm s$, are zero. We obtain

$$\overline{\lim_{n\to\infty}} \|(H_{F_n})_{I_1,\ldots,I_m} - (H_F)_{I_1,\ldots,I_m}\|$$
$$\leq \overline{\lim_{n\to\infty}} \|(H_{F_n})_{[t_{-s},t_{-s+1}),\ldots,[t_{-2},t_{-1}),[t_{-1},t_1),[t_1,t_2),\ldots,[t_{s-1},t_s)}$$
$$- (H_F)_{[t_{-s},t_{-s+1}),\ldots,[t_{-2},t_{-1}),[t_{-1},t_1),[t_1,t_2),\ldots,[t_{s-1},t_s)}\| = 0,$$

so that, in view of 3.1.14., the convergence relation $H_{F_n} \underset{n\to\infty}{\Rightarrow} H_F$ is satisfied.

2. From $H_{F_n} \underset{n\to\infty}{\Rightarrow} Q$ it follows that $F_n \underset{n\to\infty}{\Rightarrow} F$, $F \in \boldsymbol{F} \setminus \{k_{[0,+\infty)}\}$, as well as $Q = H_F$.

Every subsequence of (F_n) contains a subsequence (F_{n_m}) which is convergent with respect to ϱ_F, so that $F_{n_m} \underset{m\to\infty}{\Rightarrow} F$. Assuming that $F = k_{[0,+\infty)}$, we were able to conclude that

$$\lim_{m\to\infty} H_{F_{n_m}}\big(\Phi((0,t)) < s\big) = 0$$

for all $t > 0$ and all natural numbers s, which contradicts the weak convergence of H_{F_n} toward Q. Hence F is different from $k_{[0,+\infty)}$, and applying 1. we observe that $H_{F_{n_m}} \underset{m\to\infty}{\Rightarrow} H_F$, i.e. that $H_F = Q$. Consequently the distribution function F does not depend on the choice of the subsequence.

Each subsequence of (F_n) has a subsequence which converges toward F with respect to ϱ_F. Hence, $F_n \underset{n\to\infty}{\Rightarrow} F$. ∎

The distributions of the form H_F, $F(0) = 0$, $\lim_{x\to+\infty} F(x) = 1$, $\Delta_F = \int x \, \mathrm{d}F(x) < +\infty$ are dense in the set $\{H_G\}_{G\in\boldsymbol{F}\setminus\{k_{[0,+\infty)}\}}$. We even have

10.4.5. *To each F in $\boldsymbol{F}$ there exists a sequence (F_n) in $\boldsymbol{F}$ having the properties*

$$F_n(0) = 0, \quad \lim_{x\to+\infty} F_n(x) = 1, \quad \Delta_{F_n} < +\infty \quad \textit{for} \quad n = 1, 2, \ldots,$$
$$\Delta_{F_n} \xrightarrow[n\to\infty]{} +\infty, \qquad F_n \underset{n\to\infty}{\Rightarrow} F.$$

Proof. We set

$$F_n(t) = \begin{cases} 0 & & 0 \leq t < n^{-1} \\ (1 - n^{-1})\, F(t) & \text{for} & n^{-1} \leq t < t_n \\ 1 & & t_n \leq t \end{cases}$$

where t_n is chosen large enough to satisfy $t_n > n$ as well as $\Delta_{F_n} > n$. ∎

Immediately from 9.6.2., 10.4.5., 10.4.4., 10.3.7., 8.3.1., and 8.3.3. there follows

10.4.6. Proposition. *To each distribution function F in $\boldsymbol{F}$ being different from $k_{[0,+\infty)}$, and to each non-negative real l, there exists a well-determined*

stationary infinitely divisible distribution P on $\mathfrak{M}$ such that

$$(\tilde{P})^0 = lH_F.$$

For $\lim_{x\to+\infty} F(x) < 1$, *$P$ is a regular, for* $\lim_{x\to+\infty} F(x) = 1$ *a singular infinitely divisible distribution.*

In the special case $F(x) = 0$ for $x \geqq 0$ we have $(\tilde{P})^0 = l\delta_{\delta_0}$, i.e. $P = P_{l\mu}$. If $F(0) = 0$, $\lim_{x\to+\infty} F(x) = 1$ and $\Delta_F = +\infty$ is fulfilled, then we get the mixing infinitely divisible distributions P introduced in Section 9.6.

The investigation of the structure of the stationary infinitely divisible distributions P introduced in 10.4.6., having the property $(\tilde{P})^0 = lH_F$, can be reduced to the special case $F(0) = 0$ by the following considerations.

Let $c_1, c_2, \ldots$ be any sequence of non-negative real numbers such that $\sum_{n=1}^{\infty} c_n = 1$ and $c = \sum_{n=1}^{\infty} nc_n < +\infty$. By the set-up

$$x \frown \varkappa_{(x)} = \sum_{n=1}^{\infty} c_n \delta_{n\delta_x} \qquad (x \in R^1)$$

we get a continuous cluster field on R^1 which is of type β). For all measures H on $\mathfrak{M}$, the measure $H_\varkappa$ exists. Obviously $H_\varkappa$ is stationary if H has this property. If W is a stationary measure in $\boldsymbol{W}$ such that $0 < i_W < +\infty$, then by 4.2.2. we have $i_{W_\varkappa} = ci_W$ and $W_\varkappa$ belongs to $\boldsymbol{W}$, too. Now, we will compute the Palm distribution $(W_\varkappa)_0$ under the additional assumption that W is simple.

Using 9.3.5. we obtain

$$\left\| W_\varkappa\left(T_{y(\Phi,X_n)}\Phi \in (.) \mid \Phi(X_n) > 0\right) - (W_\varkappa)_\square \right\| \xrightarrow[n\to\infty]{} 0$$

for each sequence (X_n) of sets in $\mathfrak{R}^1$ satisfying the condition (i). On the other hand, for $n = 1, 2, \ldots$ we have

$$W_\varkappa\left(T_{y(\Phi,X_n)}\Phi \in (.) \mid \Phi(X_n) > 0\right) = \left(W\left(T_{y(\Psi,X_n)}\Psi \in (.) \mid \Psi(X_n) > 0\right)\right)_\varkappa$$

from which, by applying 9.3.3. and 4.2.6. it follows that

$$\begin{aligned}
&\overline{\lim_{n\to\infty}} \left\| W_\varkappa\left(T_{y(\Phi,X_n)}\Phi \in (.) \mid \Phi(X_n) > 0\right) - (W_0)_\varkappa \right\| \\
&= \overline{\lim_{n\to\infty}} \left\| \left(W\left(T_{y(\Psi,X_n)}\Psi \in (.) \mid \Psi(X_n) > 0\right)\right)_\varkappa - (W_0)_\varkappa \right\| \\
&\leqq \overline{\lim_{n\to\infty}} \left\| W\left(T_{y(\Psi,X_n)}\Psi \in (.) \mid \Psi(X_n) > 0\right) - W_0 \right\| = 0.
\end{aligned}$$

Finally one obtains $(W_\varkappa)_\square = (W_0)_\varkappa$. Since $i_{W_\varkappa} = ci_W$, we may now use 9.3.6. to conclude that

$$(W_0)_\varkappa = c \sum_{n=1}^{\infty} n^{-1} (W_\varkappa)_0 \left(\Phi \in (.), \Phi(\{0\}) = n\right)$$

so that we obtain

$$\begin{aligned}(W_\varkappa)_0 &= \sum_{n=1}^{\infty} (W_\varkappa)_0 \left(\Phi \in (.), \Phi(\{0\}) = n\right) \\ &= c^{-1} \sum_{n=1}^{\infty} n(W_0)_\varkappa \left(\Phi \in (.), \Phi(\{0\}) = n\right) \\ &= c^{-1} \sum_{n=1}^{\infty} nc_n(W_0)_\varkappa \left(\Phi \in (.) \mid \Phi(\{0\}) = n\right).\end{aligned}$$

We thus observe the truth of

10.4.7. Proposition. *Let $c_1, c_2, \ldots$ be a sequence of non-negative real numbers having the properties*

$$\sum_{n=1}^{\infty} c_n = 1, \qquad c = \sum_{n=1}^{\infty} nc_n < +\infty.$$

If we set

$$\varkappa_{(x)} = \sum_{n=1}^{\infty} c_n \delta_{n\delta_x} \qquad (x \in R^1)$$

then for each stationary simple measure W in $\boldsymbol{W}$ with $0 < i_W < +\infty$,

$$(W_\varkappa)_0 = c^{-1} \sum_{n=1}^{\infty} nc_n(W_0)_\varkappa \left(\Phi \in (.) \mid \Phi(\{0\}) = n\right).$$

Now let F be any distribution function in $\boldsymbol{F} \setminus \{k_{[0,+\infty)}\}$,

$$\varkappa_{(x)} = \sum_{n=1}^{\infty} \left(1 - F(0)\right) \left(F(0)\right)^{n-1} \delta_{n\delta_x} \qquad (x \in R^1),$$

and l any positive number. According to 10.4.6. let P be the stationary infinitely divisible distribution having the property that

$$(\tilde{P})^0 = \left(1 - F(0)\right) lH_{F^*}.$$

Using the "clustering theorem" 4.3.3. as well as 4.2.2., we obtain

$$\left(\widetilde{P_\varkappa}\right)^0 = \left((\tilde{P})_\varkappa\right)^0 = l\left((\tilde{P})_\varkappa\right)_0.$$

Since, together with H_{F^*}, the measure $\tilde{P}$ is also simple, by Proposition 10.4.7. and 10.4.2. we have

$$\begin{aligned}\left((\tilde{P})_\varkappa\right)_0 &= \sum_{n=1}^{\infty} n\left(1 - F(0)\right)^2 \left(F(0)\right)^{n-1} \left((\tilde{P})_0\right)_\varkappa \left(\Phi \in (.) \mid \Phi(\{0\}) = n\right) \\ &= \sum_{n=1}^{\infty} n\left(1 - F(0)\right)^2 \left(F(0)\right)^{n-1} (H_{F^*})_\varkappa \left(\Phi \in (.) \mid \Phi(\{0\}) = n\right) = H_F.\end{aligned}$$

Thus we observe the truth of

10.4.8. Proposition. *Let l be a non-negative real number, F a distribution function in $\boldsymbol{F} \setminus \{k_{[0,+\infty)}\}$ and P the stationary infinitely divisible distribution on $\mathfrak{M}$, uniquely determined by*

$$(\tilde{P})^0 = \big(1 - F(0)\big)\, lH_{F^*}.$$

Further let $\varkappa$ be the cluster field on R^1 with phase space R^1, defined by

$$\varkappa_{(x)} = \sum_{n=1}^{\infty} \big(1 - F(0)\big) \big(F(0)\big)^{n-1} \delta_{n\delta_x} \qquad (x \in R^1).$$

Under these assumptions the stationary infinitely divisible distribution $Q = P_\varkappa$ satisfies

$$(\tilde{Q})^0 = lH_F.$$

In Section 9.5. to every F in $\boldsymbol{F} \setminus \{k_{[0,+\infty)}\}$ having the properties

$$F(0) = 0, \quad \lim_{x \to +\infty} F(x) = 1, \quad \Delta_F < +\infty$$

we associated the stationary recurrent distribution P_F by the set-up $(P_F)_0 = H_F$. Now we drop the assumption $F(0) = 0$ and set

$$P_F = (P_{F^*})_\varkappa$$

with the cluster field $\varkappa$ formulated in 10.4.8. Then, in virtue of 4.2.2., we have

$$i_{P_F} = \big(1 - F(0)\big)^{-1} i_{P_{F^*}} = \big(1 - F(0)\big)^{-1} (\Delta_{F^*})^{-1} = (\Delta_F)^{-1}$$

and by 10.4.7. and 10.4.2.

$$\begin{aligned}(P_F)_0 &= \big((P_{F^*})_\varkappa\big)_0 \\ &= \big(1 - F(0)\big) \sum_{n=1}^{\infty} n\big(1 - F(0)\big) \big(F(0)\big)^{n-1} (H_{F^*})_\varkappa \big(\Phi \in (.) \mid \Phi(\{0\}) = n\big) = H_F.\end{aligned}$$

In this way to each F in $\boldsymbol{F} \setminus \{k_{[0,+\infty)}\}$ having the properties

$$\lim_{x \to +\infty} F(x) = 1, \quad \Delta_F < +\infty$$

we have associated a stationary distribution P_F which fulfils

$$(P_F)_0 = H_F, \quad P_F(\{o\}) = 0.$$

Also in the case of the afore-mentioned more general assumptions, we call P_F the distribution of the *stationary recurrent point process* associated with F.

As in the derivation of Proposition 9.6.2. one can see that for all $l \geqq 0$ and all F in $F \setminus \{k_{[0,+\infty)}\}$ satisfying $\lim_{x \to +\infty} F(x) = 1, \Delta_F < +\infty$, the stationary infinitely divisible distribution P, uniquely determined by $(\tilde{P})^0 = lH_F$, has the form $P = \mathcal{U}_{l\Delta_F P_F}$.

10.5. Infinitesimal Triangular Arrays of Stationary Recurrent Distributions

Immediately from 10.3.11. and 10.4.4. the following convergence proposition is obtained for special infinitesimal triangular arrays of stationary recurrent distributions:

10.5.1. Proposition. *Let (F_n) be a sequence of distribution functions in $\boldsymbol{F}$ having the properties*

$$\lim_{x \to +\infty} F_n(x) = 1, \quad \Delta_{F_n} = \int_0^\infty x \, \mathrm{d}F_n(x) < +\infty, \quad F_n(0) < 1 \text{ for } n = 1, 2, \ldots,$$

$$\Delta_{F_n} \xrightarrow[n \to \infty]{} +\infty .$$

If a sequence (m_n) of natural numbers satisfies the condition

$$m_n(\Delta_{F_n})^{-1} \xrightarrow[n \to \infty]{} i, \quad 0 < i < +\infty,$$

then

$$(P_{F_n})^{m_n} \underset{n \to \infty}{\Rightarrow} P, \quad i_P = i$$

if and only if (F_n) converges, with respect to ϱ_F, toward some F in $\boldsymbol{F}$ different from $k_{[0,+\infty)}$. The distribution P is infinitely divisible, and $(\tilde{P})_0 = H_F$.

By virtue of 10.4.5., Proposition 10.5.1. provides an interpretation of all infinitely divisible stationary distributions P having the property that

$$(\tilde{P})^0 = lH_F, \quad l > 0, \quad F \in \boldsymbol{F} \setminus \{k_{[0,+\infty)}\}.$$

The following example shows that the condition $i_P = i$ cannot be dropped in 10.5.1. Let

$$F_n = (1 - n^{-1})\, k_{[0,+\infty)} + n^{-1} k_{[n^2,+\infty)}; \qquad m_n = n \quad \text{for} \quad n = 1, 2, \ldots$$

Then $\lim_{t \to +\infty} F_n(t) = 1$, $F_n(0) < 1$, $\Delta_{F_n} = n$, and hence $m_n(\Delta_{F_n})^{-1} = 1$ for $n = 1, 2, \ldots$ On the other hand, $(P_{F_n})^{m_n} \underset{n \to \infty}{\Rightarrow} \delta_o$, for by 9.1.14. we have, for all $t > 0$ and sufficiently large n,

$$\begin{aligned}(P_{F_n})^{m_n} (\Phi((0, t)) > 0) &\leqq m_n P_{F_n}(\Phi((0, t)) > 0) \\ &= n P_{(F_n)^*}(\Phi((0, t)) > 0) = n P_{k_{[n^2,+\infty)}}(\Phi((0, t)) > 0) = n \frac{t}{n^2} \xrightarrow[n \to \infty]{} 0.\end{aligned}$$

Let $\mathfrak{F}$ denote the smallest σ-algebra of subsets of $\boldsymbol{F}$ with respect to which the real function

$$F \curvearrowright F(t) \qquad (F \in \boldsymbol{F})$$

is measurable for all $t \geqq 0$.

10.5.2. *The σ-algebra of the Borel sets of the metric space $[\boldsymbol{F}, \varrho_{\boldsymbol{F}}]$ coincides with $\mathfrak{F}$.*

Proof. 1. Given any continuous real function h which is defined on $[0, +\infty]$. For all F in $\boldsymbol{F}$ we obtain

$$\int h(x)\, p_F(\mathrm{d}x) = \lim_{n\to\infty} \left(\sum_{i=1}^{n^2} h\left(\frac{i-1}{n}\right) p_F\left(\left[\frac{i-1}{n}, \frac{i}{n}\right)\right) + h(n)\, p_F([n, +\infty])\right),$$

where p_F as above denotes the measure on the σ-algebra of Borel sets of $[0, +\infty]$, associated with F. Hence the real function $F \curvearrowright \int h(x)\, p_F(\mathrm{d}x)$ is measurable with respect to $\mathfrak{F}$. The σ-algebra $\mathfrak{F}$ therefore includes the σ-algebra of the Borel subsets of $\boldsymbol{F}$.

2. For all $t \geqq 0$ and all F in $\boldsymbol{F}$, the value $F(t)$ equals the limit of the sequence $\left(\int h_{t,n}(x)\, p_F(\mathrm{d}x)\right)$, where

$$h_{t,n}(x) = \begin{cases} 1 & 0 \leqq x < t \\ 1 - n(x-t) \quad \text{for} & t \leqq x < t + n^{-1} \\ 0 & t + n^{-1} \leqq x < +\infty. \end{cases}$$

Hence $F \curvearrowright F(t)$ is measurable with respect to the σ-algebra of the Borel sets, i.e. this σ-algebra includes $\mathfrak{F}$. ∎

In view of 10.4.4. and 10.5.2., the mapping $F \curvearrowright H_F$ of $\boldsymbol{F} \setminus \{k_{[0,+\infty)}\}$ into $\boldsymbol{P}$ is measurable with respect to the σ-algebras $\mathfrak{F}$ and $\mathfrak{P}$. Hence to each distribution q on $\mathfrak{F}$ satisfying the condition $q(\{k_{[0,+\infty)}\}) = 0$, the distribution $\int H_F(.)\, q(\mathrm{d}F)$ can be associated.

As a generalization of 10.4.3., we have

10.5.3. *The mapping $q \curvearrowright \int H_F(.)\, q(\mathrm{d}F)$ is one-to-one.*

Proof. 1. From $q_1\big(F(0) = 0\big) = 1 = q_2\big(F(0) = 0\big)$ and $\int H_F(.)\, q_1(\mathrm{d}F) = \int H_F(.)\, q_2(\mathrm{d}F)$ it follows that $q_1 = q_2$.

Let h be a function on $\boldsymbol{F}$ having the form $h(F) = \prod_{i=0}^{n} F(t_i)$, where an arbitrary finite sequence of positive numbers can be chosen for $t_0, \ldots, t_n$. Then

$$\begin{aligned} &\int h(F)\, q_1(\mathrm{d}F) \\ &= \int H_F\big((x_0(\Phi) - x_{-1}(\Phi)) \leqq t_0, \ldots, (x_n(\Phi) - x_{n-1}(\Phi)) \leqq t_n\big)\, q_1(\mathrm{d}F) \\ &= \int H_F\big((x_0(\Phi) - x_{-1}(\Phi)) \leqq t_0, \ldots, (x_n(\Phi) - x_{n-1}(\Phi)) \leqq t_n\big)\, q_2(\mathrm{d}F) \\ &= \int h(F)\, q_2(\mathrm{d}F). \end{aligned}$$

Let us now consider functions of the type $f(F) = \prod_{i=0}^{m} \int_0^{t_i} F(\tau)\,\mu(\mathrm{d}\tau)$, where again an arbitrary finite sequence of positive numbers can be chosen for $t_0, \ldots, t_m$. Each f is the limit approached by uniform convergence of a sequence of linear combinations of functions h. We obtain $\int f(F)\,q_1(\mathrm{d}F) = \int f(F)\,q_2(\mathrm{d}F)$. The same holds also for all finite linear combinations $g = \alpha_0 + \sum_{k=1}^{l} x_k f_k$.

The set of these g's forms a subalgebra of the algebra of all continuous real functions on the compact metric space $[\boldsymbol{F}, \varrho_{\boldsymbol{F}}]$. Obviously this subalgebra separates points in $\boldsymbol{F}$, i.e. $F_1 \neq F_2$ implies the existence of some g having the property that $g(F_1) \neq g(F_2)$. In view of Stone-Weierstrass Theorem, every continuous real function v on $\boldsymbol{F}$ is a limit approached by uniform convergence of a sequence (g_n) so that we are able to conclude that $\int v(F)\,q_1(\mathrm{d}F) = \int v(F)\,q_2(\mathrm{d}F)$, i.e. that $q_1 = q_2$.

2. From $q_1(\{k_{[0,+\infty)}\}) = 0 = q_2(\{k_{[0,+\infty)}\})$ and $\int H_F(.)\,q_1(\mathrm{d}F) = \int H_F(.)\,q_2(\mathrm{d}F)$ it follows that $q_1 = q_2$.

To each F in $\boldsymbol{F}$ we associate another element $F^!$ in $\boldsymbol{F}$ by

$$F^!(t) = \begin{cases} 0 & \text{for} \quad 0 \leqq t < 1 \\ F(t-1) & \phantom{\text{for}} \quad 1 \leqq t < +\infty, \end{cases}$$

thus obtaining a one-to-one mapping $F \curvearrowright F^!$ of $\boldsymbol{F}$ into itself, which is measurable with respect to $\mathfrak{F}$ in both directions. Every distribution q on $\mathfrak{F}$ is transformed by $F \curvearrowright F^!$ into a distribution $q^!$ on $\mathfrak{F}$ with the property that $q^!\big(F(0) = 0\big) = 1$.

Let Φ be in M^0. If $\Phi(\{0\}) = 1$, then according to the interpretation of Φ as a description of a realization of a renewal process, to Φ we can measurably associate a uniquely determined $\Phi^!$ in M^0 such that all 'lifetimes' x, including the 'lifetimes' $x = 0$, are replaced by $x + 1$. If $\Phi(\{0\}) = 1 + l$, $l \geqq 0$, for every j in $\{0, \ldots, l\}$, we may interpret $\Phi(\{0\})\,\delta_0$ as follows

$$\Phi(\{0\})\,\delta_0 = j\delta_{0-0} + \delta_0 + (l - j)\,\delta_{0+0}$$

and can again measurably associate a $\Phi^{!,j}$ in M^0 by replacing all 'lifetimes' x by $x + 1$. We set

$$K\big(\Phi, (.)\big) = \sum_{0 \leqq j \leqq \Phi(\{0\})-1} \big(\Phi(\{0\})\big)^{-1}\,\delta_{\Phi^{!,j}}$$

thus obtaining a stochastic kernel K on $[M^0, \mathfrak{M}^0]$ with phase space $[M^0, \mathfrak{M}^0]$. Obviously, for all F in $\boldsymbol{F} \setminus \{k_{[0,+\infty)}\}$,

$$\int K\big(\Phi, (.)\big)\,H_F(\mathrm{d}\Phi) = H_{F^!}.$$

From

$$\int H_F(.)\,q_1(\mathrm{d}F) = \int H_F(.)\,q_2(\mathrm{d}F)$$

it follows that

$$\int H_{F^!}(.)\,q_1(\mathrm{d}F) = \int H_{F^!}(.)\,q_2(\mathrm{d}F),$$

hence

$$\int H_F(.)\,(q_1)^!\,(\mathrm{d}F) = \int H_F(.)\,(q_2)^!\,(\mathrm{d}F),$$

from which it can be concluded by means of 1. that $(q_1)^! = (q_2)^!$. Consequently, $q_1 = q_2$. ∎

Instead of 10.4.4., we now have

10.5.4. *Let (q_n) be a sequence of distributions on $\mathfrak{F}$ having the property that $q_n(\{k_{[0,+\infty)}\}) = 0$ for $n = 1, 2, \ldots$ Then $\int H_F(.)\,q_n(\mathrm{d}F) \underset{n\to\infty}{\Rightarrow} Q$ is equivalent to the fact that the sequence (q_n) converges weakly toward a distribution q having the property $q(\{k_{[0,+\infty)}\}) = 0$ and that Q coincides with $\int H_F(.)\,q(\mathrm{d}F)$.*

Proof. 1. Let $q_n \underset{n\to\infty}{\Rightarrow} q$, $q(\{k_{[0,+\infty)}\}) = 0$. Further let h be an arbitrary continuous bounded real function defined on M. If we set

$$\int H_F(.)\,q_n(\mathrm{d}F) = H_n, \qquad \int H_F(.)\,q(\mathrm{d}F) = H,$$

then

$$\int h(\Phi)\,H_n(\mathrm{d}\Phi) = \int \Big(\int h(\Phi)\,H_F(d\Phi)\Big)\,q_n(\mathrm{d}F).$$

In view of 10.4.4., $F \curvearrowright g(F) = \int h(\Phi)\,H_F(\mathrm{d}\Phi)$ is a continuous real function defined on $\boldsymbol{F} \setminus \{k_{[0,+\infty)}\}$. Moreover, this function is bounded and can be extended to a function g defined on $\boldsymbol{F}$, almost everywhere continuous with respect to q. Thus we get (cf. Billingsley [1], Theorem 5.2.)

$$\int h(\Phi)\,H_n(\mathrm{d}\Phi) = \int g(F)\,q_n(\mathrm{d}F) \xrightarrow[n\to\infty]{} \int g(F)\,q(\mathrm{d}F) = \int h(\Phi)\,H(\mathrm{d}\Phi)$$

which means that $H_n \underset{n\to\infty}{\Rightarrow} H$.

2. Conversely, suppose that, as n increases, the sequence of the $H_n = \int H_F(.)\,q_n(\mathrm{d}F)$ converges weakly toward a distribution Q.

Since the metric space $[\boldsymbol{F}, \varrho_{\boldsymbol{F}}]$ is compact, we may choose from each subsequence a subsequence (q_{n_m}) which converges weakly toward a distribution q on $\mathfrak{F}$. Let $q(\{k_{[0,+\infty)}\}) > 0$.

For any positive number s there are positive numbers σ and ω such that

$$F \in \boldsymbol{F} \setminus \{k_{[0,+\infty)}\}, \qquad F(\sigma) \geqq 1 - \omega$$

implies

$$H_F\Big(\Phi\big((0,1)\big) \geqq s\Big) \geqq 1/2.$$

Moreover, there exists a $\delta > 0$ such that all F in the open ball $S_\delta(k_{[0,+\infty)})$ satisfy $F(\sigma) \geqq 1 - \omega$. Then we have

$$\begin{aligned}
&\underline{\lim}_{m\to\infty} \int H_F\Big(\Phi\big((0,1)\big) \geqq s\Big)\,q_{n_m}(\mathrm{d}F)\\
&\geqq 1/2\ \underline{\lim}_{m\to\infty}\ q_{n_m}\big(S_\delta(k_{[0,+\infty)})\big)\\
&\geqq 1/2\ q\big(S_\delta(k_{[0,+\infty)})\big) \geqq 1/2\ q(\{k_{[0,+\infty)}\}),
\end{aligned}$$

which contradicts the weak compactness of $\{H_{n_m}\}_{m=1,2,\ldots}$. Consequently, $q(\{k_{[0,+\infty)}\}) = 0$, so that by means of 1. we can conclude that $Q = \int H_F(.)\, q(\mathrm{d}F)$. Hence the distribution q does not depend on the choice of the subsequences.

Every subsequence of (q_n) contains a subsequence converging weakly toward q. Hence (q_n) converges weakly toward q. ∎

As an immediate consequence of 10.3.9., 10.3.10., and 10.5.4., we now obtain a generalized version of 10.5.1.:

10.5.5. Theorem. *Let* $(P_{F_{n,j}})_{\substack{n=1,2,\ldots \\ 1\leqq j\leqq m_n}}$ *be an infinitesimal triangular array of stationary recurrent distributions having the property that*

$$\sum_{1\leqq j\leqq m_n} (\varDelta_{F_{n,j}})^{-1} \xrightarrow[n\to\infty]{} i, \qquad 0 < i < +\infty.$$

Then

$$\mathop{\ast}_{1\leqq i\leqq m_n} P_{F_{n,j}} \underset{n\to\infty}{\Rightarrow} P, \qquad i_P = i,$$

is equivalent to the weak convergence of

$$\Bigl(\sum_{1\leqq j\leqq m_n} (\varDelta_{F_{n,l}})^{-1}\Bigr)^{-1} \sum_{1\leqq j\leqq m_n} (\varDelta_{F_{n,j}})^{-1}\, \delta_{F_{n,j}}$$

toward a distribution q having the property that $q(\{k_{[0,+\infty)}\}) = 0$. *The stationary distribution P is infinitely divisible, and*

$$(\tilde{P})_0 = \int H_F(.)\, q(\mathrm{d}F).$$

We shall conclude this section by showing

10.5.6. Proposition. *In 10.5.5., q runs through all distributions on $\mathfrak{F}$ having the property that* $q(\{k_{[0,+\infty)}\}) = 0$.

Proof. Let ϱ denote the Prohorov metric in the set of all distributions on the σ-algebra $\mathfrak{F}$. Further let q be an arbitrary given distribution on $\mathfrak{F}$ which satisfies the condition $q(\{k_{[0,+\infty)}\}) = 0$. We construct by rows an associated triangular array $(P_{F_{n,j}})$.

Given any natural number n, we first choose a mixture $p = \sum_{i=1}^{m} c_i \delta_{F_i}$, $F_i \in \boldsymbol{F}$ for $1 \leqq i \leqq m$, such that $\varrho(p, q) < n^{-1}$. In this case it is always possible to select coefficients of the form $c_i = r_i k^{-1}$ with natural numbers r_i and a natural number $k > n$. The proof of 10.4.5. shows that the F_i can be replaced by distribution functions L_i in $\boldsymbol{F}$ such that $L_i(0) = 0$, $\lim_{t\to+\infty} L_i(t) = 1$, as well as $\varDelta_{L_i} = rk$ with some natural number r, the distance between $\sum_{i=1}^{m} \frac{rr_i}{rk} \delta_{L_i}$ and q being still less than n^{-1}. Now the n-th row of the desired triangular array can be constructed as follows. Let $m_n = \sum_{i=1}^{m} rr_i = rk$. It is required that in each case rr_i terms $P_{F_{n,j}}$ are equal to P_{L_i}. As n increases, the above mixture converges weakly toward q. ∎

10.6. The Distributions H_q

Proposition 10.4.6. raises the problem to derive in a direct-algebraic manner the Palm character of the distributions H_F for arbitrary F in $\boldsymbol{F} \setminus \{k_{[0,+\infty)}\}$. To this end, in the special case $\lim\limits_{x\to+\infty} F(x) = 1$, one can use the symmetry operator $\widehat{\Gamma}_2$ introduced in Section 8.5. Instead of this, we prefer to demonstrate a general approach that is not restricted to the dimension number $s = 1$, and leads to an interesting class of Palm distributions. The remaining case $\lim\limits_{x\to+\infty} F(x) < 1$ will be considered in the Sections 11.5. and 11.6.

Let s be an arbitrary integer, $[K, \varrho_K]$ be an arbitrary bounded complete separable metric space and $[A, \varrho_A]$ the direct product of R^s with the mark space $[K, \varrho_K]$. Generalizing a notation of Section 9.4., we denote by V the set of all doubly infinite sequences $([h_n, l_n])_{-\infty<n<+\infty}$ of elements of A fulfilling the condition

$$\lim_{k\to\infty} \left\| \sum_{n=0}^{k} h_n \right\| = +\infty = \lim_{k\to\infty} \left\| \sum_{n=1}^{k} h_{-n} \right\|.$$

Obviously V belongs to the σ-algebra $(\mathfrak{R}^s \otimes \mathfrak{K})^{\{\ldots,-1,0,1,\ldots\}}$. As in the proof of 9.4.2. we see that the mapping

$$([h_n, l_n]) \frown \Phi = \sum_{n=1}^{\infty} \delta_{[-(h_{-1}+\cdots+h_{-n}),l_{-n}]} + \delta_{[[0,\ldots,0],l_0]} + \sum_{n=0}^{\infty} \delta_{[h_0+\cdots+h_n,l_{n+1}]}$$

of V into M is measurable with respect to the σ-algebras $(\mathfrak{R}^s \otimes \mathfrak{K})^{\{\ldots,-1,0,1,\ldots\}}$ and $\mathfrak{M}$. In this way every distribution q on $(\mathfrak{R}^s \otimes \mathfrak{K})^{\{\ldots,-1,0,1,\ldots\}}$, fulfilling $q(V) = 1$, is transformed into a distribution H_q on $\mathfrak{M}$.

Partially generalizing 9.6.1. we obtain (cf. Mori [1])

10.6.1. Theorem. *Let q be a distribution on $(\mathfrak{R}^s \otimes \mathfrak{K})^{\{\ldots,-1,0,1,\ldots\}}$ stationary with respect to time shift, with $q(V) = 1$. Then for all $l \geqq 0$ there exists exactly one singular infinitely divisible distribution P on $\mathfrak{M}$ having the property*

$$(\tilde{P})^0 = lH_q.$$

Proof. Obviously $H_q(M \setminus M^0) = 0$. We shall show, that the Campbell measure $\mathscr{C}_{H_q}$ of H_q is invariant with respect to the measurable mapping $[x, \Phi] \frown [-x, T_x\Phi]$ of $C \cap (R^s \times M^0)$ onto itself.

For all $e = ([h_n, l_n])$ in V we denote the corresponding counting measure Φ in M by $\Phi_{(e)}$. We set

$$r_k(e) = \begin{cases} h_0 + \cdots + h_k & k > -1 \\ 0 & \text{for} \quad k = -1 \\ -(h_{-1} + \cdots + h_{-k+1}) & k < -1. \end{cases}$$

For all non-negative real functions v on C being measurable with respect to $\mathfrak{C}$ we now obtain

$$\int v(x, \Phi)\, \mathscr{C}_{H_q}(\mathrm{d}[x, \Phi]) = \int \Big(\sum_{x \in R^s, \Phi^\times(\{x\})>0} \Phi^\times(\{x\})\, v(x, \Phi) \Big) H_q(\mathrm{d}\Phi)$$

$$= \int \Big(\sum_{-\infty<k<+\infty} v\big(r_{-k}(e), \Phi_{(e)}\big) \Big) q(\mathrm{d}e) = \sum_{-\infty<k<+\infty} \int v\big(r_{-k}(e), \Phi_{(e)}\big)\, q(\mathrm{d}e).$$

Since q is stationary with respect to time shift, we may continue the above chain of equalities as follows

$$= \sum_{-\infty<k<+\infty} \int v\Big(r_{-k}\big(([h_{n+k-1}, l_{n+k-1}])\big), \Phi_{(([h_{n+k-1}, l_{n+k-1}]))}\Big)\, q(\mathrm{d}e).$$

From

$$r_{-k}\big(([h_{n+k-1}, l_{n+k-1}])\big)$$

$$= \begin{cases} h_{0+k-1} + \cdots + h_{-k+k-1} = h_{k-1} + \cdots + h_{-1} & k < 1 \\ 0 & \text{for} \quad k = 1 \\ -(h_{-1+k-1} + \cdots + h_{-k+1+k-1}) = -(h_{k-2} + \cdots + h_0) & k > 1 \end{cases}$$

we get, for all integers k,

$$r_{-k}\big(([h_{n+k-1}, l_{n+k-1}])\big) = -r_{k-2}\big(([h_n, l_n])\big).$$

Consequently

$$(*) \qquad \int v(x, \Phi)\, \mathscr{C}_{H_q}(\mathrm{d}[x, \Phi]) = \sum_{-\infty<k<+\infty} \int v\big(-r_{k-2}(e), \Phi_{(([h_{n+k-1}, l_{n+k-1}]))}\big)\, q(\mathrm{d}e).$$

For $k = 1$ we have $r_{k-2}(e) = 0$ and obtain

$$\Phi_{(([h_{n+k-1}, l_{n+k-1}]))} = \Phi_{(e)} = T_{r_{k-2}(e)}\, \Phi_{(e)}.$$

For $k > 1$ we have

$$r_{k-2}(e) = h_0 + \cdots + h_{k-2}$$

and obtain

$$T_{r_{k-2}(e)}\, \Phi_{(e)} = \sum_{n=1}^{\infty} \delta_{[-(h_{-1}+\cdots+h_{-n}) - r_{k-2}(e), l_{-n}]} + \delta_{[[0,\ldots,0] - r_{k-2}(e), l_0]}$$

$$+ \sum_{n=0}^{\infty} \delta_{[h_0+\cdots+h_n - r_{k-2}(e), l_{n+1}]}$$

$$= \sum_{n=1}^{\infty} \delta_{[-(h_{-n}+\cdots+h_{-1}+h_0+\cdots+h_{k-2}), l_{-n}]} + \delta_{[-(h_0+\cdots+h_{k-2}), l_0]}$$

$$+ \sum_{n=0}^{\infty} \delta_{[h_0+\cdots+h_n - (h_0+\cdots+h_{k-2}), l_{n+1}]}$$

$$= \sum_{n=-(k-2)}^{\infty} \delta_{[-(h_{-n}+\cdots+h_{k-2}), l_{-n}]} + \delta_{[[0,\ldots,0], l_{k-1}]} + \sum_{n=k-1}^{\infty} \delta_{[h_{k-1}+\cdots+h_n, l_{n+1}]}.$$

Substituting $m = n + k - 1$, $m = n - (k - 1)$, respectively, we get

$$= \sum_{m=1}^{\infty} \delta_{[-(h_{-m+(k-1)}+\cdots+h_{-1+(k-1)}),l_{-m+(k-1)}]} + \delta_{[[0,\ldots,0],l_{0+(k-1)}]}$$

$$+ \sum_{m=0}^{\infty} \delta_{[h_{0+(k-1)}+\cdots+h_{m+(k-1)},l_{m+1+(k-1)}]} = \Phi_{(([h_{n+k-1},l_{n+k-1}]))}.$$

Thus

$$T_{r_{k-2}(e)}\Phi_{(e)} = \Phi_{(([h_{n+k-1},l_{n+k-1}]))}$$

for all $k > 1$.

By analogous computations we may derive this formula in the remaining case $k < 1$.

Substituting this formula in (*) we get

$$\int v(x, \Phi)\, \mathscr{C}_{H_q}(\mathrm{d}[x, \Phi]) = \sum_{-\infty<k<+\infty} \int v(-r_{k-2}(e), T_{r_{k-2}(e)}\Phi_{(e)})\, q(\mathrm{d}e)$$

$$= \int \sum_{-\infty<j<+\infty} v\big(-r_j(e), T_{r_j(e)}\Phi_{(e)}\big)\, q(\mathrm{d}e) = \int v(-x, T_x\Phi)\, \mathscr{C}_{H_q}(\mathrm{d}[x, \Phi]).$$

Theorem 8.6.1. implies that H_q is the Palm measure of a stationary σ-finite measure W on $\mathfrak{M}$. As $W^0 = \big(W((.) \setminus \{o\})\big)^0$, we may additionally suppose that $W(\{o\}) = 0$. As $i_{W^x} = W^0(M^0) = H_q(M^0) = 1$, the measure W belongs to $\boldsymbol{W}$. Consequently, $lW \in \boldsymbol{W}$ and

$$(lW)^0 = lW^0 = lH_q.$$

By 8.3.1. there exists exactly one stationary infinitely divisible distribution P satisfying $(\tilde{P})^0 = lH_q$. As $H_q(\chi \text{ finite}) = 0$, 8.3.3. shows that P is singular infinitely divisible. ∎

Using 10.6.1. it is possible (cf. K. Hermann [1]) to construct, in the case $s > 1$, an ergodic, but not mixing stationary infinitely divisible distribution.

We select an arbitrary element k_0 from K, an arbitrary c_0 from $R^s \setminus \{[0, \ldots, 0]\}$ and set $q = \delta_{([h_n,k_0])}$, where $h_n = c_0$ for all integers n. Theorem 10.6.1. implies the existence of a singular infinitely divisible distribution P satisfying

$$(\tilde{P})^0 = H_q = (\tilde{P})_0.$$

Obviously, $H_q(s(\chi^x) = 0) = 1$. Setting $X = \{x : x \in R^s, \|x\| < c_0\}$, we obtain $H_q(\chi^x(X - nc_0) > 0) = 1$ for all integers n. Thus, by 9.2.1. the distribution P is ergodic, but not mixing.

11. Homogeneous Cluster Fields

11.0. Introduction

A cluster field $\varkappa$ on R^s with phase space R^s is said to be homogeneous if to each translation of the producing point x there corresponds the same translation of the distribution $\varkappa_{(x)}$ of the random cluster produced by x, i.e. if the equation $\varkappa_{(x)} = \varkappa_{(y)}\big(T_{y-x}\chi \in (.)\big)$ is satisfied for all x, y in R^s. Thus a homogeneous cluster field is uniquely determined by the distribution $\varkappa_{([0,\ldots,0])}$ of the random cluster produced in the origin. Moreover, in virtue of 11.1.1., the correspondence $\varkappa \curvearrowright \varkappa_{([0,\ldots,0])}$ provides a one-to-one mapping of the set of all homogeneous cluster fields onto the set $\boldsymbol{P}$ of all distributions on $\mathfrak{M}$. The homogeneous cluster field which is uniquely associated to some Q in $\boldsymbol{P}$ is denoted by $[Q]$.

We denote by $\boldsymbol{Q}$ the set of all distributions Q on $\mathfrak{M}$ satisfying $\int Q\big(\chi(X-x) > 0\big)\,\mu(\mathrm{d}x) < +\infty$ for all bounded X in the σ-algebra $\mathfrak{R}^s$ of Borel subsets of R^s. In view of 11.2.1. we have $Q(\chi \text{ finite}) = 1$ for all Q in $\boldsymbol{Q}$. If a stationary distribution P on $\mathfrak{M}$ is different from δ_0 and $P_{[Q]}$ exists for a distribution Q in $\boldsymbol{P}$, in virtue of 11.2.4., Q belongs to $\boldsymbol{Q}$. On the other hand, by 11.2.3., the existence of $P_{[Q]}$ is ensured if P is of finite intensity and Q belongs to $\boldsymbol{Q}$.

A clustering representation $P = (P_\nu)_\varkappa$ of a distribution P in $\boldsymbol{P}$ on R^s is said to be homogeneous if ν is translation-invariant and $\varkappa$ is homogeneous, i.e. if $\nu = l\mu$, $l \geqq 0$ and $\varkappa = [Q]$, $Q \in \boldsymbol{Q}$. In view of 4.5.3., a distribution P on $\mathfrak{M}$ is infinitely divisible if and only if it has a clustering representation on R^s. For stationary distributions P, this proposition might be expected to permit some refinement to be added, such that P should be infinitely divisible if and only if there exists a homogeneous clustering representation $P = (P_{l\mu})_{[Q]}$. This is, however, not true. In virtue of $Q(\chi \text{ finite}) = 1$, we have $\widetilde{(P_{l\mu})_{[Q]}}(\chi \text{ infinite}) = \int [Q]_{(x)}(\chi \text{ infinite})\, l\mu(\mathrm{d}x) = 0$. Thus a distribution P on $\mathfrak{M}$ having a homogeneous clustering representation on R^s is stationary and regular infinitely divisible. By Theorem 11.4.3. this necessary condition is also sufficient (cf. Matthes [2], Kerstan & Matthes [1], Lee [1], and Goldman [2]). In this way we have derived a convenient characterization of the class of all distributions having the form $(P_{l\mu})_{[Q]}$, $l \geqq 0$, $Q \in \boldsymbol{Q}$. In several interesting cases, the representability of a given stationary distribution P in this form can be derived in a quite different way. In this context we refer to Hawkes & Oakes [1], Radecke [1], and Daley & Oakes [1].

To every distribution ν on the σ-algebra $\mathfrak{R}^s$ we may associate the homogeneous cluster field $[Q_\nu]$. The clustering by means of $[Q_\nu]$ causes the individual points of the realizations to be shifted independently of each other by random vectors which are distributed according to ν. As it was shown in Doob [1], all stationary Poisson distributions $P_{l\mu}$, $l \geqq 0$, are cluster-invariant with respect to $[Q_\nu]$. Then the same holds also for all mixtures $\int P_{l\mu}(.)\,\sigma(\mathrm{d}l)$, i.e. for all stationary purely random distributions. Theorem 11.10.5. says that for each non-lattice distribution ν on $\mathfrak{R}^s$, a stationary distribution P on $\mathfrak{M}$ is cluster-invariant with respect to $[Q_\nu]$, if and only if it is purely random.

As it was shown in Section 6.4., for each stationary distribution L on $\mathfrak{M}$, the sample intensity $s(\Phi) = \lim_{n\to\infty} (2n)^{-s}\,\Phi\big([-n, n)^s\big)$ of Φ exists almost everywhere with respect to L. In virtue of Theorem 11.9.5. and Theorem 11.10.4., subject to the additional assumption that $L\big(s(\Phi) < +\infty\big) = 1$, for all non-lattice distributions ν on $\mathfrak{R}^s$, the n-fold clustered distribution $\big(\ldots(L_{[Q_\nu]})\ldots\big)_{[Q_\nu]}$ i.e. the distribution $L_{[Q_{\nu^n}]}$, tends weakly toward the mixture $P = \int P_{s(\Phi)\mu}(.)\,L(\mathrm{d}\Phi)$ as n increases. Hence the limit distribution P depends only on the distribution σ_L of $s(\Phi)$ formed with respect to L.

The weak convergence of $L_{[Q_{\nu_n}]}$ towards $\int P_{l\mu}(.)\,\sigma_L(\mathrm{d}l)$ is not bound to the fact that the sequence (ν_n) has the special form (ν^n). It suffices (Theorem 11.9.5. (cf. Dobrushin [1], Stone [1], Debes, Kerstan, Liemant & Matthes [1]) that the sequence is weakly asymptotically uniformly distributed in the sense that for all x in R^s and all distributions γ on $\mathfrak{R}^s$ which are absolutely continuous with respect to the Lebesgue measure μ, the variation distance $\operatorname{Var}(\gamma * \nu_n - \delta_x * \gamma * \nu_n)$ converges towards zero as n increases.

The convergence statement indicated above remains valid with weaker assumptions on L. In fact, it is only required that there exists a non-negative real function s_L which is defined almost everywhere with respect to L and measurable with respect to $\mathfrak{M}$, such that, for all $c > 0$, the convergence relation

$$\lim_{n\to\infty} \int_{\{\Phi : s_L(\Phi) \leqq c\}} \left|(2n)^{-s}\,\Phi\big([-n, n)^s + x\big) - s_L(\Phi)\right| L(\mathrm{d}\Phi) = 0$$

is satisfied uniformly in x. We denote the set of these distributions L by $\mathbf{S}$.

Following Kallenberg [7] we associate in Section 11.3. to each distribution Q satisfying the condition $Q(\chi \text{ finite}) = 1$ the measure $\rangle Q \langle = \mathscr{C}_Q{}^T$ on $\mathfrak{R}^s \otimes \mathfrak{M}^0$. Let Q be a distribution in $\boldsymbol{Q}$ and H be a stationary measure on $\mathfrak{M}$ such that $H_{[Q]}$ exists. If we suppose now that both $H\big((.) \setminus \{o\}\big)$ and $H_{[Q]}\big((.) \setminus \{o\}\big)$ are σ-finite, then

$$(H_{[Q]})^0 = \int \left(\int k_{(.)}(\chi + T_z\Psi)\,(H^{0!})_{[Q]}\,(\mathrm{d}\Psi)\right) \rangle Q \langle\,(\mathrm{d}[z, \chi]),$$

where

$$H^{0!} = H^0\big((\Phi - \delta_{[0,\ldots,0]}) \in (.)\big).$$

11.1. The Mapping $V \rightsquigarrow [V]$

Let s be any natural number and $[K_i, \varrho_{K_i}]$, $1 \leq i \leq 2$, bounded complete separable metric spaces. We denote by $[A_i, \varrho_{A_i}]$ the direct product of R^s and $[K_i, \varrho_{K_i}]$.

A cluster field $\varkappa$ on $[A_1, \varrho_{A_1}]$ with phase space $[A_2, \varrho_{A_2}]$ is called *homogeneous* if, for all $[x, k]$ in A_1,

$$\varkappa_{([x,k])} = \varkappa_{([[0,\ldots,0],k])}\big(T_{-x}\chi \in (.)\big)$$

is satisfied. Consequently, to each homogeneous cluster field $\varkappa$ we can associate, in the form of

$$k \curvearrowright V_{(k)} = \varkappa_{([[0,\ldots,0],k])} \qquad (k \in K_1),$$

a cluster field $V_{(.)}$ on $[K_1, \varrho_{K_1}]$ with phase space $[A_2, \varrho_{A_2}]$ in a one-to-one way. Obviously $V_{(.)}$ is continuous if the cluster field $\varkappa$ is continuous.

Conversely, let $V_{(.)}$ be any cluster field on $[K_1, \varrho_{K_1}]$ with phase space $[A_2, \varrho_{A_2}]$. Setting

$$[V]_{([x,k])} = V_{(k)}\big(T_{-x}\chi \in (.)\big) \qquad ([x, k] \in A_1),$$

we obtain a mapping $[V]_{(.)}$ of A_1 into $\boldsymbol{P}_2$. For all Y in $\mathfrak{M}_2$,

$$[V]_{([x,k])}(Y) = \int k_Y(T_{-x}\chi)\, V_{(k)}(\mathrm{d}\chi)$$
$$= \int k_Y(T_{-y}\chi)\, (V_{(k)} \times \delta_x)\, (\mathrm{d}[\chi, y]).$$

Thus from 6.1.1. and the measurability of the mapping $V_{(.)}$ we can conclude that the real function $a \curvearrowright [V]_{(a)}(Y)$ is measurable with respect to the σ-algebra $\mathfrak{A}_1$. Hence $[V]_{(.)}$ is an homogeneous cluster field on $[A_1, \varrho_{A_1}]$ with phase space $[A_2, \varrho_{A_2}]$.

Now we suppose that the cluster field $V_{(.)}$ is continuous and that f is any bounded real function on M_2 which is continuous with respect to ϱ_{M_2}. Then we have

$$\int f(\chi)\, [V]_{([x,k])}(\mathrm{d}\chi) = \int f(T_{-y}\chi)\, (V_{(k)} \times \delta_x)\, (\mathrm{d}[\chi, y]).$$

The mapping $x \curvearrowright \delta_x$ is continuous with respect to ϱ_M and even with respect to the weak convergence. Therefore (cf. Billingsley [1], Theorem 3.2) $[x, k] \curvearrowright V_{(k)} \times \delta_x$ is continuous with respect to the weak convergence, and by 6.1.1. we can conclude the continuity of the mapping

$$a \curvearrowright \int f(\chi)\, [V]_{(a)}(\mathrm{d}\chi) \qquad (a \in A_1).$$

Thus, by 3.1.2. the cluster field $[V]$ is continuous.

Summarizing we observe the truth of

11.1.1. Proposition. *The correspondence* $V \curvearrowright [V]$ *provides a one-to-one mapping of the set of all cluster fields* $V_{(.)}$ *on* $[K_1, \varrho_{K_1}]$ *with phase space* $[A_2, \varrho_{A_2}]$ *onto the set of all homogeneous cluster fields on* $[A_1, \varrho_{A_1}]$ *with phase space* $[A_2, \varrho_{A_2}]$. *In this case* $[V]_{(.)}$ *is continuous if and only if* $V_{(.)}$ *has this property.*

Obviously the cluster field $\varkappa(K)$ associated with a substochastic kernel K on $[A_1, \varrho_{A_1}]$ with phase space $[A_2, \varrho_{A_2}]$ is homogeneous if and only if, for all x in R^s,

$$K([x, k], X_1 \times X_2) = K([[0, \ldots, 0], k], (X_1 - x) \times X_2) \qquad (X_1 \in \mathfrak{R}^s, X_2 \in \mathfrak{K}_2)$$

is satisfied. If we then denote by I the substochastic kernel

$$[k, Z] \frown K([[0, \ldots, 0], k], Z) \qquad (k \in K_1,\ Z \in \mathfrak{A}_2)$$

on $[K_1, \varrho_{K_1}]$ with phase space $[A_2, \varrho_{A_2}]$, we get

$$\varkappa(K) = [\varkappa(I)].$$

Now let $[K_3, \varrho_{K_3}]$ be a further bounded complete separable metric space and $[A_3, \varrho_{A_3}]$ the direct product of R^s and $[K_3, \varrho_{K_3}]$. Immediately from the definition of homogeneous cluster fields it follows that

11.1.2. *If $V_{(.)}$ is a cluster field on $[K_1, \varrho_{K_1}]$ with phase space $[A_2, \varrho_{A_2}]$ and $W_{(.)}$ a cluster field on $[K_2, \varrho_{K_2}]$ with phase space $[A_3, \varrho_{A_3}]$ then the cluster field $[W] \circ [V]$ exists if and only if the cluster field $k \frown (V_{(k)})_{[W]}$ exists, and we have*

$$[W] \circ [V] = [(V_{(.)})_{[W]}].$$

Clearly the cluster field $(V_{(.)})_{[W]}$ exists if for all k in K_1 the condition $V_{(k)}(\chi \text{ finite}) = 1$ is satisfied. If, in addition, $W_{(j)}(\chi \text{ finite}) = 1$ holds for all j in K_2, then we have $(V_{(k)})_{[W]}(\chi \text{ finite}) = 1,\ k \in K_1$. In particular, if $[K_1, \varrho_{K_1}]$ coincides with $[K_2, \varrho_{K_2}]$, for all cluster fields $V_{(.)}$ on $[K_1, \varrho_{K_1}]$ with phase space $[A_1, \varrho_{A_1}]$ having the property that $V_{(k)}(\chi \text{ finite}) = 1,\ k \in K_1$, we define inductively by

$$V_{(k)}^{[0]} = \delta_{\delta_{[[0,\ldots,0],k]}}, \qquad V_{(k)}^{[n+1]} = (V_{(k)}^{[n]})_{[V]} \qquad (k \in K_1;\ n = 0, 1, \ldots)$$

the *cluster powers* $V_{(.)}^{[n]}$ of $V_{(.)}$. Then, for all non-negative integers n, m

$$V_{(.)}^{[n+m]} = (V_{(.)}^{[n]})_{[V^{[m]}]}.$$

Obviously, for any stationary measure H on $\mathfrak{M}_1$ and any cluster field $V_{(.)}$ on $[K_1, \varrho_{K_1}]$ with phase space $[A_2, \varrho_{A_2}]$, the measure $H_{[V]}$ is also stationary if it exists.

11.1.3. Proposition. *If for a stationary distribution P on $\mathfrak{M}_1$ and a cluster field $V_{(.)}$ on $[K_1, \varrho_{K_1}]$ with phase space $[A_2, \varrho_{A_2}]$, the distribution $P_{[V]}$ exists, then we have*

$$\gamma_{P_{[V]}}(L) = \int \varrho_{V_{(l)}}(R^s \times L)\, \gamma_P(\mathrm{d}l) \qquad (L \in \mathfrak{K}_2).$$

Proof. For all natural numbers k and all Φ in M_1 we set

$$\Phi_k = \sum_{\substack{[x,j] \in A_1, \Phi(\{[x,j]\}) > 0 \\ \Phi^x(\{x + (-2,2)^s\}) \leq k}} \Phi(\{[x, j]\})\, \delta_{[x,j]}.$$

It can be shown with the method used in the first step of the proof of 6.4.2. that all mappings $\Phi \frown \Phi_k$ are measurable with respect to $\mathfrak{M}_1$. We denote by P_k the image of P with respect to $\Phi \frown \Phi_k$. Obviously, all P_k are stationary and we have

$$\sup_{k=1,2,\ldots} \varrho_{P_k}(Z) = \varrho_P(Z) \qquad (Z \in \mathfrak{A}_2)$$

and $\gamma_{P_k}(K_1) \leqq k$. From $P([V]_{(\Phi)}$ exists$) = 1$ we obtain $P_k([V]_{(\Phi)}$ exists$) = 1$ and from this the existence of all $(P_k)_{[V]}$, $k = 1, 2, \ldots$. Using 4.2.2. and 6.1.5., we now get for all L in $\mathfrak{R}_2$

$$\gamma_{P_{[V]}}(L) = \varrho_{P_{[V]}}\big([0,1)^s \times L\big) = \int \varrho_{[V](a)}\big([0,1)^s \times L\big)\, \varrho_P(\mathrm{d}a)$$

$$= \sup_{k=1,2,\ldots} \int \varrho_{[V](a)}\big([0,1)^s \times L\big)\, \varrho_{P_k}(\mathrm{d}a)$$

$$= \sup_{k=1,2,\ldots} \int \varrho_{V_{(l)}}\big(([0,1)^s - x) \times L\big)\, (\mu \times \gamma_{P_k})\, (\mathrm{d}[x, l])$$

$$= \sup_{k=1,2,\ldots} \int \Big(\int \varrho_{V_{(l)}}\big(([0,1)^s - x) \times L\big)\, \mu(\mathrm{d}x)\Big)\, \gamma_{P_k}(\mathrm{d}l)$$

$$= \sup_{k=1,2,\ldots} \int \Big(\int \mu\big([0,1)^s - x\big)\, \varrho_{V_{(l)}}\big((\mathrm{d}x) \times L\big)\Big)\, \gamma_{P_k}(\mathrm{d}l)$$

$$= \sup_{k=1,2,\ldots} \int \varrho_{V_{(l)}}(R^s \times L)\, \gamma_{P_k}(\mathrm{d}l) = \int \varrho_{V_{(l)}}(R^s \times L)\, \gamma_P(\mathrm{d}l). \quad \blacksquare$$

If a distribution P on $\mathfrak{M}_1$ is free from after-effects, this is not necessarily true for $P_{[V]}$. We have, however, (cf. Westcott [1])

11.1.4. Proposition. *If, for an ergodic or a mixing stationary distribution P on $\mathfrak{M}_1$ and a cluster field $V_{(.)}$ on $[K_1, \varrho_{K_1}]$ with phase space $[A_2, \varrho_{A_2}]$, the distribution $P_{[V]}$ exists, the latter is also ergodic or mixing, respectively.*

Proof. 1. We consider a new mark space $[K', \varrho_{K'}] = [M_2, \varrho_{M_2}]$ and write $[A', \varrho_{A'}]$ for the direct product of R^s and $[K', \varrho_{K'}]$. The correspondence

$$[x, k] \curvearrowright \varkappa_{([x,k])} = \int \delta_{\delta_{[x,\chi]}}(.)\, V_{(k)}(\mathrm{d}\chi) \qquad ([x, k] \in A_1)$$

defines an homogeneous cluster field $\varkappa$ on $[A_1, \varrho_{A_1}]$ with phase space $[A', \varrho_{A'}]$. Each marked point $[x, k]$ in A_1 is replaced by the marked point $[x, \chi]$, χ being selected at random according to $V_{(k)}$. Obviously, $P_\varkappa$ exists for all distributions P on $\mathfrak{M}_1$.

2. Let $\mathfrak{V}$ be the union of all σ-algebras ${}_{X \times K'}\mathfrak{M}'$ where X are bounded sets in $\mathfrak{R}^s$. If Y_1, Y_2 are sets in $\mathfrak{V}$, then we have, for all y in R^s with sufficiently large norm and all Φ in M_1,

$$\varkappa_{(\Phi)}(Y_1 \cap T_y Y_2) = \varkappa_{(\Phi)}(Y_1)\, \varkappa_{(\Phi)}(T_y Y_2)$$

$$= \varkappa_{(\Phi)}(Y_1)\, \varkappa_{(T_{-y}\Phi)}(Y_2).$$

Hence, for all mixing stationary distributions P on $\mathfrak{M}_1$,

$$\lim_{\|y\| \to +\infty} P_\varkappa(Y_1 \cap T_y Y_2) = \lim_{\|y\| \to +\infty} \int \varkappa_{(\Phi)}(Y_1)\, \varkappa_{(T_{-y}\Phi)}(Y_2)\, P(\mathrm{d}\Phi)$$

$$= \int \varkappa_{(\Phi)}(Y_1)\, P(\mathrm{d}\Phi) \int \varkappa_{(\Phi)}(Y_2)\, P(\mathrm{d}\Phi) = P_\varkappa(Y_1)\, P_\varkappa(Y_2).$$

The ring $\mathfrak{V}$ generates $\mathfrak{M}'$, so we can conclude that $P_\varkappa$ is mixing, too.

If P is ergodic, we get in the same way, by 6.2.9.,

$$\lim_{t\to\infty} (2t)^{-s} \int_{[-t,t)^s} P_\varkappa(Y_1 \cap T_y Y_2)\, \mu(dy) = P_\varkappa(Y_1)\, P_\varkappa(Y_2) \qquad (Y_1, Y_2 \in \mathfrak{B}).$$

Consequently, 6.2.9. also yields the ergodicity of $P_\varkappa$.

3. If P is a stationary distribution on $\mathfrak{M}_1$ such that $P_{[V]}$ exists, then the counting measure

$$\Psi^v = \sum_{a'=[x,\chi]\in A', \Psi(\{a'\})>0} \Psi(\{a'\})\, T_{-x}\chi$$

exists almost everywhere with respect to $P_\varkappa$ and is distributed according to $P_{[V]}$, if Ψ is distributed according to $P_\varkappa$. The measurable mapping $\Psi \rightsquigarrow \Psi^v$ of M' into M_2 commutes with all translation operators T_x. Hence, if $P_\varkappa$ is ergodic or mixing, respectively, this is also true for $P_{[V]}$. Thus, in view of 2., $P_{[V]}$ is ergodic or mixing, respectively, if this is true for P. ∎

The following example, due to K. Hermann, shows that the converse of 11.1.4. is not true.

Let $s = 1$, $i = 1, 2$, $K_i = \{1, 2\}$, and $F_i = k_{[i,+\infty)}$. Then the measurable mapping $\Phi \rightsquigarrow \Phi \times \delta_i$ transforms the distribution P_{F_i} into a distribution P_i on $\mathfrak{M}_1$. We consider the cluster field $V_{(j)} = {}_{[0,j)\times\{1\}}(P_{\mu\times\delta_1})$, $j = 1, 2$, on K_1 with phase space $[A_2, \varrho_{A_2}]$. Then, for the stationary non-ergodic distribution $P = 2^{-1}(P_1 + P_2)$, the distribution $P_{[V]} = P_{\mu\times\delta_1}$ is mixing.

11.2. The Set $\boldsymbol{Q}$

Now we continue the investigations of the preceding section under the additional assumption that the mark space $[K_1, \varrho_{K_1}]$ consists of only one element. For simplicity reasons we identify $[A_1, \varrho_{A_1}]$ with R^s and omit the index 1 in the corresponding expressions. Instead of cluster fields $V_{(.)}$ on $[K_1, \varrho_{K_1}]$ with phase space $[A_2, \varrho_{A_2}]$ we now have distributions Q on $\mathfrak{M}_2$.

Let $\boldsymbol{Q}$ denote the set of those distributions Q on $\mathfrak{M}_2$ which satisfy for all bounded sets X in $\mathfrak{R}^s$ the condition

$$\int Q\big(\chi^x(X - x) > 0\big)\, \mu(dx) < +\infty.$$

The question whether Q belongs to $\boldsymbol{Q}$ depends only on $(Q^x)^*$. A distribution Q on $\mathfrak{M}_2$ belongs to $\boldsymbol{Q}$ already if

$$\int Q\big(\chi^x(X_0 - x) > 0\big)\, \mu(dx) < +\infty$$

is valid for some bounded set X_0 in $\mathfrak{R}^s$ having a non-empty interior. Indeed, every bounded set X in $\mathfrak{R}^s$ can be covered by a finite sequence $X_0 + x_1, \ldots, X_0 + x_n$, so we obtain

$$\begin{aligned}\int Q\big(\chi^x(X - x) > 0\big)\, \mu(dx) &\leq \int Q\Big(\chi \Big(\bigcup_{i=1}^{n} (X_0 + x_i) - x\Big) > 0\Big)\, \mu(dx)\\ &\leq \sum_{i=1}^{n} \int Q\big(\chi^x(X_0 + x_i - x) > 0\big)\, \mu(dx)\\ &= n \int Q\big(\chi^x(X_0 - x) > 0\big)\, \mu(dx).\end{aligned}$$

For all Q in $\mathbf{Q}$, we get

$$\sum_{-\infty<m_1,\ldots,m_s<+\infty} Q\left(\chi^x\left(\underset{i=1}{\overset{s}{\times}} [m_i, m_i+1)\right) > 0\right)$$

$$\leqq \sum_{-\infty<m_1,\ldots,m_s<+\infty} \int_{\underset{i=1}{\overset{s}{\times}}[m_i,m_i+1)} Q\left(\chi^x([-1, 1)^s + y) > 0\right) \mu(\mathrm{d}y)$$

$$= \int Q\left(\chi^x([-1, 1)^s) - x\right) \mu(\mathrm{d}x) < +\infty.$$

Hence, in virtue of the Borel-Cantelli Lemma, we have

$$Q\left(\sum_{-\infty<m_1,\ldots,m_s<+\infty} \chi^x\left(\underset{i=1}{\overset{s}{\times}} [m_i, m_i+1)\right) < +\infty\right) = 1,$$

which leads to the following necessary condition:

11.2.1. *For all distributions Q in $\mathbf{Q}$, $Q(\chi$ finite$) = 1$.*

For all bounded X in $\mathfrak{R}^s$ and all distributions Q on $\mathfrak{M}_2$, we have

$$\int Q\left(\chi^x(X - x) > 0\right) \mu(\mathrm{d}x) \leqq \int \varrho_{Q^x}(X - x)\, \mu(\mathrm{d}x)$$
$$= \int \mu(X - x)\, \varrho_{Q^x}(\mathrm{d}x) = \mu(X)\, \varrho_Q(A_2),$$

which leads to the following sufficient condition:

11.2.2. *Every distribution Q on $\mathfrak{M}_2$ with the property $\varrho_Q(A_2) < +\infty$ belongs to $\mathbf{Q}$.*

If Q belongs to $\mathbf{Q}$, then so does the distribution $Q_\varkappa$ if the cluster field $\varkappa$ is defined by

$$\varkappa_{(a)} = \sum_{n=0}^{\infty} p_n \delta_{n\delta_a} \qquad (a \in A_2).$$

But $\varrho_{Q_\varkappa}(A_2)$ is infinite if $Q \neq \delta_o$ and $\sum_{n=1}^{\infty} np_n = \infty$. Consequently, the sufficient condition 11.2.2. is not necessary.

For every stationary distribution L on $\mathfrak{M}$ with finite intensity i_L and every Q in $\mathbf{Q}$, we get in virtue of 6.1.5. for all bounded sets X in $\mathfrak{R}^s$

$$\int [Q]_{(x)} \left(\chi(X \times K_2) > 0\right) \varrho_L(\mathrm{d}x) = i_L \int Q\left(\chi^x(X - x) > 0\right) \mu(\mathrm{d}x) < +\infty.$$

Recalling that each set in $\mathfrak{B}_2$ is contained in $X \times K_2$ for some bounded set X in $\mathfrak{R}^s$, we obtain immediately, by 4.2.1., the existence of $L_{[Q]}$.

If a stationary distribution P on $\mathfrak{M}$ only fulfils the weaker condition $P\left(s(\Phi) < +\infty\right) = 1$, then we set

$$P = \sum_{0 \leqq j < +\infty,\, \sigma_P([j,j+1))>0} \sigma_P\left([j, j+1)\right) P\left((.) \mid j \leqq s(\Phi) < j+1\right)$$

and apply the above considerations to the stationary distributions $P\left((.) \mid j \leqq s(\Phi) < j+1\right)$. Consequently, $P_{[Q]}$ exists in this case, too.

Hence, we observe the truth of

11.2.3. Proposition. *For all stationary distributions P on $\mathfrak{M}$ satisfying $P(s(\Phi) < +\infty) = 1$ and for all Q in $\boldsymbol{Q}$, the distribution $P_{[Q]}$ exists.*

As K. Hermann has shown, 11.2.3. has the following converse.

11.2.4. *If the clustered distribution $P_{[Q]}$ exists for a stationary distribution P on $\mathfrak{M}$ different from δ_o and a distribution Q on $\mathfrak{M}_2$, then Q belongs to $\boldsymbol{Q}$.*

Proof. Let $G = \{\ldots, -1, 0, 1, \ldots\}^s$. By virtue of

$$\Phi \curvearrowright \Phi^e = \Big(\sum_{g \in G} \Phi([0, 1)^s + g)\, \delta_g\Big)^*$$

we get a mapping of M into itself which is measurable with respect to $\mathfrak{M}$. We denote by P^e the image of P with respect to this mapping. Obviously, for all g in G,

$$P^e(\Psi(\{g\}) > 0) = P\big(\Phi([0, 1)^s + g) > 0\big) = c > 0.$$

For all Φ in M it follows from

$$\int Q(\chi^x(X - x) > 0)\, \Phi(\mathrm{d}x) < +\infty \qquad \text{for all bounded } X \text{ in } \mathfrak{R}^s$$

that

$$\int Q(\chi^x(X - x) > 0)\, \Phi^e(\mathrm{d}x) < +\infty \qquad \text{for all bounded } X \text{ in } \mathfrak{R}^s,$$

i.e., by 4.1.2., $\Phi \in M_{[Q]}$ implies $\Phi^e \in M_{[Q]}$. Therefore also $(P^e)_{[Q]}$ exists.

Let U be the distribution of an independent family $(\chi_g)_{g \in G}$, where χ_g is distributed according to $[Q]_{(g)}$. We set $W = P^e \times U$.

Now we suppose that Q does not belong to $\boldsymbol{Q}$, i.e. there exists a bounded set X_o in $\mathfrak{R}^s$ with the property $\int Q(\chi^x(X_o - x) > 0)\, \mu(\mathrm{d}x) = +\infty$. Then there is also a bounded set X in $\mathfrak{R}^s$ with $\sum_{g \in G} Q(\chi^x(X - g) > 0) = +\infty$. We set for all h in G

$$Y_h = \{[\Psi, (\chi_g)_{g \in G}] : \chi_h^x(X) > 0\},$$

$$Z_h = Y_h \cap \{[\Psi, (\chi_g)_{g \in G}] : \Psi(\{h\}) > 0\}.$$

Then we have

$$W(Z_h) = cW(Y_h),$$

$$\sum_{h \in G} W(Y_h) = +\infty = \sum_{h \in G} W(Z_h).$$

Let $n \curvearrowright g_n$ be a fixed one-to-one mapping of the set of all natural numbers onto G. Recalling that the events Y_g, $g \in G$, are independent we get

$$\lim_{n \to \infty} \frac{\sum_{i,j=1}^{n} W(Y_{g_i} \cap Y_{g_j})}{\Big(\sum_{l=1}^{n} W(Y_{g_l})\Big)^2} = \lim_{n \to \infty} \frac{\Big(\sum_{l=1}^{n} W(Y_{g_l})\Big)^2 - \sum_{l=1}^{n} \big(W(Y_{g_l})\big)^2 + \sum_{l=1}^{n} W(Y_{g_l})}{\Big(\sum_{l=1}^{n} W(Y_{g_l})\Big)^2} = 1.$$

Hence

$$\lim_{n\to\infty} \frac{\sum\limits_{i,j=1}^{n} W(Z_{g_i} \cap Z_{g_j})}{\left(\sum\limits_{l=1}^{n} W(Z_{g_l})\right)^2}$$

$$= \lim_{n\to\infty} \frac{\sum\limits_{i,j=1}^{n} P^e\big(\Psi(\{g_i\}) = 1 = \Psi(\{g_j\})\big) W(Y_{g_i} \cap Y_{g_j})}{c^2 \left(\sum\limits_{l=1}^{n} W(Y_{g_l})\right)^2}$$

$$\leqq c^{-2} \lim_{n\to\infty} \frac{\sum\limits_{i,j=1}^{n} W(Y_{g_i} \cap Y_{g_j})}{\left(\sum\limits_{l=1}^{n} W(Y_{g_l})\right)^2} = c^{-2}.$$

Taking into consideration that $\sum\limits_{i=1}^{\infty} W(Z_{g_i}) = +\infty$, we can conclude (cf. Spitzer [1], Proposition VI.26.3) that

$$W\left(\bigcap_{n=1}^{\infty} \bigcup_{m=n}^{\infty} Z_{g_m}\right) \geqq c^2.$$

Thus

$$W\left(\sum_{h\in G} \Psi(\{h\})\, \chi_h^x(X) = +\infty\right) \geqq c^2 > 0$$

which contradicts the existence of $(P^e)_{[Q]}$. Therefore Q belongs to $\boldsymbol{Q}$. ∎

In 11.2.3., the assumption $P\big(s(\Phi) < +\infty\big) = 1$ cannot be dropped. We have

11.2.5. Proposition. *Let Q be any distribution on $\mathfrak{M}$. Then the following two conditions are equivalent.*

a) *There exists an X in $\mathfrak{B}_2$ such that $Q = {}_XQ$.*

b) *$P_{[Q]}$ exists for all stationary distributions P on $\mathfrak{M}$.*

Proof. 1. First we prove that a) implies b). For any Φ in M and Z in $\mathfrak{B}_2$,

$$\int [Q]_{(x)}\big(\chi(Z) > 0\big)\, \Phi(\mathrm{d}x) = \int Q\Big(\chi\big((Z - x) \cap X\big) > 0\Big)\, \Phi(\mathrm{d}x) < +\infty.$$

Thus, 4.1.2. yields $M_{[Q]} = M$, i.e. $P_{[Q]}$ exists for all distributions P on $\mathfrak{M}$.

2. Suppose that a) is not fulfilled. Then there exists a sequence (x_i) of points in R^s such that the sets $[0, 1)^s + x_i$ $(i = 1, 2, \ldots)$ are pairwise disjoint and fulfil $Q\big(\chi^x([0, 1)^s + x_i) > 0\big) > 0$. For each natural number n, we have for the n-th convolution power of Q

$$\sum_{i=1}^{\infty} \int_{[0,1)^s + x_i} Q^n\big(\chi^x([-2, 2)^s + x) > 0\big)\, \mu(\mathrm{d}x) \geqq \sum_{i=1}^{\infty} Q^n\big(\chi^x([0, 1)^s + x_i) > 0\big).$$

But $Q^n\left(\chi^x([0,\,1)^s + x_i) > 0\right)$ tends monotonically to 1 as n increases and we get

$$\lim_{n\to\infty} \int Q^n\left(\chi^x([-2,\,2)^s + x) > 0\right) \mu(\mathrm{d}x) = +\infty.$$

Now we can choose a sequence (p_n) of positive real numbers satisfying

$$\sum_{n=1}^{\infty} p_n = 1, \qquad \sum_{n=1}^{\infty} p_n \int Q^n\left(\chi^x([-2,\,2)^s - x) > 0\right) \mu(\mathrm{d}x) = +\infty.$$

If we set $D = \sum_{n=1}^{\infty} p_n \delta_{n\delta_{[0,\ldots,0]}}$ we can therefore conclude that $\left((Q_\mu)_{[D]}\right)_{[Q]}$ does not belong to $\boldsymbol{W}_2$. Consequently, by the "clustering theorem" 4.3.3., $\left((P_\mu)_{[D]}\right)_{[Q]}$ does not exist, i.e. b) cannot be valid. ∎

Now we suppose in addition that K_2 consists of only one element. We set $[A_1, \varrho_1] = [A_2, \varrho_{A_2}] = R^s$ and omit the indices 1, 2 in the corresponding expressions.

In the set of all distributions V on $\mathfrak{M}$ with the property $V(\chi \text{ finite}) = 1$ we define by

$$[V_2, V_1] \curvearrowright (V_1)_{[V_2]}$$

an associative operation with the neutral element $\delta_{\delta_{[0,\ldots,0]}}$. The following example shows that $\boldsymbol{Q}$ is not closed with respect to this operation. Thus, the necessary condition 11.2.1. is not sufficient.

Let $s = 1$. For all natural numbers n, m, we set

$$\Psi_{n,m} = \delta_1 + \cdots + \delta_{n-1} + m\delta_n.$$

Now we choose a sequence (p_n) of positive real numbers with the properties

$$\sum_{n=1}^{\infty} p_n = 1, \qquad \sum_{n=1}^{\infty} n p_n < +\infty.$$

Then for each sequence of non-negative real numbers satisfying the condition $\sum_{n=1}^{\infty} q_n = 1$, the distribution

$$Q = \sum_{n,m=1}^{\infty} p_n q_m \delta_{\Psi_{n,m}}$$

belongs to $\boldsymbol{Q}$, since $\varrho_{Q^*}(R^1) = \sum_{n=1}^{\infty} n p_n < +\infty$; thus, by 11.2.2., we have $Q^* \in \boldsymbol{Q}$ and hence also $Q \in \boldsymbol{Q}$. We obtain

$$\begin{aligned}
&\int Q^{[2]}(\chi([0,\,1) - x) > 0)\, \mu(\mathrm{d}x) \\
&= \int \left(\int [Q]_{(\Psi)} (\chi([0,\,1) - x) > 0)\, Q(\mathrm{d}\Psi) \right) \mu(\mathrm{d}x) \\
&= \sum_{n,m=1}^{\infty} p_n q_m \int [Q]_{(\Psi_{n,m})} (\chi([0,\,1) - x) > 0)\, \mu(\mathrm{d}x)
\end{aligned}$$

$$\geqq \sum_{m=1}^{\infty} q_m \left(\sum_{n=1}^{\infty} p_n \int [Q]_{(m\delta_n)}(\chi([0,1) - x) > 0)\, \mu(\mathrm{d}x) \right)$$

$$= \sum_{m=1}^{\infty} q_m \int [Q]_{(m\delta_0)}(\chi([0,1) - x) > 0)\, \mu(\mathrm{d}x)$$

$$= \sum_{m=1}^{\infty} q_m \int Q^m(\chi([0,1) - x) > 0)\, \mu(\mathrm{d}x) = \sum_{m=1}^{\infty} q_m \left(\sum_{k=1}^{\infty} k z_{k,m} \right).$$

Here $z_{k,m}$ denotes the probability that the maximum of m independent random natural numbers which are identically distributed according to (p_n) attains the value k.

As m increases, the expectation $\sum_{k=1}^{\infty} k z_{k,m}$ tends towards $+\infty$. Thus we can choose (q_m) such that

$$\sum_{m=1}^{\infty} q_m \left(\sum_{k=1}^{\infty} k z_{k,m} \right) = +\infty$$

and hence $Q^{[2]} \notin \boldsymbol{Q}$.

To each distribution ν on $\mathfrak{R}^s$ we can associate the cluster field $[Q_\nu]$, and we have

$$[Q_\nu] = \varkappa(K) \quad \text{with} \quad K(x, X) = \nu(X - x) \qquad (x \in R^s,\ X \in \mathfrak{R}^s).$$

Obviously

$$(Q_\nu)_{[Q_\gamma]} = Q_{\nu * \gamma}$$

for all distributions γ on $\mathfrak{R}^s$. Consequently,

$$(Q_\nu)^{[n]} = Q_{\nu^n} \qquad (n = 0, 1, \ldots).$$

To finish this section we shall consider, in the special case $s = 1$, a certain class of distributions on $\mathfrak{M}$. Let F be a distribution function in $\boldsymbol{F}$ which satisfies the conditions $F(0) < 1$, $\lim_{x \to +\infty} F(x) = 1$, and let p be the distribution of a random non-negative integer. Denote by V the distribution of a sequence $\xi_1, \xi_2, \ldots$ of independent random non-negative numbers which are identically distributed according to F. The correspondence

$$[(x_n), h] \curvearrowright \Phi = \delta_{x_0} + \delta_{x_0 + x_1} + \cdots + \delta_{x_0 + \cdots + x_h}, \qquad x_0 = 0$$

provides a measurable mapping of $[0, +\infty)^{\{1,2,\ldots\}} \times \{0, 1, \ldots\}$ into M. We denote by $L_{F,p}$ the distribution of Φ with respect to $V \times p$. We have

11.2.6 *For all F in $\boldsymbol{F}$ satisfying the conditions $F(0) < 1$, $\lim_{x \to +\infty} F(x) = 1$, for all distributions p of random non-negative integers and for all $t > 0$,*

$$\int L_{F,p}(\chi((0, t] - x) > 0)\, \mu(\mathrm{d}x) = t + \left(\sum_{n=1}^{\infty} n p(\{n\}) \right) \int_0^t (1 - F(x))\, \mu(\mathrm{d}x).$$

Proof. Obviously we have

$$\begin{aligned}
&\int L_{F,p}(\chi((0,t]-x)>0)\,\mu(\mathrm{d}x)\\
&=\int_0^t L_{F,p}(\chi((0,t]-x)>0)\,\mu(\mathrm{d}x)+\int_{-\infty}^0 L_{F,p}(\chi((0,t]-x)>0)\,\mu(\mathrm{d}x)\\
&=t+\int_0^{+\infty} L_{F,p}(\chi((0,t]+x)>0)\,\mu(\mathrm{d}x).
\end{aligned}$$

Let η_i denote the sum $\xi_1+\cdots+\xi_i$ and let F_i be the distribution function associated with η_i. Then the integrand above can be transformed for non-negative x in the following way:

$$\begin{aligned}
&L_{F,p}(\chi((0,t]+x)>0)\\
&=p(\{1\})\,(F(x+t)-F(x))\\
&\quad+\sum_{j=2}^{\infty}p(\{j\})\left(\sum_{i=1}^{j-1}V(\eta_i\leqq x<\eta_{i+1}\leqq x+t)+F(x+t)-F(x)\right)\\
&=p(\{1,2,\ldots\})\,(F(x+t)-F(x))\\
&\quad+\sum_{i=1}^{\infty}p(\{i+1,i+2,\ldots\})\,V(\eta_i\leqq x<\eta_{i+1}\leqq x+t)\\
&=p(\{1,2,\ldots\})\,(F(x+t)-F(x))\\
&\quad+\sum_{i=1}^{\infty}p(\{i+1,i+2,\ldots\})\int_0^x(F(x+t-u)-F(x-u))\,\mathrm{d}F_i(u).
\end{aligned}$$

We obtain

$$\begin{aligned}
&p(\{1,2,\ldots\})\int_0^{\infty}(F(x+t)-F(x))\,\mu(\mathrm{d}x)\\
&\quad+\sum_{i=1}^{\infty}p(\{i+1,i+2,\ldots\})\int_0^{\infty}\left(\int_0^x(F(x+t-u)-F(x-u))\,\mathrm{d}F_i(u)\right)\mu(\mathrm{d}x)\\
&=p(\{1,2,\ldots\})\int_0^{\infty}(F(x+t)-F(x))\,\mu(\mathrm{d}x)\\
&\quad+\sum_{i=1}^{\infty}p(\{i+1,i+2,\ldots\})\int_0^{\infty}\left(\int_u^{\infty}(F(x+t-u)-F(x-u))\,\mu(\mathrm{d}x)\right)\mathrm{d}F_i(u)\\
&=p(\{1,2,\ldots\})\int_0^{\infty}(F(x+t)-F(x))\,\mu(\mathrm{d}x)\\
&\quad+\sum_{i=1}^{\infty}p(\{i+1,i+2,\ldots\})\left(\int_0^{\infty}\mathrm{d}F_i(u)\right)\int_0^{\infty}(F(x+t)-F(x))\,\mu(\mathrm{d}x)\\
&=\left(\sum_{i=1}^{\infty}ip(\{i\})\right)\int_0^{\infty}(F(x+t)-F(x))\,\mu(\mathrm{d}x)
\end{aligned}$$

$$= \left(\sum_{i=1}^{\infty} ip(\{i\})\right) \lim_{v\to+\infty} \left(\int_0^v F(x+t)\,\mu(\mathrm{d}x) - \int_0^v F(x)\,\mu(\mathrm{d}x)\right)$$

$$= \left(\sum_{i=1}^{\infty} ip(\{i\})\right) \lim_{v\to+\infty} \left(-\int_t^{t+v} (1-F(x))\,\mu(\mathrm{d}x) + \int_0^v (1-F(x))\,\mu(\mathrm{d}x)\right)$$

$$= \left(\sum_{i=1}^{\infty} ip(\{i\})\right) \lim_{v\to+\infty} \left(\int_0^t (1-F(x))\,\mu(\mathrm{d}x) - \int_v^{v+t} (1-F(x))\,\mu(\mathrm{d}x)\right).$$

Since $\int_v^{v+t} (1-F(x))\,\mu(\mathrm{d}x) \leqq t(1-F(v))$ we thus have

$$\int_0^{\infty} L_{F,p}(\chi((0,t]+x)>0)\,\mu(\mathrm{d}x) = \left(\sum_{i=1}^{\infty} ip(\{i\})\right) \int_0^t (1-F(x))\,\mu(\mathrm{d}x)$$

which completes the proof. ∎

On account of the relation

$$(P_{l\mu})_{[Q]}(\Phi(X)=0) = \exp\left(-\widetilde{(P_{l\mu})_{[Q]}}(\Psi(X)>0)\right)$$
$$= \exp\left(-(Q_{l\mu})_{[Q]}(\Psi(X)>0)\right) = \exp\left(-l\int Q(\chi(X-x)>0)\,\mu(\mathrm{d}x)\right)$$

which is valid for all $l \geqq 0$, $X \in \mathfrak{R}^s$ and $Q \in \boldsymbol{Q}$, we obtain by means of 11.2.6. (cf. Lewis [1], Franken & Richter [1])

11.2.7. Proposition. *Under the assumptions of 11.2.6., the distribution $L_{F,p}$ belongs to $\boldsymbol{Q}$ if and only if the expectation $\sum_{n=1}^{\infty} np(\{n\})$ is finite. In this case*

$$(P_{l\mu})_{[L_{F,p}]}(\Phi((0,t])=0)$$
$$= \exp\left(-lt - l\left(\sum_{n=1}^{\infty} np(\{n\})\right)\int_0^t (1-F(x))\,\mu(\mathrm{d}x)\right)$$

holds for all $l, t \geqq 0$.

11.3. The Mapping $\boldsymbol{Q} \rightsquigarrow \rangle\boldsymbol{Q}\langle$

In this section we use the same notations and assumptions as in the beginning of Section 11.2.

With each σ-finite measure Q on $\mathfrak{M}_2$ satisfying Q (χ finite) $= 1$ we associate by

$$\rangle Q\langle = \mathscr{C}_Q{}^T$$

a σ-finite measure $\rangle Q\langle$ on $\mathfrak{R}^s \otimes (\mathfrak{M}_2)^0$. Obviously, $\rangle Q_1\langle = \rangle Q_2\langle$ is equivalent to $\mathscr{C}_{Q_1} = \mathscr{C}_{Q_2}$ and by 5.1.10. therefore equivalent to $Q_1((.) \setminus \{o\}) = Q_2((.) \setminus \{o\})$. Thus, it follows

11.3.1. *Two distributions Q_1, Q_2 on $\mathfrak{M}_2$ with the property $Q_1(\chi$ finite$) = 1 = Q_2(\chi$ finite$)$ coincide if and only if $\rangle Q_1\langle = \rangle Q_2\langle$ is satisfied.*

By 5.1.3. we have for all X in $\mathfrak{R}^s$

$$\rangle Q\langle\,\big(X \times (M_2)^0\big) = \mathscr{C}_Q\big((X \times M_2) \cap C_2\big) = \varrho_Q(X \times K_2).$$

Thus

11.3.2. *For all σ-finite measures Q on $\mathfrak{M}_2$ satisfying $Q(\chi$ infinite$) = 0$,*

$$\rangle Q\langle\,\big((.) \times (M_2)^0\big) = \varrho_{Q^x}.$$

In the following we employ the symmetry operator $\hat{\Lambda}$ introduced in Section 8.4., which associates to each σ-finite measure H on $\mathfrak{M}_2$ satisfying the conditions

$$H(\Phi \text{ infinite}) = 0 = H\big(\Phi^x(\{[0, \ldots, 0]\}) = 0\big)$$

the measure

$$\hat{\Lambda}H = \int \Big(\big(\Phi^x(R^s)\big)^{-1} \sum_{x \in R^s, \Phi^x(\{x\})>0} \Phi^x(\{x\})\, \delta_{T_x\Phi}(.)\Big) H(\mathrm{d}\Phi).$$

This correspondence remains meaningful when we require only that $H(\{o\}) = 0$ instead of $H\big(\Phi(\{[0, \ldots, 0]\}) = 0\big) = 0$. We shall also maintain the notation $\hat{\Lambda}$ under these weaker assumptions. For the sake of simplicity, we set

$$\langle Q\rangle = \sum_{k=1}^{\infty} k\hat{\Lambda}\Big(Q\big(\chi \in (\cdot), \chi(A_2) = k\big)\Big)$$

for all σ-finite measures Q on $\mathfrak{M}_2$ with the property $Q(\chi$ finite$) = 1$. Now we supplement 11.3.2. by

11.3.3. *Under the assumptions of* 11.3.2.,

$$\rangle Q\langle\,\big(R^s \times (\cdot)\big) = \langle Q\rangle.$$

Proof. For all Y in $\mathfrak{M}_2$, we obtain

$$\begin{aligned}\rangle Q\langle\,(R^s \times Y) &= \int \Big(\int \delta_{[x, T_x\chi]}(R^s \times Y)\, \chi^x(\mathrm{d}x)\Big)\, Q(\mathrm{d}\chi)\\ &= \int \Big(\int \delta_{T_x\chi}(Y)\, \chi^x(\mathrm{d}x)\Big)\, Q(\mathrm{d}\chi) = \int \chi(A_2)\, (\hat{\Lambda}\delta_\chi)\, (Y)\, Q(\mathrm{d}\chi)\\ &= \sum_{k=1}^{\infty} k \int_{\{\chi:\, \chi(A_2)=k\}} (\hat{\Lambda}\delta_\chi)\, (Y)\, Q(\mathrm{d}\chi). \quad \blacksquare\end{aligned}$$

For the measures $\rangle Q\langle$, $Q \in \boldsymbol{Q}$, we have the following interpretation. Let Φ be a point process which is distributed according to a stationary distribution P on $\mathfrak{M}$ with intensity one. If Φ is clustered according to the homogeneous cluster field $[Q]$, we get a point process Ψ, distributed according to $P_{[Q]}$. Each point $[y, k]$ of Ψ belongs to a cluster χ generated by a point x of Φ. To $[y, k]$ we can now associate the distance $y - x$ of $[y, k]$ to the "father point" x and the translated cluster $T_y\chi$. Then the mean number of points of Ψ with positions in $[0, 1)^s$ for which $[y - x, T_y\chi]$ belongs to a measurable set Z, is $\rangle Q\langle\,(Z)$.

To be more precise, we use the modified metric $\frac{\|x - y\|}{1 + \|x - y\|}$ in R^s and get a bounded complete separable metric space which we denote by $\check{R}^s$. Moreover we denote by $[K_3, \varrho_{K_3}]$ the direct product of $\check{R}^s$ and $[(M_2)^0, \varrho_{(M_2)^0}]$ and by $[A_3, \varrho_{A_3}]$ the direct product of R^s and the mark space $[K_3, \varrho_{K_3}]$. By

$$\chi \rightsquigarrow \chi^{\circledast} = \int \delta_{[x,[x,T_x\Psi]]}(.)\ \chi^x(\mathrm{d}x)$$

we obtain a measurable mapping of M_2 into M_3. If V is a measure on $\mathfrak{M}_2$ we will denote by $V^{\circledast}$ the image of V with respect to that mapping.

Now we will give a precise description of the afore-mentioned interpretation of $\rangle Q\langle$, $Q \in \boldsymbol{Q}$.

11.3.4. *For all Q in $\boldsymbol{Q}$ and all stationary distributions P on $\mathfrak{M}$ satisfying $i_P < +\infty$ we have*

$$\gamma_{P_{[Q^{\circledast}]}} = i_P \rangle Q\langle .$$

Proof. The existence of $P_{[Q]}$ follows from 11.2.3. Hence $P_{[Q^{\circledast}]}$ exists, too. With the help of 11.1.3. we get for all Z in $\mathfrak{K}_3$

$$\begin{aligned}\gamma_{P_{[Q^{\circledast}]}}(Z) &= i_P \varrho_{Q^{\circledast}}(R^s \times Z)\\ &= i_P \int \left(\int \delta_{[x,[x,T_x\chi]]}(R^s \times Z)\ \chi^x(\mathrm{d}x)\right) Q(\mathrm{d}\chi)\\ &= i_P \int \left(\int \delta_{[x,T_x\chi]}(Z)\ \chi^x(\mathrm{d}x)\right) Q(d\chi) = i_P \mathscr{C}_Q{}^T. \ \blacksquare\end{aligned}$$

Let H be any stationary measure on $\mathfrak{M}$ such that H^0 is defined, i.e. $H\big((.) \setminus \{o\}\big)$ is σ-finite. For the sake of simplicity we denote by $H^{0!}$ the σ-finite measure $H^0\big(\Phi - \delta_{[0,\ldots,0]} \in (.)\big)$ on $\mathfrak{M}$. Let Q be any distribution in $\boldsymbol{Q}$ such that $H_{[Q]}$ exists and $H_{[Q]}\big((\cdot) \setminus \{o\}\big)$ is σ-finite. Under those assumptions we can deduce a simple expression for the Palm measure $(H_{[Q]})^0$.

First of all we state that, by the invariance of $M_{[Q]}$ with respect to all translation operators T_x, $x \in R^s$, and by 8.1.6., from $H(M \setminus M_{[Q]}) = 0$ it follows $H^0(M \setminus M_{[Q]}) = 0$. Therefore the existence of $H_{[Q]}$ implies the existence of $(H^0)_{[Q]}$ and now more than ever the existence of $(H^{0!})_{[Q]}$.

In virtue of 5.2.3. we obtain

$$\begin{aligned}\mathscr{C}_{H_{[Q]}} &= \int \mathring{\Psi}(.)\ H_{[Q]}(\mathrm{d}\Psi) = \int \left(\int \mathring{\Psi}(.)\ [Q]_{(\Phi)}\ (\mathrm{d}\Psi)\right) H(\mathrm{d}\Phi)\\ &= \int \mathscr{C}_{[Q]_{(\Phi)}}(.)\ H(\mathrm{d}\Phi) = \int \mathscr{C}_{\underset{y \in R^s, \Phi(\{y\})>0}{\circledast} ([Q]_{(y)})^{\Phi(\{y\})}}(.)\ H(\mathrm{d}\Phi)\\ &= \int \left(\int \mathscr{C}_{[Q]_{(y)}} \circledast [Q]_{(\Phi - \delta_y)}\ (.)\ \Phi(\mathrm{d}y)\right) H(\mathrm{d}\Phi).\end{aligned}$$

For each $Z \in \mathfrak{C}$, the measurability of the mappings

$$y \rightsquigarrow \mathscr{C}_{[Q]_{(y)}}(Z) \qquad (y \in R^s),$$

$$\Psi \rightsquigarrow [Q]_{(\Psi)} \qquad (\Psi \in M_{[Q]})$$

implies with the help of 5.2.2. the measurability of

$$[y, \Phi] \curvearrowright \left(\mathscr{C}_{[Q](y)} \circledast [Q]_{(\Phi-\delta_y)}\right)(Z) \qquad ([y, \Phi] \in C),$$

and, by 5.1.2. and 8.1.1., we can continue the chain of equations with

$$\begin{aligned}
&= \int (\mathscr{C}_{[Q](y)} \circledast [Q]_{(\Phi-\delta_y)})\,(.)\,\mathscr{C}_H(\mathrm{d}[y, \Phi]) \\
&= \int (\mathscr{C}_{[Q](y)} \circledast [Q]_{(T_{-y}\Phi-\delta_y)})\,(.)\,\mathscr{C}_H{}^T(\mathrm{d}[y, \Phi]) \\
&= \int \left(\mathscr{C}_{[Q](y)} \circledast [Q]_{(T_{-y}(\Phi-\delta_{[0,\ldots,0]}))}\right)(.)\,(\mu \times H^0)\,(\mathrm{d}[y, \Phi]) \\
&= \int (\mathscr{C}_{[Q](y)} \circledast [Q]_{(T_{-y}\Phi)})\,(.)\,(\mu \times H^{0!})\,(\mathrm{d}[y, \Phi]).
\end{aligned}$$

Hence, by 8.1.1.,

$$\begin{aligned}
\mu \times (H_{[Q]})^0 &= (\mathscr{C}_{H_{[Q]}})^T \\
&= \int (\mathscr{C}_{[Q](y)} \circledast [Q]_{(T_{-y}\Phi)})^T\,(.)\,(\mu \times H^{0!})\,(\mathrm{d}[y, \Phi]) \\
&= \int \left(\int \left(\int \delta_{[z, T_z(\chi+\Psi)]}(.)\,\chi^x(\mathrm{d}z)\right)([Q]_{(y)} \times [Q]_{(T_{-y}\Phi)}) \right. \\
&\qquad \left. (\mathrm{d}[\chi, \Psi])\right)(\mu \times H^{0!})\,(\mathrm{d}[y, \Phi]).
\end{aligned}$$

Now we use the homogeneity of the cluster field $[Q]$ and continue with

$$\begin{aligned}
&= \int \left(\int \left(\int \delta_{[z+y, T_z(\chi+\Psi)]}(.)\,\chi^x(\mathrm{d}z)\right)(Q \times [Q]_{(\Phi)})\,(\mathrm{d}[\chi, \Psi])\right) \\
&\qquad\qquad (\mu \times H^{0!})\,(\mathrm{d}[y, \Phi]) \\
&= \int \left(\int \left(\int \left(\int \left(\int \delta_{[z+y, T_z(\chi+\Psi)]}(.)\,\mu(\mathrm{d}y)\right)\chi^x(\mathrm{d}z)\right)[Q]_{(\Phi)}\,(\mathrm{d}\Psi)\right)H^{0!}(\mathrm{d}\Phi)\right)Q(\mathrm{d}\chi) \\
&= \mu \times \int \left(\int \left(\int \delta_{T_z(\chi+\Psi)}(.)\,\chi^x(\mathrm{d}z)\right)(H^{0!})_{[Q]}\,(\mathrm{d}\Psi)\right)Q(\mathrm{d}\chi).
\end{aligned}$$

For all Y in $\mathfrak{M}_2$ we therefore have

$$\begin{aligned}
&(H_{[Q]})^0\,(Y) \\
&= \int \left(\int \left(\int k_Y(T_z\chi + T_z\Psi)\,\chi^x(\mathrm{d}z)\right)Q(\mathrm{d}\chi)\right)(H^{0!})_{[Q]}\,(\mathrm{d}\Psi) \\
&= \int \left(\int k_Y(\chi + T_z\Psi)\,\mathscr{C}_Q{}^T(\mathrm{d}[z, \chi])\right)(H^{0!})_{[Q]}\,(\mathrm{d}\Psi).
\end{aligned}$$

Thus, we observe the truth of

11.3.5. Theorem. *Let Q be a distribution in $\boldsymbol{Q}$ and H be a stationary measure on $\mathfrak{M}$ such that $H_{[Q]}$ exists. We suppose that both $H\big((.) \setminus \{o\}\big)$ and $H_{[Q]}\big((.) \setminus \{o\}\big)$ are σ-finite. Then*

$$(H_{[Q]})^0 = \int \left(\int k_{(.)}(\chi + T_z\Psi)\,(H^{0!})_{[Q]}\,(\mathrm{d}\Psi)\right) \rangle Q \langle\,(\mathrm{d}[z, \chi]).$$

As in the end of the preceding section we now suppose that both $[A_1, \varrho_{A_1}]$ and $[A_2, \varrho_{A_2}]$ coincide with R^s.

If ν is any distribution on $\Re^s$ and P any distribution on $\mathfrak{M}$ such that $P_{[Q_\nu]}$ exists, then, immediately by 11.3.5. and

$$\rangle Q_\nu \langle = \nu \times \delta_{\delta[0,\ldots,0]}$$

it follows

$$\begin{aligned}(P_{[Q_\nu]})^0 &= \int \left(\int k_{(.)}(\delta_{[0,\ldots,0]} + T_z\Psi)\,(P^{0!})_{[Q_\nu]}\,(\mathrm{d}\Psi)\right) \nu(\mathrm{d}z) \\ &= \delta_{\delta[0,\ldots,0]} * \int (P^{0!})_{[Q_{\nu*\delta_{-z}}]}\,(.)\,\nu(\mathrm{d}z).\end{aligned}$$

Consequently (cf. Port & Stone [2]),

11.3.6. Proposition. *Let ν be a distribution on $\Re^s$. If for a stationary distribution P on $\mathfrak{M}$ the clustered distribution $P_{[Q_\nu]}$ exists then*

$$(P_{[Q_\nu]})^0 = \delta_{\delta[0,\ldots,0]} * \int (P^{0!})_{[Q_{\nu*\delta_{-z}}]}\,(.)\,\nu(\mathrm{d}z)$$

holds.

11.4. Homogeneous Clustering Representations

In this section, too, we use first of all the same assumptions and notations as in the beginning of Section 11.2.

By 4.3.3., all stationary distributions having the form $(P_{l\mu})_{[Q]}$ with $l \geqq 0$ and $Q \in \boldsymbol{Q}$ are infinitely divisible, and we have

$$\widetilde{(P_{l\mu})_{[Q]}} = (Q_{l\mu})_{[Q]}\left((.) \setminus \{o\}\right) = l \int [Q]_{(x)}\left((.) \setminus \{o\}\right) \mu(\mathrm{d}x).$$

We conceive of a *homogeneous clustering representation* of a distribution P in $\boldsymbol{P}_2$ as a clustering representation of P on R^s having the form $P = (P_{l\mu})_{[Q]}$, where $l \geqq 0$ and $Q \in \boldsymbol{Q}$. Obviously two homogeneous clustering representations $P = (P_{l_1\mu})_{[Q_1]}$, $P = (P_{l_2\mu})_{[Q_2]}$ of a distribution P in $\boldsymbol{P}_2$ being different from δ_o are equivalent if and only if $l_1 = l_2$ as well as $Q_1 = Q_2$ are satisfied.

Using a given homogeneous clustering representation of P, the Palm measure $(\tilde{P})^0$ can readily be calculated.

11.4.1. Proposition. *For every $l \geqq 0$ and every Q in $\boldsymbol{Q}$,*

$$\left(\widetilde{(P_{l\mu})_{[Q]}}\right)^0 = l\langle Q\rangle.$$

Proof. With the help of 11.3.5. and 11.3.3. we obtain

$$\begin{aligned}\left(\widetilde{(P_{l\mu})_{[Q]}}\right)^0 &= l\left((Q_\mu)_{[Q]}\right)^0 \\ &= l \int \left(\int k_{(.)}(\chi + T_z\Psi)\,\delta_o(\mathrm{d}\Psi)\right) \rangle Q \langle\,(\mathrm{d}[z,\chi]) \\ &= l \int k_{(.)}(\chi)\,\langle Q\rangle\,(\mathrm{d}\chi) = l\langle Q\rangle. \blacksquare\end{aligned}$$

Obviously a homogeneous clustering representation $P = (P_{l\mu})_{[Q]}$ of a distribution P on $\mathfrak{M}_2$ different from δ_o is reduced if and only if $Q(\{o\}) = 0$ is satisfied. In the light of the general statements made in Section 4.5., the reduction of a homogeneous clustering representation $P = (P_{l\mu})_{[Q]}$ of such a P will yield the reduced homogeneous clustering representation $P = (P_{l'\mu})_{[Q']}$, where $l' = lQ(\chi \neq o)$ and $Q' = Q\big((.) \mid \chi \neq o\big)$.

Immediately from 8.3.1. and 11.4.1. we get

11.4.2. Proposition. *Let* $l_1 > 0$, $l_2 \geqq 0$, $Q_1 \in \boldsymbol{Q}$, $Q_2 \in \boldsymbol{Q}$ and $Q_1(\{o\}) = 0 = Q_2(\{o\})$. *Subject to these assumptions,* $(P_{l_1\mu})_{[Q_1]} = (P_{l_2\mu})_{[Q_2]}$ *if and only if* $l_1 = l_2$ *as well as* $\hat{\Lambda}Q_1 = \hat{\Lambda}Q_2$ *are satisfied.*

For all $l \geqq 0$ and all Q in $\boldsymbol{Q}$ we have, in view of 11.4.1., $\big(\widetilde{(P_{l\mu})_{[Q]}}\big)^0$ (Ψ infinite) $= 0$. Hence, in virtue of 8.3.3., $P = (P_{l\mu})_{[Q]}$ is a regular infinitely divisible distribution.

Conversely, let P be an arbitrary stationary regular infinitely divisible distribution on $\mathfrak{M}_2$ being different from δ_o. In view of 8.4.2. we have $(\tilde{P})^0$ (χ infinite) $= 0$, $\hat{\Lambda}(\tilde{P})^0 = (\tilde{P})^0$ and

$$\int \left(\sum_{k=1}^{\infty} k^{-1}(\tilde{P})^0 \big(\chi^x(X - x) > 0, \chi(A_2) = k\big)\right) \mu(\mathrm{d}x) < +\infty$$

for all bounded X in $\Re^s$. We shall now show that there exists a distribution Q in $\boldsymbol{Q}$ and a real number $l > 0$ such that $(\tilde{P})^0 = l\langle Q\rangle$ is valid. First of all, by our assumptions we have

$$(\tilde{P})^0 = \sum_{k=1}^{\infty} \hat{\Lambda}\left((\tilde{P})^0 \big(\chi \in (.), \chi(A_2) = k\big)\right).$$

Using the set-up

$$Q' = \sum_{k=1}^{\infty} k^{-1}(\tilde{P})^0 \big(\chi \in (.), \chi(A_2) = k\big)$$

we obtain a measure on $\mathfrak{M}_2$ which is different from 0 and has the properties

$$Q'(\{o\}) = 0 = Q' \text{ (χ infinite)}.$$

As it was shown in the second step of the proof of 8.4.2., the measure Q' is finite, so that we may choose an $l > 0$ such that $Q = l^{-1}Q'$ is a distribution. In virtue of an afore-mentioned property of Q', the distribution Q belongs to $\boldsymbol{Q}$. Now we obtain

$$\begin{aligned}(\tilde{P})^0 &= \sum_{k=1}^{\infty} \hat{\Lambda}\left((\tilde{P})^0 \big(\chi \in (.), \chi(A_2) = k\big)\right)\\ &= l \sum_{k=1}^{\infty} k\hat{\Lambda}\left(Q\big(\chi \in (.), \chi(A_2) = k\big)\right) = l\langle Q\rangle\end{aligned}$$

and, using 11.4.1. and 8.3.1., we are able to conclude that $P = (P_{l\mu})_{[Q]}$.

Finally, if P equals δ_o, then for any Q in $\boldsymbol{Q}$ we obtain in the form of $P = (P_{0\mu})_{[Q]}$ a homogeneous clustering representation.

Summarizing, we observe the truth of

11.4.3. Theorem. *A distribution P on $\mathfrak{M}_2$ has a homogeneous clustering representation if and only if it is both a stationary and regular infinitely divisible distribution.*

By means of 11.4.2. we are able to transform every reduced homogeneous clustering representation $P = (P_{l\mu})_{[Q]}$ of a stationary infinitely divisible distribution P

on $\mathfrak{M}_2$ different from δ_0 into $P = (P_{l\mu})_{[\hat{\Lambda}Q]}$. This special reduced homogeneous clustering representation, which is uniquely determined by P, corresponds to the stochastic transition

$$c_{(\Psi)} = (\Psi(A_2))^{-1}\, \Psi((.) \times K_2) \qquad (\Psi \in M_2,\ 0 < \Psi(A_2) < +\infty).$$

(cf. the third example in Section 4.6.).

Now we consider again the non-marked case $[A_1, \varrho_{A_1}] = [A_2, \varrho_{A_2}] = R^s$. By analogy to 4.5.7., we have (cf. Redheffer [1], Khinchin [2])

11.4.4. Theorem. *For each stationary distribution P on $\mathfrak{M}$ which is different from δ_0 the following statements are equivalent:*

a) *P is free from after-effects.*
b) *P has a reduced homogeneous clustering representation $P = (P_{l\mu})_{[Q]}$ with the property that*

$$Q\big(\chi(R^s \setminus \{[0, \ldots, 0]\}) > 0\big) = 0.$$

c) *P is a regular infinitely divisible distribution and has exactly one homogeneous clustering representation $P = (P_{l\mu})_{[Q]}$ with the property that*

$$Q\big(\chi(\{[0, \ldots, 0]\}) > 0\big) = 1.$$

Proof. 1. a) implies b).

By 6.1.4. P is continuous, so that — using 1.11.7. — we are able to conclude that P is infinitely divisible. In virtue of 8.3.3. it follows that $(\tilde{P})^0\,\big(\chi^*(R^s) \neq 1\big) = 0$, i.e.

$$(\tilde{P})^0\,\big(\chi(R^s \setminus \{[0, \ldots, 0]\}) > 0\big) = 0.$$

Then, in view of 11.4.3. and 11.4.1. the afore-mentioned special reduced homogeneous clustering representation $P = (P_{l\mu})_{[Q]}$, where $\hat{\Lambda}Q = Q$, has the property that $Q\big(\chi(R^s \setminus \{[0, \ldots, 0]\}) > 0\big) = 0$.

2. b) implies c).

First of all, by 11.4.3. P is a regular infinitely divisible distribution. By assumption there exists a reduced homogeneous clustering representation $P = (P_{l\mu})_{[Q]}$ having the property that $Q\big(\chi(R^s \setminus \{[0, \ldots, 0]\}) > 0\big) = 0$. But this implies $\hat{\Lambda}Q = Q$.

Now, if $P = (P_{l'\mu})_{[Q']}$ is another homogeneous clustering representation of P which has the property that $Q'\big(\chi(\{[0, \ldots, 0]\}) > 0\big) = 1$, then by means of 11.4.2. we conclude that $l = l'$, and $Q = \hat{\Lambda}Q = \hat{\Lambda}Q'$. On the other hand, we have

$$Q'\big(\chi^*(R^s) > 1\big) = \hat{\Lambda}Q'\big(\chi^*(R^s) > 1\big) = Q\big(\chi^*(R^s) > 1\big) = 0$$

and therefore $Q'\big(\chi(R^s \setminus \{[0, \ldots, 0]\}) > 0\big) = 0$, giving $\hat{\Lambda}Q' = Q'$, and consequently $Q = Q'$.

3. c) implies a).

Suppose that P satisfies c) but is not free from after-effects, i.e. that $(\tilde{P})^0\,\big(\chi^*(R^s) > 1\big) > 0$. Hence, if $P = (P_{l\mu})_{[Q]}$ is the homogeneous clustering representation according to c), then by 11.4.1.

$$Q\big(\chi(R^s \setminus \{[0, \ldots, 0]\}) > 0\big) = Q\big(\chi^*(R^s) > 1\big) > 0.$$

We shall now show the existence of a homogeneous clustering representation $P = (P_{l\mu})_{[Q']}$ having the properties $Q' \neq Q, Q'\big(\chi(\{[0, \ldots, 0]\}) > 0\big) = 1$, thus deriving a contradiction.

Since $Q\big(\chi(\{[0, \ldots, 0]\}) < \chi(R^s)\big) > 0$ we may suppose without loss of generality that

$$Q(\chi(\{x\}) > 0 \text{ for at least one point } x \text{ in } R^s,$$
$$\text{the first coordinate of which is different from zero}) > 0.$$

In virtue of the uniqueness assertion of c), $\hat{A}Q$ coincides with Q. Thus we have

$$Q(\chi(\{x\}) > 0 \text{ for at least one point } x \text{ in } R^s,$$
$$\text{the first coordinate of which is negative}) > 0.$$

Now to each χ in M satisfying the condition $0 < \chi(R^s) < +\infty$ we associate that point $u_\chi = [(u_\chi)_1, \ldots, (u_\chi)_s]$ in $\{x : x \in R^s, \chi(\{x\}) > 0\}$ for which the following conditions are satisfied:

$$(u_\chi)_1 = \min\{\alpha_1 : \alpha_1 \in R, \chi(\{x\}) > 0 \text{ for at least one } x = [x_1, \ldots, x_s] \text{ in } R^s, \text{ where } x_1 = \alpha_1\},$$

$$(u_\chi)_2 = \min\{\alpha_2 : \alpha_2 \in R, \chi(\{x\}) > 0 \text{ for at least one } x \text{ in } R^s, \text{ where } x_1 = (u_\chi)_1, x_2 = \alpha_2\},$$

. . . .

$$(u_\chi)_s = \min\{\alpha_s : \alpha_s \in R, \chi(\{x\}) > 0 \quad \text{for} \quad x = [(u_\chi)_1, \ldots, (u_\chi)_{s-1}, \alpha_s]\}.$$

The mapping $\chi \rightarrow u_\chi$ from M into R^s is measurable with respect to $\mathfrak{M}$. This is because, in view of 5.1.5., there exists a measurable indexing relation

$$[x, \Phi] \rightarrow i(x, \Phi) \qquad \big(\Phi \in M, x \in R^s, \Phi(\{x\}) > 0\big).$$

If we denote as we did in Section 5.1., the inverse mapping of

$$[x, \Phi] \rightarrow [i(x, \Phi), \Phi] \quad \text{by} \quad [s, \Phi] \rightarrow [z_s(\Phi), \Phi],$$

then u_χ can be expressed in a measurable way by the vectors $z_s(\chi)$, $s \in I_\chi$, and hence depends measurably on χ.

Now, by 6.1.1., also the mapping $\chi \rightarrow T_{u_\chi}\chi$ from M into M is measurable with respect to the σ-algebra $\mathfrak{M}$. If we denote by Q' the distribution of $T_{u_\chi}\chi$ formed with respect to Q, then we still have $Q'\big(\chi(\{[0, \ldots, 0]\}) > 0\big) = 1$, whereas

$$Q'(\chi(\{x\}) > 0 \text{ for at least one point } x \text{ in } R^s,$$
$$\text{the first coordinate of which is negative}) = 0,$$

and hence $Q' \neq Q$. On the other hand $\hat{A}Q' = \hat{A}Q$, and hence $(P_{l\mu})_{[Q']} = (P_{l\mu})_{[Q]}$. ∎

11.5. The Distributions $H_{q,c}$

In the present section we use the same assumptions and notations as in the beginning of the Sections 11.2., 11.3., and 11.4.

Now we continue the considerations of Section 10.6. Let q be any distribution on $(\mathfrak{R}^s \otimes \mathfrak{K}_2)^{\{\ldots,-1,0,1,\ldots\}}$ stationary with respect to time shift, and c be a number in $[0, 1)$. By the set-up

$$g(\{k\}) = c^k(1 - c) \qquad (k = 0, 1, \ldots)$$

a distribution g on $\mathfrak{P}(\{0, 1, \ldots\})$ is defined, and we denote by p the direct product $g \times q \times g$. The measurable mapping

$$\left[v_1, ([h_n, l_n]), v_2\right] \curvearrowright \sum_{0<n\leqq v_2} \delta_{[-(h_{-1}+\cdots+h_{-n}),l_{-n}]} + \delta_{[[0,\ldots0],l_0]} + \sum_{0\leqq n<v_1} \delta_{[h_0+\cdots+h_n,l_{n+1}]}$$

transforms p into a distribution $H_{q,c}$ on $\mathfrak{M}^0$ satisfying $H_{q,c}(\chi \text{ finite}) = 1$. Consequently

$$H_{q,c} = \sum_{v_1,v_2=0}^{\infty} c^{v_1+v_2}(1-c)^2\, q\left(\left(\sum_{0<n\leqq v_2} \delta_{[-(h_{-1}+\cdots+h_{-n}),l_{-n}]} + \delta_{[[0,\ldots,0],l_0]} + \sum_{0\leqq n<v_1} \delta_{[h_0+\cdots+h_n,l_{n+1}]}\right) \in (.)\right).$$

We denote by $L_{q,c}$ the image of p with respect to the mapping

$$\left[v_1, ([h_n, l_n]), v_2\right] \curvearrowright \delta_{[[0,\ldots,0],l_0]} + \sum_{0\leqq n<v_1} \delta_{[h_0+\cdots+h_n,l_{n+1}]},$$

i.e. we set

$$L_{q,c} = \sum_{v=0}^{\infty} c^v(1-c)\, q\left(\left(\delta_{[[0,\ldots,0],l_0]} + \sum_{0\leqq n<v} \delta_{[h_0+\cdots+h_n,l_{n+1}]}\right) \in (.)\right).$$

By 11.2.2., $L_{q,c}$ always belongs to $\boldsymbol{Q}$. We have

11.5.1. $H_{q,c} = (1 - c)\,\langle L_{q,c}\rangle$.

Proof. First of all

$$\begin{aligned}
&(1 - c)\,\langle L_{q,c}\rangle \\
&= \sum_{v=0}^{\infty} (v+1)\, c^v(1-c)^2\, \hat{A} L_{q,c}\left(\left(\delta_{[[0,\ldots,0],l_0]} + \sum_{0\leqq n<v} \delta_{[h_0+\cdots+h_n,l_{n+1}]}\right) \in (.)\right) \\
&= \sum_{v=0}^{\infty} c^v(1-c)^2 \sum_{m=0}^{v} q\left(\left(\sum_{0<n\leqq m} \delta_{[-(h_{m-1}+\cdots+h_{m-n}),l_{m-n}]} + \delta_{[[0,\ldots,0],l_m]}\right.\right. \\
&\qquad \left.\left. + \sum_{0\leqq n<v-m} \delta_{[h_m+\cdots+h_{m+n},l_{m+n+1}]}\right) \in (.)\right).
\end{aligned}$$

Now the invariance of q with respect to time shift allows to continue with

$$= \sum_{v=0}^{\infty} c^v(1-c)^2 \sum_{\substack{v_1,v_2 \geq 0 \\ v_1+v_2=v}} q\Bigg(\Bigg(\sum_{0<n\leq v_2} \delta_{[-(h_{-1}+\cdots+h_{-n}),l_{-n}]} + \delta_{[[0,\ldots,0],l_0]} + \sum_{0\leq n<v_1} \delta_{[h_0+\cdots+h_n,l_{n+1}]}\Bigg) \in (\cdot)\Bigg)$$

$$= H_{q,c}. \blacksquare$$

By 11.4.1. it follows immediately that for all $l \geq 0$

$$\left(\widetilde{(P_{l(1-c)\mu})_{[L_{q,c}]}}\right)^0 = lH_{q,c}.$$

Bearing in mind 8.3.1. and 8.3.3., we thus observe (cf. Mori [1], K. Hermann [1])

11.5.2. Theorem. *Let q be a distribution on $(\mathfrak{R}^s \otimes \mathfrak{K}_2)^{\{\ldots,-1,0,1,\ldots\}}$ being stationary with respect to time shift, and $c \in [0, 1)$, $l \in [0, +\infty)$. Subject to these assumptions there exists exactly one regular infinitely divisible stationary distribution P on $\mathfrak{M}_2$ with the property*

$$(\tilde{P})^0 = lH_{q,c}$$

and we have

$$P = (P_{l(1-c)\mu})_{[L_{q,c}]}.$$

Under the assumptions of 10.6.1. obviously the convergence relation

$$lH_{q,c_n} \underset{n\to\infty}{\Rightarrow} lH_q$$

holds for each sequence (c_n) of numbers in $[0, 1)$ converging towards one. With the help of the continuity proposition 10.3.6., the proof of which can immediately extended to the marked case, the last convergence relation yields

$$(P_{l(1-c_n)\mu})_{[L_{q,c_n}]} \underset{n\to\infty}{\Rightarrow} P$$

with $(\tilde{P})^0 = lH_q$.

11.6. The Distributions H_F. II

Throughout this section we shall assume that the phase spaces $[A_1, \varrho_{A_1}]$, $[A_2, \varrho_{A_2}]$ coincide with R^1.

In Section 10.4. we associated with each distribution function F in $\boldsymbol{F} \setminus \{k_{[0,+\infty)}\}$ a distribution H_F on $\mathfrak{M}^0$. If $0 < \lim_{x\to+\infty} F(x) < 1$ is satisfied, then we set $F' = \left(\lim_{x\to+\infty} F(x)\right)^{-1} F$ and obtain

$$(*) \qquad H_F = H_{q_{F'}, \lim_{x\to+\infty} F(x)},$$

where $q_{F'}$, as was done in Section 10.4., denotes the distribution of a doubly infinite sequence of independent random numbers identically distributed according to F'. If we suppose that $F(x) = 0$ for all $x \geq 0$, then $(*)$ remains true if we take for F' any distribution function in $\boldsymbol{F} \setminus \{k_{[0,+\infty)}\}$ satisfying $\lim_{x\to+\infty} F'(x) = 1$.

For the sake of simplicity in what follows we shall set for all F in $\boldsymbol{F}$ satisfying $\lim_{x\to+\infty} F(x) < 1$,

$$L_F = L_{q_{F'}, \lim_{x\to+\infty} F(x)}.$$

The distribution L_F can also be constructed in the following way. Let $\xi_0, \xi_1, \ldots$ be an independent sequence of random elements in $[0, +\infty]$ identically distributed according to F. Then there exists with probability one a first index k with the property that $\xi_k = +\infty$, and the random counting measure

$$\Phi = \delta_0 + \sum_{0\leq n<k} \delta_{\xi_0+\cdots+\xi_n}$$

is distributed according to L_F.

From 11.5.2. we immediately obtain (cf. Franken, Liemant & Matthes [1], Kovalenko [1])

11.6.1. Theorem. *For all F in $\boldsymbol{F}$ having the property that $\lim_{x\to+\infty} F(x) < 1$ and all $l \geqq 0$ there exists exactly one regular infinitely divisible stationary distribution P on $\mathfrak{M}$ satisfying $(\tilde{P})^0 = lH_F$, and we have*

$$P = (P_{v\mu})_{[L_F]}, \qquad v = l\left(1 - \lim_{x\to+\infty} F(x)\right).$$

For all distribution functions F in $\boldsymbol{F} \setminus \{k_{[0,+\infty)}\}$ having the property that $\lim_{x\to+\infty} F(x) = 1$ and for all $l > 0$, the singular infinitely divisible stationary distribution P which is uniquely determined by the condition $(\tilde{P})^0 = lH_F$ is different from δ_o, so that it is not possible to find a homogeneous clustering representation of P. Referring to the fourth example of Section 4.6., however, a "causal" clustering representation of the non-stationary distribution $_{[0,+\infty)}P$ can be derived by means of a homogeneous cluster field, which can be considered a substitute for the structure assertion of Proposition 11.6.1. Indeed, in the case $\lim_{x\to+\infty} F(x) = 1$ the distribution q_F yields by the set-up

$$\Phi = \delta_0 + \sum_{n=0}^{\infty} \delta_{\xi_0+\cdots+\xi_n}$$

a distribution on $\mathfrak{M}$ which we also denote by L_F. Then we have (cf. Kovalenko [1])

11.6.2. Theorem. *For all distribution functions F in $\boldsymbol{F} \setminus \{k_{[0,+\infty)}\}$ and all $l > 0$, the stationary infinitely divisible distribution P on $\mathfrak{M}$ which is uniquely determined by the condition $(\tilde{P})^0 = lH_F$ satisfies the equation*

$$_{[0,+\infty)}P = (P_\nu)_{[L_F]}, \quad \textit{where} \quad \nu(\mathrm{d}x) = l(1 - F(x))\,\mu(\mathrm{d}x \cap [0, +\infty)).$$

Proof. 1. First of all we will suppose that $F(0) = 0$ as well as $\lim_{x\to+\infty} F(x) = 1$. In view of the statements made at the end of Section 9.6., $\widetilde{{}_{[0,+\infty)}P} = {}_{[0,+\infty)}(\tilde{P})$ is the image of a measure p with respect to the measurable mapping

$$(x_n)_{n=0,1,\ldots} \curvearrowright \sum_{n=0}^{\infty} \delta_{x_0+\cdots+x_n}$$

of $[0, +\infty)^{\{0,1,\ldots\}}$ into M, with $p = \mathop{\times}\limits_{n=0}^{\infty} p_n$, where we have to set $p_0 = \nu$ and $p_n(\mathrm{d}x) = F(\mathrm{d}x)$, $n = 1, 2, \ldots$. In other words we have

$$\widetilde{{}_{[0,+\infty)}P} = (Q_\nu)_{[L_F]}$$

which enables us, using the clustering theorem 4.3.3., to conclude that

$${}_{[0,+\infty)}P = (P_\nu)_{[L_F]}.$$

2. Now let $F(0) = 0$ as well as $\lim_{x\to+\infty} F(x) < 1$. For all natural numbers n, we set

$$F_n(x) = \max\{F(x), k_{[n,+\infty)}(x)\} \qquad (x \geqq 0)$$

thus obtaining an F_n which has the properties $F_n(0) = 0$, $\lim_{x\to+\infty} F_n(x) = 1$. Obviously $F_n \underset{n\to\infty}{\Rightarrow} F$, from which we may conclude by 10.4.4., that

$$H_{F_n} \underset{n\to\infty}{\Rightarrow} H_F.$$

For all natural numbers n, let P_n denote the stationary infinitely divisible distribution which is uniquely determined by the condition that $(\tilde{P}_n)^0 = lH_{F_n}$. By 10.3.8. we have $P_n \underset{n\to\infty}{\Rightarrow} P$, and hence, by 3.1.11.,

$${}_{[0,+\infty)}(P_n) \underset{n\to\infty}{\Rightarrow} {}_{[0,+\infty)}P.$$

In view of 1. we have, for $n = 1, 2, \ldots$,

$${}_{[0,+\infty)}(P_n) = (P_{\nu_n})_{[L_{F_n}]},$$

where

$$\nu_n(\mathrm{d}x) = l\big(1 - F_n(x)\big)\,\mu\big(\mathrm{d}x \cap [0, +\infty)\big).$$

But it follows immediately from the construction of the F_n that, for $n = 1, 2, \ldots$,

$${}_{[0,n)}\big((P_{\nu_n})_{[L_{F_n}]}\big) = {}_{[0,n)}\big((P_\nu)_{[L_F]}\big),$$

so that we obtain

$${}_{[0,+\infty)}(P_n) \Rightarrow (P_\nu)_{[L_F]},$$

which means that

$${}_{[0,+\infty)}P = (P_\nu)_{[L_F]}.$$

3. Suppose that $0 < F(0) < 1$. As in Section 10.4. we set

$$F^* = \left(1 - F(0)\right)^{-1} \left(F - F(0)\right)$$

and get for the stationary infinitely divisible distribution V, which by 10.4.6. is uniquely determined by $(\tilde{V})^0 = l\left(1 - F(0)\right) H_{F^*}$,

$$V_{[D]} = P$$

where

$$D = \sum_{n=1}^{\infty} \left(1 - F(0)\right) \left(F(0)\right)^{n-1} \delta_{n\delta_0}.$$

Hence

$$_{[0,+\infty)}P = {}_{[0,+\infty)}(V_{[D]}) = \left({}_{[0,+\infty)}V\right)_{[D]} = \left((P_{\nu'})_{[L_{F^*}]}\right)_{[D]},$$

where

$$\nu'(\mathrm{d}x) = l\left(1 - F(0)\right) \left(1 - F^*(x)\right) \mu\left(\mathrm{d}x \cap [0, +\infty)\right).$$

But

$$(L_{F^*})_{[D]} = L_F, \qquad \nu' = \nu,$$

so we get by 11.1.2.

$$_{[0,+\infty)}P = (P_\nu)_{[(L_{F^*})_{[D]}]} = (P_\nu)_{[L_F]}. \blacksquare$$

Using the additional assumption that P is simple, we may derive the following converse of Theorem 11.6.2. (cf. Fleischmann [2]):

11.6.3. Proposition. *Let P be a simple stationary distribution on $\mathfrak{M}$ which is different from δ_0. If there exists a ν in N as well as a distribution D on $\mathfrak{M}$ such that*

$$\nu = {}_{[0,+\infty)}\nu, \qquad D = {}_{[0,+\infty)}D, \qquad D\left(\chi(\{0\}) > 0\right) = 1, \qquad {}_{[0,+\infty)}P = (P_\nu)_{[D]}$$

is satisfied, then i_P is finite and, for

$$F(x) = D\left(x_0(\chi) \leqq x\right) \qquad (x \geqq 0)$$

we have

$$(\tilde{P})_0 = H_F,$$

$$\nu(\mathrm{d}x) = i_P\left(1 - F(x)\right) {}_{[0,+\infty)}\mu(\mathrm{d}x), \qquad D = L_F.$$

Proof. 1. If we use the inversion formula obtained from 8.2.9. by reflection in the origin, then for all Y in $\mathfrak{M}$ we get

$$\tilde{P}\left(\Psi \in Y, x_0(\Psi) < +\infty\right) = \int_0^{+\infty} (\tilde{P})^0 \left(-x_{-2}(\chi) > x, T_{-x}\chi \in Y\right) \mu(\mathrm{d}x),$$

where one has to set $x_{-2}(\chi) = -\infty$ if $\chi\left((-\infty, 0)\right) = 0$. Now let Z be an arbitrary set in $\mathfrak{M}^0$, and t an arbitrary positive number. Setting

$$Y = \{\Phi : \Phi \in M, x_0(\Phi) \leqq t, {}_{[0,+\infty)}(T_{x_0(\Phi)}\Phi) \in Z\}$$

we obtain

$$\tilde{P}\big(x_0(\Psi) \leqq t, {}_{[0,+\infty)}(T_{x_0(\Psi)}\Psi) \in Z\big)$$
$$= \int_0^t (\tilde{P})^0 \big(-x_{-2}(\chi) > x, {}_{[0,+\infty)}\chi \in Z\big)\, \mu(\mathrm{d}x)$$

since, for $x_{-2}(\chi) < -x < 0$, we have $x_0(T_{-x}\chi) = x$. In the special case where $Z = M^0$, we thus obtain

$$+\infty > \tilde{P}\big(x_0(\Psi) \leqq t\big) = \int_0^t (\tilde{P})^0 \big(-x_{-2}(\chi) > x\big)\, \mu(\mathrm{d}x).$$

Hence $(\tilde{P})^0 \big(-x_{-2}(\chi) > x\big)$ is finite for almost all $x \geqq 0$. In view of the monotony in x of this expression, we even have

$$(\tilde{P})^0 \big(-x_{-2}(\chi) > x\big) < +\infty \qquad (x > 0).$$

2. By Theorem 8.5.5., for all $x > 0$ we obtain

$$(\tilde{P})^0 \big(-x_{-2}(\chi) > x\big)$$
$$= (\tilde{P})^0 \Big(-x_{-2}(\chi) > x, \chi\big((-\infty, 0)\big) = +\infty = \chi\big((0, +\infty)\big)\Big)$$
$$+ (\tilde{P})^0 \left(-x_{-2}(\chi) > x, \chi \text{ finite}\right).$$

The first summand can be transformed by means of the symmetry operator $\widehat{\Gamma_1}$, and accordingly we continue the above relation as follows:

$$= (\tilde{P})^0 \Big(x_0(\chi) > x, \chi\big((-\infty, 0)\big) = +\infty = \chi\big((0, +\infty)\big)\Big)$$
$$+ (\tilde{P})^0 \Big(-x_{-2}(\chi) > x, \chi\big((-\infty, 0)\big) = 0 = \chi\big((0, +\infty)\big)\Big)$$
$$+ (\tilde{P})^0 \Big(-x_{-2}(\chi) > x, 0 < \chi\big((-\infty, 0)\big) < +\infty, \chi\big((0, +\infty)\big) = 0\Big)$$
$$+ (\tilde{P})^0 \Big(-x_{-2}(\chi) > x, \chi\big((-\infty, 0)\big) = 0, 0 < \chi\big((0, +\infty)\big) < +\infty\Big)$$
$$+ (\tilde{P})^0 \Big(-x_{-2}(\chi) > x, 0 < \chi\big((-\infty, 0)\big), \chi\big((0, +\infty)\big) < +\infty\Big).$$

The last three summands can be transformed by means of the translation operator $\widehat{\Gamma_3}$, so that we obtain

$$= (\tilde{P})^0 \Big(x_0(\chi) > x, \chi\big((-\infty, 0)\big) = +\infty = \chi\big((0, +\infty)\big)\Big)$$
$$+ (\tilde{P})^0 \Big(x_0(\chi) > x, \chi\big((-\infty, 0)\big) = 0 = \chi\big((0, +\infty)\big)\Big)$$
$$+ (\tilde{P})^0 \Big(x_0(\chi) > x, \chi\big((-\infty, 0)\big) < +\infty, \chi\big((0, +\infty)\big) = 1\Big)$$
$$+ (\tilde{P})^0 \Big(x_0(\chi) > x, 0 < \chi\big((-\infty, 0)\big) < +\infty, \chi\big((0, +\infty)\big) = 0\Big)$$
$$+ (\tilde{P})^0 \Big(x_0(\chi) > x, \chi\big((-\infty, 0)\big) < +\infty, \chi\big((0, +\infty)\big) > 1\Big)$$

$$= (\tilde{P})^0 \left(x_0(\chi) > x, \chi\big((-\infty, 0)\big) = +\infty = \chi\big((0, +\infty)\big)\right)$$
$$+ (\tilde{P})^0 \left(x_0(\chi) > x, \chi \text{ finite}\right)$$
$$= (\tilde{P})^0 \left(x_0(\chi) > x\right).$$

For all $x > 0$, we thus have

$$(\tilde{P})^0 \left(-x_{-2}(\chi) > x\right) = (\tilde{P})^0 \left(x_0(\chi) > x\right).$$

3. By Proposition 2.2.4. and the "clustering theorem" 4.3.3., we have

$$_{[0,+\infty)}\tilde{P}\big((.) \setminus \{o\}\big) = \widetilde{_{[0,+\infty)}P} = \widetilde{(P_\nu)_{[D]}} = (\tilde{P}_\nu)_{[D]} \big((.) \setminus \{o\}\big)$$
$$= (Q_\nu)_{[D]} \big((.) \setminus \{o\}\big) = (Q_\nu)_{[D]}$$

so that by 1. we obtain, for all $t > 0$ and all Z in $\mathfrak{M}^0$,

$$\int_0^t (\tilde{P})^0 \left(-x_{-2}(\chi) > x, {}_{[0,+\infty)}\chi \in Z\right) \mu(\mathrm{d}x)$$
$$= \tilde{P}\left(x_0(\Psi) \leqq t, {}_{[0,+\infty)}(T_{x_0(\Psi)}\Psi) \in Z\right)$$
$$= \nu([0, t])\, D(Z).$$

Hence, in particular,

$$\nu([0, t]) = \tilde{P}\left(x_0(\Psi) \leqq t\right)$$
$$= \int_0^t (\tilde{P})^0 \left(-x_{-2}(\chi) > x\right) \mu(\mathrm{d}x)$$

so that we obtain

$$\int_0^t (\tilde{P})^0 \left(-x_{-2}(\chi) > x, {}_{[0,+\infty)}\chi \in Z\right) \mu(\mathrm{d}x)$$
$$= D(Z)\, \nu([0, t]) = D(Z) \int_0^t (\tilde{P})^0 \left(-x_{-2}(\chi) > x\right) \mu(\mathrm{d}x).$$

Taking into consideration that in view of 1., the two integrands are finite for all $x > 0$ and, as functions of x, continuous on the right, then for all $x > 0$ and all Z in $\mathfrak{M}^0$ we obtain

$$(\tilde{P})^0 \left(-x_{-2}(\chi) > x, {}_{[0,+\infty)}\chi \in Z\right)$$
$$= D(Z)\, (\tilde{P})^0 \left(-x_{-2}(\chi) > x\right).$$

For $x \to 0 + 0$, this yields

$$(\tilde{P})^0 \left({}_{[0,+\infty)}\chi \in Z\right) = D(Z)\, (\tilde{P})^0\, (M) = D(Z)\, i_{\tilde{P}}.$$

For all sufficiently small positive x, we have $D\big(x_0(\chi) > x\big) > 0$. Setting now

$$Z_0 = \{\Psi : \Psi \in M, x_0(\Psi) > x\},$$

with the help of 1. and 2. we conclude that

$$+\infty > (\tilde{P})^0 \big(x_0(\chi) > x\big)$$
$$= (\tilde{P})^0 \left({}_{[0.+\infty)}\chi \in Z_0\right) = D\big(x_0(\chi) > x\big)\, i_{\tilde{P}}$$
$$= D\big(x_0(\chi) > x\big)\, i_P$$

and hence that

$$i_P < +\infty .$$

For all $x > 0$ and all Z in $\mathfrak{M}^0$, we thus have

$$(\tilde{P})_0 \big(-x_{-2}(\chi) > x,\, {}_{[0,+\infty)}\chi \in Z\big)$$
$$= D(Z)\,(\tilde{P})_0 \big(-x_{-2}(\chi) > x\big).$$

As $x \to 0 + 0$, this yields

$$(\tilde{P})_0 \left({}_{[0,+\infty)}\chi \in Z\right) = D(Z) \qquad (Z \in \mathfrak{M}^0).$$

Further we observe that the coordinates of the vector $[-x_{-2}(\chi),\, {}_{[0,+\infty)}\chi]$ are stochastically independent with respect to $(\tilde{P})_0$.

4. For all $x \geqq 0$, using 2. and 3. we set

$$F(x) = (\tilde{P})_0 \big(-x_{-2}(\chi) \leqq x\big) = (\tilde{P})_0 \big(x_0(\chi) \leqq x\big)$$
$$= D\big(x_0(\chi) \leqq x\big).$$

We shall now show that $(\tilde{P})_0 = H_F$ is satisfied.

The measure

$$V = (\tilde{P})_0 \big(\chi \in (.),\, \chi \text{ finite}\big)$$

is invariant under the application of the one-to-one translation operator $\widehat{\Gamma}_3$ and is uniquely determined by its values on the sets of the form

$$Y = \{\Psi : \Psi \in M^0,\, y_{-k-1}(\Psi) = +\infty,\, y_n(\Psi) \leqq s_n \quad \text{for} \quad -k \leqq n < l,$$
$$y_l(\Psi) = +\infty\}$$

$(k, l = 0, 1, \ldots;\, s_{-k}, \ldots, s_{l-1} > 0)$. Here we have to set $y_i(\Psi) = x_{i+1}(\Psi) - x_i(\Psi)$. By applying $\widehat{\Gamma}_3$ k-times, one obtains

$$V(Y) = (\tilde{P})_0 \big(-x_{-2}(\chi) = +\infty,\, y_{n+k}(\chi) \leqq s_n$$
$$\text{for} \quad -k \leqq n < l,\, y_{l+k}(\chi) = +\infty\big).$$

In view of 3., the event "$-x_{-2}(\chi) = +\infty$" is independent of ${}_{[0,+\infty)}\chi$. Thus, setting

$$(\tilde{P})_0 \big(-x_{-2}(\chi) = +\infty\big) = 1 - \lim_{x \to +\infty} F(x) = c,$$

we have

$$V(Y) = c\,V\big(y_{n+k}(\chi) \leqq s_n \quad \text{for} \quad -k \leqq n < l,\, y_{l+k}(\chi) = +\infty\big).$$

By repeated transformation by means of $\left(\widehat{\Gamma}_3\right)^{-1}$ and application of the independence property formulated in 3., we obtain

$$V(Y) = c\Big(\prod_{-k\leqq n<l} F(s_n)\Big)\,(\tilde{P})_0\,\big(x_0(\chi) = +\infty\big)$$

$$= c^2 \prod_{-k\leqq n<l} F(s_n) = H_F(\chi \in Y, \chi \text{ finite}),$$

which means that

$$(\tilde{P})_0\,\big(\chi \in (.), \chi \text{ finite}\big) = H_F\big(\chi \in (.), \chi \text{ finite}\big),$$

and hence also

$$(\tilde{P})_0\,(\chi \text{ finite}) = H_F(\chi \text{ finite}).$$

But this probability is equal to one or zero, depending on whether $\lim_{x\to+\infty} F(x) < 1$ or $\lim_{x\to+\infty} F(x) = 1$. Thus, in the first case, we have already shown by our above consideration that $(\tilde{P})_0 = H_F$.

In the second case, the distribution

$$(\tilde{P})_0 = (\tilde{P})_0\,\Big(\chi \in (.), \chi\big((-\infty, 0)\big) = +\infty = \chi\big((0, +\infty)\big)\Big)$$

is invariant with respect to the one-to-one translation operator $\widehat{\Gamma}_1$ and is uniquely determined by its values on sets of the form

$$Y = \{\Psi : \Psi \in M^0, y_n(\Psi) \leqq s_n \quad \text{for} \quad -1 \leqq n \leqq l\},$$

where $l = 1, 2, \ldots;\ s_{-1}, \ldots, s_l > 0$. By repeated transformation by means of $\left(\widehat{\Gamma}_1\right)^{-1}$ and application of the independence property formulated in 3., it follows that

$$(\tilde{P})_0\,(Y) = \prod_{-1\leqq n<l} F(s_n) = H_F(Y)$$

which again leads to

$$(\tilde{P})_0 = H_F.$$

In view of 3., we now have

$$D = {}_{[0,+\infty)}(\tilde{P})_0 = {}_{[0,+\infty)}(H_F) = L_F$$

and

$$\nu([0, t]) = \int_0^t i_P\big(1 - F(x)\big)\,\mu(\mathrm{d}x) \qquad (t \geqq 0). \ \blacksquare$$

As an immediate consequence of 11.6.2. and 11.6.3., we have

11.6.4. Theorem. *A stationary infinitely divisible distribution P on $\mathfrak{M}$ which is different from δ_0 has the form*

$$(\tilde{P})^0 = lH_F, \qquad l > 0, \qquad F \in \boldsymbol{F}, \qquad F(0) = 0$$

if and only if it is simple and ${}_{[0,+\infty)}P$ *has a "causal" clustering representation of the form*

$${}_{[0,+\infty)}P = (P_\nu)_{[D]}$$

where

$$\nu \in N, \qquad \nu = {}_{[0,+\infty)}\nu, \qquad D \in \boldsymbol{P}, \qquad D = {}_{[0,+\infty)}D, \qquad D\big(\chi(\{0\}) > 0\big) = 1.$$

Theorem 11.6.4. cannot be generalized in a way that $F(0) = 0$ is replaced by $F(0) < 1$ and the assumption that P be simple is dropped.

From

$${}_{[0,+\infty)}P = (P_\nu)_{[D]}$$

we may conclude that, for all sequences $q_1, q_2, \ldots$ satisfying the conditions $q_n \geqq 0$ for $n = 1, 2, \ldots$; $\sum_{n=1}^{\infty} q_n = 1$, setting $L = \sum_{n=1}^{\infty} q_n \delta_{n\delta_o}$,

$${}_{[0,+\infty)}(P_{[L]}) = ({}_{[0,+\infty)}P)_{[L]} = ((P_\nu)_{[D]})_{[L]} = (P_\nu)_{[D_{[L]}]}.$$

In this way it is possible to construct a stationary infinitely divisible distribution having the form $(P_\nu)_{[Q]}$, $Q = {}_{[0,+\infty)}Q$, $Q(\chi(\{0\}) > 0) = 1$, the intensity of which is infinite.

Theorem 11.6.4. shows that, as a rule, also for stationary infinitely divisible distributions P which are different from δ_o, the equivalence class of the "causal" clustering representations of ${}_{[0,+\infty)}P$ does not contain any distinguished clustering representation having a homogeneous cluster field. A corresponding statement is true also for the uniquely determined measure ν:

11.6.5. Proposition. *Let P be a stationary infinitely divisible distribution on $\mathfrak{M}$. If, for any $l > 0$ and a cluster field $\varkappa$ on $[0, +\infty)$ with phase space R^1 which satisfies the condition $\varkappa_{(x)}(\chi((-\infty, x)) = 0, \chi(\{x\}) > 0) = 1$ for all $x \geqq 0$, the relation*

$${}_{[0,+\infty)}P = (P_{l\mu})_\varkappa$$

is satisfied, then P is free from after-effects.

Proof. For all $t > 0$, we have

$$lt = l\mu((0, t]) = \tilde{P}(\Psi((0, t]) > 0).$$

Consequently, for all $t_1, t_2 > 0$,

$$\begin{aligned} l(t_1 + t_2) &= \tilde{P}(\Psi((0, t_1 + t_2]) > 0) \\ &= \tilde{P}(\Psi((0, t_1]) > 0) + \tilde{P}(\Psi((t_1, t_1 + t_2]) > 0) \\ &\quad - \tilde{P}(\Psi((0, t_1]) > 0, \Psi((t_1, t_1 + t_2]) > 0) \\ &= lt_1 + lt_2 - \tilde{P}(\Psi((0, t_1]) > 0, \Psi((t_1, t_1 + t_2]) > 0), \end{aligned}$$

i.e. we have

$$\tilde{P}(\Psi((0, t_1]) > 0, \Psi((t_1, t_1 + t_2]) > 0) = 0.$$

For all disjoint intervals I_1, I_2, we thus obtain

$$\tilde{P}(\Psi(I_1) > 0, \Psi(I_2) > 0) = 0$$

which means that

$$\tilde{P}(\Psi^*(R^1) \neq 1) = 0.$$

Now the truth of our assertion is immediately inferred from Proposition 2.2.13. ∎

11.7. The Set S

In this and the following sections we will mostly use phase spaces of the form R^s.

Immediately from 4.2.3. and 4.2.2. we get

11.7.1. *If a measure H on $\mathfrak{M}$ and a distribution Q in $\mathbf{P}$ satisfies the condition $\varrho_Q * \varrho_H \in N$ then the measure $H_{[Q]}$ exists.*

11.7.2. *If for a measure H on $\mathfrak{M}$ and a distribution Q in $\mathbf{P}$ the measure $H_{[Q]}$ exists then*

$$\varrho_{H_{[Q]}} = \varrho_Q * \varrho_H$$

holds.

A measure H on $\mathfrak{M}$ is said to be *of bounded intensity* if the inequality

$$\sup_{x \in R^s} \varrho_H(X + x) < +\infty$$

is satisfied for all bounded X in $\mathfrak{R}^s$. In virtue of 6.1.5., each stationary measure H on $\mathfrak{M}$ with the property $i_H < +\infty$ is of bounded intensity.

11.7.3. *If a measure H on $\mathfrak{M}$ is of bounded intensity, then, for all distributions Q in $\mathbf{P}$ satisfying $\varrho_Q(R^s) < +\infty$ the measure $H_{[Q]}$ exists and is again of bounded intensity.*

Proof. For all bounded X in $\mathfrak{R}^s$ and all x in R^s we get

$$(\varrho_Q * \varrho_H)\,(X + x) = (\varrho_H * \varrho_Q)\,(X + x)$$
$$= \int \varrho_H(X + x - y)\,\varrho_Q(\mathrm{d}y) \leqq \varrho_Q(R^s) \sup_{z \in R^s} \varrho_H(X + z).$$

From this follows the statement with the help of 11.7.1. and 11.7.2. ∎

Let $\mathbf{S}$ denote the set of those distributions L on $\mathfrak{M}$ which have the property that there exists a non-negative real function s_L which is defined almost everywhere with respect to L and measurable with respect to $\mathfrak{M}$, such that the convergence relation

$$\sup_{x \in R^s} \int_{\{\Phi : s_L(\Phi) \leqq c\}} \left| \frac{\Phi([-m, m)^s + x)}{(2m)^s} - s_L(\Phi) \right| L(\mathrm{d}\Phi) \xrightarrow[m \to \infty]{} 0$$

is satisfied for all $c > 0$.

For all L in $\mathbf{S}$, all $\varepsilon > 0$ and all natural numbers k we obtain

$$\varlimsup_{n \to \infty} \sup_{x \in R^s} L\left(\left|(2n)^{-s}\, \Phi([-n, n)^s + x) - s_L(\Phi)\right| > \varepsilon\right)$$
$$\leqq \varlimsup_{n \to \infty} \Big(L\big(s_L(\Phi) > k\big)$$
$$+ \sup_{x \in R^s} L\Big(s_L(\Phi) \leqq k,\ \left|(2n)^{-s}\, \Phi\big([-n, n)^s + x\big) - s_L(\Phi)\right| > \varepsilon\Big)\Big)$$

$$\leqq L\big(s_L(\Phi) > k\big)$$
$$+ \overline{\lim_{n\to\infty}} \sup_{x\in R^s} \varepsilon^{-1} \int_{\{\Phi: s_L(\Phi)\leqq k\}} \left|(2n)^{-s}\,\Phi\big([-n,n)^s + x\big) - s_L(\Phi)\right| L(\mathrm{d}\Phi)$$
$$= L\big(s_L(\Phi) > k\big).$$

Because k is arbitrary it follows

$$\lim_{n\to\infty} \sup_{x\in R^s} L\left(\left|(2n)^{-s}\,\Phi\big([-n,n)^s + x\big) - s_L(\Phi)\right| > \varepsilon\right) = 0.$$

and we observe the truth of

11.7.4. *Each distribution L in $\mathbf{S}$ belongs to $\boldsymbol{H}$, and we can choose for f_L any variant of s_L.*

In accordance with a notation in Section 7.3. we denote by σ_L the uniquely determined distribution of $s_L(\Phi)$ with respect to L. For all Borel subsets Z of $[0, +\infty)$ having the property that $\sigma_L(Z) > 0$, together with L also $L' = L\big((.) \mid s_L(\Phi) \in Z\big)$ belongs to $\mathbf{S}$. We can choose $s_{L'} = s_L$ and have $\sigma_{L'} = \sigma_L\big((.) \mid Z\big)$.

For all $c > 0$, let $\mathbf{S}^{(c)}$ denote the set of those L in $\mathbf{S}$ which satisfy the condition $\sigma_L\big([0, c]\big) = 1$.

11.7.5. *Each stationary distribution L on $\mathfrak{M}$ satisfying the condition $L\big(s(\Phi) < +\infty\big) = 1$ belongs to $\mathbf{S}$, and s_L can be set equal to the restriction of the sample intensity s on $\{\Phi : \Phi \in M,\ s(\Phi) < +\infty\}$.*

Proof. Since s is invariant and L is stationary, we have only to show that, for all $c > 0$,

$$\lim_{m\to\infty} \int_{\{\Phi: s(\Phi)\leqq c\}} \left|\frac{\Phi\big([-m,m)^s\big)}{(2m)^s} - s(\Phi)\right| L(\mathrm{d}\Phi) = 0$$

is satisfied. This statement is trivially valid if $\sigma_L([0, c]) = 0$. Otherwise it is equivalent to

$$\lim_{m\to\infty} \int \left|\frac{\Phi\big([-m,m)^s\big)}{(2m)^s} - s(\Phi)\right| L\big(\mathrm{d}\Phi \mid s(\Phi) \leqq c\big) = 0$$

and hence, in view of the estimates formed in the introduction of the sample intensity, it is a consequence of the statistical ergodic theorem 6.2.4., since $L\big((.) \mid s(\Phi) \leqq c\big)$ is stationary and, by 6.2.3., has the intensity $\int s(\Phi)$ $L\big(\mathrm{d}\Phi \mid s(\Phi) \leqq c\big) \leqq c$. ∎

The following proposition yields an interesting class of distributions in $\mathbf{S}$ being not necessarily stationary.

11.7.6. Proposition. *A Poisson distribution P_ν on $\mathfrak{M}$ belongs to $\mathbf{S}$ if and only if there exists an $l \geqq 0$ satisfying*

$$\sup_{x\in R^s} \left|(2n)^{-s}\,\nu\big([-n,n)^s + x\big) - l\right| \xrightarrow[n\to\infty]{} 0.$$

In this case $\sigma_{P_\nu} = \delta_l$ holds.

Proof. 1. For the sake of simplicity we set $[-n, n)^s + x = X_{n,x}$. If ν and l fulfil the condition formulated above we then obtain

$$\begin{aligned}
&\overline{\lim_{n\to\infty}} \sup_{x\in R^s} \int |(2n)^{-s}\, \Phi(X_{n,x}) - l|\, P_\nu(\mathrm{d}\Phi) \\
&= \overline{\lim_{n\to\infty}} \sup_{x\in R^s} \int |(2n)^{-s}\, y - l|\, \pi_{\nu(X_{n,x})}(\mathrm{d}y) \\
&\leqq \overline{\lim_{n\to\infty}} \sup_{x\in R^s} (2n)^{-s} \int |y - \nu(X_{n,x})|\, \pi_{\nu(X_{n,x})}(\mathrm{d}y) \\
&\quad + \overline{\lim_{n\to\infty}} \sup_{x\in R^s} |(2n)^{-s}\, \nu(X_{n,x}) - l| \\
&\leqq \overline{\lim_{n\to\infty}} \sup_{x\in R^s} (2n)^{-s} \Big(\int |y - \nu(X_{n,x})|^2\, \pi_{\nu(X_{n,x})}(\mathrm{d}y)\Big)^{1/2} + 0 \\
&= \Big(\overline{\lim_{n\to\infty}} \sup_{x\in R^s} (2n)^{-2s}\, \nu(X_{n,x})\Big)^{1/2} \\
&\leqq \Big(\overline{\lim_{n\to\infty}} (2n)^{-s} \sup_{x\in R^s} |(2n)^{-s}\, \nu(X_{n,x}) - l| + \overline{\lim_{n\to\infty}} (2n)^{-s}\, l\Big)^{1/2} \\
&= 0.
\end{aligned}$$

Therefore P_ν belongs to $\mathbf{S}$ and we have $\sigma_{P_\nu} = \delta_l$.

2. We suppose that, for a ν in N and a non-negative real function h which is defined almost everywhere with respect to L and measurable with respect to $\mathfrak{M}$,

$$\sup_{x\in R^s} \int_{\{\Phi:\, h(\Phi)\leqq c\}} |(2n)^{-s}\, \Phi(X_{n,x}) - h(\Phi)|\, P_\nu(\mathrm{d}\Phi) \xrightarrow[n\to\infty]{} 0$$

is satisfied for all $c > 0$. In virtue of 11.7.4. from this it follows that

$$\sup_{x\in R^s} P_\nu\big(|(2n)^{-s}\, \Phi(X_{n,x}) - h(\Phi)| > \varepsilon\big) \xrightarrow[n\to\infty]{} 0$$

for all $\varepsilon > 0$. In particular, we get the following stochastic convergence relations with respect to P_ν:

$$(2n)^{-s}\, \Phi\big([-n, n)^s + [n, \ldots, n]\big) \xrightarrow[P_\nu]{} h(\Phi),$$
$$(2n)^{-s}\, \Phi\big([-n, n)^s - [n, \ldots, n]\big) \xrightarrow[P_\nu]{} h(\Phi).$$

Because P_ν is free from after-effects, the sequence $\big((2n)^{-s}\, \Phi\big([-n, n)^s + [n, \ldots, n]\big)\big)$ is stochastically independent of the sequence $\big((2n)^{-s}\, \Phi\big([-n, n)^s - [n, \ldots, n]\big)\big)$. Hence $h(\Phi)$ is stochastically independent of $h(\Phi)$, i.e. there is an $l \geqq 0$ such that $h(\Phi)$ is equal to l almost everywhere with respect to P_ν, which implies (if we again work even in $\boldsymbol{H}$)

$$\sup_{x\in R^s} P_\nu\big(|(2n)^{-s}\, \Phi(X_{n,x}) - l| > \varepsilon\big) \xrightarrow[n\to\infty]{} 0,$$

i.e. that

$$\sup_{x \in R^s} \pi_{\nu(X_{n,x})}\big(|(2n)^{-s}\,\xi - l| > \varepsilon\big) \xrightarrow[n\to\infty]{} 0$$

for all $\varepsilon > 0$.

Let (x_n) be a sequence of elements in R^s. We write $c_n = \nu(X_{n,x_n})$. Immediately from the last convergence relation it follows that the distribution of $(2n)^{-s}\,\xi$ with respect to π_{c_n} converges weakly to δ_l, and we have

$$\exp\big(-c_n(1 - e^{-(2n)^{-s}y})\big) = \int e^{-(2n)^{-s}xy}\pi_{c_n}(\mathrm{d}x) \xrightarrow[n\to\infty]{} \int e^{-xy}\delta_l(\mathrm{d}x)$$
$$= \exp(-yl) \qquad (y \geqq 0).$$

This implies

$$c_n(1 - e^{-(2n)^{-s}}) \xrightarrow[n\to\infty]{} l,$$

i.e. that

$$(2n)^{-s}\,c_n \xrightarrow[n\to\infty]{} l.$$

Consequently

$$\sup_{x \in R^s} |(2n)^{-s}\,\nu(X_{n,x}) - l| \xrightarrow[n\to\infty]{} 0. \blacksquare$$

The proof given above shows that a Poisson distribution P_ν on $\mathfrak{M}$ belongs to $\mathbf{S}$ if and only if it belongs to $\boldsymbol{H}$.

The distributions in $\mathbf{S}$ are not necessarily of finite intensity, but we have

11.7.7. *For any $c > 0$, every L in $\mathbf{S}^{(c)}$ is of bounded intensity and we have for each sequence (t_n) of positive numbers converging towards infinity, and for each bounded set X in $\mathfrak{R}^s$ satisfying $\mu(\partial X) = 0$,*

$$\sup_{x \in R^s} \Big|t_n^{-s}\varrho_L(t_nX + x) - \mu(X)\int l\sigma_L(\mathrm{d}l)\Big| \xrightarrow[n\to\infty]{} 0.$$

Proof. Let X be any bounded set in $\mathfrak{R}^s$. Then we get for all sufficiently large natural numbers m

$$\sup_{x \in R^s} \Big|\varrho_L\big([-m, m)^s + x\big) - (2m)^s \int l\sigma_L(\mathrm{d}l)\Big|$$
$$\leqq (2m)^s \sup_{x \in R^s} \int \Big|(2m)^{-s}\,\Phi\big([-m, m)^s + x\big) - s_L(\Phi)\Big|\,L(\mathrm{d}\Phi) < +\infty,$$

hence

$$\sup_{x \in R^s} \varrho_L(X + x) \leqq \sup_{x \in R^s} \varrho_L\big([-m, m)^s + x\big) < +\infty,$$

i.e. that L is of bounded intensity.

If a sequence (t_n) of positive numbers converges to infinity then — similarly to the proof of 7.3.2. — we obtain

$$\sup_{x \in R^s} \int \Big|(2t_n)^{-s}\,\Phi\big([-t_n, t_n)^s + x\big) - s_L(\Phi)\Big|\,L(\mathrm{d}\Phi) \xrightarrow[n\to\infty]{} 0.$$

Therefore

$$\sup_{x \in R^s} \int |t_n^{-s}\Phi(t_nX_0 + x) - \mu(X_0)\, s_L(\Phi)|\, L(\mathrm{d}\Phi) \xrightarrow[n\to\infty]{} 0$$

is valid for each finite union X_0 of pairwise disjoint sets having the form $[-t, t)^s + y$, $t > 0$, $y \in R^s$. If now X is any bounded set in $\mathfrak{R}^s$ with the property $\mu(\partial X) = 0$ then we can find to given $\varepsilon > 0$ sets X_1, X_2 of the afore-mentioned form which satisfy $X_1 \subseteqq X \subseteqq X_2$ and $\mu(X_2 \setminus X_1) < \varepsilon$. So we finally obtain the convergence relation

$$\sup_{x \in R^s} \int |t_n^{-s}\Phi(t_nX + x) - \mu(X)\, s_L(\Phi)|\, L(\mathrm{d}\Phi) \xrightarrow[n\to\infty]{} 0,$$

which implies

$$\sup_{x \in R^s} \left|t_n^{-s}\varrho_L(t_nX + x) - \mu(X) \int l\sigma_L(\mathrm{d}l)\right| \xrightarrow[n\to\infty]{} 0. \blacksquare$$

Now let L be any distribution in **S**, and Q a distribution in ***P*** with the property that $\varrho_Q(R^s) < +\infty$. We set

$$L = \sum_{\substack{i=1,2,\ldots \\ \sigma_L([i-1,i))>0}} \sigma_L\big([i-1, i)\big)\, L\big((.)\,|\, i - 1 \leqq s_L(\Phi) < i\big).$$

In virtue of 11.7.7. and 11.7.3. all of the distributions

$$\big(L((.) \mid i - 1 \leqq s_L(\Phi) < i)\big)_{[Q]}, \qquad \sigma_L\big([i-1, i)\big) > 0,$$

and hence also $L_{[Q]}$ exist. We even have

11.7.8. Proposition. *For all L in* **S** *and all Q in* ***P*** *satisfying $\varrho_Q(R^s) < +\infty$, the distribution $L_{[Q]}$ exists and again belongs to* **S** *and we have*

$$\sigma_{L_{[Q]}} = \sigma_L\big(\varrho_Q(R^s)\, \xi \in (.)\big).$$

Proof. 1. We denote by $[A', \varrho_{A'}]$ the direct product of R^s and $[M, \varrho_M]$. As in the proof of 11.1.4. we define by

$$\varkappa_{(x)} = \int \delta_{\delta_{[x,\chi]}}(.)\, Q(\mathrm{d}\chi) \qquad (x \in R^s)$$

a homogeneous cluster field $\varkappa$ on R^s with phase space $[A', \varrho_{A'}]$. If ω is distributed according to $L_\varkappa$ then the random non-marked counting measure ω^x has the distribution L.

For all ω in M' and all X_1, X_2 in $\mathfrak{R}^s$ we set

$$\omega(X_1, X_2) = \int \chi(X_2 - x)\, ({}_{X_1 \times M}\omega)\, (\mathrm{d}[x, \chi])$$

and get a measurable mapping $\omega \curvearrowright \omega(X_1, X_2)$ of M' into $[0, +\infty]$. Hence, for all X_1 in $\mathfrak{R}^s$, $\omega \curvearrowright \omega\big(X_1, (.)\big)$ is a measurable mapping from M' into M. The existence of $L_{[Q]}$ means that — almost everywhere with respect to $L_\varkappa$ — the measure $\omega\big(R^s, (.)\big)$ and hence all measures $\omega\big(X_1, (.)\big)$ belong to M. Obviously, the distribution of $\omega\big(X_1, (.)\big)$ with respect to $L_\varkappa$ coincides with $({}_{X_1}L)_{[Q]}$.

2. For the sake of abbreviation we set

$$v = \varrho_Q(R^s), \quad X_{n,x} = [-n, n)^s + x \qquad (n = 1, 2, \ldots;\ x \in R^s).$$

First of all we additionally suppose that L belongs to $\mathbf{S}^{(c)}$ for some $c > 0$. Then we have by 4.2.2.

$$\begin{aligned}
&\overline{\lim_{n\to\infty}} \sup_{x\in R^s} (2n)^{-s} \int \omega(X_{n,x}, R^s \setminus X_{n,x})\, L_\varkappa(\mathrm{d}\omega) \\
&= \overline{\lim_{n\to\infty}} \sup_{x\in R^s} (2n)^{-s} \int \Psi(R^s \setminus X_{n,x})\, ({}_{X_{n,x}}L)_{[Q]}\, (\mathrm{d}\Psi) \\
&= \overline{\lim_{n\to\infty}} \sup_{x\in R^s} (2n)^{-s} \int_{X_{n,x}} \varrho_Q\big((R^s \setminus X_{n,x}) - y\big)\, \varrho_L(\mathrm{d}y) \\
&\leqq \overline{\lim_{n\to\infty}} \sup_{x\in R^s} (2n)^{-s} \int_{X_{n-2k,x}} \varrho_Q\big(R^s \setminus (X_{n-k,x} - y)\big)\, \varrho_L(\mathrm{d}y) \\
&\quad + v\, \overline{\lim_{n\to\infty}} \sup_{x\in R^s} (2n)^{-s} \big(\varrho_L(X_{n,x}) - \varrho_L(X_{n-2k,x})\big)
\end{aligned}$$

for all natural numbers k. For a given $\varepsilon > 0$ we specify k in such a manner that $\varrho_Q(R^s \setminus X_{k,0}) < \varepsilon$ is satisfied. By 11.7.7. we can continue the above chain of inequalities as follows:

$$\leqq \varrho_Q(R^s \setminus X_{k,0})\, \overline{\lim_{n\to\infty}} \sup_{x\in R^s} (2n)^{-s}\, \varrho_L(X_{n-2k,x}) + 0 \leqq \varepsilon c\,.$$

Hence

$$(*) \qquad \overline{\lim_{n\to\infty}} \sup_{x\in R^s} \int (2n)^{-s}\, \omega(X_{n,x}, R^s \setminus X_{n,x})\, L_\varkappa(\mathrm{d}\omega) = 0.$$

In a similar way we obtain

$$\begin{aligned}
&\overline{\lim_{n\to\infty}} \sup_{x\in R^s} \int (2n)^{-s}\, \omega(R^s \setminus X_{n,x}, X_{n,x})\, L_\varkappa(\mathrm{d}\omega) \\
&= \overline{\lim_{n\to\infty}} \sup_{x\in R^s} (2n)^{-s} \int \varrho_Q(X_{n,x} - y)\, \varrho_L(\mathrm{d}y \setminus X_{n,x}) \\
&= \overline{\lim_{n\to\infty}} \sup_{x\in R^s} (2n)^{-s} \int \varrho_L\big((X_{n,x} - y) \setminus X_{n,x}\big)\, \varrho_Q(\mathrm{d}y) \\
&\leqq \overline{\lim_{n\to\infty}} \sup_{x\in R^s} (2n)^{-s} \int_{X_{k,0}} \varrho_L\big((X_{n,x} - y) \setminus (X_{n-k,x} - y)\big)\, \varrho_Q(\mathrm{d}y) \\
&\quad + \varepsilon\, \overline{\lim_{n\to\infty}} \sup_{z\in R^s} (2n)^{-s}\, \varrho_L(X_{n,z}) \\
&\leqq v\, \overline{\lim_{n\to\infty}} \sup_{z\in R^s} (2n)^{-s} \big(\varrho_L(X_{n,z}) - \varrho_L(X_{n-k,z})\big) + \varepsilon c = \varepsilon c
\end{aligned}$$

and so we also have

$$(**) \qquad \overline{\lim_{n\to\infty}} \sup_{x\in R^s} \int (2n)^{-s}\, \omega(R^s \setminus X_{n,x}, X_{n,x})\, L_\varkappa(\mathrm{d}\omega) = 0.$$

3. In view of (*) and (**) we get

$$\overline{\lim_{n\to\infty}} \sup_{x\in R^s} \int |(2n)^{-s}\, \omega(R^s, X_{n,x}) - v s_L(\omega^x)|\, L_\varkappa(\mathrm{d}\omega)$$

$$\leqq \overline{\lim_{n\to\infty}} \sup_{x\in R^s} \int (2n)^{-s}\, |\omega(R^s, X_{n,x}) - \omega(X_{n,x}, R^s)|\, L_\varkappa(\mathrm{d}\omega)$$

$$+ \overline{\lim_{n\to\infty}} \sup_{x\in R^s} \int |(2n)^{-s}\, \omega(X_{n,x}, R^s) - v s_L(\omega^x)|\, L_\varkappa(\mathrm{d}\omega)$$

$$\leqq \overline{\lim_{n\to\infty}} \sup_{x\in R^s} \int (2n)^{-s} \big(\omega(R^s \setminus X_{n,x}, X_{n,x}) + \omega(X_{n,x}, R^s \setminus X_{n,x})\big)\, L_\varkappa(\mathrm{d}\omega)$$

$$+ \overline{\lim_{n\to\infty}} \sup_{x\in R^s} \int \Big(\int |(2n)^{-s}\, \omega(X_{n,x}, R^s) - v s_L(\Phi)|\, \varkappa_{(\Phi)}(\mathrm{d}\omega)\Big)\, L(\mathrm{d}\Phi)$$

$$\leqq 0 + \overline{\lim_{n\to\infty}} \sup_{x\in R^s} \int \mathrm{E} \Big| (2n)^{-s} \sum_{1\leqq l \leqq \Phi(X_{n,x})} \xi_l - (\mathrm{E}\xi_l)\, (2n)^{-s}\, \Phi(X_{n,x}) \Big|\, L(\mathrm{d}\Phi).$$

In this case we denote by $\xi_1, \xi_2, \ldots$ a sequence of random non-negative numbers distributed according to $p = Q_{R^s} \times Q_{R^s} \times \cdots$ and stochastically independent of Φ, and by E the mathematical expectation with respect to p. We can continue with

$$= \overline{\lim_{n\to\infty}} \sup_{x\in R^s} \int_{\{\Phi:\, \Phi(X_{n,x})>0\}} (2n)^{-s}\, \Phi(X_{n,x})$$
$$\mathrm{E} \Big| \big(\Phi(X_{n,x})\big)^{-1} \sum_{1\leqq l\leqq \Phi(X_{n,x})} (\xi_l - \mathrm{E}\xi_l) \Big|\, L(\mathrm{d}\Phi)$$

$$\leqq \overline{\lim_{n\to\infty}} \sup_{x\in R^s} \int_{\{\Phi:\, \Phi(X_{n,x})>0\}} s_L(\Phi)\, \mathrm{E} \Big| \big(\Phi(X_{n,x})\big)^{-1} \sum_{1\leqq l\leqq \Phi(X_{n,x})} (\xi_l - \mathrm{E}\xi_l) \Big|\, L(\mathrm{d}\Phi)$$

$$+ \mathrm{E}|\xi_1 - \mathrm{E}\xi_1| \lim_{n\to\infty} \sup_{x\in R^s} \int |(2n)^{-s}\, \Phi(X_{n,x}) - s_L(\Phi)|\, L(\mathrm{d}\Phi)$$

$$\leqq c\, \overline{\lim_{n\to\infty}} \sup_{x\in R^s} \sum_{k=1}^{\infty} L\big(\Phi(X_{n,x}) = k,\, s_L(\Phi) > 0\big)\, \mathrm{E} \Big| k^{-1} \sum_{l=1}^{k} (\xi_l - \mathrm{E}\xi_l) \Big| + 0.$$

For a given $\varepsilon > 0$ we choose an k_0 such that the inequality $\mathrm{E} \Big| k^{-1} \sum_{l=1}^{k} (\xi_l - \mathrm{E}\xi_l) \Big| < \varepsilon$ is satisfied for all $k \geqq k_0$. Now we can continue as follows:

$$\leqq c\mathrm{E}|\xi_1 - \mathrm{E}\xi_1|\, \overline{\lim_{n\to\infty}} \sup_{x\in R^s} L\big(\Phi(X_{n,x}) < k_0,\, s_L(\Phi) > 0\big) + c\varepsilon = c\varepsilon.$$

Thus we obtain

$$\lim_{n\to\infty} \sup_{x\in R^s} \int |(2n)^{-s}\, \omega(R^s, X_{n,x}) - v s_L(\omega^x)|\, L_\varkappa(\mathrm{d}\omega) = 0.$$

4. Now let L be an arbitrary distribution in **S** and $w > 0$. We consider

$$\overline{\lim_{n\to\infty}} \sup_{x\in R^s} \int_{\{\omega:\, s_L(\omega^x)\leqq w\}} |(2n)^{-s}\, \omega(R^s, X_{n,x}) - v s_L(\omega^x)|\, L_\varkappa(\mathrm{d}\omega).$$

In the case $L\big(s_L(\Phi) \leqq w\big) = 0$ this expression vanishes. Otherwise we have the estimate

$$\leqq \overline{\lim_{n\to\infty}} \sup_{x\in R^s} \int |(2n)^{-s}\, \omega(R^s, X_{n,x}) - v s_L(\omega^x)| \left(L\left(\Phi \in (.) \mid s_L(\Phi) \leqq w\right)\right)_\varkappa (\mathrm{d}\omega).$$

But because of 3. this term vanishes, too. Hence

$$\lim_{n\to\infty} \sup_{x\in R^s} \int_{\{\omega:\, s_L(\omega^x)\leqq w\}} |(2n)^{-s}\, \omega(R^s, X_{n,x}) - v s_L(\omega^x)|\; L_\varkappa(\mathrm{d}\omega) = 0.$$

Now we replace $\omega \curvearrowright v s_L(\omega^x)$ by a variant of the form $\omega\big(R^s, (.)\big) \curvearrowright f\Big(\omega\big(R^s, (.)\big)\Big)$ and get

$$\lim_{n\to\infty} \sup_{x\in R^s} \int_{\{\Psi:\, f(\Psi)\leqq vw\}} |(2n)^{-s}\, \Psi(X_{n,x}) - f(\Psi)|\; L_{[Q]}(\mathrm{d}\Psi) = 0.$$

Consequently, $L_{[Q]}$ belongs to $\boldsymbol{S}$ and $f(\Psi)$ is distributed according to $L_{[Q]}$ as $v s_L(\omega^x)$ according to $L_\varkappa$, i.e. as $v s_L(\Phi)$ according to L. ∎

11.8. Asymptotically Uniformly Distributed Sequences of Distributions on $\mathfrak{R}^s$

For all distributions ν on $\mathfrak{R}^s$ we have $[Q_\nu] = \varkappa(K)$, where $K(x, X) = \nu(X - x)$ for all $x \in R^s$, $X \in \mathfrak{R}^s$. Because of $K * \mu = \nu * \mu = \mu$ we observe with the help of 4.8.1. the truth of

11.8.1. *For all distributions ν on $\mathfrak{R}^s$ and all distributions P on $\mathfrak{M}$ having the property g_μ*

$$P_{[Q_\nu]} = P.$$

In virtue of 1.14.8. each distribution P with the property g_μ has the form $P = \int P_{l\mu}(.)\, \sigma(\mathrm{d}l)$ and is consequently stationary. The statements in Section 6.4. show that $L\big(s(\Phi) \in (.)\big) = \sigma$, i.e. $L\big(s(\Phi) < +\infty\big) = 1$ and $\sigma_P = \sigma$.

Now let L be any stationary distribution on $\mathfrak{M}$ satisfying $L\big(s(\Phi) < +\infty\big) = 1$. In view of the definition of the property g_μ, we can expect, as n increases, that $L_{[Q_{\nu_n}]}$ converges weakly towards the distribution $\int P_{l\mu}(.)\, \sigma_L(\mathrm{d}l)$, if we set, for instance,

$$\nu_n = (2n)^{-s}\, \mu\big((.) \cap [-n, n)^s\big) \qquad (n = 1, 2, \ldots).$$

A sequence (ν_n) of distributions on $\mathfrak{R}^s$ is said to be *asymptotically uniformly distributed* if, for all x in R^s, the convergence relation

$$\operatorname{Var}(\nu_n - \delta_x * \nu_n) \xrightarrow[n\to\infty]{} 0$$

is satisfied. Obviously the sequence of the $(2n)^{-s}\, \mu\big((.) \cap [-n, n)^s\big)$ has this property.

The asymptotic uniform distribution of a sequence (ν_n) is introduced here as a property "in the large", and the Lebesgue measure μ does not occur in the definition. It can, however, be characterized also by saying that the ν_n are locally approximately uniformly distributed except for sets having small ν_n-measures, in the sense that, for all bounded neighbourhoods U of $[0, \ldots, 0]$ which belong to $\mathfrak{R}^s$, and for all $\varepsilon > 0$, the convergence relation

$$\lim_{n\to\infty} \nu_n \left(\operatorname{Var}\left(\nu_n((.) \mid U + \xi) - (\mu(U))^{-1} \mu((.) \cap (U + \xi))\right) \geqq \varepsilon\right) = 0$$

is satisfied (cf. Kerstan & Matthes [6] and Herrmann, Kerstan & Liese [1]).

It turns out that the approximation of $L_{[Q_{\nu_n}]}$ towards $\int P_{l\mu}(.)\, \sigma_L(\mathrm{d}l)$ is achieved for arbitrary L in $\mathbf{S}$ and arbitrary asymptotically uniformly distributed sequences (ν_n). We have

11.8.2. Theorem. *For all distributions L in $\mathbf{S}$, all asymptotically uniformly distributed sequences (ν_n) of distributions on $\mathfrak{R}^s$ and all bounded sets X in $\mathfrak{R}^s$, the following relation holds:*

$$\sup_{x \in R^s} {}_{(X+x)}\left\| L_{[Q_{\nu_n}]} - \int P_{l\mu}(.)\, \sigma(\mathrm{d}l)\right\| \xrightarrow[n\to\infty]{} 0 .$$

The proof of 11.8.2. is based on

11.8.3. *A sequence (ν_n) of distributions on $\mathfrak{R}^s$ is asymptotically uniformly distributed if and only if all distributions γ_1, γ_2 on $\mathfrak{R}^s$ satisfy the condition*

$$\operatorname{Var}(\gamma_1 * \nu_n - \gamma_2 * \nu_n) \xrightarrow[n\to\infty]{} 0 .$$

Proof. Obviously, the condition stated above is sufficient, since we may set $\gamma_1 = \delta_{[0,\ldots,0]}$, $\gamma_2 = \delta_x$. Conversely, let (ν_n) be asymptotically uniformly distributed. For all γ_1, γ_2, we obtain

$$\overline{\lim_{n\to\infty}} \operatorname{Var}(\gamma_1 * \nu_n - \gamma_2 * \nu_n) \leqq \overline{\lim_{n\to\infty}} \left(\operatorname{Var}(\gamma_1 * \nu_n - \nu_n) + \operatorname{Var}(\gamma_2 * \nu_n - \nu_n)\right)$$

$$\leqq \overline{\lim_{n\to\infty}} \left(\operatorname{Var} \int (\delta_x * \nu_n - \nu_n)\, \gamma_1(\mathrm{d}x) + \operatorname{Var} \int (\delta_x * \nu_n - \nu_n)\, \gamma_2(\mathrm{d}x)\right)$$

$$\leqq \overline{\lim_{n\to\infty}} \int \operatorname{Var}(\delta_x * \nu_n - \nu_n)\, (\gamma_1 + \gamma_2)\, (\mathrm{d}x) .$$

The integrand $\operatorname{Var}(\delta_x * \nu_n - \nu_n)$ remains bounded and tends towards zero for each x in R^s, so that we are able to conclude that

$$\overline{\lim_{n\to\infty}} \operatorname{Var}(\gamma_1 * \nu_n - \gamma_2 * \nu_n) = 0 . \ \blacksquare$$

For every compact subset X of R^s and every $\varepsilon > 0$ there exists an m such that

$$\sup_{x \in X} \operatorname{Var}\left((2m)^{-s} \mu((.) \cap [-m, m)^s) - \delta_x * (2m)^{-s} \mu((.) \cap [-m, m)^s)\right) < \varepsilon$$

is satisfied. Then, in view of 11.8.3.

$$\overline{\lim_{n\to\infty}} \sup_{x\in X} \operatorname{Var}(\nu_n - \delta_x * \nu_n) \leqq \overline{\lim_{n\to\infty}} \operatorname{Var}\Big(\nu_n - \big((2m)^{-s}\,\mu\big((.)\cap[-m, m)^s\big)\big) * \nu_n\Big)$$

$$+ \overline{\lim_{n\to\infty}} \sup_{x\in X} \operatorname{Var}\Big((2m)^{-s}\,\mu\big((.)\cap[-m, m)^s\big) * \nu_n$$

$$- \delta_x * (2m)^{-s}\,\mu\big((.)\cap[-m, m)^s\big) * \nu_n\Big)$$

$$+ \overline{\lim_{n\to\infty}} \operatorname{Var}\Big(\delta_x * \Big(\nu_n - \big((2m)^{-s}\,\mu\big((.)\cap[-m, m)^s\big)\big) * \nu_n\Big)\Big)$$

$$< 0 + \varepsilon + 0$$

for each asymptotically uniformly distributed sequence (ν_n) of distributions on $\mathfrak{R}^s$. The number $\varepsilon > 0$ is arbitrary. We thus observe the truth of

11.8.4. Proposition. *For all asymptotically uniformly distributed sequences (ν_n) of distributions on $\mathfrak{R}^s$ and all compact subsets X of R^s,*

$$\sup_{x\in X} \operatorname{Var}(\nu_n - \delta_x * \nu_n) \xrightarrow[n\to\infty]{} 0\,.$$

A sequence (ν_n) of distributions on $\mathfrak{R}^s$ is said to be *finally absolutely continuous* if the weight $\nu_n'(R^s)$ of the part ν_n' of ν_n which is absolutely continuous with respect to μ tends towards one as n increases. We have

11.8.5. *Each asymptotically uniformly distributed sequence (ν_n) of distributions on $\mathfrak{R}^s$ is finally absolutely continuous.*

Proof. We set $\sigma = \mu\big((.)\cap[0, 1]^s\big)$. All distributions $\sigma * \nu_n$ are absolutely continuous with respect to μ. Hence

$$\operatorname{Var}(\nu_n - \sigma * \nu_n) = \operatorname{Var}(\nu_n' - \sigma * \nu_n) + (\nu_n - \nu_n')\,(R^s)\,.$$

By 11.8.3. we obtain

$$\overline{\lim_{n\to\infty}}\,(\nu_n - \nu_n')\,(R^s) \leqq \overline{\lim_{n\to\infty}} \operatorname{Var}(\nu_n - \sigma * \nu_n) = 0\,,$$

i.e. $\nu_n'(R^s) \xrightarrow[n\to\infty]{} 1$. ∎

Now, we will take up the proof of 11.8.2. But we will do that in a sharpened version which will become useful from the point of view of the next chapter.

Let $\boldsymbol{D}$ denote the set of those distributions D on $\mathfrak{M}$ for which $\int \chi(R^s)\,D(\mathrm{d}\chi) = 1$, i.e. for which ϱ_D represents a distribution on $\mathfrak{R}^s$. All distributions having the form Q_ν belong to $\boldsymbol{D}$. By 11.2.2. $\boldsymbol{D}$ is included in $\boldsymbol{Q}$. Together with D_1 and D_2, also $(D_1)_{[D_2]}$ belongs to $\boldsymbol{D}$, since by 11.7.2. we have $\varrho_{(D_1)_{[D_2]}} = \varrho_{D_2} * \varrho_{D_1}$.

For all L in $\mathbf{S}$, all distributions ν on $\mathfrak{R}^s$ and all bounded sets X in $\mathfrak{R}^s$ we obtain

$$\sup_{x\in R^s} {}_{(X+x)}\Big\|L_{[Q_\nu]} - \int P_{l\mu}(.)\,\sigma_L(\mathrm{d}l)\Big\|$$

$$\leqq \sup_{x\in R^s, D\in\boldsymbol{D}} {}_{(X+x)}\Big\|\big(L_{[Q_{\nu_n}]}\big)_{[D]} - \Big(\int P_{l\mu}(.)\,\sigma_L(\mathrm{d}l)\Big)_{[D]}\Big\|$$

for we can set $D = \delta_{\delta_{[0,\ldots,0]}}$. By 11.7.8. all distributions $(L_{[Q_\nu]})_{[D]}$ exist. We are now able to formulate the following sharpened version of 11.8.2.:

11.8.6. Theorem. *For all L in $\mathbf{S}$, all asymptotically uniformly distributed sequences (ν_n) of distributions on $\mathfrak{R}^s$, and all bounded sets X in $\mathfrak{R}^s$*

$$\sup_{x\in R^s, D\in \boldsymbol{D}} {}_{(X+x)}\left\|\left(L_{[Q_{\nu_n}]}\right)_{[D]} - \left(\int P_{l\mu}(.)\ \sigma_L(\mathrm{d}l)\right)_{[D]}\right\| \xrightarrow[n\to\infty]{} 0.$$

Proof. 1. If a distribution V on $\mathfrak{M}$ is of bounded intensity, then, for all distributions γ on $\mathfrak{R}^s$,

$$\sup_{x\in R^s, D\in \boldsymbol{D}} {}_{(X+x)}\left\|\left(V_{[Q_{\nu_n}]}\right)_{[D]} - \left(V_{[Q_{\gamma*\nu_n}]}\right)_{[D]}\right\| \xrightarrow[n\to\infty]{} 0.$$

We set

$$\nu_n = \omega_n + \alpha_n, \qquad \gamma * \nu_n = \omega_n + \beta_n \qquad (n = 1, 2, \ldots),$$

where the measures ω_n, α_n, and β_n are chosen in such a way that α_n and β_n are purely singular with respect to each other. Consequently, $\omega_n = \nu_n \wedge (\gamma * \nu_n)$, where $\wedge$ denotes the infimum with respect to the natural semi-order of measures. Since (ν_n) is asymptotically uniformly distributed,

$$(\alpha_n + \beta_n)\,(R^s) = \operatorname{Var}\,(\nu_n - \gamma * \nu_n)$$

tends — in view of 11.8.3. — towards zero as n increases. For all a in R^s and all D in $\boldsymbol{D}$, we obtain

$$[(Q_{\nu_n})_{[D]}]_{(a)} = (Q_{\delta_a*\nu_n})_{[D]} = (Q_{\delta_a*\omega_n})_{[D]} + (Q_{\delta_a*\alpha_n})_{[D]},$$
$$[(Q_{\gamma*\nu_n})_{[D]}]_{(a)} = (Q_{\delta_a*\gamma*\nu_n})_{[D]} = (Q_{\delta_a*\omega_n})_{[D]} + (Q_{\delta_a*\beta_n})_{[D]}$$

and by 11.7.3., 4.3.5. and 4.2.8. it follows for all x in R^s that

$$\begin{aligned}
&{}_{(X+x)}\left\|\left(V_{[Q_{\nu_n}]}\right)_{[D]} - \left(V_{[Q_{\gamma*\nu_n}]}\right)_{[D]}\right\| \\
&\leqq 2\int \left((Q_{\delta_a*\alpha_n})_{[D]}\left(\chi(X+x)>0\right) + (Q_{\delta_a*\beta_n})_{[D]}\left(\chi(X+x)>0\right)\right) \varrho_V(\mathrm{d}a) \\
&= 2\int (Q_{(\alpha_n+\beta_n)})_{[D]}\left(\chi(X+x-a)>0\right) \varrho_V(\mathrm{d}a) \\
&\leqq 2\int \varrho_{(Q_{(\alpha_n+\beta_n)})_{[D]}}(X+x-a)\ \varrho_V(\mathrm{d}a) \\
&= 2\int \left(\varrho_D * (\alpha_n+\beta_n)\right)(X+x-a)\ \varrho_V(\mathrm{d}a) \\
&= 2\left(\varrho_D * (\alpha_n+\beta_n) * \varrho_V\right)(X+x) = 2\left(\varrho_V * \left(\varrho_D * (\alpha_n+\beta_n)\right)\right)(X+x) \\
&= 2\int \varrho_V(X+x-a)\left(\varrho_D * (\alpha_n+\beta_n)\right)(\mathrm{d}a) \leqq 2\left(\sup_{y\in R^s} \varrho_V(X+y)\right)(\alpha_n+\beta_n)\,(R^s).
\end{aligned}$$

This yields the asserted convergence statement.

2. Let m be an arbitrary but fixed natural number. If we now set

$$X_m = [-m, m)^s, \qquad I_m = \{2m[l_1, \ldots, l_s]\}_{l_1,\ldots,l_s=0,\pm1,\pm2,\ldots},$$

then the family $(X_m + x)_{x \in I_m}$ forms a countable covering of R^s by pairwise disjoint sets. We define a measurable mapping h of R^s into itself by

$$a \curvearrowright h(a) = x, \quad \text{if} \quad a \in (X_m + x) \quad \text{and} \quad x \in I_m,$$

denoting by ω the associated deterministic cluster field $a \curvearrowright \delta_{\delta_{h(a)}}$ on R^s with phase space R^s.

For all measures H on $\mathfrak{M}$, the measure H_ω exists, H being transformed into the measure H_ω by the measurable mapping

$$\Phi \curvearrowright {}^m\Phi = \sum_{x \in I_m} \Phi(X_m + x)\, \delta_x.$$

If a measure H on $\mathfrak{M}$ is of bounded intensity, this is also true for H_ω, since

$$\sup_{x \in R^s} \varrho_{H_\omega}(Z + x) \leqq \sup_{x \in R^s} \varrho_H(Z + X_m + x)$$

for all open Z.

3. If a distribution V on $\mathfrak{M}$ is of bounded intensity, then

$$\sup_{x \in R^s, D \in \boldsymbol{D}} {}_{(X+x)}\left\|\left(V_{[Q_{\nu_n}]}\right)_{[D]} - \left((V_\omega)_{[Q_{\nu_n}]}\right)_{[D]}\right\| \xrightarrow[n \to \infty]{} 0,$$

the existence of the distributions $\left((V_\omega)_{[Q_{\nu_n}]}\right)_{[D]}$ being ensured by 2. and 11.7.3.

For every $\varepsilon > 0$ we now choose a natural number $k > 2m$ such that

$$(2k)^{-s}\, \mu(X_{k+2m} \setminus X_{k-2m}) < \varepsilon$$

is satisfied. For the purpose of abbreviation we set

$$\sigma = (2k)^{-s}\, \mu\big((.) \cap X_k\big), \qquad \sigma' = (2k)^{-s}\, \mu\big((.) \cap X_{k-2m}\big),$$
$$\sigma'' = (2k)^{-s}\, \mu\big((.) \cap X_{k+2m}\big),$$

thus obtaining

$$([D] \circ [Q_{\sigma * \nu_n}])_{(a)} = (Q_{\delta_a * \sigma * \nu_n})_{[D]} = (Q_{\delta_a * \sigma' * \nu_n})_{[D]} + (Q_{\delta_a * \sigma * \nu_n - \delta_a * \sigma' * \nu_n})_{[D]},$$

$$([D] \circ [Q_{\sigma * \nu_n}] \circ \omega)_{(a)} = (Q_{\delta_{h(a)} * \sigma * \nu_n})_{[D]} = (Q_{\delta_a * \sigma' * \nu_n})_{[D]} + (Q_{\delta_{h(a)} * \sigma * \nu_n - \delta_a * \sigma' * \nu_n})_{[D]}$$

for all a in R^s and all natural numbers n, since we have

$$o \leqq (\delta_a * \sigma * \nu_n - \delta_a * \sigma' * \nu_n),\ (\delta_{h(a)} * \sigma * \nu_n - \delta_a * \sigma' * \nu_n) \leqq \delta_a * (\sigma'' - \sigma') * \nu_n.$$

Using 4.2.8. we now conclude as in 1. that

$$\sup_{x \in R^s, D \in \boldsymbol{D}} {}_{(X+x)}\left\|\left(V_{[Q_{\sigma * \nu_n}]}\right)_{[D]} - \left((V_\omega)_{[Q_{\sigma * \nu_n}]}\right)_{[D]}\right\|$$

$$\leqq 2 \sup_{x \in R^s, D \in \boldsymbol{D}} \int 2(Q_{\delta_a * (\sigma'' - \sigma') * \nu_n})_{[D]} \left(\chi(X + x) > 0\right) \varrho_V(\mathrm{d}a)$$

$$\leqq 4(\sigma'' - \sigma')\,(R^s) \sup_{y \in R^s} \varrho_V(X + y) < 4\varepsilon \sup_{y \in R^s} \varrho_V(X + y)$$

for all natural numbers n. By means of 1. and 2., it follows that

$$\overline{\lim_{n\to\infty}} \sup_{x\in R^s, D\in \boldsymbol{D}} {}_{(X+x)}\left\|\left(V_{[Q_{\nu_n}]}\right)_{[D]} - \left((V_\omega)_{[Q_{\nu_n}]}\right)_{[D]}\right\|$$

$$\leqq \overline{\lim_{n\to\infty}} \sup_{x\in R^s, D\in \boldsymbol{D}} {}_{(X+x)}\left\|\left(V_{[Q_{\nu_n}]}\right)_{[D]} - \left(V_{[Q_{\sigma*\nu_n}]}\right)_{[D]}\right\|$$

$$+ \overline{\lim_{n\to\infty}} \sup_{x\in R^s, D\in \boldsymbol{D}} {}_{(X+x)}\left\|\left(V_{[Q_{\sigma*\nu_n}]}\right)_{[D]} - \left((V_\omega)_{[Q_{\sigma*\nu_n}]}\right)_{[D]}\right\|$$

$$+ \overline{\lim_{n\to\infty}} \sup_{x\in R^s, D\in \boldsymbol{D}} {}_{(X+x)}\left\|\left((V_\omega)_{[Q_{\nu_n}]}\right)_{[D]} - \left((V_\omega)_{[Q_{\sigma*\nu_n}]}\right)_{[D]}\right\|$$

$$\leqq 0 + 4\varepsilon \sup_{y\in R^s} \varrho_V(X+y) + 0 .$$

Since ε was arbitrary, this proves our assertion.

4. Let H be a distribution on $\mathfrak{M}$ with the property that for some $\eta > 0$

$$\varrho_H(X_m + x) \leqq \eta\mu(X_m) \quad \text{for all } x \text{ in } I_m .$$

Then obviously H is of bounded intensity. For, if $X_m + x_1, \ldots, X_m + x_k$, $x_i \in I_m$, is a finite covering of $X + (-2m, 2m)^s$ then we obtain

$$\sup_{x\in R^s} \varrho_H(X + x) \leqq k\eta\mu(X_m) .$$

We shall now prove that the inequality

$$\overline{\lim_{n\to\infty}} \sup_{x\in R^s, D\in \boldsymbol{D}} \left(H_{[Q_{\nu_n}]}\right)_{[D]} \left(\Phi(X+x) > 0\right) \leqq \eta\mu(X)$$

is valid for all bounded X in $\mathfrak{R}^s$

To show this, first of all we introduce a cluster field τ on R^s with phase space R^s by

$$\tau_{(a)} = Q_{(2m)^{-s}\mu((.)\cap(X_m+x))} \quad \text{for} \quad a \in (X_m + x), \qquad x \in I_m .$$

Setting now $H_\tau = V$, we obtain $\varrho_V \leqq \eta\mu$. On the other hand, we have $H_\omega = V_\omega$, and by means of 3. we get

$$\overline{\lim_{n\to\infty}} \sup_{x\in R^s, D\in \boldsymbol{D}} {}_{(X+x)}\left\|\left(H_{[Q_{\nu_n}]}\right)_{[D]} - \left(V_{[Q_{\nu_n}]}\right)_{[D]}\right\|$$

$$\leqq \overline{\lim_{n\to\infty}} \sup_{x\in R^s, D\in \boldsymbol{D}} {}_{(X+x)}\left\|\left(H_{[Q_{\nu_n}]}\right)_{[D]} - \left((H_\omega)_{[Q_{\nu_n}]}\right)_{[D]}\right\|$$

$$+ \overline{\lim_{n\to\infty}} \sup_{x\in R^s, D\in \boldsymbol{D}} {}_{(X+x)}\left\|\left(V_{[Q_{\nu_n}]}\right)_{[D]} - \left((V_\omega)_{[Q_{\nu_n}]}\right)_{[D]}\right\|$$

$$= 0 .$$

Consequently,

$$\begin{aligned}
&\overline{\lim_{n\to\infty}} \sup_{x\in R^s,\, D\in \boldsymbol{D}} \left(H_{[Q_{\nu_n}]}\right)_{[D]} \left(\Phi(X+x) > 0\right) \\
&= \overline{\lim_{n\to\infty}} \sup_{x\in R^s,\, D\in \boldsymbol{D}} \left(V_{[Q_{\nu_n}]}\right)_{[D]} \left(\Phi(X+x) > 0\right) \\
&\leqq \overline{\lim_{n\to\infty}} \sup_{x\in R^s,\, G\in \boldsymbol{D}} V_{[G]}\left(\Phi(X+x) > 0\right) \leqq \sup_{x\in R^s,\, G\in \boldsymbol{D}} (\varrho_G * \varrho_V)(X+x) \\
&= \sup_{x\in R^s,\, G\in \boldsymbol{D}} \int \varrho_V(X+x-a)\, \varrho_G(\mathrm{d}a) \leqq \sup_{y\in R^s} \varrho_V(X+y) \leqq \eta\mu(X).
\end{aligned}$$

5. Now let $c > 0$ and let W be an arbitrary distribution in $\boldsymbol{S}^{(c)}$. For each Φ in the domain of definition of s_W, we set

$$\Phi_m = [s_W(\Phi)\,(2m)^s] \sum_{x\in I_m} \delta_x,$$

where, as usual, $[v]$ denotes the integral part of the real number v, and we choose Φ', Φ'' in M in such a way that

$${}^m\Phi = \Phi' + \Phi_m - \Phi'', \qquad \Phi' \wedge \Phi'' = o$$

are satisfied. All of the mappings

$$\Phi \curvearrowright \Phi_m, \qquad \Phi \curvearrowright \Phi', \qquad \Phi \curvearrowright \Phi''$$

from M in M are measurable with respect to the σ-algebra $\mathfrak{M}$. We denote by W_m and $W^!$ the distributions of Φ_m and $\Phi' + \Phi''$, respectively, which are formed with respect to W. Obviously W_m depends only on σ_W and m.

In view of 11.7.7. and 2., W_ω is of bounded intensity. As in 4. we observe that W_m is also of bounded intensity. Finally, we have

$$\varrho_{W^!} \leqq \varrho_{W_\omega} + \varrho_{W_m},$$

so that $W^!$ is of bounded intensity, too. Now, if G belongs to $\boldsymbol{D}$, then by 11.7.3. the distributions

$$[G]_{(\Phi')}, \qquad [G]_{({}^m\Phi \wedge \Phi_m)}, \quad \text{and} \quad [G]_{(\Phi'')}$$

exist almost everywhere with respect to W.

The correspondence

$$\Phi \curvearrowright [G]_{(\Phi')} \times [G]_{({}^m\Phi\wedge\Phi_m)} \times [G]_{(\Phi'')}$$

provides a mapping from M into the set of all distributions on $\mathfrak{M}^3$, where — in view of 4.1.3. — the real function

$$\Phi \curvearrowright \left([G]_{(\Phi')} \times [G]_{({}^m\Phi\wedge\Phi_m)} \times [G]_{(\Phi'')}\right)(Z)$$

is measurable with respect to $\mathfrak{M}$ for all $Z = Y_1 \times Y_2 \times Y_3$ ($Y_1, Y_2, Y_3 \in \mathfrak{M}$), and hence also for arbitrary Z in $\mathfrak{M}^3$.

Obviously we have

$$^m\Phi = {}^m\Phi \wedge \Phi_m + \Phi', \qquad \Phi_m = {}^m\Phi \wedge \Phi_m + \Phi''.$$

Hence, if the random element $[\Psi_1, \Psi_2, \Psi_3]$ of the measurable space $[M^3, \mathfrak{M}^3]$ is distributed according to

$$H = \int ([G]_{(\Phi')} \times [G]_{({}^m\Phi \wedge \Phi_m)} \times [G]_{(\Phi'')}) (.) \; W(\mathrm{d}\Phi),$$

then $\Psi_1 + \Psi_2$, $\Psi_2 + \Psi_3$, and $\Psi_1 + \Psi_3$ have the distributions $(W_\omega)_{[G]}$, $(W_m)_{[G]}$, and $(W^!)_{[G]}$, respectively.

Let $x \in R^s$. Then

$$H\big({}_{X+x}(\Psi_2 + \Psi_3) \in (.), \Psi_1(X + x) = 0 = \Psi_3(X + x)\big)$$
$$= H\big({}_{X+x}(\Psi_1 + \Psi_2) \in (.), \Psi_1(X + x) = 0 = \Psi_3(X + x)\big),$$

and we obtain

$$\begin{aligned}
&{}_{(X+x)}\|(W_m)_{[G]} - (W_\omega)_{[G]}\| \\
&\leqq \big\|H\big({}_{X+x}(\Psi_2 + \Psi_3) \in (.), (\Psi_1 + \Psi_3)(X + x) > 0\big) \\
&\qquad - H\big({}_{X+x}(\Psi_1 + \Psi_2) \in (.), (\Psi_1 + \Psi_3)(X + x) > 0\big)\big\| \\
&\leqq 2H\big((\Psi_1 + \Psi_3)(X + x) > 0\big) = 2(W^!)_{[G]}\big(\Phi(X + x) > 0\big).
\end{aligned}$$

6. Now let L be a distribution in $\boldsymbol{S}^{(c)}$. We set $P = \int P_{l\mu}(.)\, \sigma_L(\mathrm{d}l)$ and, in view of $P_m = L_m$, using 3. and 5. we obtain

$$\begin{aligned}
&\overline{\lim_{n\to\infty}} \sup_{x\in R^s, D\in \boldsymbol{D}} {}_{(X+x)}\big\|\big(L_{[Q_{v_n}]}\big)_{[D]} - P_{[D]}\big\| \\
&\doteq \overline{\lim_{n\to\infty}} \sup_{x\in R^s, D\in \boldsymbol{D}} {}_{(X+x)}\big\|\big(L_{[Q_{v_n}]}\big)_{[D]} - \big(P_{[Q_{v_n}]}\big)_{[D]}\big\| \\
&\leqq \overline{\lim_{n\to\infty}} \sup_{x\in R^s, D\in \boldsymbol{D}} {}_{(X+x)}\big\|\big(L_{[Q_{v_n}]}\big)_{[D]} - \big((L_m)_{[Q_{v_n}]}\big)_{[D]}\big\| \\
&\quad + \overline{\lim_{n\to\infty}} \sup_{x\in R^s, D\in \boldsymbol{D}} {}_{(X+x)}\big\|\big(P_{[Q_{v_n}]}\big)_{[D]} - \big((P_m)_{[Q_{v_n}]}\big)_{[D]}\big\| \\
&\leqq \overline{\lim_{n\to\infty}} \sup_{x\in R^s, D\in \boldsymbol{D}} {}_{(X+x)}\big\|\big((L_\omega)_{[Q_{v_n}]}\big)_{[D]} - \big((L_m)_{[Q_{v_n}]}\big)_{[D]}\big\| \\
&\quad + \overline{\lim_{n\to\infty}} \sup_{x\in R^s, D\in \boldsymbol{D}} {}_{(X+x)}\big\|\big((P_\omega)_{[Q_{v_n}]}\big)_{[D]} - \big((P_m)_{[Q_{v_n}]}\big)_{[D]}\big\| \\
&\leqq 2\, \overline{\lim_{n\to\infty}} \sup_{x\in R^s, D\in \boldsymbol{D}} \big((L^!)_{[Q_{v_n}]}\big)_{[D]} \big(\Phi(X + x) > 0\big) \\
&\quad + 2\, \overline{\lim_{n\to\infty}} \sup_{x\in R^s, D\in \boldsymbol{D}} \big((P^!)_{[Q_{v_n}]}\big)_{[D]} \big(\Phi(X + x) > 0\big).
\end{aligned}$$

For all x in I_m we have

$$\varrho_{L^!}(X_m + x) = \int |\Phi(X_m + x) - [s_L(\Phi)\ (2m)^s]|\ L(\mathrm{d}\Phi)$$

$$\leqq \mu(X_m) \sup_{y \in R^s} \int \left| \frac{\Phi(X_m + y)}{(2m)^s} - \frac{[s_L(\Phi)\ (2m)^s]}{(2m)^s} \right| L(\mathrm{d}\Phi)$$

$$\leqq \mu(X_m) \left(\sup_{y \in R^s} \int \left| \frac{\Phi(X_m + y)}{(2m)^s} - s_L(\Phi) \right| L(\mathrm{d}\Phi) \right.$$

$$\left. + \int \left| \frac{[s_L(\Phi)\ (2m)^s]}{(2m)^s} - s_L(\Phi) \right| L(\mathrm{d}\Phi) \right)$$

$$= \mu(X_m)\ \eta_{L,m} \quad \text{(say)}.$$

Since the inequality $s_L(\Phi) \leqq c$ is satisfied almost everywhere with respect to L, as an immediate consequence of the definition of $\boldsymbol{S}^{(c)}$ and of Lebesgue's theorem the sequence $(\eta_{L,m})$ tends to zero as m increases. Accordingly we obtain

$$\varrho_{P^!}(X_m + x) \leqq \mu(X_m)\ \eta_{P,m}, \qquad \eta_{P,m} \xrightarrow[m \to \infty]{} 0.$$

Now we can continue our estimation with the help of 4. as follows:

$$\overline{\lim_{n \to \infty}} \sup_{x \in R^s, D \in \boldsymbol{D}} {}_{(X+x)}\left\| \left(L_{[Q_{v_n}]} \right)_{[D]} - P_{[D]} \right\| \leqq 2\mu(X)\ (\eta_{L,m} + \eta_{P,m}).$$

However, this yields the approximation statement 11.8.6. in the case $L \in \boldsymbol{S}^{(c)}$.

7. Now let L be any distribution in $\boldsymbol{S}$ having the property that $\sigma_L\big([c, +\infty)\big) > 0$ for all $c > 0$. Then, if c is sufficiently large, $L' = L\big((.) \mid s_L(\Phi) < c\big)$ and $L'' = L\big((.) \mid s_L(\Phi) \geqq c\big)$ belong to $\boldsymbol{S}$ and we have

$$L = \sigma_L\big([0, c)\big)\ L' + \sigma_L\big([c, +\infty)\big)\ L'',$$

$$\sigma_L = \sigma_L\big([0, c)\big)\ \sigma_{L'} + \sigma_L\big([c, +\infty)\big)\ \sigma_{L''}.$$

Consequently,

$$\overline{\lim_{n \to \infty}} \sup_{x \in R^s,\, D \in \boldsymbol{D}} {}_{(X+x)}\left\| \left(L_{[Q_{v_n}]} \right)_{[D]} - \left(\int P_{l\mu}(.)\ \sigma_L(\mathrm{d}l) \right)_{[D]} \right\|$$

$$\leqq \overline{\lim_{n \to \infty}} \sup_{x \in R^s,\, D \in \boldsymbol{D}} {}_{(X+x)}\left\| \left((L')_{[Q_{v_n}]} \right)_{[D]} - \left(\int P_{l\mu}(.)\ \sigma_{L'}(\mathrm{d}l) \right)_{[D]} \right\| + 2\sigma_L\big([c, +\infty)\big)$$

for all sufficiently large c, such that by 6. we observe the truth of the statement formulated in 11.8.6. ∎

To conclude this section, we shall consider the following simple model: For the individual particles x of a random population Φ, a velocity ξ_x at time zero is chosen at random according to a distribution ν on $\mathfrak{R}^s$. For all $t \geqq 0$, x will move uniformly with this velocity. Thus every x in Φ has the position $x + t\xi_x$ at time $t > 0$. If we denote the distribution of $t\xi_x$ by ν_t, i.e. if we set

$$\nu_t(X) = \nu(t^{-1}X) \qquad (X \in \mathfrak{R}^s),$$

then the random translation described above is represented by the homogeneous cluster field $[Q_{\nu_t}]$ and we have (cf. Breiman [1], Thedéen [1], and Stone [1])

11.8.7. Theorem. *For all L in* **S**, *all distributions ν on $\mathfrak{R}^s$ being absolutely continuous with respect to μ, and all compact subsets X of R^s,*

$$\lim_{t\to\infty} \sup_{x\in R^s} {}_{(X+x)}\left\|L_{[Q_{\nu_t}]} - \int P_{l\mu}(\cdot)\ \sigma_L(\mathrm{d}l)\right\| = 0.$$

The proof is readily derived from 11.8.2. together with the following proposition.

11.8.8. Proposition. *For each distribution ν on $\mathfrak{R}^s$ and each sequence (t_n) of positive real numbers which tends towards infinity, the following statements are equivalent:*

a) *ν is absolutely continuous with respect to μ.*
b) *(ν_{t_n}) is asymptotically uniformly distributed.*

Proof. 1. In view of 11.8.5. a) is a consequence of b).

2. Conversely, suppose that $\nu = \int_{(\cdot)} f(y)\ \mu(\mathrm{d}y)$. For all x in R^s, we obtain

$$\begin{aligned}\operatorname{Var}(\nu_{t_n} - \delta_x * \nu_{t_n}) &= \operatorname{Var}(\nu - \delta_{t_n^{-1}x} * \nu)\\ &= \int |f(z) - f(z - t_n^{-1}x)|\ \mu(\mathrm{d}z).\end{aligned}$$

As is known, for all f in the function space $L^1(\mu)$ the value $\int |f(z) - f(z-h)|\ \mu(\mathrm{d}z)$ tends towards zero as $h \to 0$. ∎

11.9. Weakly Asymptotically Uniformly Distributed Sequences of Distributions on $\mathfrak{R}^s$

A sequence (ν_n) of distributions on $\mathfrak{R}^s$ is said to be *weakly asymptotically uniformly distributed* if to every $\varepsilon > 0$ there exists a distribution γ_ε on $\mathfrak{R}^s$ such that $\gamma_\varepsilon([-\varepsilon, \varepsilon]^s) = 1$ and the sequence $(\gamma_\varepsilon * \nu_n)$ is asymptotically uniformly distributed.

Each asymptotically uniformly distributed sequence (ν_n) is weakly asymptotically uniformly distributed, since in this case we may set γ_ε equal to $\delta_{[0,\ldots,0]}$ for all $\varepsilon > 0$. On the other hand, not each of the weakly asymptotically uniformly distributed sequences is also asymptotically uniformly distributed.

As for example, the sequence of the

$$\nu_n = (2n^2)^{-s} \sum_{-n^2 \leq i_1,\ldots,i_s < n^2} \delta_{[(i_1/n),\ldots,(i_s/n)]}, \qquad n = 1, 2, \ldots,$$

is not asymptotically uniformly distributed. In fact, if at least one coordinate x_i of $x = [x_1, \ldots, x_s]$ is irrational, then we have

$$\operatorname{Var}(\nu_n - \delta_x * \nu_n) = 2 \qquad (n = 1, 2, \ldots).$$

On the other hand, (ν_n) is weakly asymptotically uniformly distributed, which can be observed by choosing for γ_ε the distribution $\varepsilon^{-s}\mu((.) \cap [0, \varepsilon]^s)$.

The above definition can be simplified as follows:

11.9.1. Proposition. *For all sequences (ν_n) of distributions on $\mathfrak{R}^s$, the following statements are equivalent:*

a) *(ν_n) is weakly asymptotically uniformly distributed.*
b) *For all distributions λ_1, λ_2 on $\mathfrak{R}^s$ which are absolutely continuous with respect to μ,*

$$\operatorname{Var}(\lambda_1 * \nu_n - \lambda_2 * \nu_n) \xrightarrow[n\to\infty]{} 0.$$

c) *For all distributions λ on $\mathfrak{R}^s$ which are absolutely continuous with respect to μ, the sequence $(\lambda * \nu_n)$ is asymptotically uniformly distributed.*

Proof. 1. a) implies b).

So let (ν_n) be weakly asymptotically uniformly distributed. For all $\varepsilon > 0$ and all distributions

$$\lambda_1 = \int_{(.)} f_1(x)\,\mu(\mathrm{d}x), \qquad \lambda_2 = \int_{(.)} f_2(x)\,\mu(\mathrm{d}x)$$

which are absolutely continuous with respect to μ, using 11.8.3. we obtain

$$\begin{aligned}
&\overline{\lim_{n\to\infty}} \operatorname{Var}(\lambda_1 * \nu_n - \lambda_2 * \nu_n)\\
&\leqq \overline{\lim_{n\to\infty}} \operatorname{Var}\big(\lambda_1 * (\gamma_\varepsilon * \nu_n) - \lambda_2 * (\gamma_\varepsilon * \nu_n)\big)\\
&\qquad + \overline{\lim_{n\to\infty}} \operatorname{Var}(\lambda_1 * \gamma_\varepsilon * \nu_n - \lambda_1 * \nu_n) + \overline{\lim_{n\to\infty}} \operatorname{Var}(\lambda_2 * \gamma_\varepsilon * \nu_n - \lambda_2 * \nu_n)\\
&\leqq 0 + \operatorname{Var}(\lambda_1 * \gamma_\varepsilon - \lambda_1) + \operatorname{Var}(\lambda_2 * \gamma_\varepsilon - \lambda_2)\\
&= \sum_{i=1}^{2} \int \left| \int f_i(x-y)\,\gamma_\varepsilon(\mathrm{d}y) - f_i(x) \right| \mu(\mathrm{d}x)\\
&= \sum_{i=1}^{2} \int \left| \int \big(f_i(x-y) - f_i(x)\big)\,\gamma_\varepsilon(\mathrm{d}y) \right| \mu(\mathrm{d}x)\\
&\leqq \sum_{i=1}^{2} \int \Big(\int |f_i(x-y) - f_i(x)|\,\gamma_\varepsilon(\mathrm{d}y)\Big)\,\mu(\mathrm{d}x)\\
&= \sum_{i=1}^{2} \int \Big(\int |f_i(x-y) - f_i(x)|\,\mu(\mathrm{d}x)\Big)\,\gamma_\varepsilon(\mathrm{d}y)\\
&\leqq \sum_{i=1}^{2} \sup_{y\in[-\varepsilon,\varepsilon]^s} \int |f_i(x-y) - f_i(x)|\,\mu(\mathrm{d}x) \xrightarrow[\varepsilon\to 0]{} 0.
\end{aligned}$$

2. b) implies c).

Suppose now that, for all λ_1, λ_2 which are absolutely continuous with respect to μ, we have $\lim_{n\to\infty} \operatorname{Var}(\lambda_1 * \nu_n - \lambda_2 * \nu_n) = 0$. For each distribution λ absolutely continuous with respect to μ, and each x in R^s, we may set $\lambda_1 = \lambda$, $\lambda_2 = \delta_x * \lambda$, which gives

$$\operatorname{Var}\big(\lambda * \nu_n - \delta_x * (\lambda * \nu_n)\big) \xrightarrow[n\to\infty]{} 0,$$

i.e. the sequence $(\lambda * \nu_n)$ is asymptotically uniformly distributed.

3. c) implies a).

This is immediately obvious. We may choose for γ_ε any distribution λ which is absolutely continuous with respect to μ and has the property that $\lambda\big([-\varepsilon, \varepsilon)^s\big) = 1$, say the uniform distribution on $[-\varepsilon, \varepsilon]^s$. ∎

Every weakly asymptotically uniformly distributed sequence shows a dispersion, i.e. we have (cf. Kerstan & Matthes [6])

11.9.2. *For each weakly asymptotically uniformly distributed sequence* (ν_n) *of distributions on* $\mathfrak{R}^s$ *and each bounded set* X *in* $\mathfrak{R}^s$,

$$\sup_{x\in R^s} \nu_n(X + x) \xrightarrow[n\to\infty]{} 0.$$

Proof. 1. It will suffice to show that the assertion is valid for asymptotically uniformly distributed sequences. Indeed, if (ν_n) is weakly asymptotically uniformly distributed, and σ the uniform distribution on $(0, 1)^s$, then for all x in R^s we have

$$\nu_n(X + x) \leqq (\sigma * \nu_n)\big((0, 1)^s + X + x\big).$$

However, like X itself, the open set $(0, 1)^s + X$ is bounded, and by 11.9.1. the sequence $(\sigma * \nu_n)$ is asymptotically uniformly distributed.

2. Now suppose that (ν_n) be asymptotically uniformly distributed. For each natural number m we may choose a finite sequence $h_1, \ldots, h_m$ of points in R^s, such that the sets $X + h_1, \ldots, X + h_m$ are pairwise disjoint. For all x in R^s and $i = 1, \ldots, m$, we obtain

$$\nu_n(X + x) \leqq \nu_n(X + x + h_i) + \operatorname{Var}(\nu_n - \delta_{h_i} * \nu_n).$$

Hence, for all x in R^s,

$$m\nu_n(X + x) \leqq \nu_n\left(\left(\bigcup_{i=1}^{m} (X + h_i)\right) + x\right) + \sum_{i=1}^{m} \operatorname{Var}(\nu_n - \delta_{h_i} * \nu_n).$$

This yields

$$\sup_{x\in R^s} \nu_n(X + x) \leqq \frac{1}{m} + \frac{1}{m} \sum_{i=1}^{m} \operatorname{Var}(\nu_n - \delta_{h_i} * \nu_n).$$

Consequently, for all natural numbers m we have

$$\varlimsup_{n\to\infty} \sup_{x\in R^s} \nu_n(X + x) \leqq \frac{1}{m}. \quad ∎$$

The following characterization of the weakly asymptotically uniformly distributed sequences can be used as starting-point for generalizations (cf. Kerstan & Debes [1], Debes, Kerstan, Liemant & Matthes [2], and Fichtner [1]).

11.9.3. Proposition. *For each sequence* (ν_n) *of distributions on* $\mathfrak{R}^s$, *the following properties are equivalent.*

a) (ν_n) *is weakly asymptotically uniformly distributed.*

b) *For all bounded* X *in* $\mathfrak{R}^s$ *and all distributions* γ *on* $\mathfrak{R}^s$,

$$\int |\nu_n(X - x) - (\gamma * \nu_n)(X - x)|\, \mu(dx) \xrightarrow[n\to\infty]{} 0.$$

c) *For all bounded sets* X *in* $\mathfrak{R}^s$ *and all distributions* γ *on* $\mathfrak{R}^s$ *we have*

$$\int_Z \big(\nu_n(X - x) - (\gamma * \nu_n)(X - x)\big)\, \mu(dx) \xrightarrow[n\to\infty]{} 0$$

for all Z *in* $\mathfrak{R}^s$.

Proof. 1. Let X be a bounded set in $\mathfrak{R}^s$ satisfying $\mu(X) > 0$ and σ the uniform distribution on $-X$. For all distributions ω on $\mathfrak{R}^s$ and all U in $\mathfrak{R}^s$ we obtain

$$\begin{aligned}(\sigma * \omega)(-U) &= \int \big(\mu(-X)\big)^{-1} \mu\big((-U - x) \cap (-X)\big)\, \omega(dx)\\ &= \big(\mu(X)\big)^{-1} \int \mu\big(U \cap (X - x)\big)\, \omega(dx) = \big(\mu(X)\big)^{-1} \int_U \omega(X - x)\, \mu(dx).\end{aligned}$$

Consequently, $\big(\mu(X)\big)^{-1} \omega\big(X - (.)\big)$ is a density function with respect to μ of the distribution $(\sigma * \omega)^\vee = (\sigma * \omega)\big(-(.)\big)$.

2. Let $\gamma, \nu_1, \nu_2, \ldots$ be any distributions on $\mathfrak{R}^s$. In view of 1, we get

$$\begin{aligned}&\int |\nu_n(X - x) - (\gamma * \nu_n)(X - x)|\, \mu(dx)\\ &= \mu(X)\, \mathrm{Var}\,\big((\sigma * \nu_n)^\vee - (\sigma * \gamma * \nu_n)^\vee\big)\\ &= \mu(X)\, \mathrm{Var}\,(\sigma * \nu_n - \sigma * \gamma * \nu_n).\end{aligned}$$

In this way we observe that b) implies a), because we can set $X = [-\varepsilon, \varepsilon]^s$, $\varepsilon > 0$.

Conversely, if (ν_n) is weakly asymptotically uniformly distributed, then the previous chain of equations yields b) under the additional assumption $\mu(X) > 0$. However, in the case of $\mu(X) = 0$ the assertion b) is always satisfied since

$$\begin{aligned}&\int |\nu_n(X - x) - (\gamma * \nu_n)(X - x)|\, \mu(dx)\\ &\leqq (\nu_n * \mu)(X) + (\gamma * \nu_n * \mu)(X) = 2\mu(X).\end{aligned}$$

Consequently a) implies b).

3. Obviously, c) is a weakening of b). Now we suppose that (ν_n) has the property c).

Let X be a bounded set in $\mathfrak{R}^s$ and γ a distribution on $\mathfrak{R}^s$. Setting

$$\varrho_m = (2m)^{-s}\, \mu\big((\cdot) \cap [-m, m)^s\big) \qquad (m = 1, 2, \ldots)$$

we get for all bounded sets B in $\mathfrak{R}^s$

$$\begin{aligned}
&\overline{\lim_{n\to\infty}} \int_B |\nu_n(X - x) - (\gamma * \nu_n)(X - x)|\, \mu(\mathrm{d}x) \\
&\leqq \overline{\lim_{n\to\infty}} \int_B |\nu_n(X - x) - (\varrho_m * \nu_n)(X - x)|\, \mu(\mathrm{d}x) \\
&\quad + \overline{\lim_{n\to\infty}} \int_B \big((\varrho_m * \nu_n)(X - x) + (\varrho_m * \gamma * \nu_n)(X - x)\big)\, \mu(\mathrm{d}x) \\
&\quad + \overline{\lim_{n\to\infty}} \int_B |(\varrho_m * \gamma * \nu_n)(X - x) - (\gamma * \nu_n)(X - x)|\, \mu(\mathrm{d}x).
\end{aligned}$$

However

$$\begin{aligned}
(\varrho_m * \tau)(X - x) &= (2m)^{-s} \int \mu\big((X - x - y) \cap [-m, m)^s\big)\, \tau(\mathrm{d}y) \\
&\leqq (2m)^{-s}\, \mu(X)
\end{aligned}$$

for all distributions τ on $\mathfrak{R}^s$. Thus we can continue the above inequality with

$$\begin{aligned}
&\leqq \overline{\lim_{n\to\infty}} \int_B \big(\nu_n(X - x) - (\varrho_m * \nu_n)(X - x)\big)\, \mu(\mathrm{d}x) + 6(2m)^{-s}\, \mu(X)\, \mu(B) \\
&\quad + \overline{\lim_{n\to\infty}} \int_B \big((\gamma * \nu_n)(X - x) - (\varrho_m * \gamma * \nu_n)(X - x)\big)\, \mu(\mathrm{d}x).
\end{aligned}$$

The first summand vanishes according to our assumption on (ν_n). Correspondingly, we get for the last summand

$$\begin{aligned}
&\overline{\lim_{n\to\infty}} \int_B \big((\gamma * \nu_n)(X - x) - (\varrho_m * \gamma * \nu_n)(X - x)\big)\, \mu(\mathrm{d}x) \\
&= \overline{\lim_{n\to\infty}} \int \Big(\int_{B+y} \big(\nu_n(X - z) - (\varrho_m * \nu_n)(X - z)\big)\, \mu(\mathrm{d}z) \Big)\, \gamma(\mathrm{d}y) = 0.
\end{aligned}$$

Summarizing we obtain for all natural numbers m

$$\overline{\lim_{n\to\infty}} \int_B |\nu_n(X - x) - (\gamma * \nu_n)(X - x)|\, \mu(\mathrm{d}x) \leqq 6(2m)^{-s}\, \mu(X)\, \mu(B),$$

i.e.

$$(*) \qquad \int_B |\nu_n(X - x) - (\gamma * \nu_n)(X - x)|\, \mu(\mathrm{d}x) \xrightarrow[n\to\infty]{} 0$$

for all distributions γ on $\mathfrak{R}^s$ and all bounded sets X, B in $\mathfrak{R}^s$.

For the sake of abbreviation we set

$$f_n(x) = \nu_n(X - x) - (\gamma * \nu_n)(X - x) \qquad (x \in R^s,\ n = 1, 2, \ldots)$$

and suppose

$$c = \overline{\lim_{n\to\infty}} \int |f_n(x)|\, \mu(\mathrm{d}x) > 0.$$

In view of (*) we obtain for all bounded B in $\mathfrak{R}^s$

$$(**) \qquad c = \overline{\lim_{n\to\infty}} \int_{R^s \setminus B} |f_n(x)|\, \mu(\mathrm{d}x).$$

We select an $\varepsilon > 0$ such that $5\varepsilon < c$. Besides there exists a natural number n_1 such that

$$\int |f_{n_1}(x)|\, \mu(\mathrm{d}x) > c - \varepsilon.$$

To n_1 there exists a bounded B_1 in $\mathfrak{R}^s$ such that

$$\int_{B_1} |f_{n_1}(x)|\, \mu(\mathrm{d}x) > c - \varepsilon, \qquad \int_{R^s \setminus B_1} |f_{n_1}(x)|\, \mu(\mathrm{d}x) < \varepsilon.$$

In virtue of (*) and (**) we can select an $n_2 > n_1$ such that

$$\int_{R^s \setminus B_1} |f_{n_2}(x)|\, \mu(\mathrm{d}x) > c - \varepsilon, \qquad \int_{B_1} |f_{n_2}(x)|\, \mu(\mathrm{d}x) < \varepsilon/2.$$

Therefore there is a bounded B_2 in $\mathfrak{R}^s$, disjoint to B_1 and satisfying

$$\int_{B_2} |f_{n_2}(x)|\, \mu(\mathrm{d}x) > c - \varepsilon, \qquad \int_{(R^s \setminus B_1) \setminus B_2} |f_{n_2}(x)|\, \mu(\mathrm{d}x) < \varepsilon/2.$$

Hence

$$\int_{R^s \setminus B_2} |f_{n_2}(x)|\, \mu(\mathrm{d}x) < \varepsilon.$$

Applying anew (*) and (**) we can select an $n_3 > n_2$ such that

$$\int_{R^s \setminus (B_1 \cup B_2)} |f_{n_3}(x)|\, \mu(\mathrm{d}x) > c - \varepsilon, \qquad \int_{B_1 \cup B_2} |f_{n_3}(x)|\, \mu(\mathrm{d}x) < \varepsilon/2,$$

etc. In this way we get a sequence $n_1 < n_2 < \cdots$ of natural numbers and a sequence $B_1, B_2, \ldots$ of pairwise disjoint bounded sets in $\mathfrak{R}^s$ such that

$$(***) \qquad \int_{B_i} |f_{n_i}(x)|\, \mu(\mathrm{d}x) > c - \varepsilon, \qquad \int_{R^s \setminus B_i} |f_{n_i}(x)|\, \mu(\mathrm{d}x) < \varepsilon$$

for all natural numbers i.

We set

$$C_i = \{x : x \in B_i, f_{n_i}(x) > 0\}, \quad C_i' = B_i \setminus C_i, \quad V = \bigcup_{i=1}^{\infty} C_i, \quad W = \bigcup_{i=1}^{\infty} C_i',$$

$$I = \left\{ i : i \in \{1, 2, \ldots\}, \int_{C_i} f_{n_i}(x)\, \mu(\mathrm{d}x) > 2^{-1}(c - \varepsilon) \right\},$$

and

$$J = \left\{ i : i \in \{1, 2, \ldots\}, -\int_{C_i'} f_{n_i}(x)\, \mu(\mathrm{d}x) > 2^{-1}(c - \varepsilon) \right\}.$$

In view of (***) we have for all i in I

$$\int_V f_{n_i}(x)\,\mu(\mathrm{d}x) = \int_{C_i} f_{n_i}(x)\,\mu(\mathrm{d}x) + \int_{V\setminus B_i} f_{n_i}(x)\,\mu(\mathrm{d}x)$$

$$\geqq 2^{-1}(c-\varepsilon) - \int_{V\setminus B_i} |f_{n_i}(x)|\,\mu(\mathrm{d}x) \geqq 2^{-1}(c-\varepsilon) - \varepsilon \geqq \varepsilon,$$

and this is a contradiction to c) if I is infinite. Otherwise (***) yields that J is infinite and we obtain in a similar way for each j in J

$$-\int_W f_{n_j}(x)\,\mu(\mathrm{d}x) \geqq \varepsilon$$

and hence a contradiction to c).

Consequently, b) is a consequence of c). ∎

The statement of 11.8.2. may become invalid if the assumption that "(ν_n) be asymptotically uniformly distributed" is replaced by the weaker assumption that "(ν_n) be weakly asymptotically uniformly distributed".

If, for instance, we set

$$\nu_n = (2n^2)^{-s} \sum_{-n^2 \leqq i_1,\ldots,i_s < n^2} \delta_{[i_1/n,\ldots,i_s/n]},$$

and $X = Y^s$, where Y be the set of irrational numbers in $[0, 1]$, as well as

$$L = \delta_{\sum_{[j_1,\ldots,j_s]\in Z^s} \delta_{[j_1,\ldots,j_s]}}$$

– here Z denotes the set of integers –, then L is in $\mathbf{S}$ and, for all natural numbers n, one obtains

$$L_{[Q_{\nu_n}]}(\Phi(X) > 0) = 0.$$

On the other hand $\sigma_L = \delta_1$, so that it is impossible that

$$\left\|\left(L_{[Q_{\nu_n}]}\right)_X - \left(\int P_{l\mu}(.)\,\sigma_L(\mathrm{d}l)\right)_X\right\| \underset{n\to\infty}{\longrightarrow} 0.$$

The following weaker convergence statement can, however, be derived:

11.9.4. Theorem. *For all L in $\mathbf{S}$, all weakly asymptotically uniformly distributed sequences (ν_n) of distributions on $\mathfrak{R}^s$, and all finite sequences $X_1, \ldots, X_k$ of bounded sets in $\mathfrak{R}^s$ having the property that $\mu(\partial X_1) = \cdots = \mu(\partial X_k) = 0$,*

$$\sup_{x\in R^s} \left\|\left(L_{[Q_{\nu_n}]}\right)_{X_1+x,\ldots,X_k+x} - \left(\int P_{l\mu}(.)\,\sigma_L(\mathrm{d}l)\right)_{X_1,\ldots,X_k}\right\| \underset{n\to\infty}{\longrightarrow} 0.$$

By 6.4.3., every continuity set X with respect to μ is also a continuity set with respect to the stationary distribution $\int P_{l\mu}(.)\,\sigma_L(\mathrm{d}l)$. Hence, using 3.1.7., the following theorem is readily derived from 11.9.4.

11.9.5. Theorem. *For all L in $\mathbf{S}$ and all weakly asymptotically uniformly distributed sequences (ν_n) of distributions on $\mathfrak{R}^s$,*

$$L_{[Q_{\nu_n}]} \underset{n\to\infty}{\Rightarrow} \int P_{l\mu}(.)\,\sigma_L(\mathrm{d}l).$$

Conversely, for all weakly asymptotically uniformly distributed sequences (ν_n) and all sequences (x_n) of points in R^s, the sequence $(\delta_{x_n} * \nu_n)$ is also weakly asymptotically distributed. Thus, Theorem 11.9.4. can easily be derived from 11.9.5., so that both theorems are equivalent.

These theorems are an immediate consequence of

11.9.6. Theorem. *For all L in $\mathbf{S}$, all weakly asymptotically uniformly distributed sequences (ν_n) of distributions on $\mathfrak{R}^s$, and all finite sequences $X_1, \ldots, X_k$ of bounded sets in $\mathfrak{R}^s$ having the property that $\mu(\partial X_1) = \cdots = \mu(\partial X_k) = 0$,*

$$\lim_{n\to\infty} \sup_{x\in R^s, D\in \boldsymbol{D}} \left\| \left((L_{[Q_{\nu_n}]})_{[D]} \right)_{X_1+x, \ldots, X_k+x} - \left(\left(\int P_{l\mu}(.)\, \sigma_L(\mathrm{d}l) \right)_{[D]} \right)_{X_1, \ldots, X_k} \right\| = 0 .$$

Proof. 1. Suppose that δ, $c > 0$ and let L be a distribution in $\boldsymbol{S}^{(c)}$, X a compact subset of R^s, and λ the uniform distribution on $[-\delta, \delta]^s$. Subject to these assumptions, to each $\eta > 0$ there exists an $\varepsilon > 0$, such that, for all distributions γ on $\mathfrak{R}^s$ satisfying the condition $\gamma([-\varepsilon, \varepsilon]^s) = 1$, the inequality

$$\sup_{x\in R^s, D\in \boldsymbol{D}} {}_{(X+x)}\left\| \left((L_{[Q_\gamma]})_{[D]} \right)_{[Q_\lambda]} - (L_{[D]})_{[Q_\lambda]} \right\| < \eta$$

is valid.

Obviously to every $\varrho > 0$ there exists an $\varepsilon > 0$ and a measure ν on $\mathfrak{R}^s$ with the properties

$$\nu(R^s) > 1 - \varrho, \qquad \delta_z * \lambda = \nu + \sigma_z, \qquad \sigma_z \in N$$

for all z in $[-\varepsilon, \varepsilon]^s$, in which case $z \curvearrowright \sigma_z$ is measurable with respect to $\mathfrak{R}$.

For all y in R^s, we have

$$\begin{aligned}
&([Q_\lambda] \circ [D] \circ [Q_\gamma])_{(y)} = \int ([Q_\lambda] \circ [D] \circ [Q_{\delta_z}])_{(y)}\, (.)\, \gamma(\mathrm{d}z) \\
&= \int ([Q_{\delta_z * \lambda}] \circ [D])_{(y)}\, (.)\, \gamma(\mathrm{d}z) = \int ([D]_{(y)})_{[Q_{\nu+\sigma_z}]}\, (.)\, \gamma(\mathrm{d}z) \\
&= \int \left(\int \mathop{\Large *}_{x\in R^s,\, \Phi(\{x\})>0} \left(\int \delta_{\delta_{x+u}}(.)\, (\nu + \sigma_z)\, (du) \right)^{\Phi(\{x\})} (.)\, [D]_{(y)}\, (\mathrm{d}\Phi) \right) (.)\, \gamma(\mathrm{d}z) \\
&= \int \mathop{\Large *}_{x\in R^s,\, \Phi(\{x\})>0} \left(\int \delta_{\delta_{x+u}}\, (.)\, \nu(\mathrm{d}u) \right)^{\Phi(\{x\})} (.)\, [D]_{(y)}\, (\mathrm{d}\Phi) + R_{(y),\gamma} \quad \text{(say)}.
\end{aligned}$$

But the intensity measure $\varrho_{y,\gamma}$ of the correcting summand $R_{(y),\gamma}$ in $\boldsymbol{E}^+$ satisfies for all Z in $\mathfrak{R}^s$

$$\begin{aligned}
\varrho_{y,\gamma}(Z) &= (\lambda * \varrho_D * \gamma - \nu * \varrho_D)\, (Z - y) \leqq (\lambda * \varrho_D * \gamma - \nu * \varrho_D)\, (R^s) \\
&= 1 - \nu(R^s) < \varrho .
\end{aligned}$$

So we obtain with the help of 4.2.8.

$$\begin{aligned}
&{}_{(X+x)}\left\| \left((L_{[Q_\gamma]})_{[D]} \right)_{[Q_\lambda]} - (L_{[D]})_{[Q_\lambda]} \right\| \\
&\leqq 2 \int (R_{(y),\gamma} + R_{(y),\delta_{[0,\ldots,0]}})\, \left(\chi(X + x) > 0 \right) \varrho_L(\mathrm{d}y)
\end{aligned}$$

$$\leqq 2 \int (\varrho_{y,\gamma} + \varrho_{y,\delta_{[0,\dots,0]}}) (X + x) \varrho_L(\mathrm{d}y)$$

$$= 2 \int \big((\lambda * \varrho_D * \gamma - \nu * \varrho_D) + (\lambda * \varrho_D * \delta_{[0,\dots,0]} - \nu * \varrho_D)\big) (X + x - y) \varrho_L(\mathrm{d}y)$$

$$\leqq 4\varrho \sup_{z \in R^s} \varrho_L(X + z).$$

Since $\varrho > 0$ is arbitrary, this leads to assertion 1.

2. Subject to the same assumptions as in 1., we have

$$\lim_{n\to\infty} \sup_{x \in R^s,\, D \in \boldsymbol{D}} {}_{(X+x)}\Big\| \big((L_{[Q_{\nu_n}]})_{[D]}\big)_{[Q\lambda]} - \big((\textstyle\int P_{l\mu}(.)\, \sigma_L(\mathrm{d}l))_{[D]}\big)_{[Q\lambda]} \Big\| = 0.$$

If we use the distributions γ_ε introduced in the definition of the weakly asymptotically uniformly distribution, then by 11.8.6. and 1. we get

$$\overline{\lim_{n\to\infty}} \sup_{x \in R^s,\, D \in \boldsymbol{D}} {}_{(X+x)}\Big\| \big((L_{[Q_{\nu_n}]})_{[D]}\big)_{[Q\lambda]} - \big((\textstyle\int P_{l\mu}(.)\, \sigma_L(\mathrm{d}l))_{[D]}\big)_{[Q\lambda]} \Big\|$$

$$\leqq \overline{\lim_{n\to\infty}} \sup_{x \in R^s,\, D \in \boldsymbol{D}} {}_{(X+x)}\Big\| \big((L_{[Q_{\nu_n}]})_{[D]}\big)_{[Q\lambda]} - \big((L_{[Q_{\gamma_\varepsilon * \nu_n}]})_{[D]}\big)_{[Q\lambda]} \Big\|$$

$$+ \overline{\lim_{n\to\infty}} \sup_{x \in R^s,\, G \in \boldsymbol{D}} {}_{(X+x)}\Big\| (L_{[Q_{\gamma_\varepsilon * \nu_n}]})_{[G]} - \big(\textstyle\int P_{l\mu}(.)\, \sigma_L(\mathrm{d}l)\big)_{[G]} \Big\|$$

$$\leqq \sup_{x \in R^s,\, G \in \boldsymbol{D}} {}_{(X+x)}\Big\| (L_{[G]})_{[Q\lambda]} - \big((L_{[Q_{\gamma_\varepsilon}]})_{[G]}\big)_{[Q\lambda]} \Big\| < \eta.$$

3. Let $X_1, \dots, X_k$ be a finite sequence of continuity sets with respect to μ. For each natural number n, we denote by $K_{i,n}$ the compact set $\{x : x \in R^s,\ x + (-1/n, 1/n)^s \subseteqq X_i\}$. Obviously, $K_{i,n}$ depends on n in a monotonically increasing way, the union of the $K_{i,n}$ giving the interior of X_i.

The sequences $(G_{i,n})$, $1 \leqq i \leqq k$, of the open sets

$$G_{i,n} = X_i + (-1/n, 1/n)^s$$

are all monotone decreasing and have the intersections $X_i \cup (\partial X_i)$.

We choose now any natural number m, setting $\lambda = (2/m)^{-s} \mu\big((.) \cap (-1/m, 1/m)^s\big)$, and obtain for all $c > 0$ and all $L \in \mathbf{S}^{(c)}$

$$\overline{\lim_{n\to\infty}} \sup_{x \in R^s,\, D \in \boldsymbol{D}} \Big\| \big((L_{[Q_{\nu_n}]})_{[D]}\big)_{X_1+x,\dots,X_k+x} - \big((\textstyle\int P_{l\mu}(.)\, \sigma_L(\mathrm{d}l))_{[D]}\big)_{X_1,\dots,X_k} \Big\|$$

$$\leqq \overline{\lim_{n\to\infty}} \sup_{x \in R^s,\, D \in \boldsymbol{D}} \Big\| \Big(\big((L_{[Q_{\nu_n}]})_{[D]}\big)_{[Q\lambda]}\Big)_{X_1+x,\dots,X_k+x} \big((L_{[Q_{\nu_n}]})_{[D]}\big)_{X_1+x,\dots,X_k+x} \Big\|$$

$$+ \overline{\lim_{n\to\infty}} \sup_{x \in R^s,\, D \in \boldsymbol{D}} \Big\| \Big(\big((L_{[Q_{\nu_n}]})_{[D]}\big)_{[Q\lambda]}\Big)_{X_1+x,\dots,X_k+x}$$

$$- \Big(\big((\textstyle\int P_{l\mu}(.)\, \sigma_L(\mathrm{d}l))_{[D]}\big)_{[Q\lambda]}\Big)_{X_1+x,\dots,X_k+x} \Big\|$$

$$+ \overline{\lim_{n\to\infty}} \sup_{D \in \boldsymbol{D}} \Big\| \Big(\big((\textstyle\int P_{l\mu}(.)\, \sigma_L(\mathrm{d}l))_{[D]}\big)_{[Q\lambda]}\Big)_{X_1,\dots,X_k} - \big((\textstyle\int P_{l\mu}(.)\, \sigma_L(\mathrm{d}l))_{[D]}\big)_{X_1,\dots,X_k} \Big\|.$$

If X is an arbitrary compact subset of R^s which includes $(X_1 \cup \cdots \cup X_k)$, then obviously the middle summand is

$$\leqq \overline{\lim_{n\to\infty}} \sup_{x\in R^s,\, D\in \boldsymbol{D}} {}_{(X+x)}\left\| \left((L_{[Q_{\nu_n}]})_{[D]}\right)_{[Q_\lambda]} - \left(\left(\int P_{l\mu}(.)\ \sigma_L(\mathrm{d}l)\right)_{[D]}\right)_{[Q_\lambda]}\right\|$$

and hence vanishes in view of 2. Thus we may continue the above chain of inequalities by means of considerations such as we made at the end of step 5. in the proof of Proposition 11.8.6., once more applying 2.:

$$\leqq 2 \overline{\lim_{n\to\infty}} \sup_{x\in R^s,\, D\in \boldsymbol{D}} \left((L_{[Q_{\nu_n}]})_{[D]}\right)_{[Q_\lambda]} \left(\Phi\left(\bigcup_{i=1}^{k} (G_{i,m} \setminus K_{i,m}) + x\right) > 0\right)$$

$$+ 2 \sup_{D\in\boldsymbol{D}} \left(\left(\int P_{l\mu}(.)\ \sigma_L(\mathrm{d}l)\right)_{[D]}\right)_{[Q_\lambda]} \left(\Phi\left(\bigcup_{i=1}^{k} (G_{i,m} \setminus K_{i,m})\right) > 0\right)$$

$$\leqq 4 \sup_{G\in\boldsymbol{D}} \left(\int P_{l\mu}(.)\ \sigma_L(\mathrm{d}l)\right)_{[G]} \left(\Phi\left(\bigcup_{i=1}^{k} (G_{i,m} \setminus K_{i,m})\right) > 0\right)$$

$$\leqq 4 \sum_{i=1}^{k} \sup_{G\in\boldsymbol{D}} \varrho_{\left(\int P_{l\mu}(.)\sigma_L(\mathrm{d}l)\right)_{[G]}}(G_{i,m} \setminus K_{i,m})$$

$$= 4 \sum_{i=1}^{k} \sup_{G\in\boldsymbol{D}} \left(\varrho_G * \left(\int l\sigma_L(\mathrm{d}l)\right) \mu\right) (G_{i,m} \setminus K_{i,m})$$

$$\leqq 4c \sum_{i=1}^{k} \mu(G_{i,m} \setminus K_{i,m}).$$

All sequences $(G_{i,m} \setminus K_{i,m})$, $1 \leqq i \leqq k$, are monotone decreasing and have intersection ∂X_i. Hence the last expression tends towards zero as m increases, and we have the statement in the case $L \in \boldsymbol{S}^{(c)}$, $c > 0$.

4. Just as in the last step in the proof of 11.8.6., we may convince ourselves that it suffices to prove the assertion of 11.9.6. for arbitrary L in $\bigcup_{c>0} \boldsymbol{S}^{(c)}$. ∎

In view of 11.8.6., the assumption that

$$\mu(\partial X_1) = \cdots = \mu(\partial X_k) = 0$$

can be dropped in 11.9.6. if the sequence (ν_n) is asymptotically uniformly distributed, since we have for $X = X_1 \cup \cdots \cup X_k$

$$\sup_{x\in R^s,\, D\in\boldsymbol{D}} \left\| \left((L_{[Q_{\nu_n}]})_{[D]}\right)_{X_1+x,\ldots,X_k+x} - \left(\left(\int P_{l\mu}(.)\ \sigma_L(\mathrm{d}l)\right)_{[D]}\right)_{X_1,\ldots,X_k} \right\|$$

$$\leqq \sup_{x\in R^s,\, D\in\boldsymbol{D}} {}_{(X+x)}\left\| (L_{[Q_{\nu_n}]})_{[D]} - \left(\int P_{l\mu}(.)\ \sigma_L(\mathrm{d}l)\right)_{[D]} \right\|.$$

Now it turns out that this sharpening of the assertion of 11.9.6., and hence also of 11.9.4., can be ensured in an other way.

11.9.7. Theorem. *If an L in $\boldsymbol{S}$ satisfies the condition $\varrho_L \leqq v\mu$ for some positive real v, then, for all weakly asymptotically uniformly distributed* *sequences* (v_n) *and all finite sequences* $X_1, \ldots, X_k$ *of bounded sets in* $\mathfrak{R}^s$,

$$\lim_{n\to\infty} \sup_{x\in R^s,\, D\in \boldsymbol{D}} \left\| \left((L_{[Q_{v_n}]})_{[D]} \right)_{X_1+x,\ldots,X_k+x} - \left(\left(\int P_{l\mu}(.)\ \sigma_L(\mathrm{d}l) \right)_{[D]} \right)_{X_1,\ldots,X_k} \right\| = 0 .$$

Proof. Let $\varepsilon > 0$. To each X_i we pick up a closed subset K_i and an open subset U_i of $\mathfrak{R}^s$ satisfying

$$K_i \subseteqq X_i \subseteqq U_i, \qquad \mu(U_i \setminus K_i) \leqq \varepsilon \qquad (i = 1, \ldots, k).$$

Moreover, we choose continuity sets Z_i with respect to μ with the property $K_i \subseteqq Z_i \subseteqq U_i$. Now we obtain

$$\overline{\lim_{n\to\infty}} \sup_{x\in R^s,\, D\in \boldsymbol{D}} \left\| \left((L_{[Q_{v_n}]})_{[D]} \right)_{X_1+x,\ldots,X_k+x} - \left(\left(\int P_{l\mu}(.)\ \sigma_L(\mathrm{d}l) \right)_{[D]} \right)_{X_1,\ldots,X_k} \right\|$$

$$\leqq \overline{\lim_{n\to\infty}} \sup_{x\in R^s,\, D\in \boldsymbol{D}} \left\| \left((L_{[Q_{v_n}]})_{[D]} \right)_{X_1+x,\ldots,X_k+x} - \left((L_{[Q_{v_n}]})_{[D]} \right)_{Z_1+x,\ldots,Z_k+x} \right\|$$

$$+ \overline{\lim_{n\to\infty}} \sup_{x\in R^s,\, D\in \boldsymbol{D}} \left\| \left((L_{[Q_{v_n}]})_{[D]} \right)_{Z_1+x,\ldots,Z_k+x} - \left(\left(\int P_{l\mu}(.)\ \sigma_L(\mathrm{d}l) \right)_{[D]} \right)_{Z_1,\ldots,Z_k} \right\|$$

$$+ \overline{\lim_{n\to\infty}} \sup_{x\in R^s,\, D\in \boldsymbol{D}} \left\| \left(\left(\int P_{l\mu}(.)\ \sigma_L(\mathrm{d}l) \right)_{[D]} \right)_{X_1,\ldots,X_k} - \left(\left(\int P_{l\mu}(.)\ \sigma_L(\mathrm{d}l) \right)_{[D]} \right)_{Z_1,\ldots,Z_k} \right\|$$

In view of 11.9.6., the middle summand vanishes so that we may continue with

$$\leqq \sup_{x\in R^s,\, G\in \boldsymbol{D}} \|(L_{[G]})_{X_1+x,\ldots,X_k+x} - (L_{[G]})_{Z_1+x,\ldots,Z_k+x}\|$$

$$+ \sup_{D\in \boldsymbol{D}} \left\| \left(\left(\int P_{l\mu}(.)\ \sigma_L(\mathrm{d}l) \right)_{[D]} \right)_{X_1,\ldots,X_k} - \left(\left(\int P_{l\mu}(.)\ \sigma_L(\mathrm{d}l) \right)_{[D]} \right)_{Z_1,\ldots,Z_k} \right\|$$

$$\leqq 2 \sup_{x\in R^s,\, G\in \boldsymbol{D}} L_{[G]} \left(\Phi\left(\bigcup_{i=1}^{k} (U_i \setminus K_i) + x \right) > 0 \right)$$

$$+ 2 \sup_{D\in \boldsymbol{D}} \left(\int P_{l\mu}(.)\ \sigma_L(\mathrm{d}l) \right)_{[D]} \left(\Phi\left(\bigcup_{i=1}^{k} (U_i \setminus K_i) \right) > 0 \right)$$

$$\leqq 2 \sup_{x\in R^s,\, G\in \boldsymbol{D}} (\varrho_G * \varrho_L) \left(\bigcup_{i=1}^{k} (U_i \setminus K_i) + x \right)$$

$$+ 2 \sup_{D\in \boldsymbol{D}} \left(\varrho_D * \left(\int l\sigma_L(\mathrm{d}l) \right) \mu \right) \left(\bigcup_{i=1}^{k} (U_i \setminus K_i) \right)$$

$$\leqq 4v\mu \left(\bigcup_{i=1}^{k} (U_i \setminus K_i) \right) \leqq 4vk\varepsilon ,$$

because immediately by the definition of $\boldsymbol{S}$ and our assumption on ϱ_L we have $\int l\sigma_L(\mathrm{d}l) \leqq v$. Since $\varepsilon > 0$ was arbitrary we obtain the statement. ∎

For all $l \geqq 0$ we denote by N_l the set of all measures ω in N which fulfil

$$\sup_{x \in R^s} \left|(2n)^{-s}\, \omega\big([-n, n)^s + x\big) - l\right| \underset{n\to\infty}{\longrightarrow} 0.$$

In virtue of 11.7.6. we have in this case $P_\omega \in \mathbf{S}$ and $\sigma_{P_\omega} = \delta_l$. If now (ν_n) is any weakly asymptotically uniformly distributed sequence of distributions on $\mathfrak{R}^s$, then 11.9.4. yields for all bounded X in $\mathfrak{R}^s$ satisfying $\mu(\partial X) = 0$,

$$\pi_{(\nu_n * \omega)(X)} = \big((P_\omega)_{[Q_{\nu_n}]}\big)_X \underset{n\to\infty}{\longrightarrow} (P_{l\mu})_X = \pi_{l\mu(X)},$$

i.e. $\nu_n * \omega(X) \underset{n\to\infty}{\longrightarrow} l\mu(X)$. By 3.2.2. from this follows $\nu_n * \omega \overset{N}{\underset{n\to\infty}{\Rightarrow}} l\mu$.

It turns out that this property of weakly asymptotically distributed sequences is a characteristic one (cf. Stone [1], Fichtner [1]).

11.9.8. Theorem. *A sequence (ν_n) of distributions on $\mathfrak{R}^s$ is weakly asymptotically uniformly distributed if and only if*

$$\nu_n * \omega \overset{N}{\underset{n\to\infty}{\Rightarrow}} l\mu$$

is satisfied for all $l \geqq 0$ and all ω in N_l.

Proof. Let γ be any distribution on $\mathfrak{R}^s$. For all Z in $\mathfrak{R}^s$ we set

$$\mu^Z = \gamma * {}_Z\mu + (\mu - {}_Z\mu)$$

and get a measure μ^Z in N. We even have $\mu^Z \in N_1$, for we have for all x in R^s and all natural numbers n

$$\begin{aligned}
&\left|(2n)^{-s}\, \mu^Z\big([-n, n)^s + x\big) - 1\right| \\
&= (2n)^{-s} \left|(\gamma * {}_Z\mu)\,\big([-n, n)^s + x\big) - {}_Z\mu\big([-n, n)^s + x\big)\right| \\
&\leqq (2n)^{-s} \int \left|{}_Z\mu\big([-n, n)^s + x - z\big) - {}_Z\mu\big([-n, n)^s + x\big)\right| \gamma(\mathrm{d}z) \\
&\leqq (2n)^{-s} \int \Big({}_Z\mu\,\Big(\big([-n, n)^s + x - z\big) \setminus \big([-n, n)^s + x\big)\Big) \\
&\qquad + {}_Z\mu\,\Big(\big([-n, n)^s + x\big) \setminus \big([-n, n)^s + x - z\big)\Big)\Big)\, \gamma(\mathrm{d}z) \\
&\leqq 2 \int (2n)^{-s}\, \mu\Big(\big([-n, n)^s - z\big) \setminus [-n, n)^s\Big)\, \gamma(dz) \underset{n\to\infty}{\longrightarrow} 0,
\end{aligned}$$

because the integrand is bounded and tends to zero as $n \to \infty$ for every z in R^s.

Now let X be any bounded set in $\mathfrak{R}^s$ which is additionally a continuity set with respect to μ. According to our assumption we have

$$(\nu_n * \mu^Z)\,(X) \underset{n\to\infty}{\longrightarrow} \mu(X),$$

i.e.

$$\nu_n * \gamma * {}_Z\mu(X) - \nu_n * {}_Z\mu(X) \underset{n\to\infty}{\longrightarrow} 0,$$

i.e.

$$\int_Z \left(\nu_n(X - x) - (\gamma * \nu_n)(X - x)\right) \mu(\mathrm{d}x) \to 0.$$

If X is not a countinuity set, then (cf. the proof of 11.9.7.) — to given $\varepsilon > 0$ — we pick up a continuity set X_ε with respect to μ satisfying $\mu\left((X \setminus X_\varepsilon) \cup (X_\varepsilon \setminus X)\right) < \varepsilon$. For each distribution τ on $\mathfrak{R}^s$ we have

$$\left|\int_Z \left(\tau(X - x) - \tau(X_\varepsilon - x)\right) \mu(\mathrm{d}x)\right| \leq \int |{}_Z\mu(X - x) - {}_Z\mu(X_\varepsilon - x)| \, \tau(\mathrm{d}x) < \varepsilon.$$

Thus

$$\overline{\lim_{n\to\infty}} \left|\int_Z \left(\nu_n(X - x) - (\gamma * \nu_n)(X - x)\right) \mu(\mathrm{d}x)\right|$$

$$< \overline{\lim_{n\to\infty}} \left|\int_Z \left(\nu_n(X_\varepsilon - x) - (\gamma * \nu_n)(X_\varepsilon - x)\right) \mu(\mathrm{d}x)\right| + 2\varepsilon = 2\varepsilon.$$

Because ε was arbitrary we get with the help of 11.9.3. that (ν_n) is weakly asymptotically uniformly distributed. ∎

Immediately from 11.7.6. and 11.9.8. we obtain the following converse of 11.9.5. (cf. Stone [1], Fichtner [1]).

11.9.9. Theorem. *A sequence (ν_n) of distributions on $\mathfrak{R}^s$ is weakly asymptotically uniformly distributed if the convergence*

$$L_{[Q_{\nu_n}]} \underset{n\to\infty}{\Rightarrow} \int P_{l\mu}(.)\, \sigma_L(\mathrm{d}l)$$

akes place for all distributions L in **S**.

11.10. Uniform-Distributional Properties of Sequences of Convolution Powers of Distributions on $\mathfrak{R}^s$

To every sequence (ν_n) of distributions on $\mathfrak{R}^s$ we associate the set $V\left((\nu_n)\right)$ of those x in R^s which satisfy the condition $\mathrm{Var}\,(\sigma * \nu_n - \delta_x * \sigma * \nu_n) \underset{n\to\infty}{\longrightarrow} 0$ for all distributions σ on $\mathfrak{R}^s$ which are absolutely continuous with respect to μ. Obviously $V\left((\nu_n)\right)$ is always a subgroup of the additive group R^s. We call it the *translation group* of (ν_n). The sequence (ν_n) is weakly asymptotically uniformly distributed if and only if $V\left((\nu_n)\right)$ coincides with R^s.

The fact that a difference $x_1 - x_2$ belongs to $V\left((\nu_n)\right)$ can also be characterized as follows:

11.10.1. *Let (ν_n) be a sequence of distributions on $\mathfrak{R}^s$. Further let $x_1, x_2 \in R^s$. Subject to these assumptions, $x_1 - x_2$ belongs to $V\left((\nu_n)\right)$ if and only if to any $\varepsilon > 0$ there exist distributions $\varrho_{1,\varepsilon}, \varrho_{2,\varepsilon}$ on $\mathfrak{R}^s$ which satisfy*

$$\varrho_{1,\varepsilon}(x_1 + [-\varepsilon, \varepsilon]^s) = 1 = \varrho_{2,\varepsilon}(x_2 + [-\varepsilon, \varepsilon]^s)$$

as well as

$$\mathrm{Var}\,(\varrho_{1,\varepsilon} * \nu_n - \varrho_{2,\varepsilon} * \nu_n) \underset{n\to\infty}{\longrightarrow} 0.$$

Proof. Let $(x_1 - x_2) \in V\big((\nu_n)\big)$ as well as $\varepsilon > 0$. Setting $\varrho = (2\varepsilon)^{\;s} \mu\big((.) \cap [-\varepsilon, \varepsilon]^s\big)$, we obtain

$$0 = \lim_{n\to\infty} \operatorname{Var}\left(\varrho * \nu_n - \delta_{x_1-x_2} * \varrho * \nu_n\right) = \lim_{n\to\infty} \operatorname{Var}\big((\delta_{x_1} * \varrho) * \nu_n - (\delta_{x_2} * \varrho) * \nu_n\big)$$

as well as

$$(\delta_{x_1} * \varrho)\,(x_1 + [-\varepsilon, \varepsilon]^s) = 1 = (\delta_{x_2} * \varrho)\,(x_2 + [-\varepsilon, \varepsilon]^s),$$

i.e. the condition stated above for $(x_1 - x_2)$ to belong to $V\big((\nu_n)\big)$ is necessary.

Conversely, suppose that this condition be satisfied for x_1, x_2. For all distributions $\sigma = \int_{(.)} f(y)\, \mu(\mathrm{d}y)$, we have

$$\lim_{x\to[0,\dots,0]} \operatorname{Var}(\sigma - \delta_x * \sigma) = \int |f(y) - f(y - x)|\, \mu(\mathrm{d}x) = 0.$$

Hence to each natural number m there exists an $\varepsilon_m > 0$, such that the inequality

$$\operatorname{Var}(\sigma - \delta_x * \sigma) < m^{-1}$$

is satisfied for all $x \in [-\varepsilon_m, \varepsilon_m]^s$. We get

$$\begin{aligned}\operatorname{Var}(\varrho_{1,\varepsilon_m} * \sigma - \delta_{x_1} * \sigma) &= \operatorname{Var}\big((\delta_{-x_1} * \varrho_{1,\varepsilon_m}) * \sigma - \sigma\big)\\ &\leqq \int \operatorname{Var}(\delta_y * \sigma - \sigma)\, (\delta_{-x_1} * \varrho_{1,\varepsilon_m})\,(\mathrm{d}y)\\ &\leqq \sup_{y\in[-\varepsilon_m,\varepsilon_m]^s} \operatorname{Var}(\delta_y * \sigma - \sigma) \leqq m^{-1}.\end{aligned}$$

Analogously, we obtain

$$\operatorname{Var}(\varrho_{2,\varepsilon_m} * \sigma - \delta_{x_2} * \sigma) \leqq m^{-1}.$$

Thus we have

$$\begin{aligned}&\overline{\lim_{n\to\infty}} \operatorname{Var}(\sigma * \nu_n - \delta_{x_2-x_1} * \sigma * \nu_n) = \overline{\lim_{n\to\infty}} \operatorname{Var}\big((\delta_{x_1} * \sigma) * \nu_n - (\delta_{x_2} * \sigma) * \nu_n\big)\\ &\leqq \operatorname{Var}(\delta_{x_1} * \sigma - \varrho_{1,\varepsilon_m} * \sigma) + \overline{\lim_{n\to\infty}} \operatorname{Var}(\varrho_{1,\varepsilon_m} * \sigma * \nu_n - \varrho_{2,\varepsilon_m} * \sigma * \nu_n)\\ &\quad + \operatorname{Var}(\delta_{x_2} * \sigma - \varrho_{2,\varepsilon_m} * \sigma)\\ &\leqq m^{-1} + 0 + m^{-1},\end{aligned}$$

and hence

$$\overline{\lim_{n\to\infty}} \operatorname{Var}(\sigma * \nu_n - \delta_{x_1-x_2} * \sigma * \nu_n) = 0,$$

which means that

$$(x_1 - x_2) \in V\big((\nu_n)\big). \blacksquare$$

Now let a be an accumulation point of $V((\nu_n))$, and ε an arbitrary positive number. We choose some x in $(a + (-\varepsilon, \varepsilon)^s) \cap V((\nu_n))$. There exists an $\eta > 0$ having the properties

$$\eta < \varepsilon, \qquad (x + [-\eta, \eta]^s) \subseteqq (a + (-\varepsilon, \varepsilon)^s).$$

By 11.10.1. there exist distributions ϱ_1, ϱ_2 on $\Re^s$ satisfying

$$\varrho_2([0, \ldots, 0] + [-\eta, \eta]^s) = 1 = \varrho_1(x + [-\eta, \eta]^s)$$

as well as

$$\operatorname{Var}(\varrho_1 * \nu_n - \varrho_2 * \nu_n) \xrightarrow[n\to\infty]{} 0.$$

By our choice of η, we have

$$\varrho_2([0, \ldots, 0] + [-\varepsilon, \varepsilon]^s) = 1 = \varrho_1(a + [-\varepsilon, \varepsilon]^s),$$

so that by 11.10.1. we are able to conclude that $(a - [0, \ldots, 0]) \in V((\nu_n))$, i.e. we have

11.10.2. *For all sequences (ν_n) of distributions on $\Re^s$, $V((\nu_n))$ is a closed set.*

Let us now determine the translation group of the sequence $\nu, \nu^2, \nu^3, \ldots$ of the convolution powers of a distribution ν on $\Re^s$. To this end we shall first introduce the following concepts: the support Θ_ν of ν is understood to be the smallest closed subset X of R^s having the property that $\nu(X) = 1$, i.e. the set of those x in R^s for which every open neighbourhood U of x satisfies the condition $\nu(U) > 0$. Further let $\mathfrak{H}_\nu$ denote the system of those closed subgroups H of R^s for which there exist elements x_H in R^s, such that $\nu(x_H + H) = 1$, i.e. that $\Theta_\nu \subseteqq x_H + H$. Obviously a closed subgroup H belongs to $\mathfrak{H}_\nu$ if and only if $\Theta_\nu - \Theta_\nu \subseteqq H$ is satisfied. From this it follows that also $H_\nu = \bigcap_{H \in \mathfrak{H}_\nu} H$ belongs to $\mathfrak{H}_\nu$. We call H_ν the *difference module* of the distribution ν.

Now we are in a position to solve the above-formulated problem as follows:

11.10.3. Theorem. *For all distributions ν on $\Re^s$, the translation group $V((\nu^n))$ coincides with the difference module H_ν.*

Proof. 1. Let $x \notin H_\nu$. Since H_ν is closed, there exists an $\varepsilon > 0$ such that $(x + [-2\varepsilon, 2\varepsilon]^s) \cap H_\nu = \emptyset$, so that we obtain

$$(x + [-\varepsilon, \varepsilon]^s + H_\nu) \cap ([-\varepsilon, \varepsilon]^s + H_\nu) = \emptyset.$$

Now let $\sigma = (2\varepsilon)^{-s} \mu((.) \cap [-\varepsilon, \varepsilon]^s)$. For $n = 1, 2, \ldots$, we get

$$(\sigma * \nu^n)(n x_{H_\nu} + H_\nu + [-\varepsilon, \varepsilon]^s) = 1,$$

$$(\delta_x * \sigma * \nu^n)(x + n x_{H_\nu} + H_\nu + [-\varepsilon, \varepsilon]^s) = 1.$$

On the other hand,

$$(x + n x_{H_\nu} + H_\nu + [-\varepsilon, \varepsilon]^s) \cap (n x_{H_\nu} + H_\nu + [-\varepsilon, \varepsilon]^s) = \emptyset$$

and hence

$$\mathrm{Var}\,(\sigma * \nu^n - \delta_x * \sigma * \nu^n) = 2 \qquad (n = 1, 2, \ldots).$$

Consequently, $x \notin V\big((\nu^n)\big)$, i.e. we have $V\big((\nu^n)\big) \subseteqq H_\nu$.

2. In order to prepare the next step of this proof, we begin by showing that the following inequality is true:

$$\sup_{\substack{0<p<1\\ n=0,1,\ldots\\ 0\leqq k\leqq n}} \binom{n}{k} p^k(1-p)^{n-k}\sqrt{p(1-p)}\,\sqrt{n+1} \leqq 1/2.$$

This will be shown in several steps.

a) For $n = 0, 1, \ldots$, we set

$$a_n = (n!)^{-1}\big(n + (1/2)\big)^{n+(1/2)}.$$

Then the sequence of the a_n/a_{n+1} decreases monotonically.

Indeed, we have

$$\frac{a_n}{a_{n+1}} = \frac{n+1}{n+\dfrac{3}{2}}\left(1 - \frac{1}{n+\dfrac{3}{2}}\right)^{n+(3/2)-1}$$

$$= \left(1 - \frac{1}{2\left(n+\dfrac{3}{2}\right)}\right)\left(1 - \frac{1}{n+\dfrac{3}{2}}\right)^{n+(3/2)-1}.$$

We shall now consider the following function f

$$f(x) = \left(1 - \frac{1}{2x}\right)\left(1 - \frac{1}{x}\right)^{x-1} \qquad \big(x \in (1, +\infty)\big).$$

We have

$$f'(x) = \left(\left(1 - \frac{1}{2x}\right)\left(\ln\left(1 - \frac{1}{x}\right) + \frac{1}{x}\right) + \frac{1}{2x^2}\right)\left(1 - \frac{1}{x}\right)^{x-1}.$$

Since

$$\left(1 - \frac{1}{2x}\right)\left(\ln\left(1 - \frac{1}{x}\right) + \frac{1}{x}\right) = \left(1 - \frac{1}{2x}\right)\sum_{k=2}^{\infty}\left(-\frac{1}{kx^k}\right)$$

$$= -\frac{1}{2x^2} + \sum_{k=3}^{\infty}\left(\frac{1}{2(k-1)} - \frac{1}{k}\right)\frac{1}{x^k}$$

we obtain

$$f'(x) = -\left(1 - \frac{1}{x}\right)^{x-1}\sum_{k=3}^{\infty}\frac{k-2}{2k(k-1)}\,\frac{1}{x^k} < 0,$$

from which the monotonicity of a_n/a_{n+1} is inferred.

b) We set

$$P_{n,k} = \sup_{0<p<1} \binom{n}{k} p^k(1-p)^{n-k} \sqrt{p(1-p)(n+1)}.$$

Then

$$P_{n,k} = \binom{n}{k} \frac{\left(k+\frac{1}{2}\right)^{k+(1/2)}}{(n+1)^{n+1}} \sqrt{n+1} \left(n-k+\frac{1}{2}\right)^{n-k+(1/2)}.$$

In fact, we have

$$\frac{d}{dx}\left(x^{k+(1/2)}(1-x)^{n-k+(1/2)}\right) = \left(k+\frac{1}{2}-(n+1)\,x\right) x^{k-(1/2)}(1-x)^{n-k-(1/2)},$$

so that the maximum on $[0, 1]$ is assumed at $p = (n+1)^{-1}\left(k+(1/2)\right)$. Hence

$$P_{n,k} = \binom{n}{k} \left(\frac{k+\frac{1}{2}}{n+1}\right)^{k+(1/2)} \left(\frac{n-k+\frac{1}{2}}{n+1}\right)^{n-k+(1/2)} \sqrt{n+1}$$

$$= \binom{n}{k} \frac{\left(k+\frac{1}{2}\right)^{k+(1/2)}}{(n+1)^{n+1}} \sqrt{n+1} \left(n-k+\frac{1}{2}\right)^{n-k+(1/2)}.$$

c) By means of a) we observe that

$$P_{n,k} = a_k \sqrt{n+1}\, \frac{a_{n-k}}{a_n} \frac{\left(n+\frac{1}{2}\right)^{n+(1/2)}}{(n+1)^{n+1}} = a_k \frac{a_{n-k}}{a_n} \left(1-\frac{1}{2(n+1)}\right)^{n+(1/2)}$$

is a monotone decreasing function of n. Hence

$$\sup_{n\geqq k} P_{n,k} = P_{k,k} = \sqrt{\frac{1}{2}} \left(1-\frac{1}{2(k+1)}\right)^{(k+1)-(1/2)}.$$

d) Let $a > 0$. Then, for $x \geqq a + (1/2)$,

$$\frac{d}{dx}\left(1-\frac{a}{x}\right)^{x-a} = \left(1-\frac{a}{x}\right)^{x-a} \left(\ln\left(1-\frac{a}{x}\right)+\frac{a}{x}\right).$$

Now, $\ln\left(1-(a/x)\right)+(a/x) < 0$, and hence $\left(1-(a/x)\right)^{x-a}$ decreases monotonically for $x \geqq a + (1/2)$. Consequently,

$$\left(1-\frac{1}{2(k+1)}\right)^{(k+1)-(1/2)}$$

decreases monotonically for $k = 0, 1, \ldots$, so that

$$\sup_{k=0,1,\ldots} P_{k,k} = P_{0,0} = \sqrt{\frac{1}{2}}\left(1 - \frac{1}{2}\right)^{1/2} = \frac{1}{2},$$

which proves the above statement.

3. Given any points x_1, x_2 in Θ_ν. For any $\varepsilon > 0$, the two probabilities $\nu(x_1 + [-\varepsilon, \varepsilon]^s)$, $\nu(x_2 + [-\varepsilon, \varepsilon]^s)$ are then positive. Thus ν can be represented as a mixture

$$\nu = p\nu\big((.) \mid x_1 + [-\varepsilon, \varepsilon]^s\big) + q\nu\big((.) \mid x_2 + [-\varepsilon, \varepsilon]^s\big) + r\gamma$$

where $p > 0$, $q > 0$, $r \geqq 0$, $p + q + r = 1$, and ε sufficiently small.

For abbreviation we set

$$\nu\big((.) \mid x_1 + [-\varepsilon, \varepsilon]^s\big) = \varrho_1, \qquad \nu\big((.) \mid x_2 + [-\varepsilon, \varepsilon]^s\big) = \varrho_2,$$

$$u = p(p + q)^{-1}, \qquad v = q(p + q)^{-1} \quad \text{and} \quad u\varrho_1 + v\varrho_2 = \varrho.$$

Then first of all we have

$$\varrho_1(x_1 + [-\varepsilon, \varepsilon]^s) = 1 = \varrho_2(x_2 + [-\varepsilon, \varepsilon]^s).$$

For $n = 0, 1, \ldots$, we get

$$\begin{aligned}
&\operatorname{Var}(\varrho_1 * \varrho^n - \varrho_2 * \varrho^n) \\
&= \operatorname{Var}\left(\sum_{k=0}^{n} \binom{n}{k} u^k v^{n-k} \varrho_1{}^{k+1} \varrho_2{}^{n-k} - \sum_{k=0}^{n} \binom{n}{k} u^k v^{n-k} \varrho_1{}^{k} \varrho_2{}^{n-k+1}\right) \\
&\leqq \sum_{k=0}^{n+1} \left|\binom{n}{k-1} u^{k-1} v^{n-(k-1)} - \binom{n}{k} u^k v^{n-k}\right|.
\end{aligned}$$

Since the numbers $\binom{n}{k} u^k v^{n-k}$ as functions of k will initially increase and then decrease, it is shown by means of 2. that, for $n = 0, 1, \ldots$,

$$\begin{aligned}
\operatorname{Var}(\varrho_1 * \varrho^n - \varrho_2 * \varrho^n) &\leqq 2 \max_{0 \leqq k \leqq n} \binom{n}{k} u^k v^{n-k} \\
&\leqq \frac{1}{\sqrt{uv}} \frac{1}{\sqrt{n + 1}} = \frac{p + q}{\sqrt{pq}} \frac{1}{\sqrt{n + 1}}.
\end{aligned}$$

Now we set $s = p + q$. We have $\nu = s\varrho + r\gamma$. For $n = 1, 2, \ldots$, we have

$$\operatorname{Var}(\varrho_1 * \nu^n - \varrho_2 * \nu^n) \leqq \sum_{k=0}^{n} \binom{n}{k} s^k r^{n-k} \operatorname{Var}(\varrho_1 * \varrho^k - \varrho_2 * \varrho^k)$$

$$\leqq \sum_{k=0}^{n} \binom{n}{k} s^k r^{n-k} \frac{s}{\sqrt{pq}} \frac{1}{\sqrt{k + 1}} = \frac{s}{\sqrt{pq}} \sum_{k=0}^{n} \sqrt{\frac{1}{k + 1}} \binom{n}{k} s^k (1 - s)^{n-k}.$$

In view of the Jensen inequality, we may continue as follows:

$$\leqq \frac{s}{\sqrt{pq}} \left(\sum_{k=0}^{n} \frac{1}{k+1} \binom{n}{k} s^k (1-s)^{n-k} \right)^{1/2}$$

$$= \frac{s}{\sqrt{pq}} \left(\frac{1}{n+1} \sum_{k=0}^{n} \binom{n+1}{k+1} s^k (1-s)^{n-k} \right)^{1/2}$$

$$= \frac{s}{\sqrt{pq}} \left(\frac{1}{(n+1)s} \sum_{k=1}^{n+1} \binom{n+1}{k} s^k (1-s)^{n+1-k} \right)^{1/2}$$

$$= \frac{s}{\sqrt{pq}} \left(\frac{1}{(n+1)s} \left(1-(1-s)^{n+1}\right) \right)^{1/2} \leqq \sqrt{\frac{p+q}{pq}} \sqrt{\frac{1}{n+1}}.$$

Thus we get

$$\operatorname{Var}(\varrho_1 * \nu^n - \varrho_2 * \nu^n) \xrightarrow[n\to\infty]{} 0,$$

so that by 11.10.1. we are able to conclude that $(x_1 - x_2) \in V((\nu^n))$. Consequently, $\Theta_\nu - \Theta_\nu \subseteqq V((\nu^n))$, which means by 11.10.2. that $H_\nu \subseteqq V((\nu^n))$. ∎

A distribution ν on $\mathfrak{R}^s$ is said to be *non-lattice* if H_ν coincides with $\mathfrak{R}^s$. Now the following statement is readily derived from 11.10.3. (cf. H. Hermann [1], Stam [1], [2], Kerstan & Matthes [3], [6], and Liese [1]):

11.10.4. Theorem. *The sequence of convolution powers of a distribution ν on $\mathfrak{R}^s$ is weakly asymptotically uniformly distributed if and only if ν is non-lattice.*

With the help of 11.8.1., 11.10.4. and 11.9.5. we obtain

11.10.5. Theorem. *For each non-lattice distribution ν on $\mathfrak{R}^s$, a stationary distribution P on $\mathfrak{M}$ is cluster-invariant with respect to $[Q_\nu]$, if and only if it has the property g_μ.*

Proof. Let P be a stationary distribution on $\mathfrak{M}$ with the property $P_{[Q_\nu]} = P$. We use the measurable mappings $\Phi \curvearrowright \Phi_k$ of M into itself and the distributions P_k which were introduced in the proof of 6.4.2. for all natural numbers k. Let V_k denote the distribution of $[\Phi_k, \Phi - \Phi_k]$ formed with respect to P. For all natural numbers k, m, n one gets

$$\begin{aligned} P\left(\Phi([0,1)^s) \geqq m\right) &= P_{[Q_{\nu^n}]}\left(\Phi([0,1)^s) \geqq m\right) \\ &= \int [Q_{\nu^n}]_{(\Phi_1+\Phi_2)} \left(\Phi([0,1)^s) \geqq m\right) V_k(\mathrm{d}[\Phi_1,\Phi_2]) \\ &\geqq \int [Q_{\nu^n}]_{(\Phi_k)} \left(\Phi([0,1)^s) \geqq m\right) P(\mathrm{d}\Phi) \\ &= (P_k)_{[Q_{\nu^n}]}\left(\Phi([0,1)^s) \geqq m\right). \end{aligned}$$

P_k is stationary and of finite intensity. Hence, by 11.7.5., 11.10.4., and 11.9.4. it follows that

$$P\left(\Phi([0,1)^s) \geqq m\right) \geqq \int P_{l\mu}\left(\Phi([0,1)^s) \geqq m\right) \sigma_{P_k}(\mathrm{d}l).$$

Because of 1.7.3., $P_{l\mu}\big(\Phi([0,1)^s) \geqq m\big) = \pi_l(\xi \geqq m)$ increases monotononsly for $l > 0$. Hence

$$P\left(\Phi([0,1)^s) \geqq m\right) \geqq \pi_l(\xi \geqq m)\, P\big(s(\Phi_k) > l\big)$$

for all $l > 0$.

The sequence of the $s(\Phi_k)$ is defined and monotonously increasing almost everywhere with respect to P. We even have

$$s(\Phi_k) \xrightarrow[k\to\infty]{} s(\Phi)$$

almost everywhere with respect to P. Otherwise the invariant set

$$Y_n = \left\{\Phi : \Phi \in M,\ \sup_{k=1,2,\ldots} s(\Phi_k) < \min\big(n, s(\Phi)\big)\right\}$$

would have a positive probability with respect to P for at least one natural number n_0. Setting now $P' = P\big((.) \mid Y_{n_0}\big)$ one obtains, by means of the theorem of B. Levi

$$\begin{aligned}\int s(\Phi)\, P'(\mathrm{d}\Phi) &= \int \Phi([0,1)^s)\, P'(\mathrm{d}\Phi) = \int \Big(\sup_{k=1,2,\ldots} \Phi_k([0,1)^s)\Big)\, P'(\mathrm{d}\Phi) \\ &= \sup_{k=1,2,\ldots} \int \Phi_k([0,1)^s)\, P'(\mathrm{d}\Phi) = \sup_{k=1,2,\ldots} \int s(\Phi_k)\, P'(\mathrm{d}\Phi) \\ &= \int \Big(\sup_{k=1,2,\ldots} s(\Phi_k)\Big)\, P'(\mathrm{d}\Phi) \leqq n_0,\end{aligned}$$

and hence

$$\int \Big(s(\Phi) - \sup_{k=1,2,\ldots} s(\Phi_k)\Big)\, P'(\mathrm{d}\Phi) = 0.$$

This implies, however, that

$$\int_{Y_{n_0}} \Big(s(\Phi) - \sup_{k=1,2,\ldots} s(\Phi_k)\Big)\, P(\mathrm{d}\Phi) = 0,$$

i.e. that

$$P\Big(\Phi \in Y_{n_0},\ s(\Phi) > \sup_{k=1,2,\ldots} s(\Phi_k)\Big) = P(Y_{n_0}) = 0,$$

which contradicts the above statement.

Hence for all $l > 0$ we obtain

$$\begin{aligned}P\left(\Phi([0,1)^s) \geqq m\right) &\geqq \pi_l(\xi \geqq m) \sup_{k=1,2,\ldots} P\big(s(\Phi_k) > l\big) \\ &\geqq \pi_l(\xi \geqq m)\, P\big(s(\Phi) > l\big) \geqq \pi_l(\xi \geqq m)\, P\big(s(\Phi) = +\infty\big).\end{aligned}$$

Consequently

$$\begin{aligned}P\left(\Phi([0,1)^s) \geqq m\right) &\geqq \lim_{l\to\infty} \pi_l(\xi \geqq m)\, P\big(s(\Phi) = +\infty\big) \\ &= P\big(s(\Phi) = +\infty\big).\end{aligned}$$

The above inequality holds for all natural numbers m. Thus it follows that

$$0 = \lim_{m\to\infty} P\left(\Phi([0, 1)^s) \geqq m\right) \geqq P\left(s(\Phi) = +\infty\right),$$

so that by 11.7.5. the distribution P belongs to $\mathbf{S}$. In view of 11.10.4. and 11.9.5. we then have $P = \int P_{l\mu}(.)\, \sigma_P(\mathrm{d}l)$, and by 6.4.12. the proof is finished. ∎

As in 7.3.3. the restriction of stationarity cannot be dropped in 11.10.5.

Let $s = 1$. We set $P = P_\varrho$, where $\varrho(\mathrm{d}x) = e^x\mu(\mathrm{d}x)$. Then, by 4.4.4., P is cluster-invariant with respect to $[Q_\nu]$ if and only if the distribution ν satisfies the integral equation $\varrho * \nu = \varrho$. If we now choose ν to be the uniform distribution on $[-c, 1-c]$, $0 < c < 1$, then $\varrho * \nu$ has the density

$$g(x) = \int_{-c}^{1-c} e^{x-y}\mu(\mathrm{d}y) = e^x(e^c - e^{c-1}).$$

For $c = \ln\left((e-1)^{-1} e\right)$, $g(x)$ coincides with e^x for all real x, i.e. we have $\varrho * \nu = \varrho$, and hence $P_{[Q_\nu]} = P$.

Consider now asymptotically uniformly distributed sequences of the form $\nu, \nu^2, \nu^3, \ldots$ As a counterpart to 11.10.4. we shall prove (cf. H. Hermann [1], Stam [1], [2], Kerstan & Matthes [3], [6], Stone [1], and Letac [1])

11.10.6. Theorem. *The sequence of convolution powers of a distribution ν on $\mathfrak{R}^s$ is asymptotically uniformly distributed if and only if at least one convolution power ν^{n_0} has a non-vanishing part absolutely continuous with respect to μ.*

Proof. 1. By 11.8.5. the condition stated for (ν^n) to be asymptotically uniformly distributed is necessary.

2. For all natural numbers n, let σ_n and ω_n denote the parts of ν^n which are absolutely continuous with respect to μ and purely singular, respectively. Now, if $\sigma_{n_0} \neq 0$ for any natural number n_0, then the sequence (ν^n) is finally absolutely continuous.

For $n = 1, 2, \ldots$, we have

$$\sigma_{n+1}(R^s) \geqq (\nu * \sigma_n)\,(R^s) = \sigma_n(R^s),$$

i.e. the sequence $\left(\sigma_n(R^s)\right)$ increases monotonously.

For $k = 0, 1, \ldots$, we obtain

$$\omega_{(k+1)n_0}(R^s) \leqq (\omega_{n_0} * \omega_{kn_0})\,(R^s) = \left(1 - \sigma_{n_0}(R^s)\right)\omega_{kn_0}(R^s),$$

where $\omega_0 = \delta_{[0,\ldots,0]}$. Thus, by complete induction

$$\omega_{kn_0}(R^s) \leqq \left(1 - \sigma_{n_0}(R^s)\right)^k,$$

and hence $\lim_{k\to\infty} \omega_{kn_0}(R^s) = 0$. Now, recalling that $\left(\omega_n(R^s)\right)$ is monotone, we conclude that $\lim_{n\to\infty} \omega_n(R^s) = 0$.

3. Subject to the assumption employed in 2., the distribution ν is non-lattice. Indeed, if this were not the case, a proper closed subgroup H of R^s would satisfy the condition $\nu^{n_0}(n_0x_H + H) = 1$ for some element x_H of R^s. This would imply that $\mu(H) = \mu(n_0x_H + H) > 0$. However, the structure of such subgroups implies that always $\mu(H) = 0$.

Hence, by 11.10.4., $V((\nu^n)) = R^s$. For all x in R^s, all natural numbers k, and all $n > k$,

$$\begin{aligned}\operatorname{Var}(\nu^n - \delta_x * \nu^n) &= \operatorname{Var}\left((\sigma_k + \omega_k) * \nu^{n-k} - \delta_x * (\sigma_k + \omega_k) * \nu^{n-k}\right)\\ &\leqq \operatorname{Var}(\sigma_k * \nu^{n-k} - \delta_x * \sigma_k * \nu^{n-k}) + 2\omega_k(R^s),\end{aligned}$$

so that we are able to conclude that

$$\overline{\lim_{n\to\infty}} \operatorname{Var}(\nu^n - \delta_x * \nu^n) \leqq 2\omega_k(R^s),$$

i.e. that, in view of 2.,

$$\operatorname{Var}(\nu^n - \delta_x * \nu^n) \xrightarrow[n\to\infty]{} 0. \blacksquare$$

We finish this section with the following alternative theorem (cf. Kerstan & Matthes [3], [6]).

11.10.7. Theorem. *Let ν be any distribution on $\mathfrak{R}^s$ and x any point in R^s. Then one of the following assertions is fulfilled:*

a) $\sup_{n=1,2,\ldots} \sqrt{n} \operatorname{Var}(\nu^n - \delta_x * \nu^n) < +\infty$,

b) $\operatorname{Var}(\nu^n - \delta_x * \nu^n) = 2 \qquad (n = 1, 2, \ldots)$.

Proof. We will suppose that b) is not true, i.e. $\operatorname{Var}(\nu^r - \delta_x * \nu^r) < 2$ for some natural number r. Then there exist distributions γ, α, β on $\mathfrak{R}^s$ and a number $p \in (0, 1]$ with the properties

$$\nu^r = p\gamma + (1-p)\,\alpha, \qquad \delta_x * \nu^r = p\gamma + (1-p)\,\beta.$$

Hence $\nu^r = p\varrho_1 + (1-p)\,\varrho_2$ with

$$\varrho_1 = 2^{-1}(\gamma + \delta_{-x} * \gamma), \qquad \varrho_2 = 2^{-1}(\alpha + \delta_{-x}*\beta).$$

Continuing in the same way as in the third step of the proof of theorem 11.10.3. we obtain for $n = 1, 2, \ldots$

$$\operatorname{Var}(\nu^{nr} - \delta_x * \nu^{nr}) = \operatorname{Var}(\nu^{nr} - \delta_{-x} * \nu^{nr})$$

$$= \operatorname{Var}\left(\sum_{k=0}^{n} \binom{n}{k} p^k(1-p)^{n-k} \left(\varrho_1{}^k * \varrho_2{}^{n-k} - (\delta_{-x} * \varrho_1{}^k) * \varrho_2{}^{n-k}\right)\right)$$

$$\leqq \sum_{k=0}^{n} \binom{n}{k} p^k(1-p)^{n-k} \operatorname{Var}(\varrho_1{}^k - \delta_{-x} * \varrho_1{}^k)$$

$$= \sum_{k=0}^{n} \binom{n}{k} p^k(1-p)^{n-k} \operatorname{Var}\left(\sum_{j=0}^{k} \binom{k}{j} 2^{-k}\gamma^k * (\delta_{-x})^j - \sum_{j=0}^{k} \binom{k}{j} 2^{-k}\gamma^k * (\delta_{-x})^{j+1}\right)$$

$$\leqq \sum_{k=0}^{n} \binom{n}{k} p^k (1-p)^{n-k} \sum_{j=0}^{k+1} \left| \binom{k}{j-1} 2^{-k} - \binom{k}{j} 2^{-k} \right|$$

$$\leqq \sum_{k=0}^{n} \binom{n}{k} p^k (1-p)^{n-k}\, 2(k+1)^{-1/2} \leqq 2(p(n+1))^{-1/2}.$$

Now let $m > r$. We set $m = nr + t$ with $0 \leqq t < r$. Then

$$\begin{aligned} \operatorname{Var}(\nu^m - \delta_x * \nu^m) &\leqq \operatorname{Var}(\nu^{nr} - \delta_x * \nu^{nr}) \\ &\leqq 2(p(n+1))^{-1/2} \leqq 2\sqrt{r}\,(p(nr+t))^{-1/2}, \end{aligned}$$

which means that

$$\sup_{m>r} \sqrt{m} \operatorname{Var}(\nu^m - \delta_x * \nu^m) \leqq 2(r/p)^{1/2} < +\infty. \blacksquare$$

In Guy Fourt [1], it has been shown that there exist non-lattice distributions ν on $\mathfrak{R}^s$ such that the two distributions ν^n, $\delta_x * \nu^n$ are purely singular with respect to each other for all x in $R^s \setminus \{[0, \ldots, 0]\}$ and all natural numbers n.

12. Spatially Homogeneous Branching Processes

12.0. Introduction

Each cluster field of the form $[Q]$, $Q\big(\chi(R^s) < +\infty\big) = 1$, can be considered as a description of a spatially homogeneous branching mechanism with discrete time. If we start with a random finite initial population χ_0 with distribution L, repeated application of the stochastic transition $\Psi \rightsquigarrow [Q]_{(\Psi)}$ yields a random sequence $\chi_0, \chi_1, \ldots$ of finite counting measures. The i-th term χ_i describes the i-th generation and has the distribution $\big(\ldots(L_{[Q]})\ldots\big)_{[Q]}$. If we are only interested in the total number of particles, i.e. in the random sequence $\chi_0(R^s)$, $\chi_1(R^s), \ldots$ of non-negative integers, we observe that it can be described by the pattern of a Galton-Watson process, L_{R^s} being the initial distribution, and Q_{R^s} the distribution governing the random number of all of a particle's descendants. Hence this coarser branching process is critical if and only if the distribution Q satisfies the condition $\int \chi(R^s)\, Q(\mathrm{d}\chi) = 1$. We denote the set of all these distributions by $\boldsymbol{D}$.

In contrast to the classical theory of Galton-Watson processes, in the spatially homogeneous branching mechanism outlined above we may also admit random infinite initial populations χ_0. Thus, the question of the possibility of a statistical equilibrium with respect to the clustering operation $\Psi \rightsquigarrow [D]_{(\Psi)}$ becomes meaningful not only in the case $D\big(\chi(R^s) = 1\big) = 1$ but for all distributions D in $\boldsymbol{D}$. A distribution D in $\boldsymbol{D}$ is called stable if there exists a stationary distribution P on $\mathfrak{M}$ which is cluster-invariant with respect to $[D]$ and has the property $0 < i_P < 1$.

This chapter is mainly devoted to the investigation of those stationary distributions P on $\mathfrak{M}$ which are cluster-invariant with respect to a given homogeneous cluster field $[D]$, $D \in \boldsymbol{D}$, i.e. of the stationary solutions P of the functional equation $P_{[D]} = P$. Here three problems arise:

a) Is it possible to express the property of stability of a distribution D in $\boldsymbol{D}$ in terms of elementary characteristics of D, at least under additional assumptions?
b) What is the structure of the set of all stationary distributions P on $\mathfrak{M}$ which are cluster-invariant with respect to $[D]$?
c) Under which assumptions on L and D does the n-fold clustered distribution $\big(\ldots(L_{[D]})\ldots\big)_{[D]}$ tend weakly for $n \to \infty$ towards a given stationary distribution which is cluster-invariant with respect to $[D]$?

In the investigation of the n-fold clustered distributions $\big(\ldots(L_{[D]})\ldots\big)_{[D]}$, $L \in \boldsymbol{S}$, the following fact should be taken into consideration: it is possible that the realizations χ_n show the tendency to be built up by larger and larger but, on the other hand, more and more rare "lumps", so that for $n \to \infty$ for each bounded Borel subset X of R^s the probability of the event $\chi_n(X) > 0$ tends towards zero, although by virtue of $\varrho_D(R^s) = 1$, the conservation of the average particle density is guaranteed (cf. especially Proposition 12.1.6.).

As was shown in Kallenberg [7] it is generally impossible to express the stability of D only in terms of ϱ_D and D_{R^s}. For distributions D in $\boldsymbol{D}$ satisfying $D(\chi(R^s) = 1) < 1$, $\int (\chi(R^s))^2 D(\mathrm{d}\chi) < +\infty$, however, the random walk corresponding to the symmetrized intensity measure ${}^0\varrho_D$ of D is transient if and only if D is stable (Theorems 12.6.4. and 12.6.6.). In this way a partial solution of problem a) can be given.

Let $\boldsymbol{C}$ denote the set of those distributions D in $\boldsymbol{D}$ which are stable and whose intensity measure ϱ_D is non-lattice. In virtue of Theorem 12.4.1. to every D in $\boldsymbol{C}$ and every $l \geqq 0$ there exists exactly one ergodic[1]) stationary distribution $]D[_{(l)}$ which is cluster-invariant with respect to $[D]$ and has the intensity l. Theorem 12.4.6. says that, for all D in $\boldsymbol{C}$, the correspondence $\sigma \rightsquigarrow \int]D[_{(l)}(.)\, \sigma(\mathrm{d}l)$ provides a one-to-one mapping of the set of all distributions σ of random non-negative numbers upon the set of all stationary distributions on $\mathfrak{M}$ which are cluster-invariant with respect to $[D]$. Thus, a partial solution of problem b) is given.

In virtue of Theorem 12.4.3. for all D in $\boldsymbol{C}$ and all L in $\boldsymbol{S}$ the n-fold clustered distribution $(\ldots(L_{[D]})\ldots)_{[D]}$ tends weakly towards the mixture $\int]D[_{(l)}(.)\, \sigma_L(\mathrm{d}l)$ for $n \to \infty$. To ensure the weak convergence towards $\int]D[_{(l)}(.)\, \sigma(\mathrm{d}l)$ we thus only need to suppose that σ_L coincides with σ. This is a partial solution of problem c).

The statements made above in connection with the problems a), b) and c) can be transferred to the case where, instead of R^s, an arbitrary locally compact Abelian topological group satisfying the second axiom of countability is used as phase space (cf. Debes, Kerstan, Liemant & Matthes [1] and Liemant [2]). This generalization is of importance even for problems concerning phase spaces R^s only, especially for the solution of problem b) for arbitrary stable distributions (cf. Kallenberg [7]). Furthermore, also at most countable mark spaces K can be permitted (cf. Warmuth [1]), whereby it is possible to describe critical age-dependent spatially homogeneous branching processes as well as critical spatially homogeneous branching processes with an at most countable set of types of particles.

If Q is subcritical, i.e. if $0 < c(Q) = \int \chi(R^s)\, Q(\mathrm{d}\chi) < 1$, for all L in $\boldsymbol{S}$ the n-fold clustered distribution L tends weakly towards δ_o. Using the contraction operator which was introduced in Section 7.3., we obtain the weak convergence of $\mathscr{K}_{(c(Q))^n}\big((\ldots(L_{[Q]})\ldots)_{[Q]}\big)$ towards a stationary distribution P, where P is of intensity $\int l\sigma_L(\mathrm{d}l)$ in the "non-explosive" case

$$\sum_{k=1}^{\infty} k \lim_{n\to\infty} \underbrace{(\ldots(Q_{[Q]})\ldots)_{[Q]}}_{(n-1)\text{-times}} \big(\chi(R^s) = k \mid \chi \neq o\big) < +\infty,$$

and coincides with δ_o in the "explosive case" (Theorem 12.5.5., a generalization to the case of finite mark spaces was given in Fleischmann & Prehn [3]).

---

[1]) These distributions $]D[_{(l)}$, $l \geqq 0$, are even mixing (cf. Fleischmann [4]).

12.1. The Distributions $]\boldsymbol{D}[_{(l)}$

In this chapter we use as a rule a phase space of the form R^s. Non-indicated mathematical objects as M, $\mathfrak{M}$ and $\boldsymbol{P}$ should always refer to this phase space.

In virtue of 11.3.2., for all D in $\boldsymbol{D}$ we have

$$\rangle D \langle\, (R^s \times M^0) = 1 = \langle D\rangle\, (M^0).$$

For the sake of abbreviation we introduce the modified distributions

$$\rangle D\langle^! = \rangle D\langle\, \big([y, \Phi - \delta_{[0,\ldots,0]}] \in (.)\big), \qquad \langle D\rangle^! = \langle D\rangle\, \big((\Phi - \delta_{[0,\ldots,0]}) \in (.)\big)$$

on the σ-algebras $\mathfrak{R}^s \otimes \mathfrak{M}$, $\mathfrak{M}$, respectively.

Proceeding in the same way as in the derivation of 11.3.5. we obtain for all H, Q in $\boldsymbol{D}$ and all Y in $\mathfrak{R}^s \otimes \mathfrak{M}^0$

$$\begin{aligned}
\rangle H_{[Q]}\langle\, (Y) &= \int (\mathscr{C}_{[Q]_{(y)}} \circledast [Q]_{(T_{-y}\Phi)})^T\, (Y)\, \rangle H\langle^!\, (\mathrm{d}[y, \Phi]) \\
&= \int \Big(\int \Big(\int \delta_{[z, T_z(\chi+\Psi)]}\, (Y)\, \chi(\mathrm{d}z)\Big)\, ([Q]_{(y)} \times [Q]_{(T_{-y}\Phi)})\, (\mathrm{d}[\chi, \Psi])\Big)\, \rangle H\langle^!\, (\mathrm{d}[y, \Phi]) \\
&= \int \Big(\int \Big(\int \Big(\int \delta_{[z+y, T_z\chi + T_z\Psi]}\, (Y)\, \chi(\mathrm{d}z)\Big)\, Q(\mathrm{d}\chi)\Big)\, [Q]_{(\Phi)}\, (\mathrm{d}\Psi)\Big)\, \rangle H\langle^!\, (\mathrm{d}[y, \Phi]) \\
&= \int \Big(\int \Big(\int k_Y([z + y, \chi + T_z\Psi])\, \rangle Q\langle\, (\mathrm{d}[z, \chi])\Big)\, [Q]_{(\Phi)}(\mathrm{d}\Psi)\Big)\, \rangle H\langle^!\, (\mathrm{d}[y, \Phi]).
\end{aligned}$$

Thus we observe the truth of (cf. Kallenberg [7])

12.1.1. *For all H, Q in $\boldsymbol{D}$ we have*

$$\rangle H_{[Q]}\langle = \int \Big(\int \Big(\int k_{(.)}([z + y, \chi + T_z\Psi])\, [Q]_{(\Phi)}(\mathrm{d}\Psi)\Big)\, \rangle H\langle^!\, (\mathrm{d}[y, \Phi])\Big)\, \rangle Q\langle\, (\mathrm{d}[z, \chi]).$$

Remembering the interpretation of $\rangle D\langle$ which was connected with 11.3.4., we may interprete 12.1.1. in the following way. Let ω be a stationary random counting measure with non-vanishing finite intensity. Clustering of ω by $[H]$ yields a random counting measure γ which is transformed in a random counting measure σ by clustering according to $[Q]$. We now choose "at random" an individual I_0 from σ, to be placed at the origin for convenience, and denote by $-z$ the position of its "father" and by χ the counting measure describing its "brothers". Furthermore, let $-z - y$ be the position of the "grandfather" I_2 of I_0, and $T_z\Phi$ be the counting measure describing the "brothers" of I_1. Then $\big[[y, \Phi], [z, \chi]\big]$ is distributed according to $\rangle H\langle^! \times \rangle Q\langle^!$. Given y, z, Φ and χ, the superposition τ of the offsprings of $T_z\Phi$, i.e. the counting measure describing the "cousins" of I_0, coincides with $Q_{(T_z\Phi)}$. Thus, we have constructed the distribution of $[z + y, \chi + \tau]$, i.e. the distribution $\rangle H_{[Q]}\langle^!$.

In virtue of 12.1.1. we obtain for all D in $\boldsymbol{D}$ and $n = 2, 3, \ldots$

$$\begin{aligned}
\rangle D^{[n]}\langle^! &= \rangle (D^{[n-1]})_{[D]}\langle^! \\
&= \int \Big(\int \Big(\int k_{(.)}([x_1 + y_2, \eta_1 + T_{x_1}\Psi_2])\, [D]_{(\Phi_2)}(\mathrm{d}\Psi_2)\Big)\, \rangle D^{[n-1]}\langle^!\, (\mathrm{d}[y_2, \Phi_2])\Big) \\
&\qquad \rangle D\langle^!\, (\mathrm{d}[x_1, \eta_1]).
\end{aligned}$$

Now, if $n > 2$, applying anew 12.1.1., we may continue with

$$= \int \Big(\int \Big(\int \Big(\int \Big(\int k_{(.)}([x_1 + x_2 + y_3, \eta_1 + T_{x_1}\Psi_2])\, [D]_{(\eta_2 + T_{x_2}\tau)}\, (\mathrm{d}\Psi_2)\Big)$$

$$[D]_{(\Phi_3)}\, (\mathrm{d}\tau)\Big)\, \rangle D^{[n-2]}\langle^!\, (\mathrm{d}[y_3, \Phi_3])\Big)\, \rangle D\langle^!(\mathrm{d}[x_2, \eta_2])\Big)\, \rangle D\langle^!\, (\mathrm{d}[x_1, \eta_1])$$

$$= \int \Big(\int \Big(\int \Big(\int \Big(\int \Big(\int k_{(.)}([x_1 + x_2 + y_3, \eta_1 + T_{x_1}\omega_2 + T_{x_1+x_2}\Psi_3])\, [D]_{(\eta_2)}\, (\mathrm{d}\omega_2)\Big)$$

$$[D]_{(\tau)}\, (\mathrm{d}\Psi_3)\Big)\, [D]_{(\Phi_3)}\, (\mathrm{d}\tau)\Big)\, \rangle D^{[n-2]}\langle^!\, (\mathrm{d}[y_3, \Phi_3])\Big)\, \rangle D\langle^!\, (\mathrm{d}[x_2, \eta_2])\Big)$$

$$\rangle D\langle^!\, (\mathrm{d}[x_1, \eta_1])$$

$$= \int \Big(\int \Big(\int \Big(\int \Big(\int k_{(.)}([x_1 + x_2 + y_3, \eta_1 + T_{x_1}\omega_2 + T_{x_1+x_2}\Psi_3])\, [D^{[2]}]_{(\Phi_3)}\, (\mathrm{d}\Psi_3)\Big)$$

$$[D]_{(\eta_2)}\, (\mathrm{d}\omega_2)\Big)\, \rangle D^{[n-2]}\langle^!\, (\mathrm{d}[y_3, \Phi_3])\Big)\, \rangle D\langle^!\, (\mathrm{d}[x_2, \eta_2])\Big)\, \rangle D\langle^!\, (\mathrm{d}[x_1, \eta_1]).$$

Continuing recursively, we thus obtain (cf. Kallenberg [7])

12.1.2. Proposition. *For all D in $\boldsymbol{D}$ we have for $n = 2, 3, \ldots$*

$$\rangle D^{[n]}\langle = \int\Big(\int k_{(.)}([x_1 + \cdots + x_n, \delta_{[0,\ldots,0]} + \eta_1 + T_{x_1}\omega_2 + \cdots + T_{x_1+\cdots+x_{n-1}}\omega_n])$$

$$\Big(\mathop{\times}_{k=2}^{n} [D^{[k-1]}]_{(\eta_k)}\Big)\, (\mathrm{d}[\omega_2, \ldots, \omega_n])\Big)\, (\rangle D\langle^! \times \cdots \times \rangle D\langle^!)\, \big(\mathrm{d}\big[[x_n, \eta_n], \ldots, [x_1, \eta_1]\big]\big).$$

By 12.1.2. the distribution $\langle D^{[n]}\rangle$, $n = 2, 3, \ldots$, may be constructed in the following way. Let $[x_1, \eta_1]$, $[x_2, \eta_2]$, ... be a random sequence of elements of $R^s \times M$ which is distributed according to $\rangle D\langle^! \times \rangle D\langle^! \times \cdots$. Given this sequence, we select $\gamma_2, \gamma_3, \ldots$ according to the distribution $[D]_{(T_{x_1}\eta_2)} \times [D^{[2]}]_{(T_{x_1+x_2}\eta_3)} \times \cdots$. Then, for $n = 2, 3, \ldots$, the counting measure

$$\lambda_n = \delta_{[0,\ldots,0]} + \eta_1 + \gamma_2 + \cdots + \gamma_n$$

is distributed according to $\langle D^{[n]}\rangle$.

The random sequence $\lambda_1 = \delta_{[0,\ldots,0]} + \eta_1, \lambda_2, \lambda_3, \ldots$ of counting measures is almost surely increasing. If $\lambda = \sup\limits_{n=1,2,\ldots} \lambda_n$ belongs to M with probability one, i.e. if

$$\lambda\big(S_k([0, \ldots, 0])\big) = 1 + \eta_1\big(S_k([0, \ldots, 0])\big) + \sum_{n=2}^{\infty} \gamma_n\big(S_k([0, \ldots, 0])\big) < +\infty \qquad (k = 1, 2, \ldots),$$

the distribution $\langle D^{[n]}\rangle$ of λ_n converges weakly towards the distribution Q of λ. Otherwise we have $\lambda\big(S_{k_0}([0, \ldots, 0])\big) = +\infty$ with positive probability for some natural number k_0, i.e.

$$\inf_{j=1,2,\ldots}\ \sup_{n=1,2,\ldots} \langle D^{[n]}\rangle \big(\chi\big(S_{k_0}([0, \ldots, 0])\big) \geqq j\big) > 0,$$

such that $\{\langle D^{[n]}\rangle\}_{n=1,2,\ldots}$ is not relatively compact with respect to $\varrho_{\boldsymbol{P}}$. In this way we obtain (cf. Kallenberg [7])

12.1.3. *For all D in $\boldsymbol{D}$, the following properties are equivalent.*

a) $\{\langle D^{[n]}\rangle\}_{n=1,2,\ldots}$ *is relatively compact with respect to* $\varrho_{\boldsymbol{P}}$.
b) $(\langle D^{[n]}\rangle)$ *converges weakly towards a distribution Q in $\boldsymbol{P}$.*

A distribution D on $\mathfrak{M}$ is called *stable* if there exists a stationary distribution P in $\boldsymbol{P}$ with the properties $0 < i_P < +\infty$, $P_{[D]} = P$. In virtue of 11.1.3., each stable distribution belongs to $\boldsymbol{D}$.

Let $\langle D^{[n]}\rangle \underset{n\to\infty}{\Rightarrow} Q$ for a distribution D in $\boldsymbol{D}$. In view of 11.4.1., for all $l \geqq 0$, we get

$$\left(\overline{(P_{l\mu})_{[D^{[n]}]}}\right)^0 = l\,\langle D^{[n]}\rangle \underset{n\to\infty}{\Rightarrow} lQ\,.$$

By 10.3.7. there exists a stationary infinitely divisible distribution $]D[_{(l)}$ satisfying $\left(\overline{]D[_{(l)}}\right)^0 = lQ$, and in virtue of 10.3.8. we obtain

$$(P_{l\mu})_{[D^{[n]}]} \underset{n\to\infty}{\Rightarrow}]D[_{(l)}\,, \qquad i_{]D[_{(l)}} = l\,.$$

By 4.7.3.

$$(P_{l\mu})_{[D^{[n+1]}]} = \left((P_{l\mu})_{[D^{[n]}]}\right)_{[D]} \underset{n\to\infty}{\Rightarrow} (]D[_{(l)})_{[D]}\,.$$

Hence $]D[_{(l)}$ is cluster-invariant with respect to $[D]$ for $l \geqq 0$.

Now, let

$$\inf_{j=1,2,\ldots}\ \sup_{n=1,2,\ldots} \langle D^{[n]}\rangle \left(\chi\big(S_{k_0}([0,\ldots,0])\big) \geqq j\right) > 0$$

for some natural number k_0. In view of 11.3.5., for any stationary distribution L in $\boldsymbol{P}$ with the properties $i_L < +\infty$, $L_{[D]} = L$, we get for $j, n = 1, 2, \ldots$

$$L^0\left(\Phi\big(S_{k_0}([0,\ldots,0])\big) \geqq j\right) \geqq i_L \langle D^{[n]}\rangle \left(\chi\big(S_{k_0}([0,\ldots,0])\big) \geqq j\right).$$

Thus

$$\begin{aligned} 0 &= \inf_{j=1,2,\ldots} L^0\left(\Phi\big(S_{k_0}([0,\ldots,0])\big) \geqq j\right) \\ &\geqq i_L \inf_{j=1,2,\ldots}\ \sup_{n=1,2,\ldots} \langle D^{[n]}\rangle \left(\chi\big(S_{k_0}([0,\ldots,0])\big) \geqq j\right) \end{aligned}$$

and hence $i_L = 0$.

Summarizing, we obtain the truth of (cf. Kallenberg [7])

12.1.4. Theorem. *For all D in $\boldsymbol{D}$, the following properties are equivalent.*

a) *D is stable.*
b) $\{\langle D^{[n]}\rangle\}_{n=1,2,\ldots}$ *is relatively compact with respect to* $\varrho_{\boldsymbol{P}}$.
c) *For all $l \geqq 0$, the sequence $\left((P_{l\mu})_{[D^{[n]}]}\right)$ converges weakly towards a stationary infinitely divisible distribution $]D[_{(l)}$ with intensity l, which is cluster-invariant with respect to $[D]$.*

From

$$\left(\widetilde{]D[_{(l)}}\right)^0 = lQ = l\left(\widetilde{]D[_{(1)}}\right)^0 \qquad (l \geqq 0)$$

it follows that

$$\widetilde{]D[_{(l)}} = l\widetilde{]D[_{(1)}} \qquad (l \geqq 0).$$

Hence the distributions $]D[_{(l)}$, $l \geqq 0$, run through the parametrized semi-group which is, according to 2.4.1., uniquely determined by $]D[_{(1)}$.

Immediately from 11.8.1. it follows that

$$]Q_\nu[_{(l)} = P_{l\mu} \qquad (l \geqq 0)$$

for all distributions ν on $\mathfrak{A}$. Hence in the case $D(\chi(R^s) = 1) = 1$ the distributions $]D[_{(l)}$, $l \geqq 0$, are regular infinitely divisible ones. On the other hand, we have (cf. Liemant [1])

12.1.5. Proposition. *For all stable D in $\boldsymbol{D}$ having the property $D(\chi(R^s) = 1) < 1$, the distributions $]D[_{(l)}$, $l \geqq 0$, are singular infinitely divisible.*

Proof. We define a stationary regular infinitely divisible distribution P by

$$\tilde{P} = \widetilde{]D[_{(l)}}\left(\Psi \in (.), \Psi \text{ finite}\right)$$

and suppose that $]D[_{(l)}$ be not a singular infinitely divisible distribution, i.e. that P be different from δ_o. Since $D(\chi \text{ finite}) = 1$, we get by 4.3.3.

$$\begin{aligned}\widetilde{P_{[D]}} &= (\tilde{P})_{[D]}\left((\cdot) \setminus \{o\}\right)\\ &= \left(\widetilde{]D[_{(l)}}\left(\Psi \in (.),\ \Psi \text{ finite}\right)\right)_{[D]}\left((.) \setminus \{o\}\right)\\ &= \left(\widetilde{]D[_{(l)}}\right)_{[D]}\left(\Phi \in (.), 0 < \Phi(R^s) < +\infty\right)\\ &= \widetilde{]D[_{(l)}}\left(\Phi \in (.), \Phi \text{ finite}\right) = \tilde{P},\end{aligned}$$

i.e. P is cluster-invariant with respect to $[D]$. According to 11.4.3., P has a reduced homogeneous clustering representation $P = (P_{v\mu})_{[H]}$ with $v > 0$, $H \in \boldsymbol{Q}$ and $H(\{o\}) = 0$. By virtue of 11.4.2., the number v is uniquely determined.

By 11.1.2. we get the homogeneous clustering representation

$$P = P_{[D]} = \left((P_{v\mu})_{[H]}\right)_{[D]} = (P_{v\mu})_{[H_{[D]}]}.$$

By assumption $D(\{o\}) > 0$ and hence $H_{[D]}(\{o\}) > 0$. Reducing the homogeneous clustering representation, we get

$$P = (P_{v'\mu})_{[H_{[D]}((.)|\chi \neq o)]}$$

where $v' = vH_{[Q]}(\chi \neq o) < v$, which contradicts, however, the statement that v is unique. ∎

Let U be any bounded open neighbourhood of the origin. For all natural numbers k we set

$$_{U,k}\Phi = \sum_{\substack{x \in R^s, \Phi(\{x\})>0 \\ \Phi(U+x) \leqq k}} \Phi(\{x\})\, \delta_x .$$

Just as in the proof of 6.4.2. we see that $\Phi \curvearrowright {}_{U,k}\Phi$ is a measurable mapping of M into M. Let $_{U,k}P$ denote the image of a distribution P with respect to this mapping. Together with P the distribution $_{U,k}P$ is stationary, too.

The following proposition (cf. Debes, Kerstan, Liemant & Matthes [1]) gives us some insight into the nature of stability.

12.1.6. Proposition. *Let U be any bounded open neighbourhood of the origin. Then, if a distribution D in $\boldsymbol{D}$ is stable,*

$$\varrho_{_{U,k}(D^{[n]})}(R^s) \xrightarrow[k\to\infty]{} 1$$

is satisfied uniformly in n. If, however, D is not stable, then

$$\varrho_{_{U,k}(D^{[n]})}(R^s) \xrightarrow[n\to\infty]{} 0$$

holds for all natural numbers k.

Proof. 1. Let $f_{U,k}$ denote the indicator function of the subset $\{[x,\chi]: [x,\chi] \in C,\ \chi(U+x) \leqq k\}$, i.e. $\{[x,\chi]: [x,\chi] \in C,\ ({}_{U,k}\chi)(\{x\}) > 0\}$ of $C = \{[x,\chi]: [x,\chi] \in R^s \times M,\ \chi(\{x\}) > 0\}$. In view of 5.1.2. we obtain for all distributions V in $\boldsymbol{D}$

$$\begin{aligned}\varrho_{_{U,k}V}(R^s) &= \int \Big(\int f_{U,k}(x,\chi)\, \chi(\mathrm{d}x)\Big)\, V(\mathrm{d}\chi)\\ &= \int f_{U,k}(x,\chi)\, \mathscr{C}_V(\mathrm{d}[x,\chi]) = \int f_{U,k}(x, T_{-x}\Psi) > \rangle V \langle\, (\mathrm{d}[x,\Psi])\\ &= \langle V\rangle \big(\Psi(U) \leqq k\big).\end{aligned}$$

2. From 1. and 12.1.4. it follows immediately that for any stable D in $\boldsymbol{D}$

$$\varrho_{_{U,k}(D^{[n]})}(R^s) \xrightarrow[k\to\infty]{} 1$$

is satisfied uniformly in n.

3. Let W be any open neighbourhood of the origin with the properties $W - W \subseteqq U$, $\mu(\partial W) = 0$. For all natural numbers n we obtain

$$\begin{aligned}&(P_\mu)_{[D^{[n]}]}\big(\Phi(W) = 0\big) = \exp\Big(-\int D^{[n]}\big(\chi(W-x) > 0\big)\, \mu(\mathrm{d}x)\Big)\\ &\leqq \exp\Big(-\int k^{-1}\, \varrho_{_{U,k}(D^{[n]})}(W-x)\, \mu(\mathrm{d}x)\Big)\\ &= \exp\Big(-k^{-1} \int \mu(W-x)\, \varrho_{_{U,k}(D^{[n]})}(\mathrm{d}x)\Big) = \exp\Big(-k^{-1}\mu(W)\, \varrho_{_{U,k}(D^{[n]})}(R^s)\Big).\end{aligned}$$

4. Let us suppose that the monotonously decreasing sequence $\Big(\langle D^{[n]}\rangle \big(\chi(U) \leqq k\big)\Big)$, i.e. $\Big(\varrho_{_{U,k}(D^{[n]})}(R^s)\Big)$, has a positive limit c_k for some natural number k.

For all natural numbers n, we set

$$H_n = n^{-1} \sum_{l=1}^{n} (P_\mu)_{[D^{[l]}]}.$$

In view of 3.2.8. there exists a subsequence (H_{n_m}) which converges weakly towards a distribution P in $\boldsymbol{P}$. By 10.3.4. and 3.1.12. the distribution P is stationary and of finite intensity. Thus, by 3. and 3.1.9. we obtain

$$P(\Phi(W) = 0) = \lim_{m\to\infty} H_{n_m}(\Phi(W) = 0) \leqq \exp\left(-k^{-1}\mu(W)\, c_k\right) < 1,$$

so that P is different from δ_o.

In virtue of 4.7.3. we get

$$(H_{n_m})_{[D]} \underset{m\to\infty}{\Rightarrow} P_{[D]}.$$

On the other hand

$$(H_n)_{[D]} = n^{-1} \sum_{l=2}^{n+1} (P_\mu)_{[D^{[l]}]},$$

and hence

$$\varrho_{\boldsymbol{P}}\left(H_n, (H_n)_{[D]}\right) \leqq \|H_n - (H_n)_{[D]}\| \underset{n\to\infty}{\longrightarrow} 0.$$

Consequently,

$$(H_{n_m})_{[D]} \underset{m\to\infty}{\Rightarrow} P,$$

which means that P is cluster-invariant with respect to $[D]$. Thus D is stable. ∎

Now let L be any stationary distribution with finite intensity. We obtain for all U, k and n

$$i_{U,k}(L_{[D^{[n]}]}) \leqq i_L \varrho_{U,k}(D^{[n]})\,(R^s).$$

Thus, if D is not stable, by 12.1.6. we get

$$i_{U,k}(L_{[D^{[n]}]}) \underset{n\to\infty}{\longrightarrow} 0.$$

Consequently, for all U, k we have

$$\begin{aligned}
&\overline{\lim_{n\to\infty}}\, L_{[D^{[n]}]}\left(\Phi(U) > 0\right)\\
&\leqq \overline{\lim_{n\to\infty}}\, L_{[D^{[n]}]}\left(({}_{U,k}\Phi)\,(U) > 0\right) + \overline{\lim_{n\to\infty}}\, L_{[D^{[n]}]}\left(\Phi(U + U) > k\right)\\
&\leqq 0 + k^{-1} i_L \mu(U + U).
\end{aligned}$$

Thus, we obtain (cf. Kallenberg [7])

12.1.7. *If a distribution D in $\boldsymbol{D}$ is not stable, for all stationary distributions L of finite intensity, we have*

$$L_{[D^{[n]}]} \underset{n\to\infty}{\Rightarrow} \delta_o.$$

In Kallenberg [7] it was proved that for all stable distributions D in $\boldsymbol{D}$ and all stationary L in $\boldsymbol{P}$ with finite intensity i_L the sequence $(L_{[D^{[n]}]})$ converges weakly towards a stationary distribution P with the properties $P_{[D]} = P$, $i_P = i_L$.

12.2. A Decomposition of the Cluster Powers $D^{[n]}$

Let D be any distribution on $\mathfrak{M}$ with the properties $D \neq \delta_o$, $D(\chi \text{ finite}) = 1$. According to our statements in Section 11.1. all cluster powers $D^{[n]}$, $n = 0, 1, \ldots$, are defined and different from δ_o. For the sake of abbreviation we set

$$z_n(D) = D^{[n]}(\chi \neq o), \qquad D^{(n)} = D^{[n]}\big((.) \mid \chi \neq o\big) \qquad (n = 0, 1, \ldots).$$

For the desired decomposition of $D^{[n]}$ it is necessary to use a more detailed description of the spatially homogeneous branching model determined by D which is different from our considerations in Section 12.0. For this we use the marking technique. Each particle from the n-th generation χ_n, $n = 1, 2, \ldots$, will get a mark so that all family connections of particles from χ_n can be read-off by using the corresponding marks. As the possible family connections grow more and more as n increases, our mark spaces depend on the generation number n.

As in Section 11.3. we denote by $\check{R}^s$ the set R^s equipped with the metric $\|x - y\| (1 + \|x - y\|)^{-1}$. We inductively define a sequence of mark spaces $[K_n, \varrho_{K_n}]$. Let $[K_1, \varrho_{K_1}]$ be the direct product of $\{1, 2, \ldots\}$ equipped as usually with the discrete metric, and $\check{R}^s$ and for all natural numbers n let $[K_{n+1}, \varrho_{K_{n+1}}]$ be the direct product of $[K_1, \varrho_{K_1}]$ and $[K_n, \varrho_{K_n}]$. Furthermore, we denote the direct product of R^s and $[K_n, \varrho_{K_n}]$ by $[A_n, \varrho_{A_n}]$ and set $[A_0, \varrho_{A_0}] = R^s$. We sign with the lower index n the objects formed with respect to the phase space $[A_n, \varrho_{A_n}]$, $n = 0, 1, \ldots$.

With the help of D we now define for each natural number n a cluster field ${}_n\varkappa$ on $[A_{n-1}, \varrho_{A_{n-1}}]$ with phase space $[A_n, \varrho_{A_n}]$. In the case $n = 1$ we set for all x in R^s

$${}_1\varkappa_{(x)} = \int \delta_{\sum\limits_{y \in R^s, \chi(\{y\})>0} \; \sum\limits_{r=1,\ldots,\chi(\{y\})} \delta_{[y,[r,x]]}}(.) \, [D]_{(x)} \, (\mathrm{d}\chi),$$

and in the case $n > 1$ we set for all $[x, k]$ in A_{n-1}

$${}_n\varkappa_{([x,k])} = \int \delta_{\sum\limits_{y \in R^s, \chi(\{y\})>0} \; \sum\limits_{r=1,\ldots,\chi(\{y\})} \delta_{[y,[r,x,k]]}}(.) \, [D]_{(x)} \, (\mathrm{d}\chi).$$

The distribution ${}_n\varkappa_{([x,k])}$ describes the immediate progeny of a particle of the $(n-1)$-th generation at the position x with mark k where in the case $n = 1$ the mark k has to be cancelled. This immediate progeny arises in two steps. First of all the particle generates a random population χ distributed according to $[D]_{(x)}$. After that all particles of χ will be signed by marks of K_n. To this first of all the particles of χ with the same position will be numbered, and then the marks consist of the so introduced number r, of the position x of the original particle, and in the case $n > 1$ additionally of its mark k.

For all natural numbers m, by

$$\Psi \curvearrowright {}_{m}\varkappa_{(\Psi)} \qquad (\Psi \in M_{m-1},\ \Psi \text{ finite})$$

transition probabilities from $[(M_{m-1})_e, (\mathfrak{M}_{m-1})_e]$ into $[(M_m)_e, (\mathfrak{M}_m)_e]$ are given which allows us to define for each x in R^s an inhomogeneous Markov process $\Psi_0 = \delta_x, \Psi_1, \Psi_2, \ldots$ with initial distribution δ_{δ_x}. We denote its distribution by $\mathscr{P}_x$. Here Ψ_n describes (for an initial particle at position x) the *reduced family tree* of the n-th generation. The counting measure Ψ_n gives the whole n-th generation with all its family connections. The additional term "reduced" has to express the fact that the branches of the total family tree which did not reach the n-th generation have been cancelled.

Obviously, all distributions $\mathscr{P}_x\big(\Psi_n \in (.)\big)$, $n = 1, 2, \ldots$, are simple. If we now omit the additional marking we get back the homogeneous cluster fields $[D^{[n]}]$, i.e.

$$\mathscr{P}_x\big((\Psi_n)^x \in (.)\big) = [D^{[n]}]_{(x)} \qquad (x \in R^s,\ n = 1, 2, \ldots).$$

Now we will sketch how to derive the measurability of the integrand appearing in the definition of ${}_1\varkappa$. Analogous conclusions are valid for $n = 2, 3, \ldots$.

We have to show that for all Y in $\mathfrak{M}_1$ the real function

$$[x, \chi] \curvearrowright \delta_{\sum\limits_{y \in R^s, \chi(\{y\})>0} \ \sum\limits_{r=1,\ldots,\chi(\{y\})} \delta_{[y,[r,x]]}}(Y)$$

is measurable with respect to $\mathfrak{R}^s \otimes \mathfrak{M}$. However, this is equivalent to the measurability of the mapping

$$[x, \chi] \curvearrowright \Phi = \sum_{y \in R^s, \chi(\{y\})>0} \ \sum_{r=1,\ldots,\chi(\{y\})} \delta_{[y,[r,x]]}$$

with respect to $\mathfrak{R}^s \otimes \mathfrak{M}$ and $\mathfrak{M}_1$, i.e. to the measurability of the real function $[x, \chi] \curvearrowright \Phi(X)$ for all X in $\mathfrak{B}_1$.

Let $(\mathfrak{Z}_n)$ be any distinguished sequence of decompositions of R^s. If we put now

$$\Phi_n = \sum_{Z \in \mathfrak{Z}_n} \ \sum_{r=1,\ldots,\chi(Z)} (z\chi)^* \times \delta_r \times \delta_x,$$

then all mappings $[x, \chi] \curvearrowright \Phi_n$ are measurable. On the other hand, for all X in $\mathfrak{B}_1$ the equation $\Phi_n(X) = \Phi(X)$ is fulfilled for all $n > n_X$. Hence all real functions $[x, \chi] \curvearrowright \Phi(X)$, $X \in \mathfrak{B}_1$, are measurable.

Let n be any natural number. Each a in A_n has the form

$$a = [x_n, r_n, x_{n-1}, r_{n-1}, \ldots, x_1, r_1, x_0].$$

It describes a particle of the n-th generation which is arising from the initial population δ_{x_0}. The ancestor of this particle at time 1 was the r_1-th particle at the position x_1 of the immediate descendants of the initial particle etc. If a further point $a' = [x_n', \ldots, x_0']$ in A_n is given, then we can recognize without difficulties the common ancestors of the particles described by a, a'. Common ancestors exist if and only if $x_0 = x_0'$. In this case we define $m(a, a')$ as the minimum of all non-negative integers k such that

$$[x_{n-k}, r_{n-k}, \ldots, x_0] = [x_{n-k}, r_{n-k}, \ldots, x_0']$$

is satisfied. Hence $0 \leqq m(a, a') \leqq n$. Now, if Ψ is a finite simple counting measure in M_n different from o such that all points of Ψ have the same last coordinate, we define by

$$m(\Psi) = \max\left\{m(a, a') : a, a' \in A_n, \Psi(\{a\}) = 1 = \Psi(\{a'\})\right\}$$

the *source time* of Ψ. For such Ψ the last common ancestor of all particles of Ψ is in the $(n - m(\Psi))$- th generation. In particular $m(\Psi) = 0$ is satisfied if and only if Ψ consists of exactly one particle.

Obviously, $m(\Psi) \leqq k$ for $k = 0, \ldots, n$ if and only if $\Psi = \Psi^*$ and $0 < \Psi(A_n) < +\infty$ holds and the counting measure Ψ' in M_{n-k} obtained by omitting the first $2k$ coordinates of all points in Ψ has the property $(\Psi')^*(A_{n-k}) = 1$. Hence $\Psi \curvearrowright m(\Psi)$ is a measurable mapping from M_n into $\{0, 1, \ldots\}$. For convenient notation we set

$$q_k^n(D) = \mathscr{P}_{[0,\ldots,0]}\big(m(\Psi_n) = k \mid \Psi_n \neq o\big) \qquad (n = 1, 2, \ldots; \ 0 \leqq k \leqq n).$$

For all k in $\{0, \ldots, n\}$ we define a mapping t_{n-k} of $\{\Psi : \Psi \in M_n, m(\Psi) \leqq k\}$ into R^s in the following way: Let $a = [x_n, r_n, \ldots, x_0]$ be any point in Ψ. Then $t_{n-k}(\Psi) = x_{n-k}$. Thus we obtain a measurable mapping from M_n into R^s. The point $t_{n-k}(\Psi)$ describes the position of the common ancestor at time $n - k$ of the n-th generation given by Ψ. We speak of the *trunk particle* at time $n - k$ of the reduced family tree Ψ. We set

$$\mathscr{P}_{[0,\ldots,0]}\big(m(\Psi_n) \leqq k, t_{n-k}(\Psi_n) \in (.)\big) = \tau_{n,k}.$$

We may consider $\tau_{n,k}(X)$, $X \in \mathfrak{R}^s$, as probability that in the $(n-k)$-th generation given by Ψ_{n-k} exactly one of the particles situated in X has descendants at time n and all other particles of the $(n-k)$-th generation have no descendants at time n. Consequently,

$$\begin{aligned}\tau_{n,k}(X) &= \int\limits_{\{\chi:\chi\neq o\}} \chi(X)\, z_k(D)\, \big(1 - z_k(D)\big)^{\chi(R^s)-1}\, D^{[n-k]}(d\chi)\\ &\leqq z_k(D) \int \chi(X)\, D^{[n-k]}(d\chi) = z_k(D)\, \varrho_{[D^{[n-k]}]}(X),\end{aligned}$$

which leads us to (cf. Fleischmann & Prehn [3])

12.2.1. *For each natural number n, all k in $\{0, 1, \ldots, n\}$, and each non-negative measurable real function f on R^s, we have*

$$\begin{aligned}\int f(x)\, \tau_{n,k}(dx) &= z_k(D) \int\limits_{\{\chi:\chi\neq o\}} \left(\int f(x)\, \chi(dx)\right) \big(1 - z_k(D)\big)^{\chi(R^s)-1}\, D^{[n-k]}(d\chi)\\ &\leqq z_k(D) \int f(x)\, (\varrho_D)^{n-k}\, (dx).\end{aligned}$$

With similar arguments as in the derivation of 12.2.1. we get for all Y in $\mathfrak{M}$

$$\begin{aligned}&\mathscr{P}_{[0,\ldots,0]}\big((\Psi_n)^x \in Y, m(\Psi_n) \leqq k\big)\\ &= \int\limits_{\{\chi:\chi\neq o\}} \left(\int \mathscr{P}_x\big((\Psi_k)^x \in Y, \Psi_k \neq o\big) \big(1 - z_k(D)\big)^{\chi(R^s)-1} \chi(dx)\right) D^{[n-k]}(d\chi)\\ &= z_k(D) \int\limits_{\{\chi:\chi\neq o\}} \left(\int [D^{(k)}]_{(x)}\, (Y)\, \chi(dx)\right) \big(1 - z_k(D)\big)^{\chi(R^s)-1}\, D^{[n-k]}(d\chi).\end{aligned}$$

With the help of 12.2.1. we can continue with

$$= \int [D^{(k)}]_{(x)} (Y) \, \tau_{n,k}(\mathrm{d}x) = (Q_{\tau_{n,k}})_{[D^{(k)}]} (Y)$$

and obtain (cf. Fleischmann & Prehn [3])

12.2.2. Proposition. *For all distributions D on $\mathfrak{M}$ satisfying $D \neq \delta_o$, $D(\chi$ finite$) = 1$, all natural numbers n, and all k in $\{0, \ldots, n\}$ we have*

$$D^{[n]} = \mathscr{P}_{[0,\ldots,0]}\big((\Psi_n)^x \in (\cdot), \Psi_n = o \quad \text{or} \quad m(\Psi_n) > k\big) + (Q_{\tau_{n,k}})_{[D^{(k)}]}.$$

12.3. Approximation Theorems in the Critical Case

Let D be any distribution in $\boldsymbol{D}$. For all $n > m > 0$, the distribution $D_{n,m} = (Q_{(\varrho_D)^m})_{[D^{[n-m]}]}$ belongs to $\boldsymbol{D}$. Obviously, $D_{n,m}$ coincides with $D^{[n]}$ if $D\big(\chi(R^s) = 1\big) = 1$, i.e. $D = Q_{\varrho_D}$ is fulfilled. It turns out that, for fixed m, the distributions $D_{n,m}$ are in the following sense asymptotically equivalent to the cluster powers $D^{[n]}$.

12.3.1. *To each D in $\boldsymbol{D}$, each $\varepsilon > 0$ and each natural number m, there is some $n_{\varepsilon,m}$ having the following property: For all $n \geqq n_{\varepsilon,m}$ there exist measures $V_{n,m}$, $V'_{n,m}$ and $V''_{n,m}$ on $\mathfrak{M}$ which satisfy the conditions*

$$D^{[n]} = V_{n,m} + V'_{n,m}, \qquad D_{n,m} = V_{n,m} + V''_{n,m},$$

$$\varrho_{V'_{n,m}}(R^s) = \varrho_{V''_{n,m}}(R^s) < \varepsilon.$$

Proof. 1. In the special case $D\big(\chi(R^s) = 1\big) = 1$ our assertion is trivial. Thus in the following we can assume that $D\big(\chi(R^s) = 1\big) < 1$ holds, and in virtue of a well-known result in the theory of critical Galton-Watson processes (cf. e.g. Athreya & Ney [1], Section 1.5.) we get $z_n(D) \xrightarrow[n\to\infty]{} 0$.

2. Let $n > m > 0$. Then by 12.2.1. we have

$$\begin{aligned} D_{n,m} &= \big(1 - z_{n-m}(D)\big)\, \delta_o + z_{n-m}(D) \int [D^{(n-m)}]_{(x)} (\cdot)\, (\varrho_D)^m (\mathrm{d}x) \\ &\geqq \int [D^{(n-m)}]_{(x)} (\cdot)\, \tau_{n,n-m}(\mathrm{d}x) = (Q_{\tau_{n,n-m}})_{[D^{(n-m)}]}. \end{aligned}$$

On the other hand, 12.2.2. yields

$$D^{[n]} \geqq (Q_{\tau_{n,n-m}})_{[D^{(n-m)}]}.$$

Thus, setting $V_{n,m} = (Q_{\tau_{n,n-m}})_{[D^{(n-m)}]}$, we obtain

$$D^{[n]} = V_{n,m} + V'_{n,m}, \qquad D_{n,m} = V_{n,m} + V''_{n,m},$$

where $V_{n,m}$, $V'_{n,m}$, $V''_{n,m}$ belong to $\boldsymbol{E}^+$.

3. We have

$$\varrho_{D^{[n]}}(R^s) = 1 = \varrho_{D_{n,m}}(R^s) \qquad (n > m > 0).$$

By 12.2.1. we obtain

$$\varrho_{V_{n,m}}(R^s) = \big(z_{n-m}(D)\big)^{-1} \tau_{n,n-m}(R^s)$$
$$= \int\limits_{\{\chi:\chi \neq o\}} \chi(R^s) \big(1 - z_{n-m}(D)\big)^{\chi(R^s)-1} D^{[m]}(\mathrm{d}\chi).$$

The integrand is monotonically increasing towards $\chi(R^s)$. Thus,

$$\lim_{n\to\infty} \varrho_{V_{n,m}}(R^s) = \int \chi(R^s)\, D^{[m]}(\mathrm{d}\chi) = 1. \blacksquare$$

Now we are ready to derive

12.3.2. *Let L belong to* **S**, *D be in* $\boldsymbol{D}$, *and let X be a bounded set in* $\mathfrak{R}^s$. *If at least one convolution power of ϱ_D has a non-vanishing part which is absolutely continuous with respect to μ, then*

$$\sup_{x\in R^s} {}_{(X+x)}\big\|L_{[D^{[n]}]} - \big(\textstyle\int P_{l\mu}(.)\, \sigma_L(\mathrm{d}l)\big)_{[D^{[n]}]}\big\| \xrightarrow[n\to\infty]{} 0.$$

Proof. At first we shall in addition assume that L be in $\mathbf{S}^{(c)}$. Now let $\varepsilon > 0$ and let m be an arbitrary natural number. Using the notation employed in 12.3.1. we get, by 4.2.8.

$$\sup_{x\in R^s} {}_{(X+x)}\big\|L_{[D^{[n]}]} - \big(L_{[Q_{(\varrho_D)^m}]}\big)_{[D^{[n-m]}]}\big\|$$
$$\leqq 2 \sup_{x\in R^s} \int \big(\varrho_{V'_{n,m}}(X + x - a) + \varrho_{V''_{n,m}}(X + x - a)\big)\, \varrho_L(\mathrm{d}a)$$
$$= 2 \sup_{x\in R^s} \int \varrho_L(X + x - a)\, (\varrho_{V'_{n,m}} + \varrho_{V''_{n,m}})\, (\mathrm{d}a)$$
$$\leqq 4\varepsilon \sup_{a\in R^s} \varrho_L(X - a)$$

for $n \geqq n_{\varepsilon,m}$. Accordingly, taking into account that $(P_{l\mu})_{[Q_\nu]} = P_{l\mu}$ and setting $P = \int P_{l\mu}(.)\, \sigma_L(\mathrm{d}l)$ we also obtain

$$\sup_{x\in R^s} {}_{(X+x)}\|P_{[D^{[n]}]} - P_{[D^{[n-m]}]}\|$$
$$\leqq 4\varepsilon \sup_{a\in R^s} \varrho_P(X - a) \leqq 4\varepsilon c \mu(X)$$

for $n \geqq n_{\varepsilon,m}$. Hence

$$\overline{\lim_{n\to\infty}} \sup_{x\in R^s} {}_{(X+x)}\|L_{[D^{[n]}]} - P_{[D^{[n]}]}\| \leqq 4\varepsilon \Big(c\, \mu(X) + \sup_{a\in R^s} \varrho_L(X - a)\Big)$$
$$+ \overline{\lim_{n\to\infty}} \sup_{x\in R^s} {}_{(X+x)}\big\|\big(L_{[Q_{(\varrho_D)^m}]}\big)_{[D^{[n-m]}]} - P_{[D^{[n-m]}]}\big\|$$
$$\leqq 4\varepsilon \Big(c\, \mu(X) + \sup_{a\in R^s} \varrho_L(X - a)\Big) + \sup_{x\in R^s, G\in \boldsymbol{D}} {}_{(X+x)}\big\|\big(L_{[Q_{(\varrho_D)^m}]}\big)_{[G]} - P_{[G]}\big\|.$$

This estimate is valid for all m. By 11.10.6. and 11.8.6. the second term tends towards zero as $m \to \infty$, and because $\varepsilon > 0$ was arbitrary, it follows that

$$\overline{\lim_{n\to\infty}} \sup_{x\in R^s} {}_{(X+x)}\left\| L_{[D^{[n]}]} - \left(\int P_{l\mu}(.)\, \sigma_L(\mathrm{d}l)\right)_{[D^{[n]}]}\right\| = 0.$$

Just as in the last step in the proof of 11.8.6. we observe that the assertion of 12.3.2. is generally valid if it holds for all L in $\bigcup_{c>0} \mathbf{S}^{(c)}$. ∎

A distribution D in $\boldsymbol{D}$ is said to be *non-lattice* if its intensity measure ϱ_D has this property, i.e. if the inequality

$$D\left(\chi\left(R^s \setminus (x_H + H)\right) > 0\right) > 0$$

is satisfied for all closed proper subgroups H of R^s and all x_H in R^s. Then in view of 11.10.4. the sequence $\left((\varrho_D)^n\right)$ is weakly asymptotically uniformly distributed, and using 11.9.6. we get

12.3.3. *For each L in $\mathbf{S}$, each non-lattice distribution D in $\boldsymbol{D}$, and each finite sequence $X_1, \ldots, X_k$ of bounded sets in $\mathfrak{R}^s$ having the property that $\mu(\partial X_1) = \cdots = \mu(\partial X_k) = 0$,*

$$\lim_{n\to\infty} \sup_{x\in R^s} \left\| (L_{[D^{[n]}]})_{X_1+x,\ldots,X_k+x} - \left(\left(\int P_{l\mu}(.)\, \sigma_L(\mathrm{d}l)\right)_{[D^{[n]}]}\right)_{X_1,\ldots,X_k}\right\| = 0.$$

Proof. As in the last step of the proof of 11.8.6. we see that it can be additionally assumed that L be in $\mathbf{S}^{(c)}$, $c > 0$.

Now let m be an arbitrary natural number. Setting $X = X_1 \cup \cdots \cup X_k$, $P = \int P_{l\mu}(.)\, \sigma_L(\mathrm{d}l)$, we obtain

$$\overline{\lim_{n\to\infty}} \sup_{x\in R^s} \left\|(L_{[D^{[n]}]})_{X_1+x,\ldots,X_k+x} - (P_{[D^{[n]}]})_{X_1,\ldots,X_k}\right\|$$

$$\leqq \overline{\lim_{n\to\infty}} \sup_{x\in R^s} \left\| (L_{[D^{[n]}]})_{X_1+x,\ldots,X_k+x} - \left((L_{[Q_{(\varrho_D)^m}]})_{[D^{[n-m]}]}\right)_{X_1+x,\ldots,X_k+x}\right\|$$

$$+ \overline{\lim_{n\to\infty}} \sup_{x\in R^s} \left\|\left((L_{[Q_{(\varrho_D)^m}]})_{[D^{[n-m]}]}\right)_{X_1+x,\ldots,X_k+x} - (P_{[D^{[n-m]}]})_{X_1,\ldots,X_k}\right\|$$

$$+ \overline{\lim_{n\to\infty}} \left\|(P_{D^{[n]}})_{X_1,\ldots,X_k} - (P_{[D^{[n-m]}]})_{X_1,\ldots,X_k}\right\|$$

$$\leqq \overline{\lim_{n\to\infty}} \sup_{x\in R^s} {}_{(X+x)}\left\|L_{[D^{[n]}]} - \left(L_{[Q_{(\varrho_D)^m}]}\right)_{[D^{[n-m]}]}\right\|$$

$$+ \sup_{x\in R^s, G\in \boldsymbol{D}} \left\|\left((L_{[Q_{(\varrho_D)^m}]})_{[G]}\right)_{X_1+x,\ldots,X_k+x} - (P_{[G]})_{X_1,\ldots,X_k}\right\|$$

$$+ \overline{\lim_{n\to\infty}} {}_X\left\|P_{[D^{[n]}]} - P_{[D^{[n-m]}]}\right\|.$$

Using 11.10.4. and 11.9.6. we see that the middle expression tends towards zero as $m \to \infty$. To a given $\varepsilon > 0$, the other terms can be estimated just as in the proof of 12.3.2. from which our statement follows. ∎

Now, if instead of 11.9.6. we use 11.9.7., the method of proof applied above can be transferred, leading us to

12.3.4. *If any L in $\boldsymbol{S}$ satisfies the condition $\varrho_L \leqq v\mu$ for a positive number v, then, for all non-lattice distributions D in $\boldsymbol{D}$ and all finite sequences $X_1, \ldots, X_k$ of bounded sets in $\mathfrak{R}^s$,*

$$\lim_{n\to\infty} \sup_{x\in R^s} \left\| (L_{[D^{[n]}]})_{X_1+x,\ldots,X_k+x} - \left(\left(\int P_{l\mu}(.)\, \sigma_L(\mathrm{d}l)\right)_{[D^{[n]}]}\right)_{X_1,\ldots,X_k} \right\| = 0 .$$

12.4. Convergence Theorems in the Critical Case

Let $\boldsymbol{C}$ denote the set of those distributions in $\boldsymbol{D}$ which are stable and non-lattice. For such D the distributions $]D[_{(l)}, l \geqq 0$, can be characterized in the following way.

12.4.1. Theorem. *For all D in $\boldsymbol{C}$ and all $l \geqq 0$, a stationary distribution P in $\boldsymbol{P}$ of intensity l coincides with $]D[_{(l)}$ if and only if it is ergodic and cluster-invariant with respect to $[D]$.*

Proof. 1. Let P be any stationary distribution with the properties

$$P_{[D]} = P, \qquad \sigma_P = \delta_l .$$

By 12.3.3. we obtain

$$\lim_{n\to\infty} \left\| P_{X_1,\ldots,X_m} - \left((P_{l\mu})_{[D^{[n]}]}\right)_{X_1,\ldots,X_m} \right\| = 0$$

for all finite sequences $X_1, \ldots, X_m$ of bounded sets in $\mathfrak{R}^s$ satisfying $\mu(\partial X_1) = \cdots = \mu(\partial X_m) = 0$, so that by means of 3.1.9. and 12.1.4. it can be concluded that $P =]D[_{(l)}$.

2. According to our assumptions there exists a stationary distribution L on $\mathfrak{M}$ which has the properties $L_{[D]} = L, 0 < i_L < +\infty$.

Just as it was done in the second step of the proof of 9.2.6. we now represent L as a mixture of conditional distributions with respect to the random variable $s(\Phi)$, thus obtaining

$$L = \int L_{(l)}(.)\, \sigma_L(\mathrm{d}l) .$$

By definition of the $L_{(l)}, l \geqq 0$, the equation $L_{(l)}\big(s(\Phi) = l\big) = 1$ holds almost everywhere with respect to σ_L, and almost all of the distributions $L_{(l)}$ are stationary.

Now let $L'_{(.)}$ be an arbitrary measurable mapping of $[0, +\infty)$ into $[\boldsymbol{P}, \mathfrak{P}]$ which has the following properties:

a) $L'_{(l)}\big(s(\Phi) = l\big) = 1$ almost everywhere with respect to σ_L,

b) $L = \int L'_{(l)}(.)\, \sigma_L(\mathrm{d}l)$.

We show that subject to these assumptions, $L'_{(.)}$ is a version of the conditional distribution of L with respect to $s(\Phi)$.

Let Y be a set in $\mathfrak{M}$, and Z a Borel subset of $[0, +\infty)$. We get

$$L\big(\Phi \in Y, s(\Phi) \in Z\big) = \int L'_{(l)}\big(\Phi \in Y, s(\Phi) \in Z\big)\, \sigma_L(\mathrm{d}l) = \int_Z L'_{(l)}(Y)\, \sigma_L(\mathrm{d}l),$$

which proves our assertion.

3. Let us now return to the starting point of 2. We have

$$L = L_{[D]} = \int (L_{(l)})_{[D]}\,(.)\, \sigma_L(\mathrm{d}l).$$

By 11.7.5., 11.7.8. and 2. it follows that $(L_{(l)})_{[D]} = L_{(l)}$ almost everywhere with respect to σ_L.

Due to $i_L > 0$, the probability $\sigma_L\big((0, +\infty)\big)$ is positive. Thus, there exists some $l_0 > 0$ for which $L_{(l_0)}$ is stationary and satisfies the conditions $(L_{(l_0)})_{[D]} = L_{(l_0)}$, $\sigma_{L_{(l_0)}} = \delta_{l_0}$. In virtue of 1., $L_{(l_0)}$ coincides with $]D[_{(l_0)}$. Hence

$$]D[_{(l_0)}\big(s(\Phi) = l_0\big) = 1.$$

In view of 6.4.10., the stationary infinitely divisible distribution $]D[_{(l_0)}$ is ergodic. By 6.4.9., we obtain

$$l_0 \widetilde{]D[_{(1)}}\big(s(\Psi) \neq 0\big) = \widetilde{]D[_{(l_0)}}\big(s(\Psi) \neq 0\big) = 0.$$

Consequently, for all $l \geqq 0$,

$$\widetilde{]D[_{(l)}}\big(s(\Psi) \neq 0\big) = l\, \widetilde{]D[_{(1)}}\big(s(\Psi) \neq 0\big) = 0.$$

Using anew 6.4.9. we derive the ergodicity of $]D[_{(l)}$ for all $l \geqq 0$. ∎

The following example shows that there exists a sequence (D_n) of distributions in $\boldsymbol{D}$ such that $(P_\mu)_{[D_n]}$ converges weakly towards a non-ergodic stationary distribution of intensity one.

Let $s = 1$ and $D_n = (1 - n^{-1})\, \delta_o + n^{-1} P_{\mu((.)\cap(0,n))}$ for $n = 1, 2, \ldots$ We have

$$\begin{aligned}
\Big(\widetilde{(P_\mu)_{[D_n]}}\Big)^0 &= \langle D_n \rangle = n^{-1} \sum_{l=1}^{\infty} e^{-n} (l!)^{-1} \langle (Q_{\mu((.)\cap(0,n))})^l \rangle \\
&= n^{-1} e^{-n} \sum_{l=1}^{\infty} (l!)^{-1}\, l \hat{\Lambda}\big((Q_{\mu((.)\cap(0,n))})^l\big) \\
&= n^{-1} e^{-n} \sum_{l=1}^{\infty} \big((l-1)!\big)^{-1} \delta_{\delta_0} * \int_0^n (Q_{\mu((.)\cap(-x,n-x))})^{l-1}\,(.)\, \mu(\mathrm{d}x) \\
&= \delta_{\delta_0} * n^{-1} \int_0^n P_{\mu((.)\cap(-x,n-x))}(.)\, \mu(\mathrm{d}x) \\
&\underset{n\to\infty}{\Rightarrow} \delta_{\delta_0} * P_\mu = (P_\mu)^0 = \Big(\widetilde{\mathscr{U}_{P_\mu}}\Big)^0.
\end{aligned}$$

Hence, by 10.3.8.,

$$(P_\mu)_{[D_n]} \underset{n\to\infty}{\Rightarrow} \mathscr{U}_{P_\mu}.$$

In virtue of 2.4.2. to each distribution σ of a random non-negative number and each D in $\mathbf{C}$ we can associate the stationary distribution $P = \int]D[_{(l)}(.)\, \sigma(dl)$. Since

$$P_{[D]} = \int (]D[_{(l)})_{[D]}(.)\, \sigma(dl) = \int]D[_{(l)}(.)\, \sigma(dl) = P,$$

P is cluster-invariant with respect to $[D]$. By 12.4.1., for all Z in $\mathfrak{R}^1$ we obtain

$$P\big(s(\Phi) \in Z\big) = \int]D[_{(l)}\big(s(\Phi) \in Z\big)\, \sigma(dl) = \int k_Z(l)\, \sigma(dl)\,,$$

i.e. $\sigma_P = \sigma$. In view of 12.3.3., we obtain for any L in $\mathbf{S}$ with the property $\sigma_L = \sigma$

$$\begin{aligned}
&\overline{\lim_{n\to\infty}} \sup_{x\in R^s} \|(L_{[D^{[n]}]})_{X_1+x,\dots,X_m+x} - P_{X_1,\dots,X_m}\| \\
&\leqq \overline{\lim_{n\to\infty}} \sup_{x\in R^s} \left\|(L_{[D^{[n]}]})_{X_1+x,\dots,X_m+x} - \Big(\big(\textstyle\int P_{l\mu}(.)\, \sigma(dl)\big)_{[D^{[n]}]}\Big)_{X_1,\dots,X_m}\right\| \\
&\quad + \overline{\lim_{n\to\infty}} \left\|P_{X_1,\dots,X_m} - \Big(\big(\textstyle\int P_{l\mu}(.)\, \sigma(dl)\big)_{[D^{[n]}]}\Big)_{X_1,\dots,X_m}\right\| \\
&= 0
\end{aligned}$$

for all finite sequences $X_1, \dots, X_m$ of bounded sets in $\mathfrak{R}^s$ satisfying $\mu(\partial X_1) = \cdots = \mu(\partial X_m) = 0$. Thus, we observe the truth of (cf. Debes, Kerstan, Liemant & Matthes [1])

12.4.2. Theorem. *For all L in $\mathbf{S}$, all D in $\mathbf{C}$, and all finite sequences $X_1, \dots, X_m$ of bounded sets in $\mathfrak{R}^s$ having the property $\mu(\partial X_1) = \cdots = \mu(\partial X_m) = 0$,*

$$\lim_{n\to\infty} \sup_{x\in R^s} \left\|(L_{[D^{[n]}]})_{X_1+x,\dots,X_m+x} - \big(\textstyle\int]D[_{(l)}(.)\, \sigma_L(dl)\big)_{X_1,\dots,X_m}\right\| = 0\,.$$

By means of 3.1.9., the following theorem is readily derived from that one above.

12.4.3. Theorem. *For all L in $\mathbf{S}$ and all D in $\mathbf{C}$,*

$$L_{[D^{[n]}]} \underset{n\to\infty}{\Rightarrow} \int]D[_{(l)}(.)\, \sigma_L(dl)\,.$$

Using now 12.3.4. instead of 12.3.3. we get (cf. Kerstan, Matthes & Prehn [1])

12.4.4. Theorem. *If any L in $\mathbf{S}$ satisfies the condition $\varrho_L \leqq v\mu$ for some $v \geqq 0$, then, for all D in $\mathbf{C}$ and all finite sequences $X_1, \dots, X_m$ of bounded sets in $\mathfrak{R}^s$,*

$$\lim_{n\to\infty} \sup_{x\in R^s} \left\|(L_{[D^{[n]}]})_{X_1+x,\dots,X_m+x} - \big(\textstyle\int]D[_{(l)}(.)\, \sigma_L(dl)\big)_{X_1,\dots,X_m}\right\| = 0\,.$$

If the sequence $\big((\varrho_D)^n\big)$ is asymptotically uniformly distributed, we may use 12.3.2. to obtain (cf. Debes, Kerstan, Liemant & Matthes [1])

12.4.5. Theorem. *Let L be in $\mathbf{S}$, D in $\mathbf{C}$, and X a bounded set in $\mathfrak{R}^s$. If at least one convolution power of ϱ_D has a non-vanishing part which is absolutely continuous with respect to μ, then*

$$\lim_{n\to\infty} \sup_{x\in R^s} {}_{(X+x)}\left\| L_{[D^{[n]}]} - \int]D[_{(l)}\,(.)\,\sigma_L(\mathrm{d}l)\right\| = 0\,.$$

Because of 12.4.3. for each D in $\mathbf{C}$ the correspondence

$$\sigma \curvearrowright P = \int]D[_{(l)}\,(.)\,\sigma(\mathrm{d}l)$$

transforms the set of all distributions σ of random non-negative numbers onto the set of all distributions P in $\mathbf{S}$ which are cluster-invariant with respect to $[D]$. This mapping is one-to-one, for we have $\sigma_P = \sigma$. Thus, all distributions in $\mathbf{S}$ which are cluster-invariant with respect to $[D]$ are stationary. It can be shown that conversely each stationary distribution on $\mathfrak{M}$ which is cluster-invariant with respect to $[D]$ belongs to $\mathbf{S}$.

12.4.6. Theorem. *For all D in $\mathbf{C}$ the correspondence*

$$\sigma \curvearrowright \int]D[_{(l)}\,(.)\,\sigma(\mathrm{d}l)$$

provides a one-to-one mapping of the set of all distributions σ of random non-negative numbers onto the set of all stationary distributions on $\mathfrak{M}$ which are cluster-invariant with respect to $[D]$.

Proof. We can literally repeat the proof of 11.10.5. if we replace $P_{l\mu}$ by $]D[_{(l)}$ and use $]D[_{(l_1)} *]D[_{(l_2)} =]D[_{(l_1+l_2)}$ instead of $P_{l_1\mu} * P_{l_2\mu} = P_{(l_1+l_2)\mu}$. ∎

In Section 11.10. it was shown by a counter-example that in 11.10.5. and therefore also in 12.4.6. the assumption of stationarity cannot be dropped. The cluster-invariant, non-stationary distribution occurring in this counter-example is not of bounded intensity. This is not by chance, for we have

12.4.7. Proposition. *If D belongs to $\mathbf{D}$, then each distribution P on $\mathfrak{M}$ which is cluster-invariant with respect to $[D]$ and is of bounded intensity is stationary.*

Proof. Given any distribution ν on $\mathfrak{R}^s$ which is absolutely continuous with respect to μ and a bounded set X in $\mathfrak{R}^s$. For all natural numbers m, n with $m < n$, we obtain

$$\overline{\lim_{n\to\infty}}\, {}_X\left\|P - (P_{[Q_\nu]})_{[D^{[n]}]}\right\| = \overline{\lim_{n\to\infty}}\, {}_X\left\|P_{[D^{[n]}]} - (P_{[Q_\nu]})_{[D^{[n]}]}\right\|$$

$$\leqq \overline{\lim_{n\to\infty}}\, {}_X\left\|P_{[D^{[n]}]} - \left(P_{[Q_{(\varrho_D)^m}]}\right)_{[D^{[n-m]}]}\right\|$$

$$+ \overline{\lim_{n\to\infty}}\, {}_X\left\|\left(P_{[Q_{(\varrho_D)^m}]}\right)_{[D^{[n-m]}]} - \left(P_{[Q_{\nu*(\varrho_D)^m}]}\right)_{[D^{[n-m]}]}\right\|$$

$$+ \overline{\lim_{n\to\infty}}\, {}_X\left\|(P_{[Q_\nu]})_{[D^{[n]}]} - \left((P_{[Q_\nu]})_{[Q_{(\varrho_D)^m}]}\right)_{[D^{[n-m]}]}\right\|\cdot$$

Using now the notations employed in 12.3.1. we obtain, by 4.2.8.

$$\overline{\lim_{n\to\infty}}\ {}_X\left\|P_{[D^{[n]}]} - \left(P_{[Q_{(\varrho_D)^m}]}\right)_{[D^{[n-m]}]}\right\|$$

$$\leqq 2 \overline{\lim_{n\to\infty}} \int \left(\varrho_{V'_{n,m}}(X-x) + \varrho_{V''_{n,m}}(X-x)\right) \varrho_P(\mathrm{d}x)$$

$$= 2 \overline{\lim_{n\to\infty}} \int \varrho_P(X-x)\left(\varrho_{V'_{n,m}} + \varrho_{V''_{n,m}}\right)(\mathrm{d}x)$$

$$\leqq 2\left(\overline{\lim_{n\to\infty}}\left(\varrho_{V'_{n,m}}(R^s) + \varrho_{V''_{n,m}}(R^s)\right)\right) \sup_{x\in R^s} \varrho_P(X-x) = 0.$$

In the same way we get

$$\overline{\lim_{n\to\infty}}\ {}_X\left\|(P_{[Q_\nu]})_{[D^{[n]}]} - \left((P_{[Q_\nu]})_{[Q_{(\varrho_D)^m}]}\right)_{[D^{[n-m]}]}\right\|$$

$$\leqq 2\left(\overline{\lim_{n\to\infty}}\left(\varrho_{V'_{n,m}}(R^s) + \varrho_{V''_{n,m}}(R^s)\right)\right) \sup_{x\in R^s} (\varrho_P * \nu)(X-x)$$

$$\leqq 2\left(\overline{\lim_{n\to\infty}}\left(\varrho_{V'_{n,m}}(R^s) + \varrho_{V''_{n,m}}(R^s)\right)\right) \sup_{x\in R^s} \varrho_P(X-x) = 0.$$

Hence,

$$\overline{\lim_{n\to\infty}}\ {}_X\|P - (P_{[Q_\nu]})_{[D^{[n]}]}\|$$

$$\leqq \overline{\lim_{m\to\infty}} \sup_{W\in\boldsymbol{D}}\ {}_X\left\|\left(P_{[Q_{(\varrho_D)^m}]}\right)_{[W]} - \left(P_{[Q_{\nu*(\varrho_D)^m}]}\right)_{[W]}\right\|.$$

Setting

$$\gamma_m = (\varrho_D)^m \wedge (\nu * (\varrho_D)^m), \qquad \gamma'_m = (\varrho_D)^m - \gamma_m, \qquad \gamma''_m = \nu * (\varrho_D)^m - \gamma_m$$

we have

$$\gamma_m(R^s) = 1 - 2^{-1} \operatorname{Var}\left((\varrho_D)^m - \nu * (\varrho_D)^m\right).$$

Since the sequence $((\varrho_D)^m)$ is weakly uniformly distributed, we may conclude that

$$\lim_{m\to\infty} \gamma'_m(R^s) = 0 = \lim_{m\to\infty} \gamma''_m(R^s).$$

Thus, by means of 4.2.8. we get

$$\overline{\lim_{m\to\infty}} \sup_{W\in\boldsymbol{D}}\ {}_X\left\|\left(P_{[Q_{(\varrho_D)^m}]}\right)_{[W]} - \left(P_{[Q_{\nu*(\varrho_D)^m}]}\right)_{[W]}\right\|$$

$$\leqq 2 \overline{\lim_{m\to\infty}} \sup_{W\in\boldsymbol{D}} \int \left(\varrho_{(Q_{\gamma'_m})_{[W]}}(X-x) + \varrho_{(Q_{\gamma''_m})_{[W]}}(X-x)\right) \varrho_P(\mathrm{d}x)$$

$$\leqq 2 \overline{\lim_{m\to\infty}} \sup_{W\in\boldsymbol{D}} \left(\varrho_{(Q_{\gamma'_m})_{[W]}}(R^s) + \varrho_{(Q_{\gamma''_m})_{[W]}}(R^s)\right) \sup_{x\in R^s} \varrho_P(X-x)$$

$$= 2\left(\overline{\lim_{m\to\infty}}\left(\gamma'_m(R^s) + \gamma''_m(R^s)\right)\right) \sup_{x\in R^s} \varrho_P(X-x),$$

since

$$(Q_{(\varrho_D)^m})_{[W]} = (Q_{\gamma_m})_{[W]} + (Q_{\gamma'_m})_{[W]}, \qquad (Q_{\nu*(\varrho_D)^m})_{[W]} = (Q_{\gamma_m})_{[W]} + (Q_{\gamma''_m})_{[W]}.$$

Consequently, for all bounded X in $\mathfrak{R}^s$,

$$_X\|P - (P_{[Q_\nu]})_{[D^{[n]}]}\| \xrightarrow[n\to\infty]{} 0$$

and hence

$$(P_{[Q_\nu]})_{[D^{[n]}]} \underset{n\to\infty}{\Rightarrow} P.$$

In view of the homogeneity of $[D^{[n]}] \circ [Q_\nu]$, it follows that

$$\left(P_{[Q_{\delta_x * \nu}]}\right)_{[D^{[n]}]} = \left((P_{[Q_\nu]})_{[D^{[n]}]}\right)_{[Q_{\delta_x}]} \underset{n\to\infty}{\Rightarrow} P_{[Q_{\delta_x}]}$$

for all x in R^s. On the other hand, also $\delta_x * \nu$ is absolutely continuous with respect to μ, hence

$$\left(P_{[Q_{\delta_x * \nu}]}\right)_{[D^{[n]}]} \underset{n\to\infty}{\Rightarrow} P.$$

Consequently, we always have $P = P_{[Q_{\delta_x}]}$, i.e. P is stationary. ∎

Obviously, the convergence assertion of 12.1.7. remains valid if the assumption $i_L < +\infty$ is replaced by $L(s(\Phi) < +\infty) = 1$. Consequently, by 12.3.3. we get the following counterpart to 12.4.3.

12.4.8. Proposition. *Let D be any distribution in $\boldsymbol{D}$ which is non-lattice and not stable. Then, for all L in* $\mathbf{S}$,

$$L_{[D^{[n]}]} \underset{n\to\infty}{\Rightarrow} \delta_o.$$

12.5. A Convergence Theorem in the Subcritical Case

For the sake of abbreviation we set

$$c(D) = \int \chi(R^s)\, D(\mathrm{d}\chi) \qquad (D \in \mathbf{P}).$$

A distribution D on $\mathfrak{M}$ is called *subcritical* if the corresponding Galton-Watson process has this property, i.e. if $0 < c(D) < 1$ holds.

Immediately from the theory of subcritical Galton-Watson processes (cf. for instance Athreya & Ney [1], Sections 1.8. and 1.11.) we obtain the following two propositions.

12.5.1. Proposition. *For all subcritical distributions D on $\mathfrak{M}$, the limits*

$$p_l{}^*(D) = \lim_{n\to\infty} D^{(n)}\left(\chi(R^s) = l\right) \qquad (l = 1, 2, \ldots)$$

exist and we have

$$\sum_{l=1}^{\infty} p_l{}^*(D) = 1.$$

For all subcritical distributions D on $\mathfrak{M}$ we set

$$v(D) = \left(\sum_{l=1}^{\infty} l p_l{}^*(D)\right)^{-1}.$$

Thus $0 \leq v(D) < +\infty$.

12.5.2. Proposition. *For all subcritical distributions D on $\mathfrak{M}$ we have*

$$z_n(D)\,\big(c(D)\big)^{-n} \xrightarrow[n\to\infty]{} v(D).$$

Using 12.5.1. it can be shown (cf. Fleischmann & Prehn [1], Zubkov [1]) that the conditional distribution of the source time tends to a limit for $n \to \infty$.

12.5.3. Proposition. *For all subcritical distributions D on $\mathfrak{M}$, the limits*

$$q_k^*(D) = \lim_{n\to\infty} q_k^n(D) \qquad (k = 0, 1, \ldots)$$

exist and we have

$$\sum_{k=0}^{\infty} q_k^*(D) = 1.$$

Proof. For $n = 0, 1, \ldots$ and $-1 \leqq z \leqq 1$ we set

$$f_n(z) = \sum_{l=0}^{\infty} z^l D^{[n]}\big(\chi(R^s) = l\big), \qquad f^*(z) = \sum_{l=1}^{\infty} z^l p_l^*(D)$$

and have (cf. for instance Athreya & Ney [1], Theorem 1.8.1.)

$$(*) \qquad f^*\big(f_n(z)\big) = \big(c(D)\big)^n f^*(z) + \Big(1 - \big(c(D)\big)^n\Big).$$

From $\lim_{n\to\infty} f_n(0) = 1$ it follows for $k = 0, 1, \ldots$

$$\big(1 - f_{n-k}(0)\big)^{-1} \Big(1 - f_k\big(f_{n-k}(0)\big)\Big) \xrightarrow[n\to\infty]{} \big(c(D)\big)^k,$$

i.e.

$$(**) \qquad \lim_{n\to\infty} \big(z_{n-k}(D)\big)^{-1} z_n(D) = \big(c(D)\big)^k.$$

By 12.2.1. we get

$$\begin{aligned}
\mathscr{P}_{[0,\ldots,0]}\big(m(\Psi_n) \leqq k\big) &= \tau_{n,k}(R^s)\\
&= z_k(D) \int_{\{\chi:\,\chi\neq o\}} \chi(R^s)\,\big(f_k(0)\big)^{\chi(R^s)-1} D^{[n-k]}(\mathrm{d}\chi)\\
&= z_k(D)\, f'_{n-k}\big(f_k(0)\big).
\end{aligned}$$

Now 12.5.1. yields

$$(***) \qquad \lim_{n\to\infty} \big(z_n(D)\big)^{-1} f_n'(z) = (f^*)'\,(z) \qquad (-1 < z < 1),$$

and together with $(**)$ we get

$$\begin{aligned}
\sum_{i=0}^{k} q_i^n(D) &= \mathscr{P}_{[0,\ldots,0]}\big(m(\Psi_n) \leqq k \mid \Psi_n \neq o\big)\\
&= z_k(D)\,\big(z_n(D)\big)^{-1} z_{n-k}(D)\big(z_{n-k}(D)\big)^{-1} f'_{n-k}\big(f_k(0)\big)\\
&\xrightarrow[n\to\infty]{} z_k(D)\,\big(c(D)\big)^{-k}\,(f^*)'\,\big(f_k(0)\big).
\end{aligned}$$

Consequently, the limits in 12.5.3. exist.

We differentiate $(*)$ at $z = f_1(0)$ and get

$$(f^*)'\left(f_{n+1}(0)\right) f_n'\left(f_1(0)\right) = \left(c(D)\right)^n (f^*)'\left(f_1(0)\right).$$

By $(**)$ and $(***)$ this yields

$$\sum_{i=0}^{\infty} q_i^*(D) = \lim_{n\to\infty} z_{n+1}(D) \left(c(D)\right)^{-(n+1)} (f^*)'\left(f_{n+1}(0)\right)$$
$$= \lim_{n\to\infty} z_{n+1}(D) \left(z_n(D)\right)^{-1} \left(c(D)\right)^{-1} z_n(D) \left(f_n'\left(f_1(0)\right)\right)^{-1} (f^*)'\left(f_1(0)\right) = 1. \blacksquare$$

Let D be any subcritical distribution on $\mathfrak{M}$. For all $n > k > 0$ we set

$$D^{n,k} = \left(1 - z_k(D)\left(c(D)\right)^{n-k}\right)\delta_o + z_k(D) \int [D^{(k)}]_{(x)}(.)\,(\varrho_D)^{n-k}(\mathrm{d}x).$$

Obviously, $D^{n,k}$ is a distribution on $\mathfrak{M}$ with the property

$$c(D^{n,k}) = \left(c(D)\right)^n = c(D^{[n]}).$$

It turns out that the distributions $D^{n,k}$ are in the following sense asymptotically equivalent to the cluster powers $D^{[n]}$ (cf. Fleischmann & Prehn [3]).

12.5.4. *To each subcritical D in $\boldsymbol{P}$ satisfying $v(D) > 0$ there exists a mapping $[n, k] \curvearrowright [U_{n,k}, U'_{n,k}, U''_{n,k}]$ of $\{[n, k] : n > k > 0\}$ into $\boldsymbol{E}^+ \times \boldsymbol{E}^+ \times \boldsymbol{E}^+$ which satisfies the conditions*

$$D^{[n]} = U_{n,k} + U'_{n,k}, \qquad D^{n,k} = U_{n,k} + U''_{n,k},$$

$$\lim_{k\to\infty}\lim_{n\to\infty} \left(c(D)\right)^{-n} \varrho_{U'_{n,k}}(R^s) = 0 = \lim_{k\to\infty}\lim_{n\to\infty} \left(c(D)\right)^{-n} \varrho_{U''_{n,k}}(R^s).$$

Proof. 1. Let $n > k > 0$. Then by 12.2.1. and 12.2.2.

$$D^{n,k} \geqq \int [D^{(k)}]_{(x)}(.)\,\tau_{n,k}(\mathrm{d}x) = (Q_{\tau_{n,k}})_{[D^{(k)}]},$$
$$D^{[n]} \geqq (Q_{\tau_{n,k}})_{[D^{(k)}]}.$$

Thus, setting $U_{n,k} = (Q_{\tau_{n,k}})_{[D^{(k)}]}$, we obtain

$$D^{[n]} = U_{n,k} + U'_{n,k}, \qquad D^{n,k} = U_{n,k} + U''_{n,k},$$

where $U_{n,k}, U'_{n,k}, U''_{n,k}$ belong to $\boldsymbol{E}^+$.

2. In view of 12.5.2. and 12.5.3. we obtain

$$\left(c(D)\right)^{-n} \varrho_{U_{n,k}}(R^s) = \left(c(D)\right)^{-n} \tau_{n,k}(R^s)\, c(D^{(k)})$$
$$= \left(c(D)\right)^{-n} z_n(D)\, \mathscr{P}_{[0,\ldots,0]}\left(m(\Psi_n) \leqq k \mid \Psi_n \neq o\right) \left(z_k(D)\right)^{-1} \left(c(D)\right)^k$$
$$\xrightarrow[n\to\infty]{} \left(v(D) \sum_{j=0}^{k} q_j^*(D)\right) \left(\left(c(D)\right)^{-k} z_k(D)\right)^{-1} \xrightarrow[k\to\infty]{} 1.$$

Thus,

$$\lim_{k\to\infty}\lim_{n\to\infty} \left(c(D)\right)^{-n} \varrho_{U'_{n,k}}(R^s) = 0 = \lim_{k\to\infty}\lim_{n\to\infty} \left(c(D)\right)^{-n} \varrho_{U''_{n,k}}(R^s). \blacksquare$$

For all subcritical distributions D on $\mathfrak{M}$ we set

$$D_* = \sum_{l=1}^{\infty} p_l^*(D)\, \delta_{l\delta_{[0,\ldots,0]}}.$$

Using the contraction operator $\mathscr{K}_c$ introduced in Section 7.3. we obtain the following convergence theorem (cf. Fleischmann & Prehn [2] and Fleischmann [3]).

12.5.5. Theorem. *For all subcritical distributions D on $\mathfrak{M}$, all distributions L in $\mathbf{S}$ and all finite sequences $X_1, \ldots, X_m$ of bounded sets in $\mathfrak{R}^s$ with the property $\mu(\partial X_1) = \cdots = \mu(\partial X_m) = 0$ we have*

$$\lim_{n\to\infty} \sup_{x\in R^s} \Big\| \big(\mathscr{K}_{(c(D))^n}(L_{[D^{[n]}]})\big)_{X_1+x,\ldots,X_m+x} - \Big(\big(\textstyle\int P_{lv(D)\mu}(.)\, \sigma_L(\mathrm{d}l)\big)_{[D_*]}\Big)_{X_1,\ldots,X_m}\Big\| = 0.$$

Proof. 1. According to 11.7.8. the distributions $L_{[D^{[n]}]}$, $n = 1, 2, \ldots$, exist.

Just as in the last step of the proof of 11.8.6. we observe that the assertion of 12.5.5. is generally valid if it holds for all L in $\bigcup_{c>0} \mathbf{S}^{(c)}$. Thus, we can additionally assume that L belongs to $\mathbf{S}^{(c)}$ for some $c > 0$.

2. We set $t_n = \big(c(D)\big)^{-n/s}$. Let $v(D) > 0$. By 4.2.8. and 12.5.4. we obtain for all bounded X in $\mathfrak{R}^s$ and all $n > k > 0$

$$\begin{aligned}
&\overline{\lim_{k\to\infty}}\ \overline{\lim_{n\to\infty}} \sup_{x\in R^s} {}_{(X+x)}\big\|\mathscr{K}_{(c(D))^n}(L_{[D^{[n]}]}) - \mathscr{K}_{(c(D))^n}(L_{[D^{n,k}]})\big\| \\
&= \overline{\lim_{k\to\infty}}\ \overline{\lim_{n\to\infty}} \sup_{y\in R^s} {}_{(t_nX+y)}\|L_{[D^{[n]}]} - L_{[D^{n,k}]}\| \\
&\leqq 2\ \overline{\lim_{k\to\infty}}\ \overline{\lim_{n\to\infty}} \sup_{y\in R^s} \int (\varrho_{U'_{n,k}} + \varrho_{U''_{n,k}})\,(t_nX + y - z)\, \varrho_L(\mathrm{d}z) \\
&\leqq 2\ \overline{\lim_{k\to\infty}}\ \overline{\lim_{n\to\infty}} \big(c(D)\big)^{-n} (\varrho_{U'_{n,k}} + \varrho_{U''_{n,k}})\,(R^s) \sup_{z\in R^s} (t_n)^{-s}\, \varrho_L(t_nX - z).
\end{aligned}$$

Let Z be any bounded set in $\mathfrak{R}^s$ satisfying $Z \supseteqq X$, $\mu(\partial Z) = 0$. In view of 12.5.4. and 11.7.7. we can continue as follows:

$$\leqq 2c\mu(Z) \lim_{k\to\infty} \lim_{n\to\infty} \big(c(D)\big)^{-n} (\varrho_{U'_{n,k}} + \varrho_{U''_{n,k}})\,(R^s) = 0.$$

3. We fix a natural number k. For all $n > k$ a substochastic kernel K_n on R^s with phase space R^s is defined by

$$K_n(y, X) = z_k(D)\,(\varrho_D)^{n-k}\,(t_nX - y) \qquad (y \in R^s, X \in \mathfrak{R}^s).$$

Then, for all L in $\mathbf{S}^{(c)}$, $c > 0$, and all finite sequences $X_1, \ldots, X_m$ of bounded sets in $\mathfrak{R}^s$ satisfying $\mu(\partial X_1) = \cdots = \mu(\partial X_m) = 0$ we have

$$\lim_{n\to\infty} \sup_{x\in R^s} \big\|(L_{\varkappa(K_n)})_{X_1+x,\ldots,X_m+x} - \big(\textstyle\int P_{ld_k\mu}(.)\, \sigma_L(\mathrm{d}l)\big)_{X_1,\ldots,X_m}\big\| = 0,$$

where $d_k = z_k(D)\,\big(c(D)\big)^{-k}$.

To prove this we use 7.4.6. Indeed, if we substitute $[A', \varrho_{A'}]$, $[A, \varrho_A]$ by R^s, $\boldsymbol{L}$ by $\{L(T_y\Psi \in (.))\}_{y \in R^s}$, $i_{L(T_y\Psi \in (.)),(\Phi)}$ by $s_L(\Phi)\, d_k\mu$ and $\mathfrak{H}$ by the ring of all bounded sets X in $\mathfrak{R}^s$ having the property $\mu(\partial X) = 0$, the above convergence statement is transformed into the statement of 7.4.6. Therefore, it remains to show only that the assumptions of 7.4.6. are satisfied after the above substitutions.

First of all we have to show that for all natural numbers n and all P in $\boldsymbol{L}$ the measure $\int K_n(x, (.))\, \varrho_P(\mathrm{d}x)$ belongs to N, i.e. that for all bounded X in $\mathfrak{R}^s$ and all y in R^s,

$$z_k(D) \int (\varrho_D)^{n-k}\, (t_n X - x - y)\, \varrho_L(\mathrm{d}x)$$
$$= z_k(D) \int \varrho_L(t_n X - x - y)\, (\varrho_D)^{n-k}\, (\mathrm{d}x) < +\infty.$$

But this follows immediately because L is, according to 11.7.7., of bounded intensity.

Next we have to show that for all H in $\mathfrak{H}$

$$\lim_{n\to\infty} \sup_{P \in \boldsymbol{L}} \int \left| \int K_n(x, H)\, \Phi(\mathrm{d}x) - i_{P,(\Phi)}(H) \right| P(\mathrm{d}\Phi) = 0,$$

i.e. that for all bounded X in $\mathfrak{R}^s$ satisfying $\mu(\partial X) = 0$

$$(*) \quad \lim_{n\to\infty} \sup_{y \in R^s} \int \left| \int z_k(D)\, (\varrho_D)^{n-k}\, (t_n X - x - y)\, \Phi(\mathrm{d}x) - d_k s_L(\Phi)\, \mu(X) \right| L(\mathrm{d}\Phi)$$
$$= 0,$$

i.e.

$$\overline{\lim_{n\to\infty}} \sup_{y \in R^s} \int \left| \int t_n^{-s} \Phi(t_n X - x - y) \left(c(D)\right)^{-(n-k)} (\varrho_D)^{n-k}\, (\mathrm{d}x) - s_L(\Phi)\, \mu(X) \right|$$
$$L(\mathrm{d}\Phi) = 0.$$

This limit superior has the upper estimate

$$\leq \overline{\lim_{n\to\infty}} \sup_{y \in R^s} \int \left(\int |t_n^{-s} \Phi(t_n X - x - y) - s_L(\Phi)\, \mu(X)|\, L(\mathrm{d}\Phi) \right)$$
$$\left((c(D))^{-1}\, \varrho_D \right)^{n-k} (\mathrm{d}x).$$

As was shown in the proof of 11.7.7.,

$$\lim_{n\to\infty} \sup_{z \in R^s} \int |t_n^{-s} \Phi(t_n X - z) - s_L(\Phi)\, \mu(X)|\, L(\mathrm{d}\Phi) = 0.$$

Thus, the upper estimate is equal to zero.

The remaining third condition

$$\lim_{n\to\infty} \sup_{P \in \boldsymbol{L}} \int \left(K_n(x, H)\right)^2 \varrho_P(\mathrm{d}x) = 0 \qquad (H \in \mathfrak{H})$$

means that for all bounded X in $\mathfrak{R}^s$ with $\mu(\partial X) = 0$

$$\lim_{n\to\infty} \sup_{y \in R^s} \int \left(z_k(D)\right)^2 \left((\varrho_D)^{n-k}\, (t_n X - x - y)\right)^2 \varrho_L(\mathrm{d}x) = 0.$$

In virtue of 11.7.7.

$$\lim_{n\to\infty} \sup_{z\in R^s} \left| t_n^{-s}\varrho_L(t_nX - z) - \mu(X) \int l\sigma_L(\mathrm{d}l) \right| = 0,$$

and we conclude that

$$\overline{\lim_{n\to\infty}} \sup_{y\in R^s} \big(z_k(D)\big)^2 \int \big((\varrho_D)^{n-k} (t_nX - x - y)\big)^2 \varrho_L(\mathrm{d}x)$$

$$\leqq \big(z_k(D)\big)^2 \big(c(D)\big)^{-k} \overline{\lim_{n\to\infty}} \sup_{y\in R^s} \int t_n^{-s}(\varrho_D)^{n-k} (t_nX - x - y) \varrho_L(\mathrm{d}x)$$

$$= \big(z_k(D)\big)^2 \big(c(D)\big)^{-k} \overline{\lim_{n\to\infty}} \sup_{y\in R^s} \int t_n^{-s}\varrho_L(t_nX - x - y) (\varrho_D)^{n-k} (\mathrm{d}x)$$

$$\leqq \big(z_k(D)\big)^2 \big(c(D)\big)^{-k} \overline{\lim_{n\to\infty}} c\mu(X) \big(c(D)\big)^{n-k} = 0.$$

4. Immediately from the definition of $D^{n,k}$ and K_n we get

$$\mathscr{K}_{(c(D))^n}(L_{[D^{n,k}]}) = \mathscr{K}_{(c(D))^n}\Big(\Big(\big(\mathscr{D}_{z_k(D)(c(D))^{n-k}}L\big)_{[Q((c(D))^{-1}\varrho_D)^{n-k}]}\Big)_{[D^{(k)}]}\Big)$$

$$= \Big(\mathscr{K}_{(c(D))^n}\Big(\big(\mathscr{D}_{z_k(D)(c(D))^{n-k}}L\big)_{[Q((c(D))^{-1}\varrho_D)^{n-k}]}\Big)\Big)_{[\mathscr{K}_{(c(D))^n}D^{(k)}]}$$

$$= (L_{\varkappa(K_n)})_{[\mathscr{K}_{c((D))^n}D^{(k)}]}.$$

For all natural numbers k we set

$$_*D^{(k)} = \int \delta_{\chi(R^s)\delta_{[0,\ldots,0]}}(.)\, D^{(k)}(\mathrm{d}\chi).$$

Next we show that for all finite sequences $X_1, \ldots, X_m$ of bounded sets in $\mathfrak{R}^s$ with the property $\mu(\partial X_1) = \cdots = \mu(\partial X_m) = 0$,

$$\lim_{n\to\infty} \sup_{x\in R^s} \Big\| \big((L_{\varkappa(K_n)})_{[\mathscr{K}_{(c(D))^n}D^{(k)}]}\big)_{X_1+x,\ldots,X_m+x}$$

$$- \Big(\big(\int P_{ld_k\mu}(.)\, \sigma_L(\mathrm{d}l)\big)_{[_*D^{(k)}]}\Big)_{X_1,\ldots,X_m} \Big\| = 0.$$

First of all for each bounded X in $\mathfrak{R}^s$ satisfying $\mu(\partial X) = 0$ by means of $(*)$ we get

$$\overline{\lim_{n\to\infty}} \sup_{z\in R^s} \Big| \int z_k(D) (\varrho_D)^{n-k} (t_nX - z - y) \varrho_L(\mathrm{d}y) - \int d_k s_L(\Phi) \mu(X) L(\mathrm{d}\Phi) \Big|$$

$$\leqq \overline{\lim_{n\to\infty}} \sup_{z\in R^s} \int \Big| \int z_k(D) (\varrho_D)^{n-k} (t_nX - z - y) \Phi(\mathrm{d}y) - d_k s_L(\Phi) \mu(X) \Big| L(\mathrm{d}\Phi) = 0,$$

i.e.

$$(**) \qquad \lim_{n\to\infty} \sup_{x\in R^s} \Big| \varrho_{L_{\varkappa(K_n)}}(X - x) - d_k\mu(X) \int l\sigma_L(\mathrm{d}l) \Big| = 0.$$

Thus, we have a "uniform convergence" of the intensity measure of $L_{\varkappa(K_n)}$ towards the intensity measure of $\int P_{d_kl\mu}(.)\, \sigma_L(\mathrm{d}l)$.

Let ε be any positive number, Z the set $[-\varepsilon, \varepsilon)^s$, ϱ_n the intensity measure of $(L_{\varkappa(K_n)})_{[(R^s\setminus Z)(\mathscr{K}_{(c(D))^n D^{(k)}})]}$, X any bounded set in $\Re^s$ and X' any bounded set in $\Re^s$ with the properties $X' \supseteqq X$, $\mu(\partial X') = 0$. By (**) we obtain

$$\overline{\lim_{n\to\infty}} \sup_{x\in R^s} \varrho_n(X - x) \leqq \overline{\lim_{n\to\infty}} \sup_{x\in R^s} \varrho_{L_{\varkappa(K_n)}}(X' - x)\, \varrho_{D^{(k)}}(R^s \setminus t_n Z)$$
$$\leqq d_k \mu(X') \int l\sigma_L(\mathrm{d}l)\, \overline{\lim_{n\to\infty}}\, \varrho_{D^{(k)}}(R^s \setminus t_n Z) = 0\,.$$

Thus, in view of 4.2.7., we get

$$_{(X+x)}\left\|(L_{\varkappa(K_n)})_{[\mathscr{K}_{(c(D))^n D^{(k)}}]} - (L_{\varkappa(K_n)})_{[z(\mathscr{K}_{(c(D))^n D^{(k)}})]}\right\|$$
$$\leqq \int {}_{(X+x)}\left\|[\mathscr{K}_{(c(D))^n}D^{(k)}]_{(y)} - \left[z\left(\mathscr{K}_{(c(D))^n}D^{(k)}\right)\right]_{(y)}\right\| \varrho_{L_{\varkappa(K_n)}}(\mathrm{d}y)$$
$$\leqq 2 \int \left[{}_{(R^s\setminus Z)}\left(\mathscr{K}_{(c(D))^n}D^{(k)}\right)\right]_{(y)} \left(\chi(X + x) > 0\right) \varrho_{L_{\varkappa(K_n)}}(\mathrm{d}y)$$
$$\leqq 2 \sup_{x\in R^s} \varrho_n(X + x) \xrightarrow[n\to\infty]{} 0\,,$$

i.e.

$$(+) \qquad \lim_{n\to\infty} \sup_{x\in R^s} {}_{(X+x)}\left\|(L_{\varkappa(K_n)})_{[\mathscr{K}_{(c(D))^n D^{(k)}}]} - (L_{\varkappa(K_n)})_{[z(\mathscr{K}_{(c(D))^n D^{(k)}})]}\right\| = 0\,.$$

Let $X_1, \ldots, X_m$ be a finite sequence of pairwise disjoint bounded sets in $\Re^s$ satisfying $\mu(\partial X_1) = \cdots = \mu(\partial X_m) = 0$. For $i = 1, \ldots, m$ let X_i' resp. X_i'' denote the union of all cubes

$$Z + 2\varepsilon z \qquad (z \in \{\ldots, -1, 0, 1, \ldots\}^s)$$

for which

$$2Z + 2\varepsilon z \subseteqq X \qquad \text{resp.} \qquad (2Z + 2\varepsilon z) \cap X \neq \emptyset$$

holds. We set

$$X' = \bigcup_{i=1}^{m} X_i', \qquad X = \bigcup_{i=1}^{m} X_i, \qquad X'' = \bigcup_{i=1}^{m} X_i'',$$

$${}_{\varepsilon,n}D^{(k)} = \int \delta_{\chi(Z)\delta_{[0,\ldots,0]}}(\cdot)\, (\mathscr{K}_{(c(D))^n}D^{(k)})\, (\mathrm{d}\chi) \qquad (n = 1, 2, \ldots),$$

and obtain

$$\left\|\left((L_{\varkappa(K_n)})_{[z(\mathscr{K}_{(c(D))^n D^{(k)}})]}\right)_{X_1+x,\ldots,X_m+x} - \left((L_{\varkappa(K_n)})_{[{}_{\varepsilon,n}D^{(k)}]}\right)_{X_1+x,\ldots,X_m+x}\right\|$$
$$\leqq \int \Big\|[z(\mathscr{K}_{(c(D))^n}D^{(k)})]_{((X''+x)\Phi)}\big)_{X_1+x,\ldots,X_m+x}$$
$$- \left([{}_{\varepsilon,n}D^{(k)}]_{((X''+x)\Phi)}\right)_{X_1+x,\ldots,X_m+x}\Big\|\, L_{\varkappa(K_n)}(\mathrm{d}\Phi)\,.$$

Taking into consideration

$$\left(\left([z(\mathscr{K}_{(c(D))^n}D^{(k)})]_{((X'+x)\Phi)}\right)_{X_1+x,\ldots,X_m+x} = \left([{}_{\varepsilon,n}D^{(k)}]_{((X'+x)\Phi)}\right)_{X_1+x,\ldots,X_m+x}\,,$$

we are able to continue the above inequality with

$$\leqq \int \Big\|\Big([z(\mathcal{K}_{(c(D))^n}D^{(k)})]_{(((X''\setminus X')+x)\Phi)}\Big)_{X_1+x,\ldots,X_m+x}$$
$$-\Big([_{\varepsilon,n}D^{(k)}]_{(((X''\setminus X')+x)\Phi)}\Big)_{X_1+x,\ldots,X_m+x}\Big\|\, L_{\varkappa(K_n)}(\mathrm{d}\Phi)$$
$$\leqq \int 2\Phi\big((X''\setminus X')+x\big)\, L_{\varkappa(K_n)}(\mathrm{d}\Phi) \leqq 2\sup_{x\in R^s} \varrho_{L_{\varkappa(K_n)}}\big((X''\setminus X')+x\big).$$

However, by (**) the last expression converges towards

$$2d_k\mu(X''\setminus X')\int l\sigma_L(\mathrm{d}l) \leqq (2cd_k)\,\mu(X''\setminus X').$$

Thus,

$$(++)\qquad \lim_{\varepsilon\to 0+0}\ \overline{\lim_{n\to\infty}}\ \sup_{x\in R^s}\Big\|\Big((L_{\varkappa(K_n)})_{[z(\mathcal{K}_{(c(D))^n}D^{(k)})]}\Big)_{X_1+x,\ldots,X_m+x}$$
$$-\Big((L_{\varkappa(K_n)})_{[\varepsilon,nD^{(k)}]}\Big)_{X_1+x,\ldots,X_m+x}\Big\| = 0.$$

Proceeding in the same way as in the proof of (+) we obtain

$$(+++)\qquad \lim_{n\to\infty}\ \sup_{x\in R^s}\ {}_{(X+x)}\|(L_{\varkappa(K_n)})_{[\varepsilon,nD^{(k)}]} - (L_{\varkappa(K_n)})_{[_*D^{(k)}]}\| = 0.$$

Because of

$$[_*D^{(k)}]_{(y)}\big(\chi(R^s\setminus\{y\})>0\big) = 0 \qquad (y\in R^s)$$

we finally obtain in using 4.2.6. and 3.

$$(++++)\qquad \overline{\lim_{n\to\infty}}\ \sup_{x\in R^s}\Big\|\Big((L_{\varkappa(K_n)})_{[_*D^{(k)}]}\Big)_{X_1+x,\ldots,X_m+x}$$
$$-\Big(\Big(\int P_{ld_k\mu}(.)\,\sigma_L(\mathrm{d}l)\Big)_{[_*D^{(k)}]}\Big)_{X_1+x,\ldots,X_m+x}\Big\|$$
$$\leqq \overline{\lim_{n\to\infty}}\ \sup_{x\in R^s}\Big\|(L_{\varkappa(K_n)})_{X_1+x,\ldots,X_m+x} - \Big(\int P_{ld_k\mu}(.)\,\sigma_L(\mathrm{d}l)\Big)_{X_1+x,\ldots,X_m+x}\Big\| = 0.$$

In view of (+), (++), (+++) and (++++), the convergence statement of 4. is proved under the additional assumption that the sets $X_1,\ldots,X_m$ are pairwise disjoint. This assumption can be easily removed by passing to a disjoint refinement of the finite sequence $X_1,\ldots,X_m$.

5. For all bounded X in $\mathfrak{R}^s$ we get

$${}_X\Big\|\Big(\int P_{ld_k\mu}(.)\,\sigma_L(\mathrm{d}l)\Big)_{[_*D^{(k)}]} - \Big(\int P_{ld_k\mu}(.)\,\sigma_L(\mathrm{d}l)\Big)_{[D_*]}\Big\|$$
$$\leqq \int {}_X\|(P_{ld_k\mu})_{[_*D^{(k)}]} - (P_{ld_k\mu})_{[D_*]}\|\,\sigma_L(\mathrm{d}l)$$
$$\leqq \int \Big(\int {}_X\|[_*D^{(k)}]_{(x)} - [D_*]_{(x)}\|\, ld_k\mu(\mathrm{d}x)\Big)\,\sigma_L(\mathrm{d}l)$$
$$\leqq cd_k\mu(X)\,\|_*D^{(k)} - D_*\|.$$

However 12.5.1., 12.5.2. yield

$$\lim_{k\to\infty}\|_*D^{(k)} - D_*\| = 0,\qquad \lim_{k\to\infty} d_k = v(D),$$

so that the above expression tends to zero for $k \to \infty$. On the other hand, by 1.12.1.,

$$\begin{aligned} &{}_X\left\|\left(\int P_{l d_k \mu}(.)\, \sigma_L(\mathrm{d}l)\right)_{[D_*]} - \left(\int P_{l v(D) \mu}(.)\, \sigma_L(\mathrm{d}l)\right)_{[D_*]}\right\| \\ &\leqq \int {}_X\|P_{l d_k \mu} - P_{l v(D) \mu}\|\, \sigma_L(\mathrm{d}l) \\ &\leqq 2 \int \operatorname{Var}\left({}_X(l d_k \mu) - {}_X(l v(D)\, \mu)\right) \sigma_L(\mathrm{d}l) \\ &\leqq 2c\mu(X)\, |d_k - v(D)|, \end{aligned}$$

which converges in the same way by 12.5.2. towards zero.

Summarizing, we obtain

$$\overline{\lim_{k\to\infty}}\ {}_X\left\|\left(\int P_{l d_k \mu}(.)\, \sigma_L(\mathrm{d}l)\right)_{[_* D^{(k)}]} - \left(\int P_{l v(D) \mu}(.)\, \sigma_L(\mathrm{d}l)\right)_{[D_*]}\right\| = 0$$

for all bounded X in $\mathfrak{R}^s$.

6. Immediately from 1., 2., 4. and 5. we obtain the convergence assertion of 12.5.5. for all subcritical distributions D with the property $v(D) > 0$.

7. Proceeding in the same way as in the derivation of 12.1.2. we see that the distributions $\left(c(D)\right)^{-n} \langle D^{[n]} \rangle^!$, $n = 1, 2, \ldots$, may be constructed in the following way. Let $[x_1, \eta_1], [x_2, \eta_2], \ldots$ be a random sequence of elements of $R^s \times M$ which is distributed according to $\left(\left(c(D)\right)^{-1} \rangle D \langle^!\right) \times \left(\left(c(D)\right)^{-1} \rangle D \langle^!\right) \times \cdots$. Given this sequence, we select $\gamma_1, \gamma_2, \cdots$. according to the distribution $[D^{[0]}]_{(\eta_1)} \times [D]_{(T_{x_1} \eta_2)} \times [D^{[2]}]_{(T_{x_1+x_2} \eta_3)} \times \ldots$. Then, for $n = 1, 2, \ldots$, the counting measure $\gamma_1 + \cdots + \gamma_n$ is distributed according to $\left(c(D)\right)^{-n} \langle D^{[n]} \rangle^!$. We denote by V the distribution of $(\gamma_n)_{n=1,2,\ldots}$. Obviously, the sequence $\left(\gamma_n(R^s)\right)$ is stochastically independent with respect to V.

We set

$$\zeta = \sup\{n : n \geqq 1,\ \gamma_n(R^s) \neq 0\}.$$

Thus $\zeta = 0$ holds if and only if all $\gamma_n(R^s)$ vanish. For all natural numbers m we obtain

$$\begin{aligned} V(\zeta \leqq m) &= V\left(0 = \gamma_{m+1}(R^s) = \gamma_{m+2}(R^s) = \cdots\right) = \prod_{i>m} V\left(\gamma_i(R^s) = 0\right) \\ &= \prod_{i>m} \left(\left(c(D)\right)^{-1} \sum_{k=1}^{\infty} D\left(\chi(R^s) = k\right) k (D^{[i-1]})^{k-1} (\Psi = o)\right) \\ &= \prod_{i>m} \left(\left(c(D)\right)^{-1} \sum_{k=1}^{\infty} D\left(\chi(R^s) = k\right) k\left(f_{i-1}(0)\right)^{k-1}\right) = \prod_{i>m} \left(c(D)\right)^{-1} f_1'\left(f_{i-1}(0)\right). \end{aligned}$$

Hence, $V(\zeta \leqq m)$ is different from zero if and only if

$$\sum_{j=1}^{\infty} \left(1 - \left(c(D)\right)^{-1} f_1'\left(f_j(0)\right)\right) < +\infty$$

holds. This in turn is equivalent to $v(D) > 0$ (cf., for instance, Athreya & Ney [1], Section 1.11.). Thus, $V(\zeta \text{ infinite}) = 1$ is equivalent to $v(D) = 0$.

8. Let $v(D) = 0$ and $L \in \mathbf{S}^{(c)}$, $c > 0$. Setting

$$L_n = \mathscr{K}_{(c(D))^n}\left(L_{[D^{[n]}]}\right) = \left(\mathscr{K}_{(c(D))^n}L\right)_{[\mathscr{K}_{(c(D))^n}D^{[n]}]}$$

we obtain for any x in R^s and any bounded neighbourhood U of the origin with the property $\mu(\partial U) = 0$

$$L_n\big(({}_{U,k}\Phi)\,(U + x) > 0\big) \leqq \varrho_{U,k(L_n)}(U + x)$$

$$\leqq \varrho_{(\mathscr{K}_{(c(D))^n}L)_{[U,k(\mathscr{K}_{(c(D))^n}D^{[n]})]}}(U + x)$$

$$= \int \left(c(D)\right)^n \varrho_{\mathscr{K}_{(c(D))^n}L}(U + x - y)\,\left(c(D)\right)^{-n}\,\varrho_{U,k(\mathscr{K}_{(c(D))^n}D^{[n]})}(\mathrm{d}y)$$

$$\leqq \Big(\sup_{x\in R^s} t_n^{-s}\varrho_L(t_nU + x)\Big)\,\left(c(D)\right)^{-n}\,\varrho_{U,k(\mathscr{K}_{(c(D))^n}D^{[n]})}(R^s).$$

By 11.7.7. the first factor converges to $\mu(U)\int l\sigma_L(\mathrm{d}l)$ for $n \to \infty$.

By 7. we have $V\left(\sum\limits_{l=1}^{\infty}\gamma_l(R^s) = +\infty\right) = 1$. Consequently, for any natural number k, analogously to the first step in the proof of 12.1.6., we get

$$\varlimsup_{n\to\infty}\left(c(D)\right)^{-n}\varrho_{U,k(\mathscr{K}_{(c(D))^n}D^{[n]})}(R^s) = \varlimsup_{n\to\infty}\left(c(D)\right)^{-n}\langle D^{[n]}\rangle^!\left(\chi(t_nU) < k\right)$$

$$= \varlimsup_{n\to\infty} V\left(\sum_{l=1}^{n}\gamma_l(t_nU) < k\right) = 0.$$

Hence,

$$\lim_{n\to\infty}\sup_{x\in R^s} L_n\big(\Phi(U + x) > 0\big)$$

$$= \inf_{k=1,2,\ldots}\varlimsup_{n\to\infty}\sup_{x\in R^s} L_n\big((\Phi - {}_{U,k}\Phi)\,(U + x) > 0\big) = 0. \blacksquare$$

12.6. Stability Conditions

By 12.1.2., for all D in $\boldsymbol{D}$ and all natural numbers n, the intensity measure of $\langle D^{[n]}\rangle$ is the sum of the intensity measures of the random counting measures $(\delta_{[0,\ldots,0]} + \eta_1), \gamma_2, \ldots, \gamma_n$. By definition, the intensity measure of the first term is $\varrho_{\langle D\rangle} = \delta_{[0,\ldots,0]} + \varrho_{\langle D\rangle^!}$. In virtue of 11.7.2., given $x_1, \ldots, x_{i-1}$, the intensity measure of γ_i, $i = 2, 3, \ldots$, is

$$\left(\varrho_{D^{[i-1]}} * \varrho_{\langle D\rangle^!}\right)\left((.) + x_1 + \cdots + x_{i-1}\right).$$

Taking into consideration that by 11.3.2. the sum $x_1 + \cdots + x_{i-1}$ is distributed according to $\varrho_{D^{[i-1]}} = (\varrho_D)^{i-1}$, we see that the intensity measure of γ_i, $i = 2, 3, \ldots$, is

$$\int \left((\varrho_D)^{i-1} * \varrho_{\langle D\rangle^!}\right)\left((.) + x\right)(\varrho_D)^{i-1}(\mathrm{d}x) = \left(\varrho_D * (\varrho_D{}^\vee)\right)^{i-1} * \varrho_{\langle D\rangle^!},$$

where $(\varrho_D{}^\vee) = \varrho_D\big(-(.)\big)$. Setting ${}^0\varrho_D = \varrho_D * (\varrho_D{}^\vee)$, we obtain

12.6.1. *For all D in $\mathbf{D}$, and all natural numbers n,*

$$\varrho_{\langle D^{[n]}\rangle} = \delta_{[0,\ldots,0]} + \left(\sum_{k=0}^{n-1} ({}^0\varrho_D)^k\right) * \varrho_{\langle D\rangle}!\,.$$

This simple expression for $\varrho_{\langle D^{[n]}\rangle}$ leads us to

12.6.2. Theorem. *For all D in $\mathbf{D}$, the following properties are equivalent*

a) *The measure* $\left(\sum_{k=0}^{\infty} ({}^0\varrho_D)^k\right) * \varrho_{\langle D\rangle}!$ *belongs to N.*

b) *D is stable, and for all $l \geqq 0$ and all bounded X in $\mathfrak{R}^s$, we have*

$$\int (\Phi(X))^2 \,]D[_{(l)}\, (\mathrm{d}\Phi) < +\infty\,.$$

Proof. Combining the derivation of 12.1.3. and 12.6.1. we see that the measure $\tau = \left(\sum_{k=0}^{\infty} ({}^0\varrho_D)^k\right) * \varrho_{\langle D\rangle}!$ belongs to N if and only if $(\langle D^{[n]}\rangle)$ tends weakly towards a distribution Q of finite intensity. Thus, by 12.1.4., the measure τ belongs to N if and only if D is stable and $\left(\widetilde{]D[_{(l)}}\right)^0$ is of finite intensity for all $l \geqq 0$.

The remaining part of the proof now follows immediately from 2.2.6. and the following lemma (cf. Mori [1] and Krickeberg [3]).

12.6.3. *For all stationary measures W on $\mathfrak{M}$ with finite intensity i_W the corresponding Palm measure W^0 is of finite intensity if and only if $\int (\Phi(X))^2\, W(\mathrm{d}\Phi) < +\infty$ is satisfied for all bounded sets X in $\mathfrak{R}^s$.*

Proof. Let X be any bounded set in $\mathfrak{R}^s$. We select a symmetric bounded open set $U \supseteq X$ and obtain

$$\mu(X) \int \Phi(X)\, W^0(\mathrm{d}\Phi) = \mu(X) \sum_{k=1}^{\infty} W^0\big(\Phi(X) \geqq k\big)$$

$$= \sum_{k=1}^{\infty} \mathscr{C}_W^{\mathrm{T}}\big(x \in X, \Phi(X) \geqq k\big) \leqq \sum_{k=1}^{\infty} \mathscr{C}_W\big(x \in U + U, \Phi(U + U) \geqq k\big)$$

$$= \sum_{k=1}^{\infty} k\mathscr{C}_W\big(x \in U + U, \Phi(U + U) = k\big).$$

By 5.1.8. we may continue as follows:

$$= \sum_{k=1}^{\infty} k^2 W\big(\Phi(U + U) = k\big) = \int \big(\Phi(U + U)\big)^2\, W(\mathrm{d}\Phi).$$

Thus ϱ_{W^0} belongs to N if $\int (\Phi(Z))^2\, W(\mathrm{d}\Phi)$ is finite for all bounded Z in $\mathfrak{R}^s$.

In the opposite direction we get

$$\int (\Phi(X))^2\, W(\mathrm{d}\Phi) = \sum_{k=1}^{\infty} \mathscr{C}_W\big(x \in X, \Phi(X) \geqq k\big)$$

$$\leqq \sum_{k=1}^{\infty} k\mathscr{C}_W^{\mathrm{T}}\big(x \in U + U, \Phi(U + U) = k\big) = \mu(U + U) \int \Phi(U + U)\, W^0(\mathrm{d}\Phi)\,. \blacksquare$$

For any distribution α on $\mathfrak{R}^s$ let q_α denote the distribution of a stochastically independent family $(\xi_i)_{-\infty<i<+\infty}$ of random elements of R^s, distributed identically according to α. The random sequence $\zeta_0 = 0$, $\zeta_1 = \xi_1$, $\zeta_2 = \xi_1 + \xi_2, \ldots$, i.e. the *random walk* corresponding to α, is called *transient*, if almost surely with respect to q_α the measure $\Phi = \sum_{n=0}^{\infty} \delta_{\zeta_n}$ belongs to M, i.e. if the measure H_{q_α} exists (cf. Section 10.6.). It turns out that the random counting measure Φ is of finite intensity in the transient case (cf. for instance Loynes [1]). Thus, (ζ_n) is transient if and only if the measure $\sum_{n=0}^{\infty} \alpha^n$ belongs to N. This property of α can be expressed in terms of the characteristic function of α (cf. for instance Loynes [1] and Port & Stone [1]).

For all D in $\boldsymbol{D}$ we have

$$\varrho_{\langle D\rangle}{}^!(R^s) + 1 = \varrho_{\langle D\rangle}(R^s) = \sum_{k=1}^{\infty} k\langle D\rangle\big(\chi(R^s) = k\big) = \sum_{k=1}^{\infty} k^2 D\big(\chi(R^s) = k\big).$$

Thus $\varrho_{\langle D\rangle}{}^!$ is finite and different from o if and only if the second moment of D_{R^s} is finite and $D\big(\chi(R^s) = 1\big) < 1$ holds.

The following theorem (cf. Liemant [2]) yields an effective criterion of "second order stability".

12.6.4. Theorem. *For all D in $\boldsymbol{D}$ satisfying*

$$D\big(\chi(R^s) = 1\big) < 1, \qquad \int \big(\chi(R^s)\big)^2 D(\mathrm{d}\chi) < +\infty,$$

the following properties are equivalent.

a) *The random walk $(\zeta_n)_{n=0,1,\ldots}$ corresponding to $\alpha = {}^0\varrho_D$ is transient.*

b) *D is stable, and for all $l \geqq 0$ and all bounded X in $\mathfrak{R}^s$ we have*

$$\int \big(\Phi(X)\big)^2\,]D[_{(l)}\, (\mathrm{d}\Phi) < +\infty.$$

Proof. 1. Let $\left(\sum_{k=0}^{\infty} \alpha^k\right) * \varrho_{\langle D\rangle}{}^! \in N$. For all bounded open neighbourhoods U, V of the origin we obtain

$$\varrho_{\langle D\rangle}{}^!(V)\left(\sum_{k=0}^{\infty} \alpha^k\right)(U) \leqq \int_V \left(\sum_{k=0}^{\infty} \alpha^k\right)(U + V - x)\, \varrho_{\langle D\rangle}{}^!(\mathrm{d}x)$$

$$\leqq \left(\left(\sum_{k=0}^{\infty} \alpha^k\right) * \varrho_{\langle D\rangle}{}^!\right)(U + V) < +\infty.$$

For all sufficiently large V we have $0 < \varrho_{\langle D\rangle}{}^!(V) < +\infty$. Hence $\left(\sum_{k=0}^{\infty} \alpha^k\right)(U) < +\infty.$

2. Let $\left(\sum_{k=0}^{\infty} \alpha^k\right) \in N$. For all bounded open neighbourhoods U of the origin and all x in R^s we get

$$\left(\sum_{k=0}^{\infty} \alpha^k\right)(U - x) = \mathrm{E}_{q_\alpha}\big(\Phi(U - x)\big)$$

$$= \sum_{n=0}^{\infty} q_\alpha(\zeta_n \in U - x, \zeta_i \notin U - x \quad \text{for} \quad 0 \leqq i < n)$$

$$\mathrm{E}_{q_\alpha}\big(\Phi(U - x) \mid \zeta_n \in U - x, \zeta_i \notin U - x \quad \text{for} \quad 0 \leqq i < n\big)$$

$$\leqq \sum_{n=0}^{\infty} q_\alpha(\zeta_n \in U - x, \zeta_i \notin U - x \quad \text{for} \quad 0 \leqq i < n)\, \mathrm{E}_{q_\alpha}\big(\Phi(U - U)\big)$$

$$\leqq \mathrm{E}_{q_\alpha}\big(\Phi(U - U)\big) = \left(\sum_{k=0}^{\infty} \alpha^k\right)(U - U).$$

Thus

$$\left(\left(\sum_{k=0}^{\infty} \alpha^k\right) * \varrho_{\langle D\rangle}!\right)(U) = \int \left(\sum_{k=0}^{\infty} \alpha^k\right)(U - x)\, \varrho_{\langle D\rangle}!(\mathrm{d}x)$$

$$\leqq \varrho_{\langle D\rangle}!(R^s) \left(\sum_{k=0}^{\infty} \alpha^k\right)(U - U) < +\infty.$$

3. The assertion of 12.6.4. now follows immediately from 12.6.2., 1. and 2. ∎

As is known (see, for instance, Petrov [1], Chapt. III, § 2 — the proof given there for $s = 1$ can be extended to the multi-dimensional case), for all distributions ν on $\mathfrak{R}^s$, the support of which is not contained in some proper linear submanifold, we have for all bounded X in $\mathfrak{R}^s$

$$\sup_{x \in R^s} \nu^n(X + x) = O(n^{-s/2}).$$

Now, if the support of ϱ_D is not contained in some linear submanifold of dimension less than three, the support of ${}^0\varrho_D$ has the same property. Thus, for all bounded X in $\mathfrak{R}^s$, we have $({}^0\varrho_D)^n(X) = O(n^{-3/2})$, and by 12.6.4. we obtain (cf. Fleischmann [1])

12.6.5. Theorem. *In the case $s \geqq 3$, a distribution D in $\boldsymbol{D}$ satisfying $\int (\chi(R^s))^2 D(\mathrm{d}\chi) < +\infty$ is stable if the support of ϱ_D is not contained in some two-dimensional linear submanifold of R^s.*

Under the condition $\int (\chi(R^s))^2 D(\mathrm{d}\chi) < +\infty$ stability is equivalent to "second order stability". This is an immediate consequence of 12.6.4. and the following necessary condition, which was proved in Kallenberg [7].

12.6.6. Theorem. *For all stable D in $\boldsymbol{D}$ satisfying $D(\chi(R^s) = 1) < 1$ the random walk $(\zeta_n)_{n=0,1,\ldots}$ corresponding to $\alpha = {}^0\varrho_D$ is transient.*

References

(= ...) Reference to English translation

Ambartzumian, R. V.

[1] On an equation for stationary point processes. (Russian; Armenian summary.) *Dokl. Akad. Nauk Armjan. SSR* **42**, 141–147 (1966).

[2] Palm distributions and superpositions of independent point processes in R^n. In *Stochastic Point Processes: Statistical Analysis, Theory and Applications* (ed. P. A. W. Lewis), Wiley, New York 1972, 626–645.

[3] Homogeneous and isotropic random point fields in the plane. (Russian.) *Math. Nachr.* **70**, 365–385 (1975).

Athreya, K. B., & Ney, P. E.

[1] *Branching Processes.* Springer-Verlag, Berlin 1972.

Belyaev, Ju. K.

[1] Limit theorems for flows subject to rarefaction. (Russian; English summary.) *Theor. Veroyatnost. i Primenen.* **8**, 175–184 (1963). (= *Theor. Probability Appl.* **8**, 165–173).

[2] New results and generalizations of problems of level crossings. (Russian.) Appendix to the Russian edition of H. Cramér and M. R. Leadbetter [1]. MIR, Moscow 1969.

Billingsley, P.

[1] *Convergence of Probability Measures.* Wiley, New York 1968.

Breiman, L.

[1] The Poisson tendency in traffic distribution. *Ann. Math. Statist.* **34**, 308–311 (1963).

Brown, M.

[1] A property of Poisson processes and its application to macroscopic equilibrium of particle systems. *Ann. Math. Statist.* **41**, 1935–1941 (1970). Corrigendum **42**, 1777.

[2] Discrimination of Poisson processes. *Ann. Math. Statist.* **42**, 773–776 (1971).

Cox, D. R.

[1] Some statistical models connected with series of events. *J. Roy. Statist. Soc. Ser. B*, **17**, 129–164 (1955).

Cox, D. R., & Lewis, P. A. W.

[1] *The Statistical Analysis of Series of Events.* Wiley, New York 1966.

Cramér, H., & Leadbetter, M. R.

[1] *Stationary and Related Stochastic Processes.* Wiley, New York 1967.

Csiszár, I.

[1] Information-type measures of difference of probability distributions. *Stud. Sci. Math. Hung.* **2**, 299–318 (1967).

Daley, D. J.

[1] Various concepts of orderliness for point processes. In *Stochastic Geometry and Analysis* (ed. D. G. Kendall and E. F. Harding). Wiley, London 1974, 148–161.

[2] Poisson and alternating renewal processes with superposition a renewal process. *Math. Nachr.* **57**, 359–369 (1973).

Daley, D. J., & Oakes, D.

[1] Random walk point processes. *Z. Wahrscheinlichkeitstheorie und verw. Gebiete* **30**, 1–16 (1974).

Daley, D. J., & Vere-Jones, D.

[1] A summary of the theory of point processes. In *Stochastic Point Processes Statistical Analysis, Theory and Applications* (ed. P. A. W. Lewis), Wiley New York 1972, 299–383.

Debes, H., Kerstan, J., Liemant, A., & Matthes, K.

[1] Verallgemeinerungen eines Satzes von Dobruschin I. *Math. Nachr.* **47**, 183–244 (1970).

[2] Verallgemeinerungen eines Satzes von Dobruschin III. *Math. Nachr.* **50**, 99–139 (1971).

Dennler, G.

[1] Unbegrenzt teilbare Verteilungsgesetze in dem Dual eines separablen nuklearen lokal-konvexen Raumes. *Math. Nachr.* **39**, 195–215 (1969).

Derman, C.

[1] Some contributions to the theory of denumerable Markov chains. *Trans. Amer. Math. Soc.* **79**, 541–555 (1955).

Dobrushin, R. L.

[1] On the Poisson law for distributions of particles in space. (Russian.) *Ukrain. Mat. Z.* **8**, 127–134 (1956).

[2] Description of a random field by means of conditional probabilities and the conditions governing its regularity. *Teor. Veroyatnost. i Primenen.* **13**, 201–229 (1968) (= *Theor. Probability Appl.* **13**, 197–224 (1968).

Doob, L. L.

[1] *Stochastic Processes.* Wiley, New York 1953.

[2] Renewal theory from the point of view of the theory of probability. *Trans. Amer. Math. Soc.* **63**, 422–438 (1948).

Dunford, N., & Schwartz, J. T.

[1] *Linear Operators. Part I: General Theory.* Interscience Publishers, New York 1958.

Feller, W.

[1] *An Introduction to Probability Theory and its Applications,* Volume I, second edition. Wiley, New York 1957.

[2] *An Introduction to Probability Theory and its Applications,* Volume II. Wiley, New York 1966.

Fichtner, K. H.

[1] Gleichverteilungseigenschaften substochastischer Kerne und zufällige Punktfolgen. *Math. Nachr.* **62**, 251—260 (1974).

[2] Charakterisierung Poissonscher zufälliger Punktfolgen und infinitesimale Verdünnungsschemata. *Math. Nachr.* **68**, 93—104 (1975).

[3] Schwache Konvergenz von unabhängigen Überlagerungen verdünnter zufälliger Punktfolgen. *Math. Nachr.* **66**, 333—341 (1975).

[4] Unabhängige Überlagerungen verschobener zufälliger Punktfolgen. *Math. Nachr.* **70**, 133—153 (1975).

Fieger, W.

[1] Zwei Verallgemeinerungen der Palmschen Formeln, *Transactions 3-rd Prague Conf. Inf. Theory Stat. Dec. Functions Random Proc.*, 107—122 (1964).

[2] Eine für beliebige Call-Prozesse geltende Verallgemeinerung der Palmschen Formeln. *Math. Scand.* **16**, 121—147 (1965).

Fleischmann, K.

[1] On the stability of spatially homogeneous cluster fields. *Trans. 7-th Prague Conf. Inf. Theory. Stat. Dec. Functions Random Proc.* 1974 (to appear).

[2] Eine Charakterisierung einer Klasse unbegrenzt teilbarer Punktprozesse. *Math. Nachr.* **70**, 177—182 (1975).

[3] Subkritische räumlich homogene Verzweigungsprozesse II. *Math. Nachr.* (to appear).

[4] Mixing properties of cluster-invariant distributions. *Litovsk Math. Sb.* (to appear).

[5] A continuity theorem for clustering. (Submitted to *Litovsk. Mat. Sb.*)

Fleischmann, K., & Prehn, U.

[1] Ein Grenzwertsatz für subkritische Verzweigungsprozesse mit endlich vielen Typen von Teilchen. *Math. Nachr.* **64**, 357—362 (1974).

[2] Subkritische räumlich homogene Verzweigungsprozesse. *Math. Nachr.* **70**, 231—250 (1975).

[3] Limit theorems for spatially homogeneous branching processes with a finite set of types II. *Math. Nachr.* (to appear).

Fleischmann, K., & Siegmund-Schultze, R.

[1] An invariance principle for reduced family trees of critical spatially homogeneous branching processes. *Serdica Bulgaricae mathematicae publicationes* (to appear).

Franken, P.

[1] Approximation durch Poissonsche Prozesse. *Math. Nachr.* **26**, 101—114 (1963).

[2] Einige Anwendungen der Theorie zufälliger Punktprozesse in der Bedienungstheorie I. *Math. Nachr.* **70**, 303—319 (1975).

Franken, P., Liemant, A., & Matthes, K.

[1] Stationäre zufällige Punktfolgen III. *Jber. Deutsch. Math.-Verein.* **67**, 183—202 (1965).

Franken, P., & Richter, G.

[1] Über eine Klasse von zufälligen Punktfolgen. *Wiss. Z. Friedrich-Schiller-Univ. Jena* **14**, 247—249 (1965).

Fremlin, D. H.

[1] A direct proof of the Matthes-Wright integral extension theorem. *J. London math. Soc., II. Ser.* **11**, 276—284 (1975).

Geman, D., & Horowitz, J.

[1] Remarks on Palm measures. *Ann. Inst. H. Poincaré* **9**, 215–232 (1973).

Georgii, H.-O.

[1] Canonical and grand canonical Gibbs states for continuum systems. *Commun. in Mathematical Physics* **48**, 31–51 (1976).

Gikhman, I. I., & Skorohod, A. V.

[1] *The Theory of Stochastic Processes I.* (Russian.) Izdat. Nauka, Moscow 1971 (= Springer-Verlag, Berlin 1974).

Gnedenko, B. V., & Kovalenko, I. N.

[1] *Introduction to Queueing Theory.* (Russian.) Izdat. Nauka, Moscow 1966 (= Israel Program for Scientific Translations, Jerusalem 1968).

Goldman, J. R.

[1] Stochastic point processes: limit theorems. *Ann. Math. Statist.* **38**, 771–779 (1967).

[2] Infinitely divisible point processes in R^n. *J. Math. Anal. Appl.* **17**, 133–146 (1967).

Grandell, J.

[1] A note on characterization and convergence of non-atomic random measures. In *Abstract of Communications* T. 1., 175–176. International Conference on Probability Theory and Mathematical Statistics at Vilnius 1973.

[2] *Doubly Stochastic Poisson Processes.* Lecture Notes in Mathematics **529**. Springer-Verlag, Berlin 1976.

Grigelionis, B.

[1] On the convergence of sums of random step processes to a Poisson process. (Russian.) *Teor. Veroyatnost. i Primenen.* 8, 189–194 (1963) (= *Theor. Probability Appl.* 8, 177–182).

[2] Superposition of integer-valued random measures. (Russian.) *Litovsk. Mat. Sb.* **6**, 359–363 (1966).

[3] On the question of convergence of sums of steplike random processes to a Poisson process (Russian; Lithuanian and English summaries.) *Litovsk. Mat. Sb.* **6**, 241–244 (1966).

Guy Fourt, M.

[1] Existence de measures á puissances singuliéres á toutes leurs translatées. *C. r. Acad. Sci., Paris, Sér. A* **274**, 648–650 (1972).

Haberland, E.

[1] Infinitely divisible stationary recurrent point processes. *Math. Nachr.* **70**, 259–264 (1975).

Halmos, P. R.

[1] *Measure Theory.* New York 1950.

Harris, T. E.

[1] Counting measures, monotone random set functions. *Z. Wahrscheinlichkeitstheorie und verw. Gebiete* **10**, 102–119 (1968).

[2] Random measures and motions of point processes. *Z. Wahrscheinlichkeitstheorie und verw. Gebiete* **18**, 85–115 (1971).

Hawkes, A. G., & Oakes, D.

[1] A cluster process representation of a self-exciting process. *J. Appl. Probability* **11**, 493—503 (1974).

Hermann, K.

[1] Konstruktion Palmscher Verteilungen mittels stationärer Folgen im R^s. Diplomarbeit, Humboldt-Universität Berlin, Sektion Mathematik, 1975.

Herrmann, H.

[1] Glättungseigenschaften der Faltung. *Wiss. Z. Friedrich-Schiller-Universität Jena* **14**, 221—234 (1965).

Herrmann, H., Kerstan, J., & Liese, F.

[1] Zweidimensionale und lokale Charakterisierung der asymptotischen Gleichverteilung. *Math. Nachr.* (to appear).

[2] o-stetige Maße und asymptotische Gleichverteilung. *Math. Nachr.* (to appear).

Herrmann, U.

[1] Ein Approximationssatz für Verteilungen stationärer zufälliger Punktfolgen. *Math. Nachr.* **30**, 377—381 (1965).

Hewitt, E., & Stromberg, K.

[1] *Real and Abstract Analysis.* Springer-Verlag, Berlin 1965.

Hilico, C.

[1] Processus ponctuels marqués stationnaires. Application à l'interaction sélective de deux processus ponctuels stationnaires. *Ann. Inst. H. Poincaré* **9**, 177—192 (1973).

Huff, B. W.

[1] On the infinite divisibility of certain discrete mixtures. *Queen's Mathematical Preprint* No. 1976—11, Ontario 1976.

Jagers, P.

[1] On the weak convergence of superpositions of point processes. *Z. Wahrscheinlichkeitstheorie und verw. Gebiete* **22**, 1—7 (1972).

[2] On Palm probabilities. *Z. Wahrscheinlichkeitstheorie und verw. Gebiete* **26**, 17—32 (1973).

[3] Aspects of random measures and point processes. In *Advances in Probability and Related Topics* **3**, 179—239. Marcel Dekker, New York 1974.

Jagers, P., & Lindvall, T.

[1] Thinning and rare events in point processes. *Z. Wahrscheinlichkeitstheorie und verw. Gebiete* **28**, 89—98 (1974).

Jiřina, M.

[1] Conditional probabilities on σ-algebras with countable basis. (Russian.) *Czechosl. math. J.* **4**, 372—380 (1954).

[2] Branching processes with measure-valued states. *Trans. 3-rd Prague Conf. Inf. Theory Stat. Dec. Functions Random Proc.* 333—357 (1964).

[3] Asymptotic behaviour of measure-valued branching processes. *Rozpravy Českosl. Akad. Věd. Ř. Mat. Přirodnich Věd.* **76**, No. 3 (1966).

Kallenberg, O.

[1] Characterization and convergence of random measures and point processes. Thesis, Chalmers institute of technology and the university of Göteborg, 1972.

[2] Characterization and convergence of random measures and point processes. *Z. Wahrscheinlichkeitstheorie und verw. Gebiete* **27**, 9—21 (1973).

[3] Characterization of continuous random processes and signed measures. *Stud. Sci. Math. Hung.* **8**, 473—477 (1973).

[4] On symmetrically distributed random measures. *Trans. Amer. Math. Soc.* **202**, 105—121 (1975).

[5] Limits of compound and thinned point processes. *J. Appl. Probability* **12**, 269—278 (1975).

[6] *Random Measures*. Akademie-Verlag, Berlin 1975, and Academic Press, London—New York—San Francisco 1976.

[7] Stability of critical cluster fields. *Math. Nachr.* **77**, 7—43 (1977).

[8] On the structure of stationary flat processes. *Z. Wahrscheinlichkeitstheorie und verw. Gebiete* **37**, 157—174 (1976).

[9] On conditional intensities of point processes. *Z. Wahrscheinlichkeitstheorie und verw. Gebiete* (to appear).

[10] A counterexample to R. Davidson's conjecture on line processes. *Math. Proc. Camb. Phil. Soc.* (to appear).

[11] On the independence of velocities in a system of non-interacting particles. *Tech. report, Math. Dept., Göteborg* 1977.

[12] On the asymptotic behavior of line processes and systems of non-interacting particles. Preprint, Göteborg 1977.

Karbe, W.

[1] Konstruktion einfacher zufälliger Punktfolgen. Diplomarbeit, Friedrich-Schiller-Universität Jena, Sektion Mathematik, 1973.

Katti, S. K.

[1] Infinite divisibility of integer-valued random variables. *Ann. Math. Statist.* **38**, 1306—1308 (1967).

Kendall, D. G.

[1] Foundations of a theory of random sets. In *Stochastic Geometry and Analysis* (ed. D. G. Kendall and E. F. Harding). Wiley, London 1974, 322—376.

Kerstan, J.

[1] Teilprozesse Poissonscher Prozesse. *Trans. 3-rd Prague Conf. Inf. Theory Stat. Dec. Functions Random Proc.* 377—403 (1964).

Kerstan, J., & Debes, H.

[1] Zufällige Punktfolgen und Markoffsche Übergangsmatrizen ohne stationäre Verteilungsgesetze. *Wiss. Z. Friedrich-Schiller-Universität Jena* **18**, 349—359 (1969).

Kerstan, J., & Matthes, K.

[1] Stationäre zufällige Punktfolgen II. *Jber. Deutsch. Math.-Verein.* **66**, 106—118 (1964).

[2] Verallgemeinerung eines Satzes von Sliwnjak. *Rev. Roumaine Math. Pures. Appl.* **9**, 811—829 (1964).

[3] Gleichverteilungseigenschaften von Faltungspotenzen auf lokalkompakten abelschen Gruppen. *Wiss. Z. Friedrich-Schiller-Universität Jena* **14**, 457—462 (1965).

[4] A generalization of the Palm-Khinchin theorem. (Russian.) *Ukrain. Mat. Z.* **17**, 29—36 (1965).

[5] Ergodische unbegrenzt teilbare stationäre zufällige Punktfolgen. *Trans. 4-th Prague Conf. Inf. Theory Stat. Dec. Functions Random Proc.* 399—415 (1967).

[6] Gleichverteilungseigenschaften von Faltungen von Verteilungsgesetzen auf lokal-kompakten abelschen Gruppen I. *Math. Nachr.* **37**, 267–312 (1968).

[7] Gleichverteilungseigenschaften von Faltungen von Verteilungsgesetzen auf lokalkompakten abelschen Gruppen II. *Math. Nachr.* **41**, 121–132 (1969).

Kerstan, J., Matthes, K., & Prehn, U.

[1] Verallgemeinerungen eines Satzes von Dobruschin II. *Math. Nachr.* **51**, 149–188 (1971).

Khinchin, A. Ya.

[1] *Mathematical Methods in the Theory of Queueing.* (Russian.) Trudy Mat. Inst. Steklov **49** (1955) (= Griffin, London 1960).

[2] Sequences of chance events without after-effects. (Russian.) *Teor. Veroyatnost. i Primenen.* **1**, 3–18 (1956) (= *Theor. Probability Appl.* **1**, 1–15).

[3] On Poisson streams of random events. (Russian.) *Teor. Veroyatnost. i Primenen.* **1**, 320–327 (1956) (= *Theor. Probability Appl.* **1**, 291–297).

Kingman, J. F. C.

[1] On doubly stochastic Poisson processes. *Proc. Cambridge Philos. Soc.* **60**, 923 to 930 (1964).

[2] Completely random measures. *Pac. J. Math.* **21**, 59–78 (1967).

König, D., & Matthes, K.

[1] Verallgemeinerung der Erlangschen Formeln I. *Math. Nachr.* **26**, 45–56 (1963).

König, D., Matthes, K., & Nawrotzki, K.

[1] *Verallgemeinerungen der Erlangschen und Engsetschen Formeln* (*Eine Methode in der Bedienungstheorie*). Akademie-Verlag, Berlin 1967.

[2] Unempfindlichkeitseigenschaften von Bedienungsprozessen. Appendix to the German edition of B. V. Gnedenko & I. N. Kovalenko [1]. Akademie-Verlag, Berlin 1971.

Kovalenko, I. N.

[1] On the class of limit distributions for a sequence of series of sums of independent renewal processes. (Russian; Lithuanian and English summaries.) *Litovsk. Mat. Sb.* **5**, 561–568 (1965).

Kozlov, O. K.

[1] Gibbsian description of point random fields (Russian). *Teor. Veroyatnost. i Primenen.* **21**, 348–365 (1976).

[2] Consistent systems of conditional distributions for random field (Russian). *Probl. Pereda. Inform.* **13**, 77–90 (1977).

Krickeberg, K.

[1] Moments of point processes. In *Stochastic Geometry and Analysis* (ed. D. G. Kendall and E. F. Harding), Wiley, London 1974, 89–113.

[2] The Cox process. *Istituto Nazionale di Alta Matematica, Symposia Mathematica* **9**, 151–167 (1972).

[3] *Lectures on point processes* (invietnamese). Institute of Mathematics Hanoi 1976.

Kummer, G., & Matthes, K.

[1] Verallgemeinerung eines Satzes von Sliwnjak II. *Rev. Roumaine Math. Pures Appl.* **15**, 845–870 (1970).

[2] Verallgemeinerung eines Satzes von Sliwnjak III. *Rev. Roumaine Math. Pures Appl.* **15**, 1631–1642 (1970).

Kurtz, T. G.

[1] Point processes and completely monotone set functions. *Z. Wahrscheinlichkeitstheorie und verw. Gebiete* **31**, 57–67 (1974).

Leadbetter, M. R.

[1] On three basic results in the theory of stationary point processes. *Proc. Amer. Math. Soc.* **19**, 151–217 (1968).

[2] On basic results of point process theory. *Proc. 6-th Berkeley Symp. Math. Statist. Probability* **3**, 449–462 (1972).

Le Cam, L.

[1] An approximation theorem for the Poisson binomial distribution. *Pacific J. Math.* **10**, 1181–1197 (1960).

Ledrappier, F.

[1] *Energie locale pour les systèmes continus.* Preprint. Paris 1976.

Lee, P. M.

[1] A structure theorem for infinitely divisible point processes. Address to I.A.S.P.S., Bern 1964 (unpublished).

[2] Infinitely divisible stochastic processes. *Z. Wahrscheinlichkeitstheorie und verw. Gebiete* **7**, 147–160 (1967).

[3] Some examples of infinitely divisible point processes. *Stud. sci. Math. Hung.* **3**, 219–224 (1968).

Lenard, A.

[1] Correlation functions and the uniqueness of the state in classical statistical mechanics. *Commun. math. Phys.* **30**, 35–44 (1973).

[2] States of classical statistical mechanical systems of infinitely many particles I. *Arch. Rational Mech. Anal.* **59**, 219–239 (1975).

[3] States of classical statistical mechanical systems of infinitely many particles II. Characterization of correlation measures. *Arch. Rational Mech. Anal.* **59**, 241–256 (1975).

Letac, G.

[1] Groupe de Stam d'une probabilité. *Ann. Inst. Henri Poincaré, Sect. B* **8**, 175–181 (1972).

Lévy, P.

[1] Sur les exponentielles de polynomes et sur l'arithmétique des produits de lois de Poisson. *Ann. sci. École norm. sup., III. Sér.* **54**, 231–292 (1937).

Lewis, P. A. W.

[1] A branching Poisson process model for the analysis of computer failure patterns. *J. Roy. Statist. Soc. Ser. B* **26**, 398–456 (1964).

Liemant, A.

[1] Invariante zufällige Punktfolgen. *Wiss. Z. Friedrich-Schiller-Universität Jena* **18**, 361–372 (1969).

[2] Verallgemeinerungen eines Satzes von Dobruschin V. *Math. Nachr.* **70**, 387–390 (1975).

Liemant, A., & Matthes, K.

[1] Verallgemeinerungen eines Satzes von Dobruschin IV. *Math. Nachr.* **59**, 311—317 (1974).

[2] Verallgemeinerungen eines Satzes von Dobruschin VI. *Math. Nachr.* **80**, 7—18 (1977).

Liese, F.

[1] Eine Abschätzung der Entropie von Faltungspotenzen und ihre Anwendung auf den Nachweis von Gleichverteilungseigenschaften. *Studia Scientiarum Mathematicarum Hungarica* **9**, 199—217 (1974).

[2] Eine informationstheoretische Bedingung für die Äquivalenz unbegrenzt teilbarer Punktprozesse. *Math. Nachr.* **70**, 183—196 (1975).

Lindvall, T.

[1] An invariance principle for thinned random measures. Chalmers university of technology and the university of Göteborg, department of mathematics, Göteborg 1974 (preprint).

Loève, M.

[1] *Probability Theory,* second edition. Van Nostrand, Princeton 1960.

Loynes, R. M.

[1] Products of independent random elements in a topological group. *Z. Wahrscheinlichkeitstheorie und verw. Gebiete* **1**, 446—455 (1963).

Matheron, G.

[1] *Random sets & integral geometry.* Wiley & Sons, London (1975).

Matthes, K.

[1] Ergodizitätseigenschaften rekurrenter Ereignisse I. *Math. Nachr.* **24**, 109—119 (1962).

[2] Unbeschränkt teilbare Verteilungsgesetze stationärer zufälliger Punktfolgen. *Wiss. Z. Hochsch. Elektrotechn. Ilmenau* **9**, 235—238 (1963).

[3] Stationäre zufällige Punktfolgen I. *Jber. Deutsch. Math.-Verein* **66**, 66—79 (1963).

[4] Eine Charakterisierung der kontinuierlichen unbegrenzt teilbaren Verteilungsgesetze zufälliger Punktfolgen. *Rev. Roumaine Math. Pures Appl.* **14**, 1121—1127 (1969).

[5] Arithmetical problems in the theory of stationary stochastic point processe. *Proceedings of the symposium to honour Jerzy Neyman* 243—252, PWN-Polish Scientific Publisher, Warszawa 1977.

Matthes, K., Warmuth, W., & Mecke, J.

[1] Bemerkungen zu einer Arbeit von Nguyen Xuan Xanh und Hans Zessin. *Math. Nachr.* (to appear).

Mecke, J.

[1] Stationäre zufällige Maße auf lokal-kompakten abelschen Gruppen. *Z. Wahrscheinlichkeitstheorie und verw. Gebiete* **9**, 36—58 (1967).

[2] Zum Problem der Zerlegbarkeit stationärer rekurrenter zufälliger Punktfolgen. *Math. Nachr.* **35**, 311—321 (1967).

[3] Eine charakteristische Eigenschaft der doppelt stochastischen Poissonschen Prozesse. *Z. Wahrscheinlichkeitstheorie und verw. Gebiete* **11**, 74—81 (1968).

[4] Verschärfung eines Satzes von McFadden. *Wiss. Z. Friedrich-Schiller-Universität Jena* **18**, 387—392 (1969).

[5] Zufällige Maße auf lokal-kompakten Hausdorffschen Räumen. *Beiträge zur Analysis* **3**, 7—30 (1972).
[6] Invarianzeigenschaften allgemeiner Palmscher Maße. *Math. Nachr.* **65**, 335—344 (1975).
[7] Eine Charakterisierung des Westcottschen Funktionals. *Math. Nachr.* **80**, 295—313, (1977)
[8] A characterization of mixed Poisson processes. *Rev. Roumaine Math. Pures Appl.* **21**, 1355—1360 (1976).

Milne, R. K.

[1] Simple proofs of some theorems on point processes. *Ann. Math. Statist.* **42**, 368—372 (1971).

Milne, R. K., & Westcott, M.

[1] Further results for Gauss-Poisson processes. *Advances in Appl. Probability* **4**, 151—176 (1972).

Mönch, G.

[1] Verallgemeinerung eines Satzes von Rényi. Dissertation, Technische Hochschule Karl-Marx-Stadt, 1970.
[2] Verallgemeinerung eines Satzes von Rényi. *Stud. Sci. Math. Hung.* **6**, 81—90 (1971).

Moran, P. A. P.

[1] A characteristic property of the Poisson distribution. *Proc. Cambridge Philos. Soc.* **48**, 206—207 (1952).
[2] A non-Markovian quasi-Poisson process. *Stud. Sci. Math. Hung.* **2**, 425—429 (1967).

Mori, T.

[1] Stationary random measures and renewal theory. *Yokohama math. J.* **23**, 31—54 (1975).

Moyal, J. E.

[1] The general theory of stochastic population processes. *Acta Math.* **108**, 1—31 (1962).
[2] Multiplicative population processes. *J. Appl. Probability* **1**, 267—283 (1964).

Nawrotzki, K.

[1] Ein Grenzwertsatz für homogene zufällige Punktfolgen. *Math. Nachr.* **24**, 201 to 217 (1962).
[2] Mischungseigenschaften stationärer unbegrenzt teilbarer zufälliger Maße. *Math. Nachr.* **38**, 97—114 (1968).
[3] Mischungseigenschaften stationärer unbegrenzt teilbarer zufälliger Distributionen. *Wiss. Z. Friedrich-Schiller-Universität Jena* **18**, 397—408 (1969).

Neveu, J.

[1] Sur la structure des processus ponctuels stationnaires. *C. R. Acad. Sci. Paris* **267 A**, 561—564 (1968).
[2] Sur les mesures de Palm de deux processus ponctuels stationnaires. *Z. Wahrscheinlichkeitstheorie und verw. Gebiete* **34**, 199—203 (1976).

Nguyen, X. X., & Zessin, H.

[1] Punktprozesse mit Wechselwirkung. *Z. Wahrscheinlichkeitstheorie und verw. Gebiete* **37**, 91–126 (1976).

[2] Integral and differential characterizations of the Gibbs Process. *Math. Nachr.* (to appear).

Ososkov, G. A.

[1] A limit theorem for flows of similar events. *Teor. Veroyatnost. i Primenen.* (Russian.) **1**, 274–282 (1956) (= *Theor. Probability Appl.* **1**, 248–255).

Oswald, H.

[1] Zur Geschwindigkeit der Gleichverteilung einer Folge von Faltungspotenzen. *Wiss. Z. Friedrich-Schiller-Universität Jena* **18**, 373–379 (1969).

Palm, C.

[1] Intensitätsschwankungen im Fernsprechverkehr. *Ericsson Technics* no. **44**, 189 pp., 1943.

Papangelou, F.

[1] The Ambrose-Kakutani theorem and the Poisson process. In *Contributions to Ergodic Theory and Probability, Springer-Verlag Lecture Notes in Mathematics* **160**, 234–240 (1970).

[2] On the Palm probabilities of processes of points and processes of lines. In *Stochastic Geometry and Analysis* (ed. D. G. Kendall and E. F. Harding). Wiley, London 1974, 114–147.

[3] The conditional intensity of general point processes and an application to line processes. *Z. Wahrscheinlichkeitstheorie und verw. Gebiete* **28**, 207–226 (1974).

[4] Point processes on spaces of flats and other homogeneous spaces. *Math. Proc. Camb. Phil. Soc.* **80**, 297–314 (1976).

Petrov, V. V.

[1] *Sums of Independent Random Variables.* (Russian.) Izdat. Nauka, Moscow 1972 (= Springer-Verlag, Berlin 1976).

Port, S. C.

[1] Equilibrium processes. *Trans. Amer. Math. Soc.* **124**, 168–184 (1966).

Port, S. C., & Stone, C. J.

[1] Potential theory of random walks on Abelian groups. *Acta Math.* **122**, 19–114 (1969).

[2] Infinite particle systems. *Trans. Amer. Math. Soc.* **178**, 307–340 (1973).

Prehn, U., & Röder, B.

[1] Limit theorems for spatially homogeneous branching processes with a finite set of types I. (Russian.) *Math. Nachr.* **80**, 37–86 (1977),

Prekopa, A.

[1] On secondary processes generated by a random point distribution of Poisson type. *Ann. Univ. Sci. Budapest Sect. Mat.* **1**, 153–170 (1958).

[2] On the spreading process. *Trans. 2-nd Prague Conf. Inf. Theory Stat. Dec. Functions Random Proc.* 521–529 (1960).

Preston, C.

[1] *Random fields.* Lecture Notes in Mathematics **534**. Springer-Verlag, Berlin 1976.

Prohorov, Yu, V.

[1] Convergence of random processes and limit theorems in probability theory. (Russian.) *Teor. Veroyatnost. i Primenen.* **1**, 177–237 (1956) (= *Theor. Probability Appl.* **1**, 157–214).

[2] Random measures on a compactum. (Russian.) *Dokl. Akad. Nauk SSSR* **138**, 53–55 (1961) (= *Soviet Math. Dokl.* **2**, 539–541).

Radecke, W.

[1] Analyse und Konstruktion stationärer markierter Punktprozesse mit vorgegebenem Entwicklungsgesetz. *Math. Nachr.* **81**, 83–167 (1978)

[2] Über eine spezielle Klasse abhängiger Verschiebungen Poissonscher Punktprozesse. *Math. Nachr.* **81**, 83–167 (1978)

Redheffer, R. M.

[1] A note on the Poisson law. *Math. Magazine* **26**, 185–188 (1953).

Rényi, A.

[1] A characterization of Poisson processes. (Hungarian.) *Magyar Tud. Akad., Mat. Kutató, Int. Közl.* **1**, 519–527 (1956).

[2] On two mathematical models of the traffic on a divided highway. *J. Appl. Probability* **1**, 311–320 (1964).

[3] Remarks on the Poisson process. *Stud. Sci. Math. Hung.* **2**, 119–123 (1967).

Ripley, B. D.

[1] Locally finite random sets: Foundations for point process theory. *The Annals of Probability* **4**, 983–994 (1976).

[2] The foundations of stochastic geometry. *The Annals of Probability* **4**, 995–998 (1976).

[3] On stationarity and superposition of point processes. *The Annals of Probability* **4**, 999–1005 (1976).

[4] The second-order analysis of stationary point processes. *J. Appl. Prob.* **13**, 255–266 (1976).

Ryll-Nardzewski, C.

[1] Remarks on processes of calls. *Proc. 4-th Berkeley Symp. Math. Statist. Probability* **2**, 455–465 (1961).

Slivnyak, I. M.

[1] Some properties of stationary flows of homogeneous random events. (Russian.) *Teor. Veroyatnost. i Primenen.* **7**, 347–352 (1962) (= *Theor. Probability Appl.* **7**, 336–341).

[2] Stationary flows of homogeneous random events. (Russian.) *Vestnik Kharkov. Gos. Univ.* **32**, 73–116 (1966).

Spitzer, F.

[1] *Principles of Random Walk,* second edition. Springer-Verlag, New York 1976.

Srivastava, R. C.

[1] On a characterization of the Poisson process. *J. Appl. Probability* **8**, 615–616 (1971).

Stam, A. J.

[1] On shifting iterated convolutions I. *Compositio math.* **17**, 268–280 (1967).

[2] On shifting iterated convolutions II. *Compositio math.* **18**, 201–228 (1967).

Stone, C.

[1] On a theorem by Dobrushin, *Ann. Math. Statist.* **39**, 1391–1401 (1968).

Strassen, V.

[1] The existence of probability measures with given marginals. *Ann. Math. Statist.* **36**, 423–439 (1965).

Szász, D.

[1] On the general branching process with continuous time parameter. *Stud. Sci. Math. Hung.* **2**, 227–247 (1967).
[2] Once more on the Poisson process. *Stud. Sci. Math. Hung.* **5**, 441–444 (1970).

Teicher, H.

[1] On the multivariate Poisson distribution. *Skand. Aktuarietidskr.* **37**, 1–9 (1954).

Thedéen, T.

[1] A note on the Poisson tendency in traffic distribution. *Ann. Math. Statist.* **35**, 1823–1824 (1964).
[2] Convergence and invariance questions for point systems in R_1 under random motion. *Ark. Mat.* **7**, 211–239 (1967).

Thiele, B.

[1] Eine Umkehrformel für einfache stationäre Punktprozesse endlicher Intensität im R^s. *Math. Nachr.* (to appear).

Tortrat, A.

[1] Processus indéfiniment divisibles et mesures aléatoires. Faculté des Sciences, Paris 1968 (preprint).
[2] Sur les mesures aléatoires dans les groupes nonabéliens. *Ann. Inst. H. Poincaré Sect. B* **5**, 31–47 (1969).

Vasil'ev, P. I., & Kovalenko, I. N.

[1] A remark on stationary streams of homogeneous events. (Russian.) *Ukrain. Mat. Z.* **16**, 374–375 (1964).

Volkonskii, V. A.

[1] An ergodic theorem on the distribution of the duration of fades. (Russian.) *Teor. Veroyatnost. i Primenen.* **5**, 357–360 (1960) (= *Theor. Probability Appl.* **5**, 323–326).

von Waldenfels, W.

[1] Charakteristische Funktionale zufälliger Maße. *Z. Wahrscheinlichkeitstheorie und verw. Gebiete* **10**, 279–283 (1968).

Warmuth, W.

[1] Kritische räumlich homogene Verzweigungsprozesse mit abzählbarer Typenmenge. *Math. Nachr.* (to appear).

Wegmann, H.

[1] Characterization of Palm distributions and finitely divisible random measures z. Wahrscheinlichkeitstheorie und verw. Gebiete **39**, 257–262 (1977).

Weiß, P.

[1] On the singularity of Poisson processes. *Math. Nachr.* (to appear).

Westcott, M.

[1] On existence and mixing results for cluster point processes. *J. Roy. Statist. Soc. Ser. B* **33**, 290—300 (1971).
[2] The probability generating functional. *J. Aust. Math. Soc.* **14**, 448—466 (1972).
[3] Some remarks on a property of the Poisson process. *Sankhyā, Ser. A* **35**, 29—34 (1973).
[4] Simple proof of a result on thinned point processes. *The Annals of Probability* **4**, 89—90 (1976).

Zitek, F.

[1] A note on a theorem of Korolyuk. (Russian.) *Czech. Math. J.* **7**, 318—319 (1957).
[2] The theory of ordinary streams. (Russian.) *Czech. Math. J.* **8**, 448—459 (1958) (= *Select. Transl. Math. Statist. Probability* **2**, 241—251).

Zubkov, A. M.

[1] Limit distributions of the distance to the nearest mutual ancestor. (Russian; English summary.) *Teor. Veroyatnost. i Primenen.* **20**, 614—623 (1975).

Index

Index of Notations

DUNN and CLARK · Applied Statistics: Analysis of Variance and Regression
ELANDT-JOHNSON · Probability Models and Statistical Methods in Genetics
FLEISS · Statistical Methods for Rates and Proportions
GIBBONS, OLKIN and SOBEL · Selecting and Ordering Populations, a New Statistical Methodology
GNANADESIKAN · Methods for Statistical Data Analysis of Multivariate Observations
GOLDBERGER · Econometric Theory
GROSS and CLARK · Survival Distributions
GROSS and HARRIS · Fundamentals of Queueing Theory
GUTTMAN, WILKS and HUNTER · Introductory Engineering Statistics, *Second Edition*
HAHN and SHAPIRO · Statistical Models in Engineering
HALD · Statistical Tables and Formulas
HALD · Statistical Theory with Engineering Applications
HARTIGAN · Clustering Algorithms
HILDEBRAND, LAING and ROSENTHAL · Prediction Analysis of Cross Classifications
HOEL · Elementary Statistics, *Fourth Edition*
HOLLANDER and WOLFE · Nonparametric Statistical Methods
HUANG · Regression and Econometric Methods
JAGERS · Branching Processes with Biological Applications
JOHNSON and KOTZ · Distributions in Statistics
Discrete Distributions
Continuous Univariate Distributions-1
Continuous Univariate Distributions-2
Continuous Multivariate Distributions
JOHNSON and KOTZ · Urn Models and Their Application
JOHNSON and LEONE · Statistics and Experimental Design: In Engineering and the Physical Sciences, Volumes I and II, *Second Edition*
KEENEY and RAIFFA · Decisions with Multiple Objectives
LANCASTER · The Chi Squared Distribution
LANCASTER · An Introduction to Medical Statistics
LEWIS · Stochastic Point Processes
McNEIL · Interactive Data Analysis
MANN, SCHAFER and SINGPURWALLA · Methods for Statistical Analysis of Reliability and Life Data
MEYER · Data Analysis for Scientists and Engineers
OTNES and ENOCHSON · Digital Time Series Analysis
PRENTER · Splines and Variational Methods
RAO and MITRA · Generalized Inverse of Matrices and Its Applications
SARD and WEINTRAUB · A Book of Splines
SEAL · Stochastic Theory of a Risk Business
SEARLE · Linear Models